José Mira   Alberto Prieto (Eds.)

Bio-Inspired Applications of Connectionism

6th International Work-Conference on
Artificial and Natural Neural Networks, IWANN 2001
Granada, Spain, June 13-15, 2001
Proceedings, Part II

 Springer

Gerhard Goos, Karlsruhe University, Germany
Juris Hartmanis, Cornell University, NY, USA
Jan van Leeuwen, Utrecht University, The Netherlands

Volume Editors

José Mira
Universidad Nacional de Educación a Distancia
Departamento de Inteligencia Artificial
Senda del Rey, s/n., 28040 Madrid, Spain
E-mail: jmira@dia.uned.es

Alberto Prieto
Universidad de Granada
Departamento de Arquitectura y Tecnología de Computadores
Campus Fuentenueva, 18071 Granada, Spain
E-mail: aprieto@atc.ugr.es

Cataloging-in-Publication Data applied for

Die Deutsche Bibliothek - CIP-Einheitsaufnahme

International Work Conference on Artificial and Natural Neural Networks <6, 2001, Granada>:
6th International Work Conference on Artificial and Natural Neural Networks ; Granada, Spain, June 13 - 15, 2001 ; proceedings / IWANN 2001. José Mira ; Alberto Prieto (ed.). - Berlin ; Heidelberg ; New York ; Barcelona ; Hong Kong ; London ; Milan ; Paris ; Singapore ; Tokyo : Springer
Pt. 2. Bio-inspired applications of connectionism. - 2001
   (Lecture notes in computer science ; Vol. 2085)
   ISBN 3-540-42237-4

CR Subject Classification (1998): F.1, F.2, I.2, G.2, I.4, I.5, J.3, J.4, J.1

ISSN 0302-9743
ISBN 3-540-42237-4 Springer-Verlag Berlin Heidelberg New York

This work is subject to copyright. All rights are reserved, whether the whole or part of the material is concerned, specifically the rights of translation, reprinting, re-use of illustrations, recitation, broadcasting, reproduction on microfilms or in any other way, and storage in data banks. Duplication of this publication or parts thereof is permitted only under the provisions of the German Copyright Law of September 9, 1965, in its current version, and permission for use must always be obtained from Springer-Verlag. Violations are liable for prosecution under the German Copyright Law.

Springer-Verlag Berlin Heidelberg New York
a member of BertelsmannSpringer Science+Business Media GmbH

http://www.springer.de

© Springer-Verlag Berlin Heidelberg 2001
Printed in Germany

Typesetting: Camera-ready by author, data conversion by Olgun Computergrafik
Printed on acid-free paper   SPIN 10839320   06/3142   5 4 3 2 1 0

Preface

Underlying most of the IWANN calls for papers is the aim to reassume some of the motivations of the groundwork stages of biocybernetics and the later bionics formulations and to try to reconsider the present value of two basic questions. The first one is: "What does neuroscience bring into computation (the new bionics)?" That is to say, how can we seek inspiration in biology? Titles such as "computational intelligence", "artificial neural nets", "genetic algorithms", "evolutionary hardware", "evolutive architectures", "embryonics", "sensory neuromorphic systems", and "emotional robotics" are representatives of the present interest in "biological electronics" (bionics).

The second question is: "What can return computation to neuroscience (the new neurocybernetics)?" That is to say, how can mathematics, electronics, computer science, and artificial intelligence help the neurobiologists to improve their experimental data modeling and to move a step forward towards the understanding of the nervous system?

Relevant here are the general philosophy of the IWANN conferences, the sustained interdisciplinary approach, and the global strategy, again and again to bring together physiologists and computer experts to consider the common and pertinent questions and the shared methods to answer these questions.

Unfortunately, we have not always been successful in the six biennial meetings from 1991. Frequently the well-known computational models of the past have been repeated and our understanding about the neural functioning of real brains is still scarce. Also the biological influence on computation has not always been used with the necessary methodological care. However IWANN 2001 constituted a new attempt to formulate new models of bio-inspired neural computation with the deeply-held conviction that the interdisciplinary way is, possibly, the most useful one.

IWANN 2001, the 6th International Work-Conference in Artificial and Natural Neural Networks, took place in Granada (Spain) June 13-15, 2001, and addressed the following topics:

1. *Foundations of connectionism.* Brain organization principles. Connectionist versus symbolic representations.
2. *Biophysical models of neurons.* Ionic channels, synaptic level, neurons, and circuits.
3. *Structural and functional models of neurons.* Analogue, digital, probabilistic, Bayesian, fuzzy, object oriented, and energy related formulations.
4. *Learning and other plasticity phenomena.* Supervised, non-supervised, and reinforcement algorithms. Biological mechanisms of adaptation and plasticity.
5. *Complex systems dynamics.* Optimization, self-organization, and cooperative processes. Evolutionary and genetic algorithms. Large scale neural models.

6. *Artificial intelligence and cognitive processes.* Knowledge modeling. Natural language understanding. Intelligent multi-agent systems. Distributed AI.
7. *Methodology for nets design.* Data analysis, task identification, and recursive hierarchical design.
8. *Nets simulation and implementation.* Development environments and editing tools. Implementation. Evolving hardware.
9. *Bio-inspired systems and engineering.* Signal processing, neural prostheses, retinomorphic systems, and other neural adaptive prosthetic devices. Molecular computing.
10. *Other applications.* Artificial vision, speech recognition, spatio-temporal planning, and scheduling. Data mining. Sources separation. Applications of ANNs in robotics, economy, internet, medicine, education, and industry.

IWANN 2001 was organized by the Universidad Nacional de Educación a Distancia, UNED (Madrid), and the Universidad de Granada, UGR (Granada), also in cooperation with IFIP (Working Group in Neural Computer Systems, WG10.6), and the Spanish RIG IEEE Neural Networks Council.

Sponsorship was obtained from the Spanish CICYT and the organizing universities (UNED and UGR).

The papers presented here correspond to talks delivered at the conference. After the evaluation process, 200 papers were accepted for oral or poster presentation, according to the recommendations of reviewers and the authors' preferences. We have organized these papers in two volumes arranged basically following the topics list included in the call for papers. The first volume, entitled "Connectionist Models of Neurons, Learning Processes, and Artificial Intelligence" is divided into four main parts and includes the contributions on:

I. Foundations of connectionism and biophysical models of neurons.
II. Structural and functional models of neurons.
III. Learning and other plasticity phenomena, and complex systems dynamics.
IV. Artificial intelligence and cognitive processes.

In the second volume, with the title, "Bio-inspired Applications of Connectionism", we have included the contributions dealing with applications. These contributions are grouped into three parts:

I. Bio-inspired systems and engineering.
II. Methodology for nets design, and nets simulation and implementation.
III. Other applications (including image processing, medical applications, robotics, data analysis, etc.).

We would like to express our sincere gratitude to the members of the organizing and program committees, in particular to F. de la Paz and J. R. Álvarez-Sánchez, to the reviewers, and to the organizers of preorganized sessions for their invaluable effort in helping with the preparation of this conference. Thanks also to the invited speakers for their effort in preparing the plenary lectures.

Last, but not least, the editors would like to thank Springer-Verlag, in particular Alfred Hofmann, for the continuous and excellent cooperative collaboration from the first IWANN in Granada (1991, LNCS 540), the successive meetings in Sitges (1993, LNCS 686), Torremolinos (1995, LNCS 930), Lanzarote (1997, LNCS 1240), Alicante (1999, LNCS 1606 and 1607), and now again in Granada.

June 2001

José Mira
Alberto Prieto

Invited Speakers

Oscar Herreras, Dept. of Research. Hospital Ramón y Cajal (Spain)
Daniel Mange, Logic Systems Laboratory, IN-Ecublens (Switzerland)
Leonardo Reyneri, Dip. Elettronica, Politecnico di Torino (Italy)
John Rinzel, Center for Neural Science, New York University (USA)

Field Editors

Igor Aizenberg, Neural Networks Technologies Ltd. (Israel)
Amparo Alonso Betanzos, University of A Coruña (Spain)
Jose Manuel Benitez Sanchez, Universidad de Granada (Spain)
Enrique Castillo Ron, Universidad de Cantabria (Spain)
Andreu Català Mallofré, Univ. Politècnica de Catalunya (Spain)
Carolina Chang, Universidad Simón Bolívar (Venezuela)
Carlos Cotta, University of Málaga (Spain)
Richard Duro, Universidade da Coruña (Spain)
Marcos Faundez-Zanuy, Univ. Politècnica de Catalunya (Spain)
Carlos Garcia Puntonet, Universidad de Granada (Spain)
Gonzalo Joya, Universidad de Málaga (Spain)
Christian Jutten, Inst. National Polytechnique de Grenoble (France)
Dario Maravall, Universidad Politécnica de Madrid (Spain)
Eduardo Sánchez, Universidad de Santiago de Compostela (Spain)
José Santos Reyes, Universidade da Coruña (Spain)
Kate Smith, Monash University (Australia)

Reviewers

Igor Aleksander, Imperial College of Sci. Tech. and Medicine (UK)
José Ramón Álvarez-Sánchez, UNED (Spain)
Shun-ichi Amari, RIKEN (Japan)
A. Bahamonde, Universidad de Oviedo en Gijón (Spain)
Senén Barro Ameneiro, Univ. Santiago de Compostela (Spain)
J. Cabestany, Universidad Politécnica de Cataluña (Spain)
Marie Cottrell, Université Paris 1 (France)
Félix de la Paz López, UNED (Spain)
Ana E. Delgado García, UNED (Spain)
Ángel P. Del Pobil, Universidad Jaime I de Castellón (Spain)
José Dorronsoro, Universidad Autónoma de Madrid (Spain)
José Manuel Ferrández, Universidad Miguel Hernandez (Spain)
Kunihiko Fukushima, Osaka Univ (Japan)
Tamas Gedeon, Murdoch University (Australia)
Karl Goser, Univ. Dortmund (Germany)
Manuel Graña Romay, Universidad Pais Vasco (Spain)
J. Hérault, Inst. N. P. Grenoble (France)
Óscar Herreras, Hospital Ramón y Cajal (Spain)
Gonzalo Joya, Universidad de Málaga (Spain)
Christian Jutten, Inst. National Polytechnique de Grenoble (France)
Shahla Keyvan, University of Missouri-Rolla (USA)
Daniel Mange, IN-Ecublens (Switzerland)
Darío Maravall, Universidad Politécnica de Madrid (Spain)
Eve Marder, Brandeis University (USA)
José Mira, UNED (Spain)
J.M. Moreno Aróstegui, Univ. Politécnica de Cataluña (Spain)
Christian W. Omlin, University of Western Cape South (Africa)
Julio Ortega Lopera, Universidad de Granada (Spain)
F.J. Pelayo, Universidad de Granada (Spain)
Franz Pichler, Johannes Kepler University (Austria)
Alberto Prieto Espinosa, Universidad de Granada (Spain)
Leonardo Maria Reyneri, Politecnico di Torino (Italy)
John Rinzel, New York University (USA)
J.V. Sánchez-Andrés, Universidad de Alicante (Spain)
Francisco Sandoval, Universidad de Málaga (Spain)
J.A. Sigüenza, Universidad Autónoma de Madrid (Spain)
M. Verleysen, Universite Catholique de Louvin (Belgium)

Table of Contents, Part II

Bio-inspired Systems and Engineering

From Embryonics to POEtic Machines 1
    D. Mange, A. Stauffer, G. Tempesti, and C. Teuscher

Design and Codesign of Neuro-Fuzzy Hardware 14
    L.M. Reyneri

A Field-Programmable Conductance Array IC
for Biological Neurons Modeling 31
    V. Douence, S. Renaud-Le Masson, S. Saïghi, and G. Le Masson

A 2-by-n Hybrid Cellular Automaton Implementation
Using a Bio-Inspired FPGA 39
    H.F. Restrepo and D. Mange

Parametric Neurocontroller for Positioning of an Anthropomorfic Finger
Based on an Oponent Driven-Tendon Transmission System 47
    J.I. Mulero, J. Feliú Batlle, and J. López Coronado

An Integration Principle for Multimodal Sensor Data Based
on Temporal Coherence of Self-Organized Patterns. 55
    E.I. Barakova

Simultaneous Parallel Processing of Object and Position
by Temporal Correlation 64
    L.F. Lago-Fernández and G. Deco

Methodology for Nets Design, Nets Simulation and Implementation

NeuSim: A Modular Neural Networks Simulator for Beowulf Clusters 72
    C.J. García Orellana, R. Gallardo Caballero,
    H.M. González Velasco, F.J. López Aligué

Curved Kernel Neural Network for Functions Approximation 80
    P. Bourret and B. Pelletier

Repeated Measures Multiple Comparison Procedures Applied
to Model Selection in Neural Networks 88
    E. Guerrero Vázquez, A. Yañez Escolano, P. Galindo Riaño,
    J. Pizarro Junquera

Extension of HUMANN for Dealing with Noise and with Classes
of Different Shape and Size: A Parametric Study 96
   P. García Báez, C.P. Suárez Araujo, and P. Fernández López

Evenet 2000: Designing and Training Arbitrary Neural Networks in Java .. 104
   E.J. González, A.F. Hamilton, L. Moreno, J.F. Sigut,
   and R.L. Marichal

Neyman-Pearson Neural Detectors 111
   D. Andina and J.L. Sanz-González

Distance between Kohonen Classes Visualization Tool to Use SOM
in Data Set Analysis and Representation 119
   P. Rousset and C. Guinot

Optimal Genetic Representation
of Complete Strictly-Layered Feedforward Neural Networks 127
   S. Raptis, S. Tzafestas, and H. Karagianni

Assessing the Noise Immunity of Radial Basis Function Neural Networks .. 136
   J.L. Bernier, J. González, A. Cañas, and J. Ortega

Analyzing Boltzmann Machine Parameters for Fast Convergence 144
   F.J. Salcedo, J. Ortega, and A. Prieto

A Penalization Criterion Based on Noise Behaviour for Model Selection ... 152
   J. Pizarro Junquera, P. Galindo Riaño, E. Guerrero Vázquez,
   and A. Yañez Escolano

Image Processing

Wood Texture Analysis by Combining the Connected Elements Histogram
and Artificial Neural Networks 160
   M.A. Patricio Guisado and D. Maravall Gómez-Allende

Dynamic Topology Networks for Colour Image Compression 168
   E. López-Rubio, J. Muñoz-Pérez, and J.A. Gómez-Ruiz

Analysis on the Viewpoint Dependency in 3-D Object Recognition
by Support Vector Machines 176
   T. Hayasaka, E. Ohnishi, S. Nakauchi, and S. Usui

A Comparative Study of Two Neural Models for Cloud Screening
of Iberian Peninsula Meteosat Images 184
   M. Macías Macías, F.J. López Aligué, A. Serrano Pérez,
   and A. Astilleros Vivas

A Growing Cell Neural Network Structure
for Off-Line Signature Recognition 192
  K. Toscano-Medina, G. Sanchez-Perez, M. Nakano-Miyatake,
  and H. Perez-Meana

ZISC-036 Neuro-processor Based Image Processing 200
  K. Madani, G. de Trémiolles, and P. Tannhof

Self-Organizing Map for Hyperspectral Image Analysis................ 208
  P. Martínez, P.L. Aguilar, R.M. Pérez, M. Linaje, J.C. Preciado,
  and A. Plaza

Classification of the Images of Gene Expression Patterns
Using Neural Networks Based on Multi-valued Neurons 219
  I. Aizenberg, E. Myasnikova, and M. Samsonova

Image Restoration Using Neural Networks 227
  S. Ghennam and K. Benmahammed

Automatic Generation of Digital Filters by NN Based Learning:
An Application on Paper Pulp Inspection 235
  P. Campoy-Cervera, D.F. Muñoz García, D. Peña,
  and J.A. Calderón-Martínez

Image Quality Enhancement for Liquid Bridge Parameter Estimation
with DTCNN.. 246
  M.A. Jaramillo, J. Álvaro Fernández, J.M. Montanero, and F. Zayas

Neural Network Based on Multi-valued Neurons: Application
in Image Recognition, Type of Blur and Blur Parameters Identification .. 254
  I. Aizenberg, N. Aizenberg, and C. Butakoff

Analyzing Wavelets Components to Perform Face Recognition 262
  P. Isasi, M. Velasco, and J. Segovia

Man-Machine Voice Interface
Using a Commercially Available Neural Chip 271
  N.J. Medraño-Marqués and B. Martín-del-Brío

Partial Classification in Speech Recognition Verification.................. 279
  G. Hernández Ábrego and I. Torres Sánchez

Speaker Recognition Using Gaussian Mixtures Model 287
  E. Simancas-Acevedo, A. Kurematsu, M. Nakano Miyatake,
  and H. Perez-Meana

A Comparative Study of ICA Filter Structures Learnt from Natural
and Urban Images 295
  C. Ziegaus and E.W. Lang

Neural Edge Detector –
A Good Mimic of Conventional One Yet Robuster against Noise 303
  K. Suzuki, I. Horiba, and N. Sugie

Neural Networks for Image Restoration from the Magnitude
of Its Fourier Transform 311
  A. Burian, J. Saarinen, and P. Kuosmanen

Medical Applications

An Automatic System for the Location of the Optic Nerve Head
from 2D Images... 319
  M. Bachiller, M. Rincón, J. Mira, and J. García-Feijó

Can ICA Help Classify Skin Cancer and Benign Lesions?................ 328
  *C. Mies, C. Bauer, G. Ackermann, W. Bäumler, C. Abels,
  C.G. Puntonet, M. Rodríguez-Alvarez, and E.W. Lang*

An Approach Fractal and Analysis of Variogram for Edge Detection
of Biomedical Images 336
  L. Hamami and N. Lassouaoui

Some Examples for Solving Clinical Problems Using Neural Networks..... 345
  A.J. Serrano, E. Soria, G. Camps, J.D. Martín, and N.V. Jiménez

Medical Images Analysis: An Application of Artificial Neural Networks
in the Diagnosis of Human Tissues 353
  *E. Restum Antonio, L. Biondi Neto, V. De Roberto Junior,
  and F. Hideo Fukuda*

Feature Selection, Ranking of Each Feature and Classification
for the Diagnosis of Community Acquired Legionella Pneumonia 361
  E. Monte, J. Solé i Casals, J.A. Fiz, and N. Sopena

Rotation-Invariant Image Association for Endoscopic Positional
Identification Using Complex-Valued Associative Memories............. 369
  H. Aoki, E. Watanabe, A. Nagata, and Y. Kosugi

A Multi Layer Perceptron Approach for Predicting and Modeling
the Dynamical Behavior of Cardiac Ventricular Repolarisation 377
  R. El Dajani, M. Miquel, and P. Rubel

Detection of Microcalcifications in Mammograms by the Combination
of a Neural Detector and Multiscale Feature Enhancement 385
  D. Andina and A. Vega-Corona

An Auto-learning System for the Classification
of Fetal Heart Rate Decelerative Patterns 393
    B. Guijarro Berdiñas, A. Alonso-Betanzos, O. Fontenla-Romero,
    O. Garcia-Dans, and N. Sánchez Maroño

Neuro-Fuzzy Nets in Medical Diagnosis: The DIAGEN Case Study
of Glaucoma 401
    E. Carmona, J. Mira, J. García Feijó, and M.G. de la Rosa

Robotics

Evolving Brain Structures for Robot Control 410
    F. Pasemann, U. Steinmetz, M. Hülse, and B. Lara

A Cuneate-Based Network and Its Application as a Spatio-Temporal Filter
in Mobile Robotics 418
    E. Sánchez, M. Mucientes, and S. Barro

An Application of Fuzzy State Automata: Motion Control
of an Hexapod Walking Machine 426
    D. Morano and L.M. Reyneri

Neural Adaptive Force Control for Compliant Robots................... 436
    N. Saadia, Y. Amirat, J. Pontnaut, and A. Ramdane-Cherif

Reactive Navigation Using Reinforment Learning Techniques
in Situations of POMDPs 444
    P. Puliti, G. Tascini, and A. Montesanto

Landmark Recognition for Autonomous Navigation
Using Odometric Information and a Network of Perceptrons 451
    J. de Lope Asiaín and D. Maravall Gómez-Allende

Topological Maps for Robot's Navigation: A Conceptual Approach 459
    F. de la Paz López, and J.R. Álvarez-Sánchez

Information Integration for Robot Learning Using Neural Fuzzy Systems .. 468
    C. Zhou, Y. Yang, and J. Kanniah

Incorporating Perception-Based Information
in Reinforcement Learning Using Computing with Words 476
    C. Zhou, Y. Yang, and X. Jia

Cellular Neural Networks for Mobile Robot Vision 484
    M. Balsi, A. Maraschini, G. Apicella, S. Luengo, J. Solsona,
    and X. Vilasís-Cardona

Learning to Predict Variable-Delay Rewards and Its Role
in Autonomous Developmental Robotics 492
    A. Pérez-Uribe and M. Courant

Robust Chromatic Identification and Tracking 500
    J. Ramírez and G. Grittani

Sequence Learning in Mobile Robots Using Avalanche Neural Networks ... 508
    G. Quero and C. Chang

Investigating Active Pattern Recognition in an Imitative Game 516
    S. Moga, P. Gaussier, and M. Quoy

Towards an On-Line Neural Conditioning Model for Mobile Robots 524
    E. Şahin

General Applications

A Thermocouple Model Based on Neural Networks 531
    N. Medraño-Marqués, R. del-Hoyo-Alonso, and B. Martín-del-Brío

Improving Biological Sequence Property Distances
Using a Genetic Algorithm 539
    O.M. Perez, F.J. Marin, and O. Trelles

Data Mining Applied to Irrigation Water Management.................. 547
    J.A. Botía, A.F. Gómez Skarmeta, M. Valdés, and A. Padilla

Classification of Specular Object Based on Statistical Learning Theory ... 555
    T.S. Yun

On the Application of Heteroassociative Morphological Memories
to Face Localization 563
    B. Raducanu and M. Graña

Early Detection and Diagnosis of Faults in an AC Motor
Using Neuro Fuzzy Techniques: FasArt+ Fuzzy k Nearest Neighbors...... 571
    J. Juez, G.I. Sainz, E.J. Moya, and J.R. Perán

Knowledge-Based Neural Networks for Modelling Time Series............ 579
    J. van Zyl and C.W. Omlin

Using Artificial Neural Network to Define Fuzzy Comparators
in FSQL with the Criterion of Some Decision-Maker.................... 587
    R. Carrasco, J. Galindo, and A. Vila

Predictive Classification for Integrated Pest Management by Clustering
in NN Output Space 595
    M. Salmerón, D. Guidotti, R. Petacchi, and L.M. Reyneri

Blind Source Separation in the Frequency Domain: A Novel Solution
to the Amplitude and the Permutation Indeterminacies 603
    A. Dapena and L. Castedo

Evaluation, Classification and Clustering with Neuro-Fuzzy Techniques
in Integrate Pest Management 611
    E. Bellei, D. Guidotti, R. Petacchi, L.M. Reyneri, and I. Rizzi

Inaccessible Parameters Monitoring in Industrial Environment:
A Neural Based Approach.. 619
    K. Madani and I. Berechet

Autoorganized Structures for Extraction of Perceptual Primitives 628
    M. Penas, M.J. Carreira, and M.G. Penedo

Real-Time Wavelet Transform for Image Processing
on the Cellular Neural Network Universal Machine 636
    V.M. Preciado

OBLIC: Classification System Using Evolutionary Algorithm 644
    J.L. Alvarez, J. Mata, and J.C. Riquelme

Design of a Pre-processing Stage for Avoiding the Dependence
on TSNR of a Neural Radar Detector................................ 652
    P. Jarabo Amores, M. Rosa Zurera, and F. López Ferreras

Foetal Age and Weight Determination
Using a Lateral Interaction Inspired Net 660
    A. Fernández-Caballero, J. Mira, F.J. Gómez, and M.A. Fernández

Inference of Stochastic Regular Languages
through Simple Recurrent Networks with Time Dealys................. 671
    G.A. Casañ and M.A. Castaño

Is Neural Network a Reliable Forecaster on Earth? A MARS Query!...... 679
    A. Abraham and D. Steinberg

Character Feature Extraction Using Polygonal Projection Sweep
(Contour Detection) 687
    R.J. Rodrigues, G.K. Vianna, and A.C.G. Thomé

Using Contextual Information
to Selectively Adjust Preprocessing Parameters 696
    P. Neskovic and L.N. Cooper

Electric Power System's Stability Assessment and Online-Provision
of Control Actions Using Self-Organizing Maps 704
    C. Leder and C. Rehtanz

Neural Networks for Contingency Evaluation and Monitoring
in Power Systems 711
    F. García-Lagos, G. Joya, F.J. Marín, and F. Sandoval

Hybrid Framework for Neuro-dynamic Programming Application
to Water Supply Networks 719
    M. Damas, M. Salmerón, J. Ortega, and G. Olivares

Classification of Disturbances in Electrical Signals
Using Neural Networks 728
    C. León, A. López, J.C. Montaño, and Í. Monedero

Neural Classification and "Traditional" Data Analysis: An Application
to Households' Living Conditions 738
    S. Ponthieux and M. Cottrell

Nonlinear Synthesis of Vowels in the LP Residual Domain
with a Regularized RBF Network 746
    E. Rank and G. Kubin

Nonlinear Vectorial Prediction with Neural Nets 754
    M. Faúndez-Zanuy

Separation of Sources Based on the Partitioning of the Space
of Observations 762
    M. Rodríguez-Álvarez, C.G. Puntonet, and I. Rojas

Adaptive ICA with Order Statistics in Multidmensional Scenarios 770
    Y. Blanco, S. Zazo, and J.M. Paez-Borrallo

Pattern Repulsion Revistited 778
    Fabian J. Theis, C. Bauer, C. Puntonet, and E.W. Lang

The Minimum Entropy and Cumulants Based Contrast Functions
for Blind Source Extraction 786
    S. Cruces, A. Cichocki, and S.-I. Amari

Feature Extraction in Digital Mammography:
An Independent Component Analysis Approach 794
    A. Koutras, I. Christoyianni, E. Dermatas, and G. Kokkinakis

Blind Source Separation in Convolutive Mixtures:
A Hybrid Approach for Colored Sources 802
    F. Abrard and Y. Deville

A Conjugate Gradient Method and Simulated Annealing
for Blind Separation of Sources 810
    R. Martín-Clemente, C.G. Puntonet, and J.I. Acha

The Problem of Overlearning in High-Order ICA Approaches:
Analysis and Solutions 818
    J. Särelä and R. Vigário

Equi-convergence Algorithm for Blind Separation of Sources with Arbitrary Distributions 826
    L.-Q. Zhang, S. Amari, and A. Cichocki

Separating Convolutive Mixtures by Mutual Information Minimization.... 834
    M. Babaie-Zadeh, C. Jutten, and K. Nayebi

Author Index 843

Table of Contents, Part I

Foundations of Connectionism and Biophysical Models of Neurons

Dendrites: The Last-Generation Computers 1
    O. Herreras, J.M. Ibarz, L. López-Aguado, and P. Varona

Homogeneity in the Electrical Activity Pattern as a Function
of Intercellular Coupling in Cell Networks 14
    E. Andreu, R. Pomares, B. Soria, and J.V. Sanchez-Andres

A Realistic Computational Model of the Local Circuitry
of the Cuneate Nucleus 21
    E. Sánchez, S. Barro, J. Mariño, and A. Canedo

Algorithmic Extraction of Morphological Statistics
from Electronic Archives of Neuroanatomy 30
    R. Scorcioni and G.A. Ascoli

What Can We Compute with Lateral Inhibition Circuits? 38
    J. Mira and A.E. Delgado

Neuronal Models with Current Inputs 47
    J. Feng

Decoding the Population Responses of Retinal Ganglions Cells
Using Information Theory... 55
    *J.M. Ferrández, M. Bongard, F. García de Quirós, J.A. Bolea,
    J. Ammermüller, R.A. Normann, and E. Fernández*

Numerical Study of Effects of Co-transmission by Substance P
and Acetylcholine on Synaptic Plasticity in Myenteric Neurons 63
    R. Miftakov and J. Christensen

Neurobiological Modeling of Bursting Response During Visual Attention .. 72
    R. Rajimehr and L. Montaser Kouhsari

Sensitivity of Simulated Striate Neurons to Cross-Like Stimuli Based
on Disinhibitory Mechanism 81
    K.A. Saltykov and I.A. Shevelev

Synchronisation Mechanisms in Neuronal Networks..................... 87
    S. Chillemi, M. Barbi, and A. Di Garbo

Detection of Oriented Repetitive Alternating Patterns in Color Images
(A Computational Model of Monkey Grating Cells) 95
    T. Lourens, H.G. Okuno, and H. Kitano

Synchronization in Brain – Assessment by Electroencephalographic Signals 108
    E. Pereda and J. Bhattacharya

Strategies for the Optimization of Large Scale Networks of Integrate
and Fire Neurons 117
    M.A. Sánchez-Montañés

Structural and Functional Models of Neurons

A Neural Network Model of Working Memory
(Processing of "What" and "Where" Information) 126
    T. Minami and T. Inui

Orientation Selectivity of Intracortical Inhibitory Cells in the Striate
Visual Cortex: A Computational Theory and a Neural Circuitry 134
    M.N. Shirazi

Interpreting Neural Networks in the Frame of the Logic of Lukasiewicz ... 142
    C. Moraga and L. Salinas

Time-Dispersive Effects in the J. Gonzalo's Research
on Cerebral Dynamics 150
    I. Gonzalo and M.A. Porras

Verifying Properties of Neural Networks 158
    P. Rodrigues, J.F. Costa, and H.T. Siegelmann

Algorithms and Implementation Architectures
for Hebbian Neural Networks 166
    J.A. Berzal and P.J. Zufiria

The Hierarchical Neuro-Fuzzy BSP Model: An Application
in Electric Load Forecasting 174
    F.J. de Souza, M.M.R. Vellasco, and M.A.C. Pacheco

The Chemical Metaphor in Neural Computation 184
    J. Barahona da Fonseca, I. Barahona da Fonseca,
    C.P. Suárez Araujo, and J. Simões da Fonseca

The General Neural-Network Paradigm for Visual Cryptography 196
    T.-W. Yue and S. Chiang

Π-DTB, Discrete Time Backpropagation with Product Units 207
    J. Santos and R.J. Duro

Neocognitron-Type Network for Recognizing Rotated and Shifted Patterns
with Reduction of Resources 215
    S. Satoh, S. Miyake, and H. Aso

Classification with Synaptic Radial Basis Units 223
    J.D. Buldain

A Randomized Hypercolumn Model and Gesture Recognition 235
    N. Tsuruta, Y. Yoshiki, and T. El. Tobely

Heterogeneous Kohonen Networks 243
    S. Negri, L.A. Belanche

Divided-Data Analysis in a Financial Case Classification
with Multi-dendritic Neural Networks 253
    J.D. Buldain

Neuro Fuzzy Systems: State-of-the-Art Modeling Techniques 269
    A. Abraham

Generating Linear Regression Rules from Neural Networks
Using Local Least Squares Approximation........................... 277
    R. Setiono

Speech Recognition Using Fuzzy Second-Order
Recurrent Neural Networks... 285
    A. Blanco, M. Delgado, M.C. Pegalajar, and I. Requena

A Measure of Noise Immunity for Functional Networks 293
    *E. Castillo, O. Fontenla-Romero, B. Guijarro-Berdiñas,
    and A. Alonso-Betanzos*

A Functional-Neural Network
for Post-Nonlinear Independent Component Analysis 301
    *O. Fontenla Romero, B. Guijarro Berdiñas,
    and A. Alonso Betanzos*

Optimal Modular Feedfroward Neural Nets Based
on Functional Network Architectures 308
    A.S. Cofiño, J.M. Gutiérrez

Optimal Transformations in Multiple Linear Regression
Using Functional Networks 316
    E. Castillo, A.S. Hadi, and B. Lacruz

Learning and Other Plasticity Phenomena, and Complex Systems Dynamics

Generalization Error and Training Error at Singularities
of Multilayer Perceptrons 325
    S.-I. Amari, T. Ozeki, and H. Park

Bistable Gradient Neural Networks: Their Computational Properties 333
    V. Chinarov and M. Menzinger

Inductive Bias in Recurrent Neural Networks 339
    S. Snyders and C.W. Omlin

Accelerating the Convergence of EM-Based Training Algorithms
for RBF Networks 347
    M. Lázaro, I. Santamaría, and C. Pantaleón

Expansive and Competitive Neural Networks 355
    J.A. Gomez-Ruiz, J. Muñoz-Perez, E. Lopez-Rubio,
    and M.A. Garcia-Bernal

Fast Function Approximation with Hierarchical Neural Networks
and Their Application to a Reinforcement Learning Agent 363
    J. Fischer, R. Breithaupt, and M. Bode

Two Dimensional Evaluation Reinforcement Learning.................... 370
    H. Okada, H. Yamakawa, and T. Omori

Comparing the Learning Processes of Cognitive Distance Learning
and Search Based Agent 378
    H. Yamakawa, Y. Miyamoto, and H. Okada

Selective Learning for Multilayer Feedforward Neural Networks 386
    A.P. Engelbrecht

Connectionist Models of Cortico-Basal Ganglia Adaptive Neural Networks
During Learning of Motor Sequential Procedures....................... 394
    J. Molina Vilaplana, J. Feliú Batlle, and J. López Coronado

Practical Consideration on Generalization Property
of Natural Gradient Learning... 402
    H. Park

Novel Training Algorithm Based on Quadratic Optimisation
Using Neural Networks 410
    G. Arulampalam and A. Bouzerdoum

Non-symmetric Support Vector Machines 418
    J. Feng

Natural Gradient Learning in NLDA Networks 427
    J.R. Dorronsoro, A. González, and C. Santa Cruz

AUTOWISARD: Unsupervised Modes for the WISARD 435
    I. Wickert and F.M.G. França

Neural Steering: Difficult and Impossible Sequential Problems
for Gradient Descent 442
    G. Milligan, M.K. Weir, and J.P. Lewis

Analysis of Scaling Exponents of Waken and Sleeping Stage in EEG 450
    J.-M. Lee, D.-J. Kim, I.-Y. Kim, and S.I. Kim

Model Based Predictive Control Using Genetic Algorithms.
Application to Greenhouses Climate Control. 457
    X. Blasco, M. Martínez, J. Senent, and J. Sanchis

Nonlinear Parametric Model Identification with Genetic Algorithms.
Application to a Thermal Process. 466
    X. Blasco, J.M. Herrero, M. Martínez, and J. Senent

A Comparison of Several Evolutionary Heuristics
for the Frequency Assignment Problem 474
    C. Cotta and J.M. Troya

GA Techniques Applied to Contour Search in Images of Bovine Livestock . 482
    H.M. González Velasco, C.J. García Orellana,
    M. Macías Macías, and M.I. Acevedo Sotoca

Richer Network Dynamics of Intrinsically Non-regular Neurons Measured
through Mutual Information... 490
    F. Rodríguez, P. Varona, R. Huerta, M.I. Rabinovich,
    and H.D.I. Abarbanel

RBF Neural Networks, Multiobjective Optimization
and Time Series Forecasting 498
    J. González, I. Rojas, H. Pomares, and J. Ortega

Evolving RBF Neural Networks 506
    V.M. Rivas, P.A. Castillo, and J.J. Merelo

Evolutionary Cellular Configurations
for Designing Feed-Forward Neural Networks Architectures 514
    G. Gutiérrez, P. Isasi, J.M. Molina, A. Sanchís, and I.M. Galván

A Recurrent Multivalued Neural Network for the N-Queens Problem...... 522
    E. Mérida, J. Muñoz, and R. Benítez

A Novel Approach to Self-Adaptation of Neuro-Fuzzy Controllers
in Real Time 530
   H. Pomares, I. Rojas, J. González, and M. Damas

Expert Mutation Operators for the Evolution
of Radial Basis Function Neural Networks 538
   J. González, I. Rojas, H. Pomares, and M. Salmerón

Studying Neural Networks of Bifurcating Recursive Processing Elements
– Quantitative Methods for Architecture Design
and Performance Analysis 546
   E. Del Moral Hernandez

Topology-Preserving Elastic Nets 554
   V. Tereshko

Optimization with Linear Constraints in the Neural Network 561
   M. Oota, N. Ishii, K. Yamauchi, and M. Nakamura

Optimizing RBF Networks with Cooperative/Competitive Evolution
of Units and Fuzzy Rules 570
   A.J. Rivera, J. Ortega, I. Rojas, and A. Prieto

Study of Chaos in a Simple Discrete Recurrence Neural Network 579
   J.D. Piñeiro, R.L. Marichal, L. Moreno, J.F. Sigut,
   and E.J. González

Genetic Algorithm versus Scatter Search
and Solving Hard MAX-W-SAT Problems 586
   H. Drias

A New Approach to Evolutionary Computation:
Segregative Genetic Algorithms (SEGA) 594
   M. Affenzeller

Evolution of Firms in Complex Worlds: Generalized *NK* Model 602
   N. Jacoby

Learning Adaptive Parameters
with Restricted Genetic Optimization Method 612
   S. Garrido and L. Moreno

Solving NP-Complete Problems with Networks
of Evolutionary Processors 621
   J. Castellanos, C. Martín-Vide, V. Mitrana, and J.M. Sempere

Using SOM for Neural Network Visualization 629
   G. Romero, P.A. Castillo, J.J. Merelo, and A. Prieto

Comparison of Supervised Self-Organizing Maps Using Euclidian
or Mahalanobis Distance in Classification Context 637
    F. Fessant, P. Aknin, L. Oukhellou, and S. Midenet

Introducing Multi-objective Optimization in Cooperative Coevolution
of Neural Networks 645
    N. García-Pedrajas, E. Sanz-Tapia, D. Ortiz-Boyer,
    and C. Hervás-Martínez

STAR - Sparsity through Automated Rejection 653
    R. Burbidge, M. Trotter, B. Buxton, and S. Holden

Ordinal Regression with K-SVCR Machines 661
    C. Angulo and A. Català

Large Margin Nearest Neighbor Classifiers 669
    S. Bermejo and J. Cabestany

Reduced Support Vector Selection by Linear Programs 677
    W.A. Fellenz

Edge Detection in Noisy Images Using the Support Vector Machines 685
    H. Gómez-Moreno, S. Maldonado-Bascón, and F. López Ferreras

Initialization in Genetic Algorithms for Constraint Satisfaction Problems.. 693
    C.R. Vela, R. Varela, and J. Puente

Evolving High-Posterior Self-Organizing Maps 701
    J. Muruzábal

Using Statistical Techniques to Predict GA Performance 709
    R. Nogueras and C. Cotta

Multilevel Genetic Algorithm for the Complete Development of ANN 717
    J. Dorado, A. Santos, and J.R. Rabuñal

Graph Based GP Applied to Dynamical Systems Modeling 725
    A.M. López, H. López, and L. Sánchez

Nonlinear System Dynamics in the Normalisation Process
of a Self-Organising Neural Network for Combinatorial Optimisation 733
    T. Kwok and K.A. Smith

Continuous Function Optimisation via Gradient Descent
on a Neural Network Approxmiation Function 741
    K.A. Smith and J.N.D. Gupta

An Evolutionary Algorithm for the Design
of Hybrid Fiber Optic-Coaxial Cable Networks in Small Urban Areas 749
    P. Cortés, F. Guerrero, D. Canca, and J.M. García

Channel Assignment for Mobile Communications
Using Stochastic Chaotic Simulated Annealing 757
    S. Li and L. Wang

Artificial Intelligence and Cognitive Processes

Seeing is Believing: Depictive Neuromodelling of Visual Awareness 765
    I. Aleksander, H. Morton, and B. Dunmall

DIAGEN-WebDB: A Connectionist Approach
to Medical Knowledge Representation and Inference 772
    J. Mira, R. Martínez, J.R. Álvarez, and A.E. Delgado

Conceptual Spaces as Voltage Maps 783
    J. Aisbett and G. Gibbon

Determining Hyper-planes to Generate Symbolic Rules 791
    G. Bologna

Automatic Symbolic Modelling of Co-evolutionarily Learned Robot Skills . 799
    A. Ledezma, A. Berlanga, and R. Aler

ANNs and the Neural Basis for General Intelligence 807
    J.G. Wallace and K. Bluff

Knowledge and Intelligence 814
    J.C. Herrero

Conjecturing the Cognitive Plausibility of an ANN Theorem-Prover 822
    I.M.O. Vilela and P.M.V. Lima

Author Index 831

From Embryonics to POEtic Machines

Daniel Mange, André Stauffer, Gianluca Tempesti, and Christof Teuscher

Logic Systems Laboratory, Swiss Federal Institute of Technology,
CH-1015 Lausanne, Switzerland
daniel.mange@epfl.ch

Abstract. The space of bio-inspired hardware can be partitioned along three axes: phylogeny, ontogeny, and epigenesis. We refer to this as the POE model. Our Embryonics (for embryonic electronics) project is situated along the ontogenetic axis of the POE model and is inspired by the processes of molecular biology and by the embryonic development of living beings.
We will describe the architecture of multicellular automata that are endowed with self-replication and self-repair properties. In the conclusion, we will present our major on-going project: a giant self-repairing electronic watch, the BioWatch, built on a new reconfigurable tissue, the electronic wall or e–wall.

1 Introduction[1]

1.1 The POE Model of Bio-Inspired Systems

The space of *bio-inspired* hardware systems can be partitioned along three major axes: *phylogeny*, *ontogeny*, and *epigenesis*; we refer to this as the *POE model* [7], [4]. Where nature is concerned, the distinction between the axes cannot be easily drawn. Indeed the definitions themselves may be subject to discussion. Sipper et al. [7] thus defined each of the above axes within the framework of the POE model as follows: the phylogenetic axis involves *evolution*, the ontogenetic axis involves the *development* of a single individual from its own genetic material, essentially without environmental interactions, and the epigenetic axis involves *learning* through environmental interactions that take place after formation of the individual. As an example, consider the following three paradigms, whose hardware implementations can be positioned along the POE axes: (P) evolutionary algorithms are the (simplified) artificial counterpart of phylogeny in nature, (O) multicellular automata are based on the concept of ontogeny, where a single mother cell gives rise, through multiple divisions, to a multicellular organism, and (E) artificial neural networks embody the epigenetic process, where the system's synaptic weights and, sometimes, its topological structure change through interactions with the environment.

---

[1] Based on the following paper: D. Mange, M. Sipper, A. Stauffer, G. Tempesti. "Toward Robust Integrated Circuits: The Embryonics Approach". *Proceedings of the IEEE*, Vol. 88, No. 4, April 2000, pp. 516-541.

This paper is a description of bio-inspired hardware systems along the ontogenetic axis of the POE model: the Embryonics project. We conclude this presentation with directions for future research, based on the POE model, i.e. new *POEtic machines*.

1.2 Embryonics = Embryonic Electronics

Our *Embryonics* project is inspired by the basic processes of molecular biology and by the embryonic development of living beings. By adopting certain features of cellular organization, and by transposing them to the two-dimensional world of integrated circuits on silicon, we will show that properties unique to the living world, such as *self-replication* and *self-repair*, can also be applied to artificial objects (integrated circuits) [8], [1]. Self-repair allows partial reconstruction in case of a minor fault, while self-replication allows complete reconstruction of the original device in case of a major fault. These two properties are particularly desirable for complex artificial systems requiring very high level of reliability.

2 A Survey of Embryonics

2.1 Biological Inspiration

The majority of living beings, with the exception of unicellular organisms such as viruses and bacteria, share three fundamental features.

1. *Multicellular organization* divides the organism into a finite number of *cells*, each realizing a unique function (neuron, muscle, intestine, etc.). The same organism can contain multiple cells of the same kind.
2. *Cellular division* is the process whereby each cell (beginning with the first cell or *zygote*) generates one or two daughter cells. During this division, all of the genetic material of the mother cell, the *genome*, is copied into the daughter cell(s).
3. *Cellular differentiation* defines the role of each cell of the organism, that is, its particular function (neuron, muscle, intestine, etc.). This specialization of the cell is obtained through the expression of part of the genome, consisting of one or more *genes*, and depends essentially on the physical position of the cell in the organism.

A consequence of these three features is that each cell is "universal", since it contains the whole of the organism's genetic material, the genome. Should a minor (wound) or major (loss of an organ) trauma occur, living organisms are thus potentially capable of self-repair (cicatrization) or self-replication (cloning or budding) [8].

2.2 The Organism's Features: Multicellular Organization, Cellular Differentiation, and Cellular Division

The environment in which our quasi-biological development occurs is imposed by the structure of electronic circuits, and consists of a finite (but arbitrarily large) two-dimensional surface of silicon. This surface is divided into rows and columns, whose intersections define the cells. Since such cells (small processors and their memory) have an identical physical structure (i.e., an identical set of logic operators and of connections), the cellular array is homogeneous. As the program in each cell (our artificial genome) is identical, only the state of the cell (i.e., the contents of its registers) can differentiate it from its neighbors.

In this Section, we first show how to implement in our artificial organisms the three fundamental features of multicellular organization, cellular differentiation, and cellular division, by using a generic and abstract six-cell example. *Multicellular organization* divides the artificial organism (ORG) into a finite number of cells (Figure 1). Each cell ($CELL$) realizes a unique function, defined by a sub-program called the *gene* of the cell and selected as a function of the values of both the horizontal (X) and the vertical (Y) coordinates (in Figure 1, the genes are labeled A to F for coordinates $X, Y = 1, 1$ to $X, Y = 3, 2$). Our final artificial genome will be divided into three main parts: the *operative genome* (OG), the *ribosomic genome* (RG), and the *polymerase genome* (PG). Let us call operative genome (OG) a program containing all the genes of an artificial organism, where each gene (A to F) is a sub-program characterized by a set of instructions and by the cell's position (coordinates $X, Y = 1, 1$ to $X, Y = 3, 2$). Figure 1 is then a graphical representation of organism ORG's operative genome.

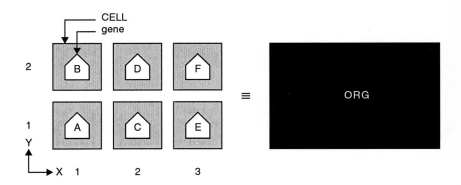

Fig. 1. Multicellular organization of a 6-cell organism ORG.

Let then each cell contain the entire operative genome OG (Figure 2a): depending on its position in the array, i.e. its place within the organism, each cell can then interpret the operative genome and extract and execute the gene which defines its function. In summary, storing the whole operative genome in each cell makes the cell universal: given the proper coordinates, it can execute any one of the genes of the operative genome and thus implement *cellular differentiation*. In our artificial organism, any cell $CELL[X,Y]$ continuously computes its coordinate X by incrementing the coordinate WX of its neighbor immediately to the west (Figure 2b). Likewise, it continuously computes its coordinate Y by incrementing the coordinate SY of its neighbor immediately to the south. Taking into consideration these computations, Figure 3 shows the final operative genome OG of the organism ORG.

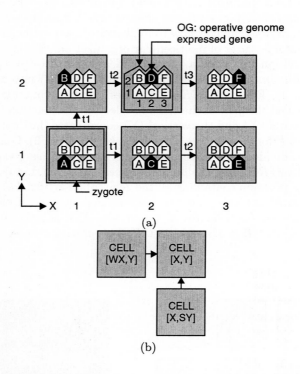

Fig. 2. Cellular differentiation and division. (a) Global organization. (b) Central cell $CELL[X,Y]$ with its west neighbor $CELL[WX,Y]$ and its south neighbor $CELL[X,SY]$; $X = WX + 1$; $Y = SY + 1$; $t1...t3$: three cellular divisions.

At startup, the first cell or *zygote* (Figure 2a), arbitrarily defined as having the coordinates $X,Y = 1,1$, holds the one and only copy of the operative genome OG. After time $t1$, the genome of the zygote (*mother* cell) is copied into the neighboring (*daughter*) cells to the east ($CELL[2,1]$) and to the north ($CELL[1,2]$). This process of *cellular division* continues until the six cells of the organism ORG are completely programmed (in our example, the farthest cell is programmed after time $t3$).

```
OG: operative genome
X = WX+1
Y = SY+1
case of X,Y:
  X,Y = 1,1: do gene A
  X,Y = 1,2: do gene B
  X,Y = 2,1: do gene C
  X,Y = 2,2: do gene D
  X,Y = 3,1: do gene E
  X,Y = 3,2: do gene F
```

Fig. 3. The operative genome OG of the organism ORG.

2.3 The Organism's Properties: Organismic Self-Replication and Organismic Self-Repair

The *self-replication* or *cloning of the organism*, i.e. the production of an exact copy of the original, rests on two assumptions.

1. There exists a sufficient number of spare cells in the array (at least six in the example of Figure 4) to contain the additional organism.
2. The calculation of the coordinates produces a cycle ($X = 1 \rightarrow 2 \rightarrow 3 \rightarrow 1...$ and $Y = 1 \rightarrow 2 \rightarrow 1...$ in Figure 4, implying $X = (WX + 1)$ modulo 3 and $Y = (SY + 1)$ modulo 2).

As the same pattern of coordinates produces the same pattern of genes, self-replication can be easily accomplished if the program of the operative genome OG, associated with the homogeneous array of cells, produces several occurrences of the basic pattern of coordinates. In our example (Figure 4), the repetition of the vertical coordinate pattern ($Y = 1 \rightarrow 2 \rightarrow 1 \rightarrow 2$) in a sufficiently large array of cells produces one copy, the *daughter organism*, of the original *mother organism*. Given a sufficiently large space, the self-replication process can be repeated for any number of specimens in the X and/or the Y axes.

In order to implement the *self-repair of the organism*, we decided to use spare cells to the right of the original organism (Figure 5). The existence of a fault is detected by a $KILL$ signal which is calculated in each cell by a built-in self-test mechanism realized at the molecular level (see Subsection 2.4 below). The state $KILL = 1$ identifies the faulty cell, and the entire column to which the faulty cell belongs is considered faulty and is deactivated (column $X = 2$ in Figure 5). All the functions (X coordinate and gene) of the cells to the right of the column $X = 1$ are shifted by one column to the right. Obviously, this process requires as many spare columns to the right of the array as there are faulty cells or columns to repair (two spare columns, tolerating two successive faulty cells, in the example of Figure 5). It also implies that the cell needs to be able to bypass the faulty column and to divert to the right all the required signals (such as the operative genome and the X coordinate, as well as the data busses).

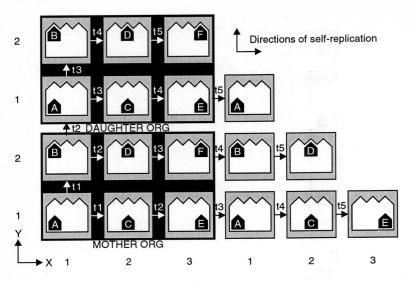

Fig. 4. Self-replication of a 6-cell organism ORG in a limited homogeneous array of 6×4 cells (situation at time $t5$ after 5 cellular divisions); $MOTHER\ ORG$ = mother organism; $DAUGHTER\ ORG$ = daughter organism.

Given a sufficient number of cells, it is obviously possible to combine self-repair in the X direction, and self-replication in both the X and Y directions.

2.4 The Cell's Features: Multimolecular Organization, Molecular Configuration, and Molecular Fault Detection

In each cell of every living being, the genome is translated sequentially by a chemical processor, the *ribosome*, to create the proteins needed for the organism's survival. The ribosome itself consists of molecules, whose description is an important part of the genome.

As mentioned, in the Embryonics project each cell is a small processor, sequentially executing the instructions of a first part of the artificial genome, the operative genome OG. The need to realize organisms of varying degrees of complexity has led us to design an artificial cell characterized by a flexible architecture, that is, itself configurable. It will therefore be implemented using a new kind of fine-grained, field-programmable gate array (FPGA).

Each element of this FPGA (consisting essentially of a multiplexer associated with a programmable connection network) is then equivalent to a *molecule*, and an appropriate number of these artificial molecules allows us to realize application-specific processors. We will call *multimolecular organization* the use of many molecules to realize one cell. The configuration string of the FPGA (that is, the information required to assign the logic function of each molecule) constitutes the second part of our artificial genome: the *ribosomic genome RG*. Fig-

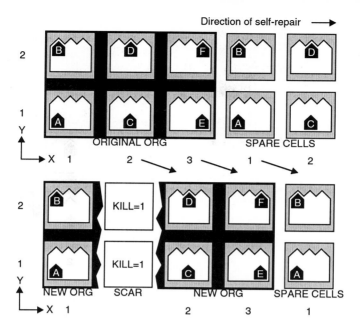

Fig. 5. Organismic self-repair.

ure 6a shows a generic and abstract example of an extremely simple cell ($CELL$) consisting of six molecules, each defined by a *molecular code* or $MOLCODE$ (*a* to *f*). The set of these six $MOLCODE$s constitutes the ribosomic genome RG of the cell.

The information contained in the ribosomic genome RG thus defines the logic function of each molecule by assigning a molecular code $MOLCODE$ to it. To obtain a functional cell, we require two additional pieces of information.

1. The *physical* position of each molecule in the cellular space.
2. The presence of one or more *spare columns*, composed of *spare molecules*, required for the self-repair described below (Subsection 2.5).

The definition of these pieces of information is the molecular configuration (Figure 6b). Their injection into the FPGA will allow:

— the creation of a border surrounding the molecules of a given cell;
— the insertion of one or more spare columns;
— the definition of the connections between the molecules, required for the propagation of the ribosomic genome RG.

The information needed for the molecular configuration (essentially, the height and width of the cell in number of molecules and the position of the spare

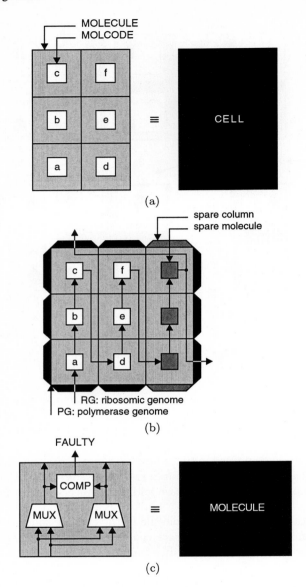

Fig. 6. The cell's features. (a) Multimolecular organization; RG: ribosomic genome: a, b, c, d, e, f. (b) Molecular configuration; PG: polymerase genome: height × width = 3 × 3; 001 = spare column. (c) Molecular fault detection; MUX: multiplexer; $COMP$: comparator.

columns) makes up the third and last part of our artificial genome: the *polymerase genome PG* [2].

Finally, it is imperative to be able to automatically detect the presence of faults at the molecular level and to relay this information to the cellular level. Moreover, if we consider that the death of a column of cells is quite expensive in terms of wasted resources, the ability to repair at least some of these faults at the molecular level (that is, without invoking the organismic self-repair mechanism) becomes highly desirable. The biological inspiration for this process derives from the DNA's double helix, the physical support of natural genomes, which provides complete redundancy of the genomic information though the presence of complementary bases in the opposing branches of the helix. By duplicating the material of each molecule (essentially the multiplexer MUX) and by continuously comparing the signals produced by each of the two copies (Figure 6c), it is possible to detect a faulty molecule and to generate a signal $FAULTY = 1$, realizing the *molecular fault detection* which will make possible cellular self-repair (described below in Subsection 2.5).

2.5 Cellular Self-Repair

The presence of spare columns, defined by the molecular configuration, and the automatic detection of faulty molecules (Subsection 2.4, Figure 6b and c) allow cellular self-repair: each faulty molecule is deactivated, isolated from the network, and replaced by a neighboring molecule, which will itself be replaced by a neighbor, and so on until a spare molecule is reached (Figure 7).

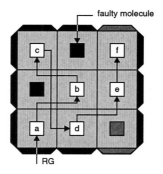

Fig. 7. Cellular self-repair.

The number of faulty molecules handled by the molecular self-repair mechanism is necessarily limited: in the example of Figure 7, we tolerate at most one faulty molecule per row. If more than one molecule is faulty in one or more rows, molecular self-repair is impossible, in which case a global signal $KILL = 1$ is generated to activate the organismic self-repair described above (Subsection 2.3 and Figure 5).

2.6 The Embryonics Landscape

The final architecture of the Embryonic project is based on four hierarchical levels of organization which, described from the bottom up, are the following (Figure 8).

1. The basic primitive of our system is the *molecule*, the element of our new FPGA, consisting essentially of a multiplexer associated with a programmable connection network. The multiplexer is duplicated to allow the detection of faults. The logic function of each molecule is defined by its molecular code or *MOLCODE*.
2. A finite set of molecules makes up a *cell*, essentially a processor with the associated memory. In a first programming step of the FPGA, the polymerase genome *PG* defines the topology of the cell, that is, its width, height, and the presence and positions of columns of spare molecules. In a second step, the ribosomic genome *RG* defines the logic function of each molecule by assigning its molecular code or *MOLCODE*.
3. A finite set of cells makes up an *organism*, an application-specific multiprocessor system. In a third and last programming step, the operative genome *OG* is copied into the memory of each cell to define the particular application executed by the organism (electronic watch, random number generator, and a Turing machine being examples we have shown to date) [2].
4. The organism can itself self-replicate, giving rise to a *population* of identical organisms, the highest level of our hierarchy.

3 Conclusion

3.1 Toward a New Reconfigurable Computing Tissue: The E–wall

Keeping in mind that our final objective is the development of very large scale integrated (VLSI) circuits capable of self-repair and self-replication, as a first step, which is the subject of this paper, we have shown that a hierarchical organization based on four levels (molecule, cell, organism, population of organisms) allows us to confront the complexity of real systems. The realization of demonstration modules at the cellular level and at the molecular level [2] demonstrates that our approach can satisfy the requirements of highly diverse artificial organisms and attain the two sought-after properties of self-repair and self-replication.

The programmable robustness of our system depends on a redundancy (spare molecules and cells) which is itself programmable. This feature is one of the main original contributions of the Embryonics project. It becomes thus possible to program (or re-program) a greater number of spare molecules and spare cells for operation in hostile environments (e.g., space exploration). A detailed mathematical analysis of the reliability of our systems is currently under way at the University of York [5], [6].

In our laboratory, the next major step in the Embryonics project is the design of the BioWatch, a complex machine which we hope to present on the occasion

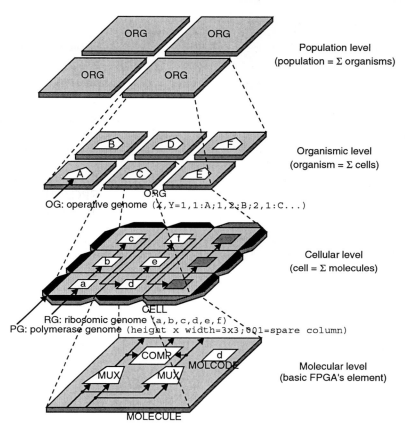

Fig. 8. The Embryonics landscape: a 4-level hierarchy.

of a cultural event which will soon take place in Switzerland. The function of the machine will be that of a self-repairing watch, counting seconds, minutes and hours. The implementation of the watch will take place in a *reconfigurable tissue*, the *electronic wall* or *e–wall* (Figure 9).

Conception, birth, growth, maturity, illness, old age, death: this is the life cycle of living beings. The proposed demonstration will stage the life cycle of the BioWatch from conception to death. Visitors will face a large wall made up of a mosaic of many thousands of transparent electronic modules, each containing a display. At rest, all the modules will be dark. A complex set of signals will then start to propagate through the space (conception) and program the modules to realize the construction of a beating electronic watch (growth). Visitors will then be invited to attempt to disable the watch: on each molecule, a push-button will allow the insertion of a fault within the module (wounding). The watch will automatically repair after each aggression (cicatrization). When the number of

Fig. 9. The BioWatch built in a new reconfigurable tissue: the e–wall (Computer graphic by E. Petraglio).

faults exceeds a critical value, the watch dies, the wall plunges once more into darkness, and the complete life cycle begins anew.

We are physically implementing the 6-digit BioWatch in our two-dimensional e–wall made up of 3200 modules. This wall is ultimately intended as a universal self-repairing and self-replicating display medium, capable of interacting "intelligently" with the user and recovering from faults. In our implementation, each molecule of the watch corresponds to a module in the wall. This module includes: (1) an input device, (2) a digital circuit, and (3) an output display [3].

The module's outer surface consists of a touch-sensitive panel which acts like a digital switch, enabling the user to kill a molecule and thereby activate the self-repair process at the molecular level (signal $FAULTY = 1$ in Figure 6c).

The module's internal circuit is a field-programmable gate array (FPGA), configured so as to realize: (1) the acknowledgment of the external (touch) input, (2) the implementation of a molecule (multiplexer with a programmable connection network), and (3) the control of the output display. This latter is a two color light-emitting diode (LED) display, made up of 64 diodes arranged as an 8×8 dot-matrix. The display allows the user to view the current molecule's state: active, faulty, spare, etc.

The configuration string of the FPGA can be considered as an ultimate part of our artificial genome, necessitated by the construction of the molecular level (Figure 8) on the shoulders of a new, lowest level: the atomic level.

3.2 Toward POEtic Machines

We presented the POE model, classifying bio-inspired hardware systems along three axes (phylogeny, ontogeny, and epigenesis), and we described our Embryonics project along the ontogenetic axis. A natural extension which suggests itself is the combination of two and ultimately all three axes, in order to attain novel bio-inspired hardware. An example of the latter would be an artificial neural network (epigenetic axis), implemented on a self-replicating and self-repairing multicellular automaton (ontogenetic axis), whose genome is subject to evolution (phylogenetic axis).

Looking (and dreaming) toward the future, one can imagine nano-scale systems becoming a reality, endowed with evolutionary, reproductive, regenerative, and learning capabilities. Such systems could give rise to novel species which will coexist alongside carbon-based organisms. This constitutes, perhaps, our ultimate challenge.

Acknowledgments

This work was supported in part by the Swiss National Foundation under grant 21-54113.98, by the Leenaards Foundation, Lausanne, Switzerland, and by the Villa Reuge, Ste-Croix, Switzerland.

References

1. D. Mange, M. Sipper, and P. Marchal. Embryonic electronics. *BioSystems*, 51(3):145–152, 1999.
2. D. Mange, M. Sipper, A. Stauffer, and G. Tempesti. Toward robust integrated circuits: The embryonics approach. *Proceedings of the IEEE*, 88(4):516–541, April 2000.
3. D. Mange, A. Stauffer, G. Tempesti, and C. Teuscher. Tissu électronique recongigurable, homogène, modulaire, infiniment extensible, à affichage électro-optique et organes d'entrée, commandé par des dispositifs logiques reprogrammables distribués. Patent pending, 2001.
4. D. Mange and M. Tomassini, editors. *Bio-Inspired Computing Machines*. Presses polytechniques et universitaires romandes, Lausanne, 1998.
5. C. Ortega and A. Tyrrell. Reliability analysis in self-repairing embryonic systems. In A. Stoica, D. Keymeulen, and J. Lohn, editors, *Proceedings of The First NASA/DOD Workshop on Evolvable Hardware*, pages 120–128, Pasadena, CA, 1999. IEEE Computer Society.
6. C. Ortega and A. Tyrrell. Self-repairing multicellular hardware: A reliability analysis. In D. Floreano, J.-D. Nicoud, and F. Mondada, editors, *Advances in Artificial Life*, Lecture Notes in Artificial Intelligence. Springer-Verlag, Berlin, 1999.
7. M. Sipper, E. Sanchez, D. Mange, M. Tomassini, A. Pérez-Uribe, and A. Stauffer. A phylogenetic, ontogenetic, and epigenetic view of bio-inspired hardware systems. *IEEE Transactions on Evolutionary Computation*, 1(1):83–97, April 1997.
8. L. Wolpert. *The Triumph of the Embryo*. Oxford University Press, New York, 1991.

Design and Codesign of Neuro-Fuzzy Hardware

L.M. Reyneri

Dipartimento di Elettronica, Politecnico di Torino
C.so Duca degli Abruzzi 24, 10129 TORINO, ITALY `reyneri@polito.it`

Abstract. This paper presents an annotated overview of existing hardware implementations of artificial neural and fuzzy systems and points out limitations, advantages and drawbacks of analog, digital, pulse stream (spiking) and other techniques. The paper also analyzes hardware performance parameters and tradeoffs, and the bottlenecks intrinsic in several implementation methodologies.

1 Introduction

In the early times of artificial neural networks and fuzzy systems, several researchers tried to exploit some of the intrinsic characteristics of such systems (like regularity, fault tolerance and adaptivity) to design efficient *hardware (***HW***) implementations* [6]-[7], which were thought to outperform other existing computing equipment in implementing neural and fuzzy systems. This, in conjunction with the powerful capabilities of neural and fuzzy systems as universal function approximators, system models and controllers, etc., gave a strong impulse to the theoretical and practical studies on *soft computing* and their HW implementations, at least until half a decade ago.

More recently, due to the ever increasing speed of general-purpose processors (like PC's and workstations), the performance increase which can be obtained from HW implementations has reduced so much that, at present, such implementations are seldom cost- (or time- or performance-) effective in comparison with commercial general-purpose processors.

A question arises naturally from those considerations, namely: *is there still a reason to design new neural or fuzzy HW devices?* and, if the answer is yes, *when an ad-hoc HW device is more effective than other commercial devices?*.

I will show that there are just *a few specific application areas* in which ad-hoc HW implementations can still offer significant benefits at a much lower cost than other off-the-shelf techniques, namely those which require either *extremely high speed* (for instance in collision detectors used by high-energy physics experiments), or *very low power consumption* (like in battery operated equipment, for instance in human pace-makers), or *increased fault tolerance* (for instance, in space applications, or to increase production yield of silicon chips), or as *building blocks* in the so-called *embedded systems* (for instance in most consumer applications) and within *systems-on-a-chip* (that is, when all parts of a system, including neuro-fuzzy, are implemented on the same chip).

Scope of this paper is to outline a set of performance parameters which are relevant to the industrial and consumer domains, which proved useful to compare implementation choices among each other and to assess if and when ad-hoc designs is more effective than existing or commercial ones.

1.1 Neuro-Fuzzy Unification and Hardware Implications

Until a few years ago, neural networks, fuzzy systems and wavelet networks were considered very different and incompatible paradigms. Manufacturers were producing interesting pieces of HW (either single devices and/or boards) targeted specifically to one predefined paradigm (usually either neural or fuzzy), selected apriori according to manufacturer's taste or experience or to market requirements. Neural networks (respectively, fuzzy systems) had to be run on dedicated neural (respectively, fuzzy) HW, while a fuzzy system could not be run on a neural HW and vice-versa. Furthermore, wavelet networks had no dedicated HW associated with them, therefore they could only be implemented as *software* (**SW**) on general-purpose computers at a limited speed.

More recently, neural networks, fuzzy systems, wavelet networks and a few other soft computing paradigms have been unified together [1] and can the at present be considered either as the different sides of the same "medal", or as different *languages* to express the same network. All neural, fuzzy, wavelet, Bayesian, linear, etc. networks can therefore be classified according to their *type* and their *order* n, as discussed in [1]. Follows a short list of most commonly used neuro-fuzzy (**NF**) systems, while other values of n and m are possible, although less frequent (note that NF systems were originally called WRBF's):

- NF-0 (namely, computing layers of order $n = 0$): perceptrons, linear systems, fuzzy I-OR implicators, Fourier and other transforms, etc.;
- NF-1 (namely, computing layers of order $n = 1$): some product fuzzy AND implicators, Reduced Coulomb Energy networks, sometimes Kohonen maps, etc.
- NF-2 (namely, computing layers of order $n = 2$): other product fuzzy AND implicators, wavelet networks, wavelet transforms, fuzzy partitions, Kohonen maps, Bayesian classifiers, etc.;
- NF-∞ (namely, computing layers of order $n = +\infty$): mini fuzzy AND implicators, Winner-takes-all, etc.;
- NORM-1/1 (namely, normalization layers of orders $n = m = 1$): input, output and inter-layer normalization, part of center-of-gravity defuzzifier, part of Bayesian classifiers, etc.;
- NORM-1/2 (namely, normalization layers of orders $n = 1$, $m = 2$): input normalization, training rule for fuzzy partitions;
- NORM-n/1 (namely, normalization layers of orders $n > 1$, $m = 1$): unsupervised training rules;
- NORM-($\pm\infty / \pm \infty$) (namely, normalization layers of orders $n = m = \pm\infty$): winner-takes-all, looser-takes-all, crisp partitions, clustering algorithms, etc.;

In the field of HW implementations, neuro-fuzzy (**NF**) unification allows to mix together the various paradigms within the same project or system and to implement all of them on similar pieces of HW (or even on a single one).

This augments significantly flexibility and re-usability of HW devices and systems, and reduces the users' "training" costs, as users need to learn only one piece of HW on which several different paradigms can be implemented, leaving the user the choice of the preferred (or the optimal) paradigm(s) to use. Conversion among the different paradigms can be made either manually or automatically in a very straightforward way [1].

1.2 Fuzzy vs. Neural Descriptions

Although it has been proven that fuzzy, neural and other soft computing paradigms can be unified into a unique paradigm, there is still a difference between the two, as regards HW optimization.

Although there is a one-to-one correspondence between (fuzzy) membership and (neural) activation functions and between fuzzy inference and neural distance computation [1], the major difference between neural (and wavelet, Bayesian, etc.) and fuzzy paradigms is that the former mostly use *additive synapses* on the first layer and weights are described numerically, while the latter mostly use *multiplicative synapses* and weights are described linguistically.

The type of description (numeric vs. linguistic) is immaterial as regards HW implementations, as an automatic conversion is available. The total number of computations is therefore different. For instance, for the first layer of a Gaussian $N \times M$ NF-2 system, where centers are on a lattice of N_C points (fuzzy sets) per input (similar considerations apply to any NF-n system):

- a fuzzy-based implementation requires computing $N \cdot N_C$ *differences, multiplications* and *non-linear functions*, plus up to $N \cdot M$ *multiplications*;
- a neural-based implementation requires computing $N \cdot M$ *differences, multiplications* and *summations*, plus M *non-linear functions*,

There are a few cases:

- *irregular networks*, where center vectors are not on a regular lattice (namely, they are on a lattice of as many point as rules, that is $N_C = M$); in this case the neural-based implementation is the simplest (independently of the language used, as one-to-one conversions are available);
- *regular networks*, where center vectors are on a regular lattice with $N_C \ll M$ points; in this case the fuzzy-based implementation is the simplest (independently of the language used, as one-to-one conversions are available);
- *irregular networks*, where center vectors are on a regular lattice with $N_C < M$ points; in this case the detailed cost of multipliers and non-linear activation functions affects the selection.

2 Taxonomy of Implementation Platforms

NF systems can be implemented in several ways. The most common ones are described in next sections. We exclude SW implementations (like DSP's, PC's, etc.), which are out of the scope of this work.

2.1 Platforms for Dedicated HW Implementations

Dedicated HW implementations are built on the following "platforms":

– *Application-specific Integrated Circuits* (**ASIC**'s) are silicon devices fully designed ad-hoc for the specific application (or class of applications), either analog or digital or pulse stream (see sect. 2.2). They include a wide variety of different solutions, which range from small to medium size networks (typically less than 10,000 synapses, sometimes up to 100,000, and offer a wide range of performance, ranging from very high speed to very low power and high performance/cost ratios (only for large productions).
ASIC's are very flexible in the design stage (as they can implement any desired function) but, once designed and manufactured, they are by far much less flexible than SW implementations, as they cannot be modified any more. Unfortunately, the design phase of a new device is often the most expensive part of a project (1-20 man-months, plus more than 100,000US\$ for masks), therefore ASIC's are justified only for large quantities or in applications where very high performance are required (see fig. 4 and fig. 5). This is the major drawback which has limited the spreading of ASIC implementations. Yet, ASIC's are coming again into favor for state-of-the-art SoC's, thanks to the availability of very powerful CAD tools, which significantly reduce design effort, risks and time-to-market. The cost of masks still has to be paid completely (or partially, for *gate-array* ASIC's).
– *Field Programmable Gate Arrays* (**FPGA**'s) are digital programmable devices, which can be configured (programmed) to implement virtually any digital system, therefore any digital NF network. An appropriate *programmer* (often very simple) is required. At present no commercial analog FPGA exists, although some prototype is currently under development.
FPGA's are as flexible as ASIC's during the design phase, but they can change their functions at any time during lifetime or operation by means of an appropriate programming sequence stored into their internal memory (either RAM or PROM).
FPGA's can be designed by means of the same tools and languages used to design ASIC's, but their programmability reduces significantly NRE (see sect. 4.5) manufacturing costs and delays, at the expense of higher RE costs, therefore they are preferred for small and medium quantities (see fig. 4 and fig. 5). Furthermore FPGA's cannot reach the ultimate performance of ASIC's, which are then preferable either for large quantities (e.g. $> 10,000$ parts/year), or for very high performance (e.g. $> 200\,\mathrm{MHz}$), or for analog and pulse stream systems. Some devices also incorporate up to a few hundreds Kb of memory, for weight and center storage.

- *HW/SW hybrid systems*, which are made of the tight interconnection of SW and digital (sometimes, analog) HW blocks. They can often offer the best performance, at the cost of an increased design effort, which can be reduced by means of HW/SW codesign techniques (see sect. 2.3, fig. 4 and fig. 5). They are often used as embedded systems in consumer applications, or within high-performance or very low power systems. They can also be implemented as SoC's (for instance, containing both micro-mechanical parts and microprocessor cores and analog circuits and high power devices).

At present there are only a few general-purpose commercial platforms for HW/SW hybrid systems, either at board level (from SIDSA, Sundance) or at chip level (from Triscend), therefore commercial applications still require ad-hoc design of the board/chip. Design effort is likely to reduce in the near future due to the increasing complexity of development tools and the spread of such systems.

2.2 Technologies for Dedicated HW Implementations

Dedicated HW systems can be designed using different technologies [2–4]:

- *analog*, mostly in either *strong inversion* [10] of *weak inversion* [4]: synaptic multipliers are often nothing but the input voltage applied to a synaptic weight's variable resistor (*transconductance multiplier*, or similar circuits). Such a circuit has either linear (strong inversion) or exponential (weak inversion) characteristic and requires only one MOS transistor, but is often difficult to design and operate.

  Summation of synaptic contributions is easy and intrinsically linear, as synaptic contributions are nearly always associated with currents, which sum up together whenever injected into the same node.

  HW-friendly non-linear activation functions are also very easy to implement, as a resistor across a simple 2-MOS amplifiers has a roughly sigmoidal transfer function between input current and output voltage.

- *bioinspired*, or *neuromorphic* [4, 11, 12, 7] is the state-of-the-art in the field of bioinspired and pulse stream implementations, as it overcomes most of their drawbacks and takes inspiration directly from biology. In practice, analog systems have characteristics similar to natural brain neurons, namely high simplicity, very small accuracy and stochasticity.

  Bioinspiration (sometimes called *opportunism*) spins off the similarities between characteristics of individual semiconductor devices and biological processes underlying biological neurons. In practice, bioinspiration goes back to the early times of artificial neural systems, when the biological neuron was the reference for every new development, but takes advantage from the latest results in neurobiology.

  Bioinspiration exploits all the benefits of analog systems (reduced size and power consumption), without being significantly affected by their limited accuracy and repeatability. Real neurons do have a very poor accuracy, even poorer than analog systems, yet they can provide very high computing power.

Several systems, like the *retina* [4, 12] and the *cochlea* [13] have been designed and manufactured, although they still have limited applications.
- *digital* [14], made of interconnected *logic gates, adders, multipliers, complex functions, look-up tables*. Digital circuits are nearly insensitive to electric noise, they can have virtually unlimited accuracy, although they require a much larger number of MOS transistors than analog circuits.
- *pulse stream* [16], which is an intermediate solution between digital and analog. They are made of a mixture of analog and digital circuits, trying to exploit the advantages of each of them [16]: digital circuits are better at transmitting signals and are less sensitive to electric noise, while analog circuits are much smaller and consume less power, but analog multipliers are often far from being ideal. It is beneficial to combine digital for signal transmission with analog for summation and hybrid techniques for multiplication. Pulse streams include (see [16] for definitions): *pulse rate modulations* (**PRM**) [14], *pulse width modulation* (**PWM**), *coherent pulse width modulation* (**CPWM**) [15], *stochastic pulse modulation* (**SPM**), *pulse delay modulations* (**PDM**) [11].

One of the major, and currently unsolved, problems of analog systems is the *long term weight storage*, beacause of *weight decay*, which completely destroys the stored value after a very short period, usually between 10 ms and 10 s (volatile weights). Several solution have been proposed to overcome that problem:
- using *fixed weights*, which are set by changing the size of transistors, resistors and capacitors. Due to the large mismatches in analog devices, fixed weights are used only in bioinspired systems, which can cope with that strong limitation.
- a special technology (*EEPROM*) can be used, where an appropriate device is used, instead of the MOS transistor to load the capacitor. Such switch does not suffer from leakage when open. The drawback is, at present, the higher cost of the technology and the limited lifetime of such devices.
- the charge on the capacitor can be periodically *refreshed* by an appropriate additional circuit. A few more transistors per synapse are enough to refresh the charge. The drawback is that weight values must be discretized in not more than 64 levels. This may sometimes create problems during on-chip training, as algorithms like back-propagation require at least 16,000 levels to operate properly [18].
- weight values can be stored on a long-term external digital memory (8-16 bits per synapse) and periodically converted to analog and transferred to the chip weight capacitor. This technique is not suited to on-chip learning, as training must modify the weights in the external memory.
- the neuro-fuzzy network is *continuously trained* by new incoming patterns, although continuous training is a very dangerous and difficult process.

2.3 Design Flows and Tools

ASIC's and FPGA's can be designed using a number of *design flows* and *tools* [2, 5], each one with different capabilities, performance, cost and time-to-market:

- *Full-Custom Approach*: is only available with ASIC's and it allows to design any circuit and optimize the size of any single transistor. It is the most commonly used approach for analog and pulse stream circuits (for which nearly no other technique exists), while it is seldom used for digital circuits, as it has a much longer time-to-market than any other tool.
- *Semi-Custom Approaches*, namely *standard-cells* and *gate arrays*; both allow to design digital circuits at the level of logic gates. It is was the most widely used method for digital circuits until a few years ago; it is still quite often used for simple circuits.
- *Silicon compilation*, which describes a digital circuit at the functional level, by means of a C-like language (usually, either *VHDL* or *Verilog*), by mans of which one can write algorithms in a rather straightforward way. Silicon compilation often makes strong use of so-called *Intellectual Properties* (**IP**'s), which are predefined, ready-to-use, complex functions.
- *Hardware/Software codesign* has recently become a valuable and efficient design method for hybrid systems [5]. Hardware/software codesign is a semiautomatic design technique to conceive, describe, design and optimize a hybrid (HW/SW) system in an uniform way, without bothering what to implement in HW and what in SW. Semi-automatic tools then allow to partition the systems into HW and SW (and in the near future into digital and analog) and to automatically design the various parts of the system.

The systems is described by means of an appropriate language (for instance, either Simulink or C or Esterel), from which the user can generate either a piece of SW code (for either a SW emulator or a DSP or an NSP) or the description of a digital system (either an ASIC or an FPGA), or any combination of both, namely the best mixture to achieve either a smaller cost, or a smaller size, or a smaller power consumption.

As HW and SW both have advantages and drawbacks, it can often be convenient to mix the two techniques together into *hybrid systems* (for instance, SoC's). Typical applications are the *embedded systems*, which are more and more pervasive in everyday life. The strength of embedded systems comes from the flexibility of SW-programmed processors augmented by the higher performance (speed and power consumption) of HW.

Unfortunately, hybrid systems are more complex to design, therefore most existing embedded systems are basically only made of SW. As long as SW offers enough performance (both speed, power and reliability), that is surely the simplest solution for any system, although it may impact cost and size. But whenever SW cannot offer the desired performance, as it often happens with neural networks (which require a much higher computational effort than many other systems), the integration of HW and SW becomes very attractive.

Until recently the design of HW/SW hybrid systems required a preliminary hand-made *partition* between HW and SW, then the design of each part independently, using any HW description language (for instance, either *VHDL* or *Verilog*). As the partition was done manually, it was based on personal experience and not on objective cost functions. A good design practice would have been to conceive several partitions, to design HW and SW parts independently, and to estimate and compare the performance of all of them. Unfortunately proper design of each partition could take several weeks or months, resulting into unaffordable design time and cost.

The advantage of HW/SW codesign is that it allows to easily build systems made of one ore more NF blocks, possibly interconnected with any other function, like integrators, derivatives, delays, memories, encoders, motor drivers, user interfaces, etc.), as required in most practical applications. I feel that HW/SW codesign is going to grow significantly, for the design of digital NF systems and, within a few years it will be used to develop most dedicated digital implementations.

A Simulink-based HW/SW codesign tool tailored specifically to NF systems is described in [5].

3 HW-Specific Design Considerations

3.1 Parallelism and Virtualization

The limitations of most general-purpose processors (like PC's, DSP's, μC's, etc.) is that they have a *sequential* processing unit, that is, only one computation can be done at a time. Therefore, $N \times M$ computing layer requires sequentially processing $N \cdot M$ computing synapses plus M neuron kernels.

On the other hand, one of the key points of most dedicated HW implementations is the high degree of parallelism available, which offers improved speed, performance and fault-tolerance (see sect. 4.1, and sect. 5). It might be said that there are very few advantages in designing a HW implementation unless it is massively parallel (see sect. 5).

Having $Q > 1$ *processing elements* (**PE**'s), either *analog* or *digital* or *optical*, increases speed and reduces computation time by up to a factor Q. In practice, the factor Q can be achieved only for very "regular" computations such as those required by NF algorithms, under the assumption that $N \cdot M \geq Q$. There is a value Q for each of the three types of PE, namely Q_S *computing synapses*, Q_F *neuron kernels* and Q_N *normalization layers*.

The highest benefits are obtained by distributing elementary operations onto a regular *lattice* of $Q_i \geq 1$ processors, preferably with the number of operations (namely, $N \cdot M$ for synapses and M for neuron kernels and normalization synapses) equal to or an integer multiple of Q_i. I define three *virtualization factors*

$$\zeta_S \triangleq \text{ceil}\left(\frac{N \cdot M}{Q_S}\right) \quad \text{and} \quad \zeta_F \triangleq \text{ceil}\left(\frac{M}{Q_F}\right) \quad \text{and} \quad \zeta_N \triangleq \text{ceil}\left(\frac{M}{Q_N}\right) \quad (1)$$

respectively, for computing synapses, neuron kernels and normalizing layers. There are a few relevant cases for synapses:

- $\zeta_S = 1$, namely one PE per synapse. These are called *fully parallel* systems, which compute the whole synaptic array in just one *cycle time* T_C, which is the time required to compute one elementary operation. Cycle time is either C_S clock cycles, for digital, or the *settling time* $\frac{k}{BW}$ for analog (where $k \geq 1$ is function of the desired accuracy, while BW is the analog bandwidth of the system), or the *response time* [16], for pulse stream implementations.

- $\zeta_S = N$, namely one PE per neuron. Each PE sequentially computes (in N consecutive cycles) the contributions of all synapses of the associated neuron. This is also called *per-neuron virtualization*.
- $\zeta_S \gg N, M$: a relatively small number of PE's which sequentially compute (in ζ_S consecutive cycles) all synaptic contributions.
- $\zeta_S = N \cdot M$, namely only one *sequential processor*. In practice, if the processor is a commercial general-purpose processor (like PC's, DSP's, μC's, etc.), this approach is merely a SW implementation.

Note that the sequential approach for activation functions ($\zeta_F = M$) is often associated either with fully sequential synapses ($\zeta_S = N \cdot M$, often SW approach) or with partially parallel synapses (e.g. $\zeta_S = M$, HW approach). I will show in sect. 4.1 that the virtualization factor of synapses should be balanced as much as possible with that of neuron kernels, to improve performance/cost ratio.

4 Performance Parameters

4.1 Sampling Time and Speed

Sampling period T_S and *system bandwidth* BW are relevant when processing time-sampled and time-continuous signals, respectively:

$$T_S \geq \begin{cases} \max_i \{\frac{k\zeta_i}{BW_i}\} & \text{for analog (usually, } \zeta_i = 1) \\ \max_i \{T_{M,i}\zeta_i\} & \text{for pulse stream (usually, } \zeta_i = 1) \\ \max_i \{T_i\zeta_i\} = \frac{\max_i \{C_i\zeta_i\}}{F_{CK}} & \text{for digital} \end{cases} \quad (2)$$

where T_i, C_i, $T_{M,i}$, ζ_i and BW_i are, respectively, *delay time* (for digital), *clock cycles per operation* (for digital), *response time* (for pulse stream [16]), *virtualization factor* (for any system) and *bandwidth* (for analog) of ith NF block (either synapses or neurons or normalization), while F_{CK} is clock frequency (for digital) and $k \geq 1$ is function of the desired accuracy (for analog).

Figure 1 compares sampling rate $\frac{1}{T_S}$ of typical analog, pulse stream and digital implementations (which depends on ζ_i ; see sect. 4.3).

As a consequence of the above formula, there is no reason to have any individual ζ_i too low (it would result into larger size); one good design goal is to *balance* virtualization factors of each block (where this is meaningful, namely mostly in digital systems):

$$C_i\zeta_i \approx \text{const.} \quad \forall i \Rightarrow \frac{\zeta_S}{\zeta_F} \approx \frac{C_F}{C_S} \quad (3)$$

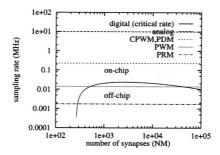

Fig. 1. Sampling rate, for typical analog and pulse stream [16] implementations, and critical sampling rate (see sect. 4.3), for digital implementations, versus number of synapses. Plot is computed for $l, l_{eq} = 6$ bits, $W_P = 0.2$ mm, $A_S = 0.45$ mm^2, $A_M = 615\,\mu$m^2, $A_P = 1,000\,\mu$m^2 (for analog), $A_P = 2,000\,\mu$m^2 (for pulse stream), $P_{AL} = 20$, $C_S = 1$, $F_{CK} = 20$ MHz, $\frac{BW}{k} = 10 MHz$.

from which the optimal number of PE's associated with the activation function is, from (1):

$$Q_F = \text{ceil}\left(\frac{C_F Q_S}{C_S N}\right)$$

As for SW implementations, speed is usually expressed in *connections per second* (**CPS**) and, when applicable, in *connection updates per second* (**CUPS**):

$$\text{CPS} \approx \frac{\sum_{i \in \text{syn}} O_i}{T_S} \qquad (4)$$

where "syn" is the set of synaptic blocks, while O_i is the *number of elementary operations* of ith NF block (either synapses or neurons or normalization).

4.2 Pin Count

Pin count is often a bottleneck of HW implementations as it significantly impacts on either size or cost or speed.

Systems with $\zeta_S = 1$ (mostly analog or pulse stream) usually have weights stored within each PE (*on-chip weight storage*), while other systems with $\zeta_S > 1$ (mostly digital, sometimes pulse stream) may have either *on-chip* or *off-chip weight storage*. With on-chip weight storage, chip pins only have to transfer input and output activations, while with off-chip weight storage chip pins also have to transfer weights.

I will analyze in the following a simple case study with just one $N \times M$ NF-n computing layer, one weight per synapse, and one data transfer through pins per clock cycle (only for digital), therefore $C_i \zeta_i$ weights are transferred at each synaptic cycle (see sect. 3.1). I also suppose that inputs, outputs and weights (for off-chip storage) are virtualized as well, to optimize pin count.

Pin count is (including P_{AL} pins for power supply and control signals):

$$N_P \approx \begin{cases} P_{AL} + N + M & \text{for analog and pulse stream,} \\ & \zeta_S = 1, \text{ non-volatile weights} \\ P_{AL} + \dfrac{l(N+M)}{C_S \zeta_S} + \log_2(C_S \zeta_S) & \text{for digital, on-chip storage} \\ P_{AL} + \dfrac{l(N+M+NM)}{C_S \zeta_S} + \log_2(C_S \zeta_S) & \text{for digital, off-chip storage} \end{cases} \quad (5)$$

where N, M and l are the number of inputs and outputs and word length, respectively, while $\log_2(C_S \zeta_S)$ are the address bits associated with inputs, outputs and weights.

Total pin count N_P is often limited, for cost reasons, to $N_P \leq P_{max}$, which usually limits ζ_S with off-chip weight storage to (from (5)):

$$\zeta_S > \frac{l(N+M+NM)}{C_S(P_{max} - P_{AL})} \quad (6)$$

The virtualization factor of neurons (ζ_F) and normalization layers (ζ_N) shall be balanced accordingly, as from (3). This bottleneck limits chip speed, for off-chip weight storage (namely for large digital networks) to (from (2), (4), (6), by noting that $\sum_i(O_i) \approx NM < (N+M+NM)$), either:

$$\text{CPS} < \frac{F_{CK}(P_{max} - P_{AL})}{l} \quad \text{or} \quad \text{CPS} < \frac{F_{mem}}{l} \quad (7)$$

where F_{mem} is memory throughput (in bits per second), for dedicated HW and for sequential processors, respectively. That bound holds independently of chips size, number of synapses and the chosen architecture (namely, either bit-serial or systolic, or parallel, etc.),

This constraint is one of the major limitation of digital devices with off-chip weight storage, as any modern commercial processor has a memory throughput F_{mem} comparable with the throughput $F_{CK}(P_{max} - P_{AL})$ of any dedicated digital device (unless very large and expensive packages are used, with $P_{max} \gg 100$ pins). The presence of internal cache in commercial processors further improves the former throughput, making dedicated HW devices with off-chip weights less effective than commercial processors (in addition, they are less flexible and more costly to design).

Whenever chip size is due to the large number of pins, the chip is called *pin-bound* (mostly with off-chip weight storage), otherwise *core-bound*, when size is mainly due to the large logic core (mostly with on-chip storage). When none of those two terms are dominant, the chip is said *dense*.

4.3 Chip Size

Chip size is given by (for one $N \times M$ NF-n computing layer, no neuron, one weight per synapse, one data transfer through pins per clock cycle, neglecting size of pins themselves:

- for $\zeta_S = 1$ (mostly analog and pulse stream, on-chip weight storage), from (5), including weight storage;

$$A_{an} \approx \max\left\{A_P NM, \left(W_P \frac{N_P}{4}\right)^2\right\}$$

where A_P is the size of one PE, while W_P is the width of each pin.
- for on-chip weight storage (mostly core-bound chips), from (1) and (2), (4):

$$A_{on} \approx A_S Q_S + A_M l NM = A_S \cdot \text{ceil}\left(\frac{C_S \text{CPS}}{F_{CK}}\right) + A_M l NM \qquad (8)$$

where A_S and A_M are the size of one PE and one bit of weight storage, respectively. In practice, chip size cannot be smaller than $\left(W_P \frac{P_{AL}}{4}\right)^2$.
- for off-chip weight storage (mostly, pin-bound chips or SW implementations on DSP's), from (7):

$$A_{off} \approx \left(W_P \frac{P_{max}}{4}\right)^2 \approx \frac{W_P^2}{16}\left(\frac{l \cdot \text{CPS}}{F_{CK}} + P_{AL}\right)^2 \qquad (9)$$

One can then choose which one, between on-chip and off-chip storage, gives the smallest area (in practice, also other considerations should be taken into account). Chip size will therefore be:

$$A \approx \min\{A_{on}, A_{off}\} \qquad (10)$$

where the first (respectively, second) term dominates for on-chip (respectively, off-chip) weight storage, as said before. Figure 2 plots chip size versus number of synapses for different values of speed. If the desired number of synapses lays on the curvilinear (respectively, horizontal) line, the chip shall be core-bound (respectively, pin-bound) and shall have on-chip (respectively, off-chip) storage.

By equating the two terms in (10) one gets the *critical speed*, which is plotted in fig. 1, in terms of *critical sampling rate* $\frac{\text{CPS}}{NM}$. While analog and pulse stream circuits (namely, all those with $\zeta_i = 1$) do have a given sampling rate, which only depends on chip size (and circuit), digital circuits (namely, all those with $\zeta_i > 1$) can trade-off area versus speed by varying ζ_i. Above (respectively, on, below) the critical speed, the chip shall be core-bound (respectively, dense, pin-bound) and shall have on-chip (respectively, on-chip, off-chip) storage. Note that, as regards critical speed, SW implementations are equivalent to off-chip storage.

It is now clear that either small or very fast networks shall be core-bound and will have on-chip storage. Those are at present the only areas in which digital HW implementations are still effective, as for off-chip storage SW implementations on commercial processors definitely offer many more advantages, making HW implementations ineffective, as discussed further in sect. 5.

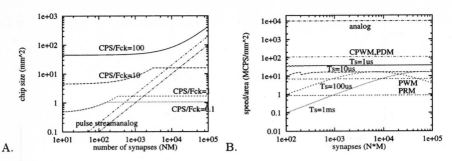

Fig. 2. A) Chip size versus number of synapses for increasing values of $\frac{CPS}{F_{CK}}$, for the same conditions of fig. 1.

Fig. 3. B) Speed/area ratio, versus number of synapses, for the same conditions of fig. 1. For digital implementations, speed also depends on the required sample period T_S (which is fixed for analog and pulse stream implementations).

4.4 Number of PE's

- For analog and pulse stream (namely, those with $\zeta_i = 1$), $Q_S = NM$.
- For digital core-bound chips (mostly on-chip weight storage), from (1), (2) and (4): $Q_S = \text{ceil}\left(\frac{C_S CPS}{F_{CK}}\right)$.
- For pin-bound chips (mostly with off-chip weight storage), by allowing all available chip size A_{off} to PE's, one gets (from (9)): $Q_S = \text{floor}\left(\frac{A_{\text{off}}}{A_S}\right)$.

4.5 Costs

There are two types of costs which are relevant, namely:

- *non-recurrent engineering* (**NRE**) costs, which include cost of design and simulation time (of chip, board and sometimes the required SW), amortization of CAD tools, fabrication of mask and test patterns, documentation, marketing, etc. NRE costs are very high for ASIC's, lower for FPGA's, very low for SW. HW/SW codesign can significantly lower NRE costs of FPGA's and at a lesser extent, of ASIC's. NRE costs of any commercial part is usually the lowest, as the manufacturer has already paid for those;
- *recurrent engineering* (**RE**) costs, which include fabrication, chip testing, packaging, assembling, system testing, commercial support, etc. RE costs of FPGA's and SW are higher that those of ASIC's. RE costs of SW can become very high if a high speed is required.

Total cost is therefore function of the production, namely:

$$\text{cost per chip} = \frac{\text{NRE}}{\text{produced parts}} + \text{RE}$$

Figure 4 and fig. 5 plot an example of costs versus production for different implementations.

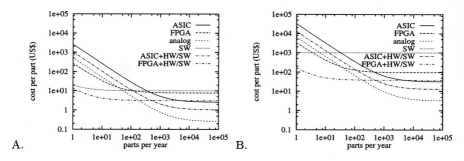

Fig. 4. A) An example of costs versus quantity per year for different implementations of a simple two-layers $32 \times 64 \times 2$ NF-0 network with on-chip training offering ≈ 1 MCPS and ≈ 0.1 MCUPS.

Fig. 5. B) Same as fig. 4, for ≈ 200 MCPS and ≈ 1 MCUPS.

4.6 Flexibility

Flexibility is often important whenever a NF system has to be applied to a set of different applications, where either the number of inputs and outputs, or the number of layers, or the size of hidden layers, or the paradigms can vary from case to case, or even only the actual weight values. The more flexible is a NF system, the more easily will adapt to a different application.

Software simulators are by far the most flexible, but they seldom offer high performance. Implementations with fixed weights (mostly analog and pulse stream, therefore ASIC's) are the least flexible, and they can be used only for large productions, when NRE costs become immaterial. Fully parallel implementation with programmable (or trainable) weights are slightly more flexible, yet a particular technology is required for analog weight storage, as indicated in sect. 2.2. At present, programmable digital circuits on FPGA's (possibly with HW/SW codesign) offers the best compromise among flexibility, cost and time-to-market, except for large productions, or for very specific requirements, as discussed in sect. 5. HW/SW codesign also improves flexibility of digital and hybrid implementations.

Partially parallel implementations offer flexible network size. In addition, bit-serial digital implementations offer flexible resolution, while certain types of PE's offer programmable NF paradigm (mainly, programmable NF order).

4.7 Time-To-Market

This is a very major concern for industrial applications, but it is less important for academic designs. Time-to-market is dominated by two components: the design phase, duration of which depends on the economical effort the designer is willing to pay, and manufacturing, the duration of which mostly depends on the chosen technology.

Analog and pulse stream ASIC's usually have the longest time-to-market, as they have to be designed, simulated at the transistor level (for different device mismatches) and tested. Overall design effort of an analog or a pulse stream NF system can range from 6 to 24 man-months and time-to-market cannot be shorter than a few months, but 6-12 months is more realistic.

On the other hand, digital systems are simpler to design and test, as they can be designed at the gate or functional level and they are insensitive to mismatches. Design effort can range from 1 to 6 man-months for ASIC's and from 0.5 to 2 man-months for FPGA's or SW, while time-to-market can range from 0.5 to 2 months for FPGA's or SW or from 1 to 6 months for ASIC's.

SW implementations usually have the shortest time-to market, thanks to their very high flexibility and to user-friendlyness of SW languages and compilers. This is definitely one of the major reasons which has ruled out the market of HW implementations.

The use of HW/SW codesign associated with FPGA's can sometimes shorten time-to-market to as low as one week, even for hybrid systems, making it competitive with SW implementations (similar time-to-market but more performance). A Simulink-based HW/SW codesign environment for NF systems is described in [5]. It allows to easily integrate NF systems together with other functional blocks taken from Simulink' standard library. At present, that is the only HW/SW codesign environment tailored to NF systems, although I foresee that other tools will soon be available.

4.8 Integration with Other Subsystems

In most practical applications, a NF system is nothing but a very small part of the whole system. It shall therefore be tightly integrated with other subsystems, very different in nature (for instance, filtering, preprocessing, user interfaces, performance evaulation, control of redundancy, etc.).

This integration may be very hard in analog and pulse stream implementations, slightly simpler in digital, while it is straightforward in SW applications. HW/SW codesign makes it nearly as straightforward in hybrid systems.

5 Concluding Remarks

Analog and pulse stream systems are preferred for large productions, very low power, very high sample rate or bandwidth, small size, possibly with fixed weights, but only when poor accuracy is sufficient (mostly, bioinspired systems).

Digital systems with on-chip memory are preferred for higher accuracy, high repeatability, low noise sensitivity, better testability, higher flexibility and compatibility with other types of preprocessing. Digital systems with off-chip memory are nearly useless, as any DSP or PC or microprocessor provides comparable

	speed	BW/sam. frequency	power	size	repeat	accur	time-to market	NRE	RE
Analog, volatile wgts	VH	VH	VL	L	L	L	VH	H	L
Analog, fixed wgts	VH	VH	VL	L	L	L	VH	H	H
P. stream, volat. wgts	H	H	L	L	L-M	L-M	VH	H	L
P. stream, fixed wgts	H	H	L	L	L-M	L-M	VH	H	H
Dig. ASIC's, off-chip	L-M	L-M	M-H	L/M†	VH	any	H	H	L
Dig. ASIC's, on-chip	H	H	M-H	H	VH	any	H	H	L
Dig. FPGA's, off-chip	L-M	L-M	H	L/M†	VH	any	L	L	M
Dig. FPGA's, on-chip	H	H	H	H	VH	any	L	L	M
SW on DSP's, PC's	L-M	L-M	H	L/M†	VH	8,16,32	VL	VL	L

Table 1. Overall performance of HW implementations: VL = very low; L = low; M = medium; H = high; VH = very high († including external weight memory).

or better performance at a smaller price, shorter time-to-market, higher availability than analog and pulse stream (see sect. 4).

Digital systems can be designed more easily, thanks to the improvements in automatic design tools. A few commercial chips are also available, most of them are like general-purpose microprocessors with an instruction set (or an internal architecture) tailored to the computation of NF systems.

On the other hand, analog systems are more difficult to be designed and can be afforded mostly for large scale productions, or for very specific applications. They are the only possible solutions in a class of problems where power consumption or high sampling frequency is a very stringent requirement.

All SW implementations and hybrid systems with FPGA's are much cheaper than any other method, for small and medium scale productions, except for very high speed (from fig. 4 and fig. 5). In particular, hybrid systems and SoC's are very suited to embedded systems, as they can offer the highest performance at the lowest price, provided that design effort is made sufficient low, as is the case with HW/SW codesign.

This is thus the answer to the opening question: for all the reasons listed above, the market of dedicated HW implementations is reducing in size, as SW implementations are becoming every day more and more attractive and cost-effective, except when ultimate performance (either size, power or speed) are required. In the latter case (very rare in practice), general-purpose HW implementations are definitely insufficient, therefore the system has to be designed ad-hoc, either in digital or in analog or pulse stream, according to the requirements (see tab. 1).

HW/SW codesign is currently making hybrid systems and SoC's more and more competitive, mainly for embedded systems, in consumer and high performance applications, therefore it is believed that in the near future this technique will overrule all the other design techniques.

Acknowledgements

This paper has been partially sponsored by Italian MADESS Project "VLSI Architectures for neuro-fuzzy systems".

References

1. L.M. Reyneri, "Unification of Neural and Wavelet Networks and Fuzzy Systems", in *IEEE Trans. on Neural Networks*, Vol. 10, no. 4, July 1999, pp. 801-814.
2. N.H. Weste, K. Eshraghian, "Principles of CMOS VLSI Design: A System Perspective", Addison Wesley (NY), 1992, ISBN 0-2012-53376-6.
3. R. Gregorian, G.C. Themes, "Analog MOS Integrated Circuits", *Wiley and sons*, NY, 1986.
4. C. Mead, "Analog VLSI and Neural Systems", *Addison Wesley*, NY, 1989, ISBN 0-201-05992-4.
5. L.M. Reyneri, M. Chiaberge, L. Lavagno, B. Pino, E. Miranda, "Simulink-based HW/SW Codesign of Embedded Neuro-Fuzzy Systems", in *Int'l Journal of Neural Systems*, Vol. 10, no. 3, June 2000, pp. 211-226.
6. R. Coggins, M. Jabri, B. Flower, S. Pickard, "A Low Power Network for On-Line Diagnosis of Heart Patients", in *IEEE MICRO*, Vol. 15, no. 3, June 1995, pp. 18-25.
7. J. Kramer, R. Sarpeshkar, C. Koch, "Pulse-Based Analog VLSI Velocity Sensor", in *IEEE Trans. Circuits and Systems - II*, Vol. 44, no. 2, February 1997, pp. 86-99.
8. E.C. Mos, J.J.L. Hoppenbrouwers, M.T. Hill, M.W. Blum, J.J.H.B. Schleipen, H. de Walt, "Optical Neuron by Use of a Laser Diode with Injection Seeding and External Optical Feedback", in *IEEE Trans. Neural Networks*, Vol. 11, no. 4, July 2000, pp. 988-996.
9. M.J. Wilcox, D.C. Thelen, "A Retina with Parallel Input and Pulsed Output, Extracting High-Resolution Information", in *IEEE Trans. Neural Networks*, Vol. 10, no. 3, May 1999, pp. 574-583.
10. K. Bohanen, "Point-to-Point Connectivity Between Neuromorphic Chips Using Address Events", in *IEEE Trans. Circuits and Systems - II*, Vol. 47, no. 5, May 2000, pp. 416-433.
11. G. Indiveri, "Neuromorphic Analog VLSI Sensor for Visual Tracking: Circuits and Application Examples", in *IEEE Trans. Circuits and Systems - II*, Vol. 46, no. 11, November 1999, pp. 1337-1347.
12. J. Lazzaro, J. Wawrzynek, A. Kramer, "Systems Technologies for Silicon Auditory Models", in *IEEE MICRO*, Vol. 14, no. 3, June 1994, pp. 7-15.
13. S. Jones, R. Meddis, S.C. Lim, A.R. Temple, "Toward a Digital Neuromorphic Pitch Extraction System", in *IEEE Trans. Neural Networks*, Vol. 11, no. 4, July 2000, pp. 979-987.
14. M. Chiaberge, E. Miranda Sologuren, L.M. Reyneri, "A Pulse Stream System for Low Power Neuro-Fuzzy Computation", on *IEEE Trans. on Circuits and Systems - I*, Vol. 42, no. 11, November 1995, pp. 946-954.
15. L.M. Reyneri, "On the Performance of Pulsed and Spiking Neurons", in *Analog Integrated Circuits and Signal Processing*, 2000 (to be published).
16. P.J. Edwards, A.F. Murray, "Fault Tolerance via Weight Noise in Analog VLSI Implementations of MLP's - A case Study with EPSILON", in *IEEE Trans. Circuits and Systems - II*, Vol. 45, no. 9, September 1998, pp. 1255-1262.
17. L.M. Reyneri e E. Filippi, "An Analysis on the Performance of Silicon Implementations of Back-propagation Algorithms for Artificial Neural Networks", on *IEEE Transactions on Computers*, Vol. 40, no. 12, December 1991, pp. 1380-1389.

A Field-Programmable Conductance Array IC for Biological Neurons Modeling

Vincent Douence [1], S. Renaud-Le Masson [1], S. Saïghi [1], and G. Le Masson [2]

[1]Laboratoire IXL, CNRS UMR 5818, ENSEIRB-Université Bordeaux 1
351 cours de la Libération, 33405 Talence, France
`renaud@ixl.u-bordeaux.fr`
[2]INSERM EPI 9914, Institut François Magendie
1 rue Camille Saint-Saens 33077 Bordeaux, France
`lemasson@bordeaux.inserm.fr`

Abstract. This paper presents the design and applications of a novel mixed analog-digital integrated circuit, that computes in real time biologically-realistic models of neurons and neural networks. The IC is organized as a programmable array of modules, which can be parameterized and interconnected like the modeled cells. On-chip memory cells dynamically store the programming data. The whole analog simulation system exploiting the circuit is described, and applications examples are reported, to validate different configurations and compare with numerical simulations.

1 Introduction

Experimentally based neuron models are now currently used in computational neuroscience for the study of biological neural networks. However, numerical simulators present a limited computation speed, and offer no direct interface with living neurons. Analog simulation is an interesting as it runs in real time and computes analog currents and voltages, like real neurons [1], [2]. Application specific integrated circuits (ASICs) designed in the IXL laboratory compute in analog mode neurons and neural networks described by conductance-based mathematical models. These artificial neurons reproduce with a high level of accuracy the neural activity, which depends on the set of model parameters. Previous chips were designed to validate the analog computation principle [3]. The model parameters were set by tuning voltage values applied on the chip pins from an external source.

We present here a novel mixed analog-digital ASIC, organized as a programmable ionic conductances array. It computes the neurons models in real time and handles on-chip parameters storage and programming, using analog dynamic and static integrated memory cells. That chip represents a new step towards the modeling in real time of more complex and biologically-realistic neural networks.

2 Modeling Principles and Circuit Architecture

The neuron electrical activity is the consequence of the flowing of ionic species through its membrane. The Hodgkin-Huxley formalism proposes an electrical equivalent schematic for the neuron membrane, where the main variable is the membrane voltage called V_{mem}, and in which varying conductances account for the ionic channels dynamic opening and closing (see figure 1). For each considered ionic specie, the associated ionic conductance is a complex variable, which depends on both variables time and V_{mem} [4]. According to the chosen complexity level, a neuron can be modeled by the interaction of only 2 ionic species (Sodium and Potassium for example in simple neurons presenting action potentials), up to 6 or 7 species for complex activity patterns. Synaptic conduction can be modeled by the same kind of mechanisms.

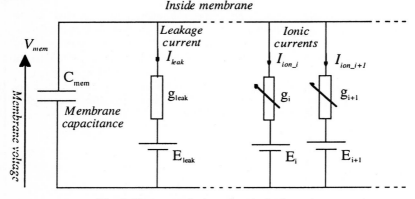

Fig. 1. Neuron equivalent electrical schematic

Analog artificial neurons are built following a structure similar to the one defined in figure 1: each chip integrates a set of modules able to compute an ionic conductance, following the Hodgkin-Huxley formalism. Additional functions expressing ionic interdependencies have been added and can be included in a neuron model (Calcium-dependence phenomenon, or regulation process).

An artificial neural network is described at three levels:
- the parameters for each ionic conductance, defined in the Hodgin-Huxley formalism equations: they are set on the chip by voltage values.
- for each individual neuron, the mapping of the ionic currents to sum on the membrane capacitance : it is defined using integrated multiplexing circuits.
- the topology of the complete neural network, also using multiplexing circuitry, to set the pre- and post-synaptic neurons in synaptic connections

The field programmable conductance array circuit ("fpca-r") integrates these three programming levels on a single chip. Its structure is presented in figure 2. Looking at figure 1, one can identify the ionic currents outputs (I_{ion}) and the membrane voltage nets (V_{mem}). Synaptic current outputs are summed on the membrane net of the post-synaptic neuron. The membrane capacitors, that represent a too large silicon area to be integrated, are external elements connected to their corresponding V_{mem} pins.

The analog core computing the neuron models has already been validated in previous ASICs [3], [5]. Improvements have been made in that version, in terms of power supply (5V instead of 10V), and biasing current (2 A instead of 50 A). The other main innovation is the implementation of analog memory cells that dynamically store voltage values representing the model parameters: we use sample-and-hold circuits, with integrated capacitors periodically refreshed, organized in a matrix structure [6]. An analog input (V_{ref}) presents successively on the array cells the parameters read in an external stack.

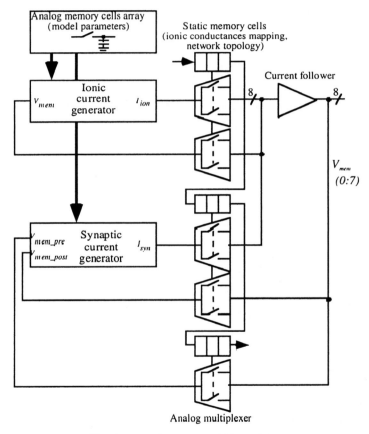

Fig. 2. Internal structure of "fpca-r".

Static digital memory cells are necessary for configuring the multiplexing circuits that define the ionic current generators mapping and the network topology. These cells are assembled in a shift register, and programmed via a serial input.

3 Circuit

The chip has been designed in a BiCMOS 0.8 micron process, and fabricated by Austria Mikro Systeme (AMS). Bipolar transistors are necessary for the integration of the sigmoïdal and the multiplication functions of the Hodgkin-Huxley equations. "fpca-r" (figure 3) has a 3186 m x 2885 m area, and comprises 10371 elements. 25% correspond to the analog core computing the neurons models. 40% are used by the dynamic memory cells, and 30% by the static ones. The last 5% are dedicated to numerical elements implementing a state machine that handles the programming and simulation sequence, and provides control and state signals for the user interface. The digital part of the circuit has been synthesized from a VHDL description, and automatically placed and routed. The analog part has been designed in "full custom" mode, with manual layout drawing.

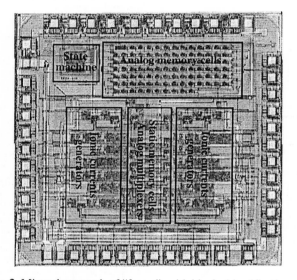

Fig. 3. Microphotograph of "fpca-r", with blocks identification.

These artificial neurons are integrated in a system that represents an analog neural simulator (figure 4). The user defines the neuron or neural network that he wants to model using an interface software running on a personal computer. The computer can act as an oscilloscope via the acquisition board. The software drives a control/acquisition board connected to a dedicated printed board, which supports the artificial neuron ASICs, an address counter, a SRAM memory circuit and a 12 bits digital-to-analog converter (DAC). When a simulation is ordered, the SRAM is filled with the models parameters values, and the state machine in "fpca-r" controls the filling of the static memory cells, to configure the neural network topology. When the process of simulation in real time starts, the V_{mem} values are computed, while the on-chip parameters memory array is continuously refreshed. The parameters are applied on the "fpca-r" chip V_{ref} input via the DAC reading the SRAM data. As shown in figure 4, the board resources (counter, SRAM, DAC) can be shared by chained "fpca-r" circuits (up to 11 chips). That resource sharing is an important point, as it allows to globally increase the number of ionic conductances available for the definition of the modeled neural network, using a single interface board.

Tests on the chip demonstrate that it works as predicted, during both the configuration and the simulation processes. The dynamic data storage happens to be well performing, and the refreshing frequency can be very slow. That property can be used to compensate the coupling problems that appear on the circuit, where glitches may randomly appear on the analog outputs (V_{mem} signals), due to the internal clock transitions. It is a common problem in mixed digital/analog integrated circuits, and should be corrected by hardening the routing during the layout process.

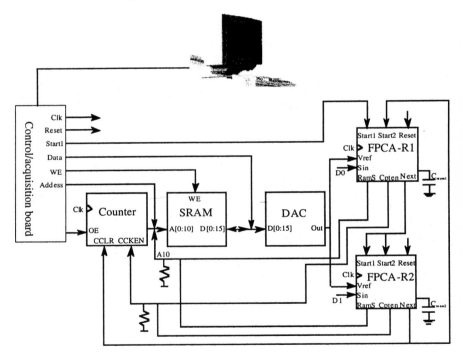

Fig. 4. Structure of the analog simulation system

Measurements presented in the next paragraph have been obtained with the system presented in figure 4, with a 25ms refreshing period. To optimize the signals variation range, artificial neurons model the biological ones with predefined x10 gains on currents and voltages.

4 Applications

We will in a first example compare the simulation results that we obtain with our analog simulator and those given by a classical numerical neural simulator (*Neuron*, [7]). The model of a two-conductances neuron, involving a Sodium and a Potassium current, is programmed on the two systems. 13 parameters are necessary to describe that neuron, following the Hodgkin-Huxley formalism. The resulting activity consists in action potentials that appear when the neuron is slightly depolarized from its resting potential by a stimulation current. Figure 6-A shows some of the model parameters, programmed on the *Neuron* interface. The others parameters are directly compiled in the model source file. The corresponding parameters for the artificial neuron are the voltage values shown in 6-B. These values are directly deduced from the biological model parameters using calibration functions, that have been specified during the "fpca-r" design process. The *Neuron* simulation result is presented in 6-C, to compare in 6-D with the measurement on the "fpca-r".

The V_{mem} waveforms present good similarities: the oscillation frequencies are identical, as well as the spike shape. The major discrepancies appear on the V_{mem} minimum and maximum values, which is a defect we often encounter in BiCMOS ASICs, due to the technological offset currents that attenuate the signal variations. Similar phenomena also exist in real neurons.

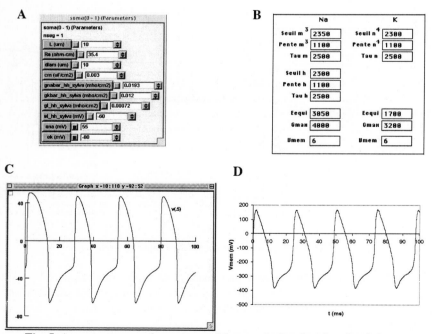

Fig. 5. A neuron two-conductances model: numerical and analog simulations.

The second example illustrates the cellular mechanisms involving Calcium ions, that have been shown to be responsible for bursting properties of neurons [8]. They are expressed by Calcium ionic currents of low amplitude and low kinetic, tuned to modulate the neuron activity in a cyclic pattern where a bursting activity periodically appears. To demonstrate the capability of "fpca-r" to treat such small amplitude currents, we implemented a simple bursting model, inspired of thalamo-cortical cells models [9]. Sodium and potassium currents generate the fast action potentials, and an additional slow Calcium currents rhythms the bursting sequence. That activity is triggered by a hyper-polarizing stimulation current I_{stim}, that can be seen in figure 6, along with the 3-conductances artificial neuron membrane potential V_{mem} measures on the "fpca-r" output.

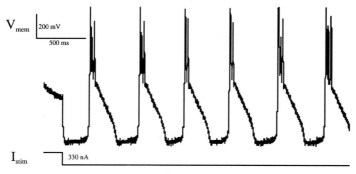

Fig. 6. Three-conductances neuron bursting activity.

Finally, we will show results of a multi-chip configuration, in which two ASICs "fpca-r" are used to model a neural network including two neurons, connected with excitatory synapses. The analog simulation system with the two ASICs is organized as shown in figure 4. On each chip are programmed the ionic currents and the synaptic current for one neuron. The two neurons of the network are tonic (as in figure 5), but their action potentials have distinct frequencies (figure 7-A). We can observe the efficiency of the synaptic currents, that synchronize the post-synaptic neuron with the pre-synaptic one (figure 7-B and 7-C).

6 Conclusion

We have presented in this paper the design and application of an integrated circuit for analog simulation in real time of biologically-realistic neuron models. The chip is organized as an programmable conductance array, with on-chip memory, and is included into a complete analog simulation system. Application configurations are reviewed, that validate the design principles. This circuit is a step towards the integration of more complex neural networks, used as a tool for computational neurosciences. Neurobiology experiments will now use "fpca-r", for the hybrid reconstruction of biological neural networks: hybrid networks are built by interconnecting in real time artificial analog neurons and living cells in an in vitro preparation. Such experiments were successfully conducted with the previous generation of circuits we designed [2], [4]. The novel functions that proposes "fpca-r" should allow us to address more advanced questions of fundamental neurobiology.

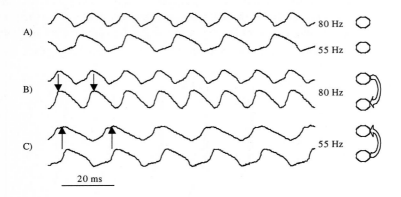

Fig. 7. Two-neurons network with excitatory synapses.

References

1. Mead, C.,: Analog VLSI and neural systems. Addison Wesley Publishing. Reading (1989).
2. Mahowald, M., Douglas, R.: A silicon neuron. Nature, Vol.345 (1991). 515-518.
3. LeMasson S., Laflaquière A., Bal, T., LeMasson, G.: Analog circuits for modeling biological networks: design and applications. IEEE transactions on biomedical engineering, Vol.46 ,n°6 (1999). 638-645.
4. Koch, C., Segev, I.,: Methods in neuronal modeling: from synapses to networks. MIT Press, Cambridge (1989).
5. Douence, V., Laflaquière, A., Bal, T., Le Masson, G.,:Analog electronics system for simulating biological neurons. Proceedings of IWANN99 in Lecture Notes in Computer Science. Vol 1606 (1999). 187-197.
6. Murray, A., Buchan, W.: A user's guide to non-volatile on-chip analogue memory. Electronics and communication engineering journal. Vol. 10, n°2 (1998). 53-63.
7. Hines, M.,: The neurosimulator NEURON . Methods in neuronal modeling, 2ème édition, Ed. C.Koch et I. Segev, MIT Press, Cambridge (1998). 129-136.
8. Meech, R., Calcium-dependent activation in nervous tissues. Annual review of biophysics and bioengineering. Vol.7 (1978). 1-18.
9. McCormick, D., Bal, T.,: Sleep and arousal: thalamocortical mechanisms. Annual Review of Neuroscience, Vol. 20 (1997). 185-215.

A 2-by-n Hybrid Cellular Automaton Implementation Using a Bio-Inspired FPGA

Hector Fabio Restrepo and Daniel Mange

Logic Systems Laboratory, Swiss Federal Institute of Technology,
IN-Ecublens, CH-1015 Lausanne, Switzerland
{HectorFabio.Restrepo, Daniel.Mange}@epfl.ch - http://lslwww.epfl.ch

Abstract. This paper presents the detailed implementation of a 2-by-n hybrid cellular automaton by using the MICTREE (for tree of micro-instructions) cell, a new kind of coarse-grained field-programmable gate array (FPGA) developed in the framework of the Embryonics project. This cell will be used for the implementation of multicellular artificial organisms with biological-like properties, i.e., capable of self-repair and self-replication.

1 Introduction

Living organisms are complex systems exhibiting a range of desirable characteristics, such as evolution, adaptation, and fault tolerance, that have proved difficult to realize using traditional engineering methodologies. The last three decades of investigations in the field of molecular biology (embryology, genetics, and immunology) has brought a clearer understanding of how living systems grow and develop. The principles used by Nature to build and maintain complex living systems are now available for the engineer to draw inspiration from [4].

In this paper, we will present a Embryonic implementation of a 2-by-n Hybrid Cellular Automaton using MICTREE cells. The MICTREE cell is used for the implementation of multicellular artificial organisms with biological-like properties, i.e., capable of self-repair and self-replication.

2 MICTREE Description

The MICTREE cell is a new kind of *coarse-grained field-programmable gate array (FPGA)*, which will be used for the implementation of multicellular artificial organisms with biological-like properties. Since such cells (small processors and their memory) have an identical physical structure, i.e., an identical set of logic operators and of connections, the cellular array is homogeneous. Only the state of the cell, that is, the content of its registers, can differentiate it from its neighbors. The MICTREE cell has been embedded into a plastic container (Figure 1a). These containers can easily be joined to obtain two-dimensional arrays as large as desired.

In all living beings, the string of characters which makes up the DNA, the *genome*, is executed sequentially by a chemical processor, the *ribosome*. Drawing inspiration from this mechanism, MICTREE is based on a *binary decision machine*(BDM) [2] (our ribosome), which sequentially executes a microprogram (our genome). In addition, the

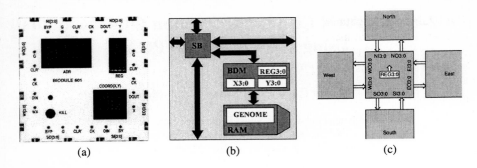

Fig. 1. *MICTRE cell. (a) BIODULE 601 demonstration module. (b) MICTREE block diagram. (c) Four neighboring cells connection diagram; REG3:0: state register.*

artificial cell is composed of a random access memory (RAM), and a communication system implemented by a switch block (SB) (Figure 1b).

The MICTREE cell sequentially executes microprograms written using the following set of instructions:

- **if** VAR **else** LABEL
- **do REG** = DATA [**on** MASK]
- **do Y** = DATA
- **nop**
- **goto** LABEL
- **do X** = DATA
- **do** VAROUT = VARIN

The state register REG and both coordinate registers are 4 bits wide (REG3:0, X3:0, and Y3:0). The 4 bits of MASK allow us to select which of the bits of REG3:0 will be affected by the assignment. By default, MASK = 1111 (all bits are affected). The variables VAROUT correspond to the four cardinal output busses (SO3:0, WO3:0, NO3:0, EO3:0), for a total of 16 bits (Figure 1c). The variables VARIN (SI3:0, WI3:0, NI3:0, EI3:0, REG3:0) correspond to the four cardinal input busses and the register REG, for a total of 20 bits. The test variables VAR include the set VARIN and the following additional variables: VAR, VARIN, WX3:0, SY3:0, and G. Where WX is the coordinate of the nearest west neighbor, SY is the coordinate of the nearest south neighbor, and G is a global variable, usually reserved for the synchronization clock.

The coordinates are transmitted from cell to cell serially, but are computed in parallel. Therefore, each cell performs a series-to-parallel conversion on the incoming coordinates WX and SY (Figure 1a), and a parallel-to-series conversion of the coordinates X and Y it computes and propagates to the east and north. The genome microprogram is also coded serially. It enters through the DIN pin(s) (Figure 1a) and is then propagated through the DOUT pin(s). The pins CK and CLR' are used for the propagation of the clock signal and to reset the binary decision machine respectively, while the signal BYP (bypass), connecting all the cells of a column, is used for self-repair.

3 The 2-by-n Hybrid Cellular Automaton

2-by-n (2-by-n arrays of cells) cellular automata (CA) are a subset of regular two-dimensional linear CA. In these automata each cell is connected only to its immediate neighbors (connectivity radius r=1), and there are three such neighbors for each cell.

The leftmost and rightmost cells are assumed to have a constant-0 input. The state Q of a cell is calculated using the cell's rule. Different cells can use different rules, making the CA hybrid [1]. Figure 2 shows an example of a 2-by-3 cellular automaton with a spare column. This 2-by-3 cellular automaton has been completely simulated, using MIC Sim [5], and implemented with our MICTREE artificial cells. Both the simulation and the physical implementation verified the theoretical results.

Fig. 2. *Two-dimensional 2-by-3 cellular automaton with a spare column.*

This CA presents two rules and the future state $Q+$ of a cell is a function of four variables (Q, QW, QE, QV)

$$QA+ = f(Q,QW,QE,QV) = QW \oplus QE \oplus QV \tag{1}$$
$$QB+ = f(Q,QW,QE,QV) = QW \oplus QE \oplus QV \oplus Q \tag{2}$$

where the operator $\oplus$ denotes the exclusive-OR function, QW, QE, and QV denote the present state of the west, east, and vertical neighbor respectively, and Q denotes the present state of the cell itself. The global state $Q1:6 = 000000$ is a fixed point of the automaton, that is, $Q1:6+ = 000000$. If we suppose that the initial state for this CA is $Q1:6 = 111111$, the remaining $2^6 - 1 = 63$ states form a cycle of maximal length, defined by the following sequence (in decimal) [1]:

$$(Q1:3,Q4:6) = \\
77 \rightarrow 43 \rightarrow 52 \rightarrow 60 \rightarrow 16 \rightarrow 46 \rightarrow 03 \rightarrow 36 \rightarrow 34 \rightarrow 11 \rightarrow 32 \rightarrow 76 \rightarrow 50 \rightarrow 45 \rightarrow 35 \rightarrow 02 \rightarrow \\
25 \rightarrow 23 \rightarrow 44 \rightarrow 26 \rightarrow 15 \rightarrow 70 \rightarrow 37 \rightarrow 27 \rightarrow 06 \rightarrow 67 \rightarrow 62 \rightarrow 33 \rightarrow 65 \rightarrow 47 \rightarrow 10 \rightarrow 21 \rightarrow \\
61 \rightarrow 05 \rightarrow 51 \rightarrow 56 \rightarrow 22 \rightarrow 57 \rightarrow 31 \rightarrow 40 \rightarrow 64 \rightarrow 54 \rightarrow 07 \rightarrow 74 \rightarrow 75 \rightarrow 66 \rightarrow 71 \rightarrow 24 \rightarrow \\
30 \rightarrow 53 \rightarrow 73 \rightarrow 01 \rightarrow 13 \rightarrow 17 \rightarrow 55 \rightarrow 14 \rightarrow 63 \rightarrow 20 \rightarrow 72 \rightarrow 12 \rightarrow 04 \rightarrow 42 \rightarrow 41 \rightarrow 77 \tag{3}$$

3.1 Genes, Coordinates, and Genome

To calculate the complete genome of this CA we need to calculate first each gene of the artificial organism (rules A and B), then its coordinates, and finally the local and global configurations.

The future state $QA+$ is represented by the Karnaugh map of Figure 3a. This Karnaugh map shows that no simplification of $QA+$ is possible (there is no *block*, i.e., no pattern formed by 2^m adjacent 0s or 1s). For a microprogrammed realization, the binary decision diagram of Figure 3b is the flowchart for the gene of rule A.

While no simplification is apparent in the Karnaugh map for $QB+$ (Figure 4a), we can identify, for $Q = 0$, a sub-map (in the form of an outlined *block of blocks*), equal to the map of function $QA+$ of Figure 3a. Transforming the complete binary decision tree

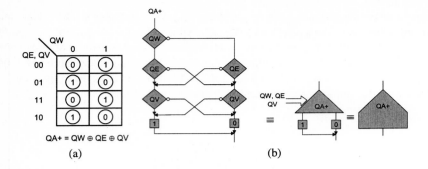

Fig. 3. *Rule A gene computation. (a) Karnaugh map. (b) Binary decision diagram and flowcharts.*

derived from Figure 4a and joining a number of identical sub-trees (the blocks *ST*1 and *ST*2, the output elements 0 and 1), we obtain the binary decision diagram of Figure 4b, which is also the flowchart for the gene of rule B. In accordance with the algebraic expression (2), we note that a part of the flowchart is identical to the flowchart for rule A ($QA+$), as shown in Figure 4b.

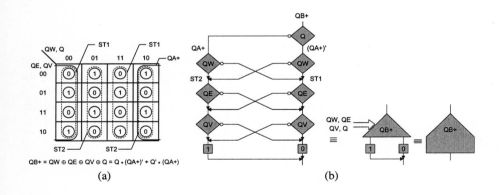

Fig. 4. *Rule B gene computation. (a) Karnaugh map. (b) Binary decision diagram and flowcharts.*

Cellular differentiation occurs, in our example, through the horizontal and vertical coordinates. From the description of Figure 2, the value of each gene can be described by a 3-variable Karnaugh map. The three variables in question are the two bits *WX1:0* of the horizontal coordinate and the bit *SY0* of the vertical coordinate (Figure 5a). The value F defines the gene of the spare cells in the column ($WX = 3$), which display a constant hexadecimal value (equal to F). The simplified binary decision tree derived from Figure 5a generates first the binary decision diagram and then the final flowchart for the operational part of the genome, **Opgenome**, which includes the sub-programs describing the genes $QA+$, $QB+$, and F (Figure 5b).

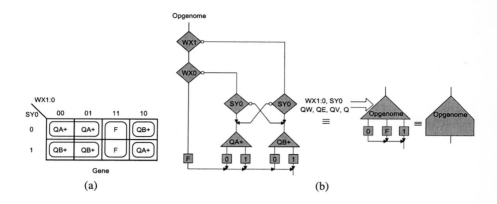

Fig. 5. *Computing the genome's operational part (sub-program Opgenome). (a) Karnaugh map. (b) Binary decision diagram and flowcharts.*

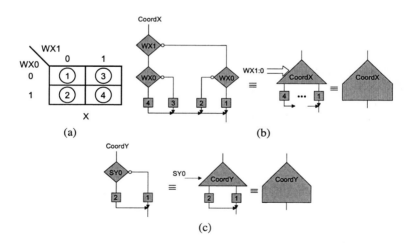

Fig. 6. *Computing the horizontal and vertical coordinates X, Y. (a) Karnaugh map for X. (b,c) Binary decision diagram and flowcharts for X and Y coordinates respectively.*

The Karnaugh map of Figure 6a expresses the horizontal coordinate X of a cell as a function of the two individual bits $WX1{:}0$ of the horizocntal coordinate WX of the west neighboring cell. The four blocks of the map define a complete binary decision tree that becomes the flowchart of the sub-program **Coord** (Figure 6b). The vertical coordinate

Y of a cell is expressed as a function of the bit $SY0$, thus the use of a Karnaugh map is not necessary. The addition of a single test element operating on $SY0$ will allow the definition of the Y coordinate. The binary decision tree, and the flowchart of the sub-program are showed in Figure 6c.

3.2 Configurations and Microprogram

A physical configuration is *global* when it is realized in all the MICTREE cells of the array, independently of the value of the coordinates (X and/or Y). The diagram of the automaton of Figure 2 leads us to choose the bit $REG0$ of the register as the state of the cell ($Q1...Q6$). From each cell, this value is sent out to the right, upper, and bottom neighbors through the east, north, and south output busses respectively ($EO0$, $NO0$, $SO0$). We thus have the global configuration of Figure 7.

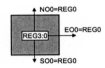

Fig. 7. *2-by-n CA global configuration.*

A physical configuration is local if it is realized by a sub-set of the MICTREE cells of the array. Such a configuration depends therefore on the value of the X and/or Y coordinates.

The diagram of Figure 2 shows such a configuration (Figure 8a), as:

- for $WX = 0, 1,$ and 2 ($X = 1, 2,$ and 3), the state $REG0$ of the cell must be sent to the cell immediately to the west through the output bus $WO0$.
- for $WX = 3$ ($X = 4$), the periodic condition, equal to the logic constant 0, must replace the state $REG0$, requiring that the rightmost cell of the array be cabled so as to assure $EI0 = 0$.

The Karnaugh map of Figure 8b describes the local configurations of $WO0$ as a function of the variables $WX1:0$ and leads to the flowchart **Localconfig** of Figure 8c. The complete genome is represented by the final flowchart **2-by-3-CA-genome** of Figure 9a. It starts with initial conditions assuring that:

- all the cells of the array are set to 1 ($REG0 = 1$), which guarantees that the initial state of the automaton will be $Q1:6 = 111111$, part of the maximal cycle, and avoids the fixed point $Q1:6 = 000000$;
- the coordinates X and Y are set to 0 ($X = 0, Y = 0$).

The microprogram then executes a double loop, controlled by the variable G (the global clock), the clock signal charged with synchronizing the cellular automaton, allowing the transition from the present state $Q1:6$ to the future state $Q1:6+$ at each rising edge of G ($G = 0 \rightarrow 1$).

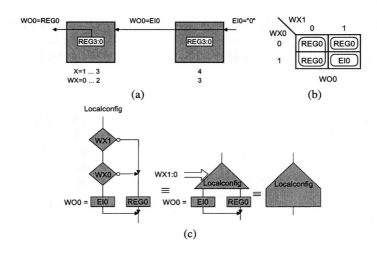

Fig. 8. *2-by-n CA local configuration. (a) Block diagram. (b) Karnaugh map. (c)* **Localconfig** *sub-program binary decision diagram and flowcharts.*

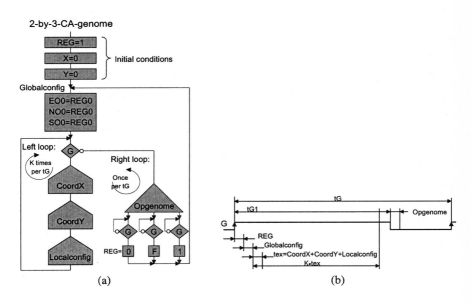

Fig. 9. *Complete genome:* **2-by-3-CA-genome** *microprogram. (a) Flowchart. (b) timing diagram; G: Global clock.*

The right-hand loop is executed once every period (of duration tG) of the global clock signal G. In this loop, the operational part of the genome (**Opgenome**) is also executed once. To assure the synchronization of all the cells, tests are performed throughout the half-period where $G = 0$, but no assignment is made until the rising edge of G

($G = 0 \rightarrow 1$), when the registers **REG** (i.e. the states of the cells) are updated and, for security, the global configuration is confirmed (**Globalconfig**).

If K is the number of used cells by row ($K = 4$ in the example), the left-hand loop must be executed at least 4 times during the half-period when $G = 1$ (Figure 9b). At the start of the microprogram, or when a repair involving a change of coordinates occurs, the coordinates are recomputed starting from the left-most cell. At least K executions of the left-hand loop are necessary to ensure that the right-most cell computes the correct coordinates. This computation occurs in the sub-programs **CoordX**, and **CoordY** which are immediately followed by the computation of the local configurations (**Localconfig**). If tex is the total execution time of the **CoordX**, **CoordY**, and **Localconfig** sub-programs, we must verify that the following inequality holds (Figure 9b):

$$K \cdot tex < tG1 \qquad (4)$$

where $tG1$ is the half-period of G when $G = 1$ (Figure 9b). The time of execution of the assignment instructions for **REG** and **Globalconfig** is negligible.

4 Conclusion

The main result of our research is to show how, thanks to the development of a new kind of coarse-grained FPGA called MICTREE, our artificial organisms (capable of self-repair and self-replication) can be easily implemented and can exploit hardware redundancy to achieve fault tolerance. The distributed automatic reconfigurability of the Embryonics approach offers considerable advantages over other reconfiguration strategies where, in most cases, a centralized agent, e.g. the operating system or a central processor, must solve the routing problem.

Diagnosis and reconfiguration functions are performed at the cellular level in the final Embryonics architecture, where each cell is itself decomposed into molecules, each molecule embedding built-in self-test [3]. No centralized agent exists. Spare elements are incorporated at different levels of an embryonic system in order to achieve resilience to faults in its constituent molecules and organisms.

References

1. K. Kattel, S. Zhang, M. Serra, and J. C. Muzio. 2-by-n Hybrid Cellular Automata with Regular Configuration: Theory and Application. *IEEE Transactions on Computers*, 48(3):285–295, March 1999.
2. D. Mange. *Microprogrammed Systems: An Introduction to Firmware Theory*. Chapman & Hall, London, 1992. (First published in French as "Systèmes microprogrammés: une introduction au magiciel", Presses Polytechniques et Universitaires Romandes, 1990).
3. D. Mange, M. Sipper, A. Stauffer, and G. Tempesti. Towards Robust Integrated Circuits: The Embryonics Approach. *Proceedings of the IEEE*, 88(4):516–541, April 2000.
4. P. Marchal, A. Tisserand, P. Nussbaum, B. Girau, and H. F. Restrepo. Array processing: A massively parallel one-chip architecture. In *Seventh International Conference on Microelectronics for Neural, Fuzzy, and Bio-Inspired Systems, MicroNeuro99*, pages 187–193, Granada, Spain, April 1999.
5. B. Wittwer. Un simulateur pour réseaux de cellules MICTREE: Documentation du projet. Rapport technique, École Polytechnique Fédérale de Lausanne, Lausanne, Suisse, juin 2000.

Parametric Neurocontroller for Positioning of an Anthropomorfic Finger Based on an Oponent Driven-Tendon Transmission System

J. I. Mulero, J. Feliú Batlle, J. López Coronado

Departamento Ingeniería de Sistemas y Automática. Universidad Politécnica de Cartagena.
Campus Muralla del Mar. C/ Doctor Fleming S/N Cartagena 30202. Murcia. Spain
juan.mulero@upct.es, Jorge.Feliu@upct.es, JLCoronado@upct.es.

Abstract. An anthropomorfic finger with a transmission system based on tendons has been proposed. This system is able to work in an agonist/antagonist mode. The main problem to control tendons proceeds from the different dimensions between the joint and tendon spaces. In order to solve this problem we propose a position controller that provides motor torques instead of joint torques as proposed in the literature. This position controller is built as a parametric neural network by using of basis functions obtained from the finger structure. This controller insure that the tracking error converges to zero and the weights of the network are bounded. Both control and weight updating has been designed by means of a Lyapunov energy function. In order to improve the computational efficient of the neural network, this has been split up into subnets to compensate inertial, coriolis/centrifugal and gravitational effects.

1 Introduction

A transmission system is reponsible of driving motion from an input to an output shaft. The objective is to get a transmission system looked like the biological tendon systems and the most proper system for this aim is the train of pulleys. The main benefit of this transmission is that allows to locate the actuator a distance far from the fingertip. This provides a reduction of the dimensions, weight and inertial effects of the fingers. One problem of the transmission systems based on pulleys and tendons is that of the different dimension of the space of joint torques (n-dimensional) and the dimension of the space of tendon force or motor torques (m-dimensional). The inputs to the dynamic model are the motor torques whereas the controller provide control actions based on the joint torques. Thus, a redundant system (m>n) is managed in order to get a total control according to the Morecki's property [6] and this implies to solve a system of equations with more number of unknown variables than equations.

$$\tau = A^T F \tag{1}$$

where A has dimension nxm with m>n, so that the problem is underspecified. A solution for this equation consists of both a homogeneous and particular solution. The homogeneous term is physically associated to the tendon forces that don't work but provide an increasing of the tension of the tendon that wraps the pulley.

These forces are obtained mathematically by means of the null space of the linear transformation. The basis vector e_i has non-zero elements in the positions 2(i-1) and 2i. This situation is that of two equal forces pulling from the end points of the tendon.

On the other hand, the particular solution generates work on the system and so that, is the most important component to take into account for being cause of motion. All the methods to solve the map of different dimension between the joint and tendon space will try to minimize the homogeneous solution, because this provides an energetic consume which is not used in the motion.

In the literature several solutions for this problem have been proposed. It can be found those that compute directy the null space and the particular solution by applyng the pseudo-inverse of Penrose and Moore [5].

Other methods use rectifiers [1],[2],[3] which are circuital components which can be implemented easily and avoid the computation of the pseudo-inverse

In order to avoid the use of a torque resolutor as seen above, a position controller has been designed providing directly the motor torques that pull from the tendons.

There are many strategies of position control based on non-linear schemes. All these schemes work in two phases or feedback loops. The inner loop carry out the dynamic compensation of the non-linearities and the external loop developes the tracking of reference trajectories.

The paper is organised in five sections. In section 2 the dynamic model of the system is proposed. The kinematic chain, tendon-driven transmission system and the whole model is described. In section 3 a description of the parametric neural controller for position is done. Then, the main simulation results are shown in section 4 and the conclusions and future works are presented in section 5.

2 Dynamic Model for the System

2.1 Articulated Mechanism

The articulated system consists of four links connected by revolute joints. The first joint causes abduction/aduction movement, whereas the other joints are orthogonals to this and develope flexion/extension of the finger. The first and second robotic joint are phisiologically associated to the metacarpophalangeal articulation (MCP), joining the metacarpus and the proximal phalange. The third robotic joint is related to the interphalangeal-proximal articulation (PIP) which joins both the proximal and middle phalange. The last robotic joint represents the behaviour of the interphalangeal-distal articulation (DIP), joining both the middle and distal phalange.

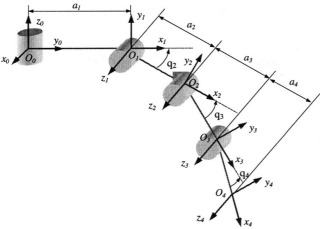

Fig. 1. Frame allocation of the links and the Denavit-Hartenberg parameters.

The allocation of the frames and the Denavit-Hartenberg parameters have been shown in the figure 1. The length a_1 is short because both the first and second joint corresponds to the MCP and his mechanical equivalence is a cardan joint.

2.2 Tendon-Driven System

We use an overactuated manipulator so that it is possible to control all the degrees of freedom as proposed by Morecki: "In order to control completely a manipulator with n degrees of freedom at least n+1 actuators are required" [5].

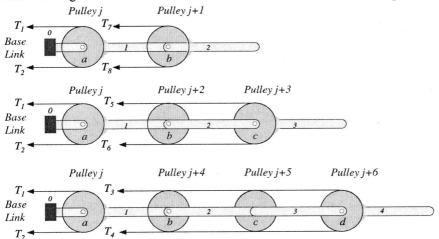

Fig. 2. Planar schematic representation for the transmission system for a finger of four degrees of freedom.

We have chosen a configuration with two tendons for each joint providing a total of m=2n tendons

In the figure 2 the planar scheme has been shown. This scheme allows to indentify the enrouting of the tendons.

2.3 Dynamic Model of the Finger

In order to describe the whole dynamics of the finger we use a vector of generalized coordinates $\Theta(t)$ that represents the angles of the articulations. Taking into account this vector, the dynamic equation of the behaviour of the kinematic chain and the transmission system is presented [5].

$$(M(\Theta)+\tilde{M})\ddot{\Theta} + \tilde{B}\dot{\Theta} + C(\Theta,\dot{\Theta})\dot{\Theta} + G(\Theta) + \tau_d = R^T B^T R_m^{-1}\tau_m = \tau \qquad (2)$$

B stands for the tendon-routing matrix, R is a diagonal matrix for the pulley radii, R_m a matrix of radii of driver pulleys, $M(\Theta)$ a matrix of inertias and τ_m the motor torques. $\tilde{M}$ and $\tilde{B}$ represent respectively the rotor inertia and the motor damping matrix reflected into the joint space. These matrices are symetric and definite-positive since $R_m^{-1} J_m R_m^{-1}$ and $R_m^{-1} B_m R_m^{-1}$ are diagonal matrices with positive eigenvalues and both $\tilde{M}$ and $\tilde{B}$ are congruent transformations.

The items in the matrix B represent the way in wich a tendon is enrouted through the pulleys in the transmission system. Hencefore, b_{ij} indicates that the tendon i is enrouted wrapping the pulley over the joint n-j+1. This is because of the fact that the joints are sorted in the next way: DIP,PIP, flexion/extension MCP and abduction/aduction MCP corresponding to the columns 1 to n. The value of b_{ij} can be −1 (clockwise turning) , 0 (non-enrouted), or +1 (counterclockwise turning) in terms of the joint turning when the motor pulls from the tendon.

3 Adaptive Controller Based on Parametric Neural Networks

Non-linear controllers carry out a compensation of non-linearities of systems and tracking of trajectories by the linearized system. First of all, the dynamic compensation can be obtained by using the property of universal approximation of neural networks. Therefore, according to the theorem of Stone-Weierstrass [9], every smooth function f(x) can be approached with a given error •(x) over a compact set • using a neural network with a large number of nodes and activation functions a(.). In the literature, neural networks with one or two hidden layers are found. Neural networks with two layers are universal approximators of non-linear functions and neural networks with one layer satisfy this property under some conditions [8]. However, multilayer neural networks are non-linear in the weights and this means a tremendous dificulty for analysis, implementation, and on-line tuning of weights. In turn, monolayer neural networks show linearity in the weights and for this reason they are very interesting to work on-line.

On the other hand, the position control system tries to carry out the stable tracking of trajectories in such way the tracking error will converge to zero ($\lim_{t\to\infty} e(t) = 0$). In fact, two kind of errors are managed in this control scheme. Firstly a computed tracking error $e(t) = \Theta_d(t) - \Theta(t)$ and secondly, a filtered tracking error $r = \dot{e} + \Lambda e$ [7], where the parameter design matrix • is definite-positive and is used to be diagonal with large positive entries, so that the system is BIBO stable.

There are a vast variety of posibilities to build a one layer neural network, in terms of the selection of base functions. So, it is possible to find neural networks such as Radial Basis Functions, Cerebellar Model Articulation Controller,... A parametric neural network has been chosen where base functions consist of trigonometric functions and products of joint velocities of the robotic finger. With this, a controller for a general class of fingers that show the axes configurations of the figure 1can be obtained.

The neurocontroller provides motor torques that pull from the tendons wrapping the pulleys in the transmission system.

$$\tau_m = \hat{f} + K_v R_m^{-1} BRr - v(t) \quad (3)$$

The control action is compound of three terms, the aproximation of the non-linear function ($\tau_{app} = \hat{f}$), the external PD-controller ($\tau_{pd} = K_v R_m^{-1} BRr$) and the robust term ($\tau_r = v(t)$) that compensates the effects of disturbances, non-knowledge dynamics and uncertainties of the model. We have assumed that there are neither external disturbances nor approximation error $\varepsilon(x) = 0$ so that the robust term is not necessary. Using a one-layer neural network for approximation of the non-linear function $f(x)$ of the system, we obtained the next controller:

$$\tau_m = \hat{W}^T \phi(x) + K_v R_m^{-1} BRr \quad (4)$$

The design of the neurocontroller is based on the idea of Lyapunov energy, This function is lower bounded and definite-positive. The variation of the Lyapunov is semidefinite negative, so that we can insure that both the error converge to zero by means of the LaSalle's theorem [7] and the weights are bounded.

The control law is defined as

$$\tau_m = \hat{W}^T \phi(x) + K_v R_m^{-1} BRr \quad (5)$$

and the weight updating law is defined as

$$\dot{\tilde{W}} = -F \tilde{W}^T \phi(x) r^T R^T B^T R_m^{-1} \quad (6)$$

4 Simulation Results

The system has been simulated in SIMULINK between 0 and 2 seconds. We use sinoidal signals to excite the system. These reference signals have amplitudes of $\frac{\pi}{3}, \frac{\pi}{4}, \frac{\pi}{5}, \frac{\pi}{6}$ radians corresponding to the angles $\theta_1, \theta_2, \theta_3, \theta_4$ respectively and frequency of 1Hz.

In figure 3 the position for the DIP joint has been shown. There is a little tracking error because we assumed that the modelization error was zero. In order to reject this error we need to add a robust term to the controller.

Initially the PD controller is responsible of controlling the finger because all the weights are initialized to zero. Then, the parametric subnets start to work, carrying out the compensation of non-linearities of the system.

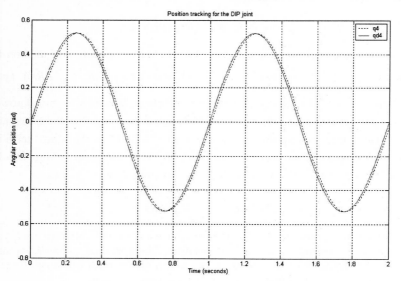

Fig. 3. Representation of both the desired actual position for the joint DIP .

In figure 4 we show the tracking errors for the four joints. These tracking errors are bounded by 0.065, and that is a 6% of error when is compared with the peak amplitude of $\frac{\pi}{3}$ corresponding to the MCP-Abduction/Aduction joint.

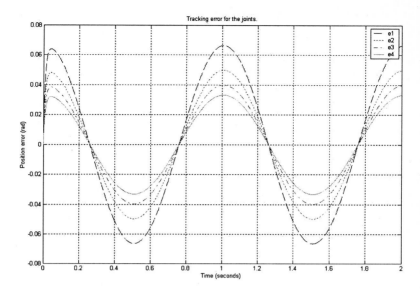

Fig. 4. Tracking errors for MCP-Abduction/Aduction, MCP Flexion/Extension, PIP and DIP joints.

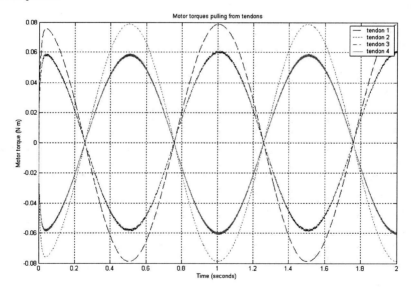

Fig. 5. Motor torques that pull from the tendon 1, 2, 3 and 4.

The motor torques that pull from the tendons 1 to 4 are shown in figure 5. The tendons 1 and 2 represents the Abduction/Aduction motion and the tendons 3 and 4 represents the Flexion/Extension for the MCP articulation. All the flexion/extension movements present a ripple since the mechanical coupling.

5 Conclusions and Future Works

A position controller built from parametric neural networks has been shown. This control scheme provides motor torques instead of joint torques, so that it is not necessary to use an aditional block to map between joint space and tendon space. On the other hand, the parametric neural network represents a controller for a general class of systems, that is index fingers. The basis functions were obtained from the structure of the finger. Nonlinear techniques based on the Lyapunov energy guarantee the convergence of the tracking error and the boundedness of the weights of the network.

In the future, other neural networks for position control will be applied. These techniques are based on static neural networks with one hidden layer using radial basis functions. The choice of static neural networks to implement the controller is that allows to discretize the input space represented exclusively by the joint position. Also it is an interesanting point to reduce the oscillations and the peak value of the tendon forces, so that we can try to filter this undesirable behaviours.

References

1. Jacobsen, S.C., Ko, H., Iversen E.K.,Davis,C.C.: Antagonist control of a tendon driven manipulator, IEEE Proc.Int. Conference on Robotics and Automation, (1989), 1334-1339.

2. Jacobsen, S.C.: The Utah/MIT hand: Work in Progress. IEEE Journal of Robotics and Automation, 3(4), (1984)

3. Lee, J.J., Tsai, L.W.: Torque Resolver Design for Tendon-Drive Manipulators, Technical Research Report. University of Maryland. TR 91-52. (1991)

4. Lee, J.J., Tsai, L.W.: Dynamic Simulation of Tendon-Driven Manipulators, Technical Research Report. University of Maryland. TR 91-53. (1991)

5. Lee, J.J.: Tendon-Driven Manipulators:Analysis, Synthesis, and Control, Thesis Report. Harvard University. (1991)

6. Tsai, L.W.: Design of Tendon-Driven Manipulators. Technical Research Report. University of Maryland. TR 95-96. (1995)

7. Li, W. and Slotine, J-J. E.. Applied Non-Linear Control. Prentice Hall, (1991)

8. Lewis, F. L. , Jagannathan, S., Yesildirek, A.: Neural Network Control of Robot Manipulators and Nonlinear Systems. Taylor & Francis, (1999)

9. Ge, S.S., Lee, T.H. and Harris, C.J.: Adaptive Neural Network Control of Robotic Manipulators. World Scientific Series in Robotics and Inteligent Systems-Vol. 19. (1998)

An Integration Principle for Multimodal Sensor Data Based on Temporal Coherence of Self-Organized Patterns

Emilia I. Barakova

GMD - Japan Research Laboratory, AIM Building 8F, 3-8-1,
Asano, Kokurakita-ku, Kitakyushu-city, 802-0001, Japan,
Work: +81 93 512 1566, Fax: +81 93 512 1588,
emilia.barakova@gmd.gr.jp, http://www.gmd.gr.jp.

Abstract. The world around us offers continuously huge amounts of information, from which living organisms can elicit the knowledge and understanding they need for survival or well-being. A fundamental cognitive feature, that makes this possible is the ability of a brain to integrate the inputs it receives from different sensory modalities into a coherent description of its surrounding environment. By analogy, artificial autonomous systems are designed to record continuously large amounts of data with various sensors. A major design problem by the last is the lack of reference of how the information from the different sensor streams can be integrated into a consistent description. This paper focuses on the development of a sinergistic integration principle, supported by the synchronization of the multimodal information streams on temporal coherence principle. The processing of the individual information streams is done by a self organizing neural algorithm, known as Neural gas algorithm. The integration itself uses a supervised learning method to allow the various information streams to interchange their knowledge as emerged experts. Two complementary data streams, recorded by exploration of autonomous robot of unprepared environments are used to simultaneously illustrate and motivate in a concrete sense the developed integration approach.

1 Motivation

The ability of a brain to integrate the inputs it receives from different sensory modalities into a consistent description of its surrounding world is its basic feature, that helps us orient in tasks with different complexity. It has been widely argued how and whether at all the integration takes place [5][6][9][10][14], and many models has been suggested therefore[3][11][12].

The integration principle, that is featured in this paper is based on the understanding, that there are two aspects of the integration process: (1) achieving a sinergistic integration of two or more sensor modalities and (2) actual combination (fusion) of the various information streams at particular moments of their processing.

The sinergistic integration relies on a hypothesis of how different percepts unify in the brain. It is based on some evidences from temporal registration and binding experiments [14]. For the actual combination the hypothesis is concreticised so that the different sources of sensory information are brought to one coherent representation. For this purpose a synchronisation on a temporal principle is proposed.

This paper focuses on an information combination method on a temporal coherence principle. The combination is made within the framework of an integration strategy proposed, and is widely intertwining with the application domain of concurrent mapping and navigation[1][5].

Two complementary data streams, recorded during the exploration of unprepared environments by an autonomous robot are used to simultaneously illustrate and motivate in a concrete sense the developed integration approach. They provide information about the movement of an autonomous robot from two perspectives: absolute - the robot movement with respect to the surrounding objects (recorded by laser range finders) and relative (recorded by the build-in gyroscope).

The neurobiological experiments have shown, that information from one type of sensors is processed separately on a certain time interval[6][14]. Accordingly, the processing of the individual data streams is done separately, by a self-organizing neural structures (neural gas algorithm in particular [8]) each. The integration of the different information streams ensues the hypothesis made, as well as the outcome of the experiments, of Triesh at al. [13] and uses a backpropagation algorithm for ensuing the different processing streams learn from each other.

This paper is organized as follows: First, an integration hypothesis chooses the scope, that the integration principle will follow. Further on the integration principle as determined by the hypothesis is narrowed down to an practically implementable approach in section 3. Simultaneously, the applications domain is briefly introduced. The flow-chart, shown in the next section follows the information transformations, which bring the information from two orthogonal data streams into a coherent description. Some results illustrate the plausibility of temporal integration principle.

2 Integration Hypotheses

It has been widely argued how the results of different processing systems come together in the brain, to give an unitary perception of the surrounding world.

Chronologically first comes the hypothesis that there are one or more areas in the brain, where integration of different processing streams physically takes place .

Neurophysiological experiments have revealed that there is not a single area in the brain to which different specialized areas uniquely connect. Instead, the brain activities, caused by perception, as well as those, related to memory experiences are simultaneously active in different, highly interconnected functionally specialized areas.

The other group of attempts to reveal the mechanisms that relate various activations is based on the hypothesis, that there is a temporal relation of operations, performed in different processing streams. A precise temporal registration of the results of this operations is possible for intervals of time bigger than one second. The brain is therefore not capable of binding together information entities from different modalities in real-time; instead, it binds the results of its own processing systems.

By far, there is not a single theory, that explains exactly how integrating takes place in the brain. Instead of trying to answer to the question *how* the integration takes place, the approach, suggested in this paper will be build on the hypothesis *why* the integration takes place.

There are variety of answers to this question. The following reasoning will suggest one, that gives a constructive basis for an integration strategy. On a level of a separate sensor modality channel, the brain operates as a self-organizing information system. It obtains inputs from various sensors and in any separate sensor modality stream it clusters the information from its inputs in a self-organizing manner into asymmetric patterns. Since every separate modality brings a different level of generality and scope of information about the external world, the information from one modality can furthermore serve as a "teacher" for the other modality.

In the static world we could use the answers we know as a teacher or expert knowledge. Instead, in a changing world routines and category judgments from the past may be inadequate or misleading. Integrating the on-line, up-date information which brings different level of generality and is sensed from different scope, can give us a key of how to adapt to the new situation and deal with it, and not to solve problems from the past in the new reality. Therefore, the information integration is the mechanism, which allows us to learn in a changing world.

This hypothesis and its preliminaries suggest that we can process separately the information from one type of sensors on a certain time interval in a self-organizing manner. Evaluating the "superiority" of a certain sensor channel to judge more generally about a specific aspect or feature of the reality, we can make it instead give the major notion about the new encountered event. The other sensor stream can tune the certainty of the information from the first stream and to enrich it with the nuances of the novelty.

3 The Integration Approach

In the previous sections of this paper an integration concept has been suggested. Here, the conceptual considerations will be brought to a concrete, technically plausible approach.

In the world of the artificial autonomous systems various sensors asynchronously provide information that has different meaning and sampling characteristics. In addition, there are not established ways of combining the information from different information sources. To achieve an actual combination of the multimodal information sources, the following arguments will be used as a starting point:

- Data, that are perceived (recorded) at the same time relate to the same situation (event).
- Processing of different data streams is done in separate modalities, followed by synchronization on a temporal principle.
- The temporal synchronization is event-based, (in contrast to fixed time interval based).

In addition, according to the conceptual considerations outlined so far, first, an event-based time intervals have to be defined. Second, the information, recorded within this intervals has to be brought to entities, that can be combined technically. And third, the actual combination has to take place.

To get a better intuition about the multimodal sensor integration approach, suggested in this paper, the application task of mapping of unknown environments for the purpose of navigation will be used.

The mapping task is to be solved by using the data, that a mobile robot records during its exploration of an unprepared environment. Figure 1 shows the experimental environment. With black points on the floor are shown some places, which are encountered by the robot as novel, and are clustered in different classes (situations), on the basis of the sensor information, as it will be outlined further on.

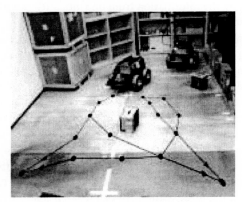

Fig. 1. Experimental environment.

One can hardly think of a group of sensors, which can imitate the consummate description of the environment, that biological systems can create. As a plausible alternative, a set of orthogonal sensors that can complement the perception of each other views on the surrounding world, can be found.

In [2] is elaborated on the relevance of the egocentric perspective of an autonomous robot in spatial modelling of previously unknown environments. The egocentric model in [2] combines two types of information: absolute and relative with respect to the robot motion.

As an absolute source of information are used the "views" that the robot perceives with a laser range finder. The individual 'view' is formed by the record of 720 samples per 360 degrees. A snapshot of a polar representation of such a record is shown at figure 2a). Snapshots are recorded at frequency of 4.7 Hz. The distances are presented in milimeters.

Sequence of such snapshots, recorded during the robot exploration and stored in a short-term memory (STM) - like manner, form a dynamical trajectory, which represents the first information stream, used for the integration. It represents the absolute perspective of the robot about its own motion (i.e. robot motion with respect to the surrounding objects). More details on dynamical trajectory formation can be found in [1]. Here only the final description of the dynamical trajectory will be given:

$$f_k(t) = (t/k)^\alpha (e^{\alpha(1-\frac{t}{k})}) \quad k = 1, ..., K \tag{1}$$

As a relative perspective of the robot is taken the information from the build-in gyroscope. It reflects the way in which the robot perceives its own motion. As a most informative is decided to be the curve, describing the angular velocity of the robot, since it reflects the changes in the direction of the trajectory of the robot and is usually associated with qualitatively novel situations in the surrounding environment, which have caused this changes.

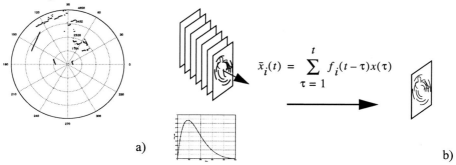

Fig. 2. a)A sensor sample; b)Dynamic trajectory formation.

The temporal synchronisation of the two information streams is performed in the following way. After exploiting the information, that the egocentric perspective that a robot can provide about its movement (based on path integration information - angular velocity data) and on landmark-type information, dynamic events have been created. This dynamic events are developed with minimal processing or interpretation of the recorded data. They contain information from two highly orthogonal sensor sources (relative and absolute).

In addition, information about the time cooccurence of the two sequences of dynamic events (i.e. two time-dependant segments of measurements or representations that happened simultaneously) is used in order to make more complete final representation.

Figure 3 illustrates the implementation of the principle of temporal synchronization over the two information streams. The first one represents the dynamic trajectory, that a robot takes during its exploration of an environment. The qualitatively different "view", that the robot observes define every new segment by this exploration (figure 3b). The duration of this segments determine the division of the other sensor data stream, that reflects the changes in the angular velocity as recorded by a gyroscope, as follows:

$$T_{ir} = \frac{f_{ia}}{f_{ir}}(T_{ia}) \tag{2}$$

Where T_{ia}, T_{ir} are correspondingly the lengths of the i-th segment of the absolute and the relative streams of data, and f_{ia}, f_{ir} are the corresponding sampling frequencies.

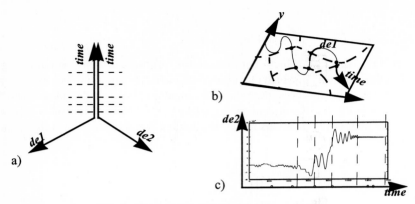

Fig. 3. The temporal synchronization principle.

This way, the synchronization of the two sensor streams is completed. Further on the data streams have to be brought to the same representational format and integrated.

4 Information Flow Chart

In this section in short will be explained the actual steps that developed approach takes to accomplish integration of multimodal information streams.

After representing the information from the laser range finders in a dynamical way, as discussed in the previous section, the first dynamical event is formed. It is prefered the term 'event' to 'feature', because feature is usually associated with some kind of processing of the underlying information, so that some of its essential properties are extracted. In this work neither processing of the information that presumes any sort of its interpretation has taken place, nor extraction of some essentials is made. Instead, the percieved information is coded as compact as possible, by using ideas from biological systems.

As mentioned in the previous section, we distinguish exploration (learning) and testing phase in our experiments. During the learning phase, a Neural gas (NG) algorithm clusters the sequences of views, recorded during the exploration of the mobile robot of previously unknown environment. The moments, when a qualitively new view is encountered, are used for a temporal division of the second dynamical stream of information, recorded by the build-in gyroscope. The distinctively new view is defined by a distance measure, which by now is empirically defined.

The velocity trajectories, recorded by the gyroscope are divided into segments on an event basis. The appearance of a new event is determined (as explained) by the absolute information stream. The so defined segments are clustered in different classes with a help of another NG network.

The testing phase takes similar operations, performed on the testing data sets. Instead of clustering, here is performed classification, according to the clusters, defined over the exploration data.

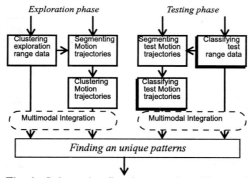

Fig. 4. Information flow by event-based integration.

For a better observability, the so described processing steps will be shown in the following information flow chart (figure 4).

5 Integration Results

To show the plausibility if the suggested integration principle, two processing strategies are compared. The first one simply combines the clustering results from the both processing streams, while the second uses the integration principle for the combination. Both strategies are tested on recognizing passed (during the exploration phase) itineraries, while operating on the test data sets.

By the processing strategy that involves integration on the basis of the developed prin-

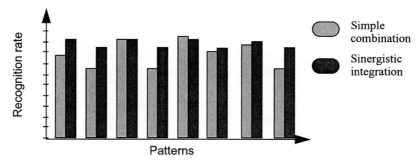

Fig. 5. Recognition rate by simple combination and by integration.

ciple, the clustering/classification results from the relative processing stream are used as a teacher for distinguishing similarly looking dynamic trajectories. After the exploration data stream is clustered and the dynamic trajectories are defined, groups with similarly looking final trajectories were distinguished. The similarities could be caused either because the robot practically never passes the same itinerary by free exploration, or because there are similarly looking places in the environment. Therefore, the velocity information, represented as a sequence of classes was used as a teacher information, so the similarly looking places are discriminated as different scenarios, while the differences in the dynamical trajectory representation, caused by slight variations of the underlying itineraries are considered as the same dynamical scenario.

6 Discussion

The elaborated integration for concurrent mapping and navigation explains the perceptual hierarchy in the following way: the knowledge about the instant movement by humans and many animals contributes a lot to the short-time navigation, while in longer time span they use their perceptions of the surrounding world. In brief analogy, the information that the robot perceives with his range sensors resembles the orientation according to the remembered views, while the velocity information has an influential similarities with the instant movement information by humans.

The important outcomes from the developed hypothesis and its implementation to the considered integration case are as follows:

- The developed method takes into account measurements from a separate sensor source on intervals that does not allow accumulation of errors which can affect the modelling process.
- There is a dynamical way of coding both: the consequent perceptions as well as the transitions between them. This is made possible also by the partially independent ways of processing of the separate information streams.
- The only interpretation of the information, contained in the data streams is made only with respect to defining the priority of the sensor judgements.

The experiments, analysed in [13] suggest, that the ways in which different modalities are integrated depend on the information cues involved, and the nature of the task. The reason, according to him that there is not an unique integration strategy developed yet is that the biological systems do not use a single immutable strategy to combine different percepts.

The hypothesis, made in this paper presumes, that the different percepts about an observed event are processed separately in different sensor modalities. Therefore they acquire different scope and generality of the information about the event they describe. As a result, the outcome of one processing stream can be used as a teacher to the other processing stream's outcome. Considering the argumentation of Triesh [13], we do not tend to show, that this hypothesis is valid for any combination of percepts. Anyway, a large range of validity of this hypothesis is expected.

The shown results reflect actual experiments that aim concurrent mapping and navigation of unknown environment. In a way the shown results are preliminary: the integration principle argues, that the both information streams can be used as teacher of each other, while the made experiments show how the second information stream, used as a teacher of the first one improves the recognition rate with respect to a simple combination of the two information streams after being processed. Our current work concentrates on the mutual learning of the two streams of each other and may need a development of a novel supervised algorithm.

References

[1] Barakova, E. I. & Zimmer, U. R.Dynamical situation and trajectory discrimination by means of clustering and accumulation of raw range measurements, Proc. of the Intl. Conf. on Advances in Intelligent Systems, Canberra, Australia, February 2000.

[2] Barakova, E.I. & Zimmer, U. R. "Global spatial modelling based on dynamics identification according to discriminated static sensations" Underwater Technology 2000, Tokyo, Japan, May 23-26 2000.

[3] Becker, S. and Hinton, G. E. (1992). A self-organizing neural network that discovers surfaces in random-dot stereograms. Nature, 355:161-163.

[4] Grossberg, S. The complementary brain :A Unifying View of Brain Specialization and Modularity.

[5] Elfes, A. "Sonar-based real-world mapping and navigation", IEEE Journal of Robotics and Automation, RA-3(3):249-265, June 1987.

[6] Eagleman, D.M. & Sejnowski, T.J. (2000) Motion integration and postdiction in visual awareness. Science. 287(5460): 2036-8.

[7] Cramer, K.C. and Mriganka Sur. Activity-dependent remodeling of connections in the mammalian visual system. Current Opinion in Neurobiology, 5:106,111, 1995.

[8] Martinetz,T.M., S.G. Berkovich, K.J. Schulten, 'Neural-Gas' Network for Vector Quantization and its Application to Time-Series Prediction, IEEE Transactions on Neural Networks, Vol. 4, No. 4, July 1993, pp. 558-569.

[9] von der Malsburg, C. The correlation theory of brain function., in Models of Neural networks II, edited by E. Domany, J.L. van Hemmen, and K. Schulten (Springer, Berlin, 1994) Chapter 2, pp. 95, 119.

[10] von der Malsburg, C. Binding in Models of Perception and Brain Function. Current Opinion in Neurobiology, 5:520, 526, 1995.

[11] McGurk, H. and MacDonald, J. (1976)., Hearing lips and seeing voices., Nature, 264:746-748.

[12] de Sa, V.R. and Dana H. Ballard, Category Learning through Multi-Modality Sensing, In Neural Computation10(5) 1998.

[13] Triesch, J. (2000). Democratic Integration: A theory of adaptive sensory integration. Technical Report NRL TR 00.1, Nat'l. Resource Lab'y. for the Study of Brain and Behavior,U. Rochester.

[14] Zeki, S. Functional specialization in the visual cortex of the rhesus monkey, Nature (Lond.)274,423-8.

Simultaneous Parallel Processing of Object and Position by Temporal Correlation

Luis F. Lago-Fernández[1,2] and Gustavo Deco[1]

[1] Siemens AG, Corporate Technology, ZT IK 4
Otto-Hahn-Ring 6, 81739 Munich, Germany
[2] E.T.S. de Ingeniería Informática,
Universidad Autónoma de Madrid, 28049 Madrid, Spain

Abstract. There is experimental evidence for the separate processing of different features that belong to a complex visual stimulus in different brain areas. The temporal correlation between neurons responding to each of these features is often thought to be the binding element. In this paper we present a neural network that separately processes objects and positions, in which the association between each stimulus and its spatial location is done by means of temporal correlation. Pools of neurons responding to the stimulus and its corresponding location tend to synchronize their responses, while other stimuli and other locations tend to activate in different time frames.

1 Introduction

One problem that is apparently easy to solve for biological agents, such as that of binding the different features that compound a complex stimulus, remains unsolved by engineers. On one hand there is strong experimental evidence for the separate processing of the different information belonging to a same stimulus in very specialized areas of the brain; on the other there is still no agreement on which kind of mechanisms the brain uses to bind all this information and associate it to a single object. One of the most plausible theories is that of temporal correlation [5]. According to it, the brain should use synchronization between all the neurons responding to the different features of a single stimulus to achieve the binding of all of them. Some experimental works have shown the existence of this kind of synchrony in cortical neurons (see [4] for a review).

In this paper we present a model that makes use of temporal correlation to achieve the binding of each object in a complex visual scene to its corresponding position. It is inspired in a previous model for visual search [2] in which two different layers of neuron pools separately process the objects and their positions. They resemble, respectively, the inferior temporal (IT) and the posterior parietal (PP) cortices in the mammalian brain. We assume the existence of competition mechanisms in both IT and PP layers, so that only one IT and one PP pools can be active at a given time. We also assume the existence of recurrent connections from IT to the input layers, so that the winning pool in IT can influence that in PP. When a particular stimulus wins in IT, the corresponding input pools are reinforced thanks to the recurrent connections. So the pools in PP responding to the stimulus position have a small advantage over the others which lead them to win the competition. We end up with a simultaneous activation of the pools responding to the object in IT and the pools responding to its position in PP.

2 The model

2.1 The mean field approach

We use population equations that describe the mean activity of a group of neurons with common functional and dynamical properties. This groups of neurons, often called pools, are constituted by a large and homogeneous population of identical spiking neurons which receive the same external input and are mutually coupled by synapses of uniform strength. The mean activity of such a pool of neurons is formally described in the framework of the mean field theory, see for example [1, 7]. Typically we have the following equation that relates the input current for the pool i to the input currents for other populations:

$$\tau_i^s \frac{dI_i(t)}{dt} = -I_i(t) + \sum_j w_{ji} F(I_j(t)) \qquad (1)$$

where τ_i^s is the synaptic time constant. The sum on the right hand side extends to all the pools that synaptically influence pool i; w_{ji} is the synaptic strength of the connection and F is the activation function, that relates the input current to the pool mean activity. F is a non-negative function that takes the value 0 below a certain threshold (minimum current needed to activate the pool) and saturates to a certain maximum due to refractoriness. Here we use the following activation function [6]:

$$F(I_j(t)) = \frac{1}{T_j - \tau_j log(1 - \frac{1}{\tau_j I_j(t)})} \qquad (2)$$

where τ_j is the membrane time constant and T_j is the refractory period.

2.2 Model architecture

The model is based on a previous model of biased competition mechanisms for visual search [2]. The biased competition hypothesis [3] states that a bias to an attended location in PP is able to balance the competition in IT in favor of the pool responding to the object at that location (object recognition). Moreover, a bias to the pool in IT that responds to a searched object can lead the competition in PP in favor of the pool that responds to its location (visual search). In this framework Deco [2] has developed a model that is able to explain the usual results of visual search experiments using a single parallel processing schema. Recurrent connections from IT to the input layers can bias the competition in PP when searching for a particular object. Also when the attention is fixated to a spatial location in PP, top-down connections from PP to the input layers bias the competition in IT favouring the object located at that position.

We have extended this work to allow simultaneous processing of more than one stimuli. When more than one object are present in the visual image, we would like to observe an oscillatory alternation of the winning pools in IT and PP, with simultaneous activation of those responding to the same object. To achieve this goal we have implemented the architecture of figures 1 and 2.

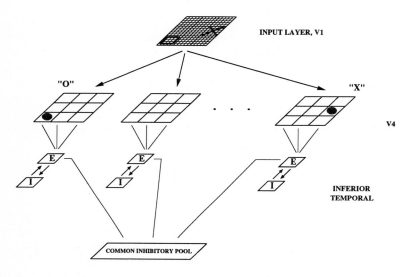

Fig. 1. Estructure of the network showing the connections reaching the IT layer. Non-directed lines represent connections in both ways.

The input layer (V1) responds to active pixels in the input image, and the V4 layer responds to simple stimuli (such as letters) at different positions. As shown in figure 1, each group of V4 pools converges on a single IT pool (E pools in the figure), and so E pools in IT respond to specific objects independently of their position. The competition in IT is mediated by a common inhibitory pool that receives connections from and sends connections to all E pools. This common inhibitory pool allows only one E pool to be active at a time.

The inhibitory pools (I) connected one to one to every E pool produce the oscillatory behavior. After a E pool has won the competition it is inhibited and a different pool can win. The final result is an oscillatory alternation of all the E pools that respond to an object in the input image.

In figure 2 we show the connections reaching the PP layer. Each PP pool responds to activation in a specific location, no matter the object that produces it. Again we have a common inhibitory pool that implements the competition between PP pools, so that only one of them can be active at a time; and local inhibition that generates the oscillations.

Now we will see the importance of the role that recurrent connections from IT to V4 play. When a pool in IT is dominating, these recurrent connections produce a small increase in the activation of the V4 pools responding to the winning object. This translates into a small bias to the PP pool that responds to the object position, which will then win its own competition. So the pools responding to the object (IT) and its position (PP) win their respective competitions simultaneously. This gives an elegant solution to the problem of binding each object to its position.

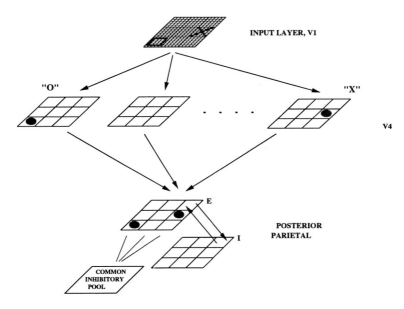

Fig. 2. Estructure of the network showing the connections reaching the PP layer. Non-directed lines represent connections in both ways.

2.3 Model equations and parameters

We now describe the neurodynamical equations that regulate the evolution of the whole system. In the recognition layer (IT), the input current for excitatory pools is given by:

$$\tau_s^E \frac{dI_i^E(t)}{dt} = -I_i^E(t) + w^{IE}F(I_i^I(t-\Delta t)) + w^{EE}F(I_i^E(t)) + w^{V_4E}\sum_j F(I_{ij}^{V_4}(t))$$

$$+ w^{CE}F(I^C(t)) + I_0 + I_{noise} \qquad (3)$$

where in the superindexes E means excitatory, I means inhibitory, C is the inhibitor mediating competition in IT and V_4 is the V4 layer; and in the subindexes i refers to "object" and j to "position". The parameters w are the strengths of the connections, and Δt in the connection from inhibitory to excitatory pools is a delay necessary for the appearance of oscillations. We have included a diffuse spontaneous background input current I_0^E and a gaussian noise term I_{noise}^E. For the inhibitory pools we have:

$$\tau_s^I \frac{dI_i^I(t)}{dt} = -I_i^I(t) + w^{EI} F(I_i^E(t)) + w^{II} F(I_i^I(t)) + I_0 \qquad (4)$$

And for the common inhibitory pool:

$$\tau_s^C \frac{dI^C(t)}{dt} = -I^C(t) + w^{EC} \sum_i F(I_i^E(t)) + w^{CC} F(I^C(t)) + I_0 + I_{noise} \qquad (5)$$

The parameters for equations 3-5 are $\tau_s^E = \tau_s^C = 3.0$, $\tau_s^I = 30.0$, $I_0 = 0.025$, $w^{EE} = 0.95$, $w^{IE} = -20.0$, $w^{V_4 E} = 0.34$, $w^{CE} = -0.5$, $w^{EI} = 1.5$, $w^{II} = w^{CC} = -0.1$, $w^{EC} = 1.5$, $\Delta t = 40.0 ms$; and the amplitude of the gaussian noise is 0.02. In the PP layer we have the following equations for the excitatory, inhibitory and common inhibitory pools:

$$\tau_s^E \frac{dI_j^E(t)}{dt} = -I_j^E(t) + w^{IE} F(I_j^I(t - \Delta t)) + w^{EE} F(I_j^E(t)) + w^{V_4 E} \sum_i F(I_{ij}^{V_4}(t))$$
$$+ w^{CE} F(I^C(t)) + I_0 + I_{noise} \qquad (6)$$

$$\tau_s^I \frac{dI_j^I(t)}{dt} = -I_j^I(t) + w^{EI} F(I_j^E(t)) + w^{II} F(I_j^I(t)) + I_0 \qquad (7)$$

$$\tau_s^C \frac{dI^C(t)}{dt} = -I^C(t) + w^{EC} \sum_j F(I_j^E(t)) + w^{CC} F(I^C(t)) + I_0 + I_{noise} \qquad (8)$$

As before, i means object and j means position. The parameters are $\tau_s^E = \tau_s^C = 3.0$, $\tau_s^I = 20.0$, $w^{EE} = 0.60$, $w^{IE} = -10.0$, $w^{V_4 E} = 0.50$, $w^{CE} = -0.1$, $w^{EI} = 1.5$, $w^{II} = w^{CC} = -0.1$, $w^{EC} = 1.0$, and $\Delta t = 40.0 ms$. In the V4 layer, the current for a pool responding to object i and position j is described by the equation:

$$\tau_s^{V_4} \frac{dI_{ij}^{V_4}(t)}{dt} = -I_{ij}^{V_4}(t) + w^{EV_4} F(I_i^E(t)) + \sum_k w_k^{V_1 V_4} F(I_k^{V_1}(t))$$
$$+ w^{V_4 V_4} F(I_{ij}^{V_4}(t)) + I_0 + I_{noise} \qquad (9)$$

where the E in the superindexes refers to excitatory pools in IT, and the sum in k extends to all the input (V1) pools. The parameters are $\tau_s^{V_4} = 3.0$, $w^{EV_4} = 0.25$, $w^{V_4 V_4} = 0.95$, and $w^{V_1 V_4}$ depends on the stimulus. Finally the equations regulating the input current in the V1 layer are:

$$\tau_s^{V_1} \frac{dI_k^{V_1}(t)}{dt} = -I_k^{V_1}(t) + w^{V_1 V_1} F(I_k^{V_1}(t)) + I_0 + I_{noise} + I_k^{bias} \qquad (10)$$

where a bias current $I_k^{bias} = 0.03$ is used to simulate the activation of the V1 pool due to the stimulus, and the rest of parameters are $\tau_s^{V_1} = 3.0$, and $w^{V_1 V_1} = 0.95$.

The activation functions for all the pools are given by equation (2), with parameters $T = 300.0$ and $\tau = 40.0$ for inhibitory pools in both IT and PP; $T = 1.0$ and $\tau = 10.0$ for excitatory pools in PP; and $T = 1.0$ and $\tau = 20.0$ for the rest of pools.

3 Results

We present the results for a simple case in which two different objects are presented in two different positions, as in figures 1 and 2. Equations 3-10 have been solved using the Euler method with a time step of 0.1 ms.

The V4 pools extract the characteristics of both objects and send the information to IT and PP, where competition takes place. In figure 3A we show the currents for the IT pools responding to each object, and in figure 3B we do the same for the PP pools responding to each active location. In both layers there is an alternation of the dominant pool and, as expected, when an object wins in IT its position also wins in PP.

To clarify the process we show in figure 3C the input current for the active V4 pools. Currents for both pools have a constant component due to the input they receive from V1. But we also observe a small increase in the current of the V4 pool each time the pool in IT responding to the same object increases its activity. This generates a small difference in the input to PP pools that balances the competition in favor of the correct position.

4 Conclusions

We have constructed a model that separately processes the objects in a visual image and their positions, and links each object to its position using temporal correlation of population activities.

This model can be easily generalized to an arbitrary number of features in the stimulus space. For each dimension in the stimulus space we can have a different processing layer. Recurrent connections among these layers would permit the simultaneous activation of the pools in each layer that respond to the particular object.

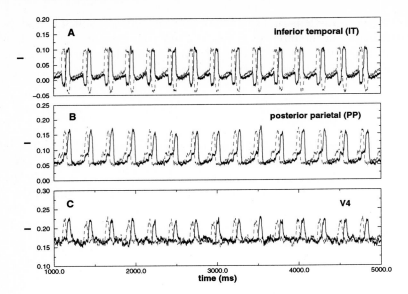

Fig. 3. Input currents in IT, PP and V4 when two different objects are presented simultaneously to the network. Solid lines: object "X"; dashed lines: object "O". A. IT excitatory pools responding to each object. B. PP excitatory pools responding to each active position. C. V4 active pools. A clear correlation is observed between pools responding to a single object.

5 Acknowledgements

We thank the Comunidad Autónoma de Madrid for FPI grant to LFL.

References

1. Abbott, L. F.: Firing rate models for neural populations. In Benhar, O., Bosio, C., Del Giudice, P., Tabet, E., eds. Neural Networks: From Biology to High-Energy Physics (ETS Editrice, Pisa, 1991) 179–196
2. Deco, G., Zihl, J.: Top-down selective visual attention: a neurodynamical approach. Vis. Cog. **8** (2001) 118–139
3. Duncan, J., Humphreys, G.: Visual search and stimulus similarity. Psychol. Rev. **96** (1989) 433–458
4. Gray, C. M.: Synchronous oscillations in neuronal systems: mechanisms and function. J. Comput. Neurosci. **1** (1994) 11–38
5. von der Malsburg, C.: The correlation theory of brain function. In Domany, E., van Hemmen, J. L., Schulten, K., eds. Models of Neural networks II (Springer, Berlin, 1994) 95–119
6. Usher, M., Niebur, E.: Modelling the temporal dynamics of IT neurons in visual search: a mechanism for top-down selective attention. J. Cog. Neurosci. **8** (1996) 311–327

7. Wilson, H. R., Cowan, J. D.: Excitatory and inhibitory interactions in localized populations of model neurons. Biophys. J. **12** (1972) 1–24

NeuSim: A Modular Neural Networks Simulator for Beowulf Clusters

Carlos J. García Orellana, Ramón Gallardo Caballero, Horacio M. González Velasco, and Francisco J. López Aligué

Dpto. de Electrónica. Facultad de Ciencias
Universidad de Extremadura
Avd. de Elvas, s/n. 06071 Badajoz - SPAIN
carlos@nernet.unex.es

Abstract We present a neural networks simulator that we have called NeuSim, designed to work in Beowulf clusters. We have been using this simulator during the last years over several architectures and soon we want to distribute it under GPL license. In this work we offer a detailed description of the simulator, as well as the performance results obtained with a Beowulf cluster of 24 nodes.

1 Introduction

As we know, the simulation of neural nets is an expensive work from the computational point of view. In the last years, the growing speed and capacity of the computers have allowed the simulation of neural networks with few neurons to be done in any standard computer, without the necessity of using workstations nor specific hardware which was so common some years ago [1,2].

However, when the objective is to work with large size neural networks and/or with biologically accurate models, we find that the simulation time and necessary memory increase a lot, so, the simulation in a conventional computer becomes very difficult. Some authors [13] work with simplified models for which they obtain results similar to the most accurate simulations. However, this is not always possible. When we need to do these simulations, parallel simulators running on a network of workstations (NOW) or on a cluster are very useful [9,11].

Our work is centered on the simulation of neural nets under Beowulf clusters [3,4]. A cluster is a collection of computers connected by a fast interconnection system. In particular, a Beowulf cluster is characterized to be built using commodity hardware (although not always). Usually, processors of the x86 family, normal memory of PC's, standard cases and fast-ethernet network adapters are used. Sometimes, hard disks are used in each node, but it is not always necessary. In general, one of the nodes acts of supervisor (master) and the other ones carry out the work (slaves).

The connection among nodes and the election of the processors are two of the critical points in the cluster design. Although it is habitual to use switched fast-ethernet (with one or several cards for node), there are other solutions that have better performance like Myrinet [10] or Gigabit ethernet, however, the cost is much bigger.

With respect to the number and type of processors, it depends on the application. In general, when the applications make intensive use of the memory it is advisable to use monoprocessors. However, in another type of applications it can be better to use SMP multiprocessors (symmetric multiprocessing). Regarding the type, if we limit ourselves to compatible x86 processors, it seems that Athlon AMD processors are a better option instead of Pentium III, not only because of cost, but also because of their best performance in floating point calculations. There is even comparatives in which the Athlon overcomes to the Pentium 4 in floating point operations [12], in spite of the excellent bandwidth with the main memory of Pentium 4.

Another fundamental component of a cluster is the operating system to install. In a Beowulf cluster the operating system is usually Linux. As it is known, Linux is a freeware version of UNIX and it is occuping a more and more important position in the world of operating systems. Without doubts, the administration of the cluster is one of the most complicated tasks, though there are Linux distributions adapted for clusters that facilitate the installation of the nodes, offer nodes supervision, common space of processes, etc. [8].

In the following section we will describe the structure and characteristic of our simulator. After this, we take into account the simulator performance, and later we will show the results, both in time of simulation and in degradation of the parallelism (efficiency), of our system with several sizes of neural nets. We will conclude with the conclusions of our work.

2 Description of the simulator

In this section, we are going to describe to the types of neuronal networks that we want to simulate and the structure of the simulator.

2.1 Neural Network models

As we have commented in the introduction, our objective is that the simulator can work with neural networks in general, but with preference neural nets of large size and with algorithms computationally expensive.

During the development of the simulator, we have carried out a modeling to facilitate the simulation of neural networks [7]. An updated object modeling diagram can be observed in the Figure 1.

From the objects that are observed in that figure, we would like to stand out the classes CONEX, CX, ACT_CONTROL and NN_SCHED.

CONEX indicates us that a connection exists between two layers, without specifying the exact form of this connection, that work is left to the objects of the class CX (connection pattern). To the class CONEX also goes bound the learning algorithm, which can be implemented by an loadable independent module.

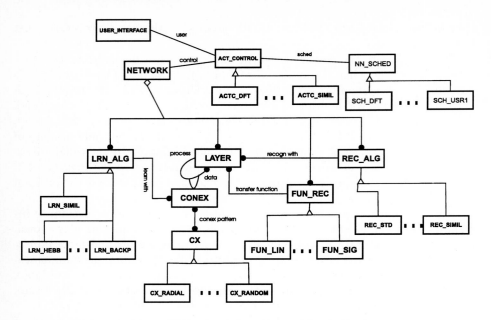

Figure 1. General diagram of the object modeling.

The class ACT_CONTROL takes charge of controlling how the simulator will do recognition and learning, for example, there are models in which one learning iteration include some recognition phases, the way how exactly this should be done is taken in charge by the objects of the class ACT_CONTROL.

The class NN_SCHED allows us to implement (through expansion modules) different ways to plan the simulation and to distribute the load among processors. By default, we use a scheduler that distributes the work of the neural net according to the power of calculation of the available processors, that is, it is not necessary to have a homogeneous cluster, this is important because when we update the cluster, it becomes heterogeneous.

In short, NeuSim is designed to work with neural nets of large size organized in rectangular layers, with connections among them in any direction (feedforward and feedback), with connection patterns, recognition and learning algorithms customizable using modules.

2.2 Simulator Structure

NeuSim is built around a client-server structure. The communication between client and server has been done using Sun RPC (Remote Procedure Call). These remote functions are used for configuring and making evolve the neural network.

For simulation system operation, we must start a daemon called '*nn_event*' in the client machine. This daemon takes the work of transmitting the events from server to client (events like: iteration end, error in simulation, progress when saving files, etc..).

Now, we are going to describe the client and server subsystems.

Client As we have commented, client part is a library to program in C language (NNLIB) under UNIX systems. This library provides functions to create objects, to delete them and to set and get their properties: i.e., to configure the neural network.

These objects can be those already defined inside of simulator or objects developed separated and incorporated to the simulator through expansion modules.

It also has functions to manage the events caused by server, to control the simulation, to see the state of the processors, etc.

Server The server is who really does the simulation, now, we are using a cluster of PC computers, connected by a switched fast-ethernet network (cluster Beowulf). We are using a Linux distribution that allows to administer in a simpler way the whole cluster (Scyld Beowulf [8]).

The server is composed by three subsystems called *postmaster*, *master* and *slave*. Next, we are going to do a description of each one of these subsystems.

- *Postmaster:* this is the start of the server. One only instance of this subsystem exists for all the simulations. As the other subsystems it is multi-thread. Their main tasks are:
  - On one hand, to wait the petition of a new simulation. When anybody wants to begin a new simulation, the postmaster executes a new instance of the master and as many slaves as nodes are available. By this way, a total independence is gotten among simulations, avoiding that a problem in one of them affects to the other ones.
  - Among the other tasks of the postmaster is to check that the simulations respond and continue active, and if it is not so, to eliminate them to liberate resources. We also have a console in which we can consult parameters and status of all simulations in course.
- *Master:* of this subsystem an instance exists for each running simulation. Each one of the master threads takes charge of doing this different tasks:
  - On one hand, to receive the messages sent by slave nodes, to control their state, and in the case of failure, to delete them from the available slaves list.
  - On the other hand, to answer to RPC request from NNLIB client, i.e. from user, and to send the events to the client.
  - Another thread takes charge of distribute and partitioning the neural net among processors (in function of its power of calculation) and of coordinating the simulation in course.
  - Also, other task is to maintain the communication with the postmaster.
  - Another of the tasks is to act as shared memory RPC server for the slaves that need some block of memory located in the master. This is a multi-thread server with the purpose of improving the performance.

- *Slave:* it is the last subsystem, its main task is to do a part of simulation (execution of the simulation modules). One instance exists in each node and for each simulation in course. Its other tasks are:
  - To maintain the communication with the master: to respond to command from it, to send results and to inform of errors.
  - Other important task of the slave is to act as shared memory server of the data that it stores. As in the server subsystem, it is multi-thread RPC server.

Finally, we would like to emphasize two aspects of the server subsystem that we think are important.

- *Distributed memory:* the form of storing and referencing a matrix (for example the neuron's weight) in the shared memory is using a list of shared memory blocks. Each element of this list makes reference to one portion of the matrix, this element of the list indicate the processor where the data are, the size and the offset regarding the beginning of the matrix. We have developed a RPC library that allows to obtain any part of the matrix from the list of blocks, to reserve new blocks, to free them, etc.
- *Modules:* the simulator extension is, together with the portability, one of the things that we have kept in mind during simulator design. The expansion modules allow to implement practically any algorithm type. In general, would be necessary to write two modules, one for the master and another for the slave, which can be loaded and discharge in a dynamic way.

3 Simulator Performance

In previous works [5,6], we have obtained expressions that estimate the simulator performance, both for the case of systems VME with shared memory and for the case of clusters of PCs (with Vxworks and Linux). The expression that estimates the time of simulation is given in (1), in which, α is inversely proportional to the power of calculation of a processor, γ is proportional to latency of communication time between processors, c is the total connections of the neural net, and p is the number of processors. The parameters α and γ can be obtained doing a non linear regression.

$$t(c,p) = \frac{\alpha \cdot c}{p} + \gamma \cdot p \qquad (1)$$

However, analyzing the experimental data, we can reach a better estimation if we suppose that the time of simulation is given by the expression (2), where κ is another parameter.

$$t(c,p) = \frac{\alpha \cdot c}{p} + \gamma \cdot p^{\kappa} \qquad (2)$$

Together with the time of simulation, it is also interesting to estimate how we are taking advantage of the addition of new processors to the simulation. To do it, we define the efficiency ($\epsilon(c,p)$) as it is shown in (3)[11].

$$\epsilon(c,p) = \frac{t(c,1)}{p \cdot t(c,p)} \times 100 \qquad (3)$$

This figure of merit indicates us in what percentage we are taking advantage of the cluster regarding the ideal case (without communication costs). Having this estimation before beginning the simulation is quite important, since this way, we will waste less resources, because if we can choose how many processors to use to obtain better efficiency, it will be possible to use the others for another simulation.

In general, we should use the processors that maintain an efficiency around 80%, although the better value of this parameter will depend on how many simulations we want to make in a simultaneous way.

4 Results

To study the performance of our simulator we have made simulations with several neural networks in our cluster, later on we have adjusted the parameters of the model shown in (2) and we have compared the real values with the estimated ones. We insist again that our objective is double, on one hand we want to check the simulator performance, and on the other, we want to study the viability of our estimation, which will allow us to decide a priori how many processors to use in a simulation.

Table 1. Simulation times and efficiency (both real and estimated) for several nets.

$c(Mc) \rightarrow$		23.8		52.5		105		310.8		1010.8	
$p \Downarrow$		$t\ (s)$	$\epsilon\ (\%)$	$t\ (s)$	$\epsilon\ (\%)$	$t\ (s)$	$\epsilon\ (\%)$	$t\ (s)$	$\epsilon\ (\%)$	$t\ (s)$	$\epsilon\ (\%)$
1	Exp.	1.033	100	1.936	100	3.699	100	—	—	—	—
	Theo.	1.041	100	1.960	100	3.640	100	10.230	100	32.626	100
2	Exp.	0.608	84.9	1.072	90.3	1.948	94.9	—	—	—	—
	Theo.	0.587	88.7	1.046	93.7	1.886	96.5	5.179	98.7	16.379	99.6
5	Exp.	0.274	75.5	0.459	84.4	0.804	92.0	2.145	95.3	—	—
	Theo.	0.290	71.8	0.474	82.7	0.810	89.9	2.127	96.1	6.607	98.8
10	Exp.	0.172	60.2	0.260	74.5	0.431	85.8	1.075	95.1	—	—
	Theo.	0.178	58.6	0.270	72.7	0.438	83.2	1.096	93.3	3.336	97.8
12	Exp.	0.145	59.5	0.213	75.6	0.370	83.3	0.915	93.1	2.805	97.0
	Theo.	0.157	55.2	0.234	69.9	0.374	81.1	0.923	92.4	2.789	97.5
15	Exp.	0.141	49.0	0.198	65.2	0.314	78.7	0.758	89.9	2.256	96.5
	Theo.	0.136	51.1	0.197	66.3	0.309	78.5	0.748	91.1	2.241	97.0
20	Exp.	0.112	46.1	0.165	58.6	0.270	68.6	0.567	90.2	1.710	95.4
	Theo.	0.113	46.1	0.159	61.7	0.243	74.9	0.572	89.4	1.692	96.4
24	Exp.	0.105	40.9	0.150	53.9	0.231	66.7	0.510	83.5	1.430	95.1
	Theo.	0.101	43.0	0.139	58.7	0.209	72.5	0.484	88.1	1.417	95.9

The cluster that we have used is built using 24 nodes with 900 MHz AMD Athlon processors and 512 Mbytes of RAM memory, connected through a switched fast-ethernet network. We have made simulations using five neural nets with 23.8, 52.5, 105, 310.8 and 1010.8 million of connections. In the Table 1 are shown the values of measured simulation time, the estimated one, as well as the real and estimated efficiency. For some of the neural nets it has not been possible to carry out the simulations with few processors because we don't have enough memory (the largest net consumes more than 4 Gbytes of RAM), for these ones, the real efficiency has been calculated using the estimated simulation time for one processor.

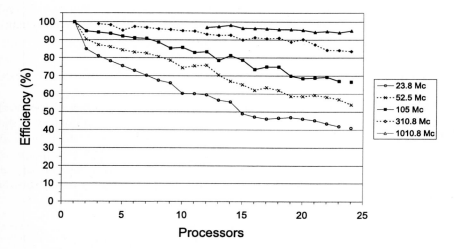

Figure 2. Evolution of efficiency with respect to processors number.

To calculate estimated results the following values have been used for the parameters of the model (2): $\alpha = 0.032$, $\gamma = 0.28$ and $\kappa = -0.45$. These values have been obtained by non linear regression.

In the figure 2 is shown the efficiency in a graphic way.

From the results in Table 1, we can observe that a good agreement exists between the real values and the estimated ones (the maximum error is around 5%). On the other hand, in figure 2, we can see how the efficiency is quite better when the neural network we are simulating grows, which is logical because the time of communication is a smaller percentage of the total time.

5 Conclusions

In the first place, we have presented a simulation system for large neural networks, designed to be used in a cluster, based on client-server structure, expandable using modules (to do new simulation types) and able to handle independent simulations from different users. As shown in results, we can conclude that the simulator performance is quite good and better when the neural network to simulate is greater.

On the other hand, we have studied the simulator performance, obtaining a good estimation of the simulation time and the efficiency. The main objective is to have an a priori estimation of the efficiency because this parameter can serve as a figure of merit to decide how many processors to use in a certain simulation, although when we use more processors the performance that we obtain is better (smaller time as is shown in results), the exploitation of the cluster is worse and since our objective is to have several simulations at once, we must try to maximize this exploitation.

Acknowledgements

This work has been supported by project 1FD970723 (financed by the FEDER and *Ministerio de Educación*).

References

1. A. Müller, A. Gunzinger and W. Guggenbühl, Fast Neural Net Simulation with DSP Processor Array, IEEE Transac. On Neural Networks, Vol. 6, No. 1, (1995).
2. U. Ramacher et al., SYNAPSE-1 – A General Purpose Neurocomputer, Siemens AG (1994).
3. D.J. Becker, T. Sterling, D. Savarese et al., Beowulf: A Parallel Workstation for Scientific Computation, Proc. International Conference on Parallel Processing (1995).
4. C. Reschke, T. Sterling, D. Ridge et al., A Desing Study of Alternative Network Topologies for the Beowulf Parallel Workstation, Procedings of High Performance and Distributed Computing (1996).
5. C.J. García Orellana, Modelado y Simulación de Grandes Redes Neuronales (in spanish), Ph. D. Thesis, University of Extremadura, (Oct. 1998).
6. C.J. García Orellana et al., Large Neural Nets Simulation under Beowulf-like Systems, Eng. App. of Bio-Inspired Artificial Neural Networks, Springer Verlag, pag. 30-39, (1999).
7. F.J. López Aligue et al., Modelling and Simulation of Large Neural Networks, Proc. Int. Joint-Conference on Neural Networks, IEEE Press Journals, (1999).
8. Scyld Beowulf. Scyld Computing Corporation, 2001. http://www.scyld.com
9. M.A. Wilson, U.S. Bhalla, J.D. Uhley and J.M. Bower. GENESIS: A system for simulating neural networks. Advances in Neural Information Processing Systems. D. Touretzky, editor. Morgan Kaufmann, San Mateo. pp. 485-492, 1989.
10. Myrinet Network Products. Myricom Inc. http://www.myrinet.com
11. P. Pacheco, M. Camperi and T. Uchino. Parallel Neurosys: A System for the Simulation of Very Large Networks of Biologically Accurate Neurons on Parallel Computers. Neurocomputing, Vol. 32-33 (1-4) (2000) pp. 1095-1102
12. Scott Wasson. Intel's Pentium 4 Processor. The Tech Report. http://www.techreport.com/reviews/2001q1/pentium4/, 2001.
13. A Omurtag, E Kaplan, B Knight and L Sirovich. A population approach to cortical dynamics with and application to orientation tuning. Network: Computation in Neural Systems. Vol. 11, No. 4, 2000.

Curved Kernel Neural Network for Functions Approximation

Paul Bourret and Bruno Pelletier

ONERA/CERT DTIM, Toulouse, France

Abstract. We propose herein a neural network based on curved kernels constituting an anisotropic family of functions and a learning rule to automatically tune the number of needed kernels to the frequency of the data in the input space. The model has been tested on two case studies of approximation problems known to be difficult and gave good results in comparison with traditional radial basis function (RBF) networks. Those examples illustrate the fact that curved kernels can locally adapt themselves to match with the observation space regularity.

1 Introduction

Interpolation of observed data has been widely studied in the past. The method of kernels has been revisited in the last ten years, see [1][2]. This method is based on the decomposition of function onto a set of decreasing function of the distance between the input vector and a set of points of the input space. These kernels are thus isotropic. Nevertheless several questions are still open nowadays. Among them we frequently have to face the following problems in practical cases.

For terrain numerical models purpose, various methods based on local regularization have been tried but a sufficiently reliable criterion allowing us to determine the regularization coefficients has not been exhibited. Especially, it is not possible with available techniques to locally tune the number of kernels with respect to the frequency of data in a part of the input space. The choice of a kernels family is still difficult even when an a priori knowledge allows us to assume that the observed phenomenon is the sum of several processes of the same class, each of these processes being close to a given kernel. This is the case for instance for the estimate of binary error rates in radio transmission. The observed error rate is the bounded sum of a few interferences and/or intermodulation processes.

When the sample of data consists of pairs of values (x_i, y_i) and when the goal is to approximate $y = f(x)$, it is commonly assumed that the y_i are noisy. But we also have to face the problem which occurs when both x_i and y_i are noisy. This is the case for the estimate of the concentration rate of a product in the sea given the solar reflectance in various parts of the spectrum.

At last, it may occur that the (x_i, y_i) are not uniformly spaced in the input space and that moreover the frequency in a part of this space is meaningfull. Thus we have looked for a learning rule and a kind of kernel with parameters that allow us to cope with the following objectives :

- To be able to use as kernels anisotropic families of functions. In fact, this amounts to use kernels with curvilinear coordinates, thus almost any shape can be approximated with an emphasis on the previously mentioned case of several processes of the same class.
- To be able to adapt the number of necessary kernels to the observed frequency of data in the input space, because this variable frequency is an important knowledge representing the fact that one variable is in fact a function of the other.

2 Model presentation

In what follows, we consider the approximation problem from a statistical viewpoint. The sample of data is seen as a set of observations of a random vector **X** and a random variable Y such that $Y = f(\mathbf{X})$ holds true. An estimate of f may be taken as the conditional mean of Y given **X**. This method requires the estimation of the joint probability density of **X** and Y.

2.1 Curved kernel

Let f be a function, X a d-dimensional random vector taking values in D, a bounded domain of $\mathcal{R}^d$ and Y a random variable such that $Y = f(X)$, defining a bounded manifold in $\mathcal{R}^{d+1}$ denoted by $\mathcal{M}$. In this case, the joint probability density of X and Y degenerates in $\delta(y - f(x))f_X(x)$ with $f_X(x)$ the marginal probability density of X. Let $M(\mathbf{x}_0, y_0) \in \mathcal{M}$ and V be a neighbourhood of M. If some knowledge about f is available that is if function f restricted to V behaves like a given function $\phi(\mathbf{x}, \mathbf{x}_0)$, the joint probability density of X and Y may be approximated using a kernel K by :

$$\frac{1}{h_x^d h_y} K\left(\frac{\mathbf{x} - \mathbf{x}_0}{h_x}\right) K\left(\frac{y - \phi(\mathbf{x}, \mathbf{x}_0)}{h_y}\right) \quad (1)$$

where the widths h_x and h_y represent a measure of the approximation confidence spread and/or embedd the fact that in the presence of measurement noise, the observed values may deviate from the manifold. When such a priori information is missing, one may use for ϕ an estimate of the first terms of the Taylor expansion of f so that the above expression becomes :

$$\frac{1}{h_x^d h_y} K\left(\frac{\mathbf{x} - \mathbf{x}_0}{h_x}\right) K\left(\frac{y - y_o - \langle \alpha, \mathbf{x} - \mathbf{x}_0 \rangle}{h_y}\right) \quad (2)$$

with α denoting the first order derivatives of f and $\langle , \rangle$ the standard scalar product. This expression has the advantage that α may be estimated from the observations by a finite difference scheme. The joint probabilty density estimate and the regression model then take the following expressions :

$$\hat{f}_{XY} = \sum_{k=1}^{M} \frac{1}{h_{x,k}^d h_{y,k}} K\left(\frac{\|\mathbf{x} - \mathbf{t}_{x,k}\|}{h_{x,k}}\right) K\left(\frac{\|y - t_{y,k} - \langle \alpha_k, \mathbf{x} - \mathbf{t}_{x,k} \rangle\|}{h_{y,k}}\right) \quad (3)$$

$$\hat{f}(x) = \frac{\sum_{k=1}^{M} \left(t_{y,k} + \langle \alpha_k, \mathbf{x} - \mathbf{t}_{x,k}\rangle\right) \frac{1}{h_{x,k}^d} K\left(\frac{\|\mathbf{x} - \mathbf{t}_{x,k}\|}{h_{x,k}}\right)}{\sum_{k=1}^{M} \frac{1}{h_{x,k}^d} K\left(\frac{\|\mathbf{x} - \mathbf{t}_{x,k}\|}{h_{x,k}}\right)} \quad (4)$$

where $\mathbf{t}$ denotes the centers of the network. The above expression reduces to the expression of generalized normalized radial basis functions (RBF) networks with ellipsoidal kernels when α equals 0. The initial idea was to measure distances respectively along the hypersurface and its normal, thus using curvilinear coordinates. But because of the introduced complexity, we kept the benefit of euclidean geometry simplicity.

2.2 Training algorithm

Network parameters optimization may be done in two different ways corresponding to functional or statistical viewpoints :

- Fitting the regression model to the data points by least square minimization for example.
- Fitting the density model to the observations which is the retained method.

As a criterion to be minimized, we used the Kullback-Liebler divergence from the true unknown probability density to the density model given by

$$D\left(f_{XY} \| \hat{f_{XY}}\right) = \int f_{XY} \log \frac{f_{XY}}{\hat{f_{XY}}} \quad (5)$$

which is a misfit measure between two probability densities. Denoting by θ the network parameters, we minimize the following quantity by a gradient descent adaptive algorithm :

$$C(\theta) = D\left(f_{XY} \| \hat{f_{XY}}\right) = E_{XY}\left[J(\theta, X, Y)\right] \quad (6)$$

where $J(\theta, X, Y) = \log \frac{f_{XY}}{\hat{f_{XY}}}$. The adaptive rule for the presentation of a pattern $(\mathbf{x}_n, y_n)$ is given by :

$$\theta_{n+1} = \theta_n + \epsilon_n \overrightarrow{grad} J(\theta, X, Y) \quad (7)$$

In what follows, we implemented the competitive version of the algorithm where only the parameters of the closer kernel to the presented pattern are updated, see [3]. For such an adaptive algorithm to be efficient, the patterns must be presented according to the true probability density f_{XY} which is unknown. We therefore need an a-priori assumption on f_{XY}. Generally, the patterns are uniformly presented, assuming the data set has been observed and is distributed according to f_{XY}. But in many practical experiments, the data set is obtained by sampling and not according to the true density f_{XY}. We therefore adopt the following scheme. Let $\mathbf{M}$ be a random vector taking values on the bounded manifod and uniformly distributed on it. Let $\Pi : \mathcal{M} \mapsto \mathcal{R}^d$ be the projection application and $\mathcal{U}$ a bounded region of $\mathcal{D}$. The marginal probability law of $\mathbf{X}$ is then :

$$p(\Pi(\mathbf{M}) \in \mathcal{U}) = p(\mathbf{M} \in f(\mathcal{U})) = \frac{\int_{f(\mathcal{U})} dS}{S_T} \quad (8)$$

where $dS = \prod_{i=1}^{d} \left[1 + \left(\frac{\partial f}{\partial x_i}\right)^2\right]^{\frac{1}{2}} dx_1...dx_d$ represents the surface element and $S_T = \int_{\mathcal{D}} dS$ is the total surface. The marginal density of random vector $\mathbf{X}$ may then be written as :

$$f_X(x) = \frac{1}{S_T} \prod_{i=1}^{d} \left[1 + \left(\frac{\partial f}{\partial x_i}\right)^2\right]^{\frac{1}{2}} \quad (9)$$

Those formula may be discretized for numerical experiments using a finite difference scheme. This gives a prior estimate of the marginal density of $\mathbf{X}$ that may be used to train the network. This assumption of uniformity represents also a smoothing a-priori.

3 Experimental results

We have tested our model on two case studies, respectively one and bi dimensional, and compared it with traditional RBF neural networks. The tested models used kernels built along hyperplanes where the coefficients were estimated from the data as shown in equation (4). The kernels centers were initially positioned on the examples randomly chosen according to the estimated prior probability law described in the previous section. This implies that the initial distribution of the kernels matchs the frequency of the data in the input space. Moreover, since the optimization method is local, one needs a good starting point for it to be effective.

3.1 One dimensional case

We approximated the function $f(x) = \sin \frac{1}{x+0.05}$ which has a varying regularity along the considered domain. The data set was composed of 1001 patterns obtained by regular sampling and normalized between 0 and 1. Our model and the RBF network both used 25 kernels. The following table shows for both models the mean absolute error, mean square error, maximum absolute error and mean absolute relative error. RBF network results are represented on figure 1 where one can see the model is accurate up to a given data regularity. Figure 2 shows the results for our model. Figure 3 represents the kernels shapes of our model and illustrates their anisotropic behaviour since they may desymetrize themselves for local adaptation.

Table 1. Comparative results for the one dimensional case.

Model	MAE	MSE	AEmax	MARE
RBF	0.025	0.0012	0.25	0.32
Curved	0.007	0.0004	0.21	0.36

3.2 Bi dimensional case.

The function to be approximated had the following expression in polar coordinates :

$$f(r, \theta) = \frac{r}{r_0^3} \exp^{-\frac{(r-r_0)^2}{2\sigma^2}} \qquad (10)$$

with $\sigma = 0.2$ and $r_0 = \exp^{0.2\theta}$. The data set was composed of 6561 patterns obtained by regular sampling in the (x, y) plane. Both models used 100 kernels. As above, the following table shows standard error criterion values for both models. The initial surface is plotted on figure 4. The approximated surfaces

Table 2. Comparative results for the bi dimensional case.

Model	MAE	MSE	AEmax	MARE
RBF	0.022	0.0023	0.73	8.22
Curved	0.013	0.0012	0.59	4.70

for the RBF and our model are respectively represented on figure 5 and 6.

4 Conclusion

We have proposed a neural network model whose kernels can locally adapt to the input space regularity with an interest in the particular case of a sum of processes of the same class where some a priori knowledge may be embedded in the kernel or estimated from the data when this latter is missing. Secondly, we have described a method for tuning the kernel distribution with respect to the frequency of the data in the input space. Experimental results show the efficiency of those techniques on approximation problems known to be difficult. Nevertheless, the problem arising when working with local approximators is the smoothness of the resulting global model. It would therefore be interesting to study a way of smoothing the joint probability density estimate. In a more general way, the results show that the estimation of some parameters of the data space intrinsic geometry is meaningfull for the construction of a model. A much more deeper insight in this intrinsic geometry would maybe provide helpful tools for model design.

References

1. Darken, C., Moody, J.: Fast learning in networks of locally-tuned processing units. Neural Computation, 1(2):281-294, 1989.
2. Poggio, T., Girosi, F.: Networks for approximation and learning. Proc. of the IEEE, Vol 78, No 9, September, pp. 1481-1497, 1990.
3. Benam, M., Tomasini, L.: Competitive and self-organizing algorithms based on the minimization of an information criterion. International Conference on Artificial Neural Networks, 1991.

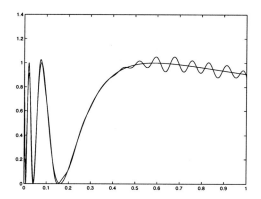

Fig. 1. Approximation by RBF neural network.

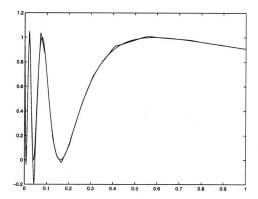

Fig. 2. Approximation by a curved kernel neural network.

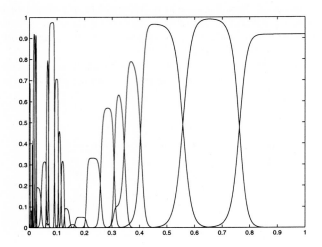

Fig. 3. Graphs of the curved kernels where we can see their adapted shapes and distribution with respect to the regularity of the data.

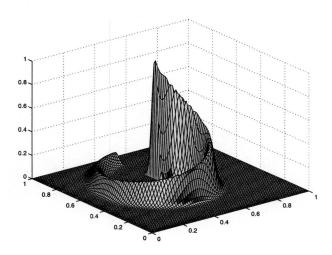

Fig. 4. Original surface.

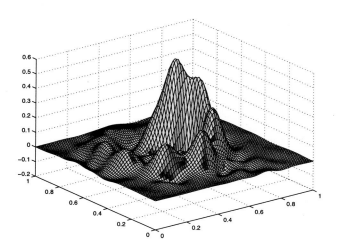

Fig. 5. Approximated surface by RBF neural network.

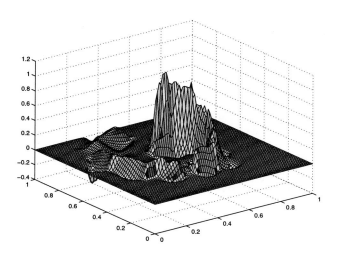

Fig. 6. Approximated surface by curved kernel neural network.

Repeated Measures Multiple Comparison Procedures Applied to Model Selection in Neural Networks

Elisa Guerrero Vázquez[1], Andrés Yañez Escolano[1], Pedro Galindo Riaño[1], Joaquín Pizarro Junquera[1]

[1] Universidad de Cádiz, Departamento de Lenguajes y Sistemas Informáticos, Grupo de Investigación *"Sistemas Inteligentes de Computación"*
C.A.S.E.M. 11510 - Puerto Real (Cádiz), Spain
{elisa.guerrero, andres.yaniez, pedro.galindo, joaquin.pizarro@uca.es}

Abstract.
One of the main research concern in neural networks is to find the appropriate network size in order to minimize the trade-off between overfitting and poor approximation. In this paper the choice among different competing models that fit to the same data set is faced when statistical methods for model comparison are applied. The study has been conducted to find a range of models that can work all the same as the cost of complexity varies. If they do not, then the generalization error estimates should be about the same among the set of models. If they do, then the estimates should be different and our job would consist on analyzing pairwise differences between the least generalization error estimate and each one of the range, in order to bound the set of models which might result in an equal performance. This method is illustrated applied to polynomial regression and RBF neural networks.

1 Introduction

In the model selection problem we must balance the complexity of a statistical model with its goodness of fit to the training data. This problem arises repeatedly in statistical estimation, machine learning, and scientific inquiry in general. Instances of model selection problem include choosing the best number of hidden nodes in a neural network, determining the right amount of pruning to be performed on a decision tree, and choosing the degree of a polynomial fit to a set of points. In each of these cases, the goal is not to minimize the error on the training data, but to minimize the resulting generalization error [8]. The model selection problem is coarsely prefigured by Occam's Razor: given two hypotheses that fit the data equally well, prefer the simpler one. For example, in the case of neural networks, oversized networks learn more rapidly since they easily monitor local minima, but exhibit poor generalization performance because of their tendency to fit data noise with the true response [11].

In a previous work [10] we proposed a resampling based multiple comparison technique where a randomized data collecting procedure was used to obtain independent error measurements. However repeated measures experimental designs offer greater statistical power relative to sample size because the procedure takes into account the fact that we have two o more observations on the same subject, and uses this information to provide a more precise estimate of experimental error [9].

This paper focuses on the use of repeated measures tests, specifically repeated measures Anova and Friedman tests, that allow to reject the hypothesis null that the models are all equal, but they do not pinpoint where the significant differences lie.

Multiple comparison procedures (MCP) are used to find out which pair or pairs of models are significantly different from each other [6].

The outline of this paper is as follows. In section 2 we briefly introduce parametric and nonparametric statistical tests, including the multiple comparison procedures that could be applied for a repeated measures design. In section 3 our strategy using statistical tests for model selection is described and illustrated in the application of RBF networks and polynomial regression.

2 Repeated Measures Tests and Post-Hoc Tests

Our objective is to consider a set of several models simultaneously, compare them and come to a decision on which to retain. Repeated measures tests allow to reject the hypothesis null that the models are all equal [2].

The application of the within-subjects ANOVA test implies the following assumptions to be satisfied: the scores in all conditions are normally distributed and each subject is sampled independently from each other subject, and, sphericity [5]. When these assumptions are violated, a non-parametric test should be used instead [3]. In our experiments we use Friedman test that is based on the median of the sample data [7].

If repeated measures Anova F test is significant and only pairwise comparisons should be made Girden [5] recommended the use of separated error terms for each comparison in order to protect the overall familywise probability of the type I error.

The approach for multiple comparisons proposed here consists on comparing only groups between which we expect differences to arise rather than comparing all pairs of treatment levels. The less tests we perform the less we have to correct the significance level, and the more power is retained. This approach is applied to the step-down sequentially rejective Bonferroni procedure [1] that is more powerful and less conservative that the singlestep Bonferroni correction.

Nemenyi test is a nonparametric multiple comparison method [12] that we will use when the repeated measures Anova assumptions are not met. This test uses rank sums instead of means.

3 Experimental Results

Our experiments can be outlined as follows:

1. Take the whole data set and create at least 30 new sets by bootstrapping [4].

2. Apply models with a degree of complexity ranging from m to n (in our experiments $m = 1$ and $n = 25$) and generate as many error measures per model as number of sets have been created in the previous step.
3. Calculate the error mean per model and select the model with minimum error mean.
4. Test repeated measures Anova assumptions
5. Apply statistical tests to check if we might reject the null hypothesis that all groups are equal:
    5.1. Repeated measures Anova test, if its assumptions are met.
    5.2. Friedman test if assumptions are not met.
6. If the null hypothesis is not rejected select the simplest model.
7. If the null hypothesis is rejected, apply a Multiple Comparison Procedure
    7.1. Sequentially rejective Bonferroni if repeated measures Anova was applied.
    7.2. Nemenyi test if Friedman test was applied.
8. Select the less complex model from the subset of models that are not significantly different from the model with minimum error mean.

In order to illustrate our strategy we conducted a range of experiments on simulated data sets. Thirty data sets Z=(X,Y) for several sample sizes were simulated according to the following experimental functions:

$$y = sin(x+2)^2 + \xi, \quad x \in (-2,+2) \tag{1}$$

$$y = -4.9389x^5 - 1.7917x^4 + 23.2778x^3 + 8.7917x^2 - 15.3380x - 6 + \xi, x \in (-1,3) \tag{2}$$

where ξ is gaussian noise.

3.1 Polynomial Fitting

We considered the problem of finding the degree N of a polynomial P(x) that better fits a set of data in a least squared sense. Polynomials with degrees ranging from 1 to 25 are used. The only aspect of the polynomials that remains to be specified is the degree.

Tables 1 and 2 show the results according to the experimental function (1) for each of the following sample sizes n = 25, 100, 500 and for each of the following N (0,0.5), N(0,1), N(0,3) and N(0,5) gaussian noise. In this case, all the experiments conducted showed that the least generalization error mean was reached with a polynomial with degree 5.

Tables 3, 4 and 5 show the results according to the experimental function (2) for each of the following sample sizes n = 250, 500, 1000 and for each of the following N (0,0.5), N(0,1), N(0,3) and N(0,5) gaussian noise.

Table 1. Polynomial Degrees (Models) with minimum test error mean, and percentage of times that occurred in our experiments.

		noise's variance			
	models (%)	0.5	1	3	5
data set size	25	**5 (93.33%)** 6 (6.67%)	4 (6.67%) **5 (90%)** 7 (3.33%)	4 (16.67%) **5 (80%)** 6 (3.33%)	4 (30%) **5 (70%)**
	50	**5 (86.67%)** 6 (10%) 7 (3.33%)	**5 (83.33%)** 6 (13.33%) 7 (3.33%)	**5 (93.33%)** 7 (3.33%) 8 (3.33%)	**5 (90%)** 6 (10%)
	100	**5 (83.33%)** 6 (3.33%) 7 (10%) 8 (3.33%)	**5 (83.33%)** 6 (13.33%) 8 (3.33%)	**5 (33.33%)** 6 (20%) 7 (3.33%) 8 (3.33%)	**5 (86.67%)** 6 (13.33%)
	500	**5 (86.67%)** 6 (6.67%)) 7 (3.33%) 10 (3.33%)	**5 (90%)** 6 (6.66%) 7 (3.33%)	**5 (96.67%)** 6 (3.33%)	**5 (80%)** 6 (6.66%) 7 (3.33%) 8 (3.33%) 9 (3.33%) 10 (3.33%)

Table 2. Selected models that are not significant different from the model with minimum test error mean (confidence level = 0.05).

		noise's variance			
	models (%)	0.5	1	3	5
data set size	25	2 (3.33%) 3 (13.33%) **4 (73.33%)** 5 (10%)	1 (10%) 3 (40%) **4 (50%)**	1 (13.33%) 2 (6.67%) **3 (60%)** 4 (20%)	1 (26.67%) 2 (23.33%) **3 (43.33%)** 4 (6.67%)
	50	4 (3.33%) **5 (96.67%)**	4 (16.67%) **5 (83.33%)**	4 (40%) **5 (60%)**	**4 (83.33%)** 5 (16.67%)
	100	**5 (100%)**	**5 (100%)**	**5 (100%)**	4 (3.33%) **5 (96.67%)**
	500	**5 (100%)**	**5 (100%)**	**5 (100%)**	**5 (100%)**

Table 3. Models with minimum generalization error mean.

		noise's variance				
	models (%)	0.1	0.25	0.5	0.75	1
data set size	250	14	14	13	12	11
	500	15	14	14	13	11
	1000	17	14	14	14	12

Table 4. Models with minimum test error mean and percentage of times that occurred in our experiments.

	models (%)	noise's variance				
		0.1	0.25	0.5	0.75	1
data set size	250	12 (40%) 13 (6.67%) **14 (60%)** 15 (13.33%) 16 (10%)	11 (16.67%) 12 (20%) 13 (16.67%) **14 (36.67%)** 15 (6.67%) 16 (3.33%)	**11 (50%)** 12 (26.67%) 13 (16.67%) 14 (6.67%)	8 (3.33%) 9 (6.67%) 10 (6.67%) **11 (60%)** 12 (6.67%) 13 (10%) 14 (3.33%) 16 (3.33%)	8 (6.67%) 9 (16.67%) 10 (26.67%) **11 (33.33%)** 12 (13.33%) 15 (3.33%)
	500	14 (33.33%) **15 (36.67%)** 16 (13.33%) 17 (10%) 18 (3.33%) 20 (3.33%)	12 (3.33%) 13 (3.33%) **14 (63.33%)** 15 (10%) 16 (16.67%) 18 (3.33%)	11 (20%) **12 (36.67%)** 13 (13.33%) 14 (20%) 15 (3.33%) 16 (6.67%)	**11 (56.67%)** 12 (6.67%) 13 (6.67%) 14 (10%) 15 (3.33%) 16 (3.33%)	9 (3.33%) 10 (3.33%) **11 (53.33%)** 12 (30%) 13 (3.33%) 14 (3.33%) 15 (3.33%)
	1000	14 (16.67%) **15 (46%)** 16 (10%) 17 (16.67%) 18 (3.33%) 20 (3.33%) 21 (3.33%)	**14 (73.33%)** 15 (16.67%) 17 (3.33%) 18 (3.33%) 19 (3.33%)	11 (3.33%) 12 (13.33%) 13 (10%) **14 (63%)** 15 (6.67%) 16 (3.33%)	11 (10%) 12 (40%) 13 (10%) **14 (33.33%)** 15 (10%) 17 (3.33%) 20 (3.33%)	**11 (53.33%)** 12 (30%) 13 (6.67%) 14 (6.67%) 18 (3.33%)

Table 5. Selected models that are not significant different from the model with minimum test error mean (confidence level = 0.05).

	models (%)	noise's variance				
		0.1	0.25	0.5	0.75	1
data set size	250	9 (6.67%) 10 (3.33%) 11 (36.67%) **12 (46.67%)** 13 (6.67%)	9 (6.67%) 10 (3.33%) **11 (90%)**	8 (46.67%) 9 (33.33%) **10 (60%)**	**8 (96.67%)** 9 (3.33%)	1 (10%) 3 (10%) 4 (3.33%) **8 (73.33%)** 9 (3.33%)
	500	**12 (53.33%)** 13 (46.67%)	**11 (100%)**	9 (20%) **10 (46.67%)** 11 (33.33%)	8 (40%) **9 (53.33%)** 10 (6.67%)	**8 (76.67%)** 9 (20%) 11 (3.33%)
	1000	**14 (100%)**	11 (3.33%) **12 (76.67%)** 13 (20%)	**11 (100%)**	9 (6.67%) 10 (16.67%) **11 (76.67%)**	8 (10%) **9 (60%)** 10 (26.67) 11 (3.33%)

3.2 Radial Basis Function Neural Networks

RBF neural network having one hidden layer for which the combination function is the Euclidean distance between the input vector and the weight vector. We use the exp activation function, so the activation of the unit is a Gaussian "bump" as a function of the inputs. The placement of the kernel functions has been accomplished using the k-means algorithm. The width of the basis functions has been set to

$$\sigma = \frac{\|max(x_i - x_j)\|}{\sqrt{2 \cdot n}} \qquad (3)$$

where n is the number of kernels.

The second layer of the network is a linear mapping from the RBF activations to the output nodes. Output weights are computed via matrix-pseudoinversion.

Tables 6, 7 and 8 show the results according to the experimental function (1) for each of the following sample sizes n = 50, 100, 500 and for each of the following N (0,0.5), N(0,3) and N(0,5) gaussian noise.

Table 6. Models with minimum generalization error mean.

	models	noise's variance		
		0.5	3	5
data set size	50	11	11	10
	100	13	12	11
	500	14	14	13

Table 7. Models with minimum test error mean and percentage of times that occurred in our experiments.

	models (%)	noise's variance		
		0.5	3	5
	50	8 (16.67%) 12 (66.67%) 13 (16.67%)	9 (20.33%) 10 (43.33%) 11 (16.67%) 13 (16.67%)	**10 (86.67%)** 11 (13.33%)
data set size	100	12 (30%) **13 (53.33%)** 15 (16.67%)	**10 (40%)** 11 (26.67%) 12 (16.67%) 13 (16.67%)	10 (20%) **11 (60%)** 12 (20%)
	500	14 (16.67%) **15 (83.33%)**	12 (36.67%) **13 (43.33%)** 15 (20%)	11 (23.33%) **12 (60%)** 13 (16.67%)

Table 8. Selected models that are not significant different from the model with minimum test error mean (confidence level = 0.05).

	models (%)	noise's variance		
		0.5	3	5
	50	6 (16.67%) 8 (20%) **9 (63.33%)**	7 (36.37%) **8 (63.33%)**	**7 (80%)** 8 (20%)
data set size	100	**9 (83.33%)** 10 (16.67%)	**8 (100%)**	7 (23.33%) **8 (76.67%)**
	500	**10 (73.33%)** 11 (26.67%)	**9 (76.67%)** 10 (23.33%)	**9 (100%)**

Tables 9, 10 and 11 show the results according to the experimental function (2) for each of the following sample sizes n = 250, 500, 1000 and for each of the following N (0,0.25), N(0,0.5) and N(0,1) gaussian noise.

Table 9. Models with minimum generalization error mean.

	models (%)	noise's variance		
		0.25	0.5	1
data set size	250	14	13	12
	500	14	12	12
	1000	17	15	13

Table 10. Models with minimum test error mean and percentage of times that occurred in our experiments.

	models (%)	noise's variance		
		0.25	0.5	1
data set size	250	12 (6.67%) 13 (33.33%) **14 (40%)** 15 (20%)	11 (33.33%) 12 (6.67%) **13 (40%)** 15 (20%)	10 (16.33%) 11 (23.33%) **12 (30.67%)** 13 (16.33%) 14 (13.33%)
	500	13 (26.67%) **14 (40%)** 15 (33.33%)	**12 (43.33%)** 14 (36.67%) 15 (20%)	**11 (33.33%)** 12 (26.67%) 13 (23.33%) 15 (16.67%)
	1000	**14 (40%)** 15 (20%) 16 (13.33%) 17 (13.33%) 18 (13.33%)	12 (13.33%) 13 (10%) **14 (63.33%)** 15 (10%) 16 (3.3%)	11 (20%) 12 (3,33%) **13 (60%)** 14 (16.67%)

Table 11. Selected models that are not significant different from the model with minimum test error mean (confidence level = 0.05).

	models (%)	noise's variance		
		0.25	0.5	1
data set size	250	10 (13.33%) 11 (20%) **12 (46.67%)** 13 (20%)	9 (40%) **10 (60%)**	5 (20%) **7 (46.67%)** 9 (23.33%) 10 (10%)
	500	**12 (46.67%)** 13 (53.33%)	10 (36.67%) 11 (63.33%)	**9 (76.67%)** 10 (23.33%)
	1000	**13 (100%)**	**11 (100%)**	9 (20%) **10 (76.67%)** 11 (3.33%)

In all the experiments, when the data set was large enough, the parametric tests were applied. Otherwise the repeated measures Anova assumptions were not satisfied and then, nonparametric test were used instead.

A systematic underfitting is observed in the method proposed. This underfitting is produced by: the use of bootstrapping techniques in the simulated data sets, and the strategy followed by our method. However, we can guarantee with high statistical reliability that the performance of the selected model is as good as the least test error mean.

4 Conclusions

In this work we have presented a model selection criterion that consists on finding the group of models that are not significant different from the model with the minimum test error mean, in order to select the model with less complexity (Occam's Razor).

The experimental results show that this criterion produces underfitting when the data set size is small but it works very well when the data set size is large enough, thus our method improves the widely used criterion of selecting the model with the least test error mean. To avoid underfitting problems another selection criterion could be applied at the cost of a larger complexity.

5 References

1. Chen, T., Seneta, E.: A stepwise rejective test procedure with strong control of familywise error rate. University of Sidney, School of Mathematics and Statistics, Research Report 99-9, March (1999)
2. Dean A., Voss, D.: Design and Analysis of Experiments. Springer-Verlag New York (1999)
3. Don Lehmkuhl, L: Nonparametric Statistics: Methods for Analyzing Data Not Meeting Assumptions Required for the Application of Parametric Tests, Journal of Prosthetics and Orthotics, 3 (8) 105-113 (1996)
4. Efron, B., Tibshirani, R.: Introduction to the Bootstrap, Chapman & Hall, (1993)
5. Girden, E. R.: Anova Repeated Measures, Sage Publications (1992)
6. Hochberg, Y., Tamhane A. C.: Multiple Comparison Procedures, Wiley (1987)
7. Hollander, M., Wolfe, D. A.: Nonparametric Statistical Methods, Wiley (1999)
8. Kearns, M., Mansour, Y.: An experimental and theorical comparison of model selection methods. Machine Learning, 27(1), (1997)
9. Minke, A.: Conducting Repeated Measures Analyses: Experimental Design Considerations, Annual Meeting of the Southwest Educational Research Association, Austin, (1997)
10. Pizarro, J., Guerrero, E., Galindo, P.: A statistical model selection strategy applied to neural networks. Proceedings of the European Symposium on Artificial Neural Networks Vol 1, pp. 55-60, Bruges (2000)
11. Vila, J.P., Wagner, V., Neveu, P.:Bayesian nonlinear model selection and neural networks: a conjugate prior approach. IEEE Transactions on neural networks, vol 11,2, march (2000)
12. Zar, J. H.: Biostatistical Analysis, Prentice Hall (1996)

Extension of HUMANN for Dealing with Noise and with Classes of Different Shape and Size: A Parametric Study

Patricio García Báez[1], Carmen Paz Suárez Araujo[2] and Pablo Fernández López

[1] Department of Statistics, Operating Research and Computation, University of La Laguna, 38071 La Laguna, Canary Islands, Spain
`pgarcia@ull.es`
[2] Department of Computer Sciences and Systems, University of Las Palmas de Gran Canaria, 35017 Las Palmas de Gran Canaria, Canary Islands, Spain
`cpsuarez@dis.ulpgc.es`

Abstract. In this paper an extension of HUMANN (hierarchical unsupervised modular adaptive neural network) is presented together with a parametric study of this network in dealing with noise and with classes of any shape and size. The study has been made based on the two most noise dependent HUMANN parameters, λ and μ, using synthesised databases (bidimensional patterns with outliers and classes with different probability density distribution). In order to evaluate the robustness of HUMANN a Monte Carlo [1] analysis was carried out using the creation of separate data in given classes. The influence of the different parameters in the recovery of these classes was then studied.

1 Introduction

For a classification process to work correctly and efficiently it will have to be able to work in environments which are non-stationary, noisy and non-linear, without prior information about the number of classes in the data. A classification scheme which has all these characteristics leads to systems based on unsupervised artificial neural networks with an important process, namely clustering.

Many clustering methods have been used in pattern recognition and signal processing problems over the years. Non neural methods such as the learning vector quantization algorithm [2], the *k*-means algorithm, the ISO-DATA [3], the Autoclass [4] and Snob [5] algorithms. There are also many neural methods such as Kohonen's SOM [6], ART2 [7], Dignet [8], and others. Most of these are significantly limited in two important aspects: *a priori* knowledge of the number of existing classes in the data and insensitivity to noise. The best non neural methods in dealing with these problems are Autoclass and Snob, and for neural methods ART2 and Dignet, though with many restrictions.

In this paper an extension of a hierarchical unsupervised modular adaptive neural network (HUMANN) [9] is presented. This extension makes it more resistant to noise and to sets of data where the classes within the sets have different distribution densities and, consequently, different sizes and different shapes for the class patterns.

HUMANN has been successfully tested in blind clustering applications with data with a high level of noise [9] [10].

2 HUMANN

HUMANN is a biologically plausible unsupervised neural modular feedforward network with an input layer and three hierarchically organised modules, with different neurodynamics, connection topologies and learning laws. The three modules are: Kohonen's s Self-Organizing Map (SOM) module, Tolerance module and Labelling module. HUMANN implements the last two stages of the general approach of the classification process, which has three stages: a) feature extraction, b) template generation, c) discrimination (labelling), in a transparent and efficient way. The feature extraction must be implemented by application-dependent pre-processing modules.

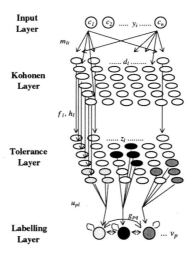

Fig. 1. HUMANN: Neural structure and topology of connections.

The SOM [6][11] module implements a nonlinear "projection" from a high-dimensional space onto a two-dimensional array. The HUMANN uses this kind of neural structure because their main feature is the formation of topology-preserving feature maps and approximation of the input probability distribution, by means a self-organizing process which can produce feature´s detectors.

The Tolerance module is the main module responsible for the robustness of HUMANN against noise. The functional layer of this module is the Tolerance layer. Its topology is a two-dimensional array which has the same dimension as the Kohonen layer and an interconnection scheme one to one with that previous layer. The main objective of this layer is to compare the fitting between the input patterns and the Kohonen detectors. If the goodness of the fit is not sufficient the pattern is regarded as an outlier and is discarded. The weights of this layer are responsible for storing the mean of the fits between the inputs and the Kohonen detectors when this neuron is the winner. This is a new concept called the "Tolerance margin". The goodness of the representation of a pattern by a detector will be a function of the ratio of the Euclidean distance between both of them and the Tolerance margin of the detector.

We present a new design of this module which extends its adaptation to noise and to different pattern distribution densities. This extension looks at the existence of SOM detectors located in regions with a low pattern density, empty regions and regions with outliers. Such detectors will be inhibited depending on the inhibition process of the ratio of neural firing as a function of the Tolerance margin of said detectors. This will represent the density of patterns in the receptive cone. This new operating scheme introduces a new neural mechanism, the adaptive firing ratio, f. This scheme is embodied in its neurodynamics, Eq. (1), and in its learning law, Eqs. (2) and (3).

$$z_l = \begin{cases} 0 & \text{if } \frac{f_l}{h_l^2} \leq \xi \text{ or } d_l \geq \lambda h_l \\ 1 - \frac{d_l}{\lambda h_l} & \text{otherwise} \end{cases} \quad (1)$$

Where h_l is the weight of the neuron l which expresses the Tolerance margin, f_l is the adaptive firing ratio of neuron l and λ and ξ two adjustable parameters which determine the degree of tolerance for the non acceptance of the input pattern.

$$\Delta h_l = \begin{cases} \beta(t)(d_c - h_c) - \eta(t)h_c & \text{if } l = c \\ -\eta(t)h_l & \text{otherwise} \end{cases} \quad (2)$$

$$\Delta f_l = \begin{cases} \beta(t)(1 - f_c) & \text{if } l = c \\ -\beta(t)f_l & \text{otherwise} \end{cases} \quad (3)$$

Where c is the winning unit of the Kohonen layer, $\beta(t)$ is the decreasing learning ratio and $\eta(t)$ is a weight decaying ratio that continuously decreases the tolerance for very infrequent patterns.

The labelling module [9] implements the discrimination task. It maps the outputs of a neural assembly belongs to the Tolerance layer which have been activated by a class, into different clusters represented by labelling units. This module present a full connection topology and a dynamic dimension, which is fitted to the number of clusters detected in the data set. This fitting is controlled by local connections between the labelling units. They are symmetrical and can be defined as structural specialisation for neurotransmission that do no produce a physiological response in the receiving cell, in the same way as silent synapses [12] in biological neural circuits.

3 Parametric Study

In this section we present an analysis of the performance of HUMANN in dealing with noise, always present in all real time data environments. The classes present in the data have different probability density distribution and, therefore, different shape and size. This analysis has been carried out using a parametric study of HUMANN. We show that HUMANN is a highly noise resistant architecture and one which should be able to build decision regions that separate classes of any shape and size.

3.1 Data Environment

The study was carried out with a data environment made up of bidimensional patterns, created using normal multivariate distributions [1] [13]. The number of real classes in the set of analysed data is three, and the patterns belonging to each class are created in accordance with the following density function:

$$P(x_i) = \frac{1}{\sqrt{2\pi}\sigma_i} e^{-\frac{1}{2\sigma_i^2}(x_i - \mu_i)^2} \quad (4)$$

Where i represents the dimension, between 1 and 2, and μ_i and σ_i respectively the average and standard deviation of the class in the dimension i.

This data environment has the two characteristics which form the basis of the performance analysis of HUMANN, namely classes with different distribution density of the patterns which comprise it and outlier type noise. Such sets of data with these characteristics will be called D^3 data and outlier data respectively. The training sets used in the analysis of HUMANN will be made up of 10,000 of these patterns.

The D^3 data have a variable σ_1 of the second class, while the remaining parameters μ_i and σ_i of all the classes are fixed, Table 1.

Table 1. Values of the parameters which define the classes of the set of data D^3

Class	μ_0	μ_1	σ_0	σ_1
1	0	0	1	1
2	8	0	1	Variable
3	0	8	1	1

With σ_1 of class 2 varying between 0.5 and 3, obtaining 26 training sets. The patterns in each class have an equal distribution probability. Later, each pattern is regarded as belonging to the class whose centroid is nearest to it in Euclidean distance.

The second type of data of our environment has different levels of outlier noise. This is created by introducing in equation (4) values of σ_0 and σ_1 9 times higher than the patterns without outliers. The outlier level of noise varies between 0% and 80%, in this way obtaining 26 training sets. The parameters of the different classes for this type of data are equal than Table 1, except that σ_1 is 1 for class 2.

In an outlier set of data the limits of each class are located in the region where the distribution density of outliers is equal to that for patterns. The allocation to each class of patterns within these limits was carried out in the same way as for D^3 data.

3.2. Results

In the analysis of the performance of HUMANN in dealing with D^3 and outlier data, the most significant HUMANN parameters were studied in noise tolerance and in capacity to build decision regions that separate classes of any shape and size: λ and ξ belonging to the Tolerance module. In the labelling layer ρ would seem to have the least effect on the correct performance of HUMANN. It must always be below 1, with a good and wide working range being noted for values between 0 and 0.3.

A study was made of the parameter intervals that would provide the best performance of HUMANN, as well as the best parametric adjustment for each situation.

In this parametric study 51 values were used for each parameter in a given range during the learning process of the network with all the training sets. For each training run 10 cycles were carried out, with all the patterns of the training set being introduced once in each cycle.

The goodness of the classification was evaluated with the labelling of each cluster created in a training run with the number of the class to which most of the patterns in the cluster belong. The success ratio was defined as the ratio between the number of patterns classified in a cluster, which has the same label as the real class of the pattern, and the total number of patterns (=10,000). The analysis of the results shows the success ratio of HUMANN as well as the number of clusters created as a function of the different sets of data used and the different values of the parameters studied.

3.2.1 Parameter λ

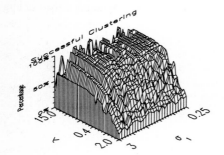 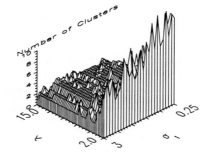

Fig. 2. Success percentage for D^3 data. $\lambda \in [2,15]$, $\xi = 0.16$

Fig. 3. Number of clusters created for D^3 data. $\lambda \in [2,15]$, $\xi = 0.16$

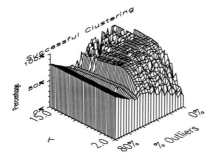 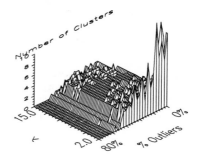

Fig. 4. Success percentage for outlier data. $\lambda \in [2,15]$, $\xi = 0.16$

Fig. 5. Number of clusters created for outlier data. $\lambda \in [2,15]$, $\xi = 0.16$

In Fig. 2 and Fig. 3 the effect of λ in the performance of HUMANN with respect to D^3 data can be seen, and in Fig. 4 and Fig. 5 with respect to outlier data. For high values of σ_l class 2 in general is expanded and the results are degraded, as its elements are regarded as outlier elements due to their low density. It can be seen how HUMANN is able to work well even with large differences of density between classes.

It can be seen from Fig. 3 how, as λ varies, the initially high number of classes decreases to the correct value. This is due to the fact that the increase in λ facilitates neuronal elimination through functional and synaptic reorganisation in the labelling layer when a higher number of neurons are fired with greater strength in the Tolerance layer. This explains why, for values of $\lambda > 12.5$, instabilities appear in the number of classes created, affecting the success percentage.

In Fig. 4 it can be seen that for $\lambda \in [6.5, 12.5]$ the network is able to classify correctly sets of data with up to 51% of outlier level. Note that if the number of classes created were zero the success percentage would coincide with the percentage of existing outliers, since all the samples would be classified as such. This is what produces the inclined plane in Fig. 4 and Fig. 8.

3.2.2 Parameter ξ

This parameter influences to a large extent the performance of the network against noise. This can be clearly seen in Fig. 8 and Fig. 9. The effect of this parameter is less notable in problems related to the shapes of the classes, Fig. 6 and Fig. 7.

For D^3 data the greatest influence of ξ can be seen when there is one class, in this case class 2, which is excessively expanded ($\sigma_l \cong 3$). If parameter ξ has a high value the class will be regarded as outlier noise, hence the reduction in the success ratio to 66% (the other two classes being well classified), Fig. 6, and the decrease in the number of clusters created from 3 to 2, Fig. 7.

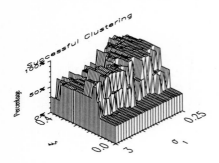 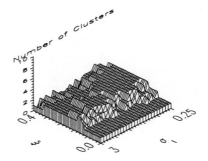

Fig. 6. Success percentage for D^3 data. $\xi \in [0,0.4]$, $\lambda = 10$

Fig. 7. Number of clusters created for D^3 data. $\xi \in [0,0.4]$, $\lambda = 10$

With low values of ξ some neurons from the tolerance layer with a receptive field of low firing density are not disregarded. This facilitates neural elimination through functional and synaptic reorganisation and can make the system detect a single cluster, with an error percentage of 33%. As this parameter increases the neurons with a receptive field of low firing density in the tolerance layer will be inhibited. This means that for excessively high values of ξ there are hardly any useful neurons, creating zero classes and achieving a success percentage equal to the percentage of outliers present (as explained above). For low values of active neurons with a receptive field of low firing density the percentage of correct firings falls while neural elimination through functional and synaptic reorganisation in the labelling layer is benefited.

From this analysis it can be deduced that parameter ξ must be set as a function of the level of outliers expected to achieve the best results.

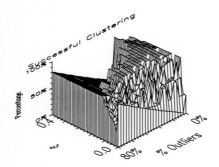 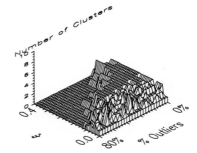

Fig. 8. Success percentage for outlier data. $\xi \in [0,0.4]$, $\lambda = 10$

Fig. 9. Number of clusters for outlier data. $\xi \in [0,0.4]$, $\lambda = 10$

4 Conclusions

A parametric study is hereby presented of an extension of HUMANN, a neural architecture conceived for classification processes with blind clustering and for greater tolerance to noise and classes with different shapes and sizes. With this extension the capacities of HUMANN are expanded to include discrimination based on the densities of the existing patterns within the input space. The Tolerance module is redesigned with the introduction of a mechanism known as 'adaptive firing ratio'. By means of a parametric study the robustness of HUMANN with respect to outlier noise is demonstrated as well as its stability in environments where the classes can have different shapes and densities of patterns. At the same time, it is possible to determine the optimum values for λ and ξ in order to achieve the best performance of HUMANN.

References

1. Afzal Upal, M.: Monte Carlo Comparison of Non-Hierarchical Unsupervised Classifiers. Master's thesis, University Of Saskatchewan (1995).
2. Makhoul, J., Roucos, S., Gish, H.: Vector quantization in speech coding. In: Proceedings of the IEEE, Vol. 73, Num. 11 (1985) 1551-1588.
3. Anderberg, M.R.: Cluster Analysis for Applications. Academic Press, New York (1973).
4. Cheesemann, P., Kelly, J., Self, M., Sutz, J., Taylor, W., Freeman, D.: Autoclass: A Bayesian clasification sustem. In: Proceedings of the Fifth International Conference on Machine Learning, San Mateo, CA (1988) 54-64.
5. Wallacem C.S., Dowe, D.L.: Intrinsic classification by MML - the Snob program. In: Proceedings of the Seventh Australian Joint Conferencie on Artificial Intelligence, Singapore (1994) 37-44.
6. Kohonen, T.: The Self-Organizing Map. In: Proceedings of IEEE, Special Issue on Neural Networks, Vol. 78, Num. 9 (1990) 1464-1480.
7. Carpenter, G.A., Grossberg, S.: ART 2: Self-Organization of Stable Category Recognition Codes for Analog Input Patterns. In: Applied Optics, Vol. 26 (1987) 4919-4930.
8. Thomopoulos, S.C.A., Bougoulias, D.K., Wann,C.D.: Dignet: An Unsupervised Learning Algorithm for Clustering and Data Fusion. In: IEEE Transactions on Aerospace and Electronic Systems, Vol. AES-31, Num. 1 (1995) 21-38.
9. García, P., Fernández, P., Suárez, C.P.: A Parametric Study of Humann in relation to the Noise. Application to the Identification of Compounds of Environmental Interest. In: Systems Analysis Modelling Simulation, in press.
10. García, P., Suárez, C.P., Rodríguez, J., Rodríguez, M.: Unsupervised Classification of Neural Spikes With a Hybrid Multilayer Artificial Neural Network. In: Journal of Neuroscience Methods, Vol. 82 (1998) 59-73.
11. Kohonen, T.: Self-Organizing Maps, 2nd ed.. Springer Series in Information Sciences, Vol. 30. Springer-Verlag, Berlin Heidelberg New York (1997).
12. Atwood, H.L., Wojtowicz, J.M.: Silent Synapses in Neural Plasticity: Current Evidence. In: Learning & Memory, Vol. 6 (1999) 542-571.
13. Afzal Upal, M., Neufeld, E.M.: Comparison of Unsupervised Classifiers. In: Proceedings of the Eleventh International Joint Conference on Artificial Intelligence, San Mateo, CA (1989) 781-787.

Evenet 2000: Designing and Training Arbitrary Neural Networks in Java

Evelio J. González, Alberto F. Hamilton, Lorenzo Moreno, José F. Sigut, Roberto L. Marichal

Department of Applied Physics. University of La Laguna.
La Laguna 38271, Tenerife, Spain.
Tel: +34 922 318286 / Fax: +34 922 319085
{evelio, alberto, lorenzo, sigut, marichal}@cyc.dfis.ull.es

Abstract. In this paper, Evenet-2000, a Java-Based neural network toolkit is presented. It is based on the representation of an arbitrary neural network as a block diagram (these blocks are, for example, summing junctions or branch points) with a set of simple manipulation rules. With this toolkit, users can easily design and train any arbitrary neural network, even time-dependent ones, avoiding the complicated calculations that the means of establishing the gradient algorithm requires when a new network architecture is designed. Evenet-2000 consists of three parts: a calculation library, a user-friendly interface and a graphic network editor with all the Java advantages: encapsulation, inheritance, powerful libraries…

1 Introduction

The means of establishing the gradient algorithm formulas when a new neural network is proposed usually requires long and complicated calculations. Evenet-2000, a Java-Based neural network toolkit, allows users to avoid this work. For this, it develops an approach presented by Wan and Beaufays [1] to derive gradient algorithms for time-dependent neural networks, using the *Signal Flow Graph theory*. This approach is based on a set of simple block diagram transformation and manipulation rules, but Evenet-2000 users do not need at all to know these rules. Moreover, the designed structure makes it not to be limited to algorithms based on the gradient. Actually, for example, algorithms based on genetic evolution are included in the tool.

The aim of this article is to present this program and its main characteristics. We think it may be an important help for neural network researchers and users, due to the increasing use of this tool in several fields (robotic systems [2], patterns classifier/detectors in biomedical signals [3]…)

2 Theoretical Base

Regarding the gradient algorithms in neural networks, several researchers [1],[4],[5] have shown that there is a reciprocal nature to the forward propagation of the states and the backward propagation of gradient terms, and this reciprocity appears in all network architectures. Based on these properties, and using the *signal flow graph theory*, they have deduced a general method for automatic determination of gradient in an arbitrary neural network. In this paper we will be interested in Wan and Beaufays approach. First step is representing the arbitrary network as a block diagram. These blocks can be summing junctions, branching points, univariate functions, multivariate functions or time-delay operators. A neuron, for example is a summing junction followed by a univariate function (sigmoid, tanh). Networks are created from these elements.

From this block diagram, an *adjoint* network can be built by reversing the flow direction in the original network, performing the following operations:

1. – Summing junctions are replaced with branching points.
2. – Branching points are replaced with summing junctions.
3. – Univariate functions are replaced with their derivatives.
4. – Multivariate functions are replaced with their Jacobians.
5. – Delay operators are replaced with advance operators.
6. – Outputs become inputs. [1]

This way, the gradient algorithm can be implemented without any chain rule expansion.

3 Description of the Tool

The fact of transforming any neural network into a block diagram suggests *object-oriented programming* as implementation method, so Java, the most popular and powerful object-oriented language, has been chosen for development of the tool. Evenet-2000 consists of three parts: a calculation library, a user-friendly interface and a graphic neural network editor. They will be described in the following sections.

3.1 Calculation Library

Evenet-2000 develops the theory shown above with object-oriented programming. Every basic element is assigned an object, knowing its preceding element and how many successors it is connected to. This way, the *adjoint* philosophy is easily implemented. When network output is required, output elements are asked for their own output. As they need their predecessor element outputs, the output elements ask them. These predecessor elements need their own predecessor element outputs, so there is propagation until input elements are reached.

Regarding the gradient calculation, when network output is given, the error vector, whose components are output element derivatives, is calculated. Once these terms have been assigned to the output elements, backpropagation starts. Every element calculates its derivative and sends it to its predecessors until input elements are reached again. In this way the gradient vector is ready to be used and no additional flow graph construction is required.

The join of these basic elements can form other network elements. For example, a *Neuron* object can be built, joining *Summing Junction, Univariate Function* and *Branching Point* objects.

But these elements are not sufficient for a complete library. They must be joined to other types of objects that implement arbitrary neural network trainings. For this, Evenet-2000 calculation library follows the UML diagram shown in Figure 1.

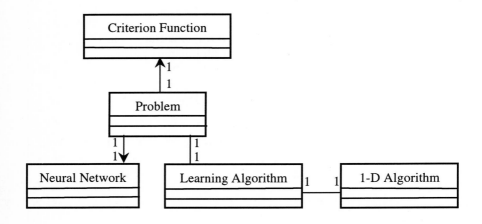

Fig. 1. Evenet-2000 Calculation Library Diagram

With this implementation, training and optimization problems have been uncoupled from the neural network structure specification. Problem object sets the chosen neural network structure, the learning algorithm and the criterion function (problem object is the connection among these objects). In this structure, the problem object sets the network inputs, asks the neural network object for the output vector. Then asks the criterion function to calculate the error vector, and finally the neural network calculates the gradient vector. Once this process has happened, the problem object sends the gradient vector to the algorithm. The algorithm calculates the new weight vector, asking the problem for the error or cost function. Unidimensional (1-D) optimisation could be required, so 1-D algorithms have been included. These steps are repeated until the design conditions are reached.

It is important to remark that algorithm classes can be applied to other optimization problems that are not based on neural networks, in the spirit of object-oriented programming. Moreover, this structure can be applied to algorithms that are not based on gradient determination. Actually, descent gradient, conjugated gradient, the Boltzmann machine, random search, genetic algorithms are available for Backpropagation (BP), Real Time Recurrent Learning (RTRL) and Adaptive Network-Based Fuzzy Inference System (ANFIS) problems. In the other hand, Evenet-2000 also has the following 1-D algorithm classes: parabolic interpolations and golden section search.

This calculation library obviously can be used independently from the rest of the program. As program classes are fully documented, users can easily design their problems and code them. For example, the following code creates and trains a 3-layers MLP (with 3, 4 and 3 neurons in each layer), with Sigmoid as activation function. Chosen learning method is Descent Gradient with Golden Search optimization.

```
/* number of layers */
int numberLayers = 3
/* vector of neurons in each layer, 3 elements */
int[] neuronsLayer = new int[number Layers];
neuronsLayer= {3,4,3};
/* activation function */
Sigmoid function= new Sigmoid();
/* creating MLP */
MLP net = new MLP (numberLayers, neuronsLayer,
      function);
/* patterns */
double[] patternVector = new double[2][6];
patternVector ={ {1,1,1,0,1,0} , {1,0,0,1,1,0} };
/* creating criterion function (cuadratic) */
Cuadratic criterionFunction= new Cuadratic();
Problem problem= new Problem(net, patternVector,
      criterionFunction);
/* initiating weight vector*/
problem.inicializateWeights();
/* creating  Algorithm object */
DescentGradient algorithm = new DescentGradient
      (problem, problem.tellMeWeights());
/* initiating derivative vector */
net.inicializateDerivative();
/* creating Golden Search object */
GoldenSearch algorithm1D = new
      GoldenSearch(algorithm);
/* joining algorithm1D to algorithm */
algorithm.tellMe1DAlgorithm(algorithm1D);
/* learning starts: desired error, number of steps,
initial learning rate */
algorithm.startLearning(0.001,1000,0.25);
/* optimal weight set */
double[] optimalSet = algorithm.tellMeOptimalSet();
/* minimal error */
```

```
double error = algorithm.tellMeError();
/* error evolution (can be used for graphical
    representation) */
double[] errorEvolution =
    algorithm.tellMeErrorEvolution();
```

As can be seen, with few code lines, users can easily design and train a neural network, without being limited to standard architectures. Moreover, as Evenet-2000 documentation is available, they can create new objects or methods through Java properties (inheritance, polymorphism,...).

3.2 User-Friendly Interface

Although the calculation library is complete and easy to use, people that are not used to object-oriented programming could not make the most of it. Because of this, the tool has been improved with a user-friendly interface. It allows training directly a MLP or a recurrent network. By way of example, a MLP case is analysed below.

When this case is selected from program main menu, a frame like the shown in Figure 2 appears. From its menu bar, desired learning method, criterion function and optimisation algorithm can be selected. Number of layers, neurons in each layer and training parameters can be easily changed through secondary frames.

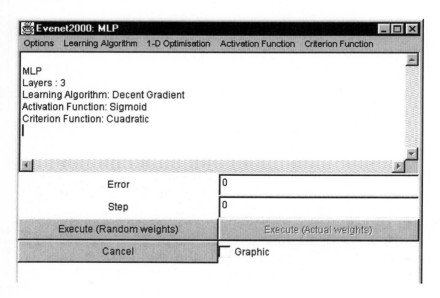

Fig. 2. Evenet-2000 MLP Frame (Fragment)

When training pairs have been loaded, users can train the network yet. Initial weight set can be fixed also. If the Graphic checkbox is selected, the program shows the training error evolution graphic after the training has finished. The error and iteration number are shown in their respective text fields, and results can be saved in a text file that can be analysed later.

3.3 Graphic Editor

Evenet-2000 user-friendly interface allows users to train MLP or recurrent networks with no code. But this interface has not taken advantage of the possibility of training a neural network with an arbitrary architecture yet. For this, a graphic editor has been included in the tool. This editor, whose frame is shown in Figure 3, can be selected from the program main menu.

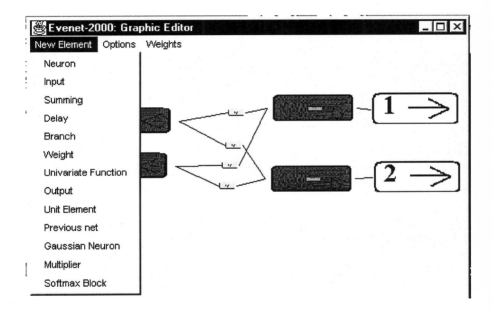

Fig. 3. Evenet-2000 Graphic Editor Frame (Fragment)

This graphic editor allows creating an arbitrary neural network, users only have to select its elements from the menu (as can be seen in Figure 3) and connect them. Actually, available elements are: neurons, summing points, delays, branch points, weights, univariate functions, inputs, outputs, unit elements (elements whose outputs are 1, useful for bias implementation), previous loaded nets, neurons whose activation is a Gaussian Function, multipliers and blocks that implements the Softmax Function. The icons can be dragged by the users. Designs can be saved in a text file and loaded as new elements (modularity), so users can develop their own neural network library. Activation function, in the case of univariate functions, can be chosen through a pop-up menu. As it was noted above, neural network designs are saved in a text file that can be modified without the help of the editor. This file can be loaded by other program window, similar to the MLP training one (described above), for its training. This way, neural networks with arbitrary architecture can be designed and trained with no code. This is obviously a great advantage for the users that do not need any complicated calculations at all.

4 Availability

Evenet2000 is available at http://www.evenet.cyc.ull.es

5 Conclusions

In the present paper, Evenet-2000, a Java-Based neural network toolkit has been presented. With this tool, users can design a neural network with arbitrary architecture and train it with no complex gradient calculation. It is based on the application of *Signal Flow Graph theory* to neural networks. According to this theory, any kind of neural network can be transformed into a block diagram. Every neural network element is assigned a Java object that can be connected to other ones.

This tool is divided into three parts: a calculation library, a user-friendly interface and a graphic editor. For the calculation library, a structure that is not limited to gradient - based algorithms has been implemented. Actually several learning algorithms (gradient descent, random search, genetic algorithms...) and 1-D optimisation methods (golden section, parabolic interpolation) have been included for three kinds of problems (BP, RTRL and ANFIS). Moreover, Evenet-2000 allows users to include easily new neural network elements and algorithms in it.

For its generality and simplicity, we think this tool is highly recommended for all those researchers interested in neural networks.

References

1. Eric A. Wan and Françoise Beaufays "Relating real-time backpropagation and backpropagation through time. An application of flow graph interrreciprocity." Neural computation, volume 6 (1994), number 2, pp. 296-306.
2. Acosta L., Marichal G.N., Moreno L., Rodrigo J.J., Hamilton A., Méndez J.A., "Robotic system based on neural network controllers". Artificial intelligence in Engineering, vol 13, number 4, pp 393-398, 1999
3. Piñeiro J.D., Marichal R.L., Moreno L., Sigut J., Estevez I, Aguilar R., Sanchez J.L. and Merino J. "Evoked potential feature detection with recurrent dynamic neural networks." Proceedings in International ICSC/IFAC Symposium on Neural Computation NC'98. Vienna Austria, 1998
4. Campolucci P., Marchegiani A., Uncini A., Piazza F., "Signal-Flow-Graph Derivation of On-line Gradient Learning Algorithms" IEEE International Conference on Neural Networks, Houston (USA), June 1997.
5. Osowski S., Herault J. "Signal Flow Graphs as an Efficient Tool for Gradient and Exact Hessian Determination". Complex Systems, Vol 9, 1995.

Neyman-Pearson Neural Detectors

Diego Andina[1] and José L. Sanz-González[1]

Universidad Politécnica de Madrid
Departamento de Señales, Sistemas y Radiocomunicaciones, E.T.S.I.
Telecomunicación, Madrid, Spain
andina@gc.ssr.upm.es

Abstract. This paper is devoted to the design of a neural alternative to binary detectors optimized in the Neyman-Pearson sense. These detectors present a configurable low probability of classifying binary symbol *1* when symbol *0* is the correct decision. This kind of error, referred in the scientific literature as *false-positive* or *false alarm probability* has a high cost in many real applications as medical Computer Aided Diagnosis or Radar and Sonar Target Detection, and the possibility of controlling its maximum value is crucial. The novelty and interest of the detector is the application of a Multilayer Perceptron instead of a classical design. Under some conditions, the Neural Detector presents a performance competitive with classical designs adding the typical advantages of Neural Networks. So, the presented Neural Detectors may be considered as an alternative to classical ones.

1 Introduction

Neural Networks can have interesting robust capabilities when applied as binary detectors. This type of networks have proved their abilities in classifying problems, and we could reduce the binary detection problem as having to decide if an input has to be classified as one of two outputs, *0* or *1*. Optimizing a Multilayer Perceptron (MLP) in the Neyman-Pearson sense means an optimization process to achieve an arbitrary low probability of classifying binary symbol *1* when symbol *0* is the correct decision. The MLP under study is optimized in that sense. The results, compared with those of the optimal detector, show the performance capabilities of the Multilayer Perceptron.

In the following sections we develop the main topics:

(a) The network structure design. Variations in performance are analyzed for different numbers of inputs and hidden layers, and different number of nodes in these hidden layers.

(b) The network training. Using the BackPropagation (BP) algorithm with momentum term, we study the training parameters: initial set of weights, threshold value for training, momentum value, and the training method. The preparation of the training set, whose key parameter is the Training Signal-to-Noise Ratio (TSNR), is discussed; also, it is shown how to find the TSNR value that maximizes the detection probability (P_d) for a given false alarm probability (P_{fa}), as is required by the Neyman-Pearson criterion.

(c) The results are analyzed, and appropriate conclusions are extracted.

2 The Neyman-Pearson Neural Detector

The detector under consideration, refered as Neyman-Pearson Neural Detector (NP-ND) is a modified envelope detector [1, 2]. The binary detection problem is reduced to decide if an input complex value (the complex envelope involving signal and noise) has to be classified as one of two outputs, 0 or 1. The need of processing complex signals with an all-real coefficient NN, requires to split off the input in its real and imaginary parts (the number of inputs doubles the number of integrated samples). The input $r(t)$ is a band-pass signal, and the complex envelope $x(t) = x_C(t) + jx_S(t)$ is sampled each T_0 seconds. Then

$$x(kT_0) = x_C(kT_0) + jx_S(kT_0), k = 1, \ldots, M. \ (j = \sqrt{-1}); \qquad (1)$$

At the Neural Network output, values in $(0,1)$ are obtained. A threshold value $T \in (0,1)$ is chosen so that output values in $(0,T)$ will be considered as binary output 0 (decision D_0) and values in $[T,1)$ will represent 1 (decision D_1).

2.1 The Multi-Layer Perceptron (MLP)

Neural Networks (NNs) have clear advantages over classical detectors as non-parametric detectors. In this case, the statistical description of the input is not available. The only information available in the design process is the performance of the detector on a group of patterns that are denominated training patterns. For this task, the BP algorithm carries out an implicit histogram of the input distribution, adapting freely to the distributions of each class.

For detection purposes, the Multi-Layer Perceptron (MLP) trained with the BackPropagation (BP) algorithm, has been found more powerful than other types of NNs [3]. While other types of NNs can learn topological maps using lateral inhibition and Hebbian learning, a MLP trained with BP can also discover topological maps. Furthermore, the MLP can be even superior to the parametric bayesian detectors when the input distribution departures from the assumptions.

2.2 The MLP Structure

Although - due to the lack of mathematical methods to calculate the dimensions of the NN - the capacities of the NN depend on the problem to be approached, for binary detection only one hidden layer seems to be necessary - empirical results for the NN show that additional hidden layers do not provide any performance improvement, on the contrary the training time is critically increased-.

So the chosen structure would be a MLP with one hidden layer, represented as $2M/N/1$ for a MLP with $2M$ input nodes, N hidden layer nodes and one output node.

Also, the number of nodes in the hidden layer cannot be established a priori. Although the exact input-output realization may imply a number of nodes up to the number of training patterns [4], this number has to be much lower for the sake of the NN generalization capability. Generally, the size is determined by test and error procedure using empirical curves [1].

The parameter Training Signal-to-Noise Ratio (TSNR) is defined as the Signal-to-Noise-Ratio used to generate the training set patterns, and is one of the key design parameters of the NN.

In the empirical curves, each one presents a knee where the most effective relation between complexity and performance is verified. After a thorough study of the values of M, N and TSNR the structure 16/8/1 has been chosen as the most efficient one.

3 The Training Algorithm

Even if the size of the NN has been precisely determined, finding the adequate weights is a difficult problem. To do so a BackPropagation (BP) algorithm is used during the training.

The dynamic of learning utilized is *cross-validation*, a method that monitorizes the generalization capabilities of the NN. This method demands to separate learning data in two sets: a training set, employed to modify the NN weights, and a testing set, which is utilized to measure its generalization capability. During the training, the performance of the NN on the training set will continue improving, but its performance on the test set will only improve until a point, beyond which it will begin to degrade, at this point the NN begins to be overtrained, loosing capacity of generalization; consequently, the training should be finished.

The Least-Mean-Squares (LMS) criterion is the most widely used for training a Multi-Layer Perceptron (MLP). However, depending on the application, other criteria can be used [6]. The following criterion functions have been analysed:

(a) Least-Mean-Squares (LMS). It has benn proved that this criterion approximates the Bayes optimal discriminant function and yields a least-squares estimate of the *a posteriori* probability of the class given in the input [7]. However, one of its drawbacks is that a least squares estimate of the probability can lead to gross errors at low probabilities.

(b) Minimum Misclassification Error (MME). This criterion minimizes the number of misclassified training samples. It approximates class boundaries directly from the criterion function, and it could perform better than LMS with less complexity networks [9].

(c) El-Jaroudi and Makhoul (JM) criterion. It is similar to the Kullback-Leibler information measure and results in superior probability estimates when compared to least squares [8].

3.1 The Training Sets

The training set has been formed by an equal number of signal plus noise patterns and only noise patterns (i.e., $P(H_0) = P(H_1)$) during the training, being respectively H_0 and H_1 the hypothesis symbol absent at the input and symbol present at the input), presented alternatively, so the desired output varies from D_0 (detecting noise) to D_1 (detecting target) in each iteration. Other choices of presenting the training pairs, as in a random sequence, are also suitable. However, changing the fifty-fifty composition of each class in the training set has resulted in degrading the performance of the resulting NP-ND.

The pattern sets are classical input models for radar detection (Marcum and Swerling) [5]. The NP-ND inputs are samples of the complex envelope of the signal. There are NNs with complex coefficients to process complex signals, but its seems more convenient to sample its *in-phase* and *quadrature* components, getting an all real coefficient net. This provides generality to this study, because the same NN can be utilized with a different preprocessing.

There are no methods that indicate the number of training patterns to be used. Assuming that the test and training patterns have the same distribution, there have been found limits to the number of training patterns necessary to achieve a given error probability: this number is approximately the number of weights divided by the desired error and makes the number of patterns necessary for training prohibitive. Empirical studies have proved an upper limit of 2,000 training patterns to be sufficient for any criterion function or input signal model.

4 Computer Results

The results obtained are based on the following set of parameters:
 Training signal-to-noise ratio: TSNR
 Input signal-to-noise ratio: SNR
 Structure: $2M/N_1/1$ ($2M$ inputs / N_1 nodes in the hidden layer / 1 output)
 Probability of detection, P_d, the probability that the detector decide D_1 under the hypothesis H_1.
 False alarm Probability P_{fa}, the probability that the detector decides D_0 under the hypothesis H_0.
 The performance of the detector is classically measured by detection curves (P_d vs. SNR, for a fixed P_{fa}) and ROC curves (Receiver Operating Characteristics curves, P_d vs. P_{fa} for a fixed SNR).

4.1 Detection curves and Training-Signal-to-Noise-Ratio (TSNR)

These curves are used to study the NP-ND performance as TSNR changes. Figure 1, shows that the detection capabilities of the NNs trained with low TSNR degrade as P_{fa} decreases, so, to achieve low values of P_{fa}, the TSNR has to be raised. However when this last value is too high P_d worsens indicating that there is an optimal value of TSNR for each P_{fa} value, that has to be empirically found.

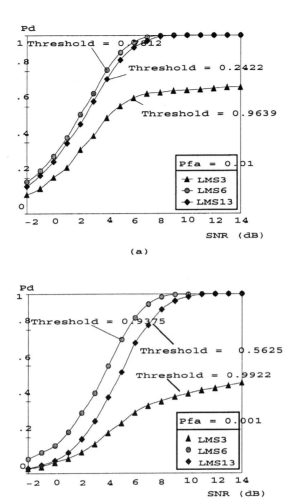

Fig. 1. Detection probability (P_d) vs. Signal-to-Noise vs. Signal-to-Noise Ratio (SNR) for a 16/8/1 MLP trained with the LMS criterion and different Training Signal-to-Noise Ratio (TSNR) in dB. For example, LMS6 means LMS criterion and TSNR=6dB. (a) $P_{fa} = 0.01$, (b) $P_{fa} = 0.001$

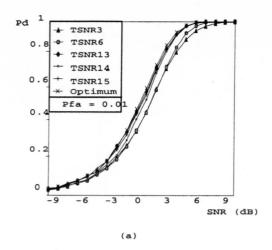

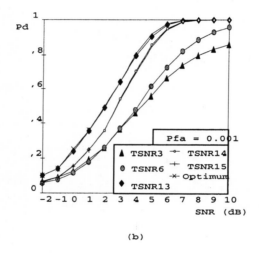

Fig. 2. Detection probability (P_d) vs. Signal-to-Noise Ratio (SNR) for a MLP of structure 16/8/1 and different Training Signal-to-Noise Ratio (TSNR) with JM criterion (a) $P_{fa} = 0.01$ and (b) $P_{fa} = 0.001$

4.2 The criterion function

A result common to all criterion functions is that rising the TSNR, the number of training iterations decreases, as the decision regions to be separated by the MLP become more different.

Concerning to the training time and convergence of the algorithm, the best results are those of the JM criterion, followed by the LMS one. The MME requires, in general, a higher number of training iterations than the others. Significant convergence improvements are obtained by using JM instead of LMS. This conclusion does not depend on the value of TSNR.

Comparing the performance through Detection Curves, also the criterion JM achieves the best results for the detection problem, fact that supports the idea suggested in [8] that the estimation of the *a posteriori* probabilities carried out by the JM criterion is more accurate than the LMS one. The MME criterion present the worst characteristics. This criterion does not adapt to the application, possibly because training the MLP on this criterion depends highly on the training threshold value, T_0. This result does not take place when criterions LMS or JM are applied. In those last cases, a value of $T_0 = 0.5$ for the training threshold always achieve good results. After the training, the threshold is adjusted by Monte Carlo tries or Importance Sampling Techniques [5] to achieve the desired maximum P_{fa} value.

4.3 Integration of samples

It can be seen that the TSNR parameter is critical in the performance of the NP-ND. If this parameter is properly chosen, the performance of the NP-ND can be quasi optimal (Figure 2).

But this excellent performance also depends on the number samples (M) used for each decision. In the NN: $2M/8/1$ if we use $M=2$ samples, or less, for each decision, the NP-ND is too small to solve the problem efficiently. If M is too high, the NP-ND is over-dimensioned, and the Neural Network curves separate from the optimum Neyman-Pearson ones (for $M=2$, 4 or 16 the best TSNR=13 dB; for $M=32$ is TSNR=6 dB). It has been observed that this behavior does not depend on the target model.

5 Conclusions

A design method to obtain optimal Neyman-Pearson detectors using a Neural Network has been presented. The resulting Detector is referred as NP-ND. The paper can be summarized in the following items:

- For the binary detection task, a Neural Network, despite of its blind learning, can reach the optimal theoretical results for the Neyman-Pearson detector for very low values of the false alarm parameter. This make the NN-PD very useful in any application where maintaining false alarm or false-positive errors under a given rate is crucial.

- A three layer MLP structure has been used, with the adequate criterion function and optimizing the relationship among signal-to-noise ratio for training (TSNR), detection probability (P_d) and false alarm probability (P_{fa}). The desired value of this last parameter is obtained calculating the appropriate output threshold by Monte Carlo tries or Importance Sampling Techniques after the training phase.

- For all experiments developed, the best TSNR depends on the criterion function and the desired desing value of P_{fa}. The lower P_{fa} value that can be achieved with the NP-ND decreases as the TSNR increases. As a general rule, the NP-ND should be trained with an intermediate TSNR (let us say 13 dB, the minimum TSNR depending on P_{fa}) and then adjust the threshold value to achieve the design requirements of P_{fa}. If this is impossible, a higher TSNR has to be used.

- El-Jaroudi and Makhoul (JM) criterion performs better than classical sum-of-square-error (LMS). It can be due to its superior a posteriori probability estimation. The training time is lower than any other criterion too.

References

1. Velazco, R., Godin, Ch., Cheynet, Ph., Torres-Alegre, S., Andina, D., Gordon, M.B.: Study of two ANN Digital Implementations of a Radar Detector Candidate to an On-Board Satellite Experiment. In: Mira, J., Sánchez-Andrés, V. (eds.): Engineering Applications of Bio-Inspired Artificial Neural Networks. Lecture Notes in Computer Science, Vol. 1607. Springer-Verlag, Berlin Heidelberg New York (1999) 615–624
2. Root, W.L.: An Introduction to the Theory of the Detection of Signals in Noise. Proc. of the IEEE, Vol 58. (1970) 610–622
3. Decatur, S.E.: Application of Neural Networks to Terrain Classification. Proc. IEEE Int. Conf. Neural Networks. (1989) 283–288
4. Hush, D.R., Horne, B.G.: Progress in Supervised Neural Networks. What's new since Lippmann?. IEEE Signal Processing Magazine (1993) 8–51.
5. Sanz-González, J.L., Andina, D.: Performance Analysis of Neural Network Detectors by Importance Sampling Techniques. Neural Processing Letters. Kluwer Academic Publishers, Netherlands. ISSN 1370-4621. **9** (1999) 257–269
6. Barnard, E., Casasent, D.: A Comparison Between Criterion Functions for Linear Classifiers, with Application to Neural Nets. IEEE Trans. Systems, Man, and Cybernetics, Vol. 19. **5** (1989) 1030–1040
7. Ruck, D.W., Rogers, S.K., Kabrisky, M., Oxley, M.E., Suter, B.W.: The Multilayer Perceptron as an Approximation to a Bayes Optimal Discriminant Function. IEEE Trans. on Neural Networks, Vol. 1. **4** (1990) 296–298
8. El-Jaroudi, A., Makhoul J.: A New Error Criterion For Posterior Probability Estimation With Neural Nets. Proc. of Int. Joint. Conf. Neural Networks, IJCNN, Vol. I. **5**, (1990) 185–192
9. Telfer, B.A., Szu, H.H., Energy Functions for Minimizing Misclassification Error With Minimum-Complexity Networks. Proc. of Int. Joint Conf. Neural Networks, IJCNN, Vol IV. (1992) 214–219

Distance between Kohonen Classes Visualization Tool to Use SOM in Data Set Analysis and Representation

Patrick Rousset[1,2] and Christiane Guinot[3]

[1] CEREQ, 10 place de la Joliette, 13474 Marseille cedex, France
[2] SAMOS, Université de Paris I, 90 rue de Tolbiac, 75013 Paris, France
[3] CE.R.I.E.S., 20 rue Victor Noir, 92521 Neuilly sur Seine cedex, France
rousset@cereq.fr
christiane.guinot@ceries-lab.com

Introduction

Representation of information given by clustering methods is of little satisfaction. Some tools able to localize classes into the input space are expected in order to provide a good visual support to the analysis of classification results. Actually, clusters are often visualized with the planes produced by factorial analysis. These representations are sometimes unsatisfying, for example when the intrinsic structure of the data is not at all linear or when the compression phenomenon generated by projections on factorial planes is very important. In the family of clustering methods, the Kohonen algorithm has the originality to organize classes considering the neighborhood structure between them [9][10][6]. It is interesting to notice that many transcription in graphical display have been conceived to optimize the visual exploitation of this neighborhood structure [5][11]. Each one helps the interpretation in a particular context. they are twinned to the Kohonen algorithm and called *Kohonen maps*. For example, one used in the following helps the interpretation of the classification from an exogenous or endogenous qualitative variable. Unfortunately, no one allows for a visualization of the data set structure in the input space. This is very regrettable when the Kohonen map makes such a folder that two classes close to each other in the input space can be far on the map. A tool that visualizes distances between all classes gives a representation of the classification structure in the input space. Such a tool is proposed in the following. As the Kohonen algorithm has the property to reveal effects of small distances also called *local distances* and the new tool is able to control big distances, this clustering method has now a large field of exploitation.

In the context of multidimensional analysis, the graphical display of all distances between classes transforms the Kohonen algorithm from a clustering technique summarizing information into a data analysis and data set representing method [1][3]. Its approach can be compared with factorial analysis. In particular it can be used to study the result of any classification c (not only the Kohonen one). To avoid any confusion, in this paper the prefix k refers to the Kohonen algorithm and c or nothing refer to the other classification. As the distance chosen for the k-algorithm and for the c-classification are coherent (Euclidean one for both, χ^2 for both, etc.), it is probable that any difference between individuals able to generate a c-class would create also a k-class. In that situation, the Kohonen map is probably more adapted to visualize the influence of local distances than a projection on a plane which is more efficient to represent big distances effects. This noticed property would reduce the very well known risk of compression induced by the association between the clustering method and a factorial data analysis. The new technique is also presented as the approximation of the data set by a non-linear surface. In a non-linear context, One can use for example a Multilayer Perceptron to do it but in that case it is not very easy to interpret the model. At the contrary, the tool that allows for the graphical display of the surface structure simplifies the interpretation and makes this method use easier than most of non-linear ones.

The Kohonen map is in the following used first as a method of classification which takes into account at the same time proximity and big distance between classes and in a second time used as a method of data analysis to represent the data set and the result of any other classification. To illustrate the possibilities of this technique, we use the example of a data set concerning the skin quality and resulting of a study realized in the C.E.R.I.E.S institute. The purpose of this study is to propose a typology of the human healthy skin out of several pertinent medical criteria. Then, the data set is build by recording presence or absence of 17 criteria on 212 volunteer women. All these variables take two values, they are considered as quantitative variables for the clustering but are also analyzed with tools that are usually applied to qualitative ones. Some studies have already been done, each one corresponding to a different approach translated by adapted distance or referring method [2][8]. All the subroutines are developed with the SAS software version 6.12. First, the Kohonen algorithm with Euclidean distance is applied to classify individuals, then is used to represent in the input space the result of a hierarchical classification making use of the Ward distance.

The Kohonen Classification and Its Associated Map

In this chapter, the Kohonen algorithm is considered only as a clustering method adapted to any distance and particularly interesting when local distances have to be taken into account. In the following, a class issued from a Kohonen classification is called *k-class* in order to exclude any confusion with a class issued from another technique. The Kohonen network presented here is a two-dimensional grid with n by n units, but the method allows for the choice of any topological organization of the network. We notice $U = n \times n$ the number of units. after the learning, the weight vectors G_u, called code vectors, represent in the input space their corresponding unit u. The delicate problem of the learning is not addressed here, it is supposed to be successfully realized. Each individual is associated to the code vector which is the closest in the input space. In this way, two individuals associated to the same code vector G_u are assigned to the same class k_u. U k-classes k_u are defined in such a way. They are represented by their corresponding code vector G_u in the input space and unit u on the network. This classification as the particularity to organize units on a chart called Kohonen map such as units neighbored on the map are corresponding to code vectors that are close in the input space. Consequently, two individuals that belong to classes referring to neighbored units are close in the input space. By contrary, to be far on the chart does not mean anything concerning the proximity. In the following, the neighborhood notion refers to the map localization. The boxes organization of figure 1 is a traditional representation of the Kohonen map in the case of a grid. The unit u_0 neighbors for the rays 0, 9 and 25 are respectively the unit u_0 itself, any unit of the square of 9 units centered on u_0 and anyone of the square of 25 units centered on u_0.

The Kohonen map allows for a representation of some characteristics of a k-class and the generalization to its neighbors. For example, the effect of an endogenous or exogenous qualitative variable Q can be visualized by including into each unit a pie reflecting the proportion of one modality occurrence in the class population, as shown in figure 1. The pie slice angle is $2 \times \pi \frac{n_{ik}}{n_{.k}}$, where n_{ik} is the number of individuals classified in the k-class k for which Q takes the i modality and $n_{.k}$ is the number of individuals that are affected to the class k. For example the pie ⬤ indicates that 33% individuals of this class have a "yellow skin". Figure 1 represents one endogenous criterion effect, "the skin is yellow or not". The Q variable is then the criterion corresponding variable. Two areas of neighbored units are emerging, one is composed of units 1,2,8 and the other one of units 28, 34, 35, 41, 42, 48, 49. Moreover an other area grouping mixed population is located in the 7, 14, 21 units area.

The main problem of such a representation is that units are localized on the map at equal distance from their neighbors while their own code vectors are not at equal distances in the input space. In this way, the intrinsic data set structure in the input space is omitted. In the skin quality sample, this translates the visual impossibility to conclude if the "yellow skin" characteristic is present in two separated areas of the data set input space or in only one. The problem of the graphical display of the data set in the input space is the aim of the following section where the proposed solutions are expected to allow for decision between one area and two.

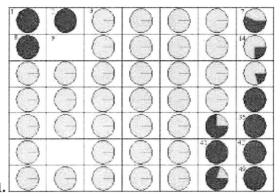

Fig. 1. Kohonen classification characterization with a qualitative endogenous or exogenous variable, numeration is going from upper-left to bottom-right. Frequency of individuals where the criterion "yellow skin" is present or not corresponding respectively to dark and light gray.

Representation of Distances between All Classes and Visualization of the Intrinsic Data Set Structure in the Input Space

In order to give an idea of the data set structure in the input space, some modest ideas have already been proposed. One visualizes the importance of the proximity between k-classes by grouping code vectors with a new classification on these U vectors [5][11]. A gray scale that is associated to each new class called *macro-class* is used to colorize the background in each unit cell as shown in figure 2. In practice, this technique groups more often code vectors located on connected areas of the map which confirms the neighbored topology. A second method represents the distance between code vectors of neighbored classes [4]. The concept consists in separating two neighbored units borders with a space which width is in proportion with the distance between their corresponding code vectors, as in figure 3. This method allows for a distinction between a population large enough to generate two macro-classes and two different populations' separation. But the information on distance is restricted to neighbored classes. When both techniques are applied to the example of the skin quality, the first method creates the figure 2 where 6 macro-classes associated to 6 level of gray are determining 6 connected areas of the map. The second one shows in figure 3 that the right border of the map (units 6, 7, 13, 14, 20, 21, 27, 28, 34, 35, 41, 42, 48, 49) is separated from the large area of the rest. But it is very difficult to see the structure of the data set and the aspect of uniformity is probably wrong. Moreover, code vectors can belong to two different macro-classes while they are close in the input space. So, it is still impossible to conclude of proximity or remoteness between individuals affected in both map opposite corners upper-left and bottom-right. In conclusion, these methods are some simple ways to give a first aspect of distances between classes but only a tool that represents any distance between two centroïds is able to give information on the input space structure.

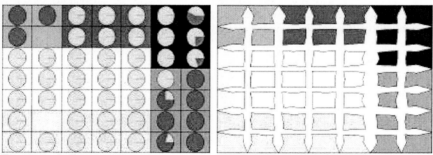

Fig. 2. and Fig. 3.

In figure2, macro-classes representation is added to previous map, (the 6 level groups of a hierarchical classification on the 49 code vectors realized with the Ward distance). Figure3 represents distance between neighbored classes centroïds.

Now the necessity is fit to represent distances between any classes, which constitutes U^2 values. U^2 is usually a number very large and these values presentation have to use the neighborhood organization in order to group the redundant information. This condition is necessary to obtain a result that can be exploited. The natural way is to describe these distances with a new Kohonen map divided in U boxes of U divisions. The principal consists in assigning on the map a level of gray to each couple of k-classes *(u, u')* in such a way that the color of the division *u'* of the *u*th box indicates the size of the distance *d(u, u')*. This visualization is a pertinent tool to understand the data set structure. Figure 4 is the map of distances applied to the skin quality sample. 4 gray levels are used , as the gray is dark as the corresponding distance is large. The first grid (upper-left) which represents distance between unit 1 and any else shows that the light units are on the upper-left corner area (its own macro class units 2, 8 and 9) or on the bottom-right one (42 and 49). Now we can conclude that the two opposite corners of the map *upper-left* and *bottom-right* are close on the input space and so any individuals whose skin has the criterion *yellow skin* are in the same space area. This last conclusion is impossible from any of the previous representations and show the power of this new tool. Moreover the properties revealed by the previous distance maps are still appearing. As example, we have previously notice that the right border is rather far from the rest, this property still appears in grid 27 and 28. But this time, the exception of the bottom-right corner which is far from the middle of the map but close to the corner upper-left is emerging. Moreover, this new map allows for an understanding of the data set structure. For example, with grids 10 and 31 we can see the opposition between the upper and bottom borders of the map. This new map is more complex to read than the previous ones but is more precise, does not introduce interpretation mistakes and makes appear properties that are invisible with the others.

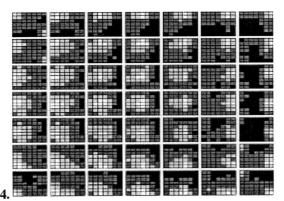

Fig. 4. Representation of any distance between centroïds, grids and divisions has the same organization than any previous Kohonen map

Any Classification Representation with a Kohonen Map

In this section, the Kohonen algorithm result is perceived as a summary of the input data set in U points of this space and the different maps presented in sections 1 and 2 as a visualization of the input data set structure. It is so natural in order to solve the initial goal to use these maps to represent any clustering method result and not the only Kohonen one. The distance choice is free which allows the use of one in coherency with the classification one. A qualitative variable Q is defined in order to affect to any individuals the index of its own class. This Q variable cartography done as in section 1 (figure 1) allows for the characterization of any part of the input space from the clusters. A similar map can be build to localize any modality of the Q variable, and so any class, in the input space. Instead of the pie of figure 1, a bar chart represents the proportion between the contingency n_{ik} of the population *composed of individuals that belong to the class k and have the modality i* and the contingency $n_{i.}$ of the population *that has the modality i as common characteristic*. As the angle of the pie is $2 \times \pi \dfrac{n_{ik}}{n_{.k}}$, the size of the bar chart is $\dfrac{n_{ik}}{n_{i.}}$ (corresponds to a k-class that contains 15% of class 1, ..., and 80% of class 6). Figure 6 is the result of this graphical technique applied to a classification in 6 classes. By cumulating contingency representation (figures 5 and 6) and distance maps information (figure 3 and 4), it is possible to have a large idea of the clusters repartition in the input space.

This method can be applied to the example of the skin quality to illustrate its possibilities to visualize for example the 6 grouping obtained at level 6 classes of a hierarchical classification making use of the Ward distance. Figures 6 and 4 show that individuals of classes 2 and 6 are in the right border and so far from the rest of the population. Class 6 is close from class 3 as it is located in the corner bottom-right. This class 3 is also present in the upper-left corner that is remained to be close from the bottom-right corner in the input space. Classes 1, 4 and 5 are constituting the main area of the map, the middle and the left side map except the upper-left corner. It looks as a large uniform population where classes 1 (bottom side) and 5 (upper side) are opposite and class 4 (middle) is the medium population. Moreover, class 3 is close from classes 1, 4, 5 and as well than from class 6 and appears also as a medium class for the data set.

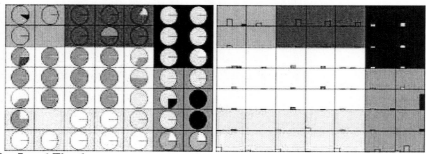

Fig. 5. and **Fig. 6.**

Contingency obtained by crossing the Kohonen k-classes with the result of a hierarchical classification. In figure 5, it is measured in proportion of the k-classes contingency. In figure 6, it is measured in proportion of the c-classes contingency and represented by the bar size.

Perspective

While code vectors are projected on a factorial plane and neighbored unit ones are linked together, the representation obtained is the kind of figure 7. The links determine parallelepipeds that generate a surface. This surface adjusts the data set by joining any U code vector to its neighbors. The data set representation with map previously presented in section 1, 2 and 3 can be considered as a graphical display of this surface and as well of any surface that adjusts the data set tying neighbored code vectors. We have then on the one hand a way to summarize the data set structure with a surface and on the other hand some tools very well adapted to non-linear structure to visualize this surface.

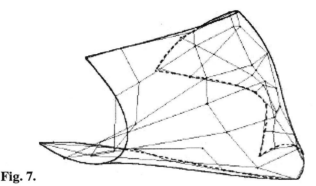

Fig. 7.

The surface generated by Kohonen classes centroïds and links between them is projected on a plane. To simplify representation, only 4 neighbors links are drawn: Map brim is overdrawn (the stippled design corresponds to the back part of the surface)

Conclusion

We have presented a method to visualize the input data set structure with the Kohonen maps that is able to substitute linear graphical displays when these one are unsatisfying. While the Kohonen map reefers to the neighborhood structure between the k-classes produced by this algorithm, a new tool that represents any distance between k-classes centroïds allows for some of properties in the input space. This new technique can be applied to a larger domain than the interpretation of the k-classes. In particular in data analysis, it looks very well adapted to some applications such as the visualization of any c-clustering result. In that context, the many charts associated to the Kohonen algorithm became also graphical displays of the data set or the c-clusters properties. For example, the one of figure 2 shows at the same time effects of any qualitative variable on the k-classification and on any clustering result. As a perspective, it can probably be used to visualize some adjustment of the data set with non-linear surfaces.

Acknowledgements

The authors wish to thank Professor Erwin Tschachler for his advice and encouragement, the CERIES team for the important contribution to the data and Mrs Annie Bouder from the CEREQ for her assistance in the manuscript.

Reference

1 Blayo F. and Demartines P. (1991). Data analysis: how to compare Kohonen neural networks to other techniques ? In *Proceedings of IWANN'91*, pages 469-476, Springer Verlag, Berlin.

2 Chavent M., Guinot C., Lechevallier Y. et Tenenhaus M. (1999). Méthodes divisives de classification et segmentation non supervisée : Recherche d'une typologie de la peau humaine saine. *Revue de Statistique Appliquée*, XLVII, 87-99.

3 Cottrell M. and Ibbou S. (1995). Multiple Correspondence Analysis of a crosstabulations matrix using the Kohonen algorithm. In *Proceedings of ESANN'95*, M. Verleysen (Eds), pages 27-32, D Facto, Bruxelles.

4 Cottrell M. and de Bodt E. (1996). A Kohonen Map Representations to Avoid Misleading Interpretations. In *Proceedings of ESANN'96*, M. Verleysen (Ed.), pages 103-110, D Facto, Bruxelles.

5 Cottrell M. and Rousset P. (1997). A powerful Tool for Analysing and Representing Multidimensional Quantitative and Qualitative Data. In *Proceedings of IWANN'97*, pages 861-871, Springer Verlag, Berlin.

6 Cottrell M., Fort J.C., Pagès G. (1998). Theorical aspects of the SOM algorithm, Neuro Computing, 21, pages 119-138

7 Cottrell M., Gaubert P., Letremy P. and Rousset P. (1999). Analyzing and representing multidimensional quantitative and qualitative data : Demographic study of the Rhône valley. The domestic consumption of the Canadian families. E. Oja and S. Kaski (Eds), pages 1-14, Elsevier, Amsterdam.

8 Guinot C, Tenenhaus M., Dubourgeat M., Le Fur I., Morizot F. et Tschachler E. (1997). Recherche d'une classification de la peau humaine saine : méthode de classification et méthode de segmentation. *Actes des XXIXe Journées de Statistique de la SFdS*, pages 429-432.

9 Kohonen T. (1993). *Self-organization and Associative Memory*. 3°ed., Springer Verlag, Berlin.

10 Kohonen T. (1995). *Self-Organizing Maps*. Springer Series in Information Sciences Vol 30, Springer Verlag, Berlin.

11 Rousset P. (1999). Application des algorithmes d'auto-organisation à la classification et à la prévision. Thesis of doctorat, University Paris I, pages 41-68.

Optimal Genetic Representation of Complete Strictly-Layered Feedforward Neural Networks

Spyros Raptis, Spyros Tzafestas, and Hermione Karagianni

Intelligent Robotics and Automation Laboratory
Division of Signals, Control and Robotics
Department of Electrical and Electronic Engineering
National Technical University of Athens
Zographou Campus, 15773, Athens, GREECE
tzafesta@softlab.ece.ntua.gr

Abstract. The automatic evolution of neural networks is both an attractive and a rewarding task. The connectivity matrix is the most common way of directly encoding a neural network for the purpose of genetic optimization. However, this representation presents several disadvantages mostly stemming from its inherent redundancy and its lack of robustness. We propose a novel representation scheme for encoding complete strictly-layered feedforward neural networks and prove that it is optimal in the sense that it utilizes the minimum possible number of bits. We argue that this scheme has a number of important advantages over the direct encoding of the connectivity matrix. It does not suffer from the curse of dimensionality, it allows only legal networks to be represented which relieves the genetic algorithm from a number of checking and rejections, and the mapping from the genotypes to phenotypes is one-to-one. Additionally, the resulting networks have a simpler structure assuring an easier learning phase.

1 Introduction

As stochastic search processes, genetic algorithms (GA's) provide no estimation on the time required to locate an adequate solution to a given problem. They use very limited (if any) a priori information on the specific problem they are addressing. In this sense, they are inadequate for on-line execution in time-sensitive problems.

On the contrary, GA's have very often been used as a high-level tool to evolve other systems that are more suited for on-line performance. Indicative examples include the definition of the input partitions or other parameters of a fuzzy system, the determination of appropriate values for the parameters of another GA, the design of an appropriate set of detectors for pattern recognition, etc. The evolutionary design of neural networks by genetic means may very well be placed in this context.

In spite of the intense research in the area of neural networks, globally applicable rules for their design are still missing and their development is still based on heuristics and rules of thumb. Design parameters include the network topology, the determination of its connectivity pattern, the selection of the neuron activation functions, the training algorithm used to calculate the weights of the connections, etc.

The design of a neural network by genetic means restates the problem in the context of an optimization process. Automatically evolving a network that is "optimal" (in the sense of a certain criterion) is particularly attractive since it offers a general methodology for the design of a "well-behaving" neural system.

The work in the field of evolving neural networks is quite extensive and certain surveys are also available (e.g. [1]). The various approaches may be roughly divided into categories based on different characteristics of the process.

Based on the genotypic representation used we can identify (a) *direct encoding schemes* (e.g. [2], [3], [4]) where the GA uses a simple and easily decodable representation of the neural network such as its connectivity matrix and (b) *indirect encoding schemes* (e.g. [5], [6], [7]) such as production rules, grammars, etc. where decoding is not so trivial. Based on the type of network being evolved, we may identify approaches for the design of (a) *feedforward networks* or (b) *recurrent networks*. The selection of the required class of neural networks is, clearly, problem-dependent. A last classification can be identified that is based on the type of network parameters being evolved. We can identify methods that (a) optimize only the *network topology*, (b) optimize the values of the *synaptic weights* and other parameters of a fixed network, or (c) optimize *both* the topology and the parameter values of the network.

In the remainder of the paper we will only consider the case of *complete strictly-layered feedforward neural networks*. A neural network is said to be *layered* if its hidden neurons are organized in layers. No connection can appear among neurons of the same layer. A network is *strictly-layered* if it is layered and the neurons of a layer can only accept inputs from neurons of the immediately preceding layer. A network is *complete* if every neuron of a layer is connected to all the neuron in the previous layer.

2 Connectivity Matrices

The connectivity matrix is the most commonly used representation of a neural network. It is a square binary matrix, C, of dimension equal to the total number of neurons in the network (including input, output and hidden neurons). Each element of the matrix, say $C(i,j)$, indicates the presence or absence of a connection from neuron i to neuron j. For a feedforward network the connectivity matrix has an upper-triangular form.

In the connectivity matrix of a feedforward neural network, the number of elements that can be non-zero is given by:

$$n_{gene} = (n+h)(m+h) - \frac{1}{2}h(h+1) \tag{1}$$

where n is the number of network inputs, m is the number of network outputs, and h is the number of hidden neurons.

The representation of neural connectivity patterns through their respective connectivity matrix contains redundant information. Thus, while the two networks displayed in Fig. 1 are functionally equivalent, their respective connectivity matrices differ. Of course, by appropriate renumbering (i.e. exchanging columns and rows) the two matrices can be made identical. The implementation of an algorithm that can transform the different versions of a matrix to a single is a relatively straightforward task.

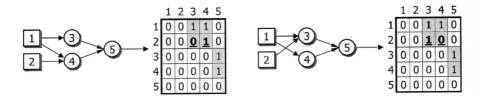

Fig. 1. Two equivalent neural networks

In the context of genetic search, a representation that relies on a direct coding of the significant bits of the connectivity matrix is bound to such a "symmetry" and the formulated mapping from the genotype domain (coded individuals) to the phenotype domain (neural networks) is, inherently, many-to-one. This is true not only for strictly layered neural networks but for any type of feedforward network as well.

For a feedforward network, the required number of bits for coding only the necessary information of the connectivity matrix is given by Eq. (1). Thus, a typical genetic representation would need to utilize n_{gene} bits.

It is important to note that random combinations of these bits do not always produce legal networks. To quantify this argument, it is interesting to get an estimate of the percentage of legal networks to the illegal ones. To this end, let us consider the simple problem of designing a feedforward neural network with two inputs ($n=2$), one output ($m=1$), and (up to) four hidden neurons ($h=4$). This gives a total of seven neurons. The dimension of the connectivity matrix of such a network will be 7×7 containing 20 significant bits (i.e. bits that can be non-zero) thus leading to 2^{20} different possible combinations. A set of 100000 such bit series were randomly produced and the respective networks were constructed and simplified. The statistics for the resulting networks are shown in Table 1.

Table 1. Statistics for 100000 neural networks of dimension 2×4×1 randomly produced using direct coding of the significant bits of the connectivity matrix

Hidden Neurons	%
0 (illegal)	11%
1	17%
2	25%
3	28%
4	19%
	100%

Num of Groups	%
0 (illegal)	11%
1	24%
2	37%
3	23%
4	5%
	100%

Num of Layers[1]	%
1	89.80%
2	9.30%
3	0.87%
4	0.03%
	100%

---

[1] Only 7% of generated networks presented layering. The percentages in this table are with respect to those layered networks.

The statistics become even more problematic for the case of more complex networks. Therefore, for realistic problems it is almost impossible to control the quality of the networks that are randomly created based on a direct coding of the significant bits of the connectivity matrix and evaluated by a genetic algorithm. Additionally, it is quite rare for a layered network to randomly come by. Thus, for networks of this type a different encoding scheme is obviously required.

From the discussion above, it is quite obvious that a different network representation scheme is necessary especially for the case of genetically evolving strictly layered neural networks.

3 The Proposed Representation Scheme

Let us consider again the case of a strictly-layered feedforward neural network of n inputs, m outputs and h hidden neurons. Then the total number of neurons on the network will be $N=n+m+h$.

Consider a string of h binary digits receiving the values "0" or "1". If we assign to the symbol "1" the meaning *"new layer"* and to the symbol "0" the meaning *"same layer"*, then we may directly express through a string of h such symbols any *complete strictly-layered feedforward neural network*.

This way, the complete neural network of Fig. 2 having $n=3$, $m=2$ and $h=7$ ($N=12$), can be represented by the simple string "1000100". In its general form, the proposed representation can be depicted as shown in the Fig. 3.

Since, the number of 1's in the string explicitly controls the number of layers in the resulting network we may directly favor "shallower" network architectures by introducing an appropriate bias during the random instantiation of such strings which assigns higher probability to 0's than to 1's. Additionally, we may impose the demand for at least one hidden layer by fixing the leftmost bit of the string to "1" and allowing only the other h-1 bits to vary. In this case, however, the number of hidden neurons in all resulting networks will be *exactly h* as discussed below.

By relaxing the requirement for at least one hidden layer (thus allowing the leftmost digit of the string to receive the value "0") and omitting all the "0" that appear in the left side of the first "1" in the string, we may directly use the same scheme to represent networks of fewer hidden nodes or even without any hidden node (string consisting entirely of 0's). Such an approach offers additional properties to the representation scheme, such as its independence from the problem dimensions, since:

$$\underbrace{\overbrace{0\,0...0}^{t}\,1\overbrace{**...*}^{k}}_{t+k+1} \equiv \underbrace{1\overbrace{**...*}^{k}}_{k+1} \tag{2}$$

4 Properties and Advantages

Compared to the direct encoding of the significant digits of the connectivity matrix, the proposed representation scheme presents a set of significant advantages.

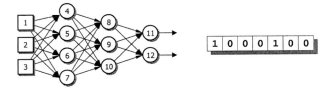

Fig. 2. The representation of the complete strictly-layered feedforward neural network of size 3×4×3×2 ($n=3$, $m=2$ and $h=7$)

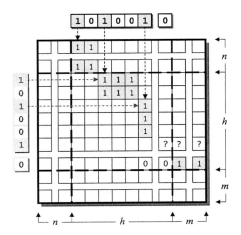

Fig. 3. The representation of a complete strictly-layered feedforward neural network

Adequacy: The direct encoding scheme of the connectivity matrix allows for the representation of legal neural networks that the proposed model cannot. Examples of such networks are networks that are not complete (i.e. networks whose neurons are not fully interconnected among successive layers) or not strictly layered (i.e. networks whose neurons may be arbitrarily interconnected, e.g. input neurons directly connected to output neurons). However, such an economy on the number of connections (as in the case of non-complete networks) or the generalized "sprawl" connectivity patterns (as in the case of networks that are not strictly layered), do not necessarily lead to more powerful or adequate networks. Not only is there no evidence that strictly-layered networks are by any means inferior to more generalized networks but, on the contrary, such networks tend to present benefits during the training phase due to their simplified structure.

Complexity: An important advantage of the proposed representation scheme is its computational simplicity: it requires $\mathbf{O}(h)$ digits as opposed to the $\mathbf{O}(h^2)$ required by the direct encoding scheme, thus avoiding the well known "curse of dimensionality" problem. So, the search space is effectively reduced and the problem of determining adequate network structures is kept tractable even for relatively large networks. For example, for the case of the network of Fig. 2, the proposed scheme requires just 7 bits (introducing a search space of 2^7 configurations) as opposed to the 62 bits required by the direct encoding scheme (leading to a search space of 2^{62} configurations)!

Correctness. The proposed scheme does not allow the representation of illegal networks configurations. So, any arbitrary string of digits corresponds to a legal network configuration. This property relieves the genetic search process from the need to perform any checking either to the randomly generated individuals or to the individuals that result from the application of genetic operators during recombination. On the contrary, the direct encoding scheme cannot guarantee the validity of the network configurations involved in the search process, rendering the extensive use of checking absolutely necessary during the initialization and recombination phases. Such a checking will result in certain individuals being rejected or "punished" by assigning particularly low fitness values during their evaluation. However, such checking and rejection can introduce severe obstacles to the search process since the genotypic resemblance (which is a main driving force of a genetic algorithm) will no longer imply phenotypic resemblance of even similarity in the assigned fitness values.

Properties of the Mapping. It can be shown that the genotypic-to-phenotypic correspondence obtained by the proposed scheme is "1-1" and "over" mapping from the domain of binary strings of appropriate length to the domain of neural network configurations. This suggests that (a) two different strings can never correspond to the same network, and (b) for *any* complete strictly-layered feedforward neural network there is a corresponding string to describe it.

5 Mathematical Proof of the Optimality of the Mapping

The main target of the present paragraph is to determine the cardinality of the set of complete strictly-layered feedforward neural networks.
In Theorem 3, we prove that the cardinality of the set of complete strictly-layered feedforward neural networks that contain *at most* h hidden neurons equals 2^h. Similarly, the cardinality of the set of strings the are composed of h binary digits is also 2^h. Thus, since the cardinality of the set of such neural networks and of the set of binary strings is the same, and given that the performed mapping is 1-1, we can prove that the proposed representation scheme is *optimal* in the sense that it is *minimal to the size of the required bits*. I.e. there is no scheme that can represent the set of complete strictly-layered feedforward neural networks using a smaller number of bits.

Let us consider a neural network containing *at most* h hidden neurons. We wish to determine the number of all its possible configurations, i.e. the number of ways h or less neurons can be partitioned to any number of hidden layers. Obviously, each layer can contain any number of neurons between 1 and h and the sum of all the neurons must not be higher that h.

Fig. 4 displays all the possible configurations of a complete strictly-layered feedforward neural network having $h=4$ or less hidden neurons.

We will first calculate the number of possible configurations of a complete strictly-layered feedforward neural network of *exactly* h hidden neurons divided in *exactly* l hidden layers where, obviously, $l \leq h$.

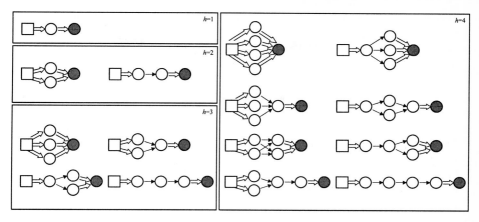

Fig. 4. Graphical representation of all the possible configurations of a complete strictly-layered feedforward neural network having *at most* h=4 hidden neurons. The subplots display all the configurations having *exactly* (a) h=1, (b) h=2, (c) h=3, and (d) h=4 neurons. The number of inputs and outputs is arbitrary. The thick arrows represent many-to-one and one-to-many connections.

Theorem 1. The number of possible configurations of a complete strictly-layered feedforward neural network that contains *exactly* h hidden neurons divided in *exactly* l hidden layers is given by:

$$N_l^h = \frac{(h-1)!}{(h-l)!(l-1)!} \tag{3}$$

Proof. The problem of determining the number of possible configurations for exactly h neurons divided in exactly l layers is equivalent to the following problem:

"Consider a binary string of h 0's. We want to determine the number of ways we can insert l-1 1's between the 0's. The insertions take place at the h-1 spaces between the zeros and each such space can only accept up to one 1."

In the above transformed version of the problem, the 0's play the role of neurons and the 1's the role of barriers separating consecutive layers.

The number possible combinations for placing r identical objects to n numbered boxes with no more than one object per box is given by:

$$C(n,r) = \frac{P(n,r)}{r!} = \frac{n!}{r!(n-r)!} \tag{4}$$

So, for r=l-1 and n=h-1, Eq. (4) takes the form:

$$N_l^h = C(h-1, l-1) = \frac{(h-1)!}{(h-l)!(l-1)!}$$

Q.E.D.■

We will proceed with the calculation of the number of possible configurations of a complete strictly-layered feedforward neural network with exactly h hidden neurons distributed in any number of hidden layers.

Theorem 2[2]. The number of possible configurations of a complete strictly-layered feedforward neural network with exactly h hidden neurons distributed in any number of hidden layers is given by:

$$N^h = 2^{h-1} \tag{5}$$

Proof. Obviously, h hidden neurons can form any number from 1 (vertical form) to h (horizontal form) hidden layers as shown in Fig. 5. So, the total number of different configurations can be calculated as the sum of the quantities given by Theorem 1 Eq. (3) for $l=1...h$, i.e.

$$N^h = \sum_{l=1}^{h} N_l^h = \sum_{l=1}^{h} \frac{(h-1)!}{(h-l)!(l-1)!} = \sum_{l=1}^{h} C(h-1,h-l) = \sum_{l=1}^{h} \binom{h-1}{h-l} = \sum_{l=1}^{h} \binom{h-1}{l-1} \tag{6}$$

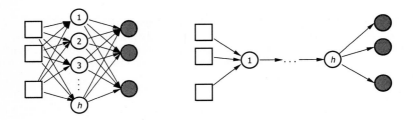

Fig. 5. Two extreme configurations of a network with h hidden neurons

It is known that:

$$\binom{h-1}{h-l} = \binom{h-1}{l-1} \Rightarrow \binom{h}{l} = \binom{h-1}{l} + \binom{h-1}{l-1} \tag{7}$$

Thus, summing from $l=1$ to $h-1$, we get:

$$\sum_{l=1}^{h-1} \binom{h-1}{l-1} = \sum_{l=1}^{h-1} \binom{h}{l} - \sum_{l=1}^{h-1} \binom{h-1}{l} \tag{8}$$

Employing the binomial expansion of $(x+y)^2$ for $x=y=1$ and $n=h$ we can arrive at:

$$\sum_{l=1}^{h} \binom{h}{l} = 2^h - 1 \text{ and } \sum_{l=1}^{h-1} \binom{h-1}{l} = 2^{h-1} - 1 \tag{9}, (10)$$

We have:

$$\sum_{l=1}^{h} \binom{h}{l} = \sum_{l=1}^{h-1} \binom{h}{l} + \binom{h}{h} \xRightarrow{Eq.(9)} 2^h - 1 = \sum_{l=1}^{h-1} \binom{h}{l} + 1 \Rightarrow \sum_{l=1}^{h-1} \binom{h}{l} = 2^h - 2 \tag{11}$$

---

[2] It is easy to see that the problem of determining N^h is equivalent to the problem of counting the additive partitions of integer h, which is a typical problem in the field of discrete mathematics.

So:

$$Eq.(8) \underset{Eq.(11)}{\overset{Eq.(10)}{\Rightarrow}} \sum_{l=1}^{h-1}\binom{h-1}{l-1} = (2^h - 2) - (2^{h-1} - 1) = 2^{h-1} - 1 \qquad (12)$$

Finally:

$$\sum_{l=1}^{h}\binom{h-1}{l-1} = \sum_{l=1}^{h-1}\binom{h-1}{l-1} + \binom{h-1}{h-1} \overset{Eq.(12)}{\Rightarrow} \sum_{l=1}^{h}\binom{h-1}{l-1} = (2^{h-1} - 2) + 1 = 2^{h-1}$$

Q.E.D.∎

The proposed scheme uses h bits for the representation networks having *up to h* hidden neurons (and so does an $h \times h$ connectivity matrix). Such networks of lower dimension can be represented by padding an appropriate number of 0's to the left of the string (or by inserting rows and columns of all zeros in a connectivity matrix). The following theorem takes these lower dimension networks into account.

Theorem 3. The number of all possible configurations of a complete strictly-layered feedforward neural network having up to h hidden nodes distributed in any number of hidden layers is given by:

$$N^{[1,h]} = 2^h \qquad (13)$$

Proof. The number of configurations of *up to h* neurons is, clearly, the sum of the number of configurations of exactly i neurons, for $i=1...h$.
Obviously, this sum is:

$$N^{[1,h]} = \sum_{i=1}^{h} N^i = \sum_{i=1}^{h} 2^{i-1} = \frac{1}{2}\sum_{i=1}^{h} 2^i = 2^h$$

Q.E.D.∎

References

1. X. Yao, "Evolving Artificial Neural Networks," *Proceedings of the IEEE*, Vol. 87, No. 9, pp. 1423-1447, September (1999)
2. T. R. J. Bossomaier and N. Snoad, "Evolution and Modularity in Neural Networks," in *Proc. IEEE Workshop on Non-Linear Signal and Image Processing*, I. Pitas, Ed., IEEE, Thessaloniki, Greece, pp. 282-292 (1992)
3. D. Dasgupta and D. R. McGregor, "Designing Application-Specific Neural Networks Using the Structured Genetic Algorithm," in *Proc. COGANN-92: Int'l Workshop on Combinations of Genetic Algorithms and Neural Networks*, L. D. Whitley and J. D. Schaffer (Eds.), IEEE Computer Society Press, June 5 (1992)
4. D. Whitley and C. Bogart, "The evolution of connectivity: pruning neural networks using genetic algorithms," in *Proc. Int'l Joint Conf. on Neural Networks*, Vol. I, pp. 134-137, Washington, DC, Lawrence Erlbaum Associates, Hillsdale, NJ (1990)
5. N. J. Radcliffe, "Genetic Set Recombination and its Application to Neural Network Topology Optimization," *Neural Computing and Applications*, Vol. 1, No. 1, pp. 67-90 (1993)
6. V. Maniezzo, "Genetic Evolution of the Network Topology and Weight Distribution of Neural Networks," *IEEE Trans. Neural Networks*, Vol. 5, No. 1, pp. 39-53, January (1994)
7. M. Mandischer, "Representation and evolution of neural networks," in R. F. Albrecht, C. R. Reeves and U. C. Steele (Eds.), *Artificial Neural Nets and Genetic Algorithms*, pp. 643-649 (1993)

Assessing the Noise Immunity
of Radial Basis Function Neural Networks

J L. Bernier, J. González, A. Cañas, J. Ortega

Dpto. Arquitectura y Tecnología de Computadores
Universidad de Granada, E18071-Granada (Spain)

Abstract. Previous works have demonstrated that Mean Squared Sensitivity (MSS) constitutes a good aproximation to the performance degradation of a MLP affected by perturbations. In the present paper, the expression of MSS for Radial Basis Function Neural Networks affected by additive noise in their inputs is obtained. This expression is experimentally validated, allowing us to propose MSS as an effective measurement of noise immunity of RBFNs.

1 Introduction

Radial Basis Function Networks (RBFNs) are one of the neural paradigms that are receiving a great interest in the present. RBFNs can be considered as universal approximators [1]. Nevertheless, the algorithms used to train them provide solutions that correspond to local-optima [1] in the space of network configurations. In this way, a modification in the parameters used during training or in the desired structure of the network would produce different configurations for the RBFN. These different configurations may present a similar respone with respect to learning, i.e., they may have a similar Mean Squared Error (MSE). However, their performance with respect to noise immunity may present great differences. In this way, two different configurations of RBFs that solve the same problem can present a similar MSE computed over the same set of input patterns, but if these patterns are affected by noise, the performance of both configurations may be degraded in a different magnitude.

In [2], it was proposed the Statistical Sensitivity as a measurement of the degradation of the performance in a Multilayer Perceptron when the weights or inputs suffer deviations. Bernier *et al.* obtained in [3,4] a better approximation of the MSE degradation of a MLP subject to perturbations, called Mean Squared Sensitivity (MSS). MSS is explicitly related with MSE degradation of a MLP when their weights or inputs are perturbed and it is computed from the statistical sensitivities of the output neurons. Thus, as MSE and MSS can be computed using the nominal values of the parameters of the trained MLP, MSS constitutes a good measurement of fault tolerance when the values of the weights change, or of noise immunity when these perturbations are related to the inputs. Moreover, MSS has been used to develop a new backpropagation algorithm [3,5] that provides weight configurations that show a better performance when the MLP is perturbed.

In this work we have applied the same methodology used in [4] to the study of RBFNs. In this way we have obtained the expression of the Mean Squared Sensitivity and shown the validity of this approximation. So, we propose to use MSS as a measurement of the degradation of performance with respect to the nominal MSE, and so providing a useful criterion to select between different RBFN configurations in order to obtain the corresponding network implementation.

The paper is organized as follows: in Section 2, the concept of statistical sensitivity is introduced, such as the perturbation model that it is used. The particular expression of statistical sensitivity for RBFs is derived in Section 3. The relationship between the statistical sensitivity and the MSE degradation is presented in Section 4, where the Mean Squared Sensitivity is defined. Section 5 is concerned to experimental results that allow us to demonstrate the validity of the obtained expressions and, finally, in Section 6 the conclussions are resumed.

2 Concept of statistical sensitivity

We are going to consider a RBF network consisting of n inputs, an unique output, and m neurons in the hidden layer. The output of this network is then computed as an averaged sum of the outputs of the m neurons, where each neuron is a radial function of the n inputs to the network:

$$y = \sum_{i=1}^{m} w_i \Phi_i = \sum_{i=1}^{m} w_i \exp\left(-\frac{\sum_{k=1}^{n}(x_k - c_{ik})^2}{r_i^2}\right) \quad (1)$$

where x_k ($k=1,...,n$) are the inputs to the network, and c_{ik} and r_i are the centers and radius of the RBF associated with neuron i, respectively.

If the inputs of the RBF are perturbed by noise, then the output y of the network is changed with respect to its nominal output. The *statistical sensitivity, S,* allows us to estimate in a quantitive way the degradation of the expected output of the RBF network when the values of the inputs change in a given amount. Statistical sensitivity is defined in [2] by the following expression:

$$S = \lim_{\sigma \to 0} \frac{\sqrt{var(\Delta y)}}{\sigma} \quad (2)$$

where σ represents the standard deviation of the changes in the inputs, and $var(\Delta y)$ is the variance of the deviation in the output (with respect to the output in the absence of perturbations) due to these changes, that can be computed as:

$$var(\Delta y) = E[(\Delta y)^2] - (E[\Delta y])^2 \quad (3)$$

where $E[\cdot]$ is the expected value of $[\cdot]$.

To compute expression (2), an additive model of input deviations is assumed that satisfies:
(a). $E[\Delta x_k] = 0$
(b). $E[(\Delta x_k)^2] = \sigma^2$
(c). $E[(\Delta x_k \Delta x_{k'})] = 0$ if $k \neq k'$

i.e. each input x_k is changed to $x_k + \delta_k$ where δ_k is a random variable with average equal to zero and varianze equal to σ^2. Moreover, perturbations of different inputs are supposed not to be statistically correlated. If the deviations of the weights of a neuron i in layer m are small enough, the corresponding deviation in the output can be approximated as :

$$\Delta y \approx \sum_{k=1}^{n} \frac{\partial y}{\partial x_k} \Delta x_k \quad (4)$$

3 The statistical sensitivity of a RBF network

Proposition 1: if $E[\Delta x_k]=0 \ \forall \ k$ then $E[\Delta y]=0$. *Proof 1:*

$$E[\Delta y] = E[\sum_{k=1}^{n} \frac{\partial y}{\partial x_k} \Delta x_k] = E[\sum_{k=1}^{n} \frac{\partial \sum_{i=1}^{m} w_i \Phi_i}{\partial x_k} \Delta x_k] = E[\sum_{k=1}^{n} \sum_{i=1}^{m} w_i \frac{\partial \Phi_i}{\partial x_k} \Delta x_k]$$

$$= E[\sum_{i=1}^{m} w_i \sum_{k=1}^{n} \frac{\partial \Phi_i}{\partial x_k} \Delta x_k] = 2 \sum_{i=1}^{m} \frac{w_i \Phi_i}{r_i^2} \sum_{k=1}^{n} (c_{ik} - x_k) E[\Delta x_k] = 0 \quad (5)$$

$\square$

Proposition 2: the statistical sensitivity to additive input perturbations of a RBF network can be expressed as:

$$S = 2\sqrt{\sum_{k=1}^{n}\left(\sum_{i=1}^{m}\frac{w_i \Phi_i}{r_i^2}(c_{ik}-x_k)\right)^2} \qquad (6)$$

Proof 2: making use of Proposition 1, $var(\Delta y) = E[(\Delta y)^2]$, that can be obtained as:

$$\begin{aligned}
E[(\Delta y)^2] &= E[(2\sum_{i=1}^{m}\frac{w_i \Phi_i}{r_i^2}\sum_{k=1}^{n}(c_{ik}-x_k)\Delta x_k)(2\sum_{j=1}^{m}\frac{w_j \Phi_j}{r_j^2}\sum_{l=1}^{n}(c_{jl}-x_l)\Delta x_l)] \\
&= 4E[\sum_{i=1}^{m}\frac{w_i \Phi_i}{r_i^2}\sum_{j=1}^{m}\frac{w_j \Phi_j}{r_j^2}\sum_{k=1}^{n}(c_{ik}-x_k)\sum_{l=1}^{n}(c_{jl}-x_l)\Delta x_k \Delta x_l] \\
&= 4\sum_{i=1}^{m}\sum_{j=1}^{m}\frac{w_i \Phi_i}{r_i^2}\frac{w_j \Phi_j}{r_j^2}\sum_{k=1}^{n}(c_{ik}-x_k)(c_{jk}-x_k)\sigma^2 \qquad (7)\\
&= 4\sigma^2 \sum_{k=1}^{n}\left(\sum_{i=1}^{m}\frac{w_i \Phi_i}{r_i^2}(c_{ik}-x_k)\right)\left(\sum_{j=1}^{m}\frac{w_j \Phi_j}{r_j^2}(c_{jk}-x_k)\right) \\
&= 4\sigma^2 \sum_{k=1}^{n}\left(\sum_{i=1}^{m}\frac{w_i \Phi_i}{r_i^2}(c_{ik}-x_k)\right)^2
\end{aligned}$$

Substituting (7) in (2) Proposition 2 is proved.

□

4 The Mean Square Sensitivity

It is usual to measure the learning performance of a RBF network with the Mean Squared Error (MSE). This error measurement is computed by a sum over a set of input patterns whose desired output is known, and its expression is the following:

$$MSE = \frac{1}{2N_p}\sum_{p=1}^{N_p}(d(p) - y(p))^2 \qquad (8)$$

where N_p is the number of input patterns considered, and $d(p)$ and $y(p)$ are the desired and obtained outputs for the input pattern p, respectively.

If the inputs of the network suffer any deviation, the MSE is altered and so, by developing expression (8) with a Taylor expansion near the nominal MSE found after trainig, MSE_0, it is obtained that:

$$MSE' = MSE_0 - \frac{1}{N_p} \sum_{p=1}^{N_p} (d(p)-y(p)) \, \Delta y(p) + \frac{1}{2N_p} \sum_{p=1}^{N_p} (\Delta y(p))^2 + 0 \qquad (9)$$

Now, if we compute the expected value of MSE' and take into account that $E[\Delta y]=0$, and that $E[(\Delta y)^2]$ can be obtained from expressions (2) and (3) as $E[(\Delta y)^2] \approx \sigma^2 S^2$, the following expression is obtained:

$$E[MSE'] = MSE_0 + \frac{\sigma^2}{2N_p} \sum_{p=1}^{N_p} (S(p))^2 \qquad (10)$$

By analogy with the definition of MSE, we define the following figure as *Mean Squared Sensitivity* (MSS):

$$MSS = \frac{1}{2N_p} \sum_{p=1}^{N_p} (S(p))^2 \qquad (11)$$

The MSS can be computed from the statistical sensitivity as expression (11) shows and, then, from the nominal values of the network parameters and inputs. In this way, combining expressions (10) and (11), the expected degradation of the MSE, E[MSE'] can be computed as:

$$E[MSE'] = MSE_0 + \sigma^2 \, MSS \qquad (12)$$

Thus, (12) shows the direct relation between the MSE degradation and the MSS. As MSE_0 and MSS can be directly computed after training, it is possible to predict the degradation of the MSE when the inputs are deviated from their nominal values into a range with standard deviation equal to σ. Moreover, as can be deduced from the expression obtained, a lower value of the MSS implies a lower value of the degradation of the MSE, so we propose using the MSS as a suitable measure of the noise immunity of RBF networks.

5 Results

In order to validate expression (12) we compared the results obtained for the E[MSE'] when the networks are subject to additive deviations in their inputs, with the predicted value obtained by using this expression. Two RBFNs were considered: a predictor of the Box-Jenkins gas furnace [6] and a predictor of the Mackey-Glass temporal series [7]. For each experiment we have perform the following steps:

```
1. Train the network
2. Compute MSE₀ and  MSS over a test set of patterns
3. Fix a value for σ
4. Repeat 100 times
        4.1 For each input pattern p and each input k of
        the test set:
           Do x_pk =x_pk + δ_pk
        4.2 Compute MSE'
5. Compute E [MSE']_experimental
6. Compute E [MSE']_predicted as MSE₀ + σ MSS
```

The gas-furnace predictor had 2 inputs, 10 RBFs in the hidden layer and 1 output neuron, while the Mackey-Glass predictor consisted of 4 inputs, 14 RBFs in the hidden layer and 1 output neuron. Table 1 shows the values of MSE_0 and MSS obtained after training with the test patterns (different from those used for training).

Table 1. MSE_o and MSS obtained after training.

	Gas-furnace	Mackey-Glass
MSE_0	0.164	$6{,}65\ 10^{-5}$
MSS	1.275	1.074

As described above, all the inputs have been deviated from their nominal values by using an additive model such that each input x_{pk} of pattern p has a value equal to $x_{pk} + \delta_{pk}$, where δ_{pk} is a random variable with standard deviation equal to σ and average equal to zero. We have considered different values for σ, and the experimental values of MSE' are averaged over 100 tests in order to compute $E[MSE']$.

Figures 1 and 2 show the degradation of the MSE for different values of σ. The values predicted and obtained experimentally are represented for the gas-furnace and the Mackey-Glass predictors, respectively. Each experimental value is plotted with its respective confidence level at 95% obtained with 100 samples. It can be observed that the predicted values of the MSE accurately fit those obtained experimentally.In this way, MSS is proposed as a measurement of noise immunity of the RBF network because the lower the MSS the lower the MSE degradation in the presence of noise. Thus, among RBF network configurations that present a similar MSE_0, the one with lower MSS provides a more stable output when its inputs are perturbed.

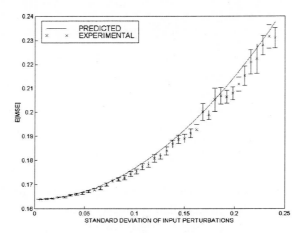

Figure 1. Predicted and experimental MSE for the predictor of the gas-furnace problem.

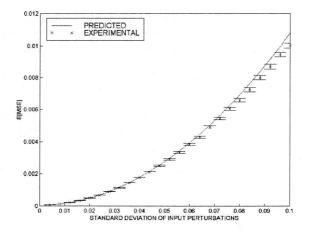

Figure 2. Predicted and experimental MSE for the predictor of the Mackey-Glass temporal series.

6 Conclusions

We have derived a quantitative measurement of noise immunity for RBF networks. This measurement, that we have called Mean Squared Sensitivity (MSS), is explicitly related to the MSE degradation in the presence of input perturbations. We have demonstrated the validity of the MSS expression by showing that a lower value of MSS implies a lower degradation of MSE. As we did in [3] for MLPs, we pretend to extend this study considering other kind of deviations (other models and over other parameters). Moreover, as MSS can be seen as a regularizer [1], it can be used to train the network to improve the final performance in the same way as it was done in [5] for MLPs.

Acknowledgements

This work has been supported in part by the projects 1FD97-0439-TEL1, CICYT TIC99-0550 and CICYT TIC2000-1348.

References

[1] S. Haykin, *Neural Networks. A Comprehensive Foundation*, second edition, Prentice Hall, 1999.
[2] J.Y. Choi, C. Choi, "Sensitivity Analysis of Multilayer Perceptron with Differentiable Activation Functions", *IEEE Trans. on Neural Networks*, vol. 3, no. 1, pp. 101-107, Jan 1992.
[3] J.L. Bernier, J. Ortega, E. Ros, I. Rojas, A. Prieto; "A Quantitative Study of Fault Tolerance, Noise Immunity and Generalization Ability of MLPs", Neural Computation, Vol. 12, pp. 2941-2964, 2000.
[4] J.L. Bernier, J. Ortega, I. Rojas, E. Ros, A. Prieto "Obtaining Fault Tolerant Multilayer Perceptrons Using an Explicit Regularization", Neural Processing Letters, Vol. 12, No.2, pp. 107-113, Oct. 2000.
[5] J.L. Bernier, J. Ortega, E. Ros, I. Rojas, A.Prieto. "A new measurement of noise immunity and generalization ability for MLPs". International Journal of Neural Systems, Vol. 9, No. 6, pp.511-522. December 1999.
[6] G.E.P Box, G.M. Jenkins, "Time Series Analysis, Forecasting and Control", 2nd ed. San Francisco, CA: Holden-Day, 1976.
[7] M.C. Mackey, L. Glass, "Oscillation and Chaos in Physiological Control Systems". Science, Vol. 197, pp. 287-289. 1977.

Analyzing Boltzmann Machine Parameters for Fast Convergence

Fco. Javier Salcedo, Julio Ortega, and Alberto Prieto

Department of Computer Architecture and Technology, University of Granada, Spain
jortega@atc.ugr.es

Abstract. The behavior of a Boltzmann Machine (BM) according to changes in the parameters that determine its convergence is experimentally analyzed to find a way to accelerate the convergence towards a solution for the given optimization problem. The graph colouring problem has been chosen as a benchmark for which convergence with quadratic time complexity has been obtained.

1 Introduction

A Boltzmann Machine (BM) is a network with binary neurons connected by symmetric weights that evolves stochastically. In the ideal case of an infinitely slow evolution, the convergence to a global optimum of the consensus function is guaranteed. However, procedures for real applications need polynomial time complexity with respect to the problem size to be of practical use.

In [4] an acceleration method for Boltzmann Machine learning is proposed using the *linear response theorem*. This method reduces computing time to $O(N^3)$, with N being the number of neurons. Other approaches try to reduce the computing time required for BM evolution, for example, by taking advantage of the parallelism offered by the neural model. A scheme for the efficient implementation of BM in coarse grain parallel computer architectures is presented in [3].

This paper shows how to improve the evolution of a sequential Boltzmann Machine (SBM), considering two possibilities: (1) the modification of the BM topology to suit it to the problem at hand; and (2) a careful determination of the parameters that control the evolution. These alternatives could reduce the time required by a given BM to provide sufficiently good solutions for combinatorial optimization problems with a polynomial time complexity.

Section 2 of this paper describes the characteristics of the BM model and considers strategies for fast convergence. Then, Section 3 illustrates the application of the above strategies to the graph colouring problem, while Section 4 shows the experimental results. Finally, some conclusions are provided in Section 5.

2 The Boltzmann Machine

As is well known [1], BM is a neural network composed of interconnected neurons with two possible states {0,1}. The network can be represented by a pseudo-graph B=(U,C), where U is the finite set of neurons and C is the set of connections between neurons. If k(u) is the state of the neuron u∈ U, neuron u is active when k(u)=1, and a connection {u,v} is active if neurons u and v are both active (k(u)*k(v)=1). Each connection has an associate number $s_{\{u,v\}}$ called the weight of the connection {u,v}. The so called *consensus function* assigns a real number to each BM state as:

$$W : K \rightarrow \Re \qquad W(k) = \sum_{\{u,v\} \in C} s_{\{u,v\}} k(u) k(v) \ . \qquad (1)$$

where K is the space of all possible BM states. The cardinality of K is $2^{|U|}$. In a sequential Boltzmann Machine (SBM), only one neuron is allowed to change its state thus contributing to the machine evolution. A change in a neuron, u, is evaluated by using expression (2), where C_u is the set of connections with neuron u at one of the extremes.

$$\Delta W_k(u) = (1 - 2k(u)) \left[\sum_{\{u,v\} \in C_u} s_{\{u,v\}} k(v) + s_{\{u,u\}} \right] \ . \qquad (2)$$

A change in a neuron state is accepted depending on the value of the activation function:

$$A_k(u,w) = \frac{1}{1 + e^{-\frac{\Delta W_k(u)}{t}}} \ . \qquad (3)$$

where t is a parameter called temperature, that decreases as the BM evolves. The value of $A_k(u,w)$ is compared at every trial with a random value between 0 and 1, which is usually uniform and independent from the global state k and temperature t. When the temperature is near 0, no state changes are accepted and then the consensus function reaches the optimum value.

To decrease the time required for BM convergence, two alternatives can be considered:

1) A reduction in the solution space to be explored by the BM.
2) An increase in the BM cooling speed.

The first strategy is related to the characteristics of the optimization problem at hand and the topology of the BM where it has been mapped. Thus, a careful design of the BM has to be made, trying to determine the machine with the lowest number of units that can be mapped to the problem at hand.

The second alternative implies adjusting the parameters that control the evolution of the BM towards a solution. An unsuitable choice of such parameters could make the convergence to an optimal solution much too slow. For each problem it it possible to find a set of variables x_1, x_2, x_3...that characterize it. Once the function is found that describes how to decrease the temperature during the BM evolution (for example a linear function), the determination of the parameters implies finding the three functions described below.

$$\begin{cases} T_i = T_i(x_1, x_2, x_3, ...) \\ T_f = T_f(x_1, x_2, x_3, ...) \\ \alpha = \alpha(x_1, x_2, x_3, ...) \end{cases} \quad (4)$$

Thus, the initial temperature T_i, final temperature T_f and the parameter to decrease the temperature α can be estimated before the evolution of the BM starts. A linear function to decrease temperature has been chosen for simplicity, as in this case only three parameters have to be determined. In a more general case, other functions can be used, although this implies determining more parameters with the form $f=f(x_1, x_2, x_3,...)$.

The following expression can be used to approximate the initial temperature T_i:

$$T_i \approx \Delta W_k(u) \ . \quad (5)$$

In (5), the initial temperature takes values near the increment in the consensus function due to a change in the state of a neuron. Once an expression for T_i in term of variables $x_1, x_2, x_3,...$, has been obtained, it is necessary to find the remaining two functions. In the next section the we show how this is done for the graph colouring problem.

3. Boltzmann Machine parameters in the Graph Colouring Problem

Given an entry graph $G=(V,E)$, where V is the set of nodes, and E the set of connections between nodes, colouring a graph requires us to find the minimum integer number, m, for which it is possible to define a function $g:V \rightarrow \{1,2,...,m\}$ such that $g(i) \neq g(j)$ for every i, j∈ V with (i,j)∈ E. The graph colouring problem is NP-Complete [5].

This is the classical formulation of the problem, but in many real applications it is more usual to have a maximum number, l, of colours or subsets to classify the nodes of graph G. An example of this kind of problem is the optimal assignment of the variables in a computer program (nodes) to the processor registers (colours) [2], taking into account that a connection (i,j) between variables i and j exists if both need to be used by the processor at the same time. In the following, we will use this latter formulation of the problem.

3.1 Decreasing the Space of Possible Solutions

The BM structure proposed to solve a graph colouring problem is a modification of the one proposed by Aarts and Korst [1]. The neurons of the BM are the cartesian product of the set of nodes, V, of the graph, and the set of colours C, V'=VxC. Each neuron (i,c) represents the node i and the color c. Thus, a color c is assigned to neuron i if in the final state, the neuron (i,c) is active. In this way, BM is defined by the graph G'.

$$G' = (V', E') = (VxC, E_h \cup E_v \cup E_b) \,. \tag{6}$$

Once BM is constructed the consensus function can be expressed as:

$$W = \sum_{\{(i,c)(i,c)\} \in E_b} \omega_{icic} x_{ic} \,. \tag{7}$$

The BM built this way saves computing time because the space of states of the BM, K, has been bounded; its cardinality has been reduced from $2^{|V|}$, of a BM with classical topology [1], to 2^l in the BM built with the procedure proposed here. An interesting graph colouring is when the number of colours, l, is less than the number of nodes |V|.

3.2 Parameters for Fast Convergence

To determine the initial temperature T_i, final temperature T_f and decreasing temperature parameter α, in a graph colouring problem, it is considered that these parameters depend on the number of colours, l, and on the complexity of the graph to be colored.

The graph complexity is indicated by a real number. If the graph to be colored, G, has N nodes and Q connections, and the maximum number of connections is the number of combinations of nodes taken two by two, the graph complexity can be defined as:

$$\chi(N, Q) = \frac{nodes \cdot connections}{maximum\ number\ of\ connections} = \frac{2Q}{N-1} \,. \tag{8}$$

Once the BM is built, the parameters that control its evolution must be determined. From (5), the initial temperature T_i is set equal to the consensus value in the initial state of the BM, divided by the number of colours, l, and the graph complexity χ. Thus, if the initial value of the consensus function is noted as W_i and the number of colours is l then:

$$T_i = \frac{W_i}{\chi l} \,. \tag{9}$$

The required initial value for consensus can be estimated from (7). As the initial state of the BM is randomly selected, it is possible to assume that there are half of the active neurons, $n_j = N/2$, in each level, and that the sum over the color weights is approximately equal to the product of the number of colours and a weight situated between the maximum and the minimum weights:

$$W_i = \sum_{j=1}^{l} n_j \omega_j \approx \frac{N}{2} \sum_{j=1}^{l} \omega_j \approx \frac{N}{2} l \frac{\omega_{max} + \omega_{min}}{2} \,. \tag{10}$$

In the case of an intermediate graph complexity ($\chi \approx N/2$), and substituting the approximated value of W_i in (10), T_i is obtained as a function of the number of colours l, $T_i = f(l)$, as $\omega_i = f(l)$.

$$T_i \approx \frac{1}{2}(\omega_{max} + \omega_{min}).\qquad(11)$$

Once the initial temperature is determined, the next step is to obtain the parameter $\alpha(\chi,l)$ and the final temperature $T_f(\chi,l)$. The range of possible values for the decrement parameter $\alpha(\chi,l)$ is divided into intervals of equal size according to the required accuracy. For a given number of colours and χ, the best assignment for α is a sigmoidal function, where $a_2 = 0.9999$, and the remaining function parameters depend on l.

$$\alpha(\chi) = \frac{a_1 - a_2}{1 + e^{\frac{x - x_0}{d}}} + a_2.\qquad(12)$$

Figure 1 shows the curves corresponding to different numbers of colours. As the number of colours increases, α grows more slowly to its maximum value. Moreover, as α grows with graph complexity, a higher evolution time is required as the graphs become more complex.

The value of the parameter T_f can be obtained once the curves for α are obtained. For each value of α obtained from curves, the corresponding final temperature is selected using the minimum value of those obtained for the same number of colours. It has been verified that, for any number of colours and graph complexity, this value for the final temperature is unique and $T_f = 10^{-4}$.

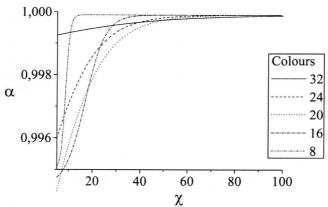

Fig. 1. Determination of the decrement parameter

4. Experimental Results

The procedures used to determine the initial temperature T_i, final temperature T_f and decreasing parameter α, in terms of graph complexity and the number of colours, can be integrated into a final version of the BM algorithm,

```
1)Built Boltzmann Machine.
2)Choose a random configuration of global state k.
3)Calculate T_i and α.
4)Choose a random neuron.
5)Calculate A_k(u,w) and a random number G.
6)IF A_k(u,w)<G THEN accept state change.
7)Reduce temperature t.
8)IF t<10^-4 THEN finish evolution ELSE GOTO 4.
```

Figure 2 shows how the convergence evolves for different numbers of colours (here, convergence is defined as the percentage of feasible graph colourings obtained from different executions of the BM). As expected, convergence grows with the number of colours, and decreases with graph complexity.

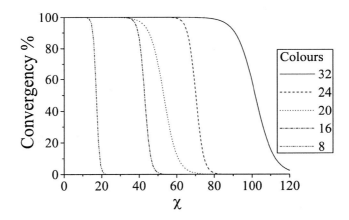

Fig. 2. Convergence obtained by a BM

Figure 3 shows how the convergence grows with the number of colours. The most important zone is that with a BM convergence higher than 95%. Using this figure, we can estimate either the minimum number of colours required to color a graph using BM, or the convergence provided by BM with a given graph and number of colours. It is possible to save computing time with this information as we can determine the expected convergence without any simulation. If this convergence is too low, the number of colours should be increased until a suitable level of convergence is reached.

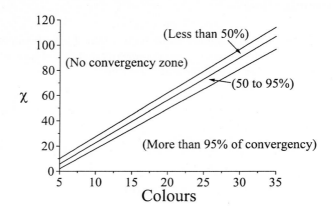

Fig. 3. Zones of convergence vs. number of colours

Figure 4 shows that the proposed BM algorithm has a quadratic time complexity. More precisely it has been calculated that this complexity is 1.54, which implies that the dominant powers are 1 and 2. This is a reasonable value because the previous BM algorithm had the following complexities in its steps:

1) $O(N^2)$
2) ... 8) $O(N)$

In fact, an algorithm with linear complexity is not possible as obtaining the BM from the graph to be colored implies $O(N^2)$.

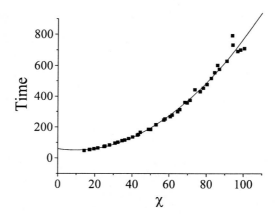

Fig. 4. Time complexity of the proposed BM algorithm for 32 colours (the curve corresponds to a quadratic function)

A graph colouring problem solved by a BM, like reference [6], required 82 seconds on a PC with an AMD K7 750 Mhz processor. The same problem has been solved using our BM in only 0.2 seconds (using the same machine).

5. Conclusions

This paper shows how it is possible to decrease the search time of a BM by: (1) a suitable definition of the BM for the problem at hand in order to reduce the number of possible search states, and (2) a suitable determination of the parameters that control the evolution of the BM.

The first alternative is not always applicable, but it is always possible to determine the parameters required to reach a significant decrease in the time required for BM evolution.

Both alternatives have been applied to graph colouring problems, and it has been shown that whenever both alternatives are simultaneously applied, the decrease in computing time can be very significant (complexities lower than $O(N^2)$ have been obtained).

Combining this method with others that explore the inherent parallelism of neural networks [3], can improve the performance of BM in the solution of combinatorial optimization problems in real applications.

Acknowledgements. Paper supported by project TIC2000-1348 (Spanish *Ministerio de Ciencia y Tecnología*)

References

1. Aarts, E.H.L., Korst, J.H.M.: Simulated Annealing and Boltzmann Machines. Wiley, New York (1989)
2. Chow, F., Hennessy, J.: Register Allocation by Priority-Based Colouring. Proceedings of the ACM SIGPLAN '84. Symposium on Compiler Construction SIGPLAN Notices, vol. 19, num. 6, (June 1984)
3. Ortega, J., and others: Parallel Coarse Grain Computing of Boltzmann Machines. Neural Processing Letters 7: 169-184, (1998)
4. Kappen, H.J., Rodríguez, F.B.: Efficient Learning in Boltzmann Machines Using Linear Response Theory. Neural Computation 10 (5): 1137-1156, (1998)
5. Garey, M. R., Johnson, D. S.: Computers and Intractability: A Guide to the Theory of NP-Completeness. Freeman (1979)
6. Zissimopoulos, V., Paschos, V. Th., Pekergin, F.:On the Approximation of NP-Complete Problems by Using the Bolzmann Machine Method: The Cases of Some Covering and Packing Problems. IEEE Transactions on Computers, vol. 40, num. 12, (December 1991)

A Penalization Criterion Based on Noise Behaviour for Model Selection

Joaquín Pizarro Junquera[1], Pedro Galindo Riaño[1], Elisa Guerrero Vázquez[1], Andrés Yañez Escolano[1]

[1] Universidad de Cadiz, Departamento de Lenguajes y Sistemas Informáticos, Grupo de Investigación *"Sistemas Inteligentes de Computación"*
C.A.S.E.M. 11510 - Puerto Real (Cádiz), Spain
{joaquin.pizarro ,pedro.galindo ,elisa.guerrero, andres.yaniez@ }uca.es

Abstract. Complexity-penalization strategies are one way to decide on the most appropriate network size in order to address the trade-off between overfitted and underfitted models. In this paper we propose a new penalty term derived from the behaviour of candidate models under noisy conditions that seems to be much more robust against catastrophic overfitting errors that standard techniques. This strategy is applied to several regression problems using polynomial functions, univariate autoregressive models and RBF neural networks. The simulation study at the end of the paper will show that the proposed criterion is extremely competitive when compared to state-of-the-art criteria.

1. Introduction

When learning a function $f:X \rightarrow Y$ from a set of training data $<x_1,y_1>, ..., <x_n,y_n>$, there is a well-known tradeoff between the size of the training sample and the complexity of the function class being considered: if the class is too complex, there is a risk of "overfitting" the training data and guessing a function that will perform poorly on future test examples. An overfitted model may be unstable in the sense that repeated samples from the same process can lead to widely differing predictions due to variability in the extraneous variables. On the other hand, if the class is too simple, underfitting occurs, and we'll get models with poor predictive ability due to the lack of detail in the model. This tradeoff can be formalized in terms of the bias/variance decomposition of the expected hypothesis error. Intuitively, we expect the variance term to increase for larger hypothesis. On the other hand, we expect the bias term to decrease as the model complexity grows up.

Under the simplest formulation of model selection, the idea is to define a set of candidate hypothesis class $H_0 \subset H_1 \subset ... \subset H_n$ and then choose the class that has the appropriate complexity for the given training data. Note, however, that for a given training sample we obtain a corresponding sequence of empirical functions $h_1 \in H_1,...,h_n \in H_n$ that achieve the *minimum observed average error* $E(error(h_i))$, and that these errors are monotonically decreasing. Therefore, choosing the function with minimum training error simply leads to choosing a function from the largest class.

The most common strategy for coping with this fact is to use some form of model selection, such us hold-out testing or complexity-penalization, to balance the tradeoff between complexity and data fit.

The first approach splits the sample data into two subsamples of size n-m and m; the first is used to fit the model and the second is used to estimate the minimum average error for the model. There is, however, a problem in deciding how many samples points, m, one should "leave out" of the first subsample. For example if we split into groups of equal size n/2 then, in the second step, we would be estimating the minimum average error for a sample of size n/2 rather than for a sample of size n (which is what we have in fact). Sample size is a critical factor in determining which model is best, on average, and our objective is to find this out for a sample of size n, not for a sample of size n/2. One could reduce m but that leave less information over to estimate the error. There are many variants of this approach, including generalized cross validation, bootstrap methods, etc.

The second strategy assigns increasing complexity values $c_0, c_1, ..., c_n$ to the successive function classes, and then chooses the hypothesis that minimizes some combination of the form $E(error(h_k)) + c_k$. The penalty term can be interpreted as postulating a particular profile for the variances as a function of model complexity [4]. If the postulated and true profiles do no match, then systematic underfitting or overfitting results, depending on whether the penalty terms are too large or too small.

2. Complexity-Penalization Strategies

In this section we describe some of the complexity-penalization selection criteria commonly used in regression modeling. Two of these, the Akaike Information Criterion(AIC) [2] and its corrected version (AICc) [5],[10] estimate the Kullback-Leibler discrepancy. The other two, the Schwarz Information Criterion (SIC) [9] and the AICu [6] were derived for their asymptotic performance properties.

AIC was the first of the Kullback-Leibler (K-L) information based model selection criteria. In his derivation Akaike makes the assumption that the true model belongs to the set of candidate models. This assumption may be unrealistic in practice, but it allows us to compute expectations for central distributions, and it also allows us to entertain the concept of overfitting. In general, AIC= -2log(*likelihood*) + 2k, where the *likelihood* is usually evaluated at the estimated parameters, and *k* is the numbers of parameters. The derivation of AIC is intended to create an estimate that is an approximation of the K-L discrepancy. In fitting a candidate model k we have

$$AIC_k = \log(\frac{SEE_k}{n}) + \frac{2(k+1)}{n} \qquad (1)$$

where n is the sample size, k is the number of parameters of the model, and SSE_k is the usual sum of squared errors.

Schwarz [9] derived SIC as an asymptotic approximation to a transformation of the Bayesian posterior probability of a candidate model. SIC is one of the most widely known and used tools in statistical model selection. Although the original derivation assumes that the observed data are independent, identically distributed, and arising from a probability distribution in the regular exponential family, SIC has traditionally been used in a much larger scope of model selection problem. In large sample settings, the fitted model favored by SIC ideally corresponds to the candidate model which is a posteriori most probable. It is defined as follows:

$$SIC_k = \log(\frac{SSE_k}{n}) + \frac{\log(n)k}{n} \qquad (2)$$

The 2(k+1) term in AIC is replaced by log(n)k in SIC, resulting in a much stronger penalty for overfitting. Many authors have shown that the small sample properties of AIC and SIC lead to overfitting. In response to this difficult other methods have been proposed.

AICc [10] is intended to correct the small-sample overfitting tendencies of AIC by estimating an approximation to E(K-L) directly rather than estimating an approximation to L-K. Hurvich and Tsai have shown that AICc does in fact outperform AIC in small samples, but that is asymptotically equivalent to AIC and therefore performs just as well in large samples.

$$AICc_k = \log(\frac{SSE_k}{n}) + \frac{n+k}{n-k-2} \qquad (3)$$

AICu [6] tries to correct the weak signal-to-noise ratio of several model selection criteria, by strengthening this ratio. It makes use of the unbiased estimate of σ^2 instead of its maximum likelihood shown below:

$$AICu_k = \log(\frac{SSE_k}{n-k}) + \frac{n+k}{n-k-2} \qquad (4)$$

Let us note that AICu has the same penalty function as AIC, but it differs by its use of the unbiased estimate of the variance.

There are several penalization theories for determining the optimal network size e.g. the NIC (Network Information Criterion) [3] which is a generalization of the AIC, the generalized final prediction error(GPE) as proposed by [7], and the Vapnik-Chervonenkis (VC) dimension[1]. NIC relies on a single well-defined minimum to the fitting function and can be unreliable when there are several local minima [8]. There is very little published computational experience of the NIC, or the GPE, and their evaluation is prohibitively expensive for large networks. For these reasons, our results on RBF's are reported on k-fold cross validation, with k=2, k=10 and directly on the generalization error.

Hence, four criteria were chosen to illustrate and compare the characteristics and performance of the proposed criterion, namely AIC, SIC, AICc and AICu.

3. Analysing Noise

Let's start by describing the model structures with which we will work, and the assumptions we will make. We define the *true model* in regression to be Y = f(X). Next we define the *general model* to be Y=f(X)+ε where Y=(y_1, y_2,...,y_m) is the vector of responses assuming that the errors are independent, identically distributed, following a gaussian distribution, $\varepsilon=N(0,\sigma)$. From the general model we only have a *random sample* of size n << m. Finally we will define the *candidate models* with respect to the general model $y_k=f_k(x)$. We will assume that the method of least squares is used to fit a model to the data. The point is that minimum average errors tend to gross underestimates of the true error in general and the degree of underestimation tends to become worse at higher complexity levels. Complexity-penalization seeks to adjust the empirical error to compensate this fact. Some methods (AIC, SIC) exhibit an overfitting tendency when the number of samples is small because the penalty term is smaller than necessary, while others adjust these tendency (AICc, AICu). This effect also appears when hold-out techniques are applied.

Our idea is to exploit the extra information which arises from the noise study to avoid many of the overfitting errors that plague standard complexity-penalization methods. This extra information attempts to estimate the variance of a function class H_k by analysing how it fits the noise distribution. The idea is illustrated using an example in which 100 random samples of size 50 are taken from the function $f(x) = -4.9389 \cdot x^5 - 1.7917 \cdot x^4 + 23.2778 \cdot x^3 + 8.7917 \cdot x^2 - 15.3389 \cdot x - 6 + N(0,3)$. We will define the candidate models to be the polynomial families from degrees 1 to 10.

Let us suppose that the noise in each data point of the sample is known. For a given model k, two polynomials are computed, one by fitting the sample $(x, y_i+\varepsilon_i)$ and another one by fitting only the noise (x, ε_i). Figure 1 shows the median experimental (sometimes called sample error or resubstitution error) and generalization errors, both for data (SE_k, GE_k) and noise (NSE_k, NGE_k). The median provides a better measure of location than the mean when there are some extremely large values, as it is the case, due to the appearance of a large number of outliers.

Let us note that in the case of an overfitted model (degrees 5 and up) $SE_k-SNE_k=0$ and $GE_k-NGE_k=0$. An inmediate conclusion is that, in this case, all the error is due to noise. If the true function does not belong to the set of candidate models, it is easily verified that SE_k and GE_k are above NSE_k and NGE_k, but the differences GE_k-NGE_k and ES_k-NSE_k can be considered constants. As a conclusion, there is an error due to data fitting and another one due to noise fitting.

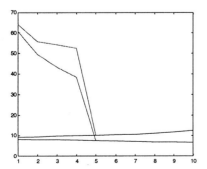 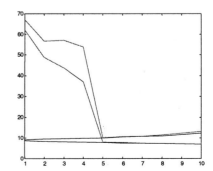

Fig. 1. Medians per model of SE, GE, NSE and NGE over 100 random sample where noise per sample is known.

Fig. 2. Medians per model of SE, GE, NSE* and NGE*, over 100 random sample where noise per sample is unknown.

Let us suppose now that we know the noise distribution, but that noise in each data point of the sample is unknown. In this case, in order to approximate NSE_k and NGE_k, a set of points from this noise distribution is generated, and a large set of noise samples of size 50 is taken (the more, the better). Then NSE_k and NGE_k from each noise sample are computed and the median NSE_k^* and NGE_k^* is taken as the error due to noise. Figure 2 shows the results. Note that $NSE_k \cong NSE_k^*$ and $NGE_k \cong NGE_k^*$.

Let us suppose now the usual situation, where the noise distribution is unknown. Figure 3 shows the median of the distributions of NSE_k^* and NGE_k^* from normal distributions $N(0,\sigma)$, where the size of the sample is 50 and $\sigma=0.5, 1, 3$ and 5.

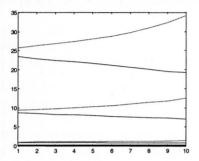
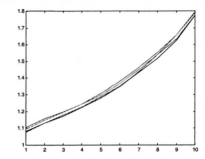

Fig. 3. NSE^* and NGE^* from 1000 random samples $N(0, \sigma)$. $\sigma = 0.5, 1, 3, 5$, size = 50

Fig. 4. Relation NGE_k^*/SNE_k^* from 1000 random samples $N(0,\sigma)$ $\sigma=0.5,1,3,5$, size=50

Figures 5 and 6 shows the distribution of NSE^*/σ and NGE^*/σ for a given complexity k=9 and different values of σ. It is interesting to note that the distributions of NSE^* and NGE^* do differ in a value which is exactly σ. This property also holds for any other complexity level. The shape of these distributions only depends on the size of the noise samples and the complexity and nature of candidate models.

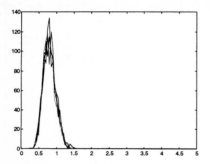
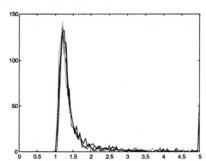

Fig. 5. Distributions of NSE_9^*/σ from 1000 random samples $N(0,\sigma)$ $\sigma=0.5,1,3,5$, size=50

Fig. 6. Distributions of NGE_9^*/σ, from 1000 random samples $N(0,\sigma)$ $\sigma=0.5,1,3,5$, size=50

In order to eliminate the dependence of the standard deviation of noise, we obtain the relation NGE_k/NSE_n, where the variance term cancels out, and, therefore, the quotient do not depend on the noise. Figure 4 shows the quotient NGE_k^*/NSE_k^* derived from 1000 random samples $N(0,\sigma)$ for four different values of σ (0.5, 1, 3, 5) and sample size equal to 50. We observe that the distributions of errors verify:

$$\frac{NSE^*_{1,k}}{\sigma^2_1} = \frac{NSE^*_{2,k}}{\sigma^2_2} = \frac{NSE^*_{3,k}}{\sigma^2_3} = \frac{NSE^*_{4,k}}{\sigma^2_4} = C_{1k} \qquad \frac{NGE^*_{1,k}}{\sigma^2_1} = \frac{NGE^*_{2,k}}{\sigma^2_2} = \frac{NGE^*_{3,k}}{\sigma^2_3} = \frac{NGE^*_{4,k}}{\sigma^2_4} = C_{2k} \qquad (5)$$

$$\frac{NGE^*_k}{NSE^*_k} = \frac{C_{2k}}{C_{1k}} = C_k \quad , \text{giving} \quad \log(NGE^*_k) = \log(NSE^*_k) + \log(C_k) \qquad (6)$$

As in any complexity-penalization method, the criterion consists of a first term, which is a measure of inaccuracy, badness of fit or bias and a second term which is a measure of complexity or penalty due to the increased unreliability for the bias in the first term. We have shown that this relation holds for noise. It also holds for data, but only when the model is overfitted. We hypothesize that this relation may be used as a criterion for model selection. Extending this formulation, we introduce two new model selection criteria, that make use of different estimates of the variance, i.e. the unbiased and the maximum likelihood estimates of σ^2, respectively, giving rise to:

$$NDIC_k = \log(\frac{SSE_k}{n}) + \log(C_k) \qquad \text{(Noise derived information criterion)} \qquad (7)$$

$$NDICu_k = \log(\frac{SSE_k}{n-k}) + \log(C_k) \qquad \text{(Noise derived information criterion, unbiased)} \qquad (8)$$

4. Experimental Results

Four different and representative problems have been simulated, polynomial fitting of a polynom, polynomial fitting of a sine function, univariate autoregressive regression and radial basis function network trained to fit a polynom. Three different sample sizes were chosen depending on the complexity of the problems to show the results from an insufficient, enough and large sample size. Different rows show the results in ascending complexity, showing how many times each model is selected.

Four criteria (AIC, SIC, AICc and AICu) were chosen for the first three problems. In these cases, the values of x were randomly generated following a uniform distribution in the interval [-2,2]. However, due to the prohibitively expensive calculation of criteria for large networks, results for RBF's are reported on k-fold cross-validation k=2,10. In this case, x_i are generated following a normal distribution N(0,1). The EG column was calculated by generating 5000 independent points from the general model, and determining the error produced for each model in all the experiments.

Table 1 shows different results in which 1000 random samples of different sizes are taken from the function $f1(x) = -4.9389 \cdot x^5 - 1.7917 \cdot x^4 + 23.2778 \cdot x^3 + 8.7917 \cdot x^2 - 15.3389 \cdot x - 6 + N(0,3)$. Table 2 shows the results obtained for the function $f2(x)=10 \cdot \sin(x+2)^2 + N(0,3)$. Table 3 shows the results obtained for the function $f3(x)= 3 \cdot x_1 + 4 \cdot x_2 - 5 \cdot x_3 + N(0,5)$ in an univariate autoregressive problem where the candidate models are $f_k(x)=a_1 \cdot x_1+...+a_k \cdot x_k$, k=1..6. Table 4 shows the results of fitting RBF neural networks to f1(x) varying the number of gaussian kernels.

Table 1. Polynomial fitting to a 5 degree polynom, f1(x), k = polynom orders = 1..10

Sample size = 100							Sample size = 25							Sample size = 15						
AIC	SIC	AICc	AICu	NDICu	NDIC	EG	AIC	SIC	AICc	AICu	NDICu	NDIC	EG	AIC	SIC	AICc	AICu	NDICu	NDIC	EG
0	0	0	0	0	0	0	0	0	0	1	1	0	11	2	2	37	113	22	4	61
0	0	0	0	0	0	0	0	0	0	0	0	0	10	0	0	19	35	10	1	64
0	0	0	0	0	0	0	0	1	1	5	3	1	1	1	12	107	149	87	50	15
0	0	0	0	0	0	0	2	4	5	9	7	5	0	23	31	108	111	116	101	0
699	958	782	905	905	776	770	519	769	904	928	927	888	668	242	375	709	580	747	816	96
103	29	97	53	51	93	135	104	84	59	39	45	71	178	78	91	17	10	17	25	67
67	10	54	24	25	59	63	94	52	15	11	11	21	87	73	62	3	2	1	3	71
65	1	39	14	14	38	16	85	38	12	5	5	12	29	93	80	0	0	0	0	19
29	1	15	3	4	16	10	82	27	3	2	1	2	13	142	100	0	0	0	0	7
37	1	13	1	1	18	6	114	25	1	0	0	0	3	346	247	0	0	0	0	0

Table 2. Polynomial fitting to f2(x)=10*sin(x+2)$_2$, k = polynom orders = 10..20

Sample size = 500							Sample size = 300							Sample size = 100						
AIC	SIC	AICc	AICu	NDICu	NDIC	EG	AIC	SIC	AICc	AICu	NDICu	NDIC	EG	AIC	SIC	AICc	AICu	NDICu	NDIC	EG
0	0	0	0	0	0	0	0	0	0	0	0	0	0	3	23	5	17	18	7	6
16	332	20	68	81	34	0	92	555	116	249	265	127	22	365	681	513	647	732	629	70
89	387	109	218	259	133	18	193	306	219	308	272	197	84	215	172	250	210	144	184	46
154	123	159	192	158	131	74	173	89	184	171	184	192	147	104	35	82	52	44	84	168
451	153	463	416	386	406	555	303	47	295	217	220	292	462	106	14	76	28	22	63	165
109	5	107	66	71	118	255	91	3	82	32	37	90	175	48	1	21	5	2	12	64
57	0	49	18	20	66	61	58	0	47	13	13	47	65	45	1	18	3	1	3	31
49	0	39	11	12	46	26	25	0	18	7	6	15	23	36	0	12	3	0	2	13
39	0	29	9	10	40	5	20	0	9	1	1	9	12	19	0	3	0	0	0	8
22	0	15	2	3	18	3	25	0	18	2	2	19	5	28	0	3	0	0	0	4
14	0	10	0	0	8	3	20	0	12	0	0	12	5	27	0	1	0	0	0	4

Table 3. Univ. Autoregressive fitting to f3(x) = 3·x1+ 4·x2 - 5·x3+ N(0,5), k=#vars = 1..6

Sample size = 100							Sample size = 25							Sample size = 15						
AIC	SIC	AICc	AICu	NDICu	NDIC	EG	AIC	SIC	AICc	AICu	NDICu	NDIC	EG	AIC	SIC	AICc	AICu	NDICu	NDIC	EG
0	0	0	0	0	0	0	0	5	1	7	5	0	0	7	25	56	138	63	18	1
0	0	0	0	0	0	0	3	7	6	10	7	3	0	22	35	65	86	64	37	5
750	957	789	910	911	795	881	673	845	853	914	878	762	802	578	694	805	736	753	669	50
129	37	123	67	63	109	81	164	91	101	57	81	132	137	126	91	46	28	60	122	62
71	5	53	20	19	56	32	77	27	22	10	17	57	41	103	72	20	12	35	76	56
50	1	35	3	7	40	6	83	25	17	2	12	46	20	164	83	8	0	25	78	26

Table 4. RBF fitting to a 5 degree polynom, f1(x), k = number of hidden units = 1..12

Sample size = 100					Sample size = 25					Sample size = 15				
2xCV	10xCV	NDICu	NDIC	EG	2xCV	10xCV	NDICu	NDIC	EG	2xCV	10xCV	NDICu	NDIC	EG
0	0	0	0	0	3	6	0	0	0	15	6	0	0	1
0	0	0	0	0	7	0	0	0	2	7	8	0	0	1
0	6	0	0	0	14	9	5	1	1	34	13	13	1	9
0	0	0	0	0	8	5	1	3	0	12	9	4	1	2
10	10	0	0	0	27	14	24	9	16	23	12	31	25	47
3	12	0	0	0	13	9	13	7	4	4	8	16	14	9
34	22	65	34	57	14	16	42	32	49	0	5	19	25	22
15	9	6	7	1	10	9	7	8	12	0	12	9	20	7
17	10	10	17	31	3	12	3	8	9	1	6	6	10	0
8	10	12	12	9	1	7	3	9	2	0	6	1	1	2
7	8	4	10	1	0	6	0	14	5	1	7	0	1	0
6	13	3	20	1	0	7	2	9	0	3	8	1	2	0

We can see from tables 1, 2 and 3 that AIC and SIC leads to small-sample overfitting problems. However, in large samples, SIC performs quite well. The corrected versions, AICc and AICu, outperform their parent criterion AIC from an overfitting perspective. The proposed criteria, NDIC and NDICu, have a similar performance in large samples, while are much better in small sample problems. This is a very important characteristic, because of the practical limitations on gathering and using data in real-world situations. Table 4 shows the results when applying RBF networks to fit the polynom f1(x) in the above explained situation, compared to k-fold cross-validation, (k=2,10) and the "real" generalization error. It is clear from the results that the criterion performs quite well in small and large sample situations, when compared to other criteria as well as to holdout schemes.

5. Conclusions

In this paper we have introduced a new criterion for model selection derived from the behaviour of candidate models to noise inputs. It is robust to overfitting problems, showing similar or better performance than other criteria reported in the literature in small sample scenarios, which is a very interesting property for real world problems. Another advantage is that it is easy to compute, and may be easily applied to neural network size determination. In fact, it has been derived independently of the nature of candidate models, achieving good results in RBF hidden neuron number determination. Further work will focus on applying this criterion to other neural network models, decision tree optimal pruning, etc.

6. References

1. Abu-Mostafa, Y. (1989), The Vapnik-Chervonenkis dimension: Information versus complexity in learning, Neural Computation 1(3), 312-317.
2. Akaike, H Information theory and an extension of the maximum likelihood principle. In B.N. Petrov and Csaki ed. 2nd Intl. Symp. Inform. Theory 267-281 , Budapest. (1973).
3. Amari, S. (1995), Learning and statistical inference, in M. A. Arbib, ed., `The Handbook of Brain Theory and Neural Networks', MIT Press, Cambridge, Massachusetts.
4. Foster, P.D. Characterizing the generalization performance of model selection strategies. Proceeding of the 14th Intl. Conf. on Machine Learning (ICML-97) Nashville, (1997).
5. Hurvich, C.M. and Tsai, C.L. Regression and time series model selection in small samples. Biometrika 76, 297-307 (1989).
6. McQuarrie A.D.R. Shumway, R.H. and Tsai C.L. The model selection criterion AICu. Statistical and Probability letters 34, 285-292. (1997).
7. Moody, J. (1992), The effective number of parameters: An analysis of generalization and regularization in nonlinear learning systems, in J. Moody et Al. eds, `Advances in Neural Information Processing Systems', Vol. 4, Morgan Kaufmann, pp. 847-854.
8. Ripley, B. (1995), Statistical ideas for selecting network architectures, Invited Presentation, Neural Information Processing Systems 8.
9. Schwarz, G. Estimating the dimension of a model. Annals of Statistics, 6, 461-515.(1978)
10. Sugiura, N. Further analysis of the data by Akaike's information criterion and the finite corrections. Communications in Statistic – Theory and Methods 7,13-26. (1978).

Wood Texture Analysis by Combining the Connected Elements Histogram and Artificial Neural Networks

M.A. Patricio Guisado [1] and D. Maravall Gómez-Allende [2]

[1] Serv. Planificación de Sistemas de la Información
[2] Departamento de Inteligencia Artificial
Universidad Politécnica de Madrid
dmaravall@fi.upm.es

Abstract. The automatic analysis of wood texture, based on a novel concept: the Frequency Histogram of Connected Elements (FHCE) is the main contribution of this work. The FHCE represents the frequency of occurrence of a random event, which not only describes the texture's gray-level distribution, but also the existing spatial dependence within the texture. The exploitation of the FHCE's shape, alongside its wavelet transform, allows the computation of excellent features for the discrimination between sound wood and defective wood; in particular, for the really hard pattern recognition problem of detecting cracks in used wood boards. A feedforward multilayer perceptron, trained with the backpropagation algorithm, is the specific ANN classifier applied for the detection and recognition of cracks in wood boards. A large digital image database, developed after an industrial project, has been used for testing purposes, attaining a success ratio far beyond those obtained with more conventional texture analysis and segmentation techniques.

1 Introduction

Texture analysis for image segmentation is one of the most critical issues in computer vision. Almost any scene taken from the real world cannot be automatically analyzed without segmenting the texture elements present in the corresponding digital images. Focusing on the automatic quality inspection of industrial products, the results attained in the segmentation phase are critical for the performance of the whole computer-based vision system. On the other hand, because of the always increasing competitivity in business and industry, quality inspection has become a key issue in any productive process. The automation of the quality inspection tasks, while maintaining at least humane-based inspection performance, is an indisputable trend. In this scenario, computer vision has witnessed a fast growth during the last decades as a serious candidate in any attempt to automate the quality inspection tasks in numerous industrial activities.

This paper emerges from a successful industrial project aimed at the complete automation of pallets inspection using computer vision techniques. Because of commercial and legal restrictions that impede a detailed description of the whole inspection system, this paper is focused on a specific technical issue: the detection and recognition of cracks in the pallet's elements. In despite of the mentioned restrictions, the paper provides a thorough perspective on the computer vision aspects at issue.

In figure 1 are displayed several examples of sound wood and defective wood –i.e. with a crack-

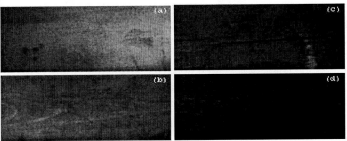

Fig. 1. Images of sound wood, (a)-(b), and defective wood, (c)-(d).

One of the main difficulties in detecting cracks in a particular piece of wood is the extreme variability that plagues the digital images of the wood boards analyzed by the computer vision system. This variability comes from the fact that the pallets arrive at the automatic inspection plant after a rather long period of use and therefore their appearance is extremely irregular. Such variability is particularly critical for the handling of false alarms –i.e. when the system "sees" an inexistent defect-.

After many attempts at applying several well-known computer vision techniques [2, 5, 6]: edge detection, line and curve extraction, region growing and histogram-based segmentation, conventional texture analysis etc. we have found that these techniques are inefficient for attaining acceptable performance as far as crack detection is concerned. Whenever we tune a particular algorithm for a better detection of cracks, the false alarm ratio increases out of control. For an efficient inspection of cracks in used wood boards we have introduced a novel texture analysis concept: what we have called the "Frequency Histogram of Connected Elements", or FHCE for short. By combining this novel concept and a multilayer perceptron neural network we have obtained excellent results in the crack detection problem. We proceed now with a detailed description of these ideas.

2 The Histogram of Connected Elements

The whole automatic inspection system has the following phases: Pallet Handling and Image Capture, Feature Extraction, Feature Selection and Recognition.

Although many cameras are employed in the inspection system, for crack detection a B&W, high resolution –1300(H) x 1030 (V)-, digital camera is used. For an optimum illumination control, the inspection system has been isolated in a big booth, totally painted in mate black. Many halogen floodlights are distributed for the entire booth in order to guarantee homogeneous illumination. The final output of this phase is –hopefully- excellent digital images of 1280x1024 pixels.

The feature extraction phase's objective is to obtain the best features for an appropriate description of the wood texture diversity. For that purpose, we have introduced a novel concept, i.e. the FHCE.

The FHCE represents the frequency of occurrence of a random event in a texture segment. The so-called "Connected Element" event not only describes the gray-level distribution of the texture but also the spatial relationship among the different gray-levels, which is extremely suitable for texture discrimination. In the sequel we very briefly describe the main technicalities underlying the FHCE.

(a) The concept of neighborhood

The neighborhood concept is central in the understanding of the FHCE concept. Let I be a digital image formed by NxM pixels: I={$x_1, x_2, ...x_{NxM}$}. If we denote by (i,j) the co-ordinates of pixel x_k, the neighborhood of pixel x is defined as follows:

$$v = \{\varphi_{i,j}, (i,j) \in I\}$$
$$\varphi_{i,j} = \{(k,l) \in I / D((k,l),(i,j)) \text{ is true}\} \quad (1)$$

where D is a predicate defined by a distance-based condition.

(b) The concept of connected element

The concept of connected element is associated to the neighborhood. We mean by a connected element:

$$C_{i,j}(T) = \{\varphi_{i,j}^{r,s} / Ng_x \subset [T-\varepsilon, T+\varepsilon], \forall x \in \varphi_{i,j}^{r,s}\} \quad (2)$$

where N_{gx} is the gray-level or brightness of pixel x_k. In other words, a connected element is any neighborhood unit such that their pixels have a gray-level close to a given gray-level T.

(c) Definition of Frequency Histogram of Connected Elements

The frequency histogram of connected elements (FHCE) is defined as:

$$H(T) = \sum_{\forall (i,j) \in I} C_{i,j}(T)$$
$$0 \leq T \leq Ng_{max} - 1 \quad (3)$$

That is to say, H(T) approaches a density function for a random event occurring in a digital image I(x,y). This event is related to the idea of connected element, which in turn is related to the pseudo-random structure of the gray-level distribution of a particular texture. Our main purpose in this paper is to demonstrate that this novel concept is an ideal tool for the automatic analysis of the textures appearing in a variety of digital images. Obviously there is no universal connected element valid for any domain application. In the design leading to the FHCE there is a critical and domain-dependent step, which is responsible for the selection of the parameters defining the optimum, connected element. Such parameters are: (1) the morphological parameter and (2) the connectivity level.

After exhaustive experimentation with a plethora of digital images of sound and defective wood boards we have selected the neighborhood function for the Connected Element event of 5x3 shape. As it can be noticed, the number of horizontal pixels is higher than the vertical pixels, the reason being the a priori knowledge available about the problem at hand. In fact, there is an empirical evidence that cracks in a piece of wood tend to present the same direction than the wood grain. As the computer vision inspection is performed in a horizontal way from the wood boards standpoint, it is easily deduced the shape of the selected neighborhood function.

To finish the connected element concept definition, it is compulsory to select the connectivity level that a particular neighborhood should possess to be considered as such. The FHCE is computed for each image portion by shifting trough all the pixels a window of the same 5x3 shape than the neighborhood. This scanning process is performed by means of a top-bottom and left-right movement and by computing at each pixel the maximum and the minimum gray level within its neighborhood. For each pixel's neighborhood it is classified as a connected element if and only if the difference between the maximum and the minimum values is small as compared with the dynamic range of the histogram in the whole window. After experimental work we have chosen a 10% ratio, which is a good compromise between wood portions in good and in bad conditions. Therefore, for a neighborhood to be labeled as connected element the following condition has to be checked: $((x_{max} - x_{min}) / (g_{max} - g_{min})) < 0.1$, where x_{min} and x_{max} are the maximum and the minimum values of the window –i.e. the dynamic range of the window's histogram-. Therefore, if a particular neighborhood possesses a gray-level variability less than ten percent of the dynamical range of the global window the corresponding pixel is a connected element and the FHCE will compute a new event with value $T = (x_{min} + x_{max})/2$. In figure 2 appears two examples of FHCE computation.

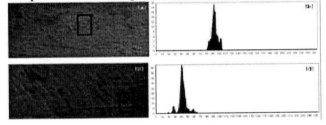

Fig. 2. Instances of FHCE for digital images of sound wood (a) and defective wood (c). Notice that the displayed FHCE correspond to the windows appearing in the digital images.

3 Discrimination Based on the FHCE Wavelet Transform

From figure 2 it can be observed the clear distinction between the FHCE for portions of sound wood and defective wood with cracks, respectively. In particular, the left side of the FHCE must be noticed; so whereas in the image corresponding to defective wood there appear side lobes, the FHCE of sound wood does not present this characteristic. Such particularity is the basic idea for the automatic recognition of cracks.

As a consequence of the quite different shape of the respective FHCE, we are in an excellent starting position to approach the task of selecting optimal features for the automatic classification of wood. We can see at the FHCE of a window corresponding to a portion of wood with a defect –a crack in our case- as a signal with two distributions. The first distribution corresponds to the connected elements that belong to the defect –i.e. the crack- and the second distribution to the connected elements belonging to the sound wood portion. On the other hand, it is quite evident that the sound wood portion has a FHCE with a single distribution. In conclusion, there is a clear evidence about the possible discrimination between sound and defective wood by using the low frequency components of the FHCE spectrum –i.e. by computing the Fourier transform of the FHCE-.However, as the FHCE of a textured region is a nonstationary function, it is advisable to work with spatio-temporal representations. The wavelet transform possesses the capacity of representing simultaneously the spatial and temporal domains of a particular function.

Broadly speaking, the wavelet transform is a mathematical tool that divides a function into a set of functions, which on their turn are escalated and shifted versions of a primitive function called the mother wavelet function. Basically, for a unidimensional discrete-time signal $f(k)$ of length $n=2^{n0}$ with n_0 being an integer, its wavelet transform can be interpreted as a decomposition of the signal into smoothed versions with respect to the frequency axis. The decomposition algorithm proposed by Mallat [3] starts with filtering the signal with a low-pass filter $h(m)$ and a high-pass filter $g(m)$, both of length M, with the purpose of its decomposition into two subwaves with a length reduction of a factor of 2. The decomposition operates as follows:

$$Hf(k) = \sum_{m=0}^{M-1} h(m) f(2k+m)$$
$$Gf(k) = \sum_{m=0}^{M-1} g(m) f(2k+m)$$
(4)

The algorithm is iteratively applied to the low-frequency sub-waves and the high-frequency components are computed and stored. The process finishes when the decomposition and filtering operations reach their top level. Therefore, if the top level is L, the discrete wavelet transform of signal $f(k)$ are: $\{Gf, GHf, GH^2f, ..., GH^Lf, GH^{L+1}f\}$ with the same length n than the original function. In principle there is an infinite number of functions that hold the condition of being mother wavelet functions, but in practice only a reduced number of them have been used. In our work we have employed the so-called Daubechies 4 function [1].

Although the wavelet transform has been intensively applied as a powerful and efficient tool in the pattern recognition discipline, it really has a clear limitation as regard to time shift. Let $g(k)$ and $f(k)$ be two generic FHCE and $WTg(u)$ and $WTf(u)$ their respective wavelets transforms; then:

$$g(k) = f(k+\tau) \text{ does not mean that } WTg(u) = WTf(u+\tau)$$
(5)

This is a significant drawback [3] affecting the usual pyramidal multiscale representation. Although in the technical literature have appeared several attempts to avoid this problem, we have chosen an alternative way based on applying a pre-processing technique for the normalization of the FHCE to be recognized. To determine whether a particular FHCE belongs to a sound wood's window or to a defective one, we restrict our analysis to the left half-side of the corresponding FHCE and afterwards the following algorithm is applied.

Step 1. Let $h(k)$ be a generic FHCE; then the following parameters are computed.:
- g_{min} : the lowest gray-level in $h(k)$ with frequency higher than zero.
- k_{max} : gray level in $h(k)$ with the highest frequency.
- max: maximum frequency in $h(k)$ that corresponds to $h(k_{max})$.

Step 2. Computation of the normalized FHCE as follows:

$$h^*(k) = \begin{cases} \dfrac{h(g_{min}+k)}{max} * 100 & for \quad 0 \le k \le k_{max} - g_{min} \\ 0 & for \quad k_{max} - g_{min} < k \le N-1 \end{cases} \quad (6)$$

In figure 3 an example of this normalization algorithm is shown.

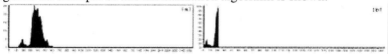

Fig. 3. The FHCE of a window corresponding to defective wood. In (b) appears the normalized window.

Finally, the feature vector is composed from the application of the transform wavelet to the normalized FHCE.

4 Crack Detection and Recognition Using a Feedforward Multilayer Perceptron

We have selected 60 boards of defective wood –i.e. with cracks- and 150 boards of sound wood, and we have run a program to automatically generate a set of training and testing samples out of these 210 boards. After thorough empirical investigation, the optimum window size chosen for the generation of digital image samples has been 30x40. Finally, for benchmark purposes, 1365 digital image samples (973 of sound wood and 392 of defect wood) have been employed. We can notice that samples corresponding to sound wood are approximately 70% of the complete set. This proportion is quite similar to that appearing in real-life inspections.

We start from 1365 feature vectors, each of them composed by 256 discriminant variables, and for feature reduction purposes a bottom-up search algorithm is applied [4]. Therefore, the best individual feature is first selected by just computing a performance index based on the Fisher discriminant ratio. The next step is to form all the possible two-elements subsets that include the best individual feature and to select the optimum subset. This incorporation process continues for feature subsets of 3,4,5, ... elements until the best performance index at a particular search level has a "slight" increase in which case the search stops. Summarizing, the algorithm finishes its search for an optimum feature subset when the increment in the discriminant ratio is lower than a certain threshold t. Through experimentation we have finally chosen a value of t=1% and we have obtained as the optimum subset of discriminant features the coefficients 17,8,3 and 16 of the FHCE wavelet transform, which is a drastic reduction of the original feature vector.

Once selected the feature vector, the next step is the classifier design. In the first place, we justify the election of an ANN classifier because our experience with a plethora of pattern recognition techniques indicates that they are best suited to deal with the problem at hand. In particular, we believe that the specific classification problem of discriminating textures of sound wood against textures of defective wood –i.e. with cracks- demands a delicate balance between generalization and particularization, and ANN are excellent classifiers as far as such dichotomy is concerned. Among the myriad of ANN methods, we have chosen the feedforward multilayer perceptron trained with the backpropagation learning algorithm. Through extensive experimentation we have finally selected the network architecture displayed in figure 4.

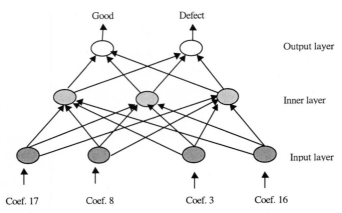

Fig. 4. The ANN architecture. Notice the four neurons of the input layer that corresponds to the selected discriminant variables.

5 Discussion of Results and Conclusions

The most exigent and reliable method for the evaluation of any automatic classifier is the *leave-one-out* policy, in which the classifier –in our case, the feedforward multilayer perceptron- is trained with all the available training samples except one, which is used for the evaluation of the classifier itself. By repeating the process with all the training samples, the average success ratio of the classifier is an excellent estimation of its future performance for new samples or cases; i.e. when it works in real-life situations. The average success ratio is computed as follows:

$$P = f\left(defect/defect\right) * f(defect) + f\left(good/good\right) * f(good) \qquad (7)$$

where $f(defect)$ and $f(good)$ are the relative frequency of a defective and a good sample, respectively; $f(defect/defect)$ and $f(good/good)$ are the a posteriori relative frequency of success for defective and good samples, respectively.

The evaluation has been carried out with different values for the maximum number of iterations and maximum deviation design parameters. In order to give just an idea of the evaluation results, table 1 displays the results obtained for maximum deviation = (0.1; 0.2; 0.3) and for several values of the maximum number of iterations. It is interesting to remark that these results correspond to the detection and recognition of cracks in wood boards, which is a really hard classification task, as it is not only important to recognize a crack, but also not to recognize a false crack. In fact, as it happens with the inspection of very complex defects, the false alarms are the real issue, rather than the defect detection.

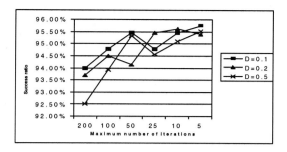

Table 1. Results of the ANN in crack detection and recognition.

It is remarkable the negative effect of high values of the maximum number of iterations parameter, which produce a clear overfitting phenomenon. The detection of this overfitting has been possible thanks to applying the leave-one-out evaluation policy. Concerning the other design parameter, i.e. the maximum deviation, it can be noticed that its influence on the ANN performance is much less important, as the difference among distinct maximum deviation values for specific number of iterations is quite small, except when the number of iterations increases significantly. The best result is obtained with a maximum deviation of 0.1 and for 5 iterations.

As a general conclusion, it can be confirmed the excellent results attained with the combination of the following elements: the novel texture analysis tool introduced in this work –i.e. the FHCE-, the wavelet transform of the FHCE to obtain an optimum subset of discriminant variables and, finally, a flexible and powerful classifier like the feedforward multilayer perceptron trained with the backpropagation learning algorithm. A success ratio higher than 95% is far beyond the performance attained using the conventional texture analysis techniques for the problem at hand – i.e. crack detection and recognition is used wood boards-. An open question is the share in the success ratio of each of the above mentioned elements: the FHCE, the selected discriminant features and the ANN.

References

1. Daubechies, I.: Ten lectures on wavelets. CBMS-NSF regional conference series in applied mathematics 61. 2nd ed. Philadelphia: SIAM, (1992).
2. Kim, C.W. and Koivo, A.J.: Hierarchical classification of surface defects on dusty wood boards. *Pattern Recognition Letters*, Vol. 7, (1994), 713-721.
3. Mallat, S.G.: A theory for multiresolution signal decomposition: the wavelet representation. *IEEE Transactions on Pattern Analysis and Machine Intelligence*. Vol. 11(7), (1989), 674-693.
4. Patricio Guisado, M.A. and Maravall Gómez-Allende, D.: Segmentation of text and graphics/image using gray-level histogram Fourier transform. Proc. of the SSPR&SPR 2000, LNCS 1876, Springer-Verlag, (2000), 757-766.
5. Pham, D.T. and Alcock, R.J.: Recent advances in intelligent inspection of woods boards. 13th Inter. Conf. on Applications of Artificial Intelligence in Engineering. Computational Mechanics Publications, Southampton, UK, (1998), 105-108.
6. Silven, O. and Kauppinen, H.: Recent developments in wood inspection. *Inter. Journal of Pattern Recognition and Artificial Intelligence*. Vol. 10(1), (1996), 83-95.

Dynamic Topology Networks for Colour Image Compression

Ezequiel López-Rubio, José Muñoz-Pérez, and José Antonio Gómez-Ruiz

Computer Science Department, University of Málaga, Campus de Teatinos s/n,
29071 Málaga, Spain
{ezeqlr, munozp, janto}@lcc.uma.es

Abstract. The Self-Organizing Dynamic Graph (SODG) is a novel unsupervised neural network that overcomes some of the limitations of the Kohonen's Self-Organizing Feature Map (SOFM) by using a dynamic topology among neurons. In this paper an application of the SODG to colour image compression is studied. A Huffman coding and the Lempel-Ziv algorithm are applied to the output of the SODG to provide considerable improvements in compression rates with respect to standard competitive learning. Furthermore, this system is shown to give mean squared errors of the reconstructed images similar to those of competitive learning. Experimental results are presented to illustrate the performance of this system.

1 Introduction

Image compression is important for many applications such as video conferencing, high definition television and facsimile systems. Vector quantization (VQ) is found to be effective for lossy image compression due to its excellent rate-distortion performance over others conventional techniques based on the codification of scalar quantities [1].

However, classic techniques for VQ are very limited due to the prohibitive computation time required. A number of competitive learning (CL) algorithms have been proposed for designing of vector quantizers. Ahalt, Krishnamurthy, Chen and Melton [2] have used algorithms based on competitive neural networks that minimise the reconstruction errors and lead to optimal or near optimal results, demonstrating their effectiveness. Likewise, Yair, Zeger and Gersho [3] have used Kohonen's learning algorithm, proposing a soft-competition algorithm for the design of optimal quantizers. Ueda and Nakano [4] prove that the standard competitive algorithm is equivalent to the Linde, Buzo and Gray [5] algorithm and they propose a new competitive algorithm based in the equi-distortion principle that allows the design of optimal quantizers. Dony and Haykin [6] show the advantages of using neural networks for vector quantization, as they are less sensitive to codebooks initialisation, lead to a smaller distortion rate and have a fast convergence rate. Cramer [7] examines much recent work regarding compression using neural networks.

The Kohonen's self-organizing neural network [8] is a realistic, although very simplified, model of the human brain. The purpose of the self-organizing feature map (SOFM) is to capture the topology and probability distribution of input data and it has been used by several authors to perform invariant pattern recognition, such as Corridoni [9], Pham [10], Subba Reddy [11] and Wang [12].

This network is based on a rigid *topology* that joins the neurons. This is not a desirable property in a self-organizing system, as Von der Marlsburg states in [13]. The Self-Organizing Dynamic Graph (SODG) [14] is a novel alternative to this network that shows a greater plasticity, while retaining the feature detection performance of the SOFM.

In this paper the SODG and competitive neural networks are used for image compression. A comparative study between the two methods is then performed, showing that the SODG algorithm leads to a smaller distortion rate and a better compression rate.

This paper is organised as follows: in Section 2 the SODG neural model is presented as a novel self-organizing network. In Section 3 a technique for the compression of colour images by SODG is presented and it is compared with standard CL compression. Finally, computational experiments and conclusions are considered in Section 4 and Section 5, respectively.

2 Self-Organizing Dynamic Graphs

Our neural model is the following. The weight vectors $\mathbf{w}_i$ are points in the input space, like in the SOFM algorithm. Nevertheless, the units are no longer joined by a static topology. Every unit i has an associated nonnegative *adjacency vector* $\mathbf{z}_i$ that reflects the degree of neighbourhood among neuron i and all the others. This means that z_{ij} is the neighbourhood between neurons i and j. We have $z_{ii}=0 \; \forall i$.

The winning neuron lookup is performed as in the SOFM:

$$i = \arg\min_{j} \left\| \mathbf{x}(t) - \mathbf{w}_j(t) \right\| \tag{1}$$

The weight vector of the winning neuron i is modified to come closer to input sample $\mathbf{x}(t)$:

$$\mathbf{w}_i(t+1) = \mathbf{w}_i(t) + (\eta(t) + \psi(t))[\mathbf{x}(t) - \mathbf{w}_i(t)] \tag{2}$$

where $\psi(t) \geq 1$ is called the *egoism parameter*, which controls how much the weight vector of the winning neuron is changed. The condition

$$\eta(t) + \psi(t) \leq 1 \quad \forall t \tag{3}$$

must be satisfied so that the modification does not move the weight vector pass the input vector. All the other neurons are modified according to their adjacency to winning neuron i:

$$\mathbf{w}_j(t+1) = \mathbf{w}_j(t) + \eta(t) z_{ij} [\mathbf{x}(t) - \mathbf{w}_j(t)] \tag{4}$$

The following condition must be satisfied in order to avoid the new weight vector to pass the input sample:

$$\eta(t) \leq 1 \quad \forall t \tag{5}$$

Von der Marlsburg [13] states that the synapses of a neuron in a self-organizing system must compete. So, a selection of the most vigorously growing synapses at the expense of the others should be performed. Note that this principle of self-organization is not considered in the SOFM. Here we introduce this principle in the network architecture by changing the strength of the synapses and by imposing the condition

$$\sum_{k=1}^{N} z_{jk}(t) = 1 \quad \forall t \ \forall j = 1,...,N \tag{6}$$

where N is the number of units.

The learning rule for the adjacency vectors of non-winning neurons is

$$\mathbf{z}_j(t+1) = \frac{1}{\sum_{k=1}^{N} y_{jk}(t)} \mathbf{y}_j(t) \quad \forall j \neq i \tag{7}$$

where

$$\mathbf{y}_j(t) = \mathbf{z}_j(t) + \frac{\|\mathbf{x}(t) - \mathbf{w}_j(t)\|}{\|\mathbf{x}(t) - \mathbf{w}_i(t)\|} \xi(t) \mathbf{u}_i, \tag{8}$$

$\mathbf{u}_i$ is a unit vector with a 1 in the ith component, $\xi(t)$ is the learning rate of the adjacency vectors and ||·|| denotes the Euclidean norm. The winning neuron does not change its adjacency vector, i. e.,

$$\mathbf{z}_i(t+1) = \mathbf{z}_i(t) \tag{9}$$

The learning rule of the adjacency vectors increases the values z_{ji}, where unit j has a weight vector close to the weight vector of winning neuron i. This reinforces the co-operation among neighbouring units.

The model presented here has a number of significant properties. Next we include three theorems, whose proofs can be found in [14].

Theorem 1 (network stability): If the input vectors are bounded, then the weight vectors are bounded.

Theorem 2 (always towards the input): Let C be a convex subset of the input space. If all the input samples lie in C, i. e.,

$$\mathbf{x}(t) \in C \quad \forall t \tag{10}$$

then every update of every weight vector $\mathbf{w}_j$ which is outside of C reduces the distance from $\mathbf{w}_j$ to C.

Theorem 3 (never go away): Let C be a convex subset of the input space. If all the input samples lie in C, i. e., condition (10) holds, and $\mathbf{w}_j(t) \in C$ then $\mathbf{w}_j(t+1) \in C$.

3 Image Compression

For image compression purposes, the original image is divided into square windows of the same size. The RGB values of the pixels of each of these windows form an input vector for the network. The compression process consists of the selection of a reduced set of representative windows $\mathbf{w}_j$ (prototypes) and the replacement of every window of the original image with the "closest" prototype, i. e., the weight vector of the winning neuron for this input vector. If we use windows with $V x V$ pixels, and the original image has $KVxLV$ pixels, then the mean quantization error per pixel E of this representation is

$$E = \frac{1}{V^2 KL} \sum_{i=1}^{KL} \min_j \left\| \mathbf{x}^i - \mathbf{w}_j \right\| \qquad (11)$$

where $\mathbf{x}^i$ is the ith window of the original image and $\mathbf{w}_j$ is the jth prototype.

For the experiments we use the image shown in Fig. 1 with 348x210 pixels (214.1 Kb in BMP format) and 256 red, green and blue values (24 bits per pixel). Therefore, we have $73080/V^2$ input patterns.

The compression has been performed by two algorithms, the SODG and the standard competitive learning. The network's initial weights have been randomly selected from all input patterns.

Fig. 1. Original figure to be compressed

3.1. Image Compression with Standard Competitive Learning

The standard competitive learning has been used in the following way. First, the network is trained with all the KL windows of the original image as input patterns. Then the final weight vectors are used as prototypes. Every window of the original image is substituted by the index of its winning neuron (closest prototype). In addition, it is necessary to store the prototype vectors in the compressed image file. The original (uncompressed) image size S_u, in bits, is

$$S_u = V^2 BKL \qquad (12)$$

where B is the number of bits per pixel (uncompressed). Typically $B=24$ or $B=32$. The compressed image size using competitive learning S_{CL} is given by

$$S_{CL} = V^2 BN + ceil(\log_2 N) KL \qquad (13)$$

where N is the number of units, and the function *ceil* rounds towards $+\infty$. The compression rate R_{CL} is obtained as

$$R_{CL} = \frac{S_{CL}}{S_u} = \frac{N}{KL} + \frac{ceil(\log_2 N) KL}{V^2 BKL} \qquad (14)$$

3.2. Image Compression with SODG

When the SODG is used to obtain prototype windows for image compression, we can use the topology information provided by the network to improve the compression rate. Neighbour windows in the original image are usually similar. This means that their prototype windows are frequently very close in the input space. Note that the SODG computes the adjacency vectors which tell us what units are close to every unit. So, for every neuron i, we can order the other units of the network by their vicinity to i by using the adjacency vector $\mathbf{z}_i$. That is, if z_{ij} is the kth greatest component of $\mathbf{z}_i$, the the unit j is in the kth place in the list of the unit i.

Hence we have a list for every unit i that shows which prototype windows that are similar to prototype window i. When we use standard competitive learning to compress image data, the prototype windows are typically coded with their index. So, the compressed image is a sequence of prototype indices. Now we can use the above mentioned lists to perform a *relative coding* of the prototype windows with respect to their predecessors in the sequence. This means that a prototype is no longer coded by its index, but by its position k in the list of its predecessor prototype. If we build the window sequence from left to right, this means that every window is coded with respect to the window on its left. Note that if the neighbour prototype is the same as the current, we write a zero in the sequence ($k=0$).

A table must be included in the compressed file to decode these numbers k into actual prototype indexes. In order to reduce the size of this table, we may store only the P closest neigbours of every unit. Then the value $k=P+1$ means that the current window is represented by a prototype which is not in the P nearest neighbours of the left prototype. Then we need to store in the compressed file the actual prototype index of this window.

The advantage of this strategy is that most times the numbers k that form the new sequence have a low value. So, if we perform a Huffman coding of this sequence, a compression will be achieved, since the data has some redundancy. Different codings have been used in image compression. An early work on this issue can be found in [15].

Then a Huffman tree is built from the absolute frequencies f_k of the numbers k in the new sequence, $k=0, 1, ..., P+1$. This tree assigns a code of b_k bits to the number k. Then the overall compressed image size (in bits) S_{SODG} is given by

$$S_{SODG} = V^2 BN + ceil(\log_2 N)(f_{P+1} + NP) + \sum_{i=0}^{P+1} f_i b_i \qquad (15)$$

So the compression rate R_{SODG} is computed as

$$R_{SODG} = \frac{S_{SODG}}{S_u} = \frac{N}{KL} + \frac{ceil(\log_2 N)(f_{P+1} + NP) + \sum_{i=0}^{P+1} f_i b_i}{V^2 BKL} \qquad (16)$$

Note that the Huffman coding does not reduce nor increase the distortion. It only improves the compression rate.

If we compare equations (14) and (16) we can see that the difference between the compression rates of the two methods depends on the frequencies f_i. If the values of f_i are high for low i, then SODG will outperform CL. This depends on the redundancy of the windows of the original image.

We have found that the final output of this coding is very redundant yet. So, we have used a sliding-window version of the Lempel-Ziv algorithm (see [16] and [17]) to provide further compression. Since the Lempel-Ziv algorithm is loseless, the distortion remains the same. If the final file size (in bits) is F, we define the final compression rate R'_{SODG} as

$$R'_{SODG} = \frac{F}{S_u} \qquad (17)$$

4 Experimental Results

Computational experiments have been carried out to compare the performance of the compression system proposed with standard CL. The image of Fig. 1 has been used for this purpose. We have run $4KL$ iterations of each method for different values of V and N. For CL simulations we have chosen a linear decay of the learning rate η, with an initial value $\eta_o=1$, and a final value near to zero. For SODG simulations we have chosen a constant learning rate $\eta=0.01$, a constant egoism parameter $\psi=0.99$, and $P=8$. Figures 2 and 3 show the results with $N=256$ and two different window sizes.

Method	V=1, N=256	V=1, N=1024	V=3, N=256	V=3, N=1024
CL	3.0969	1.7648	1.5288	1.2388
SODG	4.3972	2.9617	1.7586	1.6090

Table 1. Mean quantization errors per pixel, E.

Table 1 summarizes the mean quantization errors per pixel E obtained with both methods, for the values of V and N we have considered. We can see that both methods achieve similar results. Table 2 shows the inverse compression rates. We can see that the SODG method has the best compression rate in all cases. This means that the Huffman coding of the image sequence and the Lempel-Ziv algorithm always outperform the fixed coding used with CL.

	$V=1, N=256$	$V=1, N=1024$	$V=3, N=256$	$V=3, N=1024$
R_{CL}^{-1}	2.9688 to 1	2.3219 to 1	14.5848 to 1	5.8003 to 1
R_{SODG}^{-1}	4.5316 to 1	2.9527 to 1	15.7981 to 1	5.0536 to 1
R'_{SODG}^{-1}	13.0337 to 1	8.2629 to 1	38.8861 to 1	11.9516 to 1

Table 2. Inverse compression rates, R_{CL}^{-1}, R_{SODG}^{-1} and R'_{SODG}^{-1}.

Fig. 2. Compressed images using CL (left) and SODG (right) with $V=1$ and $N=256$

Fig. 3. Compressed images using CL (left) and SODG (right) with $V=3$ and $N=256$

5 Summary

A new image compression system has been presented, which is based in the Self-Organizing Dynamic Graph (SODG) neural network. The dynamic topology information obtained by the SODG has been used to build a Huffman coding of the image data. This is aimed to reduce the compressed image size.

Furthermore, computational experiments have been carried out. The experimental results show that Huffman coding leads to very significant improvements in compression rates with respect to standard competitive learning (CL). Nevertheless, this enhancement does not affect the distortion of the compressed images.

References

1. Gray, R.M.: Vector Quantization. IEEE ASSP Magazine 1 (1980) 4-29.
2. Ahalt, S.C., Krishnamurphy, A.K., Chen, P. and Melton, D.E.: Competitive Learning Algorithms for Vector Quantization. Neural Networks 3 (1990) 277-290.
3. Yair, E., Zeger, K. and Gersho, A.: Competitive Learning and Soft Competition for Vector Quantizer Design. IEEE Trans. Signal Processing 40 (1992), No. 2, 294-308.
4. Ueda, N. and Nakano, R.: A New Competitive Learning Approach Based on an Equidistortion Principle for Designing Optimal Vector Quantizers. Neural Networks 7 (1994), No. 8, 1211-1227.
5. Linde, Y., Buzo, A. and Gray, R.M.: An Algorithm for Vector Quantizer Design. IEEE Trans. On Communications 28 (1980), No. 1, 84-95.
6. Dony, R.D. and Haykin, S.: Neural Networks Approaches to Image Compression. Proceedings of the IEEE 83 (1995), No. 2, 288-303.
7. Cramer, C.: Neural Networks for Image and Video Compression: A Review. European Journal of Operational Research 108 (1998), 266-282.
8. Kohonen, T.: The Self-Organizing Map. Proceedings of the IEEE 78 (1990), 1464-1480.
9. Corridoni, J. M., Del Bimbo, A., and Landi, L.: 3D Object classification using multi-object Kohonen networks. Pattern Recognition 29 (1996), 919-935.
10. Pham, D. T. and Bayro-Corrochano, E. J.: Self-organizing neural-network-based pattern clustering method with fuzzy outputs. Pattern Recognition 27 (1994), 1103-1110.
11. Subba Reddy, N. V. and Nagabhushan, P.: A three-dimensional neural network model for unconstrained handwritten numeral recognition: a new approach. Pattern Recognition 31 (1998), 511-516.
12. Wang, S. S. and Lin., W.G.: A new self-organizing neural model for invariant pattern recognition. Pattern Recognition 29 (1996), 677-687.
13. Von der Malsburg, C.: Network self-organization. In Zornetzer, S.F., Davis J.L. and Lau C. (eds.): An Introduction to Neural and Electronic Networks. Academic Press, Inc. San Diego, CA (1990), 421-432.
14. López-Rubio, E., Muñoz-Pérez, J. and Gómez-Ruiz, J.A.: Self-Organizing Dynamic Graphs. Proceedings of the International Conference on Neural Networks and Applications 2001 (NNA'01), 24-28. N. Mastorakis (Ed.), World Scientific and Engineering Society Press.
15. Comstock, D. and Gobson, J.: Hamming coding of DCT compressed images over noisy channels. IEEE Transactions on Communications 32 (1984), 856-861.
16. Wyner, A.D. and Ziv, J.: The sliding-window Lempel-Ziv algorithm is asymptotically optimal. Proceedings of the IEEE 82 (1994), 872-877.
17. Wyner, A.D. and Wyner, A.J.: Improved redundancy of a version of the Lempel-Ziv algorithm. IEEE Transactions on Information Theory, 35 (1995), 723-731.

Analysis on the Viewpoint Dependency in 3-D Object Recognition by Support Vector Machines

Taichi Hayasaka, Eiichi Ohnishi, Shigeki Nakauchi, and Shiro Usui

Department of Information and Computer Sciences,
Toyohashi University of Technology,
1-1 Hibarigaoka, Tempaku, Toyohashi 441-8580, Japan
{hayasaka,ohnishi,naka,usui}@bpel.ics.tut.ac.jp

Abstract. In 3-D object recognition in human, the recognition performance across viewpoint changes is divided into 2 types: *viewpoint-dependent* and *viewpoint-invariant*. We analyzed the viewpoint dependency of objects under the theory of *image-based* object representation in human brain (Poggio & Edelman 1990, Tarr 1995) using Support Vector Machines (Vapnik 1995). We suggest from such computational approach that the features of object images between different viewpoints are major factors for human performance in 3-D object recognition.

1 Introduction

How can we recognize 3-dimensional (3-D) objects given that only 2-dimensional (2-D) information from a specific viewpoint are received on our retinae? If the object is rotated in depth or picture plane, or our viewpoint is moved, its retinal image will be changed. However, it is possible to hold the consistency of the object in our brain.

The earliest model for reconstruction of the 3-D scene from 2-D images is based on the volumetric-part representation in brain (Marr & Nishihara 1978, Biederman 1987). Recently, however, psychological, computational, and even neurophysiological studies (e.g., Bülthoff & Edelman 1992, Logothetis, et al. 1994, Poggio & Edelman 1990, Tarr 1995, Ullman & Barsi 1991) support another model for the recognition mechanism, called *image-based* model. In this model, 3-D object space is approximated by a finite set of 2-D images (or their arbitrary mathematical transformations) from different viewpoints, and in the recognition or discrimination task the visual similarity of stored 2-D information is referred in human brain.

The structural-description model was built on the evidence of the ability that recognizing familiar 3-D objects seems typically *invariant* against changes of the views. On the other hands, many psychological experiments showed the opposite results, i.e., the evidence of *viewpoint dependency* of performance for recognizing unfamiliar 3-D objects across viewpoints (e.g., Johnstone & Hayes 2000, Tarr 1995, Tarr, et al. 1997). In order to explain the reason why the opposite results are observed, we try to measure the viewpoint dependency in 3-D object recognition by the computational approach based on the image-based model. A mathematical model called Support

Vector Machines (SVM; Vapnik 1995) is applied for this problem, because its computing process is quite similar to the image-based model. We show that learning results of SVM correspond to the evidence from psychological experiments, and discuss about the factors that play an important role for recognizing or discriminating 3-D objects.

2 Models for Representation of 3-D Objects in Brain

2.1 Structural-description model

Since typical human performance in 3-D object recognition is invariant against variation of images, e.g., viewpoints, positions, illumination, size, reflection, etc., Marr and Nishihara (1978) suggested that the features for reconstruction of 3-D objects should be robust for such external conditions. The structural-description model is based on the above idea, i.e., a 3-D object is represented by several decomposed volumetric parts and their relationship.

One of the most important studies related to the structural-description model is the Recognition-By-Component (RBC) model by Biederman (1987). He sophisticated Marr and Nishihara's idea by introducing Geometrical Ion (GEON), which are basic volumetric parts, such as a cone, cylinder, pyramid, prism, etc. Applying the RBC model, 3-D objects are represented by GEON types and their positions, and the recognition or discrimination of them is executed by referring such symbolic information. Some psychological evidence for the RBC model has been obtained (e.g., Biederman & Gerhardstein 1993). The computational mechanism by using artificial neural networks has also been proposed by Hummel & Biederman (1992).

2.2 Image-based model

The idea of image-based model is rather easier; a 3-D object is represented by a set of its images from different viewpoints in brain. The recognition mechanism is based on *visual similarity*. That is, if the features of retinal images and the stored images in brain are similar, humans will recognize that they are images of the same object.

Compare to the structural-description model, object recognition performance estimated by the image-based model seems vulnerable to the change of viewpoints or other view conditions. As a matter of fact, several psychological experiments showed the sensitivity to recognize unfamiliar 3-D objects was variant across viewpoints, (e.g., Bülthoff & Edelman 1992, Johnstone & Hayes 2000, Tarr 1995, Tarr, et al. 1997). From single-cell recording in a neurophysiological study (Logothetis, et al. 1994), the mechanism of image-based representation of 3-D objects was found in inferior temporal cortex of monkeys. Moreover, computational approaches also support the realization of object recognition mechanism by image-based model (e.g., Poggio & Edelman 1990, Ullman & Barsi 1991).

2.3 Conditions for viewpoint invariance

Based on the RBC theory, Biederman & Gerhardstein (1993) discussed that unfamiliar 3-D objects that caused viewpoint-dependent performance in 3-D object recognition were not satisfied the following 3 conditions for viewpoint invariance: (1) *Objects must be decomposed into GEON parts*, (2) *Each object must include different set of GEON parts*, (3) *The set of GEON parts must be visible from different viewpoints*. However, several results for viewpoint-dependent performance have been observed by psychological experiments using 3-D objects satisfied with these conditions (e.g., Tarr et al. 1997, Hayward & Tarr 1997).

Although a few theories for connecting both performances have been obtained by the image-based model (e.g., Poggio & Edelman 1990, Tarr, et al. 1997), they have not been verified by computational approach yet. In this study, we modify the above conditions of viewpoint invariance for the image-based model, as follows:

1. *Each object's view must include different set of distinct parts.*
2. *The views of the distinct part set must be changed little in arbitrary viewpoint range.*

The first and second conditions can be regarded as the *distinctness* of images between objects and the *similarity* of individual object's images across viewpoint changes, respectively. We suppose that the combination of 2 factors decide the viewpoint dependency in 3-D object recognition performance, and analyze those by using Support Vector Machines (SVM; Vapnik 1995).

3 Object Discrimination by Support Vector Machines

Problem formulation in this study is as follows: We try to discriminate a *target* 3-D object from other *distractors* only by a set of their images from a finite number of viewpoints. SVM (Vapnik 1995) is one of the powerful statistical models to solve such a problem. If we assign the object label to +1 as a target object and −1 as distractors, and denote the images of all objects from every viewpoint by $x_1, ..., x_N$, the discrimination function is represented by a linear combination of kernels, where Gaussian functions are usually used:

$$\text{sgn}\left(\sum_j c_j \cdot \exp\left(-\gamma \|x - x_j\|^2\right)\right), \quad |c_j| \le \alpha < \infty,$$

here, c_j is a coefficient and $x_j \in \{x_1, ..., x_N\}$ plays a role of a parameter vector of the Gaussian kernels. γ is a hyper-parameter to control the width of Gaussian kernels. At the initial state of SVM, the number of kernels is equal to the number of images, i.e., it discriminates the objects accurately by using all views. By quadratic programming, SVM can choose effective kernel functions, namely the images x_j, in order to discriminate objects within a tolerance of some errors.

Since the remained x_j can be regarded as the parameters that support the rule for discrimination, they are called *Support Vector* (SV). From a statistical point of view, the more complex the rule is, the more SVs will be necessary. That is, the number of SVs will be increased if the object discrimination task becomes difficult.

In the previous section, we discussed about two factors to decide the viewpoint dependency of 3-D object recognition, namely *similarity* across viewpoints and *distinctness* between a target object and distractors. From the above-mentioned reason, the distinctness may be reflected to the number of SVs to achieve no discrimination errors. Furthermore, when the images from different viewpoints are changed a little, SVs that correspond to the images of a target object will not be required so many. Therefore, we can analyze the similarity by the number of SVs for a target object. Based on these conjectures, we examine by numerical simulations whether object features in a view space decide the viewpoint dependency. The proposition is as follows:

Proposition: *The number of SVs for discriminating "viewpoint-dependent" objects is more than "viewpoint-invariant" objects.*

4 Numerical Experiments

In this study, the input images of SVM are stimuli used in 2 psychological experiments. The paradigm of both experiments is a sequential-matching task; after two images are displayed sequentially and instantaneously (less than 500msec), subjects are instructed to answer whether these images are of the same object or not. The view angle between the images can be different, and views of each distractor are shown with some probability in the second presentation. The time and rates responded correctly are measured and analyzed with viewpoints.

4.1 Experiment 1

Stimuli of the first experiment are used in Johnstone & Hayes (2000); *unary-* and *binary-discriminative* objects, shown in Fig.1. Objects in Fig.1(a) have different volumetric parts but their configurations of parts are similar. On the other hand, although primitive parts of all objects in Fig.1(b) are the same, the spatial relations between them are arranged.

In the sequential-matching task of Johnstone & Hayes (2000), the second presentation image was the view from 0, 45, 90, or 135 degrees rotation in depth from the first image. The mean response time for unary-discriminative objects was almost constant (around 800msec) across viewpoint changes, but increased (from 1000 to 1400msec) for binary-discriminative ones. Mean error rates for binary-discriminative objects were higher (from 10% to 40%).

We verify those results by a discrimination task using SVM. Protocols of numerical simulation are as follows: We used SVM with Gaussian kernels and applied it to the object discrimination task for each object group (unary- and binary-discriminative). Number of objects in each group was 6 as shown in Fig.1(a) and (b), respectively. One object became a target by terns, and others were distractors. We made grayscaled snapshots (160×120 pixels) for each object from the viewpoints from 0 to 180 degrees with 5 degree intervals[1]. Therefore, the number of input images for SVM (initial value of the number of SVs) was equal to 216. Learning SVM was executed by RHBNC Support Vector Machine software (Saunders, et al. 1998) for each target object and group. Each kernel receives all grayscaled pixel values in an image, and computes the output by the difference from the SV images. The upper bound of absolute coefficient values of SVs was set to 10. Since it is difficult to determine the value of hyper-parameter γ, we chose its suitable range so as to achieve no discrimination errors.

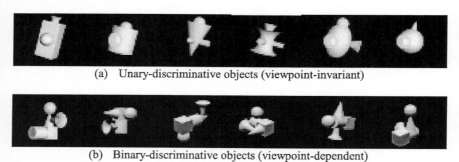

(a) Unary-discriminative objects (viewpoint-invariant)

(b) Binary-discriminative objects (viewpoint-dependent)

Fig. 1. Snapshots of stimuli in Johnstone & Hayes (2000).

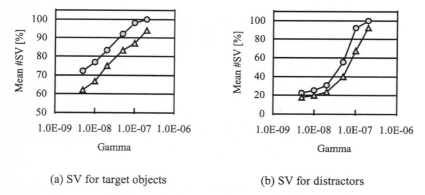

(a) SV for target objects   (b) SV for distractors

Fig. 2. Mean number of SVs in terms of 6 objects (triangles: unary, circles: binary).

Fig.2(a) and (b) show the percentage of mean number of SVs that are images of target objects and distractors, respectively. Triangle symbols show the results for unary-discriminative objects and circle symbols for binary-discriminative ones. The horizontal axes indicate the value of hyper-parameter γ. In both Fig.2(a) and (b), the

---

[1] We used POV-Ray, which is freely-available software for rendering (http://www.povray.org/).

Analysis on the Viewpoint Dependency in 3-D Object Recognition   181

number of SVs for unary-discriminative objects is less than the number for binary-discriminative ones for all γ. This result corresponds to the viewpoint dependency of object groups obtained by psychological experiments in Johnstone & Hayes (2000). Compared to the objects with prominent features in Fig.1(a), some of the objects in Fig.1(b) seem less robust for the change of their views by rotations. Since the difference of *distinctness* from distractors across viewpoints between both object groups is a little in Fig.2(b), the *similarity* of a target object across viewpoints could be a main factor to influence the viewpoint dependency in this experiment.

4.2 Experiment 2

In this experiment we used the stimuli in Tarr, et al. (1997) as shown in Fig.3, which is the case that the spatial relationship is the same for all object groups. The objects in Fig.3(a) are similar to the paperclip-like objects in Bülthoff & Edelman (1992), but one or more distinct parts (we can say they are GEON in this case) are inserted to the objects in Fig.3(b) and (c). Biederman & Gerhardstein (1993) showed that the performance to discriminate the objects in Fig.3(b) was viewpoint-invariant.

The second presentation images in the sequential-matching task of Tarr, et al. (1997) were the view from 0, ±30, ±60, or ±90 degrees rotation in depth from the first image. The results was that the reaction time and rates to correctly answer were viewpoint-dependent for the paperclip objects with 0 or 3 GEON parts, but invariant for the objects with 1 GEON part. Although the objects in Fig.3(c) satisfy the conditions for viewpoint invariance (Biederman & Gerhardstein 1993), the performance was opposite.

(a) Paperclip objects without GEON parts (viewpoint-dependent)

(b) Paperclip objects with 1 GEON part (viewpoint-invariant)

(c) Paperclip objects with 3 GEON parts (viewpoint-dependent)

Fig. 3. Snapshots of Paperclip-like object stimuli in Tarr, et al. (1997)

We try to interpret these results from the image-based approach using SVM. The protocol of numerical simulation is similar to Experiment 1, except that the number of objects in each group was 4 as shown in Fig.3. The percentage of mean number of distractor SVs is shown in Fig.4. Circle symbols denote the results for paperclip-like objects without GEON parts, diamond symbols for 1 GEON part, and triangle symbols for 3 GEON parts, respectively. The results of the numerical experiments agree with the psychological evidence for all γ, i.e., the less viewpoint dependency the objects have, the less kernels are required in order to discriminate them.

In this simulation, the number of SVs for target objects was equal to the initial value for all cases, because the objects in each group might have similar structure in which the shapes change big even by small difference of viewpoints. From Fig.4, one of the key features to make easy to discriminate objects is whether distinct parts in the objects are *unique* or not in this case. The number of distinct parts could control the *quality* of object features, and it would be reflected in the distinctness of images between a target object and distractors.

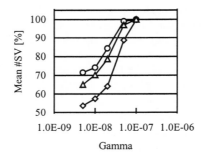

Fig. 4. Mean number of distractor SVs in terms of 4 objects (circles: no GEON parts, diamonds: 1 GEON part, triangles: 3 GEON parts).

5 Conclusions

In this study, the relationship between 3-D object images and the viewpoint dependency of performance in an object discrimination task is analyzed by using Support Vector Machines. Since the simulation results correspond to the psychological evidence, we suggest that image-based representation of 3-D objects in human brain would play an important role to recognize or discriminate them. Although Tjan & Legge (1998) have proposed another measure of viewpoint dependency based on the image-based approach, its computational cost is too expensive to analyze results.

As pointed out by Tarr & Bülthoff (1998), however, recognition mechanism based only on the image-based model is not enough to explain every human performance. We need to connect other viewpoint-invariant mechanism with image-based model in order to develop a robust model for object recognition, because our brain has an ability to recognize familiar objects from ordinary viewpoints.

Acknowledgement We thank Mr. Yoshiyuki Suzuki at Biological and Physiological Engineering Laboratory in Toyohashi University of Technology for his marvelous work.

References

Biederman, I.: Recognition-by-components: A theory of human image understanding. Psychological Review **94** (1987) 115 – 147

Biederman, I, Gerhardstein, P.C.: Recognition depth-rotated objects: Evidence and conditions for three-dimensional viewpoint invariance. J. Experimental Psychology: Human Perception and Performance **19** (1993) 1162 – 1182

Bülthoff, H.H., Edelman, S.: Psychophysical support for a two-dimensional view interpolation theory of object recognition. Proc. Natl. Acad. Sci. USA **89** (1992) 60 – 64

Hayward W.G., Tarr, M.J.: Testing conditions for viewpoint invariance in object recognition. J. Experimental Psychology: Human Perception and Performance **23** (1997) 1511 – 1521

Hummel, J.E., Biederman, I.: Dynamic binding in a neural network for shape recognition. Psychological Review **99** (1992) 480 – 517

Johnstone, M.B., Hayes, A.: An experimental comparison of viewpoint-specific and viewpoint-independent models of object representation. The Quarterly Journal of Experimental Psychology **53A** (2000) 792 – 824

Logothetis, N.K., Pauls, J., Bülthoff, H.H., Poggio, T.: View-dependent object recognition by monkeys. Current Biology **4** (1994) 401 – 414

Marr, D., Nishihara, H.K.: Representation and recognition of the spatial organization of three-dimensional shapes. Proc. R. Soc. Lond. B **200** (1978) 269 – 294

Poggio, T., Edelman, S.: A network that learns to recognize three-dimensional objects. Nature **343** (1990) 263 – 266

Saunders, C., Stitson, M.O., Weston, J., Bottou, L., Schoelkopf, B., Smola, A.: Support Vector Machine – Reference manual. Technical Report CSD-TR-98-03, Department of Computer Science, Royal Holloway, University of London, Egham, UK (1998)

Tarr, M.J.: Rotating objects to recognize them: A case study on the role of viewpoint dependency in the recognition of three-dimensional objects. Psychonomic Bulletin & Review **2** (1995) 55 – 82

Tarr, M.J., Bülthoff, H.H.: Image-based object recognition in man, monkey, and machine. Cognition **67** (1987) 1 – 20

Tarr, M.J., Bülthoff, H.H., Zabinski, M., Blanz,V.: To what extent do unique parts influence recognition across changes in viewpoint? Psychological Science **8** (1997) 282 – 289

Tjan, B.S., Legge, G.E.: The viewpoint complexity of an object recognition task. Vision Research **38** (1998) 2335 – 2350

Ullman, S., Barsi, R.: Recognition by linear combination of models. IEEE Trans. Pattern Analysis and Machine Intelligence **13** (1991) 992 – 1006

Vapnik, V.N.: The Nature of Statistical Learning Theory. Springer-Verlag, New York (1995)

A Comparative Study of Two Neural Models for Cloud Screening of Iberian Peninsula Meteosat Images

Miguel Macías Macías[1], F. Javier López Aligué[1], Antonio Serrano Pérez[2], Antonio Astilleros Vivas[2]

[1] Departamento de Electrónica e Ingeniería Electromecánica - Universidad de Extremadura,
Avda. de Elvas s/n, 06071Badajoz, Spain.
(miguel, javier)@nernet.unex.es
[2] Departamento de Física - Universidad de Extremadura, Avda. de Elvas s/n, 06071Badajoz,
Spain.
(asp,aavivas)@unex.es

Abstract. In this work we make a comparative study of the results obtained in the automatic interpretation of the Iberian Peninsula Meteosat images by means of neural networks techniques, in particular, multi-layer perceptrons and self organizing maps. The interpretation of these images implies their segmentation in the classes SEA (S), LAND (L), LOW CLOUDS (C_L), MIDDLE CLOUDS (C_M), HIGH CLOUDS (C_H) and CLOUDS WITH VERTICAL GROWTH (C_V).

1 Introduction

The cloud cover constitutes one of most important factors in the earth radiation balance, and it plays a fundamental role in the climate modulation. In spite of it, this cloud cover has not been observed suitably and, therefore, characterized, until the arrival of the satellites at the end of the 70. This complete characterization of the cloud cover implies its segmentation based on the shape and height of the clouds that make it up and it provides a great information about the dynamics associated to the air masses.

These reasons make the detection and estimation of the cloud cover from satellites images, referred as cloud screening or cloud segmentation, represent a very important problem in meteorology. For the resolution of this problem, several techniques have been used in the last two decades. The first attempts of classification were done using statistical methods [1][2][3], and threshold techniques [4], later, these techniques have opened the way for neural networks classification methods. The use of neural network was promoted by Welch [5], and since then, unsupervised SOM [6] and SOM+LVQ [7] and supervised methods PNN [8], BP [9], [10] have been used.

In most of previously mentioned references, the classification is made from images that come from the Advanced Very High Resolution Radiometer (AVHRR) on board polar satellites NOAA. The spatial resolution of these satellites is better than the one of the geostationary satellites, allowing a more accurate description of the cloud cover. Nevertheless, their poor temporal resolution (one image every 6 hours) makes them not to be suitable for cloud analysis studies, whose evolution implies temporal scales of the order of few hours. Then, geostationary satellites are interesting because they can offer the suitable temporal resolution although their space resolution is worse.

In the following section we will describe the methodology followed in this work, from the classification of the prototypes to the learning algorithms of the different neural models we have used. In this section we also specify the techniques used for the selection of the best model. In section 3 we describe the classification results obtained by each of the networks selected for each neural model, and in section 4 the conclusions and future lines which can be deduced from the present work are described.

2 Methodology

In this paper we used images from the geostationary satellite Meteosat. This satellite gives multi-spectral data in two wavelength bands, visible and infrared. The subjective interpretation of these images by several experts in Meteorology suggests to consider the following classes: SEA (S), LAND (L), LOW CLOUDS (C_L), MIDDLE CLOUDS (C_M), HIGH CLOUDS (C_H) and CLOUDS WITH VERTICAL GROWTH (C_V).

Once we have defined the classification problem and the algorithms we are going to use to carry it out, the following step in all classification problem is the incorporation of the prior knowledge for the definition of the characteristic vector representative of the input patterns. This prior knowledge also determines the possible modifications over the characteristic vector before being introduced in the neural network, or in other words, the definition of the preprocessing stage. In this sense, there have been many works about feature selection for cloud screening [11], [12].

For the classification we will use neural networks representatives of the two kinds of learning supervised and unsupervised: multi-layer perceptrons and self-organizing maps. Then, we will compare the classification results obtained by Cross Validation [13].

For the learning of the neural models, several experts in meteorology selected a large set of prototypes. These prototypes are grouped into rectangular zones, of such form that, each of these rectangular zones contains prototypes belonging to the same class. For the prototypes selection a specific plug-in for the image-processing program GIMP was created.

In order to compare the classification results obtained with the two neural models, we divided the set of prototypes into a training set, a validation set and a test set. Training set is used for the learning of the different neural networks and for each neural model, we choose the neural network with the best classification results on the validation set. Therefore, we will obtain neural networks with good generalization performance. Finally, we select the optimal model for our application based on the classification results on the test set obtained by the previously selected neural networks for each model. The different simulations were done on a linux machine with the freeware neural networks simulation program SNNS (Stuttgart Neural Network Simulator).

2.1 Preprocessing Stage

Our final objective is the definition of a segmentation system of images that are received from different hours of the day and from different days of the year. Therefore, satellite data must be corrected by the preprocessing stage to obtain physical magnitudes characteristic of clouds and independent of the measurement process.

From the infrared channel, we obtained brightness temperature information normalized of the aging and transfer function effects of the radiometer. From the visible channel we obtained albedo data independent of the radiometer aging effects, the Sun-Earth distance at the image acquisition date and the zenithal angle of the sun at the image acquisition date and for the longitude and latitude of the pixel that is being considered.

As the result of the preprocessing stage, we obtained a four dimension characteristic vector $X=(x_i)$ for each pixel in the image. The characteristic vector components are the brightness temperature and albedo of the pixel at issue, and the typical deviations of these variables calculated over its eight first neighbors. Finally, components of the characteristic vector are linearly transformed so that their values could not differ significantly. The characteristic vector is also normalized if network SOM is used.

2.2 SOM+LVQ

For the organizing of the Kohonen feature map, we started from a set of neurons distributed in a two-dimensional grid. The map is topologically arranged by using the algorithms described in [14]. According to these algorithms, the adaptation process of the weights $W_j = (w_{ij})$ of the j-th neuron of the Kohonen map for an input pattern $X=(x_i)$ is carried out according to the equation (1). In this equation N_C represents the neighborhood of the neuron whose weight vector is more similar to the input vector X (winning neuron) denoted with subscript c. If the weight and input vectors are normalized, this similarity measurement among them can be made by the scalar product of both vectors. The coefficient d_j represents the distance euclidean between the weights vectors W_j and W_c of the j-th and the winning neurons, r(t) represents the radius of the neighborhood N_C and h(t) controls the weights rate of change. To enforce the clustering process r(t) and h(t) values are decreased over time.

$$\Delta w_{ij}^{(t)} = \begin{cases} h(t) e^{-\left(\frac{d_j}{r(t)}\right)^2} (x_i(t) - w_{ij}(t)) & \text{if } j \in N_C \\ 0 & \text{if } j \notin N_C \end{cases} \quad (1)$$

The organizing of the map is made by repeating the weights adaptation process described above over the input patterns X of the training set. When the adaptation process has finished, a labeled of the map is carried out, again, the prototypes of the training set are used. Because one neuron of the map can win for prototypes belonging to different classes, we adopted the rule of the maximum, so one neuron will be considered representative of a class when it win for the prototypes of this class mainly. In the same way, if one neuron is not winning for any prototype or for equal number of prototypes of two different classes, it is considered representative of a new class, the indecision class (I).

In order to improve the classification results, the weights are fine-tuning by means of the Learning Vector Quantization LVQ algorithm described in [15]. This algorithm works according to the equation (2) where with the subscript c we denote to the winning neuron for the prototype $X=(x_i)$ of the training set, with C_C to the label of the winning neuron and with C_X to the class of the prototype.

$$\Delta w_{ic}^{(t)} = \begin{cases} +\alpha(t)(x_i(t) - w_{ic}(t)) & \text{if } C_X = C_c \\ -\alpha(t)(x_i(t) - w_{ic}(t)) & \text{if } C_X \neq C_c \end{cases} \quad (2)$$

The value of alpha, the number of iterations of the fine-tuning process and the map size are chosen based on the classification results over the validation set to obtain SOM+LVQ networks with good generalization, that is, suitable for making good predictions for new inputs.

2.3 Multi-layer Perceptron

For the training of the multi-layer perceptrons, we used the Resilient Backpropagation RProp algorithm described in [16]. Basically this algorithm is a local adaptive learning scheme performing supervised batch learning in multi-layer perceptrons and is distinguished by considering only the sign of the summed gradient information over all patterns of the training set to indicate the direction of the weight update. According to the equation (3), the size of the weight change is determined by a weight-specific, the so-called 'update-value' $\Delta_{ij}^{(t)}$. The second step of RProp learning is to determine the new update-values based on a sign-dependent adaptation process according to the equation (4).

$$\Delta w_{ij}^{(t)} = \begin{cases} -\Delta_{ij}^{(t)} & \text{if } \dfrac{\partial E^{(t)}}{\partial w_{ij}} > 0 \\ +\Delta_{ij}^{(t)} & \text{if } \dfrac{\partial E^{(t)}}{\partial w_{ij}} < 0 \\ 0 & \text{else} \end{cases} \quad (3)$$

Every time the partial derivative of the corresponding weight w_{ij} changes its sign, which indicates that the last update was too big and the algorithm has jumped over a local minimum, the update-value is decreased. If the derivative retains its sign, the update-value is slightly increased in order to accelerate convergence in shallow regions.

$$\Delta_{ij}^{(t)} = \begin{cases} \eta^{+}\Delta_{ij}^{(t-1)} & \text{if } \dfrac{\partial E^{(t-1)}}{\partial w_{ij}} \dfrac{\partial E^{(t)}}{\partial w_{ij}} > 0 \\ \eta^{-}\Delta_{ij}^{(t-1)} & \text{if } \dfrac{\partial E^{(t-1)}}{\partial w_{ij}} \dfrac{\partial E^{(t)}}{\partial w_{ij}} < 0 \quad 0 < \eta^{-} < 1 < \eta^{+} \\ \Delta_{ij}^{(t-1)} & \text{else} \end{cases} \quad (4)$$

For obtaining an optimal neural network with good generalization, we started from a multi-layer perceptron with very few neurons. This perceptron was trained on the training set by means of the algorithm described above. The learning process stops when the number of misclassifications obtained on the validation set goes through a minimum. After we repeat the process by increasing the network size and we choose the new network as the optimal if the number of misclassifications over the validation set is lower than the previous one.

3 Results

In order to implement the processes described above, the experts in Meteorology selected 1264 prototypes, 751 for the training set, 251 for the validation set and 262 for the test set. The prototype selection was made from the peninsula Iberian METEOSAT images corresponding to the years 1995-1998.

Table 1. Percent of success, mistakes and indecisions of the 4x20x6 multi-layer perceptron over the training, validation and test sets of prototyopes.

Multi-layer perceptron	Training			Validation			Test		
	Suc.	Mis.	Ind.	Suc.	Mis.	Ind.	Suc.	Mis.	Ind.
L	100	0	0	97.9	0	2.1	100	0	0
S	100	0	0	100	0	0	100	0	0
C_L	98.6	0.7	0.7	100	0	0	100	0	0
C_M	98.1	1	0.9	96.4	1.8	1.8	100	0	0
C_H	100	0	0	100	0	0	86	9.3	4.7
C_V	100	0	0	100	0	0	67.5	32.5	0

For the training of the multi-layer perceptrons, we initialized its weights randomly in the interval [-1,1] and we chose the following values for the parameters: $\eta^{+}=1.2$, η^{-}

=0.5, $\Delta_{ij}^{(0)}$=0.2. The best results on the validation set were obtained to the 252 iterations of the learning process for a multi-layer perceptron of 4, 20 and 6 neurons. The results of classification offered by this network for the three prototypes set can be observed in the table (1).

The best results for the SOM algorithm was obtained by a 6x6 map of neurons trained during 100000 iterations on the training set. The values for the neighborhood radius and the weights rate of change were r(t+1)=0.99999997r(t) and h(t)=0.99999997h(t) with r(0)=6 and h(0)=0.9. In accordance with cross validation algorithm fine-tuning went on for twenty iterations with α(t+1)=0.997α(t) and α(0)=0.39. The classification results on the three sets of prototypes can be observed in the table (2). On the other hand, we can observe the labeled of the map after the process of fine-tuning in the figure (1).

Table 2. Percent of success, mistakes and indecisions of the 6x6 Self Organizing Map over the training, validation and test sets of prototyopes.

Self Organizing Map	Training			Validation			Test		
	Suc.	Mis.	Ind.	Suc.	Mis.	Ind.	Suc.	Mis.	Ind.
L	98.5	1.5	0	81.2	8.3	10.4	100	0	0
S	100	0	0	100	0	0	100	0	0
C_L	100	0	0	100	0	0	100	0	0
C_M	79.8	20.2	0	87.5	10.7	1.8	66.7	33.3	0
C_H	92	8	0	100	0	0	83.7	14	2.3
C_V	95.6	4.4	0	80.5	19.5	0	95	5	0

Figure 1. Classifying Map

C_L	I	L	L	I	S
C_L	I	I	L	L	L
C_L	C_L	C_L	C_H	C_H	C_H
C_M	C_L	C_L	C_M	C_H	C_H
C_M	C_M	C_V	C_M	C_M	C_V
C_V	C_V	C_V	C_V	C_V	I

4 Conclusions and Future Lines

The classification results over the test set (see tables 1 and 2) of the two networks representative of the neural models are quite similar. Perhaps the most important aspect we can extract from these results is that the greatest number of misclassification is made on different classes in each case C_M for SOM and C_V for multi-layer perceptron.

These results show that both neural models are suitable for the classification problem at issue and they leave the field open to the study of the joint use of the two neural networks by means of which is known as a committee of networks [17].

Another key piece of the future work implies the validation of our neural network by means of the classification results on cloud data gathered in the SYNOP databases. These data are obtained from ground observations in several weather stations.

Besides, in order to improve the classification results, the selection of new components for the characteristic vector as well as a main component analysis of these parameters can be needed.

Acknowledgements

Thanks are due to EUMESAT for kindly providing the Meteosat data. This study was supported by the Junta de Extremadura under project IPR00C015.

References

1. Debois, M., Seze, G., and Szejwach, G.: Automatic Classification of Clouds on Meteosat Imagery: Application to High-Level Clouds. Journal of Applied Meteorology, 21 (1982) 401-412.
2. Key, J., Barry, R. G.: Cloud Cover Analysis with Arctic AVHRR Data, 1. Cloud Detection. Journal of Geophysical Research, 94(D15), December 20 (1989) 18521-18535.
3. Key J.: Cloud Cover Analysis with Arctic Advanced Very High Resolution Radiometer Data, 2. Cassification with Spectral and Textural Measures. Journal of Geophysical Research, 95(D6) May 20 (1990) 7661-7675.
4. Karlsson, K. G., Liljas E.: The SMHI Model for Cloud and Precipitation Analysis from Multispectral AVHRR Data. Technical Report 10, Swedish Meteorological and Hydrological Institute, August (1990).
5. Welch, R.M., Sengupta, S.K., Goroch, A.K., Rapindra, P., Rangaraj N. and Navar, M.S.: Polar Cloud and Surface Classification Using AVHRR Imagery: An Intercomparison of Methods. Journal of Applied Meteorology, 31, 1992, 405-419.
6. Livarinen, J., Valkealahti, K., Visa, A., Simula, O.: Feature Selection with Self-Organizing Feature Map. In International Conference on Artificial Neural Networks, Sorrento, Italy, May (1994) 26-29.
7. Visa, A., Valkealahti, K., Livarinen, J., Simula, O.: Experiences from Operational Cloud Classifier Based on Self-Organising Map. In SPIE Vol. 2243 Applications of Artificial Neural Networks V, volume 2243, Orlando, Florida, April 5-8 (1994) 484-495.
8. Bankert, R. L.: Cloud Classification of AVHRR Imagery in Maritime Regions Using a Probabilistic Neural Network. Journal of Applied. Meteorology, 33, (1994) 909-918.
9. Yhann, Stephan R., Simpson, James J., Application of Neural Networks to AVHRR Cloud Segmentation. IEEE transactions on geoscience and remote sensing, vol 33, No. 3, May (1995).
10. Lee J., Weger R. C., Sengupta S. K. And Welch R.M.: A Neural Network Approach to Cloud Classification. IEEE Transactions on Geoscience and Remote Sensing, 28(5), (1990) 846-855.
11. Welch, R.M., Sengupta, S.K., and Chen, D.W.: Cloud Field Classification Based Upon High Spatial Resolution Textural Features. Journal of Geophysical Research, Vol. 93, D10, (1988), 12663-12681.

12. Aha, D. W., and Bankert, R. L.: A Comparative Evaluation of Sequential Feature Selection Algorithms. Artificial Intelligence and Statistics V., D. Fisher and J. H. Lenz, editors. Springer-Verlag, New York, 1996.
13. Stone, M.: Cross-Validatory Choice and Assessment of Statistical predictions. Journal of the Royal Statistical Society, B 36 (1), (1974),111-147.
14. Kohonen, T.: The Self-Organizing Map. In Proceedings of the IEEE, Vol. 78, No.9, September (1990).
15. Kohonen, T.: Learning Vector Quantization for Pattern Recognition. Helsinki University of Technology, Department of Technical Physics, Laboratory of Computer and Information Science, Report TKK-F A601, (1986).
16. M. Riedmiller, M., Braun, L.: A Direct Adaptive Method for Faster Backpropagation Learning: The RPROP Algorithm. In Proceedings of the IEEE International Conference on Neural Networks 1993 (ICNN 93), 1993.
17. M. P. Perrone, and L.N., Cooper.: When networks disagree: ensemble methods for hybrid neural networks. In R. J. Mammone (Ed.), Artificial Neural Networks for Speech and Vision, pp. 126-142. London: Cahpman & Hall., 1993

A Growing Cell Neural Network Structure for Off-Line Signature Recognition

K. Toscano-Medina, G. Sanchez-Perez, M. Nakano-Miyatake, and H. Perez-Meana

SEPI ESIME Culhuacan
National Polytechnic Institute
Av. Santa Ana, San Francisco Culhuacan
04430 Mexico D.F., MEXICO
hmpm@calmecac.esimecu.ipn.mx

Abstract. The signature recognition is a topic of intensive research due to its great importance, among others, in the financial system. However it does not exist yet an enough reliable method for signature recognition and verification, especially in the forgeries detection. This paper presents an off-line signature recognition using features extracted from the off-line signature and an array of five growing cell neural network. The proposed system was evaluated using 950 signatures of 19 different persons. Experimental results show that proposed system achieves a fairly good recognition rate with a relatively low computational complexity

1. Introduction

The signature recognition is a topic of intensive research during the last several years due to the great importance it has, among others, in the financial system. However it does not exist yet an enough reliable recognition method, especially in the forgeries detection [1]-[3]. The signature recognition task has several problems that hinder its identification. Among them, the fact of that the signature is an image and then it does not have meaning in itself, and the fact that in many cases only the printed signature is available without any information about the order in which it was done.

For the signature recognition exist, basically, two approaches i.e. the on-line recognition (or dynamical recognition) and the off-line recognition (or static recognition). To accomplish the dynamical recognition it can be used the dynamic information content in the signature process, i.e., the initial and final points of the signature, the speed and intensity of each part of the signature, etc [2], besides the information in the image itself. The static recognition, on the other hand, only uses the information available in the captured image. In the second case, there is no way of knowing precisely the initial and final points of the signature. Therefore the dynamical recognition is less difficult because there are more decision elements available, in comparison to the static recognition. However considering the practical applications of both of them, the static recognition would have more applications, or in the worst case the same number than, the dynamical one. Thus it is clear that the static signature verification continues to be of great interest to the scientific community, especially considering the enormous financial impact of the automatic signature verification in check and other official documents.

Usually the researchers use the type I and type II error rates to evaluate verification systems. The type I error is when a system recognize an original signature like forged one and type II error is when the system recognize a forged signature like a original. The type II error rate is very important because it express the percentage of counterfeit signatures (forgeries) that have been accepted. Minimizing this rate often involves an increase in the number of type I error (rejections of a genuine signature).

During the last years have been developed several systems, which are able to simulate the human brain behavior in the realization of some specific jobs. Two of the most important paradigms used until now, for this purpose, are the Neural Networks and the Artificial Intelligence. Both of them are primary tools for the development of systems capable to perform tasks usually carried out easily by the human brain such as: Handwritten characters, voice, face and signature recognition as well as several other biometrics applications that have had relatively success during the last several years.

One of the greatest advances in signature verification is the increasing use of neural networks. The neural networks have found their way into identity verification systems and are now used in signature segmentation, static and dynamic signature verification, etc.

The advantages of neural networks are that they can be trained to recognize different patterns from their characteristics. Thus that they could be use to classify signatures as original or forged through a retraining process based on recent signatures. Their primary disadvantage is often the large number of specimens required to ensure that the network does in fact learn.

In the static recognition, the feature extraction is a very important factor and even basic, because the operation of the system is very dependent on the features used for identification. Because of that, it has been reported several methods for feature extraction such as those reported in the reference [2].

This paper proposes an static signature recognition system which consists of a feature extraction module and an array of 5 multilayer growing cell neural networks operating in parallel whose outputs signals are feed to another two-layers neural network which is used to take the final decision. The proposed system presents a fairly good operation performance with a relatively low computational complexity.

2. Proposed System

The proposed system consists mainly of feature extraction and identification stages. The feature extraction stage extracts, from the printed signature, the upper, lower, left and right signature envelopes and some statistic parameters used for identification. These parameters are then feed to the identification stage, which consists of five three-layers growing cells neural networks operating in parallel, whose outputs are then feed to a sparse two-layers neural network to take the final decision. The five three-layers networks are trained to identify the signature under analysis, using one of the five different extracted features.

Once these networks have been trained, the two-layers network is trained using the outputs of the above mentioned five three-layer nets to take the final decision.

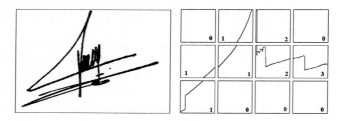

Fig. 1. Signature image together with its upper envelope divided in 12 parts

2.1 Feature Extraction Module

Once the signature has been digitized, the system proceeds to the extraction of upper, lower, left and right envelopes of the signature under analysis. After envelope extraction is carried out, each extracted envelope is divided into 12 parts subsequently to each of them it is assigned an integer value that represents the characteristics of that part of the envelope. Figure 1 shows a given signature together with its upper envelope divided in 12 parts and theirs assigned integer values according to its characteristics. These values are invariant to changes on the size and translation of the signature, as well as relatively invariant to some degree of rotation of signature. Then the system is able to properly recognize the signature even with significant changes in the size and translation of the signature, even if it has some rotation degree.

The criterion taken for assign an integer value to each of 12 parts of the signature envelope is the following:
0   If there is not any pixels inside the square,
1   If there is some pixels inside the square but the information does not form a valley, or a peak,
2   If the pixels found inside the square form a peak,
3   If the pixels found inside the square form a valley.

In addition to the envelope features extracted before, the system computes some statistic parameters to characterize the fine structure found in the signature. Thus the system obtains the values of Kurtosis, that indicates what flattened is the distribution probability with respect to a normal distribution. Also the system computes the value of Skewness that denotes the lack of symmetry of the probability distribution. The relative value between Kurtosis and Skewness, and the relative value between vertical projection and horizontal projection of the signature under analysis are also computed. [1], [2], [4].

2.2 Signature Recognition

As mentioned before, the proposed system uses 5 three-layers Growing Cell Neural Network Structure (Fig. 2) for processing the envelope features and statistic parameters. For envelope features processing the networks have 12 input data. On the other hand the network used for processing the statistic parameters has 8 inputs. The number of output neurons was the same for the five nets, which corresponds to the number of signatures to be identified. Finally, the sparse neural network used for final decision has 5N inputs, which corresponds to the outputs of the 5 multilayer perceptron, and N outputs corresponding to the number of signature to be identified.

Here firstly the 5 three layers networks are independently trained, and after these have converged the last one is trained. All of them using the proposed learning algorithm described below.

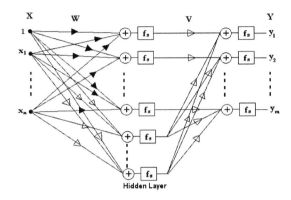

Fig. 2. Growing cell neural network used for signature recognition

The proposed neural network structure is trained using a modified backpropagation algorithm. At the beginning the weight vectors of the m ADALINE networks with nonlinear activation function are updated whose outputs are giving by [5]:

$$y_K = f_S(X^T W_K) \quad k = 1, 2, \dots m \tag{1}$$

where

$$\mathbf{X} = [1, x_1, x_2, \dots x_n]^T, \quad \mathbf{W}_k = [w_{1,k}, w_{2,k}, w_{3,k}, \dots w_{N,k}] \tag{2}$$

Subsequently their weight vectors are updated using the backpropagation algorithm without hidden layer, as follow:

$$w_{jk} = w_{jk} + \mu e_k x_j \tag{3}$$

$$e_k = d_k - y_k \tag{4}$$

is the output error and d_K is the desired output. This process is repeated until convergence is achieved. Next a recognition test is performed. If the recognition rate is satisfactory, the learning process is stopped. If it is not, a hidden neuron is added and the already estimated weights are frozen. In this stage the hidden layer defined as:

$$\mathbf{U} = [u_1, u_2, \ldots, u_m, u_{m+1}]^T \tag{5}$$

where

$$u_k = y_k \qquad r = 1, 2, \ldots, m \tag{6}$$

$$u_{m+1} = f_s(\mathbf{X}^T \mathbf{W}_{m+1}) \tag{7}$$

Now the output of the new neural network will be given by:

$$y_k = f_s(\mathbf{U}^T \mathbf{V}_k) \tag{8}$$

where

$$\mathbf{V} = \begin{bmatrix} v_{11} & 0 & 0 & - & -. & 0 \\ 0 & v_{22} & 0 & - & - & 0 \\ 0 & 0 & v_{33} & 0 & - & 0 \\ 0 & 0 & - & - & - & 0 \\ - & - & - & - & - & - \\ v_{m+1,1} & v_{m+1,2} & v_{m+1,3} & - & - & v_{m+1,n} \end{bmatrix} \tag{9}$$

Here the input weight vector $\mathbf{W}_K$, $k= 1,\ldots,m$, remains constant and only the input weight vector $\mathbf{W}_{m+1}$ and the weight matrix $\mathbf{V}$, given by (9), must be updated using the backpropagation algorithm as follow:

$$E_p = \frac{1}{2} \sum_{k=1}^{m} e_k^2 \tag{10}$$

$$v_{i,j}(n) = v_{i,j}(n-1) - \mu \frac{\partial E_p}{\partial v_{i,j}} \tag{11}$$

where

$$\frac{\partial E_p}{\partial v_{i,j}} = (d_j - y_j) f_s'(\mathbf{U}^T \mathbf{V}_i) u_i \tag{12}$$

On the other hand, the weights vector $\mathbf{W}_{M+1}$ is updated as follow:

$$w_{k,M+1}(n) = w_{k,M+1}(n-1) - \mu \frac{\partial E_p}{\partial w_{k,M+1}} \tag{13}$$

where

$$\frac{\partial E_p}{\partial w_{k,m+1}} = f'\left(\sum_{i=1}^{m+1} x_i w_{i,m+1}\right) x_k \sum_{j=1}^{m} e_k f_s'(p_j) v_{m+1,j} \tag{14}$$

and

$$p_j = u_{m+1} v_{m+1,j} \tag{15}$$

After convergence is achieved, a recognition test is performed again, and if the recognition rate is not good enough yet, a new hidden neuron is added, freezing all the already estimated weights. Then the rth stage a new hidden layer and weight vectors are defined as:

$$\mathbf{U} = [u_1, u_2, \ldots, u_m, u_{m+1}, u_{m+r}] \tag{16}$$

$$\mathbf{W}_{m+2} = [w_{1,m+r}, w_{2,m+r}, \ldots, w_{n,m+r}]^T \tag{17}$$

$$\mathbf{V} = \begin{bmatrix} v_{11} & 0 & 0 & - & -. & 0 \\ 0 & v_{22} & 0 & - & - & 0 \\ - & - & - & - & - & - \\ v_{m+1,1} & v_{m+1,2} & v_{m+1,3} & - & - & v_{m+1,m} \\ - & - & - & - & - & - \\ v_{m+r,1} & v_{m+r,2} & v_{m+r,3} & - & - & v_{m+r,n} \end{bmatrix} \tag{18}$$

and the process given by eqs. (8) – (18) is repeated again until a reasonable good recognition rate is achieved.

3. Evaluation Results

To evaluate the proposed system, it was generated a data-base that consists of 950 signatures, (50 signatures of 19 persons). 665 signatures (35 signatures of 19 different persons) are used to train the proposed system, 285 signatures are then used for evaluation of the proposed system. Table 1 and 2 show the type I error rate of proposed system using only four features and all features, respectively, extracted from the signature. The evaluation was carried out with signatures used during training.

	Type I Error	Indefinite
Upper, Lower, Left, Statistic Pram.	4.7%	3.3%
Upper, Lower, Left, Right	0.13%	14.6%
Upper, Lower, Right, Statistic Pram.	4.6%	3.8%
Upper, Left, Right, Statistic Pram.	4.3%	2.9%
Lower, Right, Left, Statistic Pram.	5.7%	2.9%

Table 1. Type I error rate with a combination of four features with signatures used during training.

	Type I Error	Indefinite
Global Verification	2.8%	2.5%

Table 2. Global type I error rate with all features and signatures used during training.

Table 3 shows the global type I error rate of the proposed system, with signatures not used during training process.

	Type I Error	Indefinite
Global Verification	6.8%	5.5%

Table 3. Global type error I of proposed structure with sigantures nor used during training.

A very important feature of a signature recognition system is its ability to detect forged signatures. In order to test the performance of proposed system in this situation was build a database of forged signatures. Table 4 shows the global type II error rate of proposed system.

	Type II Error	Indefinite
Global Verification	1.2%	1.5%

Table 4. Global Type II error rate

The evaluation results show that approximately 2.8% of Type I error when the signature used for training were also used for testing, 6.8% of Type I error obtained when the data used for training are different from those used for testing. The most important factors for this recognition rate are the upper envelope, the right envelope and statistic parameters of the signature. These characteristics contribute more to the recognition than the lower and left envelopes of the signature, because the last two envelopes have less information as shown in table 1. The proposed system also shows good performance when it is required to detect forged signatures as shown in table 4.

Fig. 3. Original and forged signature used to evaluate the proposed system. In upper part are shown 4 original signatures, while in the lower part are shown 4 forgeries of these signatures.

Figure 3 shows, in its lower part some of the forged signatures together with the original ones, upper part, used to evaluate the proposed system. All of these signatures were classified as forgeries despite the great similarity existent with the original ones.

4. Conclusions

This paper proposed a signature recognition system based on the feature extraction of 4 signature envelopes and statistic parameters, which are processed using an array of five growing cell neural networks. Computer evaluation results show that the proposed system provides a good recognition rate when the same database is used for training and testing, as well as when both database are different. The system also shows good ability to detect forged signatures. In addition to this, the proposed system is insensitive to the size and the position of the signature, that characteristic makes it attractive for practical applications.

Reference
[1] R. Bajaj and S. Chaudhury, "Signature Verification Using Multiple Neural Classifiers," Pattern Recognition, vol. 30, No. 1, Pag. 1-7, 1997.
[2] F. Leclerc and R. Plamondon, "Automatic Signature Verification: The State of the Art 1989-1993," International Journal of Pattern Recognition and Artificial Intelligence Vol. 8, No. 3 , Pag. 643-660, 1994.
[3] M. Ammar, "Progress in Verification of Skillfully Pattern Recognition and Artificial Intelligence," vol. 5, No. 1 & 2, pag. 337-351, 1991.
[4] K. Toscano M., G. Sánchez P., M. Nakano M. y H. Pérez M., "Off-Line Signature Recognition Using Feature Extraction and Multilayer Neural Networks," To appear in The Journal of Telecommunications and Radio Engineering, 2001.
[5] G. Sánchez P., K. Toscano M., Nakano M. y H. Pérez M., "Growing Cell Neural Network Structure with Backpropagation Learning Algorithm," To appear in The Journal of Telecommunications and Radio Engineering, 2001.

ZISC-036 Neuro-processor Based Image Processing

Kurosh Madani[1], Ghislain de Trémiolles[2], and Pascal Tannhof[2]

[1] Intelligence in Instrumentation and Systems Lab. (I2S)
SENART Institute of Technology - University PARIS XII
Avenue Pierre POINT, F-77127 LIEUSAINT - FRANCE

madani@univ-paris12.fr

[2] IBM France
Laboratoire d'Etude et de Developpement
224, boulevard John Kennedy
F-91105 Corbeil Essonnes Cedex, France

Abstract: This paper deals with neural based image processing and developed solutions using the ZISC-036 neuro-processor, an IBM hardware processor which implements the Restricted Coulomb Energy algorithm (RCE) and the K-Nearest Neighbor algorithm (KNN). The developed neural based techniques have been applied for image enhancement in order to restore old movies (noise reduction, focus correction, etc.), to improve digital television images, or to treat images which require adaptive processing. Experimental results, validating the exposed concepts, have been reported showing quantitative and qualitative improvement as well as the efficiency of our solutions.

1. Introduction

Even though the efficiency of Artificial Neural Networks (ANN) based techniques for solutions of an ever increasing range of problems has already been confirmed [1][2][3], their effective use in industrial world is still trifling comparing to their academic success: very few papers deal with real applications of this kind of technology. The reticence of industrial world to use more massively the neural based approaches is probably due to a number of open questions (unsolved problems) concerning the design of ANN based solution (choice of the neural model, network's topology, etc.), or to the some parameters adjustment difficulty in such models (as: learning rate in Back-Propagation or region of influence in RBF, etc.). But, this caginess is also related to the scarcity of market available and easy usable hardware implementation of such models needing to overcome execution speed constraints (related to productivity). Even if a large number of research programs, during last two decades, has concerned the implementation of ANNs, unfortunately, the most of realized hardware implementations of ANNs were presented without user interface, making them unusable.

Since 1995, a few number of industrial realizations of neural models implementations appeared on standard market among which ZISC-036 (Zero Instruction Set Computer) an IBM hardware processor which implements the Restricted Coulomb Energy algorithm (RCE) and the K-Nearest Neighbor algorithm (KNN). The very few number of functions necessary to control the ZISC-036 and integration of an incremental learning algorithm make this circuit very easy to program in order to develop applications.

This paper deals with neural based image processing and developed solutions using the ZISC-036 neuro-processor. The developed ZISC-036 based neural image processing technique has been applied for image enhancement in order to restore old movies (noise reduction, focus correction, etc.), to improve digital television, or to handle images which require adaptive processing (medical images, spatial images, special effects, etc.).

The paper is organized as follow: The section 2 presents the RCE-KNN neural model and describes the basic properties of the ZISC-036 component. In section 3, we present and discuss the developed ZISC-036 based image processing technique. Section 4 reports experimental results validating the proposed approach. In the last section we give the conclusion.

2. RCE-KNN Neural Network and It's IBM ZISC-036 Hardware Implementation

The RCE (Restricted Coulomb Energy) like neural models belong to the class of "evolutionary" learning strategy based ANNs. That means that the neural network's structure is completed during the learning process. Generally, such kind of ANNs include three layers: an input layer, a hidden layer and an output layer. It is the hidden layer which is modified during the learning phase. The output layer represents a set of categories associated to the input data. Connections between hidden and output layers are established dynamically during the learning phase. As presented in [2] and [4], RCE's algorithm (for classification) consists of an output mapping of an n-dimensional space by prototypes which are characterized by their own category and threshold. The goal of this threshold is to activate or not activate the neighboring neuron.

The goal of the learning phase is to partition the input space by prototype where each prototype is associated with a category and an influence field, a part of the input space around the prototype where generalization is possible. When a prototype is memorized, thresholds of neighboring neurons are adjusted to avoid conflict between neurons. The neural network's response is obtained from relation R1 where C_j represents a "category", $I = \begin{bmatrix} i_0 & i_1 & \ldots & i_n \end{bmatrix}^T$ is the input vector, $P^j = \begin{bmatrix} p_0^j & p_1^j & \ldots & p_n^j \end{bmatrix}^T$ represents the "prototype" memorized (learned) thanks to connections of the neuron j to the hidden layer, and λ_j the influence field associated to neuron j (or Region Of Influence – ROI).

$$C_j = 1 \quad IF \quad dist(I, P^j) < \lambda_j \quad (1)$$

Figure 1 shows the learning mechanism and associated feature space mapping in such neural model in the case of a 2-D feature space.

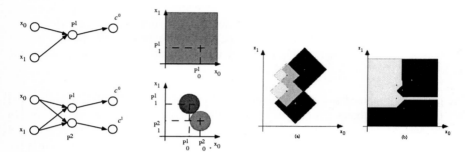

Fig. 1. Example of neurons connections and feature space mapping in a 2-D space.

Fig. 2. Example of an input mapping in a 2-D space: ROI (a) and 1-NN (b) using norm L1 in the case of the ZISC-036 RCE.

The IBM ZISC-036 is a parallel neural processor based on the RCE and KNN algorithms. Each chip is capable of performing up to 250 000 recognitions per second ([5], [6], [7] and [8]). Thanks to the integration of an incremental learning algorithm, this circuit is very easy to program in order to develop applications; a very few number of functions (about ten functions) are necessary to control it. Each ZISC-036 chip is composed of 36 neurons ([6] and [7]). This chip is fully cascadable which allows the use of as many neurons as the user needs (a PCI card has been developed with a capacity of 684 neurons) [7] and [8]. A neuron is an element which is able to:

- memorize a prototype (64 components coded on 8 bits), the associated category (14 bits), a threshold (14 bits), a context (7 bits),
- compute the distance, based on the selected norm (norm L1 given by relation 2 or LSUP given by relation 3) between its memorized prototype and the input vector (the distance is coded on 14 bits),

$$L1 : dist = \sum_{i=0}^{n} |V_i - P_i| \quad (2) \quad \text{and} \quad LSUP : dist = \max_{i=0...n} |V_i - P_i| \quad (3)$$

were P represents the memorized prototype and V is the input pattern.
- compare the distance to its threshold,
- communicate with other neurons (in order to find the minimum distance, category, etc.),
- adjust its threshold (during learning phase).

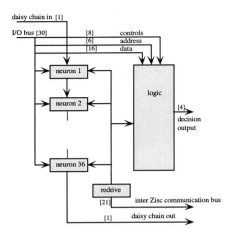
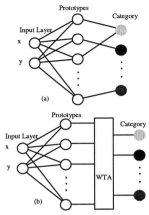

Fig. 3. IBM ZISC-036 neuron block diagram.

Fig. 4. Block diagrams of network topology for the ROI (a) and the KNN (b) operation modes.

Beside the two possible distance computing modes, ZISC implements two different operational modes. The first one operates according to RCE–adjustment where the region of influence of each connected neuron could be defined to map a given feature space. The second mode, which we will call here "KNN" mode, operates according to the WTA (Winner Takes All) principle. In this operational mode, neurons regions of influence are not used. Figure 2(a) shows that due to the close proximity of neighboring neurons, the ROI are adjusted while figure 2(b) is an example of the KNN operational mode (in the case of the figure a 1-NN has been considered). Figure 3 shows the ZISC-036 block diagram.

3. ZISC-036 Based Image Processing

The principle of our technique is very simple: association of some output category (correction, pixel value, luminance, etc.) to an input pattern which is a sub-image of the treated image. The presented principle is based on an image's physics phenomena which states that when looking at an image through a small window, there exist several kinds of shapes that no one can ever see due to their proximity and high gradient [9]. The number of existing shapes that can be seen with the human eye is limited. ZISC-036 is used to learn as many shapes as possible that could exist in an image, and then to replace inconsistent points by the middle value of the closest memorized example. The learning phase consists of memorizing small blocks of an image and associating to each the value of the middle pixel as a category. Figure 5 illustrates a simple example of the used technique. The 64 components input vector of ZISC-036 allows different possibilities to consider such sub-images: from 2x2 to 8x8. However, the most appropriated blocs are 3x3, 5x5 and 7x7.

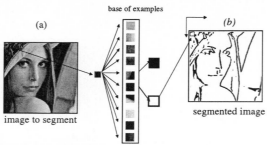

Fig. 5. Example of segmentation of an image using the above mentioned principle.

The patterns presented as learning inputs (which should be learned by the neural network) are selected according to a learning criterion. We call "learning criterion" the strategy (manner, order of presented patterns to be learned, etc.) which will be used to learn a set of patterns representing a given object (image, signal, information, etc.). The learning criterion could be based on a randomly selection of presented patterns or be based on some deterministic rules. Even if the randomly selection of patterns to be learned seems to be the most easy strategy to implement, this learning strategy doesn't lead to the best internal representation (feature space mapping). On the other hand, the used learning criterion will also lead to a different internal representation of the learned pattern. Generally, the used learning criterions are based on local difference evaluation between patterns parameters. Relation (4) gives an example of used learning criterions. In this relation, V_l^k represents the l-th component of the input vector V^k, P_l^j represents the l-th component of the j-th memorized prototype, C^k represents the category value associated to the input vector V^k, C^j is the category value associated to the memorized prototype P^j and, α and β are real coefficients adjusted empirically.

$$Learn\_Crit(V^k) = \alpha \sum_l |V_l^k - P_l^j| + \beta |C^k - C^j| \quad (4)$$

The learning strategy is based of following technique : a random example from the learning data-base is chosen and the learning criteria for that example is calculated. If the value of the learning criteria is greater than the threshold, then a neuron is engaged, and an index (the number of neurons used) is increased. If the learning criteria is less than the threshold, no neuron is engaged. After each iteration, the aforementioned threshold is decreased. Once the index reaches the desired level the learning phase is terminated. As it has been mentioned, the choice of the coefficients α and β is linked to the application and not previously defined. Numerous simulations were performed to adjust the optimal values. In order to smooth the resulting image, it is necessary to use a combination of the closest neighbors. Several solutions for combining the obtained values exist : combination of the distance of the blocks and the corresponding pixel value, the mean distance, etc. The used function, which gives good results, has been defined by relation (5) :

$$v = \frac{\dfrac{1}{0.01d_1^2+1}v_1 + \dfrac{1}{0.01d_2^2+1}v_2}{\dfrac{1}{0.01d_1^2+1} + \dfrac{1}{0.01d_2^2+1}} \quad (5)$$

where v is the estimation of the pixel value for the input block, v_1 is the pixel value associated with the closest memorized example, v_2 is the pixel value associated with the second closest memorized example, and d_1 and d_2 are the distances between the input block and the first and the second closest memorized examples respectively.

4. Experimental Results

On the basis of the ZISC-036 based neural image processing technique presented in section 3, we have developed applications relative to noise reduction, image enhancement and color images restoration in order to restore old movies, to improve digital television, to handle images which require adaptive processing (medical images, spatial images, special effects, etc.). The figure 6 shows experimental results concerning image enhancement. The images 6-a and 6-b have been used for the system training, and image 6-d has been obtained from the degraded image 6-c.

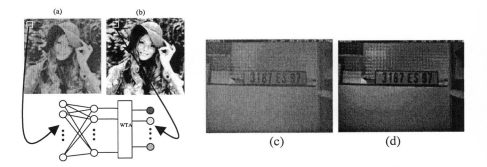

Fig. 6. Experimental results relative to image enhancement application. Images (a) and (b) have been used for learning phase. Image (d) has been obtained in generalization phase from unlearned degraded image (c).

The figure 7 shows experimental results relative to a noise reduction application using the same technique and results obtained with conventional techniques [9]. The noise added to the original image was a pulse noise. The image used for learning was the same used for the previous experiment.

Fig. 7. Experimental results relative to a noise reduction application. (a) noisy image, (b) ZISC-036 technique, (c) Histogram Equalization technique, (d) H.E. + a median filter.

	cameraman		
	ZISC036 $\alpha=1/2n$, $\beta=1/2$	Histogram Equalization	H.E. + Median Filter
3x3 (n=9)	5.41 %	18.26 %	11.47 %
5x5 (n=25)	**4.12 %**		**10.51 %**
7x7 (n=49)	4.54 %		10.79 %
7x1 (n=7)	7.45 %		10.65 %

Table 1: Comparison of the mean error per image per pixel for ZISC-036 based technique and conventional techniques (histogram equalization and median filter).

Fig. 8. Experimental results relative to color images reconstruction. (a) image used for the learning. (b) image used for generalization test.

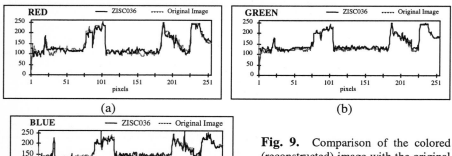

Fig. 9. Comparison of the colored (reconstructed) image with the original image in generalization phase. Red (a), Green (b) and blue (c).

5. Conclusion

In this article, we have focused our purpose on neural based solutions developed for noise reduction, image enhancement and color correction using the ZISC-036 neuro-processor, an IBM hardware processor which implements the Restricted Coulomb Energy algorithm (RCE) and the K-Nearest Neighbor algorithm (KNN). The principle of our technique is very simple: association of some output category (correction, pixel value, luminance, etc.) to an input pattern which is a sub-image of the treated image.

Obtained experimental results and a quantitative comparative study show efficiency of our solution. Furthermore, this system is easy to implement, and can be improved by the number and choice of memorized examples. Continued work on this system concerns improving the learning phase in order to construct a more robust data base by choosing the most representative patterns and the problem of defining the best adaptive error.

References

[1] Reyneri L.M.: Weighted Radial Basis Functions for Improved Pattern Recognition and Signal Processing. In: Neural Processing Letters, Vol. 2, No. 3, pp 2-6, May 1995.
[2] Mercier G., Madani K., Duchesne S.: A Specific Algorithm for A Restricted Coulomb Energy Network used as a Dynamical Process Identifier. In: RI_95, Academy Institute of Sciences, St. Petersburg, Russia, May 1995.
[3] Robert David, Erin Williams, Ghislain de Trémiolles, Pascal Tannhof, Noise reduction and image enhancement using A hardware implementation of Artificial neural networks, VI-DYNN'98 - Virtual Intelligence - Dynamic Neural Networks Stockholm - Sweden - June 22-26, 1998.
[4] Eide A., Lindblad Th., Lindsey C.S., Minerskjöld M., Sekhviaidze G. and Székely G.: An Implementation of the Zero Instruction Set Computer (ZISC-036) on a PC/ISA-bus Card. In: 1994 WNN/FNN in Washington D.C., December 1994.
[5] ZISC-036 Data Book, IBM Microelectronics, November 1994.
[6] ZISC/ISA ACCELERATOR card for PC, User Manual, IBM France, February 1995.
[7] ZISC (Zero Instruction Set Computer), http://www.fr.ibm.com/france/cdlab/zisc.htm, IBM France, January 1998.
[8] G. De Tremiolles, Contribution à l'étude théorique des modèles neuromimétiques et à leur validation expérimentale: mise en œuvre d'applications industrielles ("Contribution to the theoretical study of neuromimetic models and to their experimental validation: use in industrial applications"), Ph.D. Report, University of PARIS XII – Val de Marne, March 1998.
[9] Lawrence S., Lee Giles C., Chung Tsoi A., Back A., "Face Recognition: A Convolutional Neural Network Approach", IEEE Transactions on Neural Networks, Special Issue on Neural Networks and Pattern Recognition, 1997.
[10] Gonzalez R., Woods R., "Digital Image Processing", Addison-Wesley, USA, 1993.

Self-Organizing Map for Hyperspectral Image Analysis

P. Martínez[1], P.L. Aguilar[1], R.M. Pérez[1], M. Linaje[1], J.C. Preciado[1], A. Plaza[1]

[1]Departamento de Informática, Universidad de Extremadura, Avda. de la Universidad s/n, 10071 Cáceres, SPAIN
{pablomar, paguilar, rosapere, mlinaje, jcpreciado, aplaza}@unex.es

Abstract. In this paper we present a neural network methodology used for classifying an hyperspectral image referencied as Indian Pines. The network parameters (learning and neighborhood function) are adjusted using a test battery generated from the image, selecting the values that give the best robutness and discrimination capacity. The availity of ground truth allows us to intoduce a new stadistical measure to quantify the resulting classification accuracy. The results of this methodology show an accuracy of 80% in the classification.

1. Introduction

The use of hyperspectral imaging sensor data to study the Earth's surface is based on the capability of such sensors to provide high resolution spectra, on a per pixels basis, along with the image data.

Hyperspectral sensor provides a large number of narrow bands that provides sensor capabilities to recognize narrow absorption band, like the laboratory measurements. This capability can be used to classify and determine Earth's surface constituents signatures from the hyperspectral information provided by the sensor. Conventional algorithms use these signatures to classify and/or determine the abundances of a composite pixel spectrum [1].

This problem is known as *Hyperspectral Image Analysis*. Highly successful results have been obtained following this approach [2],[3], particularly in the field of geology were exposed lithographys have been mapped based on specific mineral hyperspectral reflectance signatures [4].

The great size of t he hyperspectral images is one of it principal disadvantages, usually 30 times larger than a Landsat TM image of the same spatial size.

The algorithms for conventional multi-spectral sensors can't be used with hyperspectral images by the following reasons:
- High dimensionality of the images.
- Great size of the image (hundred of bands, thousands of pixels)
- Sub pixel analysis possibilities.

- High discrimination to resolve classes
- Difficulties to use training data.

Un-supervised clustering is a problem with application in many areas. Given a set of N data points in a feature space R_D of D dimensions $(x_1, x_2, x_3 ..... x_D)$, we wish to characterize the data as belonging to K cluster where K, must be obtained from the data.

Parra et al. [5] assume linear combinations of reflectance spectra with some additive normal sensor noise and they derive a probabilistic maximum a posteriori framework for analyzing hyperspectral data. The material reflectance characteristics are not know a priori, so this is the problem of unsupervised linear unmixing. The incorporation of different prior information (possitivity and normalization of abundances) leads to a family of algoritms. In [5] the constrained independent component analisis (ICA) for the noise-free case is used.

The clustering is based on some distance metrics and one of the most usefully unsupervised algorithms is the Self-Organizing Neural Network or Self Organizing Map (S.O.M) proposed by T. Kohonen [6].

The present work exploits the possibility of using a Self-Organizing Neural Network to analyze the hyperspectral images. Some reason for using this neural network model, in hyperspectral analysis has been described by Bruske and Merényi [7]:
- To avoid the need to degrade the data
- To provide speed (when implemented in hardware as massively parallel algorithm.
- To surpass conventional classification algorithm performance.
- Good performance for large real life task.

The Self-Organizing neural network has the advantage that obtains by competitive procedures the class prototypes. This ability can be used for the classification of the hyperspectral images. We need tools that match up the intricacy of hyperspectral data. We propose a neural network for processing the hyperspectral information for each pixel. The neural model consists on Self-Organizing Neural Network. This net has an input neuron for each image channel. The output neurons number is related with the characteristics of the image and must be carefully optimized according some metric. Different distances and learning functions are used to obtain a better class prototypes extraction.

The result discussion also includes the influence of the following parameters in the Neural network performance:
- Neighborhood function
- Learning function
- Noise contamination in the spectra.
- Output layer geometry

2. Data

The hyperspectral unmixing algorithms proposed in this work have been tested using the public domain Indian Pines hyperspectral dataset, which has been previously used in many different studies. This image was obtained from the AVIRIS imaging spectrometer at Northern Indiana on June 12, 1992 from a NASA ER2 flight at high altitude with ground pixel resolution of 17 meters. The whole dataset comprises 145x145 pixels and 220 bands of sensor radiance without atmospheric correction. It contains two thirds of agriculture (some of the crops are in early stages of growth with low coverage), and one third of forest, two highways, a rail lane and some houses. Ground truth determines sixteen different classes (not mutually exclusive). Water absorption bands (104-108, 150-163 and 220) were removed [8], obtaining a 200 band spectrum at each pixel. In order to reduce the time of training and testing, we have selected a subscene of the complete Indian Pines dataset, which is depicted in Fig 1.

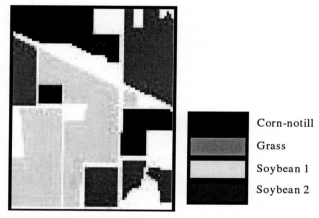

Fig. 1. A subset of the Indian Pines hyperspectral dataset with ground truth.

3. Topology of the Proposed Neural Network

The Self-Organizing Map (SOM) is based on a competitive learning that leads to the construction of topologic maps representing class prototypes. In order to understand the topology of the proposed neural network, we firstly need to define some basic concepts. A neuron is an information-processing unit. Neurons are connected by synapses or connecting links, each of them characterized by a weight. Specifically, a signal x_j at the input of synapse j connected to neuron k is multiplied by the synaptic weight w_{kj} [9]. A neural network is a set of neurons organized in the form of layers. In the simplest form, an input layer projects onto an output layer of neurons. If the input layer has N units and he output layer has M units, each unit in the output layer owns N weights associated to the connections that come from the input layer, so that the set of neural weights are organized in the form of a two-dimensional lattice (W_{MxN}). Our proposed network architecture is depicted in Fig. 2 [10],[11]. In our case, N corresponds to the number of channels of the hyperspectral image and M is the number of classes or prototypes to be extracted by the network. M must be carefully selected according with some metric (we will insist on this issue later on in the paper).

There are feedforward connections from the input layer to the output layer and self-feedback and lateral feedback connections in the output layer. These two types of local connections serve two different purposes. In the classification phase, the input signals x are projected on the feature space by the feedforward connections W, each neuron produces a selective response to input signals. In the learning phase, lateral and feedback output layer connections produce excitatory or inhibitory effects depending on the distance from the neuron to the winner neuron [12],[13]. These weigths are used to determine the W_i classification prototype por each neuron.

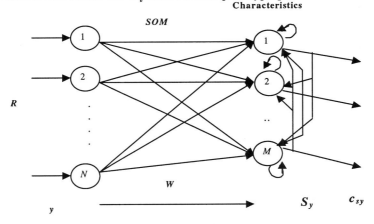

Fig. 2. SOM neural network topology including weight matriz W and learning lateral and feedback conections.

4. Training Algorithm

There are five basic steps involved in the training algorithm. These steps are repeated until the topological map is completely formed:

a) Initialisation. Choose random values for the initial weight vectors $w_i(0)$, i = 1,2,...,M. It is desirable to keep the magnitude of the weights small.

b) Sampling. Choose an input pattern x(n) belonging to a set of learning patterns or references R. The selection is done randomly.

c) Similarity Matching. Find the best-matching (winning) neuron i* at time t, using the minimum-distance criterion, as shown in the following equation:

$$i^*[x(n)] = \min_{j} dist\{x(n), w_j(t)\} \qquad j = 1, 2, \cdots, M \qquad (1)$$

where **dist (i*,i)** is the euclidean distance.

d) Learning. Adjust the synaptic weight vectors of all neurons, using the update formula (2), where η(t) is a learning-rate parameter, and γ is a Gaussian neighborhood function centred around the winning neuron. The size of the neighborhood is determined by a parameter σ(t) (see equation 3).

$$w_i(t+1) = w_i(t) + \eta(t)\gamma(t, i, i^*[x(n)])(x(n) - w_i(t)) \qquad (2)$$

From the different choice for the selection of the involved parameters, taking into acount the studies made in [13], we have choose the following:

$$\eta(t) = \frac{1}{t} \qquad \gamma(t, i, i^*[x(n)]) = e^{\frac{dist(i^*,i)^2}{\sigma(t)}} \qquad \sigma(t) = \left(\frac{\sigma_0}{t}\right)^2 \qquad (3)$$

where σ_o is the initial width that changes in the results that we present later.

e) Continue from step b) until no noticeable changes in the weight space are observed, or until the maximum convergence time is achieved.

In order to analyze a hyperspectral image using this algorithm, the network must be trained with hyperspectral signatures obtained directly from the image. The weights initially associated with each output layer neuron contain the hyperspectral signatures of some carefully selected pixels on the image (according with their spatial distribution).

5 Neighborhood and Learning Function Selection

We have realized a lot of experiments with a predetermined training set, to establish the optimum values to the neighborhood and the learning parameters for the rest of the studies.

Each training set has a fixed number of signatures with different sets generated as follows: the first set contains ten reference signatures free of noise; the second set has twenty signatures those ten and ten new signatures corresponding to noise version of them, one for each reference, the third contained thirty signatures (two noise signatures for each reference); the last set had a hundred of signatures (ten noise signatures for each reference). These sets have been used to train ten different networks.

Hundred signatures that don't belong to the training set have formed the validation set.

5.1. Learning function

We have analyzed five different learning functions given in 4:

$$a)\ \eta(t) = \frac{1}{t} \qquad b)\ \eta(t) = \frac{1}{\sqrt{t}}$$

$$c)\ \eta(t) = \eta_1\left(1 - \frac{t}{\eta_2}\right) \qquad d)\ \eta(t) = \eta_1 e^{-\frac{t}{\eta_2}} \qquad (4)$$

$$e)\ \eta(t) = \eta_1\left(\frac{\eta_1}{\eta_2}\right)^{\frac{t}{t_{max}}}$$

Where η_1 is the initial learning value (normally equal to 0,9 in c) and d) and 0,8 in e), η_2 is the final learning value (0,08 in e) or the number of iterations in c) and d)) and t_{max} is the number of iterations.

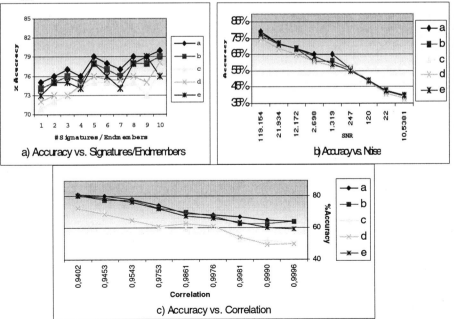

Fig. 3. Learning parameter study

214   P. Martinez et al.

The Fig. 3 shows the obtained results for three types of studies; it represents the accuracy versus the endmembers number (figure a.), the Signal to Noise Ratio defined as SNR=E(signal)/E(noise),(figure b.) and the correlation between endmembers (figure c.). We can observe that the better results are obtained by the equation 4.a.

5.2. Neighborhood function

We have analyzed different neighborhood functions given in 5.

$$\sigma(t,i,i*[x(n)]) = \begin{cases} a) & e^{\left[\frac{-d_L(i,i*[x(n)])^2}{2\sigma(t)^2}\right]}, & \sigma(t) = \frac{\sigma_0}{t} \\ b) & \frac{sen\left(\left[\frac{-d_L(i,i*[x(n)])}{\sigma(t)^2}\right]\right)}{d_L(i,i*[x(n)])} \sigma(t), & \sigma(t) = \frac{\sigma_0}{t} \\ c) & e^{\left[\frac{-d_L(i,i*[x(n)])^2}{\sigma^2(t)}\right]}, & \sigma(t) = \sigma_0 \left(\frac{\sigma_1}{\sigma_0}\right)^{\frac{t}{t_{max}}} \end{cases} \quad (5)$$

Where σ_1 is the initial neighborhood value (normally equal to middle of map), σ_2 is the final neighborhood value (0,08 in c) and t_{max} is the number of iterations.

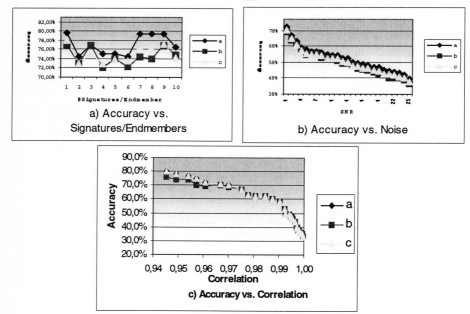

Fig. 4. Neighborhood parameter study

The figure 4 shows the obtained results for the same types of studies used in the learning function. We can observe in fig. 4 the obtained results for the same types of studies used in the learning function; the better results are obtained by the equation 5.a.)

The previous studies justify the choice of the used functions in the next section.

6. Results and Discussion

We have applied our proposed neural network to real hyperspectral data, described in section 2. Since there are several parameters involved in the training algorithm (described in section 4); in this section we analyze the influence of those parameters in the process of class prototype extraction. In particular, the parameters that we consider in the present study are the number of iterations until convergence of the neural network is reached, the size of neighborhood function γ centered around the winning neuron, which is determined by $\sigma(t)$, and the number of neurons in the output layer of the neural network.

The experiment is performed as follows. We train the network with all the hyperspectral signatures of the image. During the learning stage, we go through all the pixels of the image starting from a random pixel that is different in each of the iterations. Once class prototypes have been extracted, each pixel is classified and the confusion matrix [14] is obtained; this matrix allows us visualize the winner neuron density for each class and can be used as one different aproach of the Umatrix [15],[16].

The characteristics of the confusion matrix provide us with a comprehensive visualization of distribution in N-dimensional space, and may indicate the accuracy of the classification. Since each column corresponds to an output neuron, if in the same column, high values for different classes are presents, the overall accuracy of the classification should be low. In order to measure the degree of accuracy of the classification, we propose the following metric based on the topology of the confusion matrix:

$$E_i = 100 \frac{X_{mi}}{X_{si}} \qquad X_{mi} = \max_i (X_{ij}) \qquad X_{si} = \sum_j X_{ij} \qquad (6)$$

X_{mi} is the maximum value for a column of the confusion matrix and X_{si} is the sum of all the values in that column. E_i provides information about the capacity of each neuron to discriminate between the classes, and can be averaged for all the neurons in the network, providing a general measure about the accuracy of the classification.

Next, some results obtained for the hyperspectral data described in section 2 are provided. In the experiment, we have considered 16 neurons in the output layer, $\sigma_o = 2$ and 100 iterations. Fig. 5 shows the resulting classification provided by the neural network along with a greyscale representation of the confusion matrix obtained.

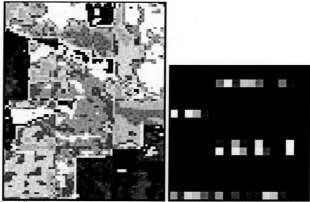

Fig. 5. Resulting classification and confusion matrix for the Indian Pines dataset considering 100 iterations, $\sigma_o = 2$ and 16 neurons in the output layer.

A favorable result would be obtained if neurons activate exclusively for a particular class, discriminating this class from the others. In the confusion matrix, this can be graphically expressed as a row for which several columns present high values. In Fig. 5 we can appreciate this situation at four different rows, indicating that there are four major classes in the image. The fact that column values overlap indicates an inaccurate classification. Another indicator of the quality of the classification is the continuity of bright zones in the confusion matrix. In this experiment, the topology of the resulting classes is not preserved since we can appreciate several discontinuities in the learnt classes. The overall accuracy of the classification obtained in this experiment was 60% according to the measure provided in equation 6. These results are obtained when we have signatures that don't belong to any class, if these signatures are not included, the accuracy increases to 80%.

One more detailed discussion of these results can be obtained in [17], from these results we can conclude two suggestion to modify the SOM learning algorithm:

1. Use the confusion matrix in the learning phase to avoid the misclassification problems associated with several winner neurons for the same learning vector. The unsupervised confusion matrix CM will be one MxM matrix and can be filled increasing for each learning vector, whose winner neuron is I, the value of $CM(I,j) = CM(I,j) + S_j(\mathbf{X}_k)$. The high values of $CM(I,j)$ should be for I=j, in the other hand must be avoided by one hard competition between I and j neurons.
2. Take into account for the hyperspectral classification the high order statistics [8] collected during the learning phase, one possibility is to store different SOM neural networks parameters for each pair of neurons I,j:
    a. $\gamma_{I,j}$
    b. $\eta_{I,j}$
    c. $\sigma(t)_{I,j}$ $\sigma_{o,I,j}$

7 Conclusions

We have presented a new approach to unsupervised classification of hyperspectral images using a Self Organizing Map. The overall performance of the method has been tested by its application to real hyperspectral data. The availability of ground truth allows us to introduce a new statistical measure to quantify the accuracy of the resulting classification. Since the training stage of the neural network incorporates several parameters, we have studied the influence of some of these parameters on the final result.

Acknowledgements

We would like to thank Applied Information Sciences Branch for their help and support during our visits to NASA/Goddard Space Flight Center. Fundings from Junta de Extremadura (PRI Program, IDUAP Grant) and European Community (LFR Program, TEITORS Grant) are also gratefully acknowledged.

References

[1] Greeen, R.O., Editor, AVIRIS Earth Science Workshop Proceedings, 1988-2000. Available at http://makalu.jpl.nasa.gov/
[2] Ifarraguerri, A.; Chang, C.-I. "Multispectral and Hyperspectral Image Analysis with Convex Cones". *IEEE Trans. Geoscience and Remote Sensing*, Vol. 37 Issue 2 Part 1, March 1999, 756–770.
[3] Jimenez, L.O., Morales-More ll, A., Creus, A., "Classification of Hyperdimensional Data Based on Feature and Decision Fusion Approaches Using Projection Pursuit, Majority Voting, and Neural Networks ", IEEE Trans. Geoscience and Remote Sensing , Vol. 37 Issue 3 Part 1, May 1999,. 1360–1366.
[4] Richardson L and Kruse F.A., "Identification and Classification of mixed phytoplankton assemblages using AVIRIS image derived spectra" *Summaries of the VIII JPL Airborne Earth Science Workshop* . (1999), 339-347
[5] Parra L, Spence C., Sadja P, Ziehe A., Müller K-R, "Unmixing Hypersepectral Data"Proc. *Advances in Neural Information Processing System 12, Proc. Of the 1999 Conference,* MIT Press (1999) 942-948
[6] Kohonen, T., *Self-Organizing Maps* (2 nd ed.) , Springer Series in Information Science. (1997)
[7] Bruske, J and Merényi, E.: "Estimating the Intrinsic Dimensionality of Hyperspectral Images ". *Proc. European Symposium on Artificial Neural Network, ESANN'99* ,Bruges, Belgium, (1999) , 105-110 .
[8] Taudjin S and Landgrebe D., Classification of High Dimensional Data with Limited Training Samples, Doctoral Thesis, School of Elelectrical Engineering and Computer Science, Purdue University. (1998).
[9] Martínez P., Pérez R.M., Aguilar P.L., Bachiller, P. and Diaz, P., "A Neuronal Tool for AVIRIS Hyperspectral Unmixing", *Summaries of the VIII JPL Airborne Earth Science Workshop* , JPL/NASA, (1999), 281-286.
[10] Aguilar, P.L., Pérez, R.M., Martínez, P., Bachiller, P., Merchán, A., "Spectra Evaluation and Recognition in the Mixture Problem Using SOFM

Algorithm", *Proc. International Symposium on Engineering of Intelligent Systems, (EIS'98)*, Vol. 2 , (1998), 118-124.
[11] Aguilar, P.L., Plaza, A., Martínez, P., Pérez, R.M., "Endmember Extraction by a Self-Organizing Neural Network on Hyperspectral Images", *Proc. International Conference on Automation, Robotics and Computer Vision*, Nanyang Technological Institute, Singapore, (2000).
[12] Aguilar P.L , Martínez, P., Pérez R.M., Hormigo, A., "Abundance Extractions from AVIRIS Images Using a Self Organizing Neural Network", *Summaries of the IX JPL Airborne Earth Science Workshop*, JPL/NASA (2000), 281-286,.
[13] Aguilar P.L., "Cuantificación de Firmas Hiperespectrales Usando Mapas Autoorganizativos", Ph. D. Thesis, Escuela Politécnica, Universidad de Extremadura, (Chapter 5), 2000.
[14] Chuvieco E., Elementos de Teledetección Espacial, Ed Rialp (2000)
[15] Ultsch, A. and Siemon, H.P., "Kohonen's Self Organizing Feature Maps for Exploratory Data Analysis", *Proc. ICNN'90 International Neural Network Conference* (1990), 305-308.
[16] Antonille, S. and Gualtieri, J.A., "Visualizing Clusters in High-Dimensional Data with a Kohonen Self Organizing Map". *Summaries of the IX JPL Airborne Earth Science Workshop*, JPL/NASA (2000) 281-286.
[17] Martínez, P, Gualtieri, J.A., Aguilar, P.L., Pérez R, Linaje M, Preciado J.C. , Plaza A., "Hyperspectral image classification using self-organizing map", *Summaries of the X JPL Airborne Earth Science Workshop*, in press, JPL/NASA, 2001.

Classification of the Images of Gene Expression Patterns Using Neural Networks Based on Multi-valued Neurons

Igor Aizenberg[1], Ekaterina Myasnikova[2] and Maria Samsonova[2]

[1] Neural Networks Technologies (NNT) Ltd., 155 Bialik st., Ramat-Gan, 52523 Israel,
igora@nnt-group.com; igora@netvision.net.il (IA)
[2] Institute of High Performance Computing and Data Bases ,118 Fontanka emb.,
St.Petersburg, 198005 Russia,
myasnikova@fn.csa.ru (EM), samson@fn.csa.ru (MS),

Abstract. Multi-valued neurons (MVN) are the neural processing elements with complex-valued weights and high functionality. It is possible to implement an arbitrary mapping described by partial-defined multiple-valued function on the single MVN. The MVN-based neural networks are applied to temporal classification of images of gene expression patterns, obtained by confocal scanning microscopy.

1 Introduction

Ability of neural networks to accumulate knowledge about objects and processes using learning algorithms makes their application in pattern recognition very promising and attractive [1]. In particular, different kinds of neural networks are successfully used for solving the image recognition problem [2].

Neural networks based on multi-valued neurons have been introduced in [3] and then developed in [4-9]. A comprehensive observation of multi-valued neurons theory, their learning and applications is given in [6]. Multi-valued neural element (MVN) is based on the ideas of multiple-valued threshold logic [6]. Its main properties are ability to implement arbitrary mapping between inputs and output described by partially defined multiple-valued function, quickly converging learning algorithms based on simple linear learning rules and complex-valued internal arithmetic.

Several kinds of MVN-based neural networks have been proposed for solving the image recognition problems. Different models of associative memory have been considered in [3, 4, 6, 7, 8]. An approach to image recognition, which will be used here, has been introduced in [5] and then also considered and developed in [6, 9]. This approach is based on the following. Since it is always difficult to present the image description, which then could be used for the learning, in some formal way, a nice solution is objectification of the image presentation using some objective procedure. Jump from the image representation in spatial domain to the representation in frequency domain is a good way to this objectification. Nature of this data presentation is clear: since in frequency domain the signal energy is concentrated in a small number of the low frequency part spectral coefficients, it makes possible to use exactly these coefficients as objective description of the signal. Taking into account that multi-valued neuron operates with the complex-valued data, it is natural to use Fourier transform basis for decorrelation of the images that have to be recognized.

In our study MVN-based neural networks are applied for classifying the objects, which vary continuously over time. A peculiarity of this type of data, which makes their classification especially difficult, is the impossibility to subdivide the dataset unambiguously into a certain number of discrete well-defined classes. Here we proceed from the preliminary classification, which was performed manually on the basis of visual inspection of objects and hence was somewhat arbitrary. A temporal class was operationally defined as a set of objects indistinguishable to a human observer, but that in turn suggested that assigning the borders of these classes was also quite arbitrary. Therefore we cannot ever say to what extent the recognition results reflect accuracy of the automatic classification, the hand classification, or both.

In this paper we consider the dataset of images of gene expression patterns, obtained by confocal scanning microscopy [10]. This is a new promising approach for acquisition of quantitative data on gene expression at the resolution of a single cell. Gene expression data are of crucial importance for elucidation of mechanisms of cell functioning, as well as for the early diagnosis of many diseases.

2 Description of the Data

We perform temporal classification of images of genes expression patterns controlling segmentation in the fruit fly *Drosophila*, which is a model organism for molecular biology studies. Like all other insects, the body of the *Drosophila* is made up of repeated units called segments. During the process of segment determination a fly embryo consists of a roughly prolate spheroid of about 5000 nuclei. Genes that act to determine segments are expressed in patterns that become more spatially refined over time. One can view each gene's expression pattern as a collection of "domains" (stripes), each of which is a region of expression containing one maximum (Fig 1). In the experiments gene expression was recorded by confocal scanning of embryos stained with fluorescence tagged antibodies. The obtained images were subjected to image segmentation procedure to obtain the data in terms of nuclear location [11] and then rescaled to remove a nonspecific background signal. In the processed image the nuclei are presented by single pixels with a fluorescence intensity proportional to the average value of gene expression in the respective nucleus.

Human observers classify the developmental stage of an embryo by careful study of its pattern, since each stripe possesses its own features at any stage of an embryo development. In such a way 809 embryos were subdivided into 8 temporal classes [12] (their representatives are shown in Fig.1). Each embryo was allocated to one of the temporal classes on the basis of thorough and extensive visual inspection of the expression pattern of the *eve* gene, which is highly dynamic. We selected embryos for scanning without regard for age, so we expect our dataset to be uniformly distributed in time. The 8 classes were approximately equally populated.

The evolution of the *eve* expression patterns during the segment determination are presented in Fig.1. Time classes 1, 2, and 3 do not have seven well-defined stripes and the number and location of stripes changes rapidly.

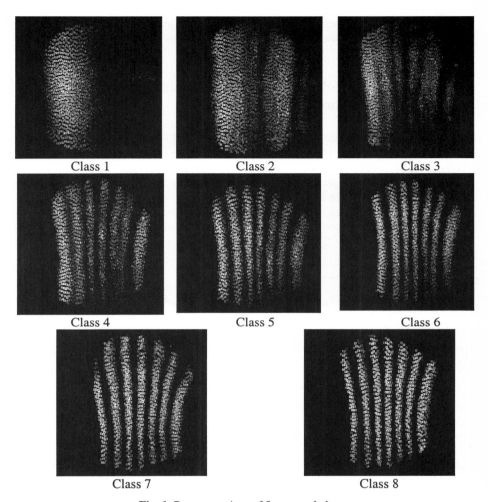

Fig. 1. Representatives of 8 temporal classes

The remaining groups (classes 4 to 8) do have seven well-defined stripes. After all the stripes are clearly visible their intensities increase in the posterior portion of the embryo. By the end of the period, all stripes have reached maximum and equal intensity and maximum sharpness. Time classes 1 to 3 could be grouped according to the number of individual stripes, which have formed. The remaining groups (classes 4, 5, 6, 7 and 8) all have seven well-defined stripes and were classified by features of the overall pattern.

In our preliminary study we performed the time classification of expression patterns for the 1D data extracted from the central 10% of an embryo along the horizontal midline. We applied the Discriminant function analysis using a fixed set of features for classification [13]. This method turned to be not very reliable and therefore the results were regarded just as preliminary. To improve the recognition we need not only to extend our study to two-dimensional data but to apply a more advanced classification method which will make an automated choice of the features responsible for the distribution of objects over the classes.

3 Multi-valued Neuron and MVN-Based Neural Network for Image Recognition

Multi-valued neuron (MVN) has been introduced in [3]. The learning algorithms for MVN have been presented in [3, 4]. The most comprehensive observation of MVN, its theoretical aspects, learning and properties is presented in [6].

We would like to remind here some key moments of the MVN theory (mathematical model of the MVN and its learning).

MVN [3, 6] performs a mapping between n inputs and single output. The performed mapping is described by multiple-valued (k-valued) function of n variables $f(x_1,...,x_n)$ via their representation through $n+1$ complex-valued weights $w_0, w_1, ..., w_n$:

$$f(x_1,...,x_n) = P(w_0 + w_1 x_1 + ... + w_n x_n), \qquad (1)$$

where $x_1,...,x_n$ are neuron's inputs or in other words variables, on which performed function depends (values of the function and of variables are also coded by complex numbers which are k-th roots of unity: $\varepsilon^j = \exp(i2\pi j/k)$, $j \in \{0, k-1\}$, i is an imaginary unity. In other words, values of the k-valued logic are represented as k-th roots of unity: $j \to \varepsilon^j$). P is the activation function of the neuron:

$$P(z) = \exp(i2\pi j/k), \text{ if } 2\pi j/k \le \arg(z) < 2\pi (j+1)/k, \qquad (2)$$

where $j=0, 1, ..., k-1$ are values of the k-valued logic, $z = w_0 + w_1 x_1 + ... + w_n x_n$ is the weighted sum, $arg(z)$ is the argument of the complex number z. So, if z belongs to the j-th sector, on which the complex plane is divided by (2), neuron's output is equal to ε^j, (Fig. 2).

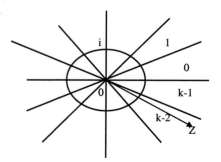

Fig. 2. Definition of the MVN activation function

MVN has some wonderful properties that make it much more powerful than traditional artificial neurons. The representation (1)–(2) makes possible implementation of the input/output mappings described by arbitrary partially defined multiple-valued functions. Such a possibility to implement arbitrary mappings on the single neuron gives an opportunity to develop networks not to perform complicate mappings, but definitely to solve the complicate applied problems.

Another important property of the MVN is simplicity of its learning. Theoretical aspects of the learning, which are based on the motion within the unit circle, have been considered in [6]. If the desired output of MVN on some element from the learning set is equal to ε^q then the weighted sum should get into the sector number q (see Fig 2). But if the actual output is equal to ε^s then the weighted sum has gotten into sector number s. A learning rule should correct the weights to move the weighted sum from the sector number s to the sector number q. We use here the following correction rule for MVN learning, which has been proposed in [6]:

$$W_{m+1} = W_m + C_m(\varepsilon^q - \varepsilon^s)\overline{X}, \qquad (3)$$

where W_m and W_{m+1} are current and next weighting vectors, $\overline{X}$ is a vector of the neuron's input signals (complex-conjugated), ε is a primitive k^{th} root of unity (k is chosen from (2)), C_m is a scale coefficient, q is a number of the desired sector on the complex plane, s is the number of the sector, to which the actual value of the weighted sum has gotten, n is a number of neuron inputs. The learning algorithm based on the rule (3) is very quickly converging. It is possible to find such a value of k in (2) that (1) will be true for given function f, which describes the mapping between neuron's inputs and output [6].

Let us consider N classes of objects, which are presented by images of $n \times m$ pixels. The problem is formulated in the following way: we have to create recognition system based on the neural network, which makes possible successful classification of the objects by fast learning on the minimal number of representatives from all classes. MVN-based neural network and methods of its organization have been proposed as such a system in [5]. This model has been also considered in [6, 9]. Frequency domain representation of data is very important for this model. All 2D images are replaced by their Fourier spectrums. Moreover, taking into account that spectrum phase contains more information about the object presented by signal than amplitude [14], only spectrum phase is used for objects representation.

In terms of neural networks to classify object we have to train a neural network with the learning set containing the spectra phase of representatives of our classes. Then the weights obtained by learning will be used for classification of unknown objects.

Representation of the recognized objects by phases of Fourier spectra coefficients is idle for MVN-based neural network. Since inputs and outputs of MVN are the complex numbers (moreover they are roots of unity), it is natural to use phases as these inputs. To put the inputs into the correspondence with value of k (value of k-valued logic or number of sectors on the complex plane) in (2), the procedure based on (2) should be applied instead of normalization:

$P(phase) = \exp(i2\pi j/k)$, if $2\pi j/k \leq phaze < 2\pi (j+1)/k$. (4)

MVN-based single-layer neural network, which contains the same number of neurons as the number of classes we have to classify has been considered in [5, 6, 9]. Each neuron has to recognize pattern belonging only to its class and to reject any pattern from any other class. This system has been successfully tested on example of face recognition [9]. The best results have been obtained experimentally, when for classification of the pattern as belonging to the given class the first $l=k/2$ sectors on the complex plane (see (2)) have been reserved. Other $k/2$ sectors correspond to the rejected patterns.

4 Recognition and Classification. Simulation Results

To solve the classification problem we used the MVN-based neural network described above. Since the images from neighbor classes are often similar to each other, organization of the network has been modified in the following "dichotomic" way.

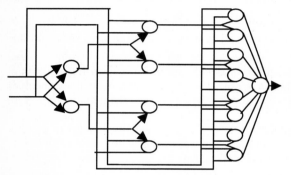

Three-layered network has been used. Two neurons of the first layer classify patterns belonging to the classes 1-4 and 5-8, respectively. Four neurons of the second layer classify pattern belonging to the classes 1-2, 3-4, 5-6 and 7-8, respectively. Finally, 8 neurons of the third layer classify pattern belonging to the corresponding particular classes (Fig. 3), and the single neuron of output layer analyses the results coming from the neurons of the third layer.

Fig. 3. Neural network for solving the problem

To create representative learning sets for all the neurons, we used images that have been *a priori* correctly classified as typical for the particular class. This classification is based on objective biological view. Since it is natural to teach the neural network using maximally objective data, definitely the typical images have been included to the learning set (from 28 up to 35 images per class) and other images have been used for testing. All the neurons have been taught using learning algorithm based on the rule (3). Convergence of the learning has been gotten quickly (10-15 seconds per neuron using software implementation on Intel Pentium-III-700 processor). The results of testing are given in the Tables 1-3 for the layers 1-3, respectively. The experimental results that are illustrated by Tables 1-3 have been obtained for $k=2048$ in (2) (2048-valued logic) and inputs taken from the Fourier phases coefficients corresponding to the frequencies from 1 to 5 (60 neuron inputs).

Table 1. The results of testing for the 1st layer of neurons.

Class #	1	2	3	4	5	6	7	8
Number of Images	75	61	99	53	76	123	98	84
Recognized	75 100%	60 98.3%	95 96%	40 75.4%	61 80.2%	119 96.7%	97 98.9%	84 100%
Unrecognized	0	1	4	5	6	4	1	0
Misclassified	0	0	0	8	9	0	0	0

Table 2. The results of testing for the 2d layer of neurons.

Class #	1	2	3	4	5	6	7	8
Number of Images	75	41	66	84	114	87	63	84
Recognized	70 93.3%	32 78%	53 80%	77 91.6%	113 99.1%	65 74.7%	59 93.6%	82 97.6%
Unrecognized	1	0	0	2	1	0	1	2
Misclassified	4	9	13	5	0	22	3	0

Table 3. The results of testing for the 3d layer of neurons (Final classification results).

Class #	1	2	3	4	5	6	7	8
Number of Images	75	41	66	53	76	87	63	50
Recognized	61 81.3%	30 73.2%	49 74.2%	45 84.9%	59 77.6%	61 70.1%	46 73%	38 76%
Unrecognized	3	0	0	1	1	2	1	0
Misclassified	11	11	17	7	16	24	16	12

5. Conclusions

We have performed a temporal classification of *Drosophila* embryos on the basis of images of their gene expression patterns, obtained by confocal microscopy. The specificity of the data varying continuously over time causes certain difficulties in their classification. In our preliminary study all embryos were subdivided into eight temporal classes by visual inspection, but due to uniformity of distribution of images over the whole observation period, the classes and their borders we assigned somewhat arbitrary. The inaccuracies of the preliminary hand classification may considerably complicate the recognition problem.

The MVN-based neural network applied to classification of the images showed good results. High accuracy of classification confirms its efficiency for image recognition. It was shown that frequency domain representation of images is highly effective for their description with a recognition purpose. It is natural and effective to use phases of the Fourier spectral coefficients as inputs of MVN-based neural network.

References

1. Haykin S. *Neural Networks. A Comprehensive Foundation.* Macmillan College Publishing Company, New York, 1994.
2. Leondes C.T. (Ed.) *Image Processing and Pattern Recognition.* Academic Press, San Diego/London/Boston/New York/Sydney/Tokyo/Toronto, 1997.
3. Aizenberg N.N., Aizenberg I.N "CNN based on multi-valued neuron as a model of associative memory for gray-scale images", *Proc. of the 2-d IEEE International Workshop on Cellular Neural Networks and their Applications, Munich, October 12-14* (1992), pp. 36-41, 1992.
4. Aizenberg N.N., Aizenberg I.N.. Krivosheev G.A. "Multi-Valued Neurons: Learning, Networks, Application to Image Recognition and Extrapolation of Temporal Series", *Lecture Notes in Computer Science*, Vol. 930, (J.Mira, F.Sandoval - Eds.), Springer-Verlag, 389-395, 1985.
5. Aizenberg I.N. "Neural networks based on multi-valued and universal binary neurons: theory, application to image processing and recognition", *Lecture Notes in Computer Sciense*, vol. 1625 (B. Reusch - Ed.), Springer-Verlag, 1999, pp. 306-316.
6. Aizenberg I.N., Aizenberg N.N., Vandewalle J. *"Multi-valued and universal binary neurons: theory, learning, applications"*, Kluwer Academic Publishers, Boston/Dordrecht/London, 2000.
7. Jankowski S., Lozowski A., Zurada M. "Complex-Valued Multistate Neural Associative Memory", *IEEE Trans. on Neural Networks*, Vol. 7, (1996), 1491-1496.
8. Aoki H., Kosugi Y. "An Image Storage System Using Complex-Valued Associative Memory", *Proceedings of the 15th International Conference on Pattern Recognition, Barcelona, Spain September 3-8, 2000*, IEEE Computer Society Press, vol. 2, pp. 626-629, 2000.
9. Aizenberg I.N., Aizenberg N.N. "Pattern recognition using neural network based on multi-valued neurons", *Lecture Notes in Computer Sciense*, vol. 1607-II (J.Mira, J.V.Sanches-Andres - Eds.), Springer-Verlag, pp. 383-392, 1999.
10. D.Kosman, J.Reinitz and D.H.Sharp "Automated Assay of Gene Expression at Cellular Resolution", *Proceedings of the 1998 Pacific Symposium on Biocomputing*, World Scientific Press, Singapore, pp. 6-17, 1997.
11. D.Kosman, S.Small and J.Reinitz "Rapid Preparation of a Panel of Polyclonal Antibodies to *Drosophila* Segmentation Proteins", *Development, Genes, and Evolution*, **208**, pp. 290-294, 1998.
12. E.Myasnikova, A.Samsonova, K.Kozlov, M.Samsonova and J.Reinitz " Registration of the Expression Patterns of Drosophila Segmentation Genes by Two Independent Methods", *Bioinformatics,* **17**, pp.3-12, 2001.
13. E.Myasnikova, D.Kosman, J.Reinitz and M.Samsonova " A Method for the Spatio-temporal Registration of the Expression Patterns of Drosophila Segmentation Genes ", *Proc. of the 1st Conf. on Modelling and Simulation In Biology, Medicine and Biomedical Engineering*, ESIEE, Noisy-le-Grand, France, pp.63-67,1999.
14. Oppenheim A.V. and. Lim S.J "The importance of phase in signals", *Proc. IEEE*, Vol. 69, (1981), pp. 529-541.

Image Restoration Using Neural Networks

Souheila Ghennam and Khier Benmahammed

Faculty of Engineering Sciences, Department of Electronic, Ferhat Abbas University,
Setif, Algeria
ghsouhi@caramail.com, and khierben@ieee.org

Abstract. For resolving a restoration problem of degraded and noisy image, we investigate the Hopfield neural network, and we employ the eliminating highest error EHE criterion in intention to improving performances of network. Moreover, with the purpose to make a better restoration, we take in consideration a human perception in restoration process. To do this, we introduce an adaptive regularization scheme, with contribution of a local statistical analysis, to assigning each pixel one regularization parameter regarding to its spatial activity. Due to various values of regularization parameter, this scheme permit us expanding the one network to a network of network NON, which we subsequently elucidate its analogy with the human visual system, a cortex.

1 Introduction

In majority of restoration methods, image is considered as a stationary process, which is incorrect since image can contain edges, textures and flat regions. Thus such methods can only produce a sub-optimal solution. We recall that these methods deal with a minimization of cost function based on a mean square error MSE, which reflects a power of signal error and has a little relationship to human visual perception, seeing that all pixels are given equal property regardless to theirs natures. In this paper, in order to produce a best solution for image restoration problem, we introduce some features of visual perception in restoration process.

Degraded and noisy image is frequently formulated by a linear equation: [1][2]

$$g = H \times f + n \qquad (1)$$

where g, f and n are $MN \times 1$ vectors representing respectively a degraded and original images of $M \times N$ size both and an additive gaussian noise. H is a Toeplitz matrix of $MN \times MN$ size representing a degradation mask $h(i,j)$. Knowing g and H, image restoration seeks into recovering original image by minimizing a quadratic cost function:[1][2]

$$\begin{aligned} J(f) &= \frac{1}{2} \left\| g - H\widehat{f} \right\|^2 + \frac{1}{2} \lambda \left\| C\widehat{f} \right\|^2 \\ &= \frac{1}{2} \widehat{f}^T \left(H^T H + \lambda C^T C \right) \widehat{f} - \left(H^T g \right)^T \widehat{f} \end{aligned} \qquad (2)$$

where C is a $MN \times MN$ matrix associated to a high pass operator $c(i,j)$, and λ is a parameter, which regularize the contribution of smoothing $\frac{1}{2}\lambda \left\|C\widehat{f}\right\|^2$ into inversion process $\frac{1}{2}\left\|g - H\widehat{f}\right\|^2$. In (2) the regularization parameter λ is the same for all pixels, so the image is considered as a stationary process. However, when humans observe the differences between two images, they do not give consideration to the differences in individual pixel level values. Instead humans are concerned with matching edges, flat regions and textures between the two images. In this paper, taking in count visual perception, we consider image as a set of dissimilar regions in statistically sense and those we will treat with different strategies or parameters. To do this, we use an adaptive regularization scheme, with contribution of statistical analysis, to assigning each pixel one regularization parameter according to its spatial activity among its neighborhood pixels.

We investigate the Hopfield network, which is a finite neural model for optimization problem such as image restoration. In addition, by dealing with the regularization scheme, a network will own a tremendous of sub-matrix of weighs instead one matrix, each one corresponds to a sub-network of neurons those the task is to restore pixels of same spatial activity. Hence, we arrive to construct a network of networks, which share same structure with human cortex or a part of it.

The paper is organized as follows: in section 2, image restoration by Hopfield network is presented and the EHE criterion is related. In section 3, adaptive regularization scheme is introduced with contribution of statistical analysis in computing different regularization parameters values. In section 4, we expand a Hopfield network to network of networks, and we give its analogy with the human cortex. The tests and results are reported and discussed in section 5. Finally, section 6 concludes the paper.

2 Image Restoration Using Hopfield Model and EHE Criterion

2.1 Image Restoration using Hopfield Model

The Hopfield model network is constituted by one layer of neurons completely or partially interconnected. Each neurone i is characterized by a bias b_i, an activation function φ, and a weight interconnection w_{ij} between him and any another neurone j. For dealing with gray image, a neurone state is a discrete value between 0 to 255 [3]. The fitting training mode for restoration problem is a sequential updating mode, seeing that network energy at such mode: [3]

$$E^S(t) = -\frac{1}{2}\sum_{i=1}^{MN}\sum_{j=1}^{MN} w_{ij}v_i(t)v_j(t) - \sum_{i=1}^{MN} b_i v_j(t)$$
$$= -\frac{1}{2}v^T(t)Wv(t) - b^T v(t) \tag{3}$$

is identical to $J(\widehat{f})$. Minimizing $E^S(t)$ occurs minimizing $J(\widehat{f})$, and thus the fixed state of network is the solution of restoration problem. A network parameters are determined by just comparison between (2) and (3): a bias vector $b = H^T g$, a state vector $v = f$, and a weights matrix $W = -\left(HH^T + \lambda CC^T\right)$. And the training of network is proceeded by applying at input $u_i(t)$ of each neurone i a negative energy gradient function according to vector state $v(t)$ such:

$$u_i(t) = -\nabla E_i^S(t) = \left(\frac{\partial E^S(t)}{\partial v(t)}\right)_i = \sum_{j=1}^{MN} w_{ij} v_j(t) + b_i \quad (4)$$

next the neurone is updated in relation with a following generalized updating rule GUR: [3][4][5]
GUR

$$vi(t) = \varphi\left(v_i(t) + \Delta v_i(t)\right) \in [1, 255] \quad (5)$$

with

$$\Delta v_i(t) = \Psi(u_i(t)) = \Psi\left(\sum_{j=1}^{MN} w_{ij} v_j(t) + b_i\right) \quad (6)$$

where

$$\Psi(x) = \begin{cases} +1 & x > +\theta_i \\ 0 & -\theta_i \leq x \leq +\theta_i \\ -1 & x < -\theta_i \end{cases} \quad (7)$$

and

$$\varphi(x) = \begin{cases} 255 & x > 255 \\ x & 0 \leq x \leq 255 \\ 0 & x < 0 \end{cases} \quad (8)$$

□

with a threshold θ_i is given by: $\theta_i = \frac{1}{2} w_{ii}$, and where w_{ii} are a diagonal elements of weights matrix W, which should be symmetric to guarantees convergence of algorithm. [3][4][5]

2.2 EHE Criterion in Image Restoration Based Hopfield Network

Processing with GUR end up to one amongst tremendous minima of energy function, whom network can reach. This minimum does not always represent the best solution of restoration. This can be reproached to executing fashion of processing. Indeed, image is skimmed through from top to bottom, without giving consideration to energy gradient values at crossed neurones. And usually neurones when gradient energy is high are treated before those when it is low, so the lasts neurones go far theirs fixed states after be so nearest them. In order to correctly converge to a fixed state, we employ the EHE criterion [5], which aims to minimize highest error at each iteration, by helping network in selection of neurones to be updated. This criterion is applied in actual network by treating neurones in decreasing order according of gradient values. To do this, at each iteration is defined a set of neurones: [5]

$$T(t) = \{i \ / \ u_i < -\theta_i \ or \ u_i > +\theta_i\} \tag{9}$$

those afterwards are rearranged in decreasing order and updated according to GUR.

3 Adaptive Regularization

3.1 Adaptive Regularization Scheme

In order to elaborate an adaptive regularization scheme in actual network, the weights are varied for implementing different values of λ according to different spatial activities in image. To substantiate this issue, we first generalize the quadratic model (2), as:

$$J(f) = \frac{1}{2} \left\| g - H\widehat{f} \right\|^2 + \frac{1}{2} \left\| \Lambda C\widehat{f} \right\|^2 \tag{10}$$

where $\Lambda = \begin{bmatrix} \sqrt{\lambda_1} & 0 & \cdots & 0 \\ 0 & \sqrt{\lambda_2} & & 0 \\ \vdots & & \ddots & \vdots \\ 0 & 0 & \cdots & \sqrt{\lambda_{MN}} \end{bmatrix}$ is a diagonal regularization matrix and where λ_i is a regularization parameter associated to pixel i. By relating (10) to the network energy formula (3), the weights between neurons i and j is given by:

$$w_{ij} = \sum_{p=1}^{MN} h_{pi} h_{pj} - \lambda_i \sum_{p=1}^{MN} c_{pi} c_{pj} \tag{11}$$

where h_{pq} and c_{pq} are respectively the elements of H and C matrices.

The new weights matrix W lacks symmetry that is an indispensable condition for convergence. To turn away this inconvenient, W is replaced by a symmetric equivalent version $\overline{W}$, such: $\overline{W} = \dfrac{W + W^T}{2}$. Note that $\overline{W}$ is symmetric since: $\overline{W}^T = \left(\dfrac{W + W^T}{2}\right)^T = \dfrac{W + W^T}{2} = \overline{W}$, and energy function remains the same, as: $E = -\frac{1}{2}\widehat{f}^T\overline{W}\widehat{f} - b^T\widehat{f} = -\frac{1}{2}\widehat{f}^T\dfrac{W + W^T}{2}\widehat{f} - b^T\widehat{f} = -\frac{1}{4}\widehat{f}^TW\widehat{f} - \frac{1}{4}\widehat{f}^TW^T\widehat{f} - b^T\widehat{f}$
$= -\frac{1}{4}\widehat{f}^TW\widehat{f} - \left(\frac{1}{4}\widehat{f}^TW\widehat{f}\right)^T - b^T\widehat{f} = -\frac{1}{2}\widehat{f}^TW\widehat{f} - b^T\widehat{f}$

3.2 Statistical Analysis

Human eye perceives degradation much in edges and texture regions, nevertheless it can't clearly detect degradation in flat regions. These remarks encouraged us the use of local statistical analysis for computing appropriate λ for each pixel. This analysis is more suitable because it take more consideration to the spatial activity of pixel inside its neighborhoods, taking in part rest of image which gives nothing information to considered pixel. In addition, seeing that human eye is more able to discern noise increase in regions of low variance than of high variance. Hence implicit form to computing λ_i is therefore a log-linear function: $\lambda_i = \alpha \cdot \log(S_i) + \beta$, where S_i is a local variance of a set of $P \times P$ pixels centered into considered pixel i.

4 Network of Network: NON

In image, numerous pixels may have same spatial activity; we will assign them to a same homogeneous area in statistically sense. Hence, image will be setting into K homogeneous areas, and by assigning each pixel one λ_i, we consequently assign each homogenous area $k(=\overline{1...K})$ one λ_k. We begin by rearranging image by indexing pixels of homogeneous area, we so obtain a new image vector $\overline{f}$, and (10) can be written as:

$$J(f) = \frac{1}{2}\left\|\overline{g} - \overline{H}\overline{f}\right\|^2 + \frac{1}{2}\left\|\overline{\Lambda C}\overline{f}\right\|^2 \quad (12)$$

where : $\overline{f} = \begin{bmatrix} \widehat{f}_1 \\ \vdots \\ \widehat{f}_K \end{bmatrix}$, $\overline{H} = \begin{bmatrix} H_1 \\ \vdots \\ H_K \end{bmatrix}$, $\overline{C} = \begin{bmatrix} C_1 \\ \vdots \\ C_K \end{bmatrix}$, $\overline{g} = \begin{bmatrix} g_1 \\ \vdots \\ g_K \end{bmatrix}$ and $\overline{\Lambda} = \begin{bmatrix} \Lambda_1 & 0 & \\ 0 & \ddots & 0 \\ & 0 & \Lambda_K \end{bmatrix}$ with $\widehat{f}_k$, H_k, Λ_k, C_k and g_k, for $k = \overline{1...K}$, are respectively a sub-matrices of $\widehat{f}$, $\overline{H}$, $\overline{\Lambda}$, $\overline{C}$ and $\overline{g}$, associated to k^{th} homogeneous area. And with $\Lambda_k = \lambda_k \cdot I$, where I is an identical matrix.

We define: $H_k = \begin{bmatrix} H_{k1} & \cdots & H_{kk} & \cdots & H_{kK} \end{bmatrix}$ and $C_k = \begin{bmatrix} C_{k1} & \cdots & C_{kk} & \cdots & C_{kK} \end{bmatrix}$ for $k = \overline{1...K}$; (12) becomes:

$$J(f) = -\frac{1}{2}\sum_{k=1}^{K}\left\{\left[\overline{f}^T\left(H_k^T H_k + \lambda_k C_k^T C_k\right)\overline{f} + 2g_k^T H_k \overline{f}\right]\right\} + \|\overline{g}\|^2$$

$$J(f) = -\frac{1}{2}\sum_{k=1}^{K}\left\{\widehat{f}_k^T\left(H_{kk}^T H_{kk} + \lambda_k C_{kk}^T C_{kk}\right)\widehat{f}_k \right.$$

$$\left. + \sum_{\substack{l=1 \\ l \neq k}}^{K}\sum_{\substack{m=1 \\ m \neq k}}^{K} \widehat{f}_l^T\left(H_{kl}^T H_{km} + \lambda_k C_{kl}^T C_{km}\right)\widehat{f}_m + 2g_k^T H_k \overline{f}\right\} + \|\overline{g}\|^2 \quad (13)$$

where $\left(H_{kk}^T H_{kk} + \lambda_k C_{kk}^T C_{kk}\right)$ denote intra-connections in k^{th} homogeneous area, while $\left(H_{kl}^T H_{km} + \lambda_k C_{kl}^T C_{km}\right)$ correspond to inter-contributions of areas l and m to area k. Thus, equation (13) represents an expansion of the one actual network into network of networks NON [6]. Additionally to architecture of NON, the operating with regard to human visual perception allows to network a processing that is like the human cortex or a part of it. Before elucidating this analogy basing on (13) and Fig. 1, we give a basic description of the cortex; it is constituted by a set of clusters of neurons, each cluster is trained to decipher some visual information. Neuron cells in same cluster are linked with synapses of identical chemical features, which permit theme to decode visual information with same fashion. The neurons of one cluster are related to neurons of other clusters with different synapses[6]. Now, in our network, by representing a pixel with a neuron, homogeneous area is mapped to a cluster of similar neurons, which are trained with same weights in order to restore pixels of same spatial activity. Moreover, since pixels of different spatial activities are depended with themselves, the neurons of one cluster are linked with neurons of other clusters. Therefore, the aim to restore image with regard to human visual perception, leads us to resulting to neuronal architecture that potentially imitates the human visual system.

5 Results and Discussion

In actual work we report tests proceeded on image (Fig. 2.a) of 128×128 size, deformed with a gaussian blur and distorted by an additive noise of $28db$SNR (Fig. 2.b). To restore image (Fig. 2.b), the following tests are made:

Test 1. Hopfield neural network with one regularization parameter is kept the same for all neurons and it is calculated to adapting the global image structure, and the processing filters. [7]

Test 2. At this time we use the adaptive regularization. The local statistic is applied by computing variance into pixel with 3×3 neighborhoods..

In each test, the network is trained in first time in simple sequential mode and in second time in sequential mode under EHE criterion. During each test is calculated the SNR improvement, and execution terminates when SNR variation become unchangeable.

It seems that expansion of Hopfield network to network of networks, via using the adaptive regularization scheme, allows to network a finest restoration comparatively to whene it is used one constant parameter regularization (Fig. 2)(Table 1), even that it is computed to adapting the global image structure and processing filters. In addition operating with EHE criterion get better results seeing improvement in SNR of $12.24db$, and visually enhancement in test image (Fig. 2).

6 Conclusion

In this paper, we are introducing some features of human visual perception in image restoration process, via using an adaptive regularization scheme. We end

Table 1. Improvement in SNR

Tests	ΔSNR(db) Sequential Mode	Sequential Mode with EHE criterion
1	5.82	5.96
2	10.90	12.24

up at a network of networks NON architecture that imitates a human visual system. Moreover, we are showing the best results that the NON can perform especially when it is conjugated with the EHE criterion, hence we can say that it bear the potential of a natural visual system to recovering lost information.

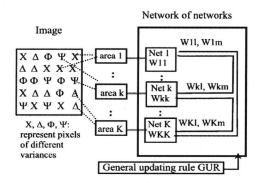

Fig. 1. Structure of network of networks NON

References

1. Gonzalez, R.C.,Woods, R.E.: Digital Image Processing. Addisson-Wesley (1992)
2. Banham, M.R., Katsaggelos, A.K.: Digital Image Restoration. IEEE Signal Processing Magazine. (March 1997) 24-41
3. Paik, J.K., Katsaggelos, A.K.: Image Restoration Using Modified Hopfield Neural network. IEEE Trans. on Image Processing, Vol. 1. (January 1992) 49-63
4. Sun, Y.: A Generalized Updating Rule For Modified Hopfield Neural Network For Quadratic Optimization. Neurocomputing. **19** (1998) 133-143
5. Sun, Y.: Hopfield Neural Network Based Algorithms For Image Restoration and Reconstruction - Part I: Algorithms and Simulation. IEEE Trans. on Signal Processing, Vol. 48. **7** (July 2000) 2119-2131
6. Kang, M. G., Katsaggelos, A. K.: General Choice of the Regularization Functional in Regularized Image Restoration. IEEE Trans. on Image Processing, Vol. 4. **5** (May 1995)
7. Anderson, J.A., Sutton, J.P.: A network of Networks: Computation and Neurobiology. World Congress of Neural Network, Vol. 1.(1995) 561-568

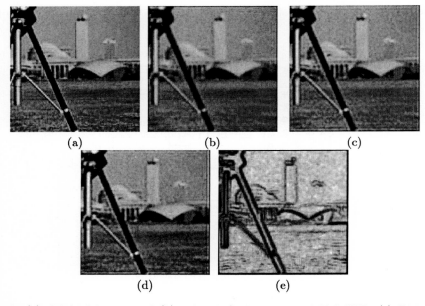

Fig. 2. (a) Original image, and (b) noisy degraded image of 28dbSNR. (c) Restored image by one Hopfield network trained in sequential mode under EHE criterion, the improvement in SNR is of 5.96db . (d) Restored image by network of networks trained in sequential mode under EHE criterion, and with adaptive regularization via local variances image in (e) where white level represents highest variance and black level represents lowest variance. The improvement in SNR is of 12.24DB

Automatic Generation of Digital Filters by NN Based Learning: An Application on Paper Pulp Inspection

Pascual Campoy-Cervera[1], David F. Muñoz-García[1],
Daniel Peña[1], and José A. Calderón-Martínez[1,2]

[1] Department of Automatic Control, Electrical Engineering
and Industrial Computing
Universidad Politecnica de Madrid
C/ Jose Gutierrez Abascal 2, 28006 Madrid, Spain
{campoy, dmunoz, dpena}@disam.upm.es
http://www.disam.upm.es/vison/
[2] Instituto Tecnologico de Aguascalientes, SEP, CONACYT, México
acaldero@disam.upm.es

Abstract. This paper presents an implementation of a digital filtering inspection system applied on a paper pulp sheet production process. The automation of the inspection phase is particularly critical during this process and its solution is highly complex. The system is based on neural network learning, allowing a compromise between resolution and processing speed. The experimental results demonstrating the use of this algorithm for the visual detection of defects in images obtained from a real factory environment are presented. These results show that the developed learning method generates filters that fulfil the required inspection standard.

1 Introduction

1.1 Motivation

This research is motivated by the desire to improve the inspection process at a Paper Pulp Plant in three aspects:

- Automatize the inspection process
- Reduce the time spent in such activity and increase the inspection frequency, therefore the inspection reliability
- Improve the overall productivity ratio by an on line connection to the production central computer
- Increase the quality of the inspection itself

Automated vision inspection systems are the most versatile non-contact inspection systems. They are fast and remarkably accurate, making possible 100% inspection in a modern manufacturing environment. As the level of automation increases in a manufacturing process, the greater is the need for some form of automatic quality monitoring and inspection of the parts that are being produced.

Automatic quality monitoring systems are used to determine the acceptance or rejection of a part or a specific production lot before parts are feed into the next process, they are also used to check the calibration of fixtures and the condition of cutting tools. Without automatic quality monitoring systems in nowadays manufacturing, effectiveness on automation would be questionable. Therefore, an automated inspection system is one of the most important factors for the successful operation of automated manufacturing systems.

There have been many approaches in texture defect detection. Conners et al. [3] proposed texture analysis methods to detect defects in lumber wood automatically. Dewaele et al. [4] used signal processing methods to detect point and line defects in texture images. Lee et al. [5] have recently used neural networks to classify defects through energy and entropy features of steel images. In this study an efficient method based on digital filters obtained from neural networks learning is presented.

The quality control process performed at a paper pulp factory has many stages, like the procurement of the pulp whiteness degree, brightness, defects presence, etc. The objective of these controls is to determine if the product is suitable for business.

2 Defect Detection on Paper Pulp Using ENCE Standard

Impurities classification must be performed according to the normalized sizes from ENCE 404 Standard [6], shown on Table 1 with the appearance percentages for each defect. This standard is more restrictive than the International TAPPI Test Method T 213 om-89 ("Dirt in Pulp"), based on reflective light, and the ISO 7213 ("Pulps Sampling for testing").

Once the real object size and position is determined, a class or defect type is obtained by comparing its size with those in Table 1.

Table 1. Defect types frequency

Defect type	Area (mm)	Appearance (pitch) %	Area (pitch) %	Appearance (shave) %	Area (shave) %
-	smaller than 0.04	-	-	-	-
P	0.04-0.1	44%	13%	9%	1%
C	0.1-0.3	38%	29%	33%	8%
B	0.3-0.8	15%	35%	30%	22%
A	bigger than 0.8	3%	23%	29%	69%

2.1 Previous Manual Method

The human inspector used to place a sheet of paper pulp on a lean transparent crystal surface (500 x 500 mm), with back lighting and searches for defects. On many occasions defects were not found because they were not located on the surface, and since light could be refracted when hitting impurities in internal layers, the aspect and size of the defect could be modified. To solve this problem, the human inspector used to scratch the defect found and once he/she visualized the defect shape, compared it with a template and classified it with the ENCE 404 standard [6].

2.2 Automatic Method

Due to the production line speed (i.e. maximum of 2,100 mm/s), the use of back lighting according to ENCE 404 standard [6], and the resolution needed, TDI (time delay integration) technology is used for image acquisition. The automatic system right now used is shown in Fig. 1. It can be seen the paper going through, the lighting, the camera and a friendly-user system screen.

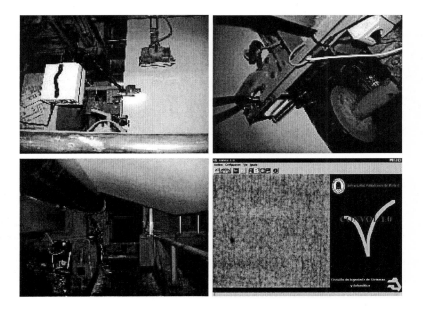

Fig. 1. Automatic system at the plant

3 Digital Filtering and the MLP Neural Net

3.1 Neural Networks

The Multi Layer Perceptron (MLP) can be considered as the most classical and spread out neural net model, either for recognition tasks or other applications. Its operation can be interpreted as a classification of vectors into two classes (i.e. 1 and 0). The frontier in that case is linear and is defined by the perpendicular hyper plane to the weight vector and is located at the distance given from the origin to the perceptron internal threshold, getting as output several values depending on where the characteristic vector of the new image's segmented object is positioned in the hyper plane [1]. When a specific combination of perceptron outputs wants to be assigned to a class and the characteristic vector to be considered as belonging to a different class, it only has to be used an additional perceptron that has as inputs the outputs of the preceding perceptrons; obtaining in this way an MLP structure [2].

3.2 Image Processing by an MLP

From the equation of the process carried out by the perceptron:

$$Y_j = \sigma \left(\sum_{i=1}^{n} W_{ij} * X_i + \omega_{0j} \right) \quad (1)$$

many conclusions would come up. W_{ij} is the jth neuron weight considering the ith input vector element. This equation reminds a functioning equation of a digital filter over an image, where a convolution of these two elements takes place, then a comparison with a threshold of value B_j and finally a classification into two classes (i.e. 1 and 0) is carried out by the sigma function.

A spatial domain is a set of pixels that make up an image; functions for image processing the spatial domain can be represented as:

$$g(x,y) = T[f(x,y)] \quad (2)$$

Where $f(x,y)$ is the input image, $g(x,y)$ is the processed image and T is an f operator, defined over the neighbor pixels of (x,y). The neighborhood of (x,y) can be defined as a squared or rectangular sub-image located at (x,y), which center is moved over the pixels and an operator is applied on each one of them obtaining a $g(x,y)$ value for every position. Techniques that use this matrices are named Mask or Filter Processing, and implies to carry out a sum of the multiplication of the coefficients by the grey levels delimited by the mask.

In both cases (i.e. Perceptron or Filter Processing) there is a product sum. But the Bj element seems to be the difference since it appears in the neuron's equation as well as in the subsequent transformation with the F function. Bj is the slant in the MLP. Every input to the neuron is evaluated with its weights and then by the transformation of the activation functions. With the presence of the slant, the input space could be divided into a less restrictive manner. What it is done indeed, is giving the neuron the possibility of varying the activity threshold instead of fixing it to 0. In conclusion, the slant in the image processing field is equivalent to a typical thresholding. Therefore, the utilization of MLP's in image processing is highly recommended because of its great advantages. Since finding an adequate sampling space means that obtaining a filter is just applying the correspondent training; getting a greater degree of automation.

4 Synthetic Sample Data for Training

4.1 The Process

The most important defects on paper pulp is the appearance of impurities which do not respond to certain pattern, and those arise from different origins. So that, the defects generated does not show fixed schemes in their arrangement. This random arrangement allows to considerate performing the defects sampling during the neural net training in a more efficient way. Since the defects found show an isotropic characteristic, it can be used for sampling, possibly not the full defect image but pixel groups at the same location.

When training the neuron, the main objective is to best characterize the defects aspect. That is why the sampling has to be as exhaustive as possible in order to obtain a mask where most of the main representative characteristics of the defect type are considered. Getting a great deal of samples is good but it is bound by the appearance of defects in the paper pulp, so it is limited by the sampling items analyzed. Therefore, it is interesting to take advantage of the isotropy shown by the defects, in order for a sampling item to produce many samples to train the neural net. In Figure 2, pixels crossed by the lines make up groups that allow characterizing the defects aspect. If one of those lines is taken as an example, the group of pixels would be like the ones shown at the bottom of Figure 2.

As it is observed, this sample contains most of the information that can be obtained from the whole image; but now there are four sub-samples. In this way not only the shortage of sampling items for training is solved, but also the training time is diminished by redesigning the neural net. Since reducing the sample size makes to pass from a matrix to a subset of a matrix, making the weights to be calculated in the neural net to be reduced. Determining the type of a sub-sample to choose is a problem; since anomalous behaviors can arise. For example, if diagonals from the image are taken, a deformation of the defects can be obtained.

These differences would cause errors while training, provoking a miss learning of the defects characteristics. Therefore the image sampling criteria is very important.

4.2 Octants Sampling

A solution is sampling the image by using octants (right of Fig. 2), so the defects shape would not be lost when digitalized. Adequate turns and symmetries have

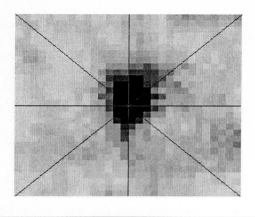 

Fig. 2. Image with a defect, a line crossing a group of pixels and an octant sample

to be taken into account in order for the samples to seem coming from the same reality.

Every defect is sampled in octants an these are sequenced and stored in an array that would finally hold the neural net training units. Introducing octants in the net as sample units, will produce an octant as output that would be used to build the filter.

5 Filter Design for Individual Patterns

5.1 How the Filters Are Obtained

The process starts getting a set of images from manually inspected paper pulp sheets, with results already clearly classified. This image taking includes samples of all different defect types, so that the morphology variety for each type is gathered, as well as samples without defects. Once the different samples are gathered they are used for training the neurons, obtaining throughout synthetic generation (Fig. 3) a set of intelligent digital filters. The CONVOL program performs a convolution of the obtained filters and the image analyzed.

5.2 Normalization

During the image gathering, conditions may change and could happen that images consecutively obtained at the same physical coordinates may have different characteristics. Those differences may be due to many causes, such as using devices (e.g. cameras, lamps, etc.) with different frequencies, or dirt in the environment, etc. Therefore, since situations like these may provoke great differences in the grey levels the images are normalized. The algorithm used is based on the adjustment of a grey level distribution to a Gaussian distribution, with mean equals to 0 and typical deviation of 1.

$$N' = \frac{N - M}{\sigma} \tag{3}$$

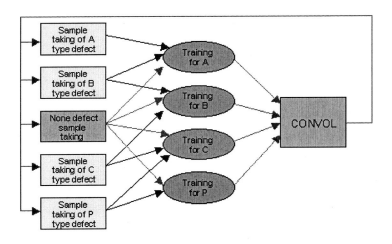

Fig. 3. Filter generation

5.3 Training the Neuron

In order for a neuron to learn a specific type of defect, it has to be trained using that type of defect and none defect samples. The number of defective samples has been fixed to 20, using them 4 times, so that the total number of defective samples used is 80. None defect samples used is 100; including defects of different type to the one the neuron is learning. For example, when training for A type defect recognition, 20 B type defects are included (as none defect samples); 20 C type when training for B type recognition; 20 P type when training for C type recognition; and only none defect samples when training for P type defect recognition. Figure 3 illustrates these combinations and Figure 4 shows the relation between defect sizes and epochs during the training. As a result the weights of the trained neuron is obtained, which conveniently arranged as a matrix would shape a digital filter, like those shown in Figure 7.

6 Classification of Several Patterns

A filter used to identify a certain type of defect is able to always detect it when appears; although it is more difficult for defect types that look alike. It

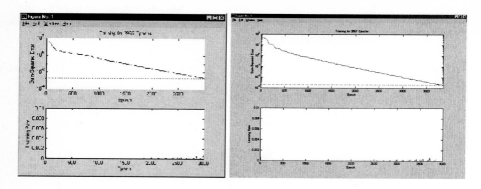

Fig. 4. Training error versus epoch for defect types B and P

means that there could be defects classified in higher categories. A solution to this problem is implementing a multi-layer filtering, making the results from the convolution to pass through another neuron that evaluates how close to some type the defect is.

This filter will conveniently weigh up the outputs taken from every two close type filters. Figure 5 illustrates how by the convolution of masks P and C, the second layer filter clearly separates the elements not well classified during the first pass. Similar processes are performed for C and B filters, and for B and A filters too.

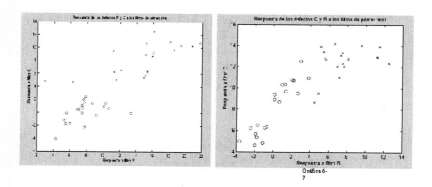

Fig. 5. Treshold for P-C and C-B type defects

7 Automatic Sample Centering

Training a neural network means to attempt it to learn only those characteristics of what it is learning. Therefore, it is very important to limit what it has to learn. Images with defective zones or zones with defective characteristics at different positions are usually found. Figure 6 shows images like these, and for analysis purpose a mark has been placed at the same coordinates in each image.

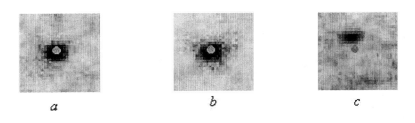

Fig. 6. Defective zones at different positions

In cases a and b the marks are at defective zones, while in c the defective zone is shifted and at the marked position there are no defective characteristics. Training the network with the first two images would cause a correct learning, but feeding it with the third one could not be convenient. In consequence, for an adequate learning the defective zones in an image should be always at the same position. So that, centering the defect in the sample image is quite important.

The solution procedure developed consists of: first the training defective images are chosen by the user as the ones with the most centered defects, then the mass centers of these images are calculated, the network is trained obtaining a temporal mask which afterwards experiments a convolution with larger images (one pixel around), obtaining finally an appropriate filter.

8 Results

All training images were taken by the developed INSPULP system installed at the factory. Two criteria were used during the training phase; for Criterion 1, the "not defective" samples used were indeed not defective ones, and for Criterion 2, the "not defective" samples were a mixture of not defective and defective samples smaller than the ones the filter was created for. Once the filters were obtained, through the previously explained process, they were tested with defective and not defective samples. A program named CONVOL is used to visualize the resulting convolutions from every defect type, as well as the original image.

At the time this report is being developed the filters have been applied at the plant, obtaining the following results:

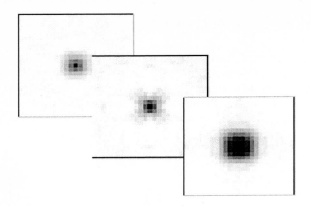

Fig. 7. Designed filters by learning

8.1 Image Acquisition and Processing

The ENCE 404 Standard [6] establishes that an area of 500x500 mm^2 from a paper pulp sheet sample of 800x800 mm^2 taken from one of every seven lots of eight 410 sheets bales, must be inspected.

Table 2. Methods comparison

$Method$	$Inspected$ $area(mm^2)$	$\%$ $Inspected$	$Speed$ (sec)
$Manual$	500x500	0.0017	$n.a.$
$Automatic$	100x800	0.12	2.6

It can be observed from Table 2 that the automatic method studied on this paper is quite superior to the manual method, based on the percentage of inspected area. The percentage of area automatically inspected has been obtained considering that when the line is at its maximum speed of 2,100 mm/s a total area of 3,000x2,100 mm^2 of paper pulp is passing through. The filters actually used during the process are those shown in figure 7. Processing data obtained for this report is listed below:

- Line speed: 92.3 m/min (i.e. 1.54 m/s)
- Image size: 496x492 $pixels$
- Filter size: 30x30 $pixels$
- Resolution: 10 pixels/mm (2x2 pixels for the defect of 0.04 mm^2)
- Image gray level: 256

9 Conclusions and Future Work

The inspection ENCE 404 Standard [6] has been completely satisfied. The system's Neural Network based architecture is robust enough under changing production parameters as well as under changing quality norms. Once the system has been tested on-line in one of the production lines, it is ready to be applied on others. In order for the system to be faster and more compact a different camera has been proposed for the following automatic inspection stations. Further work still has to be done for the system's protection, maintenance, mobility and flexibility for new lines with different textures.

References

1. Campoy, P.: *Nuevas Tendencias sobre Control Avanzado: Redes Neuronales*, pp. 37-53, Fundacion Repsol Publicaciones, 1999.
2. Campoy, P.: *Sistemas Inteligentes en la Inspeccin Visual*, "Industria XXI" Revista ETSII, Número 0, Universidad Politcnica de Madrid, pp. 23-33, 2000.
3. Conners, R. *Identifying and locating surface defects in wood*, IEEE Transactions PAMI 5 (6) (1983), pp 573-583.
4. Dewaele, P., Van Gool, L., and Oosterlinchk, A. *Texture Inspection with self-adaptive convolution filters*, Proceedings of the Ninth International Conference on Pattern Recognition, 1, 1988, pp 14-17.
5. Lee, C.S., Choi, C.H., Choi, J.Y., Kim, Y.K. and Choi, S.H., *Feature extraction algorithm based on adaptive wavelet packet for surface detect classification*, IEEE International Conference on Image Processing, 1996, pp. 673-675.
6. Normas ENCE: *Technical Specification of the Product*, Empresa Nacional de Celulosas, S.A., 1994.

Image Quality Enhancement for Liquid Bridge Parameter Estimation with DTCNN

Miguel A. Jaramillo, J. Álvaro Fernández, José M. Montanero, Fernando Zayas

Dpto. Electrónica e Ingeniería Electromecánica. Escuela de Ingenierías Industriales.
Universidad de Extremadura. Avda. de Elvas s/n. 06071 Badajoz SPAIN.
e-mail: {miguel, jalvarof}@unex.es

Abstract. This work present the use of a neural structure to augment the quality of noisy images of liquid bridges to obtain a clear representation of its border in order to determine the acceleration that it is suffering. The used network is a three layers Discrete Time Cellular Neural Network in which the last one performs the contour highlighting through the adaptive definition of the gain and threshold of their output functions. Then an easy algorithm extracts a curve from the border.

1 Introduction

The use of neural networks as image processing tools has become a growing field of research in the last years. This interest provides form the fact that learning and recognition tasks are usually performed over visual patterns. In this way, when an object is to be recognized or classified the image where it is included may be corrupted with noise so that its recognition or classification will be difficult. In addition, these objects may not appear clear enough in the picture to be easily recognized. So a preprocessing of the image in order to reject the noise or to increase the contrast may improve the system performance. On the other hand, if we could extract some characteristics of the objects, as edges or corners, its recognition would be improved as they could be defined in a more general way that provides the network with more robust learning and recognition capabilities. In this way the definition of neural networks with image processing capabilities will provide a powerful tool that will improve the system performance in pattern recognition tasks. All these issues have then encouraged the study of the image processing capabilities of neural networks, developing structures and models that provide a quality enhancement of images or extract some properties of the objects present in them.

In this field the neural model that has achieved a highest development is the so-called Cellular Neural Network (CNN) [1], [2]. It is a special case of the Hopfield model in which every neuron is only connected with their surrounding neighbors. These connections are the same for every cell, defining a repetitive structure usually named as "cloning template". This repetitive synaptic scheme represents the main feature of the model, providing a local processing of the input signal that makes it specially suited to be used in image processing, what has become the main application of CNNs. So this network provides a convolution of the input image with a window that define the processing task to be performed. In this way the CNN acts as a classical spatial filter whose effect may be modified with the inclusion of a feedback defined with a "cloning template" similar to that previously mentioned. Since the result of these convolutions is further processed by the nonlinear neural output, the neural network provides an added capacity to the convolution process. Finally a multilayer structure may be defined providing the network with a more sophisticated processing capability that allows a sequential treatment of images. So complex image processing tasks may be defined combining several layers that perform different task over the results provided by the previous one.

The combination of all these characteristics in an only structure define the CNN as a very powerful tool for image processing. An adequate definition of the overall structure, templates and parameters of the output function, will provide a very effective treatment of images where the traditional tools don't provide so good results.

In this paper we develop a such CNN structure to enhance the quality of the image of a liquid bride to obtain a profile of its contour as sharp as possible. This profile will then be used to calculate the parameters that define the bridge shape in order to obtain the accelerations acting on it. In this way a liquid bridge may be used as a precision accelerometer [3].

The work is organized as follows: Section 2 analyzes the liquid bridge equations, Section 3 describes the neural network structure and section 4 presents the results of applying this structure to the bridge contour detection.

2 Liquid Bridge Structure

Liquid bridges are volumes of liquid held between solids by surface tension forces. They occur in both natural and technological situations and have been studied for practical reasons and for basic scientific interest. The bridge consists of an isothermal drop of liquid of volume V^* held by surface tension forces between two parallel, coaxial solid disks of radii $R_o(1-H)$ and $R_o(1+H)$ (with R_o being the mean value of both disks radii) separated a distance L as shown in Fig. 1. We will assume that an acceleration g is applied in the axial direction. The bridge is surrounded by another fluid with different density, so that $\Delta\rho$ is their the density difference and σ the surface tension. With these condition the bridge shape is axisymmetric and is defined by a function $R(z)$ that measures the distance between a point in the interface and the axial axis.

The bridge shape may be described by the definition of the set of parameters {V, Λ, H, B} where $V \equiv V^*/(\pi R_o^2 L)$ is the reduced volume, $\Lambda \equiv L/(2 R_o)$ is the slenderness and $B \equiv \Delta\rho\, g\, R_o^2/\sigma$ is the so-called Bond number. The shape of the boundary of the liquid bridge, $R(z)$, is obtained from the Young-Laplace equation that describes the equilibrium conditions between the surface tension and the inertial forces. So a function defining the bridge contour $R(V, \Lambda, H, B; z)$ is obtained.

This equation may be solved for a set of values of parameters {V, Λ, H, B} providing a certain contour of the liquid bridge. On the other hand if we have the liquid bridge contour instead of its parameters we can obtain them from the minimization of an error function (for instance least squares) of the difference between the points defining the bridge contour and those obtained from the solution of $R(V, \Lambda, H, B; z)$ for a certain set of values of {V, Λ, H, B} [4]. If the liquid bridge is to be used as an accelerometer the value of the acceleration is directly obtained from the Bond number. In this way the more accurate the contour is obtained the more precisely the value of the acceleration will be. Nevertheless as the parameters are obtained from the minimization of the error function previously mentioned some small failures in the contour profile will have little effect in the parameter estimation providing a very robust method that gives very accurate values for the acceleration.

So far we have considered that only one axial acceleration is acting on the liquid bridge but it may also be assume that another one may acts on it in a normal direction [5]. If this is the case the bridge shape will not be axisymetrical and two different edges will be considered. In such a case a function $R(V, \Lambda, H, B; z, \theta)$ must be assumed, in which θ is the angle around the axis. Although this is a more sophisticated function than that for one only acceleration, the procedure to adjust the parameter set $\{V, \Lambda, H, B\}$ is the same, taking into account that instead of one only Bond number B we will have two of them: one for each acceleration.

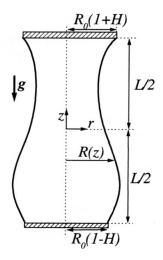

Fig. 1. Liquid bridge contour

3 The DTCNN with an Adaptive Output Function

CNNs are a particular case of the Hopfield net in which the cell connectivity expands only to those pixels or neurons surrounding the one considered. The equation defining the cell activity is [1]:

$$C\frac{dz_{ij}(t)}{dt} = -\frac{1}{R_x}z_{ij}(t) + \sum_{C(k,l)\in N_r(i,j)} A(i,j;k,l)y_{kl}(t) \qquad (1)$$
$$+ \sum_{C(k,l)\in N_r(i,j)} B(i,j;k,l)u_{kl} + I$$

In this equation $N_r(i,j)$ define the neighborhood of each neuron while A(i,j;k,l) and B(i,j;k,l) are the connecting weights with other neurons and input pixels respectively. They are the same for each cell. Usually the output function is a piecewise linear one, but we will assume a sigmoid one that is better adapted for the gray level image processing that will be performed in this work. As it can be seen (1) is a continuous function well adapted for its implementation in VLSI circuits, but as we will perform a computer simulation of the network, a discrete one will be preferred. So (1) will be substituted by [6]:

$$z_{ij}(t) = \sum_{C(k,l)\in N_r(i,j)} A(i,j;k,l)y_{kl}(t) + \sum_{C(k,l)\in N_r(i,j)} B(i,j;k,l)u_{kl} \qquad (2)$$

The threshold I is dropped to be accounted in the sigmoid output function. So (2) represents the result of the application of the convolution window B(i,j;k,l) to the input image and the convolution window A(i,j;k,l) to the network output. Usually, in image processing tasks, no feedback is included so that (2) is left as:

$$z_{ij}(t) = \sum_{C(k,l)\in N_r(i,j)} B(i,j;k,l)u_{kl} \qquad (3)$$

In this way the sigmoid output function has the form:

$$y_{ij}(t) = \sigma(z_{ij}(t)) = \frac{1}{1+\exp\{-s_{ij}(z_{ij}(t)-T_{ij})\}} \qquad (4)$$

In [7] and [8] a procedure was presented for the adaptive control of s_{ij} and T_{ij} in each cell. An appropriate definition of these two parameters will provide the sigmoid function with the capability of adapting its output range to the gray mean deviation in the neighborhood of each neuron. So we can assume that T_{ij} represents the mean gray value in that area while s_{ij} is related to its mean deviation. In this way an expansion of the gray range is accomplished as it can be seen in Fig. 2. To obtain those parameters mask sizes greater than the usual 3x3 must be assumed. Template dimensions as 9x9 or 11x11 will usually be assumed.

To use this adaptive output function we define a processing layer where it is to be used. Two other layers must be included between it and image to provided values for every s_{ij} and T_{ij}. The first one will give T_{ij} assuming that their output function parameters are: $s=4$, $T=0.5$ and B(i,j;k,l)= $1/m^2$. The second one provides the mean deviation, now assuming that $s=4$, $T=0.5$ and a new definition for the convolution product (3) is provided through the use of the extended absolute value function defined in [9]:

$$abs_\alpha(x) = (x^2 + \varepsilon^2)^{\alpha/2}, \quad \alpha \in [1, \infty), \varepsilon > 0 \quad (5)$$

Assuming $\alpha=1$ and $x=u-m$, with u the value of a pixel and m the mean value of its neighborhood, (5) has the form:

$$abs_1(u_{kl} - m_{kl}) = ((u_{kl} - m_{kl})^2 + \varepsilon^2)^{1/2} \quad (6)$$

With this expression (3) has the form:

$$z_{ij} = \sum_{C(k,l) \in N_r(i,j)} \frac{1}{m^2} |u_{kl} - m_{kl}| \quad (7)$$

So the second layer provides the mean value of the deviations from the mean of the gray levels in a pixel neighborhood. Now the output of this second layer ($d_{ij} = \sigma(z_{ij})$) will define s_{ij} in the processing one as $s_{ij} = \delta / d_{ij}$, where $\delta=2$ is assumed to obtain a better performance of the net [8].

Once the output function parameters of the output layer have been defined only the form of "cloning template" $B(i,j;k,l)$ must be specified. Although any kind of high pass filter may be used when a contrast enhancement is performed we have assumed a direct process of every pixel (i. e. $B(i,j;k,l)=[1]$) to avoid the amplification of the noise present in the images.

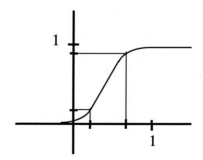

Fig. 2. Local expansion of the gray scale dynamic range

4 Contour Highlighting in Liquid Bridges

The main problem one can finds when detecting the liquid bridge contour is a bad definition of its shape due to the diffractive nature of the perception of the interface between both liquids. As we can see in Fig. 3 there isn't a clear transition from the bridge to the fluid surrounding it. So it is very difficult to represent the contour with a curve that accurately defines the bridge shape. Here the proposed model will provide a sharper representation of the bridge contour providing the use of a simple algorithm to obtain the curve representing the contour. The algorithm runs as follows. First of all the four corners of the bridge must be marked. Beginning with the two upper ones (it will be equally possible to begin with the lower ones) the three pixels immediately under it are considered, i. e., if the pixel considered has indexes (i,j) we will take those with (i+1,j-1), (i+1,j) and (i+1,j+1). The pixel with a lower gray level will be assumed as belonging to the contour. If they have the same level the central one will be taken. This procedure will be repeated until the lower corner is reached. As we have begun with two different points, two different curves will be obtained. If we are dealing with bridges with axial accelerations only one of them will be provided, but if the acceleration is normal to the axis or there are normal and axial ones acting on the bridge they both must be obtained.

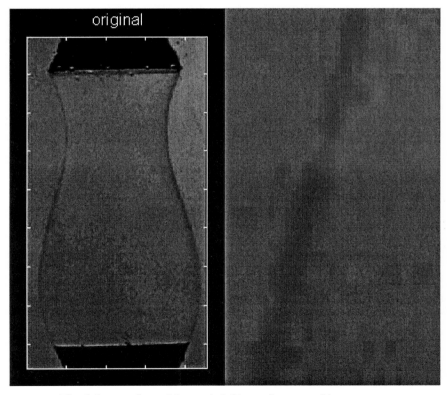

Fig. 3. Image of an axisimmetric bridge and a zoom of its contour.

We can see in Fig 4. the bridge after it has been processed and the resulting contour. As it can be noted, the contour is clearer than in the original image because a thinner and sharper profile is obtained. So the obtaining of a curve that represents this contour is now easier than from the original image. A very precise representation of it is then obtained.

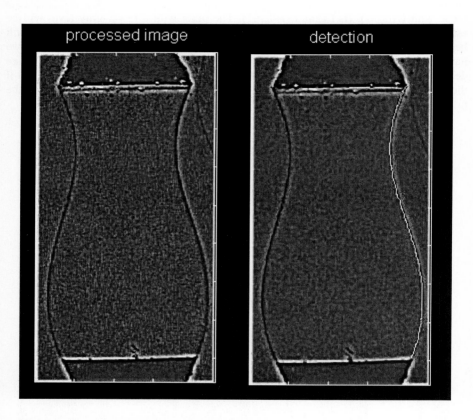

Fig. 4. The bridge contour after processing by the Neural Network and the curve obtained as the bridge contour. As an axisimmetric bridge is presented only one of the contours is obtained.

References

[1] L. O. Chua, L. Yang. "Cellular Neural Networks: Theory". IEEE Trans. on Circuits and Systems. Vol. 35, No. 10. October 1988. pp. 1257-1272.
[2] L. O. Chua, L. Yang. "Cellular Neural Networks: Applications". IEEE Trans. on Circuits and Systems. Vol. 35, No. 10. October 1988. pp. 1273-1290.
[3] J. M. Montanero, G. Cabezas, J. Acero, F. Zayas. Using rotating Liquid Bridges as Accelerometers. Microgravity Science and Technology. In Press.

[4] F. J. Acero, G. Cabezas, J. M: Montanero, F. Zayas. Método para la Medida del Contorno de Puentes Líquidos. Anales de Ingeniería Mecánica, Vol 13, Decembre 2000, pp. 1401-1406.

[5] F. Zayas, J. I. D. Alexander, J. Meseger, J.-F. Ramus. On the Stability Limits of Long Nonaxisimetric Cylindrical Liquid Bridges. Physics of Fluids, Vol. 12, No. 5, may 2000, pp.979-985.

[6] H. Harrer. Multiple Layer Discrete-Time Cellular Neural Networks Using Time-Variant Templates. IEEE Trans on Circuits and Systems II. Vol. 40, No. 3, March 1993, pp. 191-199

[7] M. A Jaramillo-Morán,. J. A. Fernández Muñoz. Adaptive Adjustment of the CNN Output Function to Obtain Contrast Enhancement. IWANN'99. Lectures Notes in Computer Science 1607. pp: 412-421. Springer Verlag, 1999.

[8] M. A Jaramillo-Morán,. J. A. Fernández Muñoz, E. Martínez de Salazar. Mejora de Contraste en Imágenes Mediante el Control Adaptativo de la Función de Salida Neuronal. XX Jornadas de Automática. Salamanca (España). September 1999.

[9] R. Dogaru, K. R. Crounse, L. O. Chua. "An Extended Class of Synaptic Operators with Applications for Efficient VLSI Implementation of Cellular Neural Networks". IEEE Transactions on Circuits and Systems, Vol. 45, No. 7, July 1998, pp.745-755

Neural Network Based on Multi-valued Neurons: Application in Image Recognition, Type of Blur and Blur Parameters Identification

Igor Aizenberg, Naum Aizenberg and Constantine Butakoff

Neural Networks Technologies Ltd. (Israel)
NNT Ltd., 155 Bialik str., Ramat-Gan, 52523 Israel
igora@netvision.net.il and igora@nnt-group.com (IA)
nauma@netvision.net.il (NA)
cbutakoff@nnt-group.com and cbutakoff@yahoo.com (CB)

Abstract. Some important ideas of image recognition using neural network based on multi-valued neurons are being developed in this paper. We are going to discuss the recognition of color images, distortion (blur) types, distortion parameters and recognition of images with distorted training set.

1. Introduction

Idea of image recognition using neural networks based on multi-valued neurons has been proposed several years ago [1]. During this short time the proposed approach has been developed [2-7].

A multi-valued neural element (MVN) is based on the ideas of multiple-valued threshold logic [3]. A comprehensive observation of multi-valued neurons, their learning and applications can be also found in [3].

Different kinds of networks that are based on MVN have been proposed [1-6]. Successful application of these networks to simulation of the associative memory [1, 3, 5, 6], image recognition and segmentation [1-3, 7], time-series prediction [1, 3] confirms their high efficiency. Highly effective quickly converged learning algorithms for MVN and neural networks based on them have been elaborated [3].

We will concentrate here on the development of image recognition approach using single-layered MVN-based neural network [2, 3, 7]. This approach is based on the following background: 1) high functionality of multi-valued neurons and quick convergence of the learning algorithm for them; 2) well-known fact about concentration of the signal energy in the low-frequency part of orthogonal spectra [8]. Because of the fact that inputs are taken from the Fourier spectrum, and FFT algorithm, which is used for Fourier spectra calculation, exists only for the signals with a length equal to power of 2, recognition of images of arbitrary sizes will be discussed. We are also going to discuss the recognition of color images.

On the other hand we will consider a problem of type of blur recognition, which is very important for image restoration, because it is impossible to get an appropriate restoration result until a blur type is unknown [9]. Blur parameters identification will be also considered.

2. Multi-valued Neurons and their Learning

Multi-valued neuron (MVN) is deeply considered in [3].
MVN performs a mapping between n inputs and single output. The mapping is described by multiple-valued (k-valued) function of n variables $f(x_1,...,x_n)$ via their representation through $n+1$ complex-valued weights $w_0, w_1, ..., w_n$:

$$f(x_1, ..., x_n) = P(w_0 + w_1 x_1 + ... + w_n x_n) \quad (1)$$

where $x_1, ..., x_n$ are variables, on which performed function depends. Values of the function and of the variables are k-th roots of unity: $\varepsilon^j = \exp(i 2\pi j/k)$, $j \in \{0, k-1\}$, i is an imaginary unity. P is the activation function of the neuron:

$$P(z) = \exp(i 2\pi j/k), \text{ if } 2\pi j/k \leq \arg(z) < 2\pi (j+1)/k \quad (2)$$

where $j=0, 1, ..., k-1$ are values of the k-valued logic, $z = w_0 + w_1 x_1 + ... + w_n x_n$ is the weighted sum, $arg(z)$ is the argument of the complex number z.

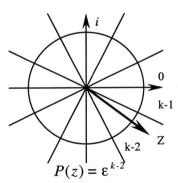

Fig. 1. Definition of the MVN activation function

For MVN, which performs a mapping described by k-valued function, we have exactly k domains. Geometrically they are the sectors on the complex plane (Fig. 1). If the desired output of MVN on some element from the learning set is equal to ε^q then the weighted sum should be exactly in the sector number q. But if the actual output is equal to ε^s, it means that the weighted sum is in the sector number s. A learning rule should correct the weights to move the weighted sum from the sector number s to the sector number q. The following correction rule for learning of the MVN has been proposed [1, 3]:

$$W_{m+1} = W_m + C_m (\varepsilon^q - \varepsilon^s) \overline{X} \quad (3)$$

where W_m and W_{m+1} are current and next weighting vectors, $\overline{X}$ is the complex-conjugated vector of the neuron's input signals, C_m is the scale coefficient.

Learning algorithm based on the rule (3) is very quickly converging.

3. Image Recognition and Blur Identification Using Single-Layered MVN-Based Neural Network

3.1 MVN-Based Neural Network for Pattern Recognition

We will use here a single-layer MVN-based neural network, which contains the same number of neurons as a number of classes we have to classify [2]. Each neuron has to recognize pattern belonging to its class and to reject any pattern from any other class.

We will use frequency domain data representation to make description of data more objective [10]. In the terms of neural networks to classify object we have to train a neural network with the learning set, which contains the spectral coefficients of the representatives of our classes.

We will use two different models for the frequency domain representation of our data. The first one is supposed to use phases of the low-frequency part of Fourier transformation coefficients and the second one is supposed to use amplitudes. Since phases of the Fourier spectrum contain more information about the object presented by signal [8], it is natural to use them as inputs for the object recognition. On the other hand, the Fourier spectrum amplitude contains more information about the signal properties (existence of noise, blur, etc.), which means that it is possible to use it for identification the existence of blur, its type and parameters.

While recognizing, output values $0,..., k/2-1$ of the i-th neuron correspond to classification of object as belonging to i-th class. Output values $l,..., k-1$ correspond to classification of object as rejected for the given neuron and class (k is taken from (2)).

3.2 Image Recognition

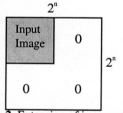

Fig. 2. Extension of image with sizes not equal to power of 2

Recognition of the gray-scale images with sizes equal to power of 2, using MVN-based neural network, has been already considered [2, 3, 7]. Let us consider, how the presented approach may be effectively applied to images with arbitrary sizes and to color images.

To operate with images of arbitrary sizes the image has to be padded with zeroes until it's square with side length equal to power of 2 (Fig.2). This solution is more appropriate than, for example, even extension of the input image. Such an extension involves appearance of the different false details that are a preventing factor for recognition because of their significant influence on the Fourier spectrum.

The simplest way to use the presented approach for color image recognition is transformation of color images to the gray-scale ones. Then a presented technique may be applied to the gray-scale images. To do thus, we extract a luminosity component from RGB color image by the following well-known [11] transformation:

$$Y = 0.299R + 0.588G + 0.113B \qquad (4)$$

where Y is a luminosity (brightness) and R, G, B are color components, respectively.

3.3 Type of Blur and Blur Parameters Recognition

To restore the blurred image (for example, using Tikhonov's algorithm), it is very important to know a type of the distortion and parameters of the corresponding distorting operator [9]. Otherwise it is impossible to obtain appropriate restoration results. A natural way to identify a type of blur or its parameters is to analyze amplitude of the image Fourier spectrum. Any blur specifically distorts the spectrum amplitude (Fig. 3).

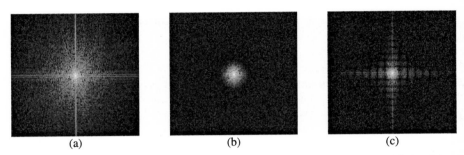

Fig. 3. Influence of the blur on Fourier spectra amplitude: (a) – spectrum amplitude of the original image; (b) – spectrum amplitude of the same image image with Gaussian blur; (c) – spectrum amplitude of the same image with rectangular blur

It is clearly visible that different blurs involve significantly different distortions of the Fourier spectrum amplitude. It means that it is possible to use these amplitudes for identification of blur type and its parameters.

3.4 Inputs' Extraction from the Representatives

For both cases, i.e. for image recognition and blur recognition we use the same idea of inputs' extraction. The Fourier Transform is applied to image and modulo or phase is extracted. Fig. 4 and Fig. 5 show how the inputs are extracted. The arrows show the order of extraction. The extracted numbers have to be normalized by the rule that follows from (2). Suppose B is the value for normalization and k is the number of sectors (2). If B is the value from Fourier phase, then we take it and find the sector, where complex number with argument B belongs to, and take its number. We will refer to it as S_B. In case B is from modulo, we look for the number of sector, where $ln(B)$ belongs to. Let A be the result of normalization. Then $A = e^{i\frac{2\pi}{k}S_B}$. This way every input vector will look like $(A_1, A_2, ..., A_N)$, where N- is the number of inputs we've chosen to take, and $A_i = e^{i\frac{2\pi}{k}S_{B_i}}$, $i=1,...,N$. B_i are the values from spectrum.

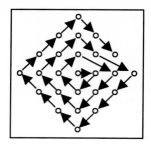 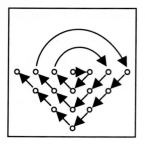

Fig. 4. Exraction from phase   Fig. 5. Extraction from modulo

4. Simulation Results

4.1 Image Recognition

Fig. 6. Some images from the learning set (color 320x240)

Fig. 7. Some images from the testing set

We would like to consider the abilities of the recognition algorithm on the face recognition example. For image recognition 10 classes[1] were taken (some examples are presented in Fig. 5, all images are color in original). Fig. 7 shows some of the images that took part in the testing (testing set contains total of 143 images for all 10 classes). The learning set contained 20 images per class (some of the images from the learning set have been created artificially using rotations, shifts, etc.). For inputs we

---

[1] The database used here for the face recognition has been created and passed to authors for experimental work by the Israeli company Video Domain Technologies Ltd.

used phases corresponding to a few (6,7 and 8, respectively) low frequencies of the Fourier spectrum. The results are summarized in the table below.

# of frq.	Parameters	Results
6	84 coefficients, 128 sectors	97.9% recognized, 0.6% not recognized, 1.3% misclassified (of 143 images)
7	144 coefficients, 128 sectors	95.1% recognized, 4.1% not recognized, 0.6% misclassified (of 143 images)
8	180 coefficients, 512 sectors	95.8% recognized, 3.4% not recognized, 0.6% misclassified (of 143 images)

4.2 Type and Parameters of Blur Recognition

For testing abilities of the recognition algorithm, to classify a type of blur and blur parameters, we used two kinds of blur: Gaussian and rectangular ones. The abilities of the presented method to identify blur parameters have been tested for different values of Gaussian blur variation and rectangular blur parameters. We also tested recognition of blur and its parameters on the noisy images. Recognition of blur parameters is very important from the point of view of further image restoration. Even minimal mistake in parameters estimation involves impossibility of satisfactory image restoration because all known restoration algorithms use blur parameters as main information, which is needed for the restoration [9]. So for testing purposes images from the AR Face Database[2] were taken, then they were corrupted by noise and blur with different parameters.

For blur identification 140 images have been taken. Only 20 images have been used in the learning set for rectangular blur and 20 for the Gaussian one. The rest 100 images have been used for testing. Fifty them have been corrupted by Gaussian blur, another fifty have been corrupted by rectangular blur. Only for 1 blurred image a type of blur has not been recognized. It means that we've got 99% successful recognition. Then 100 images with added Gaussian noise with dispersion varying from 5% to 30% of clean image dispersion were tested. This time a blur type has not been recognized for 4 images. It means that for this case we've got 96% successful recognition.

For blur parameters recognition 25 classes with different rectangular blurs without Gaussian noise and 25 classes with Gaussian noise added were created. Steps for rectangular blur parameters are equal to 2 pixels in both directions (i.e. 1x1, 1x3, ..., 1x9, 3x1, ..., 3x9, ... 9x9). The Gaussian noise was taken with the same dispersion as described above. In both cases these 25 classes contain 500 images in the testing set (20 per class), and 12 images per class in the learning set, which is 300 images.

Among the not noisy images only for 3 of 300 ones the blur parameters have not been identified, which leaves us with 99.4% of successfully identified parameters. The result with the noisy images is following: for 5 images of 300 ones the blur parameters have not been identified and for 1 of 300 they have been misclassified. It gives 99.2% of successfully recognized parameters.

To test the Gaussian blur parameters' recognition 5 classes were created where Gaussian blur with radiuses 1, 2, 3, 4 and 5 were used to corrupt the images. For the

---

[2] AR Face Database can be found at http://rvl1.ecn.purdue.edu/~aleix/aleix_face_DB.html

not noisy images a learning set contained 22 images per class, which is 110 in total, and testing set contained 10 images per class, which is 50 images in total. As the result, only 4 images were misclassified (they belong to the Gaussian blur class with radius of 5), which is 92% of successful recognition.

Then a set of noisy images with the same Gaussian blur parameters has been generated with the Gaussian noise as described above. There were total of 250 images, 50 per class. The results are following: for 7 images blur parameter has not been identified, and for 25 images blur parameter has been misidentified. So it gives us 87.2% of correctly classified images.

The tables below show a few images that took part in the experiments.

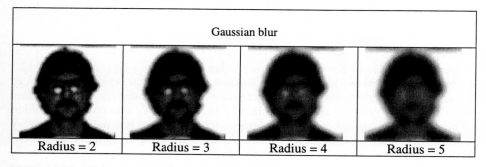

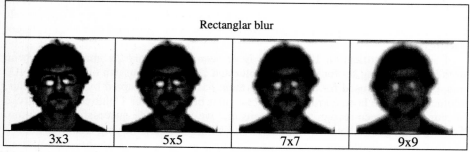

5. Conclusions and Future Work

Neural network based on multi-valued neurons, which has been considered in the paper, is highly effective, when it comes to image recognition and image properties recognition. Some very important and new solutions have been developed in this paper. The original approaches to color image recognition have been found, also as a way to work with the images of arbitrary sizes (not only of ones equal to power of two). The new idea of inputs' selection introduced in this paper gives great results for recognition, and it doesn't matter whether it is image recognition or blur identification. The ability to identify a type of blur and blur parameters (even on noisy images) is very useful, because if the parameters of distortion are known, then the original image could be restored.

Acknowledgements

This work is completely supported by the company Neural Networks Technologies Ltd. (Israel).

References

1. Aizenberg N.N., Aizenberg I.N.. Krivosheev G.A. "Multi-Valued Neurons: Learning, Networks, Application to Image Recognition and Extrapolation of Temporal Series", *Lecture Notes in Computer Science*, Vol. 930, (J.Mira, F.Sandoval - Eds.), Springer-Verlag, pp. 389-395, 1995.
2. I.N.Aizenberg, N.N.Aizenberg "Pattern Recognition Using Neural Network Based on Multi-Valued Neurons", *Lecture Notes in Computer Sciense*, Vol. 1607-II (J.Mira, J.V.Sanches-Andres - Eds.), Springer-Verlag, pp. 383-392, 1999.
3. I.N.Aizenberg, N.N.Aizenberg, J.Vandewalle *Multi-Valued and Universal Binary Neurons: Theory, Learning, Applications*, Kluwer Academic Publishers, Boston/Dordrecht/London, 2000.
4. I.Aizenberg, N.Aizenberg, C.Butakoff, E.Farberov "Image Recognition on the Neural Network Based on Multi-Valued Neurons", *Proceedings of the 15th International Conference on Pattern Recognition, Barcelona, Spain September 3-8, 2000*, IEEE Computer Society Press , **2**, pp. 993-996, 2000.
5. H.Aoki, Y.Kosugi "An Image Storage System Using Complex-Valued Associative Memory", *Proceedings of the 15th International Conference on Pattern Recognition, Barcelona, Spain September 3-8, 2000*, IEEE Computer Society Press , **2**, pp. 626-629, 2000.
6. S.Jankowski, A.Lozowski, M.Zurada "Complex-Valued Multistate Neural Associative Memory", *IEEE Trans. on Neural Networks*, **7**, pp.1491-1496, 1996.
7. S.Lawrence, C. Lee Giles, Ah Chung Tsoi and A.D.Back "Face Rocognition: A Convolutional Neural-Network Approach", *IEEE Trans. on Neural Networks*, **8**, pp. 98-113, 1997.
8. A.V.Oppenheim and S.J.Lim "The Importance of Phase in Signals", *Proceedings IEEE*, **69**, pp. 529-541, 1981.
9. O.P.Milyukova "On Justification of Image Model", *SPIE Proceedings*, **3348**, pp283-289, 1998.
10. N.Ahmed, K.R.Rao *Orthogonal Transforms for Digital Signal Processing*, Springer, 1975.
11. W.K.Pratt *Digital Image Processing*, John Wiley & Sons, N.Y., 1978.

Analyzing Wavelets Components to Perform Face Recognition

Pedro Isasi[1], Manuel Velasco[1], and Javier Segovia[2]

[1] Departamento de Informtica
Universidad Carlos III
Leganes, Madrid.
isasi@gaia.uc3m.es

[2] Departamento de Lenguajes y Sistemas Informticos
Facultad de Informtica U.P.M.
Boadilla del Monte, Madrid.

Abstract. Face recognition is a very difficult task in real environments. In those cases a good preprocessing of the images is needed to keep the images invariant to translations, scales, luminosity, shape, aspect, rotation, noise, etc... Wavelet transformation have been probed to be a good preprocessing method for many task. However, not all the coefficients of a wavelet transform have the information needed for a classification method to be efficient. This work introduce a method to select the most appropriate coefficients for a wavelet transform to allow an unsupervised neural network to well classify a set of complex faces.

1 Introduction

Most sensitivity analysis methods using neural nets have been based almost exclusively on the use of supervised net models. In such cases, the method seek to study how a specific input affects net efficiency [3]. Having detected the inputs (X_i), which hardly affect learning, they are removed from the system and learning is repeated. After the number of inputs has been reduced, if the neural net performs similarly in terms of learning and generalisation as it did with all the inputs, the method will be said to have been successful.

This paper presents a new sensitivity analysis method based on non-supervised neural nets and more specifically on Grosberg's ART models [5, 1] (ArtSen).

In this method, some inputs are gradually removed as and when their varying influence is detected. Therefore, the process is gradual and continuous, as the inputs are received in the training phase. This means that the sudden adjustment of an input does not interfere in the sensitivity analysis process (although it does have an influence on the final result), and this can output results constantly as the relationships between inputs are detected, without having to retrain.

The following sections describe the method of non-supervised sensitivity analysis and its application for determining critical parameters in an image classification task.

2 Sensitivity Analysis using ART models (ArtSen)

The presented method is based on the *ARTMap* model, and is illustrated in figure 1.

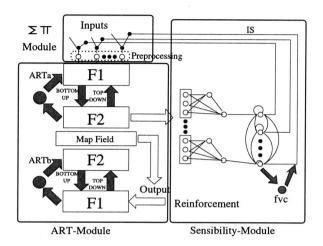

Fig. 1. ArtSen model scheme

The model is composed of mainly three modules: a first module ($\Sigma\Pi$ connectionS) where weighted inputs are received, are processed (if necessary), and an output is produced, which will be the input for the second module (*ARTMap* module). In the *ARTMap* module, the signal is propagated as usual in this model. Finally, sensitivity analysis will be carried out in the Entropy Model (*EM*) producing outputs, called Influence Signals (*IS*). These outputs biunivocally determine the influence level of each input signal and are used to weight the inputs in the next time interval

$\Sigma\Pi$ Module The first module of the system consists of a set of $\Sigma\Pi$ connections [4, 12] in which each cluster receives two signals, one from the input stimulus X_i and the other, the output signal, from the entropy model IS_i acting as weights. The generic equation for the $\Sigma\Pi$ cluster is:

$$X'_j = \Pi_{i=1}^{k} W_{ji} IS_i X_i \qquad (1)$$

The output produced by the $\Sigma\Pi$ cluster, the X'_i, will be input to a signal preprocessing module whether or not warranted by the problem to be solved.

Art Module The output of the first module is input to and propagated through an *ARTMap* module [2] (which can be replaced by an *ART*2 model [1] if there is no reinforcement signal). The module clusters the input signals unsupervised or by reinforcement. Once the net has been trained, the sensitivity analysis will involve determining the influence level of each input signal in the clustering process. The BottomUp connections (Z_{ij}) are used as input for the last module of the model, the entropy model.

Entropy Module A diagram of the architecture of this model is shown in figure 2. Figure 2 shows four layers: three Feed Forward layers, and one competitive layer. The main purposes of this module are:

1. To measure the influence of each system input to the on the ArtMap model.
2. To make this measure depending on the $ARTMap$ learning phase.

The first task is carried out in the first three layers of the net and the second one in the fourth layer.

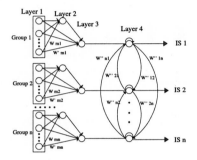

Fig. 2. Entropy Module

The first layer consists of n groups of cells, where n is the dimension of the X_i' input. Each group (G_i) will get m inputs, where m is related to the number of categories generated in the F2 layer of the Art model, (figure 3):

$$G_i = \{Z_{ij}, \forall j \in F_2\} \qquad (2)$$

where Z_{ij} refers to the connection between the ith cell of F_1 and the jth cell of F_2 of the $ARTMap$ net Art_a model (figure 3). The F_1 layer directly receives the Art net input. Therefore, each cell of this layer is related to one of the input lines of the X_i stimulus.

The second and third layers are each composed of a cell, connected as shown in figure 2. The output of this last layer has the same dimension as the $F1$ layer of Art_a, which also corresponds to the dimension of the X_i input. This output directly addresses the next competitive layer. The equations that govern the behavior of the module are:

$$\tau_s \frac{\partial S_{ij}^1}{\partial t} = -S_{ij}^1 + f_1(Z_{ij}) \qquad (3)$$

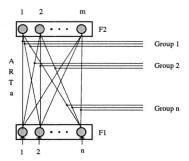

Fig. 3. Connection between the input of the Entropy Model and the BottomUp connections of the ArtMap net

$$\tau_s \frac{\partial S_i^2}{\partial t} = -S_i^2 + f_2(\Sigma_{j=1}^M W_{ji} S_{ji}^1) \quad (4)$$

$$\tau_s \frac{\partial S_i^3}{\partial t} = -S_i^3 + f_3(\Sigma_{j=1}^M W\prime_{ji}(S_{ji}^1 - S_i^2)^2) \quad (5)$$

Where:
S_{ij}^1 is the output of the ith cell of the jth group of the first layer of the entropy module, f_1, f_2 y f_3 are sigmoidal functions, Z_{ij} is the jth input of the ith group of the entropy module, M is the number of categories created so far in Art_a (F_2 dimension, coinciding with the number of neural groups), S_i^2 is the output of the ith cell of the second layer of the entropy module, W_{ji} is the weight of the connection between the jth cell of the first layer and the ith cell of the second layer. A constant value $W_{ji} = \frac{1}{M}$ can be selected, S_i^3 is the output of the ith cell of the third layer of the entropy module, W_{ji} is the weight of the connection between the jth cell of the first layer and the ith cell of the third layer. A constant value $W_{ji} = \frac{1}{M}$ can be selected, s is a temporal constant

Equation 2 shows how S_i^2 works as a threshold to compare the signals of the input groups of the entropy module, in absolute terms, thus outputting the variances of these signals, grouped by the input of F_1. This variance, S_i^3, acts as the input for the competitive layer of the entropy module, where the dynamics of the model is defined by:

$$\tau_c \frac{\partial IS_i}{\partial t} = -IS_i + f_c(S_i^3 + IS_i - \Sigma_{j=1, j/=i}^N W\prime\prime_{ij} IS_j \delta(t)) \quad (6)$$

Where:
IS_i is the activation value of the units, $W\prime\prime_{ij}$ are the connection weight of the ith cell of the third layer with the jth cell of the fourth layer, n is the dimension of the fourth layer, $\delta(t)$ is a positive value lower than 1, cell competitiveness rating of the cells, f_c is a sigmoidal threshold function, and τ_c is a temporal constant.

Values for $\tau_c \gg \tau_s$ should be used that assure that the short-term variations of S_i^3 do not strongly affect IS_i.

$\delta(t)$ introduces what has been called **competitive vigilance factor (cvf)** in the competitive layer, where the higher the value, the higher the competition. Its variation in time will be governed by the equation:

$$\delta(t) = 1 - \frac{1}{e^{\frac{t}{\tau}}} \qquad (7)$$

3 Experimental validation

The experiment chosen to validate the system was automatic face recognition. This problem has some special features that makes it suitable for use in system validation [9, 8]. It is a very complicated task involving a wide variety of factors, including translations, scales, luminosity, shape, aspect, etc. [11, 10].

The experiment involves classifying a set of 300 images of the faces of 20 different people.

Two kind of experiments are presented, those provided by the Pattern Recognition System without performing Sensitivity Analysis (Type A), and those provided by the system with the Sensitivity Analysis Process previous to the pattern selections (Type B).

The preprocessing is the same for both experiments and it is very simple. First, the images are compressed to 64x64 pixels ([3]), where each pixel represents a grey level between 0 and 255, totalling 4096 points. These images are filtered by means of a wavelet transformation into two dimensions, using the Daubechie 4 wavelet, outputting a 64x64 coefficient matrix. This matrix will be the input of the neural net, which will classify these images according to their similarity.

The reason for using 2D wavelet data processing before inputting data to the neural net is that most of the useful information in the images is located in certain transformation coefficients, particularly the low frequency coefficients. This is easily demonstrable by rebuilding the images using the inverse transformation from these coefficients only. Rebuilding is nearly perfect, albeit less perfect when fewer coefficients are selected. However, if the images are rebuilt by randomly selecting the same number of coefficients or using any other criterion, then the rebuilding process is not possible.

3.1 Type A experiments

In a first stage the patterns results from the wavelet transformations are directly introduced to the net. Each pattern has 4096 coefficients (64x64), i. e., the experiment is developed with all the information available. Table 1 presents the results from this experiment. First column shows the vigilance factor. Second column shows the number of categories generated by the net. Third column presents the number of categories classified perfectly. Fourth column indicates the percentage of global success. Finally, last column shows the typical deviation of the success by category.

Vigilance Factor	Number of Categories Generated	Number of Categories with 100% Success Rate	Global Percentage of Success	Typical Deviation of the Success Rate by Category
0.999900	19	17	68%	20.69
0.999910	22	19	67%	20.42
0.999920	27	25	68%	18.64
0.999930	35	32	76%	15.68
0.999940	45	40	84%	11.76
0.999950	64	62	97%	6.22
0.999960	97	95	98%	4.10
0.999970	133	130	99%	4.94

Table 1. Experiment with 4096 coefficients

In the next stages, the low frequency properties of the wavelet transformation are applied, obtaining sub-patterns with 1024 coefficients (32x32), removing 75% of the total information.

In a third stage 256 sub-patterns (16x16) are used, and so on, with a fourth stage (64 coefficients), and a fifth stage (16 coefficients). In this last stage only 0.39% of the information is used.

Table 2 shows partial information of these stages. A new column, with the number of coefficients, is included in the table.

Number of Coefficients	Vigilance Factor	Number of Categories Generated	Number of Categories with 100% Success Rate	Global Percentage of Success	Typical Deviation of the Success Rate by Category
1024	0,999950	52	48	92%	8,09
1024	0,999960	73	71	97%	5,56
256	0,999960	45	41	94%	6,99
256	0,999970	75	72	99%	2,33
64	0,999970	42	38	88%	10,62
64	0,999980	67	64	99%	1,47
16	0,999990	54	48	93%	8,15
16	0,999995	102	99	99%	4,02

Table 2. Experiment with rest of coefficients

To increase the vigilance factor implies that the number of categories and the success rate are higher. We consider a categorization as valid when there is a 95% of success at least and the number of categories generated is lower or equal than 40. This value is two times the number of different people.

In the first experiment (4096 coefficients) to get a 97% success 64 categories are generated, value very high for the purpose of the method. Same occurs for experiment 2, 4, and 5. However, for experiment 3 (256 coefficients), we can observe an acceptable categorization, with a 94% success and 45 categories.

3.2 Type B experiments

These experiments add the Sensitivity Analysis to make the process faster and more efficient. The patterns which are result from the wavelet transformation are processed by a module. This module has two main processes. First process weights the significance of each component. Second process transforms again the patterns building another sub-patterns only with the relevant components. The relevant components chosen by the method depends on the value for the competitive vigilance factor. In this case increasing this facto the number of coefficients selected by the method are: 171, 124, 72, 45, 37, 21, 14, 13 and 10.

Therefore the experiment is developed again with the components calculated by the process. Table 3 indicates partial results of the Type B Experiments, in the same way as was shown in type A experiments.

Number of Coefficients	Vigilance Factor	Number of Categories	Number of Categories with 100%	Global Percentage	Typical Deviation of the Success Rate
171	0,999950	42	37	86%	11,49
171	0,999960	51	48	95%	7,98
171	0,999970	77	75	99%	3,48
124	0,999960	47	43	95%	6,01
124	0,999970	66	64	98%	4,95
72	0,999970	50	46	93%	8,65
72	0,999980	85	83	99%	2,07
45	0,999970	41	36	95%	6,82
45	0,999980	59	55	97%	4,79
37	0,999980	57	52	95%	6,72
37	0,999990	111	109	99%	2,90
21	0,999980	43	37	88%	12,20
21	0,999990	78	77	99%	1,13
14	0,999990	59	53	91%	10,86
14	0,999995	109	105	99%	5,32
13	0,999990	56	49	92%	11,16
13	0,999995	108	103	98%	4,68
10	0,999990	44	37	92%	11,33
10	0,999995	82	74	96%	10,18

Table 3. Results from type B experiments

Considering the same principles as in Type A Experiments (95% success rate and 40 categories) we can observe that in first experiment (171 coefficients) 51 categories are generated for a 95% success. This can be considered as acceptable, as well as in second experiment (47 categories for a 95% success). In the fourth experiment (45 coefficients) a good categorization is observed (41 categories for a 95% success), but decreasing this number (37, 21, 14, 13, and 10 categories) no acceptable categorization is showed. Therefore, an optimum decreasing of the number of categories is demonstrated. Under this value, results begin to be worse.

Another method efficiency test has been run on a time series prediction problem [7]. The method was completely validated against a classical sensitivity analysis applied to a multilayer neural net. The efficiency of the proposed model with noisy signals was also verified in this problem.

4 Conclusions

The possibility of continuous learning, as in ART models, is highly important when dealing with some sensitivity analysis problems. In situations where the influence of the signals changes over time it is essential to have a method capable of picking up these variations, as and when they occur at a reasonable computational cost, and to be able to automatically determine when the variation took place.

The new sensitivity analysis presented here has these features. As the input changes, if the weights of the categories change or new categories are produced, these changes are automatically included in the sensitivity analysis calculations for the next input. The results vary due to the new situation.

As shown in the experimentation presented, the generality level of a system with inputs removed by the model is not reduced and, in some cases is increased due to the elimination of the noisy signals.

The analysis can be carried out in a supervised or nonsupervised manner or by reinforcement using this method by merely selecting the respective ART model.

References

1. A. Carpenter and S. Grossberg. Art2: Stable self-organization of pattern recognition code for analog input patterns. *Applied Optics*, 26:4919–4930, 1987.
2. A. Carpenter, S. Grossberg, and J. Reynolds. Artmap: Supervised real-time learning and classification of nonstationary data by a self-organizing neural network. *Neural Networks*, 4:565–588, 1991.
3. A. Carpintero, J. Castellanos, J. Rios, and J. Segovia. Automatic face recognition for access control. In *Proceedings of 1993 IJCNN*, pages 1289–1292. International Joint Conference on Neural Networks, 1993.
4. J.A. Feldman and D.H. Ballard. Connectionist models and their properties. *Cognitive science*, 6:205–254, 1982. Reprinted in Anderson and Rosenberg-1988.
5. S. Grossberg. Competitive learning: From interactive activation to adaptive resonance theory. *Cognitive science*, 11:23–63, 1987.
6. P. Isasi. *Sistema Neuronal Sensible al Entorno con Arquitectura Autoorganizada*. PhD thesis, Facultad de Inform'atica, Universidad Polit'ecnica, Madrid, 1994.

7. P. Isasi and J. Segovia. Sensitivity analysis in art models: An application in forecasting. In *XV International Symposium on Forecasting*, Toronto, Canada, June 1995.
8. Shen Jun, Shen Wei, H. J. Sun, and J. Y. Yang. Fuzzy neural nets with nonsymmetric membership functions and applications in signal processing and image analysis. *Signal Processing*, 80(6):965–983, June 2000.
9. Hongbong Kim and Kwanghee Nam. Object recognition of one-dof tools by a back-propagation neural net. *IEEE Transactions on Neural Networks*, 6(2):484–487, March 1995.
10. Sheng Liu, R. Olivia, Chih-Ping Wei, and Paul Jen-Hwa Hu. Neural net learning for intelligent patient-image retrieval. *IEEE Expert*, 13(1):49–57, February 1998.
11. K. Nezis and G. Vosniakos. Recognizing 2d shape features using a neural network and heuristics. *Computer-Aided Design*, 29(7):523–539, July 1997.
12. D. Rumelhart and McClelland. *Parallel Distributed Processing:Explorations in the Microstructure of Cognition*, volume II. MIT Press., Cambridge, 1986.

Man-Machine Voice Interface Using a Commercially Available Neural Chip

Nicolás J. Medrano-Marqués, Bonifacio Martín-del-Brío

Dept. Ingeniería Electrónica y Comunicaciones, Universidad de Zaragoza. Campus Plaza
San Francisco s/n
E-50009 Zaragoza, Spain
{nmedrano, nenet}@posta.unizar.es

Abstract. Speech recognition is a common application area for artificial neural networks. Although most of the speech recognition systems are implemented as a complex software running in a conventional computer, nowadays there are commercially available neural chips for speech recognition targeted to embedded applications. A prototype for man-computer interface based on one of these neural chips is shown; the developed circuit controls a computer mouse with speech commands. This voice-activated mouse is 'plug & play', in the sense that it can be directly connected to a mouse port without requiring any special software. The technique proposed can be easily applied to other computer input peripherals, such as keyboards or joysticks.

1 Introduction. Speech Recognition in Embedded Systems

The use of computer programs, such as text editors or spreadsheets, requires standard input peripherals (keyboard, mouse,...). However, these standard interfaces could not be appropriate for people with motor diseases or can be uncomfortable for a conventional user in some situations, as in multimedia presentations. Computer interfaces based on speech recognition can be a solution in these cases. Moreover, we think that speech recognition is the next direction for man-machine interfaces, because after many years of intensive research this technology is already available, with good performances and at competitive cost.

The processing power of present personal computers has allowed the development of sophisticated software based systems for speech recognition, such as the well known program ViaVoice, from IBM. In this kind of voice interfaces a sound card receives and processes voice commands, and a complex software (based on hidden Markov models, artificial neural networks or both techniques [1-3]) translates speech to text, or voice to computer commands. This solution requires a powerful hardware (a computer with a sound card) and the installation of a special purpose (and very complex) software.

A new approach for speech recognition is the use of dedicated, low-cost, hardware. The development in recent years of new techniques for speech recognition based on artificial neural networks and new digital signal processing algorithms, has allowed the commercialization of several solutions for speech recognition targeted to embedded systems. These new solutions can be implemented as software routines executed on standard microcontrollers or DSP processors (products from Frontier Design, Criterion Software Ltd. and Domain Dynamics Technologies), or as a dedicated piece of hardware, i.e., low cost integrated circuits for speech recognition (ISD Inc., STMicroelectronics and Cortologic Gmbh). Products commercialized by the American company Sensory Inc. are a good example of both cases [4].

These new solutions for speech recognition in embedded systems (where memory occupation is very critical) are frequently based on artificial neural networks or other new signal processing techniques (as far as we know, only ISD Inc. makes use of hidden Markov models, the usual tool in computer based speech recognition).

Due to the low cost of these new approaches (dedicated chips or software routines for standard processors), speech recognition technologies are nowadays included in consumer products, as cellular phones (Sensory and Toshiba cooperates in the development of totally hands-free telephone devices), oscilloscopes (Infinium Series from Agilent Technologies), toys, and in a near future will be included in electrical appliances and computer peripherals.

In this work we show one application of VoiceDirec™, a speech recognition chip from Sensory Inc. [4, 5]. It integrates in a QFP-64 package a standard 8-bit microcontroller, and a neural network for speech recognition that can be easily trained for recognizing up to 15 voice commands. We have developed a low-cost peripheral which acts as a voice-activated computer mouse: a VoiceDirect chip recognizes speech commands ('up', 'down', 'left', 'rigth',...) and an electronic circuits interfaces with the mouse port. Therefore, the voice-activated computer interface developed works with standard operating systems (as Windows) and standard programs (MS-Word, PowerPoint,...), without requiring any additional software (as a standard computer mouse, makes use of the mouse software driver already installed).

The paper is organized as follows. In Section 2 the operation of a standard mouse is described. Section 3 introduces the speech recognition technology of Sensory Inc., based on artificial neural network implemented in a low cost integrated circuit, which can be easily trained for the recognition of a set of speech commands. Section 4 presents the hardware prototype that we have developed, which acts as a computer mouse controlled by verbal commands, and that uses the standard mouse software driver already installed in the computer (i.e., it is 'plug & play', because does not require any special software). Finally, some conclusions of our work are provided and future developments are discussed.

2 Computer Mouse Operation

As it is well-known, the operation of a computer mouse is controlled in two different ways: by moving the mouse through a plain surface and by clicking the mouse buttons. This standard computer interface provides access to program menus, executes software commands, etc.

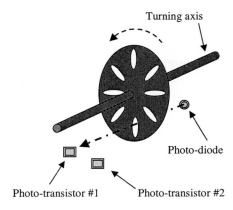

Fig. 1. Computer mouse movement sensing mechanism. When the wheel turns right to left, phototransistor 1 receives the light from the emitter before than 2, and vice-versa

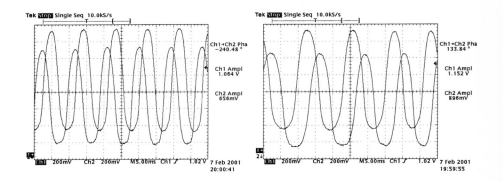

Fig. 2. Computer mouse photosensor responses: (a) the mouse moves left to right, (b) the mouse moves right to left

A computer mouse includes four photosensors, two for registering up-down movements and the other two for left-right. These sensors receive infrared light from two different emitters (one for each movement direction, Fig. 1). For instance, when a left-right movement is executed the wheel of Fig. 1 rotates, and the corresponding two sensors receive infrared light pulses (due to the holes in the wheel).

In addition, a wheel's hole aligns with only one emitter and one sensor at a time; thus, it is possible to determine if the movement is left-to-right or right-to-left, according to which of both sensors receives the light the first one. Figure 2a shows real signals recorded from two infrared sensors when the mouse movement is left-to-right; Figure 2b shows the same signals when the movement is right-to-left. Furthermore, the number of times the transistors receive light from the diode per unit of time determines the mouse speed. Note that a similar explanation can be provided for up-down movements.

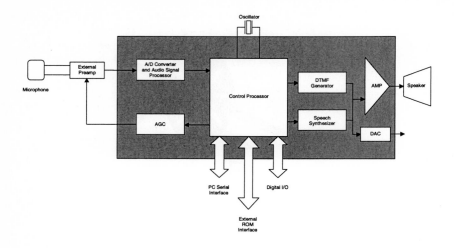

Fig. 3. Block diagram of the Voice Direct™ chip (from Sensory Inc.)

A computer mouse also includes several buttons. When a mouse button is pressed it is possible to deploy a window menu and select an option, or select an icon and drag it along the screen, to see and modify some properties or to execute the associated software with a fast double click.

Electric signals from infrared sensors and mouse buttons are sent to the mouse controller chip through several lines (up-down and left-right movements have two input pins each one, left and right buttons have independent input pins as well). The mouse controller integrated circuit translates this information to digital signals that are transmitted to the computer through a serial mouse port. In this way, every time an input line modifies its value, the mouse controller chip transmits the information to the computer in order to execute the corresponding mouse instruction.

3 Neural Integrated Circuit for Speech Recognition

Our goal is designing and building a prototype for a hardware interface between speech commands and mouse activity, appropriate for multimedia presentations or for people with motor diseases. For this purpose, a small, low cost speaker-dependent speech recognition chip from Sensory Inc., called VoiceDirect™, has been used [4, 5] (its block diagram can be seen in Fig. 3).

VoiceDirect makes use of a sophisticated neural network to recognize trained words or phrases (in any language) with greater than 99% accuracy. In training phase VoiceDirect stores the templates corresponding to voice commands in an 8KB EEPROM (up to 60 words-phrases). Then, in recall phase the neural network compares the present speech pattern with previously stored template patterns, mapping spoken commands to system control functions.

The details of the Sensory neural network for speech recognition are company trade secrets. They involve preprocessing of the raw acoustic signal into a rate and distortion-independent representation (speech templates), that are fed into the neural network integrated in the chip (it performs nonlinear Bayesian classifications). Training data consists of a large set of 300-600 voice samples, and cross-validation techniques are used (learning one command requires about one second). One of the advantages of using a neural network is that it eliminates the need for extensive RAM storage and extensive signal processing.

The VoiceDirect chip (Fig. 3) integrates in a QFP-64 package an 8-bit microcontroller (including the neural network for speech recognition), A/D converter, audio signal processor, DTMF generator, speech sintetizer, D/A converter (10 bits) and an amplifier. Its instruction set is loosely based on Intel's 8051. An external preamplifier filters and amplifies the voice commands coming from an electret microphone, and an 8 Kbytes serial EEPROM stores the speech templates (commands). Finally, an oscillator provides a 14.32 MHz clock for the voice chip operation.

The chip must be trained first with a set of voice commands from a speaker (one pin controls the learning or recall operation). Then, when it recognises a voice command turns high one or two lines of an 8 pin output port for one second (Table 1). Correctly trained, the chip can achieve 99% recognition accuracies for isolated word recognition (speaker dependent), even for high steady background noise (80 dB).

VoiceDirect has two possible working modes: in the external host mode the system receives the control instructions from an external microcontroller, being able to learn up to 60 speech commands; in the stand-alone mode the system operates without an external host and can learn up to 15 speech commands. In this work the stand-alone mode has been used. The whole system (VoiceDirect, pre-amplifier and EEPROM) operates at 5.0 volts with a typical supply current of 30 mA.

VoiceDirect is targeted to speaker dependent and isolated word recognition; nevertheless, Sensory has products for speaker independent and continuous listening [4]. Sensory commercialize low cost chips (at a price lower than $3 in some cases) and software routines for standard processors, suitable for use in applications as phone dialing, speaker verification, automotive, PDA, peripherals and toys.

4 Prototype for a Computer Mouse Voice-Controlled

The prototype for a voice-controller computer mouse we have developed makes use of a VoiceDirect integrated circuit (Fig. 3). In Figure 4 we show the block diagram of the circuit developed. In order to adapt the output signals provided for the speech recognition chip (through an output port) to execute standard mouse commands, an electronic circuit has been designed and implemented that would be included inside the computer mouse. We have used a Genius mouse for developing the prototype. The power supply and clock signal for our circuit are obtained from the Genius mouse controller chip; due to the low current level supplied by the mouse serial port, low power CMOS technology is used in the prototype.

Table 1. VoiceDirect chip's output port lines activated by every voice command

Command 1	Output 1	Command 9	Output 8 & Output 1
Command 2	Output 2	Command 10	Output 8 & Output 2
Command 3	Output 3	Command 11	Output 8 & Output 3
Command 4	Output 4	Command 12	Output 8 & Output 4
Command 5	Output 5	Command 13	Output 8 & Output 5
Command 6	Output 6	Command 14	Output 8 & Output 6
Command 7	Output 7	Command 15	Output 8 & Output 7
Command 8	Output 8		

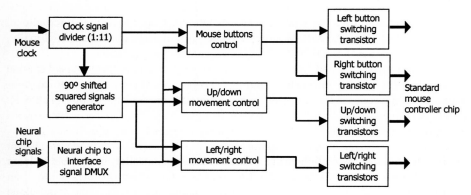

Fig. 4. Electronic circuit block diagram for the prototype of speech-to-mouse interface developed

Pins number 5 and 3 (GND and -V) from a standard PC computer serial port provide the supply voltage for the electronic interface. The Genius mouse controller chip forces pin 3 voltage to -5 volts; therefore, there is a voltage difference equal to 5 volts between both terminals. The clock signal is obtained from pin #4 of the Genius mouse controller chip. This pin provides a 32.78 kHz clock signal that is divided by means of an 11 bit counter; this clock is used for the mouse movement signals generation (Fig. 5).

The electronic circuit developed translates speech recognition chip outputs (some output port lines activates for one second, Table 1) to the following Genius mouse controller input signals: Two 90° shifted squared signals for left-right movement connected to pins #12 and #13 (Fig. 5), and for up-down (going to pins #14 and #15); single and double click signals for the left button connected to pin #11, and a single click signal for the right button (to pin #9). Table 2 summarizes relations between the VoiceDirect outputs and the corresponding mouse actions (and speech commands).

Figure 4 shows a block diagram for the circuit that carries out the signal translation from the neural chip to the mouse standard circuit (basically, this circuit replaces mechanical and analog parts of a standard mouse, Fig. 1, with a hardware voice interface connected to the conventional mouse controller chip). It consists of three 1-to-2 demultiplexers that translate the two bit signals from the voice recognition chip to 1 bit signals. Then, one circuit block (Fig. 4), consisting of logic gates and flip-flops, controls the up-down movement, and other similar block the left-right; button actions are executed by means of flip-flops. The connection between the electronic circuit developed and the conventional mouse circuit is carried out by means of PMOS transistors, operating as switches (these transistors bypass the switches and the emitter-photo-sensor electronics involved in the mouse actions of the standard computer mouse used in developing the prototype). The whole circuit works with a mean supply current lower than 1 mA, it has been implemented and is fully operative; its detailed schematic has not been included here due to space limitations.

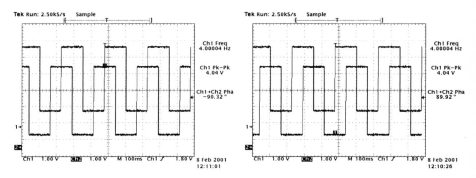

Fig. 5. Two squared signals 90° shifted representing (a) left-to-right movement and (b) right-to-left movement. Similar signals control the up-down cursor movement on the screen

Table 2. Relation between VoiceDirect outputs and mouse actions (corresponding to the speech commands the neural chip has been trained with)

Output 1	*Left*	Outputs 8 & 1	*Right*
Output 2	*Up-left*	Outputs 8 & 2	*Press left. button*
Output 3	*Down-left*	Outputs 8 & 3	*Press right button*
Output 4	*Up*	Outputs 8 & 4	*Stop movement*
Output 5	*Down*	Outputs 8 & 5	*Free buttons*
Output 6	*Down-right*	Outputs 8 & 6	*Double click left*
Output 7	*Up-right*	Outputs 8 & 7	*(Not used)*
Output 8	*(Not used)*		

5   Discussion and Future Work

In this paper a prototype for a simple and low-cost speech recognition computer interface has been designed and implemented by using the standard mouse port. It consist of a low-cost small voice recognition chip incorporated into a standard computer mouse, being its output signals converted to the appropriate input signals for the mouse controller by means of a digital electronic circuit. A prototype for the digital circuit has been implemented in low power discrete CMOS technology, which is biased by using the same supply lines used by the mouse controller chip. The prototype is fully operative.

We are now working in the implementation of the electronic circuit prototype following two different approaches, by using a low-cost microcontroller (a Motorola M68HC705J1A), and by using programmable logic devices [6] of low power consumption (such as those of ATMEL [7]). In the future we will implement the circuit in an ASIC. Any of these techniques will allow us to include all the hardware (voice recognition chip, interface circuit and mouse hardware) in the space available in a standard computer mouse, and at a very low cost and low power consumption.

Moreover, the use of an external microcontroller such as a Motorola 68HC05 family member allows the application of the neural chip in the external host mode (instead of in standalone). Several additional advantages can be obtained from this new operation mode in relation to the prototype developed, as availability of an extended spoken command set (up to 60 commands) and a lower response time.

Finally, remember that since a standard mouse controller chip sends commands to the computer, the proposed solution can be used with the standard software available (word processors, spreadsheets, multimedia software, etc.), without requiring any extra software or drivers (the standard mouse driver installed in the computer is enough). This voice interface could be easily applied to other computer input peripherals, such as joysticks, or in multimedia applications and presentations.

6 Acknowledgements

This work has been supported by the Diputación General de Aragón. Project number P82/98.

References

1. Haykin, S.: Neural Networks, A Comprehensive Foundation. 2nd edition. Prentice-Hall, New Jersey (1999)
2. Rojas, R.: Neural Networks, a Systematic Introduction.. Springer-Verlag, Berlin Heidelberg New York (1996)
3. Kohonen, T.: The 'neural' phonetic typewriter. IEEE Computer, 21 (1988) 11-22
4. Sensory Inc.: Web page http://www.sensoryinc.com/
5. Sensory Inc.: Voice Direct Data Book. Sensory Incorporated (1998)
6. Hayes, J.P.: Introduction to Digital Logic Design. Addison-Wesley (1993)
7. Atmel: CMOS Data Book (1994). Web page: http://www.atmel.com.

Partial Classification in Speech Recognition Verification

Gustavo Hernández Ábrego[1] and Israel Torres Sánchez[2]

[1] Spoken Language Technology, Sony U.S. Research Labs.
3300 Zanker Road MS/SJ1B5, San Jose CA 95134, USA
e-mail: gustavo@slt.sel.sony.com
[2] Signal Theory and Communications Department,
Universitat Politècnica de Catalunya
Jordi Girona 1-3, Campus Nord D-5, Barcelona 08034, Spain.

Abstract. Due to speech recognition imperfections, recognition results need to be verified before being used in real-life applications. Here we present two perspectives for recognition verification: direct classification and partial classification based on confidence measures. Linear classifiers, decision trees and perceptrons are used here as direct classifiers. On the other hand, we compute confidence measures through several methods, being MLP's and evolutionary fuzzy systems the best performing ones. Experimentation with three types of speech input reveals that higher correct verification rates can be achieved when verification is based on confidence measures. Moreover, classification rates can be improved when verification does not have to deal with "uncertain" examples, which are not classified. Partial classification represents a trade-off between verification accuracy and the number of recognition results verified.

1 Hypotheses Verification in Speech Recognition

In speech technology, recognition results are far from being perfect. In real-life applications, it is expected that the word hypotheses extracted from the speech signal effectively represent what the speaker said. Utterance verification is the systematic procedure that validates or rejects recognition hypotheses given some evidence of their correctness [5]. In a previous work [1], we presented confidence measures (CM's) as a feasible means to discriminate between correct and incorrect recognition hypotheses. Multi-layer perceptrons (MLP) and evolutionary fuzzy systems demonstrated being the best calculation tools for CM's. Here we want to evaluate the efficiency of CM's as the basis of recognition verification. We understand that there are some decisions harder than orders. Instead of making a wrong decision in verification, we prefer to leave out of the classification those recognition hypotheses with no clear evidence of their correctness.

2 Experimental Framework

Experimental work has been carried out using Spanish Speechdat [4] as developing and testing database. This is a database collected through the fixed telephone

	isolated	continuous	keywords
voc. size	500	59	30
speakers	989	995	993
words	989	9405	1485
false alarms	172	377	1132
accuracy	82.61 %	95.23 %	93.80 %

Table 1. Configuration, false alarms and accuracy rate of the recognition tasks tested.

network, sampled at 8 kHz and recorded under several acoustic environments. Speech was parameterized with mel-cepstrum coefficients. First and second order differential parameters plus the differential energy were employed. The recognition system utilizes Gaussian semi-continuous hidden Markov models (HMM's) for acoustic modeling. Close to 1000 speakers have been selected for each of the training and testing sets. There is no speaker overlapping between sets. To cover some of the possible frameworks where CM's are relevant, our experimentation includes the following recognition tasks:

1. Isolated words: each speaker says the name of a Spanish city.
2. Continuous speech: speakers say prompted time phrases. The average number of words per phrase is 9.4.
3. Keyword spotting: phrases containing embedded keywords are uttered by the speakers. Each sentence may contain between 1 to 4 predefined keywords.

The overall goal is to validate the recognition results that the recognizer produces when dealing with each of the experimentation tasks. The recognizer is tailored to be application independent. For acoustic modeling, it is based on high-performance sub-lexical phonetic units: Demiphones [3]. Words are built from Demiphones following canonical transcription rules. Words are then combined into sentences by means of specific language models (LM) for each application. Isolated words use null grammar so all city names have equal probability. Continuous speech recognition in "Time" is conducted by a finite-state automata. Keyword spotting has an stochastic trigram as LM. For this task, keywords are represented by Demiphones and "the rest of speech" words (filler words) by a network of phonemes [1]. Recognition accuracy, as well as the number of false alarms, and the configuration of the test sets for each of the recognition tasks is shown in table 1.

The number of false alarms is the summation of insertions and substitutions. Without any validation of the recognition hypotheses, to retrieve results with the given accuracy rate would imply to accept such amount of false alarms.

Generation of confidence measures passes through the calculation of features for each detected word. Then, the features of every recognition candidate are compiled into a CM for each word hypothesis. Besides of being used for CM generation, recognition features can also be used to train a direct classifier

3 Confidence Measures Computation

Through careful study of the nature of the speech recognition process [2], we have formulated three features as the basis of our experimentation.

3.1 Recognition Features

Our first feature is the *Likelihood Score Ratio* (LSR). For its calculation, the likelihood score of the recognition hypothesis is normalized by the score of an alternative recognizer:

$$LSR = \log L(\boldsymbol{X}|\Lambda_p) - \log L(\boldsymbol{X}|\Lambda_a). \tag{1}$$

$\boldsymbol{X}$ is the vector of acoustic features related to the actual input utterance. Λ_p and Λ_a are the sets of HMM's of the "principal" and "alternative" recognition networks respectively. There are two recognizers involved in our calculation of CM's: the principal, from which the hypotheses are taken, and the alternative, a reference used as *second opinion* [2]. Due to its unconstrained (and inaccurate) nature, it only works as a phone recognizer, the alternative network is capable to detect any sort of speech event, although its results cannot be considered as recognition hypotheses. To use its score as normalization factor helps to verify the presence of an acoustic event in the input utterance.

Our second feature is what we call *Sequence Alignment Score* (SAS). This feature is intended to express the resemblance of two independent recognition hypotheses [2]. Its calculation is done by comparing both (principal and alternative) decoded strings through time alignment. The computation of this feature is inspired by the idea of looking for confirmation in a second opinion [2] which is a natural procedure for human decision-taking. If opinions of both recognizer coincide, one corroborates the other. The comparison of two recognition hypotheses is made by calculating a "confusion" score between the "principal" and "alternative" phone sequences. In order to have a fair comparison, several weighting procedures have to be used. References [1] explain in detail the calculation process of this feature. It is interesting to notice that with this procedure we are comparing the main recognition result, i.e. the actual words, and not only a by-product of it such as the recognition likelihood score.

Our third feature, that we call *Relative Speaking Rate* (RSR), is conceived to handle the insertion and deletion errors in continuous speech recognition. Within a given utterance, it may be expected that the speaking rate is maintained by the speaker between certain margins. Whenever in the recognition hypothesis appears an abrupt change of speaking rate, insertions or deletions can be suspected. RSR is calculated according to:

$$RSR = \log \frac{N_{i,f}/t_{i,f}}{N_T/t_T}. \tag{2}$$

Being N_T is the total number of speech units found in the whole hypothesis (phrase) and $N_{i,f}$ are the units detected in the time interval $t_{i,f}$ where the actual recognition hypotheses (e.g. word) is located. t_T is the duration of the whole recognition hypothesis.

3.2 Feature Combination

To build CM's in a "feature-compilation" fashion has recently become a common procedure. Among the several combination methods tried, neural networks [5] and fuzzy logic systems [2] present the best performance. Here we used an MLP trained under the Levengberg-Marquardt paradigm. Several repetitive training steps were made and the best performing system was kept, as in a multi-start training. In spite of its good performance, MLP's lack of generalization capabilities. Hence, the same MLP cannot be efficiently used for all the speech application tested. In order to gain in flexibility, fuzzy inference systems (FIS) were used as feature compilers. A Sugeno-type FIS was chosen due to its good behavior as classifier and its simplicity. One principal disadvantage that fuzzy logic systems present compared to neural networks is the need of expert knowledge for their design. To alleviate this drawback, based on the illustrative work of Shi et al [6], we implemented an evolutionary procedure (founded on Genetic Algorithms) to train the parameters of the fuzzy system from learning examples. The goal of the training is to maximize the performance of CM's as correctness predictors. In a result classification task, performance can be evaluated through ROC (receiver operation characteristics) curves. The tradeoff between the two kinds of classification errors (false alarms and false rejections) is graphically represented in a ROC curve. A useful summary for ROC's is the normalized area below the curve, that indicates the average level of correct detections for the whole range of operating points (the normalized area of an ideal classifier is 1). Our evolution procedure uses the area below the ROC as cost function. For the feature compiler system, the inputs are the three features extracted from recognition and the output is the fuzzy value (between 0 and 1) of confidence. The population for the evolutionary procedure can be initialized randomly (learning from scratch) or with a working system that is to be optimized. With this procedure we have finely tuned the parameters of the fuzzy systems presented in [1].

4 Direct Classifiers for Verification

A version of the ROC curves, that shows the trade-off between false alarms (FA) and false rejections (FR) as a function of the decision threshold, is shown in figure 1. Total classification error is also shown as a function of the decision thresholds. Its minimum represents the optimal operating point for a correct/incorrect classifier. So, by comparing the CM value of a given recognized word against the threshold value, verification or rejection is easily performed.

One useful characteristic of CM's is that they allow the threshold to be set according to the application purposes. For instance, if an application requires of maximum confidence, threshold can be set to high CM values. The hypotheses surpassing the threshold most likely will be correct but there will be several other correct hypotheses that will be rejected. A straightforward means to evaluate the performance of a classifier is the joint classification rate defined as:

$$R(\tau_l, \tau_u) = \frac{|\mathcal{P}|+|\mathcal{Q}|}{|\mathcal{U}|+|\mathcal{V}|} \qquad \forall \quad \tau_l \leq \tau_u \qquad (3)$$

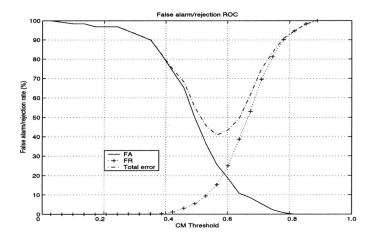

Fig. 1. Plot of FA, FR and total error as a function of the threshold for CM.

$$\mathcal{U} = \{w_i : CM(w_i) > \tau_u\}, \qquad \mathcal{P} = \{w_i \in \mathcal{U} : \text{correct}\}$$
$$\mathcal{V} = \{w_i : CM(w_i) \leq \tau_l\}, \qquad \mathcal{Q} = \{w_i \in \mathcal{V} : \text{incorrect}\}$$

being τ_l and τ_u the lower and upper thresholds, respectively, and $CM(w_i)$ the confidence value for the recognition hypothesis w_i.

In spite of its versatility, we still want to compare the CM approach against direct classification, that does not compute CM's as a pre-requisite for validation. The big difference between both perspectives is that thresholds are directly calculated at training time in direct classification whereas thresholds are set only after CM's have been calculated. What follows is a brief description of the direct classifiers we have used:

1. **Linear classifier**: a Bayesian classifier can be used as direct recognition verifier. It defines a hyperplane that divides the feature space into two regions: correct and incorrect. This scheme is well suited for linear separable classes. The coefficients used in the linear combination of our features are calculated from the covariances and means of the correct and incorrect training examples.
2. **Perceptrons**: if a hard limiting transfer function is used in a perceptron, a binary classifier is built. In contrast to the linear classifier, a multi-layer perceptron can define non-linear and even discontinuous decision hyper-surfaces in the feature space. The parameters of the perceptrons (weights and biases) are calculated through back-propagation training.
3. **Decision trees**: which represent a systematic procedure for decision taking based in consecutive thresholding operations. Decision trees are well suited for multi-class categorization and non-linear classification. The depth of the tree and the threshold values at every branching node can be calculated through a training step that adjusts the parameters of the tree by optimizing the entropy of the resulting classes.

5 Confidence Measures for Partial Classification

According to the total classification error curve in figure 1, the wisest selection for the verification threshold setting is the minimum error point. We will use this ad-hoc setting in the testing of our classifier based on CM's.

Even when the optimal decision point is selected as verification threshold, there is a considerable number of classification errors. Though, it can be seen in figure 1 that the FA and FR curves present high rates at opposite extremes. In order to take profit of the high classification rates of the system, we propose to use more than one threshold. The lower threshold is intended to optimize the classification of recognition errors and to improve the rejection of false alarms. The upper threshold is thought to optimize the classification of correct hypotheses while avoiding the acceptance of false alarms. As a consequence of this split classification, the CM range of values will be divided into three regions: correct, incorrect and "uncertain" results. The "uncertain" CM's are the ones that lie between both thresholds and that are not as low to be rejected nor as high to be accepted. The recognition examples associated to the uncertain CM's might be either rejected or re-prompted by the system. The natural consequence of multiple thresholding is incomplete classification. Not every recognition result can be labeled correct/incorrect as in direct classification. However, it is expected that the classification rates for multiple thresholds justify the incompleteness of classification. The trade-off between classification rate and total amount of hypotheses classified may be represented by the following linear combination:

$$T(\tau_l, \tau_u) = R(\tau_l, \tau_u) + w_S S(\tau_l, \tau_u) \quad \forall \quad \tau_l \leq \tau_u \tag{4}$$

which is a function of the lower, τ_l, and upper, τ_u, thresholds. $R(\tau_l, \tau_u)$ is the joint classification rate for the correct and incorrect classes according to (3). $S(\tau_l, \tau_u)$ is the amount of examples classified. Naturally, $\tau_l = \tau_u$ for complete classification. For an efficient classifier, the goal is to find the τ_l and τ_u that maximize $T(\tau_l, \tau_u)$. w_S is the relevance factor given to classification completeness. According to the relevance given to w_S, different maxima, and thresholds, in equation (4) are found.

6 Experimental Results and Discussion

Firstly, we contrast the classification performance of direct classifiers with the CM classifiers when complete classification is done. For the databases described in table 1, three direct classification methods are used: linear classifier, perceptron and decision trees. Classification from CM's are based on three calculation methods: MLP, fuzzy and evolutionary fuzzy. Joint classification rates for incorrect and incorrect classes are given. Results are shown in table 2.

Among direct classifiers, there are several behaviors. Decision trees are not capable to perform at a fair level. The linear classifier performs quite well for the keywords application revealing that the correct and incorrect classes are linearly separable in this application. In the other two applications, the best verifier is

	isolated	continuous	keywords
Linear	0.8059	0.8147	**0.8172**
Perceptron	**0.8746**	**0.8392**	0.7128
Tree	0.7452	0.6893	0.6139
MLP	0.8787	**0.9607**	**0.8290**
Fuzzy	0.8756	0.9599	0.8176
Ev. Fuzzy	**0.8797**	0.9605	0.8223

Table 2. Classification rates for different speech inputs. Direct classifiers are compared against CM ones. Complete classification is performed.

the one based on neural networks. The non-linear classification properties of multiple perceptrons produce higher rates. CM based verification systems show stable behavior. The verifiers based on evolutionary fuzzy systems and MLP's perform quite well and outperform the fuzzy based system that, maybe because its generalization properties, lacks of precision. It is noticeable the higher classification rates of CM systems over direct classifiers in continuous speech results. Nevertheless, the classification rates for isolated and, particularly, keywords application are far from optimal. Our partial classification approach may help to improve such verification results.

The results for partial classification are given in table 3. Extreme values of w_S were tried in equation (4) resulting in different amounts of classification, S (given in percentage), and recognition rates, R, for each type of speech tested.

	w_S	isolated		continuous		keywords	
		% S	R	% S	R	% S	R
MLP	1	100.0	0.8787	100.0	0.9607	100.0	0.8290
	1/4	83.5	0.9346	96.8	0.9724	79.6	0.8882
	1/10	62.3	0.9675	93.7	0.9777	50.9	0.9389
Fuzzy	1	100.0	0.8756	100.0	0.9599	100.0	0.8176
	1/4	79.2	0.9438	96.2	0.9734	70.8	0.8948
	1/10	64.9	0.9688	91.9	0.9796	2.1	1.0000
Ev. Fuzzy	1	100.0	0.8797	100.0	0.9605	100.0	0.8223
	1/4	**89.5**	**0.9164**	**97.0**	**0.9707**	**85.0**	**0.8657**
	1/10	57.5	0.9736	91.8	0.9792	1.1	1.000

Table 3. Classification rates (R) and percentage of the examples classified (%S) for the different speech types using partial classification.

For the $w_S = 1$ case, verification thresholds are equal, classification is complete and classification percentage is 100%. In such case, classification rates are equal to the ones in table 2. As w_S decreases, so it does the classification percentage but classification rates increase. Under extreme conditions, classification rates might be perfect but the classification percentage will be too small, less than 5%. A better trade-off between classification rate and percentage may be achieved using $w_S = 1/4$. By leaving a moderate number of recognition results out of the classification, remarkable improvement in the classification rates is obtained. This is particularly true for the verifier based on CM's calculated through evolutionary fuzzy systems. For instance, in the isolated words application, there is an important improvement (reducing the classification error more than 25%) by leaving a roughly 10% out of the classification. This same condition is also observed in this system for the keywords application where, at the expense of leaving a 15% out of the classification, the error classification rate is reduced 20%. In general, a wise use of partial classification clearly outperform any direct classifier in all the applications tested. A reasonable loss of classification might well justify the verification improvement.

7 Summary and Concluding Remarks

This paper has addressed the question of speech recognition results verification. We have tested several approaches to improve results classification rates. We have demonstrated the advantage of partial classification, inspired in real life decision-taking, over purely mathematics-based classifiers. CM's demonstrated to be an efficient means to drive utterance verification. Neural networks and evolutionary fuzzy systems showed to be efficient methods to calculate CM's. With partial classification based on CM's, recognition verification improved remarkably but at the expense of leaving some recognition results out of the classification. However, we have demonstrated that the classification loss may be justified by the increment in verification performance

References

1. G. Hernández-Ábrego and J. B. Mariño. Fuzzy reasoning in confidence evaluation of speech recognition. In *Proceedings of WISP'99*, pages 221–226, Budapest, September 1999. IEEE.
2. G. Hernández-Ábrego and J. B. Mariño. A second opinion approach for speech recognition verification. In *Proceedings of the VIII SNRFAI*, volume I, pages 85–92, Bilbao, May 1999.
3. J. B. Mariño, A. Nogueiras, and A. Bonafonte. The Demiphone: an efficient subword unit for continuos speech recognition. In *Proceedings of EUROSPEECH'97*, volume III, pages 1215–1218, Rhodes, September 1997.
4. A. Moreno and R. Winsky. Spanish fixed network speech corpus. Technical report, SpeechDat Project LRE-63314, 1997.
5. T. Schaaf and T. Kemp. Confidence measures for spontaneous speech recognition. In *Proceedings of 1997 ICASSP*, volume II, pages 875–878, Munich, April 1997.
6. Y. Shi, R. Eberhart, and Y. Chen. Implementation of evolutionary fuzzy systems. *IEEE Transactions on fuzzy systems*, 7(2):109–119, April 1999.

Speaker Recognition Using Gaussian Mixtures Models

Eric Simancas-Acevedo[2], Akira Kurematsu[1], Mariko Nakano Miyatake[2], and Hector Perez-Meana[2]

[1] The University of Electro-Communications
Chofu-shi, Tokyo, Japan
[2] National Polytechnic Institute of Mexico
Mexico City Mexico

Abstract. Control access to secret or personal information by using the speaker voice transmitted by long distance communication systems, such as the telephone system, requires accuracy and robustness of the identification or identity verification system, since the speech signal is distorted during the transmission process. Taking in consideration these requirements, a robust text independent speaker identifications system is proposed in which the speaker features are extracted using the Lineal Prediction Cepstral Coefficients (LPCEPSTRAL) and the Gaussian Mixture Models, which provides the features distribution and estimates the optimum model for each speaker, is used for identification. The proposed system, was evaluate using a data-base of 80 different speakers, with a pronoun phrase of 3-5s and digits in Japanese language stored during 4 months. Evaluation results show that proposed system achieves more than 90% of recognition rate.

1. Introduction

Several methods have been proposed for speaker identification and speaker verification such as the Vectors Quantization (VQ) and Dynamic Time Warping (DTW), classified as statistical methods, who use templates of small dimension instead of the routines used in the computing of the Lineal Prediction Coefficients (LPC). These methods have shown to be accurate in both text-dependent (TD) and text-independent (TI) identification when short-term utterances are used, however they are not appropriate enough, specially, in TI speaker recognition applications where long-term utterances must be used. Exist also the stochastic methods that have replaced to the statistics models in applications where it is necessary to use long-terms utterances. Among them we have the Hidden Markov Models (HMM), which assign one Model to each speaker representing all his/her features. Then, using these features during the training and test stages, then it compares them to find the model with the minimum distance, performing in this way the recognition task. The Gaussian Mixture Models (GMM), are similar to HMM with the difference that the GMM omit the information time implicit in the HMM [3]. This means that the GMM has only one state and it does not need of transition time from one state to other, as required in the HMM. So, the GMM uses only a unique Gaussian Distribution Matrix to represent all the speakers in the system. In addition, it is not necessary to use all components of the covariance matrix, because all Gaussian components are acting together to model the overall probability density function. Then full covariance matrix is not necessary even if the features are not statistically independent. This is because, to take all the components of the covariance matrix, is equivalent to take the main diagonal of the covariance matrix from each model of each speakers [1].

The Network Neural (ANN) is a recently developed for speaker recognition whose target is to build an artificial system that tries to emulate the human brain, by training one individual model for representing all speakers feature. Furthermore, ANN can be used for many applications in the speaker recognition field. Here the ANN are mainly used in speaker identification and recognition systems where they have shown efficiency in the classification of the features and good performance in the identification. In addition, the ANN uses few amounts of parameters to carry out the recognition. In contrast the ANN have the disadvantage that if one more speaker is added to the recognition system, the ANN need to estimate all its parameters again, and for this reason its use has been limited.

This paper proposes a speaker recognition based on GMM which has the capacity of representing broad acoustic classes with its individual Gaussian components and show a smooth approximation to represent long-terms distribution of the observation samples from a given speaker. These facts make the system robust, and avoid the problem of degradation of the speech signal when it pass through some kind of communication systems, such as the telephone system [1].

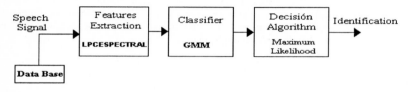

Fig. 1. Proposed Speaker Recognition System

2. Proposed Speaker Recognition System

The figure 1 shows the proposed identification system which consists of a feature extraction, a classification and a decision stages. In the feature extraction stage, the LP cepstral coefficients (LPCEPSTRAL) are calculated from a data-base of 80 speakers. They are then used as parameters in the Initialization step of the GMM model, during the classification stage, as well as in the decision stage in which a Maximum Likelihood (ML) estimation is used.

2.1. Features Extraction

A good performance of any pattern recognition system strongly depends on the features extraction stage, because appropriate information captured in a suitable form and size is very important for a good discrimination in the classification stage.

One of the most efficient representations of the speaker characteristics, regarding its performance on speaker recognition applications, have been the linear prediction coefficients (LPC) of speech signal, which have shown their effectiveness for speaker identification applications [1]. This is because the structure of the vocal tract of a person can be represented satisfactorily by using their linear prediction coefficients. Together with the LPC, other robust features that have shown a fairly good performance to build robust speaker recognition systems are the cepstral coefficients [2]. Taking this in account, in this paper, LP Cepstral Coefficients LPCEPSTRAL are used for speaker feature extraction, because, in addition, they have also proved to be robust enough to avoid distortion of the signal transmitted through conventional communication channels. Additionally, these features can be obtained very fast (about twice time faster in comparison with the mel-cepstrum) producing great competitive results. Furthermore, they computation is relatively simple just using a simple recursion after the Lineal Prediction Coefficients (LPC) were estimated. Then because the LPCEPSTRAL coefficients are derived from the LPC, they should be derived before the LPCEPSTRAL estimation.

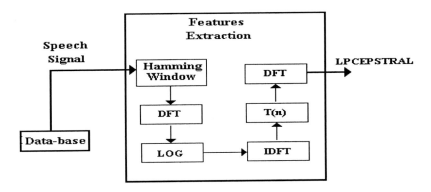

Fig. 2. LPCEPSTRAL estimation method.

Consider the LPC which can be estimated, using a linear filter, as follows

$$\hat{S}(n) = \sum_{i=1}^{P} a_i S(n-i), \qquad (1)$$

where $\hat{S}(n)$ is the estimated signal in time n, P is the filter order and a is the filter coefficients vector, usually estimated by using the Levinson Durbin algorithm. Then, once the LPC have been estimated, the LPCEPSTRAL can be obtained as follows

$$c_n = -a_n + \frac{1}{n} \sum_{i=1}^{n-1} (n-i) a_i c_{n-i}, \quad n>0 \qquad (2)$$

where C_n are the LPCEPSTRAL coefficients, $a_0 = 1$ and i <P. In this application, the LPCEPSTRAL were extracted for each one of the used phrases and digits pronounced by 80 different speakers. An alternative way for LPCEPSTRAL estimation is shown in Fig. 2. Here, firstly the speech signals is took from a data-base. Subsequently, it is

divided in several segments using Hamming window of 20ms length with 50% overlap. Next in the feature extraction stage, the LPCepstral coefficients (LPCEPSTRAL) were computed by using, either the eqs.(1) and (2) with P=16, or the procedure shown in Fig. 2, to obtain the speech features representing the speakers characteristics, where the Hamming window is given by

$$w(n) = h(n) = \begin{cases} 0.54 - 0.46 \cos\left(\dfrac{2\pi n}{N-1}\right) & \text{for } 0 \leq n \leq N-1 \\ 0 & \text{otherwise} \end{cases} \quad (3)$$

and T(n)

$$T(n) = \begin{cases} 1 + h\sin\left(\dfrac{n\pi}{L}\right) & \text{for } n = 1,2,3,\ldots L \\ 0 & \text{for } n \leq 0, n > L \end{cases} \quad (4)$$

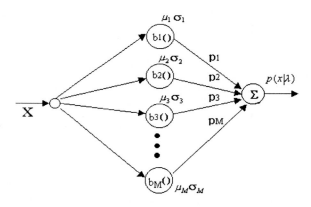

Fig. 3. Gaussian Mixture Model

2.2. Classification Stage

The classification stage uses the Gaussian Mixture Model (GMM) shown in Fig. 3, which provides a model of the main speakers voice sound. This Model is similar to Hidden Markov Model (HMM) with the difference that the GMM only has a state with a Mixture Gaussian distribution that represents different acoustic classes, and ignores the temporal information of the acoustic observation sequence. Here, only one model is building for each speaker that represents all his/her features using only the diagonal of the Covariance Matrix. This representation result in a few amount of parameters, which makes it possible that the GMM model can be used in Text-Independent (TI) speaker recognition problems.

In the GMM model the features distributions of the speech signal are modeled for each speaker by using the sum of the Gaussian weights distribution density of the speech signal of the speaker in the following form

$$p(\overline{\mathbf{x}}|\lambda) = \sum_{i=1}^{M} p_i b_i(\overline{\mathbf{x}}), \text{ with } \sum_{i=1}^{M} p_i = 1 \quad (5)$$

where $\mathbf{x}$ is a random vector of D-dimension, λ is the speaker model, p_i are the mixture weights, $b_i(\mathbf{x})$ are the density components, that is formed by the mean μ and covariance matrix σ_i to $i = 1,2,3,....M$., and each density component is a D-Variate-Gaussian distribution of the form

$$b_i(\mathbf{x}) = \frac{1}{(2\pi)^{D/2}|\sigma|^{1/2}} \exp\left\{-\frac{1}{2}(\mathbf{x}-\mu_i)'\sigma_i^{-1}(\mathbf{x}-\mu_i)\right\} \quad (6)$$

The mean vector, μ_i, variance matrix, σ_i, and mixture weights p_i of all the density components, determines the complete Gaussian Mixture Density, $\lambda=\{\mu,\sigma,p\}$, used to represent the speaker model. To obtain an optimum model representing each speaker we need to calculate a good estimation of the GMM parameters. To do that, a very efficient method is the Maximum-Likelihood Estimation (ML) approach. Here after the data to be used in training are obtained, the method maximizes the Likelihood of GMM, where for a given of T vector used for training, $\mathbf{X}=(\mathbf{x}_1,\mathbf{x}_2,...\mathbf{x}_T)$, the likelihood of GMM can be written as

$$p(\mathbf{X}|\lambda) = \prod_{t=1}^{T} p(\mathbf{x}|\lambda) \quad (7)$$

However it is a no-linear function of the parameters of the speaker model, λ, so that is not possible to make a maximization directly. Then to estimate the parameters of ML it must be used an iterative algorithm called Baum-Welch algorithm. The Baum-Welch algorithm, which is the same algorithm used by HMM to estimate its parameters has the same basic principle of the Expectation-Maximization (EM) algorithm, whose main idea is as follows: Beginning with an initial model, λ, a new model $\bar{\lambda}$ is estimated such that $P(X|\bar{\lambda}) \geq P(X|\lambda)$. Next, we make the model parameters λ equal to those estimated in the actual stage, $\bar{\lambda}$, so that, the new model become the initial model for the next iteration, and so on. Then during the estimation of the GMM parameters, to obtain an optimum model for each speaker, the parameters μ_i, σ_i and p_i should be estimated iteratively until convergence is achieved. The initial condition $p(i|x_t,\lambda)$ is obtained using the Viterbi algorithm [5]. Subsequently, the parameter $\bar{p}$, the Mean vector $\bar{\mu}$ and Variance σ_i^2 required in the subsequent iterations are given by

$$\overline{p}_i = \frac{1}{T} = \sum_{t=1}^{T} p(i|\mathbf{x}_t, \lambda) \tag{8}$$

$$\overline{\mu}_i = \frac{\sum_{t=1}^{T} p(i|\mathbf{x}_t, \lambda)\mathbf{x}_t}{\sum_{t=1}^{T} p(i|\mathbf{x}, \lambda)} \tag{9}$$

$$\overline{\sigma}_i^2 = \frac{\sum_{t=1}^{T} p(i|\mathbf{x}_t, \lambda)\mathbf{x}_t}{\sum_{t=1}^{T} p(i|\mathbf{x}_t, \lambda)} - \overline{\mu}_i^2 \tag{10}$$

where the likelihood a *posteriori* to the i-th class is given by

$$p(i|\overline{x}_t, \lambda) = \frac{p_i b_i(\overline{x}_t)}{\sum_{k=1}^{M} p_k b_k(\overline{x}_t)} \tag{11}$$

This process is repeated until convergence is achieved. The variables that need to be considered, the order of the Mixes and the model parameters previous to the Maximization of the likelihood of GMM, can be different depending on the application. In this research, we choose the Mixes of order 8 and 16, with the values of initial model parameters calculated using the Viterbi training.[5].

2.3. Decision Algorithm

After the models GMM for each speaker have been estimated, the target is to find the model which have the maximum likelihood a posteriori for a observation sequence. Usually

$$\hat{S} = \arg \max_{1 \leq k \leq S} \Pr(\lambda_k|X) = \arg \max_{1 \leq k \leq S} \frac{\Pr(X|\lambda_k)\Pr(\lambda_k)}{p(X)}, \tag{12}$$

where eq. (12) is given by the Bayes rule. Then assuming that the speaker are equally probable and noting that p(x) is the same one for all models of the speakers, the classifiers rule is reduced to

$$\hat{S} = \arg \max_{1 \leq k \leq S} \Pr(X|\lambda_k), \tag{13}$$

and then, from the above described algorithm, the estimated of the speaker identity becomes

$$\hat{S} = \arg \max_{1 \leq k \leq S} \sum_{t=1}^{T} \log p(\overline{x}_k|\lambda_k) \tag{14}$$

where $p(\mathbf{x}_k|\lambda_k)$ represents the Gaussian Mixture density given by eq. (5).

3. System Evaluation Results

This Recognition System was evaluated using a data-base of 80 speakers, provided by the KDD Corporation of Japan, which contain 10 different lists of 50 different contexts of 3-5sec duration and the digit numbers. Both in Japanese language, whose utterances were stored in real condition via telephone with sampling frequency of 8 KHz. In this research we use the speech of 80 speakers of 4 months, for each month the speaker stored the digits and 25 phrases, remaining the other utterances of 2 months to further evaluations of the system. Firstly, the system was evaluated with the data-base of 80 different Speakers, which have 10 different Text stored of about 3-5ms and digits en Japanese Language. Using a Mixes order equal to 8, the global recognition rate was equal to 88.01%. Next, the system was evaluated using separately phrases and digits. When the data-base is used separately the recognition rate improves. In his situation the recognition rate was 96.61% when only phrases are used, and 90.14% when only digits are used.

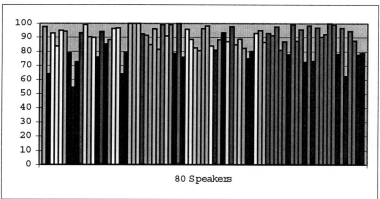

Fig. 4. Performance recognition for each speaker using phrases and digits.

Figure 4 shows the recognition performance obtained for each one of the 80 speaker individually. Here we can see that, even if there are 4 speakers with a recognition rate lower than 70%, most of them presents a recognition higher than 80%.

Next the system was trained and evaluated separately using the data-base divided into 4 groups of 20 speakers each one. Each group is then used independently for training and testing the system, obtaining in this case the following results: For the first group of 20 speakers the recognition rate was 92.15%, for the second group the recognition rate was 93.40%, for the third group it was 92.83% while for the last one the recognition rate was 87.56%. Finally, the system was trained and evaluated using all 80 speakers with a Mixes order of 16. In this situation the recognition rate was 92.87%.

4. Conclusions

This paper proposed a text-independent speaker recognition system based on the GMM model. We shown that Training the Diagonal Gaussian Model GMM with LPCEPSTRAL, the speaker identification system become robust against the noise in the communication channels that distorts the speech, since the speech data used for evaluation were stored in real conditions, and against the number of claimers in the data-base. In addition, we can see that there exist variation of the transcriptions between phrases and digits, so that we can use them to train the system independently. Several issues must still be considered such as to normalize the rate of the best likelihood, as well as to add more features such as the pitch. The addition of a verification module and evaluate the system with data not used during training or with false speakers.

Acknowledgements

This research was supported by The University of Electro-Communications of Tokyo Japan, though The Japanese University Study in Science and Technology (JUSST) program.

References

[1] Douglas A. Reynolds, "Robust Text-Independent Speaker Identification Using Gaussian Mixture Speaker Models", IEEE Transactions on Speech and Audio Processing, Vol. 3, No.1, January 1995.
[2] Richard J. Mammone, Xiaoyu Zhang, Ravi P. Ramachandran, "Robust Speaker Recognition", IEEE Signal Processing Magazine, September 1996.
[3] Douglas O' Shaughnessy, "Sppech Communication",
[4] Lawrence Rabiner , Biing-Hwang Juang, "Fundamentals Of Speech Recognition", Prentice Hall, New Jersey, 1993.
[5] Joseph Picone, "Signal Modeling Techniques In Speech Recognition", Procceding of the IEEE", Jun 3, 1993.
[6] Steve Young, D.Kershaw, J.Odell, D.Ollason, V.Valtchev, P.Woodland, "The HTK Book (for HTK Version 3.0)", Microsoft Corporation, July, 2000.
[6] E. Simancas, M. Nakano Miyatake and H. Perez.Meana, "Speaker Verification Using Pitch and Melspec Information", To appear in The Journal of Telecommunications and Radio Engineering, 2001.

A Comparative Study of ICA Filter Structures Learnt from Natural and Urban Images

Ch. Ziegaus*, E. W. Lang

Institute of Biophysics, University of Regensburg, 93040 Regensburg, Germany
email:elmar.lang@biologie.uni-regensburg.de

Abstract. The Neural-JADE and various other ICA algorithms are applied to natural and urban image ensembles to learn appropriate filter structures. The latter are shown to be represented quantitatively by Gabor and Haar wavelets in case of natural and urban image stimuli, respectively. A quantitative comparison concerning various filter characteristics demonstrates the influence of various score functions upon the resulting filter structures. Quantitative comparison will be made also with neurophysiological characteristics of these structures.

1 Introduction

The visual system is supposed to be optimally adapted to an efficient information processing of the incoming data stream. Functional structures in the early visual pathway should thus be optimally adapted to the statistical structure of the visual patterns [15],[17],[13]. This information processing strategy can be implemented using a recent data analysis technique called *Independent Component Analysis* (ICA). Various ICA algorithms have been proposed recently. Most of them formulate some objective function to be optimized. One of the major difficulties thereby is the need for an *a priori* knowledge of the source densities which is not available, however. The latter enters almost any ICA learning rule via a *score function* $\Phi(\hat{s}_m) = \frac{\partial \log P_m(\hat{s}_m)}{\partial \hat{s}_m}$. The most flexible approach to solve this problem is an adaptive kernel-based source density $P(s)$ estimation [9]. An alternative is to resort to exact algebraic approaches like the JADE [3] algorithm which do not suffer from this problem. If applied to high-dimensional data like images, JADE is, however, computationally too demanding to be practical. Realizing a neural implementation of the JADE algorithm, Neural-JADE [18], this unsatisfactory situation can be remedied.

2 Theory

Neural-JADE: The *Joint Approximative Diagonalization of Eigenmatrices* algorithm (JADE) [3] is an algebraic approach to perform *Independent Component Analysis*. It is based on fourth order cumulants of the data $Cum(z_i, z_j, z_k, z_l)$

---
* Financial support by the Clausen-Simon-Foundation is greatfully acknowledged

with the kurtosis $\kappa_i = Cum(z_i z_i z_i z_i)$ being the corresponding autocumulant. Associated with these cumulants is a fourth order signal space (FOSS) which defines the range of all mappings $[\mathbf{Q}_z(\mathbf{M})]_{ij} = \sum_{k,l=0}^{m-1} Cum(z_i, z_j, z_k, z_l) M_{kl}$ of $m \times m$ - dimensional signal matrices $\mathbf{M}$ onto corresponding *cumulant matrices* $\mathbf{Q}_z(\mathbf{M})$. The eigenmatrix decomposition of the 4-th order cumulant leads to m^2 symmetric matrices $\mathbf{M}^{(p)}$. While all eigenmatrices of the fourth-order cumulant can be jointly diagonalized by an efficient algebraic procedure [3], their algebraic determination becomes rather cumbersome with high dimensional data. Using an extension of Oja's learning algorithm for PCA to higher order neurons [16] we recently showed [18] that a learning algorithm can be derived which adaptively estimates the eigenmatrices of the FOSS. The corresponding weight update rule in case of m output neurons reads [18]

$$\Delta w_{ij}^{(p)} = \eta y^{(p)}(z_i z_j - \delta_{ij}) - \eta y^{(p)} \sum_q w_{ij}^{(q)}(y^{(q)} - Tr(\mathbf{W}^q)) \qquad (1)$$

where $w_{ij}^{(p)}$ denotes the weight matrix of output neuron p connecting to input neurons i and j. The new learning rule thus provides an efficient means to determine the eigenmatrices of the 4-th order cumulant tensor in a neural implementation.

NN-based ICA: Learning appropriate filter structures amounts to determinig a demixing matrix $\mathbf{W}$ whose output $y = \mathbf{W}z$ reconstructs the source signals *s* except for an arbitrary scaling and permutation. Various ICA algorithms in addition to the Neural-JADE have been used, all of which can be implemented on a neural network. They basically use the following learning rule which in a consistently orthogonal model reads

$$\Delta \mathbf{W} = \eta \{\mathbf{I}_n - yy^T + y\Phi^T(y) - \Phi(y)y^T\}\mathbf{W} \qquad (2)$$

Different versions of this basic learning rule result from different approximations to the score function $\Phi(y)$. Possible choices considered in this investigation are a) *nonadaptive activation functions* [2] like a hyperbolic tangent or a logistic function (denoted NN(nat) and NN(nal), respectively), b) *semi-adaptive score functions* [8] (denoted NN(sa)) and c) *adaptive score functions* using either a Gram-Charlie expansion [7] (denoted NN(GCh)) or a kernel-based density estimation [9] (denoted NN(KBDE)) of the unknown source density. Furthermore *non-linear PCA* (nlPCA(x^3)) [10] with a cubic activation function has been considered also.

3 Results

3.1 Haar and Gabor Wavelets

A discrete wavelet transform is given by $f(x) = \sum_{n,m \in \mathbf{Z}} \langle f(x)|\Psi_{n,m}(x)\rangle \Psi_{n,m}(x)$ with $\Psi_{n,m}(x) = a_0^{-n/2} \Psi\left(a_0^{-n} x - b_0 m\right)$ and with a_0, b_0 denoting a scaling and

translation (sampling rate) parameter, respectively.
Haar wavelets: The Haar basis wavelet is given by

$$\Psi(x) = \begin{cases} 1 & : \quad 0 \leq x < 0.5 \\ -1 & : \quad 0.5 \leq x < 1 \\ 0 & : \quad \text{sonst} \end{cases} \quad (3)$$

With $a_0 = 2$ and $b_0 = 1$ an orthonormal basis system $\{\Psi_{n,m}(x)\}$ may be obtained and two-dimensional Haar wavelets can then be constructed easily on a tensor product basis.
Gabor wavelets: A two-dimensional Gabor basis wavelet may be chosen as $(r = (x,y))$

$$\Psi(r) = \frac{1}{\sqrt{\pi}} \exp\left\{-\frac{1}{2}(r^2)\right\} \exp\{i(k_0 r)\} \quad (4)$$

Any two-dimensional Gabor wavelet can then be obtained via scaling, translation and rotation. In the following the special choice $a_0 = 2, b_0 = 1$ will be considered. The Gabor wavelets thus obtained may be further characterized by the following parameters (Δ_i represents the widths of the Gaussian envelopes along i): a) *spatial frequency band width* measured in octaves $\Delta k = \log_2\left(\frac{k_0 + \Delta_{k_x}}{k_0 - \Delta_{k_x}}\right)$ with $k_0 = \sqrt{k_{x0}^2 + k_{y0}^2}$, b) *orientation tuning band width* measured in degrees $\Delta\theta = \arctan\left(\frac{\Delta_{k_y}}{k_0}\right)$, c) *aspect ratio* $\lambda = \frac{\Delta_x}{\Delta_y} = \frac{\Delta_{k_y}}{\Delta_{k_x}}$. Using the aspect ratio one further obtaines the relation $\Delta\theta = \arctan\left(\lambda\frac{2^{\Delta k}-1}{2^{\Delta k}+1}\right)$. Hence, the aspect ratio may be considered a compromise between a spatial frequency and an orientation specificity.

3.2 Evaluation of the learnt filter structures

Image ensembles have been analyzed by extracting at random patches of size 12×12 pixels as input vectors to the neural nets. This corresponds to 144 different filter structures to be obtained. Not all of them could, however, be fitted to Gabor or Haar wavelets, respectively. With urban images only about half of the filter structures indeed corresponded to Haar wavelets indicating an undercomplete situation.
Natural Image Ensembles and Gabor Wavelets: Instead of graphically presenting the filter structures obtained, their relevant Gabor wavelet parameters will be summarized in the following. Filter structures can be characterized in the spatial domain through their localization $r_0 = (x_0, y_0)$, the widths of their Gaussian envelopes (Δ^x, Δ^y) and their orientation θ. Corresponding information in the spatial frequency domain is given through $k_0 = (k_{x_0}, k_{y_0})$ or $|k_0| = k_0 = \sqrt{k_{x_0}^2 + k_{y_0}^2}$ and $\Delta^{k_x}, \Delta^{k_y}$. Filters are visualized in the spatial domain by bars and in the spatial frequency domain by ellipsoids (Fig. 1,2). Results obtained with different NN-ICA algorithms are very similar. Filters spread over the spatial domain quite regularly, are well localized and do not exhibit any

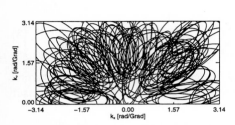 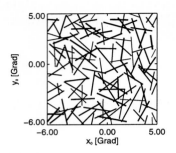

Fig. 1. Filter structures obtained with the NN(nat)-ICA algorithm *left*) in the spatial frequency domain and *right*) in the spatial domain

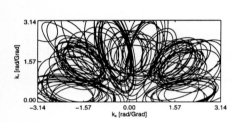 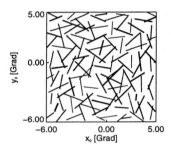

Fig. 2. Filter structures obtained with the NN(KBDE)-ICA algorithm *left*) in the spatial frequency domain and *right*) in the spatial domain

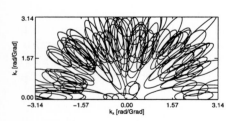 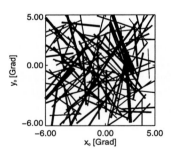

Fig. 3. Filter structures obtained with the Neural-JADE(32)-ICA algorithm *left*) in the spatial frequency domain and *right*) in the spatial domain

preferred orientation. This behaviour translates into a corresponding coverage of the spatial frequency domain with strongly overlapping filter ellipsoids. The distribution clearly exhibits a preferred spatial frequency which becomes even more pronounced in case of the Neural-JADE and the nlPCA-algorithms. Also with the latter algorithms filters become less well localized in the spatial domain, hence show a less good coverage of the spatial frequency domain either. The main axes of the ellipsoids all point towards the origin indicating that the direction of motion of the plane waves runs parallel to the short axis of the ellipsoids in accord with the information theoretic uncertainty principle $\Delta^{k_x}\Delta^x = \frac{1}{2}$ with $\Delta^x = \frac{\sigma_x}{\sqrt{2}}$. Further characterization of the filter structures can be given through corresponding distributions of the Gabor filter parameters: a) *spatial frequency bandwidths* Δk, b) *orientation bandwidths* $\Delta \theta$ and c) the *aspect ratios* λ.

Concerning spatial frequency bandwidths all NN-ICA algorithms yield almost identical distributions with maxima indicating a preferred bandwidth of one octave roughly, while Neural-JADE and nlPCA exhibited significantly narrower distributions with a preferred spatial frequency bandwidth shifted to about 0.5 octaves only (Fig.3). Similar results hold with orientation tuning bandwidth distributions, which peak at about 35° with NN-ICA algorithms and shift to 25° with Neural-JADE and nlPCA (Fig.4). Again distributions are somewhat narrower with the latter two algorithms. Quite the contrary is true in case of aspect ratios where distributions are narrower with the NN-ICA algorithms, peaking at $\lambda = 2$, than with the Neural-JADE and nlPCA algorithms (Fig.5). The latter show much broader distributions with maxima at $\lambda = 3$. These results clearly indicate that fourth order statistics alone are not sufficient to extract appropriate filter structures from the stimulus ensembles. Given the data analysis

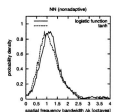
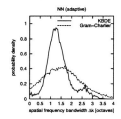
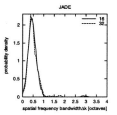

Fig. 4. Spatial frequency bandwidth distributions obtained with various ICA algorithms

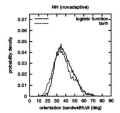
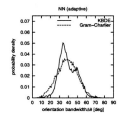
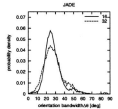

Fig. 5. Orientation bandwidth distributions obtained with various ICA algorithms

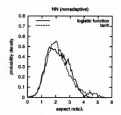 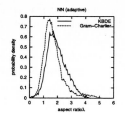 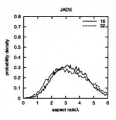

Fig. 6. Aspect ratio distributions obtained with various ICA algorithms

aspect of ICA these filters represent nothing but the columns of the demixing matrix necessary to extract the unknown source signals. Their statistics can be evaluated further by considering the activity distributions resulting given the filter structures learnt. Presenting about 5000 stimuli consisting of natural or urban image patches of size 12 × 12, corresponding output activity ensembles can be generated. With properly prewhitened and normalized sensor signals the 3-rd and 4-th order autocumulants represent *skewness* und *curtosis* of the estimated source signal distributions. With natural images source signal densities are highly symmetric (Skewness $E\{K_3\} = 0.0 \pm 0.2$) but display large values of the curtosis ($E\{K_4\} = 5.9 \pm 1.5$) indicating a rather sparse representation of the images by the output activities. The statistical independence of the estimated sources measured by their mutual information is rather hopeless to determine with high dimensional data(Dim $\approx$ 100). Hence two-dimensional joint distributions of randomly chosen pairs of output activities, their marginal counterparts and the corresponding Kullback-Leibler distances have been calculated using KBDE. Corresponding results are for the entropy H, $H = 2.06 \pm 0.03$ and for the mutual information $MI = 0.026 \pm 0.002$ with no significant differences between different algorithms. If source signals are statistically independent their mutual information vanishes as does the corresponding pairwise mutual information. The reverse is not true in general, however.

Urban image ensembles and Haar wavelets: Corresponding results have been obtained while analyzing the parameters of the Haar wavelets used to quantitatively fit the learned filter structures in case of an urban image ensemble.
Contrary to the natural image ensemble all ICA algorithms indicate some skewness ($E\{K_3\} = 0.1 \pm 0.70$) of the source distributions in case of the urban image ensemble. An even more pronounced distinction is revealed by the curtosis ($E\{K_4\} = 18 \pm 12$) which in all cases is roughly three times as large as with the natural image ensemble indicating an extremely sparse representation. This strong sparseness certainly results from the high specificity of the Haar wavelets used to quantitatively fit the learned filter structures. Again the mutual information $MI = 0.033 \pm 0.005$ between pairs of output activities almost vanishes which would be the case if the sources were indeed statistically independent. The somewhat smaller value of the entropy $H = 1.85 \pm 0.07$ is in accord with the larger curtosis of the source signal densities as the input vectors have been

prewhitened and normalized to unit variance in both image ensembles. Again no significant differences arise from different algorithms.

4 Discussion

Comparison of Gabor filter parameters with receptive field structures:
From a biological point of view the simulated filter structures might resemble receptive field structures of cortical simple cells. Hence, the filter structures and related parameters obtained with the natural image ensemble will be compared with experimental results collected with Macaque monkeys mainly.
Experimental spatial frequency bandwidths [5] typically range from 0.4 to 2.5 octaves with a maximum at $1.0 - 1.5$ octaves. Similar results have been reported from cats also [4]. Related results characterizing the Gabor wavelet filters obtained with the various ICA algorithms are collected in Table 1. It is obvious that learning algorithms which rely on fourth-order statistics only (Neural-JADE and nlPCA(x^3)) yield unrealistically narrow spatial frequency bandwidths with distributions peaking at too small bandwidths of 0.5 octaves. Orientation tuning

Table 1. Summary of spatial frequency (Δk), orientation specificity ($\Delta \Theta$) bandwidths and aspect ratios λ obtained with various ICA algorithms

ICA	Δk	$\Delta \theta$	λ	ICA	Δk	$\Delta \theta$	λ
NN(nal)	1.22	38.43	2.15	NN(GCh)	1.63	38.52	1.70
NN(nat)	1.15	37.11	2.23	JADE(16)	0.48	27.24	3.37
NN(sa)	1.16	37.19	2.19	JADE(32)	0.48	28.09	3.37
NN(KBDE)	1.37	39.70	1.95	nlPCA(x^3)	0.50	21.34	2.54

curves of cortical simple cells have been investigated by [6] and [14]. Experimental results show a large spread of orientation bandwidths ranging from 4 deg up to 50 deg. The distributions are also skewed towards larger bandwidths with an average bandwidth of 32 deg. Corresponding results from cats show a much sharper tuning with bandwidths centered around 17 deg. A comparison with simulation results demonstrates good agreement of the NN-algorithms with Macaque results. Algorithms relying on 4-th order statistics only seem to agree better to results obtained with cats. Movshon [12] and De Valois [6] also observed a strong correlation between spatial frequency and orientation bandwidths which can be substantiated by calculating the aspect ratio λ using a Gabor wavelet description of simulated filter structures. Daugman [4] proposed an aspect ratio $\lambda = 0.5$ to account for results obtained with cats. Corresponding results obtained with monkeys can only be accounted for if $\lambda = 2$. While NN-ICA algorithms yield results well represented with an aspect ratio of $\lambda = 2$, those algorithms using 4-th order statistics only afford an aspect ratio of $\lambda = 3$ in accord with the narrower and more elongated filter structures obtained.

In summary the filter structures obtained from various ICA algorithms using natural images as training stimuli resemble in many important parameters receptive field structures determined experimentally. This may suggest that the underlying principles of redundancy reduction, redundancy transformation and sparse coding are indeed the leading principles behind visual information processing. Though the emerging structures were only driven by the statistical properties of the natural image stimuli, realistic results are only obtained with ICA algorithms which account for all higher order statistics in some unspecified way.

References

1. C.L.Bajaj, G.Zhuang (1997) *An efficient Algorithm and implementation of Haar wavelets*, Technical Report Department of Computer Sciences, Purdue University, West Lafayette, USA
2. A.J.Bell, T.J.Sejnowski (1995) *An information-maximization approach to blind separation and blind deconvolution*, Neural Computation, 7:1129-1159
3. J.-F.Cardoso, A.Souloumiac (1993) *Blind beamforming for non-gaussian signals*, IEEE Proceedings - Part F, 140(6):362-370
4. J.G.Daugman (1985) *Uncertainty relation for resolution in space, spatial frequency and orientation optimized by two-dimensional visual cortical filters*, J.Opt.Soc.A, 2(7):1160-1169
5. R.L.De Valois, D.G.Albrecht, L.G.Thorell (1982a) *Spatial frequency selectivity of cells in Macaque visual cortex*, Vision Research, 22:545-559
6. R.L.De Valois, E.W.Yund, N.Hepler (1982b) *The orientation and direction selectivity of cells in Macaque visual cortex*, Vision Research, 22:531-544
7. G.Deco, D.Obradovic (1996) *An information theoretic approach to neural computing*, Springer Verlag, New York
8. M.Girolami, C.Fyfe (1996) *Negentropy and kurtosis as projection pursuit indices provide generalized ICA algorithms*, NIPS'96, Aspen, Colorado, 249-266
9. M.Habl, Ch.Bauer, Ch.Ziegaus, E.W.Lang (2000) *Analyzing brain tumor related EEG signals with ICA algorithms*, in H.Malmgren, M.Borga, L.Niklasson, eds., Persepctives in Neural Computing: Artifical Neural Networks in Medicine and Biology, pp. 131-136, Springer-Verlag Berlin
10. J.Karhunen, J.Joutsensalo (1994) *Representation and separation of signal using nonlinear PCA type learning*, Neural Networks, 7:113-127
11. T.S.Lee (1996) *Image Representation Using 2D Gabor Wavelets*, IEEE Trans. Pattern Analysis and Machine Intelligence 18:959-971
12. J.A.Movshon (1979) *Two-dimensional spatial frequency tuning of cat striate cortical neurons*, Society of Neuroscience (Abstracts), 9th Annual Meeting
13. B.A.Olshausen, D.J.Field (1997) *Sparse coding with an overcomplete basis set: A strategy employed by V1?*, Vision Research, 37
14. A.J.Parker, M.J.Hawken (1988) *Two-dimensional spatial structure of receptive fields in monkey striate cortex*, J.Opt.Soc.A, 5(4):598-605
15. D.L.Ruderman (1994) *The statistics of natural images*, Network, 5:517-548
16. J.G.Taylor, S.Coombes (1993) *Learning higher order correlations*, Neural Networks 6:423-427
17. Ch.Ziegaus, E.W.Lang (1998) *Statistical Invariances in Artificial, Natural and Urban Images*, Z.Naturforsch. 53a:1009-1021
18. Ch.Ziegaus, E.W.Lang (1999) *Neural Implementation of the JADE - Algorithm*, Lecture Notes in Computer Science 1607:487-497

Neural Edge Detector – A Good Mimic of Conventional One Yet Robuster against Noise*

Kenji Suzuki[1], Isao Horiba[1], and Noboru Sugie[2]

[1] Faculty of Information Science and Technology, Aichi Prefectural University,
Nagakute, Aichi, 480-1198 Japan
{k-suzuki, horiba}@ist.aichi-pu.ac.jp
[2] Faculty of Science and Technology, Meijo University,
Nagoya 468-8502, Japan
sugie@meijo-u.ac.jp

Abstract. This paper describes a new edge detector using a multilayer neural network, called a neural edge detector (NED), and its capacity for edge detection against noise. The NED is a supervised edge detector: the NED acquires the function of a desired edge detector through training. The experiments to acquire the functions of the conventional edge detectors were performed. The experimental results have demonstrated that the NED is a good mimic for the conventional edge detectors, moreover robuster against noise: the NED can detect the similar edges to those detected by the conventional edge detector; the NED is robuster against noise than the original one is.

1 Introduction

Edge detection is one of the most fundamental operations in image analysis. There are two major requirements for edge detection algorithms. One is (1) robust detection against noise, because the objects in an image obtained from real world scenes are generally buried in noise. The other is (2) to detect the desired edges suitable for its application. The desired edges for a certain application differ from those for another one. If we employ the active contour models such as the SNAKES [1] to extract contours, they function relatively successfully using the edges detected as thick curves rather than thin ones.

Many studies on the requirement (1) have been performed so far, and a lot of edge detectors have been proposed [2]–[5]. Most studies have focused their attention on detecting edges from noisy images; less attention has been given to the requirement (2). In an attempt to detect edges from the images with a large amount of noise, insufficient results may be obtained: not only edges but also noises are detected; the edges are detected discontinuously. Therefore, how to detect the desired edges suitable for its application from the image with a large amount of noise has remained a serious issue.

Recently, nonlinear filters based on a multilayer neural network (NN), called neural filters (NFs), have been studied. The NFs can acquire the function of a

---

* This work was supported in part by the Ministry of Education, Science, Sports and Culture of Japan, by the Kayamori Foundation of Informational Science Advancement, and by the Okawa Foundation for Information and Telecommunications.

desired filter through training. It has been reported in [6]–[8] that the performance on noise reduction is excellent. Therefore, there is a possibility that the desired edges can be detected robustly against noise by applying the NFs to edge detection.

This paper describes a new edge detector using a multilayer NN, called a neural edge detector (NED), and its capacity for edge detection against noise. The NED is a supervised edge detector: through training the NED with a set of input images together with the desired edge magnitudes, it acquires the function of a desired edge detector [9]. The experiments to acquire the functions of the conventional edge detectors were performed. By comparative evaluation, it is shown that the performances of the NED on the function acquisition and robustness against noise.

2 Neural Edge Detector

2.1 Edge Detection Problem

In general, an image obtained from real world scenes is corrupted by noise. Assuming the noise is quantum noise, the noisy image can be represented by

$$g(x,y) = f(x,y) + N(\sigma) \tag{1}$$

where $f(x,y)$ denotes a noiseless image, $N(\sigma)$ white Gaussian noise when its standard deviation $\sigma = K_N \sqrt{f(x,y)}$, and K_N a parameter determining the amount of noise. The desired edge magnitude can be calculated from the noiseless image as follows:

$$f_E(x,y) = \varphi\{f(x,y)\} \tag{2}$$

where φ denotes an operator calculating the desired edge magnitude.

The edge detection is a technique to find the operation transforming the noisy image into the desired edge magnitude. Therefore, the edge detection can be formulated as

$$\hat{f}_E(x,y) = \vartheta\{g(x,y)\} \tag{3}$$

where $\hat{f}_E(x,y)$ denotes an estimate for the desired edge magnitude, and $\vartheta(\cdot)$ the operator realizing the edge detection.

In the edge detection, the edge localization process, a process to find the exact location of an edge, is important as well as the estimation of edge magnitude. However, if the edge magnitude can be estimated accurately, the exact location of the edge can be calculated easily. Furthermore, the edge magnitude, rather than its location, is useful in many applications. Therefore, we treat the two separately: the NED estimates the edge magnitude, and then an existing edge localization process may be performed accordingly.

2.2 Architecture of the Neural Edge Detector

In the NED, the edge detection is treated as the convolution with the multilayer NN in which the activation functions of the units in the input, hidden and output layers are an identity, a sigmoid and an identity functions, respectively. We employ an identity function instead of ordinarily used sigmoid one as the activation function of the unit in the output layer, because the characteristics of the NN becomes better in the application to continuous mapping issues such as image processing. The pixel values in a local window R_S are input to the NED. The output of the NED is represented by

$$\hat{f}_E(x,y) = G_M \cdot NN\left[\{g(x-i,y-j)/G_M | i,j \in R_S\}\right] \quad (4)$$

where G_M denotes a normalization factor, and $NN(\cdot)$ the output of the multilayer NN. The universal approximation property of the multilayer NN [10] promises the capability of the NED.

2.3 Training of the Neural Edge Detector

In the NED, the edge detection is treated as the optimizing problem to find the parameters of the NN where the square error between the desired edge magnitude and its estimate is minimized, i.e., the following cost function is minimized through training:

$$E = \frac{1}{P} \sum_p (f_E{}^p/G_M - \hat{f}_E{}^p/G_M)^2 \quad (5)$$

where p is a pattern number, $f_E{}^p$ the pth pattern in the desired teaching edge magnitude, $\hat{f}_E{}^p$ the pth pattern in the estimated edge magnitude, and P the number of patterns. The NED is trained by the modified back-propagation algorithm in [11], and the structure is determined by the method in [12]–[14].

The input noisy image and the teaching edge magnitudes used for training are made from a noiseless image by using Eqs. (1) and (2), respectively. We can select the teacher edge detector as $\varphi(\cdot)$ in Eq. (2). After training the NED with these images, it is expected that the NED can detect the desired edge magnitudes, like the teacher one does, even from noisy images.

2.4 Related work

Applications of the NNs to edge detection have been studied so far. They can be classified into four broad classes: 1) edge detectors based on the cellular NNs [15]–[18]; 2) those based on the multilayer NN [19, 20]; 3) those based on the self-organizing maps [21, 22]; 4) those based on the Hopfield networks [23, 24]. Most of the edge detectors are unsupervised ones, so they can not necessarily satisfy the requirement (2). The NNs in the class 2) are supervised ones. However, they were used as classifiers to classify the pixels into classes such as boundary or not. Therefore, they can not deal with analogue values such as edge magnitude. The supervised edge detector based on a multilayer NN, can deal with analogue values like the NED, has not been proposed as yet.

In addition, the edge detectors in [15]–[17] in the class 1) are based on interesting novel neural processing elements, called the multi-valued and universal binary neurons [25]. The CNNs based on them can realize many techniques in image processing and pattern recognition, including edge detection. Since the CNNs is a locally connected network, they are essentially suitable for hardware implementation.

3 Experiments

3.1 Conventional Edge Detectors

The conventional edge detectors can be classified into three broad classes: 1) gradient-based edge detectors; 2) algorithms dealing with the edge detection as the optimal filtering problem; 3) model-based edge detectors. The better-known representative in the class 1) is the Sobel filter [2]; those in the class 2) are the Marr-Hildreth operator [4] and the Canny edge detector [5]; that in the class 3) is the Hueckel operator [3].

3.2 Training the Neural Edge Detector

To clarify the performance of the NED on acquiring the functions of different edge detectors, various teaching edge magnitudes were made by changing the teacher edge detector. The Sobel filter, the Marr-Hildreth operator, the Canny edge detector and the Hueckel operator were used as the teacher edge detectors. These edge detectors have the following features: the edge magnitude detected by the Sobel filter, the Marr-Hildreth operator, the Canny edge detector, and the Hueckel operator are natural thick, roundish, thin, and thin but a little strong, respectively. The images used for training are shown in Figs. 1(b) and 2(a). These were made from the original Columbia image (size: 480×480 pixels; gray scale: 256 levels) shown in Fig. 1(a), where K_N was set to 1.2% of the maximum level of the gray scale.

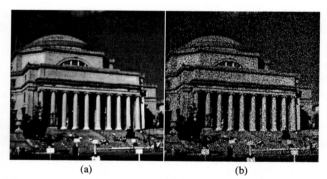

Fig. 1. (a) Original noiseless image and (b) noisy input image (Columbia).

The training set was made by sampling 10,000 points at random from the images. Three-layered NED was adopted, because it has been proved that any continuous mapping can be realized by three-layered NNs [10]. The trainings was performed on 200,000 epochs, and converged with the errors of 1.05% $\sim$ 2.82%. Then, a method for designing the structure in [12]–[14] was applied to the trained NEDs. The optimal structure was obtained: the numbers of units in the input and hidden layers became 17 $\sim$ 50 and 14 $\sim$ 18, respectively. The remaining units in the input layer were within a square region consisting of 11×11 pixels.

3.3 Results of Edge Detection

The detected edge magnitudes are shown in Fig. 2(b). The features of the edge magnitudes detected by the NEDs agree with those of the teaching edge magnitudes. In the results of the NED trained to acquire the function of the Sobel filter, the edge magnitudes are detected as thick curves, like the teacher edge detector does. In the results of the NED for the Marr-Hildreth operator, the detected edge magnitudes are thin, like that detected by the teacher. For the Canny edge detector, the detected edge magnitudes are thin, similar to the teaching one. For the Hueckel operator, the edge magnitudes of the columns in the figure are stronger, like the teaching ones.

The edge magnitudes detected by the original conventional edge detectors for noisy images are shown in Fig. 2(c). In all of the results, a lot of fine noises remain. In the result of the Sobel filter, a large amount of noise is detected mistakenly. In the result of the Marr-Hildreth operator, a lot of roundish noises ramain. In the results of the Canny edge detector and the Hueckel operator, the continuity of the edge magnitudes are spoiled. All of the detected edge magnitudes are different from the desired teaching ones. In contrast to them, in the results of the NED, there are few noises. The detected edge magnitudes are continuous and clear, and similar to the desired teaching ones.

The CPU execution time of each edge detector on a workstation (UltraSPARC-II 300MHz made by Sun Microsystems) was measured. The execution times in second are as follows: the Sobel filter: 0.15, the Marr-Hildreth operator: 2.34, the Canny edge detector: 2.92, the Hueckel operator: 42.38, and the NED: 6.43.

In order to evaluate the generalization ability, the edge detectors were applied to test images, i.e., non-training images. The images are omitted due to the limitation of space. These detected edge magnitudes showed the same features as those in the case of the training images. These results indicate that the NED can detect satisfactory edge magnitudes, similar to the desired ones, even in the case of the test images.

The mean absolute error (MAE) between the desired edge magnitude and the detected edge magnitude was adopted as a metric for evaluation of the performance. The results of evaluation are shown in Table 1. Although the NEDs are trained with only the Columbia image, the MAE's of the NED are the smallest of all even in the case of the test images. These results lead to a conclusion that the performance of the NED is the highest of all in terms of similarity to the desired edge magnitudes.

4 Conclusions

This paper describes a new edge detector using a multilayer NN, called a neural edge detector (NED), and its capacity for edge detection against noise. The NED is a supervised edge detector: the NED acquires the function of a desired edge detector through training. The experiments to acquire the functions of the conventional edge detectors, the Sobel operator, the Marr-Hildreth operator, the Canny edge detector and the Hueckel operator, were performed. The experimental results have demonstrated that the NED is a good mimic for the conventional

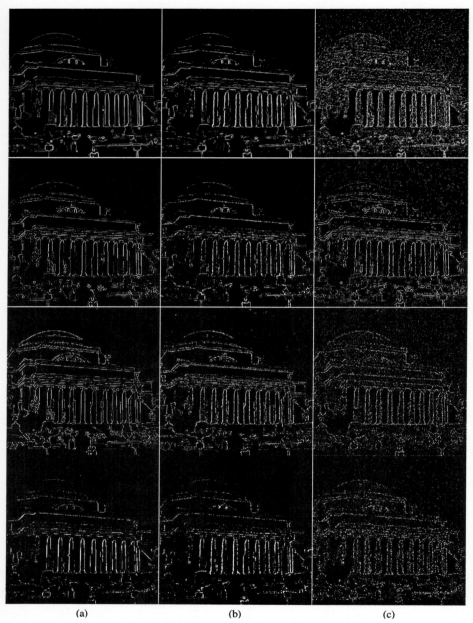

Fig. 2. Comparison of the edge magnitudes detected by the NED with those detected by the conventional edge detectors. (a) Teaching edge magnitudes made from the noiseless image by performing each teacher edge detector. From top to bottom: Sobel filter, Marr-Hildreth operator, Canny edge detector and Hueckel operator. (b) and (c) are the detected edge magnitudes for the noisy image: (b) Neural edge detector; (c) Original edge detector.

Table 1. Quantitative evaluation of the performance using the error between the desired edge magnitude and the detected edge magnitude (Training image: Columbia)

Desired one Image	SNR_I	MAE [%]							
		SF	NED_{SF}	MHO	NED_{MHO}	CED	NED_{CED}	HO	NED_{HO}
Columbia	9.41	17.0	10.1	4.3	3.6	22.5	14.4	6.4	3.5
Lena	4.42	14.7	10.4	3.7	2.4	15.8	11.0	4.9	2.3
Man	6.44	15.7	12.6	4.4	2.8	21.4	12.5	5.4	2.9
Crowd	8.07	16.6	12.0	6.1	4.0	23.4	15.0	6.4	3.5
Airplane	4.18	19.9	11.0	3.6	2.6	20.2	11.3	6.7	3.5
Lake	8.74	19.5	12.6	4.6	3.7	24.8	14.3	6.9	4.0
Bridge	7.50	22.0	19.8	7.1	4.9	34.6	16.6	7.2	4.1
Woman	4.37	15.4	11.9	2.5	2.0	19.4	11.7	5.1	2.8
Couple	5.45	18.3	15.0	4.3	3.2	26.2	13.9	6.1	3.4
Girl	4.85	10.5	8.6	2.8	1.8	13.6	10.8	3.8	2.0
Car	5.26	14.4	10.2	3.2	2.1	18.1	11.5	4.9	2.5
House	6.72	19.8	13.2	4.6	3.4	24.0	13.7	6.6	3.7
Mean	6.28	16.99	12.27	4.26	3.04	21.99	13.05	5.87	3.19

SNR_I: signal-to-noise ratio of the input image in dB. SF: Sobel filter; MHO: Marr-Hildreth operator; CED: Canny edge detector; HO: Hueckel operator; NED: Neural edge detector, the suffix means each NED's teacher edge detector.

edge detectors, moreover robuster against noise: the NED can detect the similar edges to those detected by the conventional edge detector; the NED is robuster against noise than the original one is.

We plan in the near future to perform an analytical study on the NED.

References

1. Kass, M., Witkin, A. and Terzopoulos, D.: Snakes: Active contour models, *Int. Journal Computer Vision*, Vol. 1, No. 3, pp. 321–331 (1988).
2. Duda, R. and Hart, P.: *Pattern Classification and Scene Analysis*, Wiley, pp. 267–272 (1971).
3. Hueckel, M.: A local visual operator which recognizes edges and lines, *Journal of ACM*, Vol. 20, No. 4, pp. 634–647 (1973).
4. Marr, D. and Hildreth, E.: Theory of edge detection, *Proc. Royal Soc. London*, Vol. B207, pp. 187–21 (1980).
5. Canny, J.: A computational approach to edge detection, *IEEE Trans. Pattern Analysis & Machine Intelligence*, Vol. 8, pp. 679–698 (1986).
6. Yin, L., Astola, J. and Neuvo, Y.: A new class of nonlinear filters - neural filters, *IEEE Trans. Signal Processing*, Vol. 41, No. 3, pp. 1201–1222 (1993).
7. Zhang, Z. and Ansari, N.: Structure and properties of generalized adaptive neural filters for signal enhancement, *IEEE Trans. Neural Networks*, Vol. 7, No. 4, pp. 857–868 (1996).
8. Suzuki, K., Horiba, I. and Sugie, N.: Efficient approximation of a neural filter for quantum noise removal in X-ray images, *Neural Networks for Signal Processing IX*, IEEE Press, pp. 370–379 (1999).

9. Suzuki, K., Horiba, I. and Sugie, N.: Edge detection from noisy images using a neural edge detector, *Neural Networks for Signal Processing X*, IEEE Press, pp. 487–496 (2000).
10. Funahashi, K.: On the approximate realization of continuous mappings by neural networks, *Neural Networks*, Vol. 2, pp. 183–192 (1989).
11. Suzuki, K., Horiba, I., Ikegaya, K. and Nanki, M.: Recognition of coronary arterial stenosis using neural network on DSA system, *Systems and Computers in Japan*, Vol. 26, No. 8, pp. 66–74 (1995).
12. Suzuki, K., Horiba, I. and Sugie, N.: Designing the optimal structure of a neural filter, *Neural Networks for Signal Processing VIII*, IEEE Press, pp. 323–332 (1998).
13. Suzuki, K., Horiba, I. and Sugie, N.: An approach to synthesize filters with reduced structures using a neural network, *Quantum Information II*, World Scientific Pub., pp. 205–218 (2000).
14. Suzuki, K., Horiba, I. and Sugie, N.: A simple neural network pruning algorithm with application to filter synthesis, *Neural Processing Letters*, Vol. 13, No. 1, p. 12 pages (2001).
15. Aizenberg, I.: Processing of noisy and small-detailed gray-scale image using cellular neural networks, *J. Electronic Imaging*, Vol. 6, No. 3, pp. 272–285 (1997).
16. Aizenberg, I., Aizenberg, N. and Vandewalle, J.: Precise edge detection: representation by Boolean functions, implementations on the CNN, *Proc. IEEE Int. Workshop Cellular Neural Networks & Their Appli.*, London, pp. 301–306 (1998).
17. Aizenberg, I., Aizenberg, N., Bregin, T., Butakov, C. and Farberov, E.: Image processing using cellular neural networks based on multi-valued and universal binary neurons, *Neural Networks for Signal Processing X*, IEEE Press, pp. 557–566 (2000).
18. Rekeczky, C., Roska, T. and Ushida, A.: CNN-based difference-controlled adaptive nonlinear image filters, *Int. J. Circuit Theory & Appli.*, Vol. 26, pp. 375–423 (1998).
19. Lu, S. and Shen, J.: Artificial neural networks for boundary extraction, *Proc. IEEE Int. Conf. Sys., Man and Cybernetics*, Vol. 3, pp. 2270–2275 (1996).
20. He, Z. and Siyal, M.: Edge detection with BP neural networks, *Proc. Int. Conf. Signal Processing*, Vol. 2, London, UK, pp. 1382–1384 (1998).
21. Nagai, H., Miyanaga, Y. and Tochinai, K.: An edge detection by using self-organization, *Proc. IEEE ICASSP*, pp. 2749–2752 (1998).
22. Toivanen, P., Ansamaki, J., Leppajarvi, S. and Parkkinen, J.: Edge detection of multispectral images using the 1-D self organizing map, *Proc. ICANN*, Skovde Sweden, pp. 737–742 (1998).
23. Bhuiyan, M., Sato, M., Fujimoto, H. and Iwata, A.: Edge detection by neural network with line process, *Proc. IJCNN*, pp. 1223–1226 (1993).
24. Iwata, H., Agui, T. and Nagahashi, H.: Boundary detection of color images using neural networks, *Proc. IEEE ICNN*, pp. 1426–1429 (1995).
25. Aizenberg, I., Aizenberg, N. and Vandewalle, J.: *Multi-Valued and Universal Binary Neurons –Theory, Learning and Applications–*, Kluwer Academic Pub. (2000).

Neural Networks for Image Restoration from the Magnitude of Its Fourier Transform

Adrian Burian, Jukka Saarinen, and Pauli Kuosmanen

Tampere University of Technology,
PO Box 553, FIN-33101, Tampere, Finland
burian, jukkas, pqo@cs.tut.fi

Abstract. In this paper the problem of image restoration from its Fourier spectrum magnitude is shown to be NP-complete. We propose the use of recurrent neural networks for solving the problem. The neural network incorporates the constants related to the real and imaginary parts of the image spectrum. The solution is provided by the steady state of the neural network, then is verified and eventually improved with the iterative Fourier transform algorithm. The obtained simulation results demonstrate the high efficiency of the proposed approach.

1 Introduction

The problem of reconstructing an image from only the spectrum magnitude emerges when the phase of the image is apparently lost or impractical to measure. Phase information has a fundamental importance in many two-dimensional (2-D) signal processing problems. Different applications, such as astronomy, electron microscopy, x-ray crystallography, wavefront sensing, optical imaging, requires the reconstruction of an image known to have a compact support, from its Fourier transform magnitude. The image is reconstructed if the missing Fourier phase is recovered - phase retrieval. For 2-D spatially limited non-negative objects, characterized by analytic spectrum, the solution is unique. Since the Fourier transform of the autocorrelation of the object brightness function is equivalent to the square modulus of the Fourier transform of the brightness function, an image of an object can be reconstructed if the phase retrieval problem is solved.

In the case of 1-D signals the phase retrieval problem has an unique solution only for minimum-phase signals, otherwise the problem is 'ill-posed' [2]. For non-minimum phase signals, there are infinitely many possible functions that are consistent with the given information. But, if the function and the Fourier transform are 2-D, essentially only one function is consistent with the given information. This surprisingly result is intimately linked with the Gauss' Fundamental Theorem of Algebra. The Fundamental Theorem of Algebra says that any 1-D polynomial can be factorized into a product of first order polynomials. This means that 1-D signals can not be specified uniquely from its Fourier transform magnitudes to within trivial ambiguities. For 2-D or higher dimensional signals it is proved that the set of 2-D or higher dimensional reducible polynomials has zero measure [6]. In other words, all Z-transforms of multidimensional M-D, M≥ 2, signals are irreducible, and M-D signals are uniquely specified from the Fourier transform magnitudes (to within translation, signal and space reversal ambiguities).

In [9] it was shown that the 1-D phase retrieval problem is NP-complete. For proving NP-completeness the authors used the restriction technique. They shown that the partition problem, which is a well known NP-complete problem [7], is actually a special case of 1-D phase retrieval problem. In this paper we will consider the NP-completeness of 2-D phase retrieval.

Because of its importance, the phase retrieval problem has attracted many research efforts, and a number of algorithms have been proposed in the literature. The most common one is to use an iterative Fourier transform (IFT) algorithm, which alternates between spatial and transform domains [6]. Iterative algorithms are currently one of the most effective approaches to solving a number of difficult signal recovery problems. These algorithms usually stagnate, failing to converge to a solution, and suffers also from computational complexity. In [13] the problem was formulated as a linear system of equations, while in [12] a window function was used. In [1] a maximum entropy method was considered, and the solution of the considered phase problem was obtained numerically, by using steepest-descent method. We extended this approach to a Hopfield-type maximum entropy neural network implementation in [3]. Usually, the used methods are combined with the IFT algorithm in order to improve the probability of success of phase retrieval [8].

In this paper we consider 2-D phase retrieval using a new architecture of a neural networks. The restoration of an image from its Fourier transform magnitude is regarded as a nonlinear optimization problem with a large number of variables. A recurrent neural network is used to solve directly the corresponding real and imaginary constrained system of equations, without the use of the maximum entropy concept. In this way the complexity increase given by the maximum entropy method is avoided, and a more straightforward hardware realization is obtained. Also, there is no need for setting a time variant design parameter (temperature). The use of neural networks in solving optimization problems has been already proposed [5], enabling us to solve these problems in real time, due to the massively parallel operations of the computing units.

2 Complexity of 2-D Phase Retrieval

The Fourier modulus $|X_0(n,k)|$ of a real and nonnegative object $x_0(m,l)$ of size $N \times N$ is considered to be the given data. The reconstruction problem is to find a real and nonnegative solution such that $|X(n,k)| = |X_0(n,k)|$. The 2-D unitary discrete Fourier transform (DFT) pair is defined as:

$$X(n,k) = \frac{1}{N} \sum_{m=0}^{N-1} \sum_{l=0}^{N-1} x(m,l) e^{-j2\pi(\frac{mn}{N} + \frac{lk}{N})}, \tag{1}$$

$$x(m,l) = \frac{1}{N} \sum_{n=0}^{N-1} \sum_{k=0}^{N-1} X(n,k) e^{j2\pi(\frac{mn}{N} + \frac{lk}{N})}. \quad (2)$$

Given the modulus $|X(n,k)| = |X_0(n,k)|$, the reconstruction of the image is equivalent to the reconstruction of the Fourier phase $\phi(n,k)$ from the magnitude. The phase retrieval problem is equivalent with the autocorrelation retrieval:

$$r_0(m,l) = \text{IDFT}\left[|X_0(n,k)|^2\right]. \quad (3)$$

In order to solve the problem, the usual approach is to find an iterative solution $x(m,l)$ such that $|X(n,k)| -> |X_0(n,k)|$. An important question that is needed to be raised is that whether the obtained solution with $|X(n,k)| -> |X_0(n,k)|$ will lead to $x(m,l) -> x_0(m,l)$. Unfortunately, a direct and complete answer to this question does not exist yet [8]. Instead, the question of the uniqueness of the solution of 2-D phase retrieval problem was thoroughly investigated. It has been established that the solution of a 2-D phase retrieval problem, if it exist, is typically unique to within a few trivial ambiguities - change of sign, shift in position, reversal of orientation of the solution. Previous reported results shown that the 2-D phase retrieval problems are usually solvable and unique [11], [12], [13]. Since the IFT algorithm is the best among all available techniques, the problem becomes to help IFT to overcome stagnation.

Essentially, the IFT algorithm's iterations tends to minimize:

$$E_{IFT} = \frac{1}{4N^2} \sum_{n=0}^{2N-1} \sum_{k=0}^{2N-1} \left[|X(n,k)| - |X_0(n,k)|\right]^2, \quad (4)$$

where the DFT is taken over a $2N \times 2N$ image but $x(m,l)$ is zero outside the window $N \times N$ in order to avoid aliasing in the computation of $|X(n,k)|$. The squared Euclidian distance between x and x_0 is given by:

$$E_{SED} = \|x - x_0\|_2^2 = \sum_{m=0}^{N-1} \sum_{l=0}^{N-1} [x(m,l) - x_0(m,l)]^2. \quad (5)$$

By using the Parseval's theorem we obtain:

$$E_{SED} = \frac{1}{4N^2} \sum_{n=0}^{2N-1} \sum_{k=0}^{2N-1} \left[|X(n,k) - X_0(n,k)|\right]^2. \quad (6)$$

We have that $E_{IFT} \leq E_{SED}$, which means that the error function E_{IFT} does not reflect the true error between $x(m,l)$ and $x_0(m,l)$.

For implementation purposes we separate the real and imaginary parts on both sides of equation (1). Let us denote by M_{nk} the known spectral magnitudes, and by a_{ml}^{nk} and b_{ml}^{nk} the constants related to DFT:

$$a_{ml}^{nk} = \cos(2\pi(mn + lk)/N)/N, \quad (7)$$

$$b_{ml}^{nk} = -\sin(2\pi(mn + lk)/N)/N. \quad (8)$$

The unknown phase is denoted by ϕ_{nk}. The unknown nonnegative object $x(m,l)$ is denoted by x_{ml}. The following equations are obtained:

$$M_{nk} \cos \phi_{nk} = \sum_m \sum_l x_{ml} a_{ml}^{nk}, \qquad (9)$$

$$M_{nk} \sin \phi_{nk} = \sum_m \sum_l x_{ml} b_{ml}^{nk}. \qquad (10)$$

Let us consider the following variables:

$$c_{nk} = 1 + \cos \phi_{nk}; \; s_{nk} = 1 + \sin \phi_{nk}. \qquad (11)$$

We made a translation into the positive domain in order to obtain a more uniform approach for all the unknowns.

We define the following error metric:

$$E_{RI} = \frac{1}{4N^2} \left\| \begin{bmatrix} \mathrm{Re}(X - X_0)^T \\ \mathrm{Im}(X - X_0)^T \end{bmatrix} \right\|_2^2. \qquad (12)$$

According to the triangle inequality we have that $E_{RI} \geq E_{SED}$. Let us denote by $\mathbf{A} \in R^{N^2 \times N^2}$ and $\mathbf{B} \in R^{N^2 \times N^2}$ the matrices that contains the constants given by relations (7) and (8) respectively, arranged on lines after nk and on columns after ml. It is easy to verify that $\mathbf{A} = \mathbf{A}^T$ and $\mathbf{B} = \mathbf{B}^T$. The column vector $\mathbf{M} \in R^{N^2}$ contains all the known magnitudes. The column vectors $\mathbf{C} \in R^{N^2}$, $\mathbf{S} \in R^{N^2}$ and $\mathbf{x} \in R^{N^2}$ contains all the unknowns, corresponding to the notations given above. The error metric (13) can be rewritten as:

$$E_{RI} = \frac{1}{4N^2} \left\| \begin{bmatrix} \mathbf{Ax} - \mathrm{diag}(\mathbf{M}) \cdot \mathbf{C} + \mathbf{M} \\ \mathbf{Bx} - \mathrm{diag}(\mathbf{M}) \cdot \mathbf{S} + \mathbf{M} \end{bmatrix} \right\|_2^2, \qquad (13)$$

where by $\mathrm{diag}(\mathbf{M})$ was denoted the matrix obtained by putting on the main diagonal the elements of vector $\mathbf{M}$.

In the sequel we establish that, for the 2-D phase retrieval problem, we cannot hope to find an algorithm that is efficient for all types of images. In other words, that our problem is NP-complete. So, the best we can hope for, are algorithms that are efficient most of the time, but may be very inefficient for some worst cases. In order to illustrate why the problem is NP-complete in the 2-D case, we will use the NP-completeness of the 1-D phase retrieval and the fact that the 2-D phase retrieval can be reformulated as 1-D phase retrieval with bands of zeros in it [4]. If we denote by $X(y,z)$ the 2-D Z-transform of the real image $x(m,l)$ and with $R(y,z)$ the 2-D Z-transform of its given 2-D autocorrelation (3) the 2-D phase retrieval can be written as follows:

$$X(y,z)X(1/y,1/z) = R(y,z). \qquad (14)$$

For a real image we have:

$$X(1/n, 1/k) = X^*(1/n^*, 1/k^*). \qquad (15)$$

By setting $y = Z^N$ the 2-D phase retrieval can be rewritten as the following 1-D phase retrieval problem:

$$X(Z^N, z)X(1/Z^N, 1/z) = R(Z^N, z), \qquad (16)$$

where the zero-padded 2-D sequences $x(m,n)$ and $r(m,n)$ have been laid out row by row to produce the 1-D sequence. In fact, the 2-D DFT is a 1-D DFT of rows followed by a 1-D DFT of columns. The first DFT preserves the band of zeros and part of the support of the image for the second DFT, whose magnitudes are the given 2-D DFT magnitudes. Each of these 1-D problems are solved separately in lieu of a single combined problem. But all these 1-D problems are NP-complete [9]. Our problem is then a collection of different NP-complete problems. The 2-D phase retrieval is even more difficult because an arbitrary phase factor for each of the 1-D component problems must be determined in order to preserve the remaining support for the 2-D problem. So, we can conclude that also the 2-D phase retrieval problem is NP-complete.

3 The used Neural Network Model

In this paper we will solve the 2-D phase retrieval problem by incorporating the above relations into the structure of a neural network. The obtained neural network architecture is presented in Figure 1. By using the above notations, the 2-D phase retrieval problem is regarded as the following nonlinear constrained optimization problem:

$$\text{minimize } E_{RI}, \qquad (17)$$

subject to

$$x_{ml}, c_{nk}, s_{nk} \geq 0; \qquad (18)$$

with phase normalization:

$$(c_{nk} - 1)^2 + (s_{nk} - 1)^2 = 1. \qquad (19)$$

The phase normalization condition is obtained with the block ϕ which implements the nonlinear function:

$$\phi(x, y) = 1 + \frac{x - 1}{\sqrt{(x - 1)^2 + (y - 1)^2}}. \qquad (20)$$

The time evolution of the neural network is given by the following relations:

$$\mathbf{x}^{k+1} = \mathbf{x}^k - p \cdot \mathbf{A} \left[\mathbf{A}\mathbf{x}^k - \text{diag}(\mathbf{M}) \cdot \mathbf{C}^k + \mathbf{M} \right] - p \cdot \mathbf{B} \left[\mathbf{B}\mathbf{x}^k - \text{diag}(\mathbf{M}) \cdot \mathbf{S}^k + \mathbf{M} \right]; \qquad (21)$$

$$\mathbf{C}^{k+1} = \mathbf{C}^k + p \cdot \text{diag}(\mathbf{M}) \left[\mathbf{A}\mathbf{x}^k - \text{diag}(\mathbf{M}) \cdot \mathbf{C}^k + \mathbf{M} \right]; \qquad (22)$$

$$\mathbf{S}^{k+1} = \mathbf{S}^k + p \cdot \text{diag}(\mathbf{M}) \left[\mathbf{B}\mathbf{x}^k - \text{diag}(\mathbf{M}) \cdot \mathbf{S}^k + \mathbf{M} \right], \qquad (23)$$

where $p > 0$ is the constant learning step parameter.

The steady state of the neural network with the nonlinear constraints satisfied, will provide the solution for the 2-D phase retrieval problem. In order to improve the probability of success of the phase retrieval, the IFT algorithm is applied, with the obtained image as starting point. In this way we are trying to avoid stagnation by providing a better initial guess and enabling a higher likelihood of arriving at a global minimum. To use only the IFT algorithm is not a good idea, because of the unpredictability of its convergence. In [10] a potential source of error in the numerical implementation of the algorithm is discussed. In [11] it was noted that the IFT algorithm might stagnate at some ambiguous image, which may be similar or distinctly dissimilar in appearance to the true object.

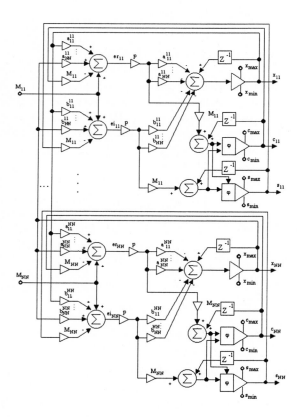

Fig. 1. The architecture of the used recurrent neural network.

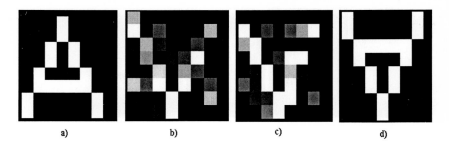

Fig. 2. Example of binary image reconstruction of a 8 × 8 pixel resolution text letter: a) original image; b)c) reconstructed images after 250 iterations; d) obtained solution after 500 iterations.

4 Simulation results

To illustrate the performance of the proposed neural network two different images will be used. For obtaining the results we used a similar procedure like in [8], and we obtained comparable performances. Because randomization is commonly used for initialization of phase retrieval algorithms, all the outputs of the neural network were randomly initialized. For the given plots we used 10 different random initial guesses and we shown only the best 2 obtained results. The first used image is a binary one shown in Figure 2a). Figures 2b) and 2c) shows the reconstructed images after 250 iterations for the best 2 obtained results. In both cases the obtained final result was the one shown in Figure 2d), with an error $E_{RI} = 0.0008$. It is obvious that this image is the reversal version of the original one. We used $p = 0.001$. We repeated the same procedure using a 16 × 16 real image with $p = 0.0001$. The best two results for this image after 2000 iterations are shown in Figure 3d)e), with the corresponding partial results in Figure 3b)c). The obtained $E_{RI} = 0.0294, 0.0468$, respectively. In all cases the last 50 iterations were IFT. The small size of the used images is justified by the high dimensionality of the involved matrices. But the obtained results are also valid for bigger sizes, which are not a problem for the hardware implementations. The power of the neural network implementation lies in its high parallelism.

5 Conclusions

In this paper, we have addressed the image recovery from its Fourier transform magnitude by using neural networks. This problem is regarded as a nonlinear optimization problem and can be considered somehow to be blind. If an ambiguity exist, the recovery process can converge to it. No way has been found to distinguish between the true solution and its ambiguous counterparts. We shown that the problem is NP-complete. The use of neural networks can be beneficially for speeding-up the process of solving the problem. The simulations we made demonstrate that the use of neural networks ensures reliable convergence to the solution.

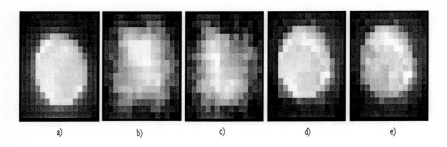

a) b) c) d) e)

Fig. 3. Example of image reconstruction for a 16 × 16 pixel resolution image: a) original image; b)c) reconstructed images after 1000 iterations; d)e) obtained solutions after 2000 iterations.

References

1. Bajkova, A.T.: Phase Retrieval Algorithm based on Maximum Entropy Method. In: Proc. of Int. Symp. on Signals, Systems and Electronics. (1995) 307–309
2. Burian, A., Kuosmanen, P., Rusu, C.: 1-D Non-Minimum Phase Retrieval by Gain Differences. In: Proc. of IEEE Int. Conf. on Electronics, Circuits and Systems, Vol. 3. Pafos, Cyprus, (1999) 1001–1005
3. Burian, A., Kuosmanen, P., Saarinen, J., Rusu, C.: Two Dimensional Phase Retrieval using Neural Networks. In: IEEE International Workshop on Neural Networks for Signal Processing. Sydney, December 2000, 652–662
4. Cantakeris, N.: Magnitude-only reconstruction of 2-D sequences with finite regions of support. In: IEEE Trans. Acoust., Speech and Signal Processing, Vol. 31. (1983) 1256–1262
5. Cichocki, A., Unbehauen, R.: Neural Networks for Optimization and Signal Processing. John Wiley & Sons, (1993)
6. Fienup, J.R.: Phase retrieval algorithms: a comparison. In: Applied Optics, Vol. 21, No. 15. (1982) 2758–2769
7. Garey, M.R., Johnson, D.S.: Computers and Intractability. A guide to the theory of NP-completeness. Bell Telephone Laboratories, Inc., (1979)
8. Mou-yan, Z., Unbehauen, R., Methods for Reconstruction of 2-D Sequences from Fourier Transform Magnitude. In: IEEE Trans. Image Processing, Vol. 6, No. 2. (1997) 222–233
9. Sahinoglou, Z., Cabrera, S.D., On Phase Retrieval of Finite-Length Sequences Using the Initial Time Sample. In: IEEE Trans. Circuits and Systems, Vol. 38, No. 8. August (1991) 954–958
10. Sauz J.L., Huang, T.S., Wu, T.F.: A note on iterative Fourier transform phase reconstruction from magnitude. In: IEEE Trans. on Acoustic, Speech, and Signal Processing, Vol. 32, No. 6. (1984) 1251–1254
11. Seldin, J.H., Fienup, J.R.: Numerical investigation of the uniqueness of phase retrieval. In: Journal Opt. Soc. Amer., Vol. 7, No. 1. (1990) 3–13
12. Wooshik, K.: Two-Dimensional Phase Retrieval using a Window Function. In: Proc. of IEEE Int. Conf. on Acoust., Speech and Signal Processing, Vol. 3. (1999) 1625–1628
13. Yagle A.E., Bell A.E.: One- and Two-Dimensional Minimum and Nonminimum Phase Retrieval by Solving Linear Systems of Equations. In: IEEE Trans. on Signal Processing, Vol. 47, No. 11. (1999) 2978–2989

An Automatic System for the Location of the Optic Nerve Head from 2D Images

Margarita Bachiller[1], Mariano Rincón[1], José Mira[1], Julián García-Feijó[2]

[1] Dpto. de Inteligencia Artificial, Facultad de Ciencias, UNED, Paseo Senda del Rey s/n
28040, Madrid, Spain
{marga,mrincon,jmira}@dia.uned.es
[2] Dpto. de Oftalmología Hospital Clínico San Carlos,Instituto de Investigaciones
Oftalmológicas Ramón Castroviejo, Universidad Complutense, Madrid, Spain
mherrerad@sego.es

Abstract. In this paper, we present a vision system in order to improve the diagnosis process of patients with glaucoma. The more accurate mean of assessing is to study the optic nerve head directly from ocular fundus images. Due to the complexity of the problem, our approach decomposes it into simpler subtasks until the primitive level is reached. An example of operationalization of each primitive is shown. Besides a number of experiments were performed in order to detect the papilla contour on real medical images that confirm the proposed operationalization. The main advantages of the system designed are the elimination of the subjectivity that exists in the process of identifying the objects that are present in the ocular fundus image and the full automatization of the process.

1. Introduction

Glaucoma is one of the major causes of preventable blindness in the developed world [1]. In order to combat the disease it is essential for it to be detected as early as possible so that the therapy acts efficiently. Moreover, patient follow-up is essential for assessing the progress of the disease. Glaucoma induces nerve damage to the optic nerve head (ONH) via increased pressure in the ocular fluid. A first assessment of the patient's eye situation related to glaucoma relies on three main measures: (1) Status of the intraocular pressure ; (2) Status of the ONH and; (3) Status of the visual field. The more accurate mean of assessing the treatment would be to assess the ONH directly. This is the most sensitive and definite indicator of glaucoma in its early stages and can also give a quantification of the treatment's results as the disease progresses [2].

As glaucoma progresses, the morphologic characteristics of the optic disk changes. This change can be evaluated by direct ophthalmoscopy or by several measuring procedures which determine the value of interest parameters such as cup-disk diameter ratio, cup-disk area ratio, neuroretinal rim area of the eye, etc. The main advantage of these measuring procedures lies in the disposal of more robust measurements independent of the human expert which examine to the patient.

The process starts with the acquisition of observables by means of different techniques: digital images, stereo images, confocal laser tomographies, ultrasound images, etc. The image obtained is then examined to identify each of the regions that exist in the ONH and, on the basis of those regions, to establish the set of parameters that enable the papilla to be classified as normal, doubtful or pathological.

The more precise measurement of all parts of eye are obtained with a scanning laser system (The Heidelberg Retina Tomograph) [10]. However, the analysis process requires the definition of the optic disk margin, in other words, it is a semiautomatic system where is necessary that the human expert controls the process [8]. In addition, this system is only used in the research field due to it is extremely expensive. Systems using simpler instrumentation [3, 4] have not yet solved the problem satisfactorily but since an ophthalmologist can recognize glaucoma from an image of the optic nerve it should be possible to design an automatic system that resolves the problem.

Specifically, the 2D image is the easiest information to obtain without the need for expensive equipment. Nonetheless, digital processing of the image is a difficult task, particularly because of the variability of the images in terms of the gray levels, the morphology of the elements sought and the existence of particular features in different patients, which may lead to an erroneous interpretation. However this case study demonstrates how, by using a knowledge-based system, it is possible to extract the information from the image that is needed by the expert to diagnose glaucoma.

The bibliography contains a considerable number of case studies devoted to creating an automatic system that detects the optic disk, although it has to be said that, to date, no entirely satisfactory results have been obtained. Cox and Wood in [3] presented a semi-automatic method in which, on the basis of digital ocular fundus images, the expert has to input a series of points with a high probability of belonging to the contour of each element. These points are used by the algorithm to determine the real contours by analyzing the gray levels. Later, Morris and Wood [6] presented a totally automatic method which determined the contour of the papilla using the properties of the gray-level gradient. However, in [7] they used the semi-automatic method again, which produced more reliable results.

On the basis of the case study presented by Lee and Brady [4] in which active contours were used to detect the contour of the ONH, Morris and Donnison improve on their results in [5] by eliminating the blood vessels that were hindering detection of the contour by incorporating discontinuities in their trajectory. In both studies, the image was pre-processed to increase the difference between the retina region and the optic disk before commencing the process of searching for the contour. Nevertheless, their research focuses solely on extracting the limit of the papilla, and provides no solution for detecting the cup. Moreover, these methods require quality images and the intervention of an expert; that is to say, the systems presented are not fully automatic.

The purpose of this research is to develop a fully automatic system which obtains a computable ONH model using only 2D ocular fundus images as input information, so that its morphology can then be analyzed and its status assessed for the diagnosis of glaucoma. This will require detailed inspection of the ONH 2D images of both a patient's eyes.

The article is organized in the following way. Section 2 describes how to address the task which, because of its complexity, has to be broken down into simpler subtasks. Section 3 explains the operationalization of these sub-tasks while Section 4 contains experimental results of its application in the localization of the papilla. Finally, the findings obtained in this work are set out in Section 5.

2. Task Decomposition

The aim is to obtain a computable model of the structure of the domain. As mentioned above, because of the complexity of this task, it is broken down into simpler sub-tasks until the primitives level is reached. The usefulness of breaking down a problem into generic tasks and sub-tasks has been demonstrated in recent decades [11,12,13,14]. The main advantage is that it has given an insight into the problem of knowledge engineering as a modeling activity, in which there is an initial structure, depending on the task, to which the domain knowledge has to be attached by assigning that knowledge to roles into the task. This clarifies the use of the domain knowledge in the task, which facilitates and steers the dialogue with the domain expert during the knowledge acquisition stage.

The process of analyzing the problem calls, in the first place, for a definition of the elements and relations that appear in an ocular fundus image. On a first approach two independent regions can be distinguished, normally differentiated by their gray levels: the ONH, characterized by the fact that it is almost circular in shape and a pale color, and the retina, which surrounds the former and is characterized by the fact that it is darker in color. The continuity of these two regions is disturbed by the presence of the blood vessels which have the darkest color in the ocular fundus. In turn, the ONH is divided into two parts: the cup, which includes the palest region (pallor), and the neuroretinal rim. Clearly there may be other elements there as well, depending on the patient, such as hemorrhages and atrophies. In this case study these last elements have not be taken into consideration owing to the fact that, in the case of the hemorrhages, their prevalence is minimal, and, in the case of the atrophies, their contribution to the diagnosis of glaucoma is limited.

Occasionally, the image portrays noise, which may make the case study focus on an erroneous region. For this reason, it is essential to find the particular zone of the ocular fundus with a high probability of belonging to the ONH. Thus, the first task that the vision system has to perform is to locate the zone that is the subject of the research or region of interest (ROI).

Sometimes it is impossible to define a set of invariant visual characteristics which enable each of the objects present in an image to be identified independently, however there is a domain model which relates the objects featured. In these cases a strategy based on the domain model is all-important for identifying the objects in the image. This happens frequently in medical image interpretation, where the visual characteristics (shape, color, texture, position, etc.) of the objects are hard to describe in precise quantitative terms, although there is a structural model which relates the domain objects. The use of this model helps reduce the complexity of the solution and in some cases is the only means of achieving it.

The localization of the objects featured in the image calls for a strategy based on progressive refinement in the localization of all the objects and the possibility of reinterpreting regions of the image in the event of inconsistencies with the domain model. The method proposed for performing this task, therefore, comprises two stages:

1) *To propose* an assignation of image regions to each of the domain objects, using the domain knowledge to do so (visual features of the objects and structural relationships between objects).
2) *To revise* this assignation of regions by actions designed to eliminate discrepancies with the domain model.

Figure 1 gives a structured picture of the breakdown of the task into sub-tasks and primitive inferences:

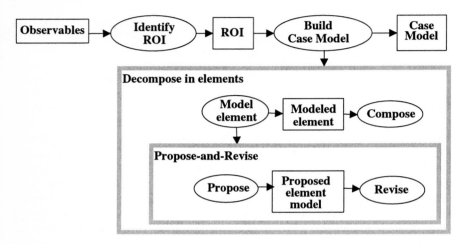

Figure 1. Task decomposition

3. Subtasks Description

The operationalization of the sub-tasks described in the previous section is explained below.

3.1 Identify Region of Interest (ROI)

As described above, it is essential to find in the image the particular region of the ocular fundus that, with a high degree of probability, belongs to ONH. A previous case study demonstrated that depending on which RGB image channel was selected it was possible to detect part of the papilla or part of the cup. This prompted a separate study of each color channel. When the image obtained through the red channel is used, global thresholding enables the papilla mainly to be located, coupled with any noise that may be present in that channel. On the other hand, when the green channel is analyzed using the same method, the pallor is the element detected, as well as the noise present in that channel.

As the noises of both channels seldom coincide, the result obtained after intersecting the elements detected by each channel enables the ocular fundus region which belongs to the ONH to be located correctly and, consequently, enables the noise in the image to be eliminated. Finally, from that region a reduced image is defined as a subset of the original image.

The main advantage of this subtask is the elimination of errors caused by the use of images corrupted by noise. In addition to this advantage, the computation time of the rest of the process that the vision system has to perform is significantly decreased, by reducing the size of the image that has to be analyzed.

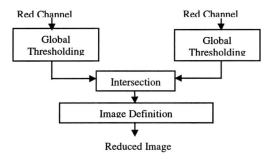

Figure 2: Flow Diagram of *Identify ROI* Subtask

3.2 Build Up Case Model

The process of analysis of the ONH involves detection of the contours that define the neuroretinal rim, since they separate the three main ocular fundus elements: retina, neuroretinal rim and cup. Image segmentation is the process that divides up the image into these three structures. In order to evaluate the ONH accurately, the segmentation process must also be very accurate. The following is a description of an operationalization of the *model element* subtask which can be applied both to locating the papilla and to the cup.

Propose subtask
Global thresholding based on the histogram is the simplest technique for performing the segmentation and the method employed in [9]. However, it is very difficult to select an adequate threshold value owing to the great variability of eye types and, consequently, the variety of histogram profiles. The process starts by selecting a threshold value equal to the average value between the first two maximums with the greatest degree of intensity. On the basis of the region detected and taking into account the a priori knowledge of the shape, the results can be improved through a procedure that alters the threshold value accordingly until the region detected is closest to the shape considered. Specifically, the process described bellow is for the particular case in which the shape sought is a circumference.

As said before the process starts by thresholding the image with a selected value. It produces a binary image formed by n independent objects with an area over and above a minimum value. A study is performed on the standard deviation of each of these objects from the shape chosen in order to select the best candidate for the region sought. Thus:

Let O_i be the i_esimo object with i = 1,2,...n, (x_{ci}, y_{ci}) its centroid, m the number of pixels of its contour and (x_j, y_j) the coordinate of the pixel j with j = 1,2,...m. Then the standard deviation d_i is given by:

$$d_i = \frac{\sum_{j=1}^{m}\sqrt{(x_j - x_{ci})^2 + (y_j - y_{ci})^2} - M}{n}$$

where M is the median of the distances from the center of gravity to each of the contour points.

If the standard deviation is lower than a particular value which is calculated experimentally, the current threshold decreases and the previous process is repeated. In another case, the threshold that produced the least deviation was selected.

Revise subtask

In order to obtain more precise results, it is necessary to revise the contour obtained during the propose subtask. On each of the pixels previously detected, the gray-level similarity with its neighboring pixels is studied. Specifically, a reduced image around the pixel under study is selected and the gradients associated with each pixel in that image are determined. It can logically be assumed that if the current pixel belongs to the contour sought, it should have the highest gradient value. Otherwise it implies that this is not the best choice and it is then altered.

During the process of modifying a contour pixel, it is subject for a force F. Let P_t and P_{t+1} be the pose of pixel j in two consecutive moments. Let $\vec{n}_j$ be the unitary vector whose direction is equal to the line between the pixel j and the centroid (x_c, y_c). Let P_c be the pixel with more high gradient following the direction of $\vec{n}_j$. The discrete version of the equation is:

$$P_{t+1} = P_t + k \cdot F_j$$

where k is a constant which is calculated experimentally and F_j is the force of pixel j defined by:

$$F_j = \|P_t P_c\| \cdot \vec{n}_j$$

being $\|P_t P_c\|$ the distance between P_t and P_c.

4. Implementation and Results

Algorithms described above were applied to a set of images of the ONH. The images analyzed were color images stored in RGB format and were processed at a resolution of 450x450 pixels.

4.1 Identify Region of Interest

The first problem that occurs during the thresholding process used in the Identify ROI subtask is how to choose the right threshold value. For any image obtained from one of the color channels there are, in principal, various regions with different gray-level values. Thus, for example, from the image obtained with the red channel, three regions are visible: retina, papilla and blood vessels, so that in an ideal histogram there should be three local maximums. However, in most cases we have a flater histogram.

Consequently, the choice of threshold is not obvious. The process employed uses an adaptative threshold so that starting with a threshold value equal to the maximum intensity of the gray level in the image, it is modified according to the size of the objects detected. This strategy involves the detection of objects in the retina region which will be eliminated in the next stage by the process of intersecting the results obtained by both channels. Figure 3 shows the results of applying the algorithm.

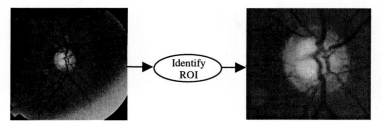

Figura 3. Results of Identify ROI subtask

The previous method was tested on a set of 100 images and a 95% accuracy rate was obtained. The failures occurred in too-dark images in which even the expert was unable to establish the ONH limits.

4.2 Build Up Case Model

In this article we have presented the results obtained when applying the algorithms described to detecting the papilla.

Propose subtask
As mentioned above, the image obtained through the red channel is, in principal, the most appropriate image for detecting the papilla. The main problem with the algorithm proposed in the previous section lies in making the right choice of threshold value from the information provided by the histogram.

There are two types of histogram profile. In the first there is a clearly differentiated maximum associated with one of the highest values of the intensity. In these cases the choice of threshold is obvious. However this histogram profile is presented by saturated images in which case the assignation of the threshold to that first maximum will detect, during the thresholding process, regions of the retina as belonging to the papilla. On the basis of the ocular fundus knowledge it is possible to tabulate the maximum area of the papilla, which will enable saturated images to be rejected as during the detection process the resulting area will be larger than the maximum area. In these cases the image used should be rejected and the image obtained should then be studied in other color channels. In the other type of profile there is no differentiated maximum, so it is necessary to select the threshold approximately between the first two relative maximums.

Figure 4 shows the results obtained from the method described in section 3.1. The results obtained in the figure on the right are insufficient and it makes necessary to revise the contour proposed in this first approach.

Revise subtask

In one of the operationalizations of this subtask, the contour detected during *propose* subtask is altered by applying the local algorithm described in section 3.2. The results obtained are shown in figure 5.

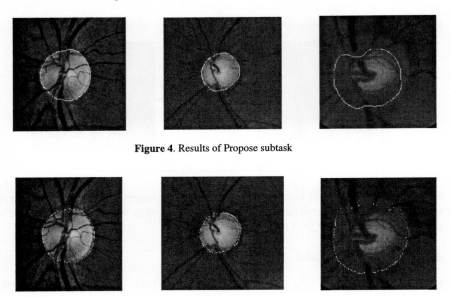

Figure 4. Results of Propose subtask

Figure 5. Results of Revise subtask

5. Conclusion

Designing a vision system capable of evaluate the ONH is no easy task, due principally to the large variety of eye fundi and, consequently, to the difficulty of defining a pattern that is valid in every case.

In the past, the systems designed have called for the intervention of the medical expert during the contour detection process. This article sets out the bases for the design of an automatic system that enables each of the ONH elements to be located fully automatically. This design has been possible as a result of prior analysis of the task, enabling it to be broken down into simpler subtasks. These subtasks can be implemented directly using current image-processing techniques and application-specific domain-knowledge and their composition has enabled an initially complex problem to be solved. To validate the method proposed for digital image processing, experimental results have been presented.

The system designed represents a considerable improvement in the diagnosis process of glaucoma as in the first place it avoids subjective input by the doctor when defining the different parts of the ONH, and provides a tool that enables a rapid comparison to be made during the follow-up of a patient's disease. It also has the following advantages: (1) accurate measurements; (2) new type of measurements; (3) reliable measurements; (4) possibility of monitoring the disease; (5) automation of the measurement for population screening.

Acknowledgments

We would like to thank to the Opthtalmology Service of Miguel Servet Hospital and to the Oftalmology Department of the San Carlos Clinic Hospital for letting us using their images in this study. We would like to thank the CICYT for its support, through the DIAGEN project (TIC-97-0604), in the context of which this case study has been carried out.

References

[1] Quigley, H.A. and Vitale, S.: Models of open-angle glaucoma prevalence and incidence in the United States. Investigate Ophthalmology & Visual Science, 38(1): PP83-91, 1997.
[2] Caprioli, J.: Clinical evaluation of the optic nerve in glaucoma. Tr Am Ophtel Soc. 1994 XCII.
[3] Cox, M. J. , Wood, I. C. J.: Computer-Assisted Optic Nerve Head Assessment. Ophthal. Physiol. Opt., vol 11, pp 27-35. 1991.
[4] Lee, S., Brady, J. M.: Integrated Stereo and Photometric Stereo to Monitor the development of Glaucoma. Proceedings of the Brithish Machine Vision Conference, pp 193-198. 1990.
[5] Morris, D. T. , Donnison, C.: Identifying the Neuroretinal Rim Boundary using Dynamic Contours. Imagen and Vision Computing, vol 17, pp169-174. 1999.
[6] Morris, D. T., Cox, M. J., Wood, I. C. J.: Automated Extraction of the Optic Nerve Head Rim. American Association of Optometrists, Annual Conference , Boston.1993.
[7] Morris, T., Wood, I.: Automated Extraction of the Optic Nerve Head. American Academy of Optometrists, biennial European meeting in Amsterdam. 1994.
[8] Quantitative Three-Dimensional Imaging of the Posterior Segment with the Heidelberg Retina Tomograph. ARTÍCULO INTERNET.
[9] Rosenthal, A. R., Falconer, D. G. and Barrett, P.: Digital measurement of pallor-disc ratio. Arch. Ophthalmol.98,2017-2031. 1980.
[10] Nasemann, J.E.R., Burk, O.W.: Scanning Laser Ophthalmology and Tomography, pp 161-264. 1990.
[11] Clancey: Heuristic classification. Artificial Intelligence 27. 1985. pp- 289-350.
[12] Chandrasekaran B.: Design problem solving: A task analysis. AI Magazine. Winter 1990. pp 59-71.
[13] Schreiber G., Wielinga B., de Hoog R., Akkermans H. and Vande Velde W.: CommonKads: A comprehensive methodology for KBS development. IEEE Expert. December 1994, pp 28-37.
[14] Schreiber G., Akkermans H. , Anjewierden A. , de Hoog R., Shadbolt N., Van de Velde W. and Wielinga B. Knowledge Engineering and Management: The CommonKADS Methodology. The MIT Press. 1999.

Can ICA Help Classify Skin Cancer and Benign Lesions?

Ch. Mies[1], Ch. Bauer[1], G. Ackermann[2], W. Bäumler[2], C. Abels[2],
C.G. Puntonet[3], M.R. Alvarez[3], and E.W. Lang[1]

[1] Institute of Biophysics, University of Regensburg,
D-93040 Regensburg, Germany
[2] Department of Dermatology, University Hospital,
D-93042 Regensburg, Germany
[3] Departamento de Arquitectura y Tecnologia de Computadores
E.T.S. Ingenieria Informatica, Universidad de Granada, E-18071 Granada, Spain

Abstract. Various neural network models for the identification and classification of different skin lesions from ALA-induced fluorescence images are presented. After different image preprocessing steps, eigenimages and independent base images are extracted using PCA and ICA, respectively. In order to extract local information in the images rather than global features, Generative Topographic Mapping is added to cluster patches of the images first and then extract local features by ICA (local ICA). These components are used to distinguish skin cancer from benign lesions. An average classification rate of 70% is obtained, which considerably exceeds the rate achieved by an experienced physician.

1 Introduction

Many kinds of biomedical data, such as fMRI, EEG or optical imaging data, form a challenge to any data processing software, due to their high dimensionality. Low dimensional representations of these signals play key role for solving many computational problems. Therefore, Principal Component Analysis (PCA) has been used in the past to provide practically useful and compact representations. Futhermore, PCA was successfully applied to classification tasks such as face recognition. [TP91]. However, one major drawback of PCA is its global, orthogonal representation, which often cannot extract the intrinsic information of highdimensional data.

Independent Component Analysis (ICA) is a generalization of PCA, which decorrelates higher order moments of the input in addition to the second order moments. In a task such as image recognition much of the important information is contained in the higher order statistics of the image. Hence ICA might be able to extract local feature-like structures of objects, such as fluoresence images of skin lesions. Bartlett [BLS98] recently demonstrated that ICA outperformed the face recognition performance of PCA.

Finally, local ICA was proposed by Karhunen *et al* [KM99] to take advantage of localized features in highdimensional data. Using Generative Topographic

Mapping, multivariate data are first split up into various clusters, followed by a local feature extraction using ICA within these clusters.

Here, we intend to classify different skin lesions, e.g. basal cell carcinoma, actinic keratosis or psoriasis plaques through their fluorescence images. Even an experienced physician is unable to distinguish the malignant from the benign lesions when only fluorescence images are taken. Note, that for sake of simplicity we will just denote the diseases as *malignant*, though basal cell carcinoma is a skin cancer but actinic keratosis is considered a premalignant condition only.

After reducing the dimension of the data by coarse graining, the images were transformed via serveral transfer functions which are standard methods in image processing. The goal was to stress either contrast or smoothness of the images. Thus, we generated different ensembles of images for each cell type.

2 Calculation of Eigenimages Using PCA

Principal component analysis is a well known method for feature extraction but it uses second order statistics only. The scheme is based on an information theoretic approach which decomposes images into a small set of orthogonal feature images called *eigenimages*. Recognition is performed by projecting a new image into the subspace spanned by the eigenimages and then classifying it by comparing its position in image space with the position of known images.

Assume the image vector $\mathbf{x} = [x(1),\ldots,x(N^2)]^\top$ to have zero-mean, i.e. $E\{\mathbf{x}\} = 0$. Then the project a of the image vector $\mathbf{x}$ onto the vector $\mathbf{u} = [u(1),\ldots,u(N^2)]^\top$ is given by $a = \mathbf{x}^\top \mathbf{u} = \mathbf{u}^\top \mathbf{x}$ and the variance of the projection can be calculated according to

$$\sigma^2 = E\{a^2\} = \mathbf{u}^\top E\{\mathbf{x}\mathbf{x}^\top\}\mathbf{u} = \mathbf{u}^\top \mathbf{C} \mathbf{u} \qquad (1)$$

where $\mathbf{C}$ is the symmetrical correlation matrix. Consider now m images $\mathbf{x}_i$ to form the columns of the $N^2 \times m$ dimensional images matrix $\mathbf{X} = (\mathbf{x}_1,\ldots,\mathbf{x}_m)$. Thus the covariance matrix is simply calculated by

$$\mathbf{C} = \frac{1}{m}\sum_{i=1}^{m}\mathbf{x}_i\mathbf{x}_i^\top = \mathbf{X}\mathbf{X}^\top,$$

where $\mathbf{C}$ is now a $N^2 \times N^2$ matrix with N^2 being the dimension of an image. Extending equ. 1 to a complete orthonormal basesystem $\mathbf{U} = (\mathbf{u}_1,\ldots,\mathbf{u}_m)$ and projecting the images into the PCA-subspace, the following eigenvalue problem results:

$$\mathbf{X}^\top \mathbf{X} \mathbf{u}_i = \Sigma \mathbf{u}_i \qquad (2)$$

with $\Sigma = \mathrm{diag}[\sigma_1^2,\ldots,\sigma_n^2]$ being the variance matrix.

Solving the eigenvalue problem for large matrices (i.e. for image we have $128^2 \times 128^2$ matrices) prooves computationally very demanding and thus Turkland introduced the following dimension reducing technique [TP91]: Consider the equation

$$\mathbf{X}^\top \mathbf{X} \mathbf{v}_i = \mu \mathbf{v}_i. \qquad (3)$$

Premultiplying eqn.3 with $\mathbf{X}$ and using eqn.2, the first m eigenvectors of the system can be calculated according to

$$\mathbf{u}_l = \sum_{i=1}^{m} v_{li}\mathbf{x}_i, \quad l = 1, \ldots, m. \tag{4}$$

With this analysis the number of calculations are greatly reduced from the order of the number of pixels N^2 in the images to the order of the number of images m in the training set.

Finally, a new image is transformed into its eigenimage components by the operation $\omega_i^k = \mathbf{u}_k^\top \mathbf{x}_i$ with $k = 1, \ldots, m' \le m$ and a subsequent backprojection according to

$$\mathbf{x}_i^f = \sum_{k=1}^{m'} \omega_i^k \mathbf{u}_k.$$

For classification the minimal reconstruction error for all images is calculated by

$$\epsilon_i = \|\mathbf{x}_i - \mathbf{x}_i^f\|, \quad i = 1, \ldots, m. \tag{5}$$

3 Extraction of Independent Base Images by ICA

Independent Component Analysis (ICA) was proposed to extract independent source signals, $\mathbf{s} = \{s_1(t), \ldots, s_m(t)\}$ after they are linearly mixed by an unknown matrix $\mathbf{A}$. Nothing is generally known about the sources or the mixing process except that there are L different recorded mixtures, $\mathbf{x} = \{x_1(t), \ldots, x_n(t)\} = \mathbf{As}$. The aim is to estimate the mixing matrix $\mathbf{A}$ by means of another matrix $\mathbf{W}$ such that the output vectors $\mathbf{y}(t) = \mathbf{Wx}(t)$ coincide with the original sources $\mathbf{s}(t)$ except for a scale factor and a permutation, i.e. $\mathbf{WA} = \mathbf{PD}$ where $\mathbf{P}$ is a permutation matrix and $\mathbf{D}$ is a diagonal matrix.

Here we use the Bell-Sejnowski learning algorithm with the natural gradient learning rule [YA97]

$$\mathbf{W}(t+1) = \mathbf{W}(t) + \eta \left(\mathbf{I} - \phi(\mathbf{y})\mathbf{y}^\top\right)\mathbf{W}(t), \tag{6}$$

where $\phi(\mathbf{y})$ denotes a sigmoidal transferfunction of the network output activities.

The images comprise the rows of the input matrix $\mathbf{X}$. With the input images in the rows of $\mathbf{X}$, the ICA outputs in the rows of $\mathbf{WX} = \mathbf{U}$ are also images and provide a set of independet basis images. In order to have control over the number of independent components the algorithm was performed on the first m principal component vectors of the image set.

Let $\mathbf{P}_m$ denote the matrix containing the principal components in its columns. For the projection into the m-dimensional PCA subspace, we first calculate $\mathbf{R}_m = \mathbf{XP}_m$ and reconstruct the images according to $\mathbf{X}^{\text{rec}} = \mathbf{R}_m \mathbf{P}_m^\top$ with a minimum squared error approximation of $\mathbf{X}$. The ICA algorithm then calculates a matrix $\mathbf{W}$ such, that $\mathbf{WP}_m^\top = \mathbf{U}$. Therefore we get

$$\mathbf{X}^{\text{rec}} = \mathbf{R}_m \mathbf{W}^{-1} \mathbf{U} \quad \text{and} \quad \mathbf{B} = \mathbf{R}_m \mathbf{W}^{-1}. \tag{7}$$

A representation for test images is achieved by using the principal component representation based on the training images to obtain $\mathbf{R}^{\text{test}} = \mathbf{X}^{\text{test}}\mathbf{P}_m$ and then computing

$$\mathbf{B}^{\text{test}} = \mathbf{R}^{\text{test}}\mathbf{W}^{-1}.$$

Image recognition performance was evaluated for the coefficient vectors $\mathbf{b}$ by the nearest neighbour algorithm. Coefficient vectors in the test set were assigned the class label of the training set with the most similar angle, as evaluated by the cosine

$$d = \frac{\mathbf{b}^{\text{test}}\mathbf{b}^{\text{train}}}{\|\mathbf{b}^{\text{test}}\| \, \|\mathbf{b}^{\text{train}}\|}. \tag{8}$$

4 Local Independent Component Analysis Using Clustering

In standard ICA (as shown above), a linear data model is used for a global description of the data. This can only provide a crude approximation for nonlinear data distributions. To overcome this deficiency here a clustering algorithm is used first, which is responsible for an overall coarse nonlinear representation of the underlying data [KM99]. Then linear ICA on each cluster is used for describing local features of the data.

Assume that the instantaneous mixtures

$$\mathbf{x}(t) = \mathbf{g}(\mathbf{s}(t))$$

of the sources $\mathbf{s}(t)$ are observed, where $\mathbf{g} : \mathbb{R}^m \to \mathbb{R}^m$ is an unknown nonlinear mixing function. The nonlinear ICA problem now consists in finding an inverse mapping $\mathbf{f} : \mathbb{R}^m \to \mathbb{R}^m$ which gives an estimate of the independent components according to $\mathbf{y}(t) = \mathbf{f}(\mathbf{x}(t))$.

The local ICA method then consists of the following steps [KM99]:

1. Cluster the data into Γ clusters. Denote the set of data vectors $\mathbf{x}^{(\gamma)}$ belonging to the γth cluster by S_γ.
2. Compute the mean vector m_γ of each cluster S_γ and subtract it from the vectors of S_γ.
3. Calculate the ICA vectors for each cluster

Instead of using SOM [MBA01] to cluster the data, here this preprocessing step is accomplished using *Generative Topographic Mapping*, which was first introduced by Bishop *et al* [BSW97]. Thereby for the nonlinear parametric mapping $\mathbf{f}(\mathbf{x}, \mathbf{W}) = [f_1(\mathbf{x}, \mathbf{W}), \ldots, f_n(\mathbf{x}, \mathbf{W})]^\top$ choose a linear combination of a set of functions

$$f_i(\mathbf{x}, \mathbf{W}) = \sum_{d=1}^{D} \phi_d(\mathbf{x}) w_{di}, \quad 0 < i \leq n \tag{9}$$

with the basis functions $\phi_d(\mathbf{x})$ being a combination of Gaussian, linear and fixed basis functions.

The basic idea then lies in maximizing the log-likelihood of the data. Consider a set of data points $\mathcal{M}_\mathbf{y} = \{\mathbf{y}^{(1)}, \ldots, \mathbf{y}^{(L)}\}$, the likelihood function can be calculated according to

$$\mathcal{L} = \prod_{l=1}^{L} p\left(\mathbf{y}^{(l)}|\mathbf{W}, \beta\right) = \prod_{l=1}^{L} \left[\frac{1}{K} \sum_{k=1}^{K} p\left(\mathbf{y}^{(l)}|\mathbf{x}^{(k)}, \mathbf{W}, \beta\right)\right] \quad (10)$$

and subsequently the log-likelihood function is given by

$$\ell = \sum_{l=1}^{L} \log\left(\frac{1}{K} \sum_{k=1}^{K} p\left(\mathbf{y}^{(l)}|\mathbf{x}^{(k)}, \mathbf{W}, \beta\right)\right). \quad (11)$$

where β^{-1} denotes the variance of $\mathbf{x}$ given $\mathbf{s}$ and $\mathbf{W}$. Setting the derivatives of the log-likelihood to zero, the update rules for the neural network can be calculated. Defining $\mathbf{G}$ to be a $K \times K$ diagonal matrix with entries $g_{kk} = \sum_{l=1}^{L} r_{kl}$, using $\mathbf{\Phi} = (\phi_{dk})_{0<d\leq D, 0<k\leq K} = \phi_d(\mathbf{x}^{(k)})$ and writing eqn.9 in matrix notation, i.e. $\mathbf{Y} = \mathbf{\Phi}\mathbf{W}$, the updated weight matrix can be derived by

$$\mathbf{\Phi}^\top \mathbf{G} \mathbf{\Phi} \mathbf{W} = \mathbf{\Phi}^\top \mathbf{R} \mathbf{T}.$$

Using a weight regularisation $\lambda \mathbf{I}_D$ with $\mathbf{I}_D$ being the identity matrix of the dimension D and λ a scaling factor we finally get

$$\mathbf{W} = \left(\mathbf{\Phi}^\top \mathbf{G} \mathbf{\Phi} + \lambda \mathbf{I}_D\right)^{-1} \mathbf{\Phi}^\top \mathbf{R} \mathbf{T}.$$

By analogy the derivate of the log-likelihood function w. r. t. β yields the new value for β:

$$\frac{1}{\beta} = \frac{1}{Ln} \sum_{l=1}^{L} \sum_{k=1}^{K} r_{kl} \|\mathbf{f}(\mathbf{x}^{(k)}, \hat{\mathbf{W}}) - \mathbf{y}^{(l)}\|^2,$$

with $\hat{\mathbf{W}}$ being yet the updated weight matrix $\mathbf{W}$.

After convergence of the algorithm, the matrix $\mathbf{W}$ learnt by the network, represents the parameters of the mapping. Thus the hidden variables $\mathbf{x}$ can be computed by inverting the mapping $\mathbf{f}(\mathbf{x}, \mathbf{W})$ according to

$$\mathbf{y} = \mathbf{W}^{-1}\mathbf{x}.$$

5 Experimental Results

For two different malignant lesions (basal cell carcinoma, actinic keratosis) and psoriasis plaques fifty fluorenscence images each were recorded with an original size of 768 × 572 pixels using a conventional CCD camera. The gray levels ranged from 0 (white) to 255 (black). After centering the pictures and reducing them to 512 × 512, coarse graining was applied by a factor four, resulting in three times 50 images with size 128 × 128 pixels. This step was necessary to minimize the computational complexity.

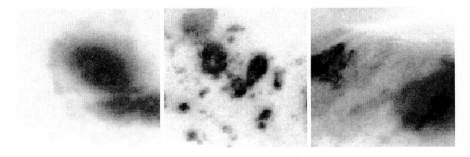

Fig. 1. Three typical fluorescence images of psoriasis, actinic keratosis and a basal cell carcinoma.

In Fig. 1, three typical recordings for psoriasis, actinic keratosis and a basal cell carcinoma can be seen. Even an experienced dermatologist cannot distinguish basal cell carcinoma, actinic keratosis or a psoriatic plaque on the basis of fluorescence images alone.

5.1 Lesion Classification Using PCA

First eigenimages were calculated for each of the three types according to eqn. 4. Then, the test and the training images were projected onto the principal components $\mathbf{a}_i = \mathbf{U}^\mathsf{T} \mathbf{x}_i$ and the euclidian distances ϵ_{ij} calculated according to

$$\epsilon_{ij} = \|\mathbf{a}_i^{\mathrm{train}} - \mathbf{a}_j^{\mathrm{test}}\|$$

The lesions were then classified according to the smallest ϵ_{ij}. Although various sorts of preprocessing steps were performed (histogram equalization and contrast manipulations via various non-linear transfer functions) the avarage classification rate did not vary to a great extend: While for psoriatic plaques with 82% a rather good recognition rate is achieved, actinic keratosis and basal cell carcinoma were hardly distinguishable (68% and 54%, respectively). Thus, the information contained in higher order statistics might be necessary.

5.2 Lesion Classification Using ICA

By analogy to the PCA, independent base images were extracted from the database of the images. Thus the original images can be understood as a linear combination (mixture) of independent source images. Unlike in PCA, the independent base images are not orthogonal and thus classification had to be accomplished using eqn.8. Using Java-code, the calculation required six hours on a 650 Pentium III operated with Linux.

In comparison to PCA, a slight improvement can be realised: While the avarage recognition rate for psoriasis or actinic keratosis increased by more than 4% (87% or 72%, respectively), the classification rate for basal cell carcinoma did not rise considerably (55%).

5.3 Lesion Classification Using Local ICA

Finally, local ICA was performed using GTM to first cluster patches of the images and then locally extract features to classify the data. As the size of the spatial structures in the images plays an important role, several cluster sizes were evaluated.

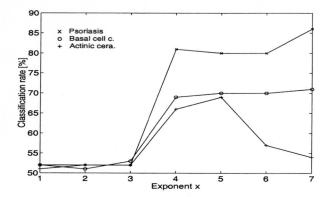

Fig. 2. Recognition rate for the different skin lesions depending on cluster size. For the cluster size the exponent to the base 2 is given on the x-axis.

In Fig. 2 the recognition rate for various patch cluster sizes is displayed. It is worth emphasising, that at least a cluster size of $2^8 = 256$ pixels is needed to achieve a reasonable rate. Although the recognition performance for psoriasis could not be ameliorated any further, the rate for basal cell carcinoma or actinic keratosis increased considerably compared to the results obtained with normal ICA.

It has to be mentioned, that for basal cell carcinoma, the recognition rate decreases with cluster sizes larger than $2^{10} = 1024$ pixels. This might be due to their inherent structure.

6 Conclusions

A new approach to classify various types of skin lesions has been presented. We tried to extract relevant features from a set of ALA-induced flourescence images of malignant and benign skin lesions to classify unknown test images. Therefore, PCA, ICA and local ICA were used in a comparative study. The results, shown in Table 1, underline the importance of higher order statistics in recognition tasks, as much information seems to be coded in higher correlations. Clustering the data in a preprocessing step using GTM, improved the overall recognition rate considerably, although it slightly decreased for psoriasis. It has to be emphasised, that a correct classification strongly depended on the cluster

Table 1. Recognition comparison for the three different algorithms used. For local ICA, the results of the best cluster size are shown.

algorithm	psoriasis	basal c. carcinoma	actinic cerat.
PCA	82	68	54
ICA	87	72	55
local ICA	81	69	66

size. Nevertheless, applying ICA to fluorescence images allows the distinction of benign or malignant skin lesions.

References

[BLS98] M.S. Bartlett, H. Martin Lades, Terrence J. Sejnowski. *Independent component representations for face recognition.* Proceedings of the SPIE Symposium on Electronic Imaging: Science and technology; Conference on Human Vision and Electronic III, San Jose, California, 1998

[BSW97] C.M. Bishop, M. Svensen, C.K.I. Williams. *A Principle Alternative to the Self Organizing Map*
Advances in Neural Information Processing Systems, volume 9, 354-360, MIT Press, 1997b

[KM99] J. Karhunen, S. Malaroiu. *Local independent component analysis using clustering.* International Workshop on Independent Component Analysis, Aussois, France, 1999

[MBA01] C. Mies, C. Bauer, G. Ackermann, W. Bäumler, C. Abels, R. M. Szeimies, E. W. Lang. *Classification of Skin Cancer And Benign Lesions Using Independent Component Analysis.* Proceedings of ISI, Dubai, 2001

[PK97] P. Pajunen, J. Karhunen. *A Maximum Likelihood Approach to Nonlinear Blind Source Separation.* Proceedings of the Int. Conf. on Artificial Neural Networks (ICANN'97), Lausanne, Switzerland, 1997

[TP91] M. Turk, A. Pentland. Eigenfaces for Recognition. Journal of Cognitive Neuroscience, 3:71-86, 1991

[YA97] H. H. Yang, S. Amari. *Adaptive On-Line Learning Algorithms for Blind Separation - Maximum Entropy and Minimum Mutual Information.* Neural Computation, 1997

An Approach Fractal and Analysis of Variogram for Edge Detection of Biomedical Images

L. Hamami and N. Lassouaoui

Polytechnic National School
Department of electronics. Signal & Communications Laboratory.
B.P. 182 16200 El-Harrach Alger
Fax: (213) 2 52 29 73

E-mail: L_HAMAMI@hotmail.com

Abstract. In this work, we study a fractal approach of edge detection. This approach is based on the evaluation of the local fractal dimension (in every pixel of the image) by using Gabor filtering. Gabor Filters use several parameters, as: radial frequency ρ and angular frequency θ. As we will see the choice of these parameters influence directly on the edge detection. Our contribution is using a mathematical tool said variogram that is going to guide us in the selection of the angular frequency θ. The method is based on exploitation of local indications that permits to affirm the existence of edge in an image on a direction θ. Results of edge detection are better since there is extraction in privileged directions of the image.

Key words: Gabor filtering, local fractal dimension, variogram, edge detection, biomedical images.

1. Introduction

Studies showed that the man is capable to recognize an object by simple observation of its edge. The edge detection causes since several years, many works of research. It permits to reduce the complexity of images by a simple description of the present objects in the stage. The main information is essentially contained in the edge of the image. Edge delimits an object and permits to distinguish between the different regions in an image. It is generally defined like an abrupt variation of gray level in an image.

The goal of this article is to present an approach fractal of edge detection. This approach appraises the local fractal dimension by using the filtering of Gabor. This filtering requires different parameters of entrance as: the angular frequencies θ. As we will see it farther, the result of detection essentially depends of θ. Our contribution is to have use a mathematical tool said variogram, that permits us to appraise θ according to the image, of goal to have a better edge detection. Our method is called « vario-spectral ».

In the section2 we present a fractal algorithm of edge detection, we give results of application. The section3 is dedicated to our vario-spectral approach; the theory and results will be given. To illustrate some results, we use the simple images of shape of segments as image 2des1.bmp. But the application is essentially made on biomedical images.

2. The Fractal Approach of Edge Detection

It has been shown that the power spectrum of a fractal surface $P(k_x, k_y)$ is given by this relation:

$$P(k_x, k_y) = \frac{1}{((k_x^2 + k_y^2)^{1/2})^\beta} = \frac{1}{\varphi^\beta} \qquad (1)$$

where
$\beta > 0$,
k_x, k_y : the two spatial frequencies coordinates in the directions x and y respectively,
ρ : the radial frequency, and is equal to: $(k_x^2 + k_y^2)^{1/2}$,
θ : is the angular frequency $\theta = \arctan(k_y / k_x)$,

Since it is admitted that images approximate the fractals surfaces, (indeed, an image is considered like a random fractal [1]), we can use the previous equation to calculate the fractal dimension of an image. Indeed the exponent β is in relation with the fractal dimension D [1] :

$$\beta = 8 - 2D \qquad (2)$$

Thus, the fractal dimension D can be calculated from the exponent $-\beta$ of the equation of the power spectrum. It is sufficient to represent the power spectrum of a log-log scale; then the curve in law of power becomes a straight line with a slope equivalent to $-\beta$. For an image the procedure gives a global value of fractal dimension. But in our case, we interest to the evaluation of the local dimension, that is to say a value of fractal dimension in every pixel of the image.

For it, we present a method based on the convolution of the image by the filter of Gabor. This filter convolute, is a bandpass filter, of which relation is:

$$h_{k_x, k_y}(x, y) = \frac{1}{\sqrt{\pi \sigma}} \exp[-(x^2 + y^2)/2\sigma^2] \exp[2\pi j(k_x x + k_y y)] \qquad (3)$$

Where   x,y : the spatial coordinates.
         k_x, k_y : coordinates of the frequency plan.

This filter samples the power spectrum to different frequencies for every point of the image. The algorithm of edge detection is as follows:

Algorithm

1. To choose N filters $h_{\rho_1,\theta}$, $h_{\rho_2,\theta}$,..., $h_{\rho_N,\theta}$ with θ fixed.
2. Convolute the image with the N filters, to sample the power spectrum to N frequencies for every point of the image.
3. To take the square of the amplitude of the filter outputs. It will give N points for every point of the image.
4. To find the power law-curves that best approximate the N points, for every point of the image. To express the exponent of the curve of the best linear approximation by $-\beta(x,y)$.
5. To calculate $D(x,y)$ for every point of the image from (2).
6. To quantify in levels of gray between 0 and 255.

The fig.1, shows the result of the application of this algorithm on a numeric image. According to images of dimensions (see fig.1), we note that edges are heightened. It is owed to the fact that limits between the homogeneous regions are a physical profile that doesn't correspond well in a fractal model. So, when points that belong to these limits between two regions of the image are examined, the approximation between the fractal model and data of the image is weak. This fact that limits are raised more that no fractals surfaces of intensities provides the means to detect points of the image that can be well edge. It ensues that when the fractal dimension of a region covering the limit between two homogeneous regions is calculated, a gotten value is lower to topological dimension, that is to say 2. Therefore, when values of fractals measurements are quantified in levels of gray, edges will be seen in black and the remainder of the regions will have a particular gray nuance.

Choice of parameters of the filter
The choice of σ

While using a fixed σ, all filters h_{k_x,k_y} will have the same width. The exchange between the frequency localization and the spatial localization is determined by the choice of σ.

If $|\phi(x,y,k_x,k_y)|$ represents the localized Fourier transform resulting of the convolution of the image $I(x,y)$ with filter h_{k_x,k_y}, then:

- A big value of σ has for result of measures of $|\phi(x,y,k_x,k_y)|^2$ that is not localized very well in the plan (x,y) but are good evaluations of the power spectrum.

- A small value of σ has for results of measures of $|\phi(x,y,k_x,k_y)|^2$ that are localized very well in the plan (x,y) but are weak appreciation of the power spectrum.

One can see it in the fig.2, it is to note that when σ decreases, the localization in the plan (x,y) grows as discussed already. (For more of details on the theory of the filter of Gabor see[1].)

The choice of radial frequency

It is necessary to adjust the radial frequencies to cover the image in the plan of frequencies. A possible solution is to calculate the frequency specter of the image by Fast Fourier and to represent the frequency coordinates k_x, k_y in the radial plan (ρ,θ). For a given θ, we determine ρ_{min} and ρ_{max} and we sample this interval in N values to generate the N Gabor filters. The result is then a juxtaposition of images of edge in different directions. (see fig.3)

The choice of θ

We applied the Gabor filtering algorithm on the same image, for three different values of the angular frequencies θ, we find the results of the fig.4.
The found results reflect an important characteristic of this algorithm to know its character isotropic. Indeed, if the power spectrum is considered as being isotropic it depends only that of ρ and no of θ. If this last is anisotropic, measures of fractal dimension in different directions will be different, reflecting anisotropic character of the specter. For a point of edge given, the spatial frequency that corresponds him has a perpendicular direction to the one of the edge (in the direction of the gradient). It is what explains the fact that the gotten edges for θ=0° and θ=45° and θ=90° are the perpendicular edge to these three directions. A second application of this algorithm would consist therefore in using him like selective extractor of edge in a direction data.
To get a global algorithm of edge extraction us propose to apply the Gabor filtering algorithm on the same image for N values of θ, and the result of the detection is the sum of N images gotten. The detection will be much better that N is big.

We note that the choice of the parameter θ is very important, and for a better detection of edges, it is necessary to apply the algorithm on a big number of angular frequencies θ, but the cost of count becomes especially very important in time of execution. We tried to use other means that are going to guide the choice of the good angular frequency for a better extraction of edges. For it, we developed a method based on the exploitation of local indication permitting to affirm such edge existence in the image by the analysis of variogram, and we called this method "vario-spectral." The theory and the application of this method are given in the following section.

3. The Vario-Spectral Method

The variogram is introduced as being a function answering to requisite qualities and to be a good indicator of texture; the theoretical properties and application of the variogram will be exploited.

3.1. General Definition of the Variogram

By definition, a variogram appraises a difference between two positions of the same function: F(x), measured in x and the same function F(x+h) measured in x+h. We suppose that the distance separating these two positions is weak and that it exists a structural dependence between them [3]. This evaluation generally operates while calculating a quadratic function (4) of the type:

$$(F(x)-F(x+h))^2 \qquad \forall x \qquad (4)$$

When h varies, this mathematical hope is not other that the prompt variogram, that we note $2\gamma(h)$. Thus, if F(x) is the random function, then we call intrinsic function of F(x), the follow function:

$$\gamma(h) = \frac{1}{2}E\left[(F(x+h)-F(x))^2\right] \qquad (5)$$

The behavior of the variogram to the origin has an analogous significance to the one of the faster fall of the autocorrelation function around its maximum. It indicates the relative speed with which the phenomena decorrelated with him self. The variogram present the properties interesting, such as:

- A periodic texture drags a variogram that will have the oscillations of the same period.
- Landings in the variogram explain that exists several different level textures.
- Besides, the pace of the variogram, it permits to take in evidence the directions in the image, and this property that we exploit in our algorithm.

3.2. Count of the Variogram

The expression of the variogram of a domain D_Ω of image of size NxN in the direction θ according to the distance θ is the following:

$$\gamma^\theta(h) = \frac{1}{(D_\Omega)}\sum_{x=n_1+1}^{N}\sum_{y=n_2+1}^{N}(I(x,y)-I(x-n_1,y-n_2))^2 \qquad (6)$$

Where I(x,y) represents the image in gray level and NxN the size of the under image.

From this formulation of the variogram, we can calculate the directional variogram, in the four main directions ($\gamma^{0°}$, $\gamma^{45°}$, $\gamma^{90°}$, $\gamma^{135°}$).

So the variogram in the horizontal direction noted γ^H (h) for θ=0° is given by following expression:

$$\gamma^H(h) = \frac{1}{(N-h)N}\sum_{x=h+1}^{N}\sum_{y=1}^{N}(I(x,y)-I(x-h,y))^2 \qquad (7)$$

The variogram in the direction of the diagonal 45° noted $\gamma^{DP}(h)$ for $\theta=45°$ is given:

$$\gamma^{DP}(h) = \frac{1}{(N-h)^2} \sum_{x=h+1}^{N} \sum_{y=h+1}^{N} (I(x,y) - I(x-h, y-h))^2 \qquad (8)$$

The variogram in the vertical direction noted $\gamma^{V}(h)$ for $\theta=90°$ is given by:

$$\gamma^{V}(h) = \frac{1}{(N-h)N} \sum_{x=1}^{N} \sum_{y=h+1}^{N} (I(x,y) - I(x, y-h))^2 \qquad (9)$$

The variogram in the diagonal direction 135° noted $\gamma^{DS}(h)$ for $\theta=135°$ is given by:

$$\gamma^{DS}(h) = \frac{1}{(N-h)^2} \sum_{x=N-h-1}^{N} \sum_{y=h+1}^{N} (I(x,y) - I(N-x-h, y-h))^2 \qquad (10)$$

While comparing curves of variograms (the axis of abscissas represents the values of the distance h and the axis of the coordinates represents the amplitude of the variogram) that correspond to the four directions, we can detect the orientation of the present objects in the stage, by simple comparison of amplitudes of variograms [3]. Of the fact, that the privileged direction of a motive or an object in an image is the one of which the amplitude of the variogram around the origin is weakest. [3]

3.3. Vario-Spectral Algorithm

Our algorithm of the detection is given by the fig.5. The algorithm of detection is the gabor filtering algorithm seen in the previous paragraph, but the angular frequencies θ are appraised by the count of variograms, and then applied to filters of Gabor.

The evaluation of angular frequency makes him self as follows:
- We divide the original image in several under images of NxN size.
- On each under image, we calculate the variogram corresponding to the four directions.
- We look the privileged direction corresponding in each under image. What gives us a matrix of directions.
- We calculate the histogram of directions for all the image (i.e. The function that gives the frequency of apparition of every direction in all image).
- The privileged direction of the image is the one having the biggest frequency.

If we want to apply two values of θ, then the second privileged direction is the one having a nearest histogram of the biggest frequency.

Very Important Remark:
We saw in results of previous section, that edge gotten by fractal approach is those in the perpendicular direction to θ. Therefore, when we finds a privileged direction θ, we applied $\theta+\pi/2$ to the algorithm of detection to have edges in the direction θ.

4. Application

The results, of application the Vario-spectral algorithm on different images, are given in fig.6.

We divided the image of size 256x256 in under-images of size 16x16, what gives 256 under-images, the variograms are calculated on every under-image, and then, we look the privileged directions while calculating the histogram of directions for all the image.

5. Conclusion

In this work, we have seen a method of count of the local fractal dimension in all image based on the Gabor filtering. The global dimension of all image can be gotten by the same algorithm, and this while taking the mean of all local measurements. If the power spectrum is isotropy then the fractal dimension will depend only of some radial frequencies ρ and no of θ, but if the power spectrum is anisotropy then the fractal dimension will depend of ρ and of θ and the fractals dimensions measured will be different in the different directions. To have better results, it is necessary to apply the algorithm for several values of the θ, but the time of count will be very big.

In the goal to improve the algorithm, we are proposed a method based on the count of variograms; this tool permits to estimate the directions present in the image, what permits us to know the privileged directions, i.e., directions where edges are most present, and results are better since there is extraction of edge in privileged directions of the image.

References

[1] C.J. Burdett and M. Desai, Localized fractal dimension measurement in digital mammographic images, *SPIE*, 2094, 1993, 141-151.
[2] J.M. Blackledge and E. Fowler, Fractal Dimension Segmentation of Synthetic Aperture Radar Images, *Image Processing and its Applications*, 1992, 445-449.
[3] S. Soltane, J. Claude Angue, Sélection d'opérateurs directionnels basée sur les variogrammes, *Revue Internationale des Technologies Avancées*, 11, 1999, 14-22.
[4] B. Mandelbrot, *Fractals Form, Chance, And Dimension*, W.H. Freeman And Company, 1977.
[5] Yuxin Liu and Yandan Li, Image Feature Extraction and Segmentation using Fractal Dimension, *IEEE Information Communications and Signal Processing*, 1997, 975-979.
[6] A. Arneodo, F.Argoul, E.Bacry, J. Elezgaray, J-F. Muzy, *Ondelettes, multifractales et turbulences de l'ADN aux croissances cristallines*, Arts et Sciences, 1995.
[7] Ph. Bolon, J-M. Chassery, J-P. Cocquerez, D. Demigny, C. Graffigne, A. Montanvert, S. Philipp, R. Zéboudj, J. Zérubia, *Analyse d'images : filtrage et segmentation*, Masson, 1995.
[8] R.C. Gonzalez, *Digital Image Processing*, 2nd Edition, Addison-Wesley, 1987.
[9] B. Mandelbrot, *Les objets fractals*, Flammarion, 1995.

Fig. 1. measurements local fractals dimensions with used 5 filters for θ=90°, σ=2, (a) original image, (b) ρ_1=18, ρ_2=18.8, ρ_3=19, ρ_4=19.5, ρ_5=20. (c) ρ_1=17, ρ_2=17.3, ρ_3=17.5, ρ_4=17.7, ρ_5=18.

Fig. 2. Localization of edge for different values of σ, with θ=90°, ρ_1=20, ρ_2=20.2, ρ_3=20.5, ρ_4=20.7, ρ_5=21.

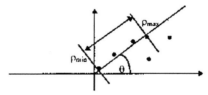

Fig. 3. Localization of radial frequencies.

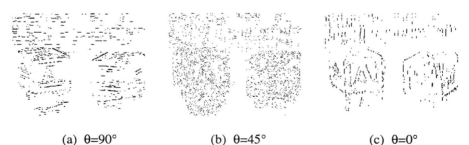

(a) θ=90°   (b) θ=45°   (c) θ=0°

Fig. 4. Edge detection for different angular frequencies with σ=2, ρ_1=20, ρ_2=20.2, ρ_3=20.5, ρ_4=20.7, ρ_5=21.

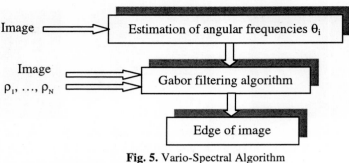

Fig. 5. Vario-Spectral Algorithm

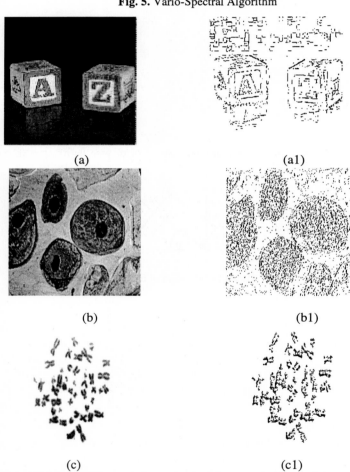

Fig. 6. Result of Application of vario-spectral algorithm, with σ=2, ρ1=20, ρ2=20.2, ρ3=20.5, ρ4=20.7, ρ5=21. (a), (b) and (c) are the original images.
(a1) the privileged directions are 90° and 0°.
(b1) the privileged directions are 90° and 45°.
(c1) the privileged directions are 90° and 0°.

Some Examples for Solving Clinical Problems Using Neural Networks

A.J. Serrano[1], E. Soria[1], G. Camps[1], J.D. Martín[1], and N.V. Jiménez[2]

[1] Grupo de Procesado Digital de Señales,
Dpto. Ingeniería Electrónica, Universidad de Valencia, Spain.
antonio.j.serrano@uv.es

[2] Dpto. Farmacia y Tecnología Farmacéutica, Universidad de Valencia. Spain.
Servicio de Farmacia del Hospital Dr. Peset, Valencia, Spain.
victor.jimenez@uv.es

Abstract. In this paper neural networks are presented for solving some pharmaceutical problems. We have predicted and prevented patients with potential risk of post-Chemotherapy Emesis and potentially intoxicated patients treated with Digoxin. Neural networks have been also used for predicting Cyclosporine A concentration and Erythropoietin concentrations. Several neural networks (multilayer perceptron for classification tasks and Elman and FIR networks for prediction) and classical methods have been used. Results show how neural networks are very suitable tools for classification and prediction tasks, outperforming the classical methods[1]. In a neural approach it is not strictly necessary to assume a specific relationship between variables and no previous knowledge of the problem is either necessary. These features allow the user better generalization performance than using classical methods. Several software applications have been developed in order to improve clinical outcomes and reduce costs to the Health Care System.

1. Introduction

Monitoring of patients treated with drugs is an essential task nowadays. Thus prediction of the appropriate drug administration level or prevention of certain risk of intoxication continues to be an important aspect in clinical decisions. In fact, the health care team in a hospital decides daily the dose to administer by assessment of the patient's factors. This decision-making can be aided with mathematical models and usually classical or statistical methods are used. However, reactions that occur in the human body are much more complex than those simulated by theoretic equations, which often do not meet the underlying hypothesis. Reality is always difficult to understand and we cannot hide under a simple model to explain a complex world. There are many difficulties involved in modeling the human body. If we described it at the systems level, we would find ourselves dealing with a non-linear time-dependent system with a large number of input variables. If we were experts in systems theory rather than health science specialists, it would never occur to us to apply a linear model to such a complex system. However, a review of the literature on

---

[1] This work has been partially supported by the FEDER project 1FD97-0935

models in this field shows basically two linear models: logistic regression for classification models and multivariate analysis for predictions problems.

Let's imagine that we do not know what system we are dealing with, i.e. the human body, and look for example, at two of the properties of linear systems:

- The sensitivity of the model's outputs vs. inputs is always of the same type.
- When faced with a linear combination of certain inputs, a linear system presents as output the linear combination of the outputs corresponding to those inputs.

Let's now compare these properties with what we would find if we applied linear methods to health science problems.

- In the case, for example, of administering drugs, the effect of a drug is not always the same, and yet a linear model would always consider the same effect. In our example, duplicating the dose would not necessary duplicate the effect.
- A given combination of inputs does not always have the same effect on outputs. Within the same protocol not all treatments make use of the same proportion of the drugs even though all of them pursue the same effect.

We now focus on a model that seems to be non-linear, i.e. the logistic regression. This mathematical model estimates the probability of belonging to a different class (e.g. healthy patients) by means of the following expression:

$$P(\vec{x}) = \frac{1}{1+e^{-(w_0+w_1 \cdot x_1+w_2 \cdot x_2+........+w_n \cdot x_n)}} \quad (1)$$

where x_k are the variables (weight, height, number of dose given, etc.) and w_k are the parameters of the model. Eq. (1) would lead us to think that the model is non-linear. We use this model to classify an input vector **x** in a given class according to whether this probability is or not above a certain value usually called threshold:

$$P(\vec{x}) = \frac{1}{1+e^{-(w_0+w_1 \cdot x_1+w_2 \cdot x_2+.......+w_n \cdot x_n)}} > k \Rightarrow \left(w_0+w_1 \cdot x_1+w_2 \cdot x_2+.......+w_n \cdot x_n\right) = \ln\left(\frac{k}{1-k}\right) \quad (2)$$

A linear condition concerns the input variables. We are assuming simple relationships between the inputs variables and, therefore, trying to model a complex system. This approach will usually fail or proportionate poor results.

Other not linear models are commonly used, but all of them have the same drawback: they require *a priori* hypothesis about the system. The criticism that these systems come in for is obvious. It is that if we want to extract information about a system, wouldn't it better no to employ any *a priori* assumption? On the other side, neural networks are non-linear regression models in which no previous knowledge is needed and it is not strictly necessary to assume a specific relationship between variables. These features allow the user better generalization performance than using classical methods.

In this paper we present a four-year project focused on the application of classical and neural approaches for solving crucial pharmaceutical problems. Several software applications have been also developed in order to improve clinical outcomes and reduce costs to the Health Care System.

In the next section we will describe a series of applications in which neural networks has been used to solve some classification and prediction problems. We will end with a brief discussion containing concluding remarks and a proposal for further work.

2. Clinical Applications

2.1. Classification of Patients with Potential Risk of Digoxin Intoxication

Digoxin is a drug that does not reduce the mortality rate among patients with congestive heart failure but is considered useful in controlling the symptoms involved in this pathology. Due to its narrow therapeutical range, commonly accepted between 0.8 and 2 ng/mL, up to 10% of the patients treated with it can suffer toxic effects. A patient is usually considered at risk for digoxin poisoning when the blood concentration of the drug is above the therapeutic range, i.e. 2 ng/mL [1].

The data used to develop the multilayer perceptron (MLP) [2] that makes it possible to identify patients at risk of digoxin intoxication corresponds to 257 patients monitored by the Pharmacy Unit of the Hospital Universitario Dr. Peset in Valencia, Spain. The best model obtained is a neural network with sensitivity (Se) and specificity (Sp) of more than 80% in the generalization set[2]. It is made up of 14 input variables (including physiological and treatment variables), nine neurons in the hidden layer and one output node that distinguishes patients with risk of been intoxicated (PC≥2ng/mL).

Table 1. Results for the Digoxin problem for the training and the validation sets. Patients are clustered in the table according the Test +/-, which represents the output of the model and the actual state of the patient (PC). Two thirds of the patients are used to build the model and the rest to validate it.

	Training		Validation	
	PC≥2ng/ml	PC<2ng/ml	PC≥2ng/ml	PC<2ng/ml
Test +	39	16	12	6
Test -	0	117	3	64
	Se=100%	Sp=88%	Se=80%	Sp=91%

A software for monitoring the clinical evolution of the patients and for predicting and preventing the symptoms of digitalis toxicity was developed [3]. This tool was developed in Visual Basic [4] and consists of several forms where the data are collected and introduced to the model. The results are showed in another screen along with the recommendations for the specialist in each case. A module for database

---

[2] Sensitivity is the correct classification percentage on intoxicated patients. Specificity is the correct classification percentage on non-intoxicated patients.

management and help files and tips for correct use of the program are also included. In Fig. 1 some windows of the software are shown.

Fig. 1. Two windows of the software PreTox-DGX.

2.2. Classification of Patients According to the Degree of Postchemotherapy Emetic Protection with Cysplatin

++Anti-neoplastic therapy forms part of most cancer treatments, in combination with surgery or radiotherapy. As a result of the toxicity of the drugs they are given cancer patients have all kinds of negative reactions. Emesis (vomiting and nausea) is, from the patient's point of view, the worst side effect of chemotherapy [5]. Treatment with cysplatin is especially difficult because of this drug's high emetogenic capacity in a great many patients.

A questionnaire on vomiting and nausea run on patients treated with cysplatin from April, 1996, to March, 1998, at the Hospital Universitario Dr. Peset and it was used to develop a neural network for predicting, on the basis of certain characteristics of the patients and the treatment, emesis during the 24h after antineoplastic chemotherapy, which is the highest risk period [6]. A multilayer perceptron was developed using 212 patterns and then validated using another 107. The best results obtained in the training set were characterized by 85% sensitivity and 96% specificity (88% success rate). In the validation set specificity was 73% and sensitivity 85% (79% success rate). This model was introduced into a web that also contained a series of elements to help with the model and the problem. In Fig. 2 a sample of the web site is shown.

2.3. Prediction of Erythropoietin (EPO) Dose

EPO is administered externally to increase hemoglobin (Hb) or hematocrit (Ht) in anemic patients. In general, the goal is to hold Hb levels at about 12 g/dL and/or Ht at 35%[7]. The final goal is to provide a mathematical model for predicting the EPO dose that guarantees the appropriate Hb level in every patient.

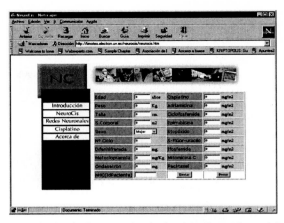

Fig. 2. Form for de data entering in the risk of emesis model.

The patients used to obtain and validate the model belong to Valencia's Nefroclub Center. The population used to generate the model comprises 77 patients (495 patterns) and another 33 patients (174 patterns) were used to validate the models. In accordance to international directives, the results are expressed as a percentage of success rates when the prediction error is below 0.5 g/dL. The Table 2 gives the results for the two best models; the corresponding architecture is indicated in brackets [8].

Table 2. Results for the EPO prediction problem.

	MLP (16×8×1)		ELMAN (16×3×1) Three Recurrent neurons	
	Success Rate	Error (g/dL)	Success Rate	Error (g/dL)
Training	97.4%	0.15	97.6%	0.15
Validation	98.3%	0.15	97.7%	0.15

The MLP model obtained was implemented in a user-friendly software that consists of a main screen on which the prediction appears after data from the preceding month are filled in, and the possibility of changing the dose of EPO depending on how the dose affects the Hb level [9]. Once the dose has been decided on, it must be stored. In Fig. 3 a capture of this tool is shown.

2.4. Prediction of Cyclosporine (CyA) Dose

CyA is generally considered a critical dose drug. Its narrow therapeutical range is an important issue in the clinical management of transplant patients, whereas underdosing may result in graft loss and overdosing causes kidney damage, increases opportunistic infections, systolic and diastolic pressure, and cholesterol [10]. Moreover, the pharmacokinetic behavior of CyA presents a substantial inter- and intra-individual variability [11].

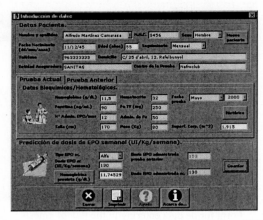

Fig. 3. Main window of the application for predicting EPO dose (PRED-EPO 1.02).

Thirty-two renal allograft recipients treated in the Nephrology Service of the Hospital Universitario Dr. Peset were included in this study. Time series for every patient are of different length ranging from 13 to 22 samples. Two thirds of the population was used for training the models and the rest for their validation. Each *pattern* or example was constituted by the present and past values of the variables described.

A comparison of the performance of all the models (Multilayer Perceptron, FIR network and Elman network[3]) is shown in Table 3 [12]. We have used the mean prediction error (ME) as a measure of bias and the root mean squared error (RMSE) and the correlation coefficient (r) as a measure of precision between the actual and the predicted signal.

3. Discussion

The present report justifies the use of neural networks in pharmacy and offers four real applications. Results have always been improved when using these models in comparison to the early stages when linear models were the first, easy though limited approaches attempted.

In Fig. 4 the best models' performance is shown for some validation patients.

At present these tools are being tested as aids in decision-making via software programs and web pages. Initial results indicate that neural networks could be used to reach a better understanding of the body's reactions to medication and, consequently, to improve treatment of patients. We are at present working on rules and knowledge extraction from trained networks. We expect that an understanding of accurate models will provide us with information about which variables are important and which meaningless, and about treatments and suitable protocols for the posology individualization and the identification and prevention of toxic effects in the patients.

---

[3] ARMA models were also attempted but results were poor due to the non-uniform sampling, the inherent complexity of the series and their short length.

Table 3. Models comparison for concentration prediction in the validation set.

*FIR Topology: (Number of Inputs × Hidden Neurons × Outputs, Order of FIR synapses in each layer)

Performance	r	ME (ng/mL) (95% CI)	RMSE (ng/mL) (95% CI)
MLP ($t_d = 2$)	0.746	1.286 (0.696, 1.876)	58.53 (-12.19, 129.25)
FIR* (7×2×10×1,1:1:1)	0.759	4.687 (4.177, 5.196)	52.34 (-71.10, 175.78)
ELMAN (RN=3)	0.750	2.528 (1.173, 3.883)	55.701 (-147.85, 259.255)

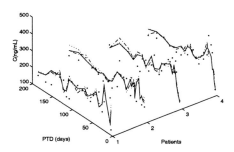

Fig. 4. Best models for the prediction of the serum concentration of CyA.

References

1. Jiménez, N. V, Albert, A., Soria, E., Camps, G, Serrano, A. J: Prediction of Digoxin Plasma Potentially Toxic Levels by Using a Neural Network Model, ASHP Midyear Clinical Meeting. Orlando, Florida U.S.A (2000)
2. Haykin, S: Neural Networks. Ed. Macmillan, (1994) 1-44
3. Camps, G., Soria, E., Jiménez, N. V.: Artificial Neural Networks for the Classification of Potentially Intoxicated Patients Treated with Digoxin, World Congress on Medical Physics and Biomedical Engineering, (2000)
4. Visual Basic 5. Edición Especial. Prentice-Hall. (1997).
5. Aapro, MS: Treatment of cancer therapy related emesis. Handbook of chemotherapy in clinical oncology. Jersey: Scientific Communication (1993) 488-92.
6. Catalán, J.L.: "Programas de emesis y sistemas informáticos para su predicción". XX Curso de terapia intravenosa y nutrición artificial. Valencia (1998)

7. Balhman,J.: Schoter KH, Scigalla P, et al. "Morbidity and mortality in hemodialysis patiens with and without erythropoietin treatment: A controled study". Contrib Nephrol, Vol. 76, (1984) 250-256.
8. Elman, J.L.: Finding structure in time, Vol 14. Cognitive science (1990), 179-211
9. Martín,J.D., Soria, E., Jiménez, N.V., Camps, G., Serrano, A.J., Pérez-Ruixo, J.: Nonlinear prediction of rhEPO by using Neural Networks. World Congress on Medical Physics and Biomedical Engineering, Chicago, (2000).
10. Belitsky: Neoral used in the renal transplant recipient, Transplantation Proceedings, Suppl. Review., Vol. 32, no. 3ª, (2000) S10-S19
11. Lindholm, A.: Factors influencing the pharmacokinetics of cyclosporine in man, Therapeutic Drug Monitoring, Vol. 13, no. 6, (1991) 465-477
12. Wan, E.A: Finite Impulse Response Neural Networks with Applications in Time Series Prediction, Master's Thesis". Department of Electrical Engineering. Standford University, (1993)

Medical Images Analysis:
An Application of Artificial Neural Networks in the Diagnosis of Human Tissues

Prof. M.Sc. Elias Restum Antonio [1], Prof. M.Sc. Luiz Biondi Neto [2],
Prof. M.Sc. Vincenzo Junior [3], Prof. M.Sc. Fernando Hideo Fukuda [4]

[1, 3, 4] Departamento de Ciências Exatas e Tecnologia, Universidade Veiga de Almeida
[2] Departamento de Eletrônica e Telecomunicações, UERJ
sosscan@br.homeshopping.com.br [1] ,lbiondi@embratel.net.br [2],
droberto@uninet.com.br[3], fukuda@unisys.com.br [4]

Abstract.

This article presents an Artificial Neural Networks (ANN) application in the image diagnosis process, by the tissues densities obtained in Computerized Tomography (CT) exams and related to Cerebral Vascular Accidents (CVAs). Among the usually analyzed aspects are the density, the form, the size and the location of these characteristic aspects of the image. As said by specialists in this area, the most relevant attribute is the analysis of the tissues densities. Considering this fact, our paper will investigate neurological pathologies in Computerized Tomography based in the tissues densities of the tomographic images.

The images to be diagnosed are digitalized, and then pre-processed, receiving an adequate mathematical treatment to be used as ANN training patterns and tests.

Keywords: Diagnosis of CVA, Diagnosis for Image, Automatic Diagnosis of CVAs and ANNs.

1. Introduction

Equipments such as Computerized Tomography (CT), Magnetic Resonance (MR), Ultrasound (US), Simple (XR) and Computerized X-Rays (RXC) produce as result, images printed in films or special papers that will be analyzed and diagnosed by specialists.

In medicine, the pre-diagnosis or even the emergency diagnosis can be improved considering the speed in the supply of the final diagnostic, through the automation of the identification processes and diagnosis of the abnormalities

In this work an application of ANN is presented, concerning the pattern recognition problem, related to diagnoses associated to images obtained by CT, in the neurological area, specifically in cerebral vascular accidents (AVC).

Because they are effective in non-linear models, as well as in situations where a group of rules cannot be easily formulated (as it happens in the analysis of images), the use of ANN appears as an interesting tool and that deserves to be investigated in cases where image processing is present.

Figure 1.1 refers to the process of conventional X-Ray. It can be seen that a source emits X-Rays in conical shape, which is driven to a certain size, and this emission is called incident radiation.

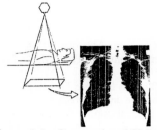

Figure 1.1 - Conventional X-Ray

The X-Rays irradiate the patient and suffer attenuation of the internal anatomy. The resulting radiation, called transmitted radiation, is then passed to a X-Ray film. The great limitation is the superposition of the anatomical structures in the film, which can hide important details, as structures that differ slightly in density, as tumors and the tissue that involves these tumors. Another difficulty is that many anatomical areas are simply inaccessible due to the bone structures that involve these areas, besides the fact that conventional X-Ray just offers qualitative information, fact that demands subjective interpretation [5].

1.1. Computerized Tomography and Diagnosis of CVA

Derived from the greek word "tomo" that means slice and "graphy" that means to write, the tomography uses the same basic principles of the conventional x-ray, with the objective of creating an anatomical representation based on the amount of X-Ray attenuation.

The compilation of the multiple vision angles of the complete rotation that will provide necessary data to the reconstruction of the slice, as shown in the figure 1.2, will be shown in a monitor and later on printed in a radiographic film.

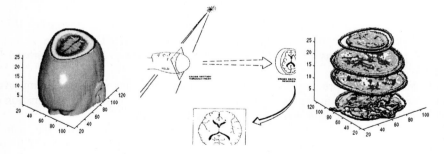

Figure 1.2 - Construction of a CT image

An important example of image diagnosis and object of our investigation is the computerized tomography of the brain, which can reveal important information about the cerebral vascular accident (CVA).

CVA is characterized by a located neurological alteration, originated from a vascular pathological process, having two groups of different problems: the thrombosis or the cerebral clot that takes to the cerebral infarct (blood absence) and the hemorrhage [6].

Our main objective is to produce a system that can distinguish a normal tomography (patient 31), of a characteristic tomography of a hemorrhagic CVA (patient 25), or a bloodless CVA (patient 39). Figure 1.3 illustrates the differences among the three investigated cases.

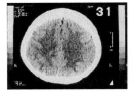 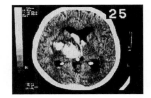 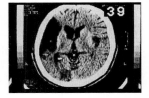

1.3 - CT image of the brain

The image analysis is considered one of the most important stages in the process of diagnosis of diseases and it should be made considering the following aspects: density, shape, size and location, where the most decisive in the diagnosis is the density which is used in our research. Thus, the first observed aspect is a difference of densities (gray intensities), characteristic that generally stands out when an image is analyzed. The presence of different shades of gray can establish what is normal and what is considered abnormal, and then take in consideration the shape, the size and the location of this varying density of the normal pattern, which was previously established.

1.2. Artificial Neural Networks

ANNs are computer programs whose objective is to learn a specific knowledge. In that way, differently from the traditional programs, which are executed directly, ANNs have two different processing phases. The first training call is characterized by the presentation of patterns trying to map a certain knowledge that the net should learn. As the neural processing is a numeric algorithm, it is necessary to transform the radiological images as well as the differentiated diagnoses emitted by the specialists (ANN training patterns) in numeric matrixes, easily interpreted by the ANN [10].

After the training process, where the knowledge acquisition process takes place, the net is ready to be executed in the second phase or execution phase. In that stage, testing patterns will be presented to the ANN, also under numeric form, searching for a diagnosis and that, usually, were not part of the training process. As it was stored in the training stage, the network will be able to compare the knowledge on the pathologies of interest and to present in the output, also under numeric form, the pathology associated to the tested image.

The Perceptron, credited to ROSENBLATT, F. (1958), was the first learning machine with possibility of supervised training. It is a network that can just be presented with one layer of neurons or with multiple layers, "multi-layer Perceptron" (MLP). Rosenblatt developed an algorithm adjusting the weights by the minimization of the sum-squared error using the method of the decreasing gradient. In that way the error ej=Tj-Oj is obtained by the difference between these values, using Oj after the activation function. Therefore Oj = F(NETj) = htan(γNETj) where γ is a positive scalar that represents the inclination of the function.

The sum-squared error is given by:

$$E = \frac{1}{2}(e_j^{(k)})^2 = \frac{1}{2}(T_j^{(k)} - O_j^{(k)})^2$$

The instant gradient is given by the differential related to the weights

$$\nabla_w E = \frac{\partial E}{\partial w_{ji}^{(k)}}$$

Applying the chain rule

$$\frac{\partial E}{\partial w_{ji}^{(k)}} = \frac{\partial E}{\partial e_j^{(k)}} \frac{\partial e_j^{(k)}}{\partial w_{ji}^{(k)}} = \frac{\partial E}{\partial e_j^{(k)}} \left(\frac{\partial e_j^{(k)}}{\partial NET_j^{(k)}} \frac{\partial NET_j^{(k)}}{\partial w_{ji}^{(k)}} \right)$$

Finally
$$\frac{\partial e_j^{(k)}}{\partial NET_j^{(k)}} = -\frac{\partial O_j^{(k)}}{\partial NET_j^{(k)}} = -\gamma[1 - (O_j^{(k)})^2]$$

$$\frac{\partial E}{\partial w_{ji}^{(k)}} = \frac{\partial E}{\partial e_j^{(k)}} \frac{\partial e_j^{(k)}}{\partial w_{ji}^{(k)}} = \frac{\partial E}{\partial e_j^{(k)}} \frac{\partial e_j^{(k)}}{\partial NET_j^{(k)}} \frac{\partial NET_j^{(k)}}{\partial w_{ji}^{(k)}}$$

$$\frac{\partial E}{\partial w_{ji}^{(k)}} = -e_j^{(k)} \gamma[1 - (O_j^{(k)})^2] x_i^{(k)}$$

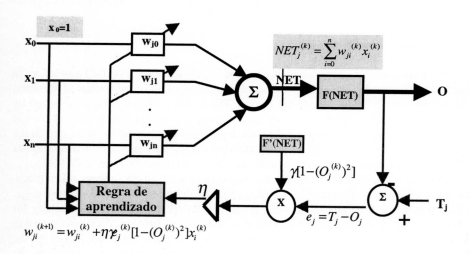

Figure 1.5 – Perceptrons Training Improvement

The Perceptron can present multiple layers (MLP) [13]. In that case an algorithm similar to the developed, and called Back-propagation, is used in the training process. The error signal should allow the error propagation to the previous layers (back propagation) until the first one [1].

2. Image Pre-processing

The block diagram presented in figure 2.1 shows the several stages of the image transformation until it gets to the representative matrixes of the training patterns, where P represents the pre-processed image and the target T represents the diagnoses presented by the specialists, numerically. Initially the file (* .bmp, * .gif, * .pcx, * .jpg, * .tif) corresponding to the original tomographic image is loaded. Then, the image passes by two successive transformations; the first creates two matrixes: a matrix corresponding to the indexed image and another matrix called color map (red, green and blue), indexed by the first one. The second transformation removes shade and saturation information of the original image, retaining intensity information, representing the black color for the intensity 0 and the white color for the intensity 1.

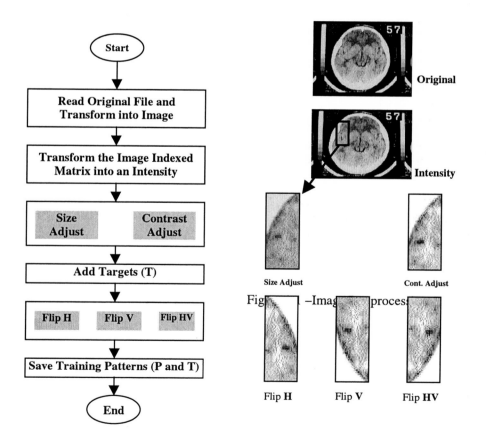

Fig. – Image processing

The following equations synthesize the problem solution for the case of continuous and discreet two variables functions and they represent the necessary mathematical base for the adequate treatment of the tomographic images to be used with ANNs. If f(x,y) is continuous and able to be integrated and F(u,v) is also able to be integrated, then the Fourrier Transform and its inverse exist and are given by:

$$\Im\{f(x.y)\} = F(u,v) = \int\int_{-\infty}^{\infty} f(x,y)\exp[-j2\pi(ux+vy)]dxdy$$

$$\Im^{-1}\{F(u,v)\} = f(x.y) = \int\int_{-\infty}^{\infty} F(u,v)\exp[j2\pi(ux+vy)]dudv$$

Where u and v are frequency domain variables.

Now lets suppose that the function f (x) can be arranged in this sequence

$$\{f(x_0), f(x_0 + \Delta x), f(x_0 + 2\Delta x), \ldots, f(x_0 + [N-1]\Delta x)\}$$

Where N represents the number of patterns and Δx the space among them.

Thus, the Fourier Transform can be represented as follows:

$$F(u,v) = \frac{1}{MN}\sum_{x=0}^{M-1}\sum_{y=0}^{N-1} f(x,y)\exp[-j2\pi(\frac{ux}{M}+\frac{vy}{N})]$$

where $u = 0,1,2,3, \ldots, M-1$ e $v = 0,1,2,3,\ldots, N-1$

$$f(x,y) = \sum_{u=0}^{M-1}\sum_{v=0}^{N-1} F(u,v)\exp[j2\pi(\frac{ux}{M}+\frac{vy}{N})]$$

where $x = 0,1,2,3, \ldots, M-1$ e $y = 0,1,2,3,\ldots, N-1$

Since the necessary transformations were made, the image can still be cut out extracting its most important characteristics and through interpolation techniques (nearest neighbor interpolation, bilinear interpolation and bicubic interpolation), the size of the image is redefined. It is also possible improving the definition of the pre-processed image through contrast adjustment. A simple algorithm is shown in Figure 2.2. In that way, the intermediate shades of gray will be better distinguished.

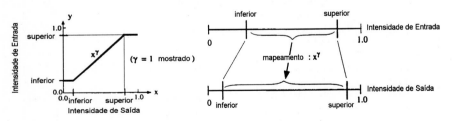

With the mapping P →T it is possible to train the ANN. With the objective of creating different patterns of a same image, increasing the diversity of training patterns, as well as to facilitate the treatment of symmetrical images, the possibility of flipping lines and columns of matrixes P and T that compose the patterns presented to the ANN is foreseen.

3. Artificial Neural Network Modeling

The number of neurons in the input layer corresponds to the number of pixels (area elements) of the image after pre-processing, and that was resized to guarantee the size standardization of the training patterns. These images are formed by small rectangles of 70 by 52, generating a total of 3640 pixels per image.

The network was trained with 320 patterns referring to the cut out of standardized size images. The activation function used for all the neurons that compose the network was the sigmoid. The optimization method chosen in our research was the Gradient and its variations. In the investigated case we used an algorithm that allowed working with an adaptative learning rate and momentum coefficient.

For the network with architecture 3640/57/1 several training parameters were tested, arranged in lots. The ones that presented the best performance were the following: error tolerance = 0.01; sampling frequency = 10; maximum number of cycles = 5000; learning rate = 0.5; momentum coefficient = 0.99;

4. Results

Studies with about 180 tests, with images that were not part of the training process were accomplished. Comparing the obtained results of all tests with the expected targets it was verified that the percentile error didn't surpass 1%. All the images used in the studies received the same treatment given to the images that were part of the ANN training. All the situations were analyzed by medical specialists whose technical opinions of the considered most significant cases, are found in table 3.1, as follows.

Table 3.1 – Results obtained by the ANN x Medical Diagnosis

Case	Tomographic Image	Target Value (expected)	Obtained Value	Error %	Diagnose (by a especialist)
1		0.5	0.4931	- 1.38	NORMAL
2		0.5	0.5012	0.24	NORMAL
3		0.0	0.0009	~ 0.0	BLOODLESS CVA
4		0.0	0.0003	~ 0.0	BLOODLESS CVA
5		1.0	0.9991	-0.19	HEMORRHAGIC CVA
6		1.0	0.9914	- 0.86	HEMORRHAGIC CVA

5. Conclusions

It can be noticed, through the tests accomplished in the studies, that the identification method of medical images patterns showed efficiency, always presenting high success rates compatible with the complexity of the neurological images tested and validated by specialists, as shown in table 3.1.

From the performance analysis of the investigated system it is possible to say that, with an adequate ANN training, it can also be used in the following applications: characterization among the structures surrounded by the cerebral liquid and the spaces (emptiness) originated from sulks, fissures, etc, differentiating them from the rest of the cerebral parenchyma; differentiation among areas full of vases and areas with decreased circulation; detection of clots, hemorrhages, infarcts and cerebral calcifications, difficult of being detected by the conventional radiological processes; definition of the components of the Turkish cell of the sphenoid; detection of focalized abscesses and parasitic lesions; alterations in the cranial vault.

As an evolution of our research it is important to point out that it would be quite interesting if, besides the pathologies identification, performed by ANNs, we could use a Specialist System (SS) to present the diagnoses under symbolic form. Thus, we would have a Hybrid System (ANNs + SS) where the ANN would identify the pathology and SS would diagnose the problem, supporting the radiologist's decision.

Bibliographical References

[1] Simon Haykin, *Neural Networks, a comprehensive foundation*, Prentice-Hall, EUA, 1994; [2] David E. Rumelhart, James L. McClelland and The PDP Research Group, *Parallel Distributed Processing*, Volume I, The MIT Press, 1986; [3] GE Medical Systems, *The Physics of Computerized Tomography*, General Electric Company, EUA, 1991; [4] GE Medical Systems, *Sytec 3000 Introduction & Prerequisite Study Package*, General Electric Company, EUA, 1992; [5] Anne G. Osborn, *Diagnostic Neuroradiology*, Mosby-Yearbook, Inc., EUA, 1994; [6] Krupp, Marcus Abraham, *Current Diagnosis & Treatment*, Lange Medical Publications, USA, 1982; [8] Richard Arnold Johnson, *Applied Multivariate Statistical Analysis*, Prentice-Hall, EUA, 1998; [9] Betty Trujillo Montoya, *Atlas Básico de Tomografia Axial Computadorizada*, Quimica Schering Colombiana, Colombia, 1990; [10] Biondi L. N., Estellita Lins Et Al, Neuro-DEA: Novo Paradigma para Determinação da Eficiência Relativa de Unidades Tomadoras de Decisão, *9° Congresso da Associação Portuguesa de Investigação Operacional – APDIO*, n ° 9, pp. 114, 2000; [11] *The Value and Importance of an Imaging Standard*, Radiological Society of North America, Department of Informatics, EUA, 1997; [12] Cichocki A. And Bargiela A., "*Neural Networks for Solving Linear Inequality Systems*", Journal of Parallel Computing, http:// www.bip.riken.go.jp/absl, 1996; [13] David, M. Skapura, *Building Neural Networks*, Addison-Wesley Publishing Company, New York, 1996; [14] Jacek M. Zurada, *Introduction to Artificial Neural Systems*, West Publishing Company, New York, 1992; [15] Rosenblatt F., *Principles of Neurodynamics*, Spartan Editions, New York, 1962; [16] Rafael C. Gonzales and Paul Wintz, *Digital Image Processing*, Addison-Wesley Publishing Company,New York,1987.

Feature Selection, Ranking of Each Feature and Classification for the Diagnosis of Community Acquired Legionella Pneumonia

Enrique Monte[1], Jordi Solé i Casals[2], Jose Antonio Fiz[3], Nieves Sopena[4]

[1]Dpt.TSC.Universitat Politécnica de Catalunya Barcelona.Spain
[2]Signal Processing Group, Universitat de Vic
[3]Pneumology dpt, [4]Infectious Diseases Unit ,Hospital U. Germans Trias i Pujol of Badalona.
E-mail:enric@gps.tsc.upc.es

Abstract. Diagnosis of community acquired legionella pneumonia (CALP) is currently performed by means of laboratory techniques which may delay diagnosis several hours. To determine whether ANN can categorize CALP and non-legionella community-acquired pneumonia (NLCAP) and be standard for use by clinicians, we prospectively studied 203 patients with community-acquired pneumonia (CAP) diagnosed by laboratory tests. Twenty one clinical and analytical variables were recorded to train a neural net with two classes (LCAP or NLCAP class). In this paper we deal with the problem of diagnosis, feature selection, and ranking of the features as a function of their classification importance, and the design of a classifier the criteria of maximizing the ROC (Receiving operating characteristics) area, which gives a good trade-off between true positives and false negatives. In order to guarantee the validity of the statistics; the train-validation-test databases were rotated by the jackknife technique, and a multistarting procedure was done in order to make the system insensitive to local maxima.

1 Introduction

In this paper we deal with the problem of designing a classifier for the diagnosis of Community-acquired pneumonia (CAP), and giving an objective ranking of the importance of the features for classification. The usual tool for determining the importance of a feature is based on the explained variance associated with each feature This supposes that the multivariate distribution is gaussian, which is no the case when part of the features are qualitative and can take only a limited number of values . In the case of the diagnosis of CAP, most of the features are qualitative, and thus, the use of the variance for explaining the importance of a given feature can be questioned. In this paper instead we use the recognition rate as the criteria for deciding the importance of each features for classification. Another problem which is dealt in this paper is the selection of a the best subset of features for discriminating the CAP. There are several techniques for finding a good subset d_s of features for classification from a

---

This research was supported by the CICYT Spanish research project TIC98-0683

given set *d*. The exhaustive search procedure takes 2^d steps to explore all possible subsets of a given set of size *d*. If the size of the subset is fixed, there are heuristic search techniques, such as the *branch and bound* , which is efficient for obtaining a subset of a given size, without a combinatorial explosion in the number of steps. This method has the property that is faster than exhaustive search, and yields the subset that maximizes the search criteria, but one must fix the size of the subset from the beginning, and this size might not be the best possible. For this paper we decided to select as search strategy the sequential backward selection, which although does not guarantee that the obtained subset corresponds to the global maximum of the criteria, it gives the eliminated feature at each stage of the search. In the case that there is no strong overlap of the classes in the feature space the methods gives the same solution as the exhaustive search in $O(d^2)$ steps. Thus in the process, this search method gives a ranking of the importance of each feature, and the best classifier for the given subset of features. Best in the sense of giving a high recognition rate, (with a low false alarm rate) in the validation and test databases. In order to make the conclusions independent of the given elements in each database (i.e. test, validation and train), we followed the jackknife technique, which consisted in the rotation of the items to be classified through the databases. The problem of local minima of the networks was dealt by the multistarting technique.

Community-acquired pneumonia (CAP) is a common condition throughout the world (1). In Europe, *Streptococcus pneumoniae* is the most frequently identified pathogen, followed by different microorganisms according to the series reviewed. Although Legionella species has been considered as a rare cause of CAP, more recent series have implicated this pathogen in 8% - 15% of the cases (2,3,4). Diagnosis of Legionella pneumonia is commonly performed retrospectively by serology study or tardy by isolation in special culture mediums from clinical samples. Recently, urinary antigen identification (5) has demonstrated high sensitivity (specially if the urine is previously concentrated) (6), and high specificity for detecting *Legionella pneumophila* serogroup 1. This technique is currently being used by several hospitals in many countries to investigate Legionella outbreaks or less frequently as a routine method in the emergency department (7). In spite of its known efficacy, laboratory results are delayed for hours, it is expensive and moreover not all medical centers can perform this technique. However, in patients with CAP, antibiotic therapy cannot be delayed and treatment decisions following different guidelines must be made (8,9). Artificial neural networks (ANN) are tools which have been successfully applied in different medical aspects such as radiologic, clinical and physiologic pattern recognition (10,11,12,13).

2 Methods

Subjects: From May 1994 to June 1999 we prospectively studied 203 patients with laboratory tests diagnosing CAP. Patients over the age of 14 years with acute symptoms consistent with pneumonia and infiltrate on chest radiography at the time of hospital admission or within 24 h were included in the study. Patients with some of the following criteria were excluded: discharge from the hospital less than 10 days before the onset of symptoms, suspicion of bronchoaspiration, obstructive pneumonia, or pulmonary tuberculosis. Residence in a nursing home, HIV infection, or pharmacologic immunosuppression were not criteria for exclusion. Sixty-two patients had CALP, 141 patients were diagnosed as having community-acquired pneumonia by another bacterial etiology (NLCAP).

Variables studied: For analysis we considered the most predictive variables (table I) according to a previous study (14): (1) demographic variables: age, sex ; (2) risk factors: smoking, chronic alcoholism; underlying disease, including COPD, HIV infection, solid or hematologic neoplasm; pharmacologic immunosuppression (steroids and chemotherapy); (3) variables of clinical presentation: cough, expectoration, thoracic pain, headache, confusion, diarrhea and antibiotic treatment with beta-lactamic drugs prior to pneumonia; (4) analytical data: sodium (Na), creatine kinase (CK) and glutamic oxaloacetic transaminase (GOT).

Definition of the Variables: A patient was considered a smoker if he had smoked>1 pack/day within the last 5 years. Alcoholism was defined as consumption of > 80 gr. of alcohol per day within the same period. Steroid use refered to treatment with >60 mg of prednisone per day over more than 2 weeks in the last month or >5 mg/d for more than the previous 3 weeks. Immunosuppresion refered to treatment with cytotoxic drugs not including steroid therapy.

Organization Objectives and of the System

We developed the system with two objectives in mind, the first was to provide a set of features selected as a function of their classification potential taking into account the possible interactions between them, and to give a ranking of their importance. This criteria is more realistic than the traditional, based on univariate analysis were using the Student's t test for comparing quantitative variables and the chi square test for qualitative variables, which can be criticized on the basis that it does not take into account possible interactions between features, and also because the hypothesis that assume about the data are not necessary true (i.e., the form of the distribution). Another criticism to this methodology is that the selection of features is done as a function of the level of significance which is indirectly related with the real objective; which is to maximize the compromise between correct classification and false alarms. In order to maximize this compromise explicitly, we took as performance measure of the classifier, the Receiving Operating Characteristic (ROC), which is a measure of this compromise. The ROC chart gives the value of the % of correct classification and % of false alarms for a given threshold for the decision. If the output of the classifier is greater than the threshold, then the class is decided to be present, on the other case, the decision is that the class is not present, thus, the ROC curve gives the dependency of the correct classification and false alarms for the values of the threshold between 0 and 1. In order to have a measure of performance of the classifier independent of the threshold, we took the area under the ROC curve. When this area is 1 (or 100%), the classifier works perfectly independently of the threshold, when the area is 1/2 (50%), the classifier gives the worst performance. For values lower than 1/2 (> 50%), the inversion of the decision rule, improves the performance, which then is transformed into new roc area=100-old roc area.

Thus we took as a classifier a multilayer perceptron, which can approximate an optimal Bayes classifier, and makes few assumptions on the form of the underlying distribution of the data. The structure of the multilayer perceptron (i.e. kind of non-linearties, and number of hidden nodes), was explored in a series of preliminary experiments, in order to establish the general structure. The criteria for selecting the general structure was the generalization capacity of the classifier, using a rotational scheme for the database (i.e. jackknife). The form of the structure was one that used as non-linearity the sigmoid function, and a range of nodes between 5 and 10. This result is consistent with the worst case bound on the generalization capabilities of a network. The rule says that for a fraction of mistakes ε the number of weights of a net should be $W \cong N*\varepsilon$. In our case, for a generalization rate of 90% ($\varepsilon=0.1$), we need about $W \cong 30$ weights. The actual number of weights in the experiments followed a range between was 210 (when all the features were present) and 72 (for the best subset), thus as the classification problem of CAP/NLCAP has a much simpler structure than in the case Baum's study, we can consider that 2 or 3 times the number of parameters of the threshold can guarantee the generalization properties. Also we used for training the multilayer perceptron a conjugate gradient method, which has better convergence properties when the input size of the network is high. This fact was confirmed empirically, and the performance of the network trained by means of a conjugate gradient method was always, (in mean) several points better (in ROC area) than the performance when the training was done by means of the backpropagation with momentum or trained by a second order algorithm such as the Levenberg-Marquart. In order to make the results as much independent of the gradient search as possible, all experiments were done by the multistarting procedure, i.e., 30 and the selection criteria was the ROC area on the validation test.

For the feature selection, we decided to use the *sequential backward elimination*, which is an algorithm, that at a given stage, computes the recognition rate for all the features present at that stage leaving one out. Once this procedure has been done, a new list is created by selecting the subset of features that maximized the ROC area, and the following stage repeats this procedure with this new list. Thus at each stage the selected list consists of the initial list of features minus the feature that when is eliminated the ROC area is maximized (a multistart procedure was done at this point) . As can be seen in figure 1, we obtain the list of the eliminated features that increase the recognition rate until the a maximum is reached, after this maximum is reached, we obtain the list of the features that contribute the maximum to the recognition rate.

An alternative to the *sequential backward elimination* is the *sequential forward selection*, which we did not consider because we wanted to take into account the possible interactions between features, in order to make a ranking. The *sequential forward selection*, begins with a list that consists of one feature and at each stage introduces the feature that best increases the performance. Our hypothesis is that in the *backward elimination*, if there is an interaction between features that increases the performance index, the are maintained during the maximum number of stages. In the *forward selection*, depending on the order in which the features are added into the best list, a good combination of features might be lost. This effect does not seem to happen in the backward elimination. The backward elimination is robust to the drop combinations of features that interact depending on the order that the lists are explores at each stage. In table I, we present the best selected lists in three different experiments (selected as representatives of a set 10 experiments) , and in table II the ROC area obtained in each experiment. The results show that the sensivity of the system to a different order on the explored features at each stage is low. It is significant that the differences on the variables present at the best subsets of each experiment are related to the last variables to be eliminated in the experiments where these variables is not present in the best list. This can be explained by the fact that at the last two stages before the maximum of the ROC area, the results are within the confidence margin at 95%, (which was of the order or ± 2.5%). If this is taken into account, the ranking of features obtained in different experiments can be considered consistent, and almost independent of the realization.

Description of the Organization of the Database for the Experiments

The database was divided into three groups (training, validation and testing), and in order to reduce the bias of the estimation of system we used the jackknife technique, with four combinations of the database. This was done by assigning individuals to each base, so that each base was representative of the whole time interval during which the data was collected. A result of this criteria was that each partition of the database was equilibrated in the sense that all had the same proportion of CALP vs. NLCAP. The criteria for defining the training, validation and testing databases was first to organize the database on the basis of the date of the pneumonia, and then to select and alternatively assign individuals to one of four groups. By assigning individuals to each sub-database in this way we guaranteed that no database was biased due to the seasonallity of the illness.

3 Results

The study consisted on two phases, a preliminary phase were the potential value features were selected by means of on univariate analysis and a multivariate analysis which selected the best subset of features and yielded the classifier. The main result of the second phase is presented in figure 1 where we show the sequence of eliminated features, as a function of the ROC area. The sequence can be divide in three parts,
- the ascending part, where features that are discarded improve the recognition rate, and thus give us an indication of the least important features by ranking.
- the descending part, where the order in which features are discarded corresponds to their importance to the recognition rate.
- the part around the maximum that corresponds to the maximum ROC area minus the confidence margin (2%), which indicate the uncertainty about the optimal features for classifying the CAP. This uncertainty is reflected in the subsets that are found if the experiment is repeated, changing the order of the features at each stage of the backward selection algorithm, as shown in table I.

As was expected, the system was sensitive to local maxima of the ROC area. Nevertheless this sensitivity was small. As can be seen in table I, the selected subset in three different experiments was similar, and the elements that were eliminated just before reaching this maxima or just afterward are common to the three experiments. Also as can be seen in table II, the ROC area associated with the different databases is approximately the same for the three experiments. This three experiments were presented as representative of a group of 10 experiments, and we took two samples that were homogeneous on the ROC area and one outlayer in the test results. Although the test results varied from one experiment to another, the order of elimination of variables was quite similar. The high ROC area associated with the validation database, can be explained by the fact, that the criteria for deciding the feature to be eliminated at each stage of the search procedure, was the generalization result on the validation database. Also we used the multistarting technique for selecting a net independent of the local minima of error, and the criteria was also the generalization ability with the validation database. The test database was not used at any point during the whole trained selection procedure.

Discussion

LCAP pneumonia has a high prevalence among CAP pneumonias being considered as the third highest cause of CAP (18, 4). In most studies early diagnosis of LCAP is difficult because other NLP pneumonias may present with a similar clinical picture and microbiologic CAP tests are not always available or have only retrospective value. In this paper we have presented a tool that might help the diagnosis of LCAP, which has two advantages, it gives a ranking of the importance of the features for classification, and also a classifier that discriminates the LCAP from the NLP.

In a previous study (14) univariate analysis showed that CAP by LP was more frequent in middle-aged, healthy male patients than CAP by another etiology. Moreover, the lack of response to previous β-lactamic drugs, headache, diarrhea, severe hyponatremia, and elevation in serum CK levels on presentation were more frequent in NLCAP by LCAP, while cough, expectoration, and thoracic pain were more frequent in NLCAP by another bacterial etiology. However, in a previous study, multivariate logistic regression analysis only demonstrated significance for underlying disease, diarrhea and elevation in the CK level, although univariate analysis was similar to the present study. We concluded that some clinical and laboratory parameters may be useful to discriminate between LCAP and NLCP, although with a low sensitivity. Based on this study, we have selected the variables which were more relevant in the sample study.

We used a neural network which yields a good sensitivity and false alarm rate, using a maximally discriminating subset of features. This classification system is inexpensive and is easy to perform. The specificity of the antigenuria test by RIA or ELISA is better, but has a lower sensitivity, only detects LP serogroup 1 and needs the support of sophisticated laboratory techniques (6). In contrast, only three parameters of our system (NA, GOT, CK) are routine laboratory techniques. The lower sensitivity and/or time consuming tests of sputum culture and direct immunofluorescence using monoclonal antibodies make these techniques relatively useful in the emergency department. On the other hand serology is a valuable diagnostic test exclusively for epidemiological purposes. As a classification technique the neural networks are a promising tool, since they may model complex interactions and arbitrary distributions in multivariable data. As a precedent to our study, other studies have been performed in different medical areas, such as the diagnosis of interstitial lung diseases (10), for the analysis of ventilation-perfusion lung scans (12), or in the survival prediction of breast cancer (19). Other studies of breast cancer, concluded that neural nets had better ROC s than radiologic criteria (11). The use of committee nets have also been used in the diagnosis of hepatoma (10). Further uses include prospective validation in the identification of acute myocardial infarction (20). In conclusion, this tool can be used as a complementary aid in medical tasks, due to its simplicity (uses only clinical and common laboratory tests), accuracy, and immediate results.

References

1. M. Woodhead. Community-acquired pneumonia guidelines. An international comparison. A view from Europe. Chest 1998 .
2. R Blanquer, R Borrás, D Auffal, P Morales, R Menendez, I Subias, L Herrero, J Redon, J Pascual. Aetiology of community acquired pneumonia in Valencia, Spain: a multicentre prospective study. Thorax 1991.
3. J Aubertin, F Dabis, J Fleurette, N Bornstein, R Salomon, E Brottier, J Brune, P Vincent, J Migueres, A Jover, C Boutin. Prevalence of legionellosis among adults: A study of community-acquired pneumonia in France. Infection 1987.
4. N Sopena, M Sabria, ML Pedro-Botet, JM Montero, L Matas, J Dominguez, JM Modol, P Tudela, V Ausina, M Foz. Prospective study of Community-Acquired Pneumonia of Bacterial Etiology in Adults Eur J Microbiol Inf Dis (in press).
5. M. E. Aguero-Rosenfeld, P.H. Edelstein, Retrospective evaluation of the Du Pont radioimmunoassay kit for detection of Legionella pneumophila serogroup 1 antigenuria in humans. Journal of clinical microbiology. 1988.
6. JA Dominguez, JM Manterolas, R Blavia, N Sopena, FJ Belda, E Padilla, M Gimenez, M Sabrià, J Morera, V Ausina. Detection of Legionella pneumophila Serogroup 1 Antigen Antigen in Nonconcentrated Urine and Urine Concentrated by Selective Ultrafiltration. Journal Clinical Microbiology 1996
7. A.E. Fiore, J.P. Nuorti, O.S.Levine, A Marx, A.C. Weltman, S. Yeager, R.F. Benson, J Prucker, P.H. Edelstein, P Green, Sh R. Zaki, B.S. Fields, J.C. Butler. Epidemic Legionnaires´ Disease two decades later: Old Sources, new diagnostic methods. Clinical infectious diseases. 1998.
8. M. Woodhead. Management of pneumonia in the outpatient setting. Seminars in respiratory infections. 1998.
9. M.S. Niederman. Community-acquired pneumonia. A North American perspective. Chest 1998.
10. N Asada, K. Doi, H. MacMahon, S.M. Montner, M. L. Giger, Ch. Abe, Y. Wu. Potential usefulness of an artificial neural network for differential diagnosis of interstitial lung diseases. Pilot study. Radiology 1990.
11. J.Baker, P.Kornguth, J.Lo, M.Williford, C.Floyd, Breast Cancer: prediction with artificail neural networks based on BI-RADS standadized lexicon. Radiology 1995.
12. J.A. Scott. Neural network analysis of ventilation-perfusion lung scans. Radiology 1993.
13. El-Solh AA, MJ Mador,, E. Ten-Brock, DW Shucard, M Abul-Khoudoud, BJ Grant. Validity of neural network in sleep apnea. Sleep 1999.
14. N Sopena. M Sabrià-Leal, M.L Pedro-Botet, E padilla, J Dominguez, J Morera, P Tudela. Comparative study of the clinical presentation of legionella pneumonia and other commnunity-acquired pneumonias. Chest 1998
15. CH.M. Bishop. Neural Networks for pattern recognition. Clarendon press. Oxford. 1995 .
16. B. Parmanto, P. Munro, Diagnosis of hepatoma by committee. NIPS 94 Post Conference workshop: Neural Network Applications in Medicine, December 1994.

17. . Lippman, L Kukolich, D. Shahian. Predicting the risk of complications in coronary arthery bypass operations using neural networks, in G. Tesauro, D. Touretzky and T. Leen (editors) Advances in Neural Infomation Processing Systems 7, Cambridge, MA: MIT Press.
18. G.D. Fang, M Fine, J Orloff. New and emerging etiology for community acquired pneumonia with applications for therapy: prospective multicenter study of 359 cases. Medicine 1990.
19. H.Burke,D.Rosen, P.Goodman, Comparing the prediction accuracy of artificial neural networks and other statistical models for breast cancer survival. NIPS*94 Post-Conference Workshop: Neural Network Applications in Medicine, Vail, Colorado, USA (2-3 December 1994).
20. W. Baxt, J Skora. Prospective validation of artificial neural network trained to identify acute myocardial infarction. The Lancet 1996.

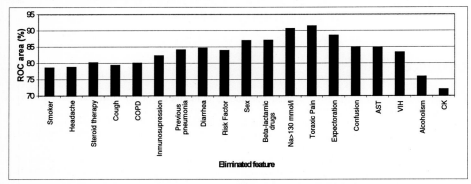

Fig. 1. Sequence of eliminated features as a function of their influence on the ROC area.

Table I Examples on the dependency on the solution with the order of search and the local minima of the neural networks at each step.

List of redundant features (Sorted by stage of elimination)			List of the Best Features		
Experiment 1	Experiment 2	Experiment 3	Experiment 1	Experiment 2	Experiment 3
Sex (m/f)	Sex (m/f)	Sex (m/f)	Age	Age	Age
Risk factor	Risk factor	Risk factor	HIV	HIV	HIV
Smoker	Smoker	Smoker	Alcoholism	Alcoholism	Alcoholism
COPD	COPD	COPD	Immunosuppression	expectoration	Immunosuppression
Previous pneumoniae	Immunosuppression	Steroid therapy	Steroid therapy	Confusion	Previous pneumoniae
Beta-lactamic drugs	Steroid therapy	Beta-lactamic drugs	expectoration	AST	Cough
Cough	Previous pneumoniae	expectoration	Confusion	CK	Confusion
Thoracic pain	Beta-lactamic drugs	Thoracic pain	CK		Na>130 mmol/l
Headache	Cough	Headache			
Diarrhea	Thoracic pain	Diarrhea			
Na>130 mmol/l	Headache	AST			
AST	Diarrhea	CK			
	Na>130 mmol/l				

Table II. ROC area on each database

	Experiment 1	Experiment 2	Experiment 3
Train	83.6 %	80.9	95.1%
Validation	90.5 %	91.4 %	92.3 %
Test	75.7	80.9	70.3%

Rotation-Invariant Image Association for Endoscopic Positional Identification Using Complex-Valued Associative Memories

Hiroyuki Aoki[1], Eiju Watanabe[2], Atsushi Nagata[3], and Yukio Kosugi[3]

[1] Department of Electronic Engineering, Tokyo National College of Technology
Hachiouji 193-0997, Japan
aoki@tokyo-ct.ac.jp
[2] Department of Neurosurgery, Tokyo Metropolitan Police Hospital
Tokyo 102-8161, Japan
eiju-ind@umin.ac.jp
[3] Frontier Collaborative Research Center, Tokyo Institute of Technology
Yokohama 226-8503, Japan
nagata@pms.titech.ac.jp
kosugi@pms.titech.ac.jp

Abstract. In order to solve the problem for identifying the tip position of fiber scopes under the endoscopic surgery, we propose an image association system. The proposed system is realized by combining complex-valued associative memories with a 2-dimensional discrete Fourier transform, and it is capable of rotation-invariant image association. Simulation results using surface images observed through endoscope are also presented.

1 Introduction

During endoscopic surgery, a surgeon should get an exact orientation during operation. However, sometimes localization of the endoscope tip is not an easy task even for experienced surgeons. For guiding a rigid-type endoscope, we may be able to utilize the surgical navigation systems [1], [2]; unfortunately, in the case of flexible fiber scopes into the intracranial ventricles, surface images (**Fig.1**.) observed through the endoscope itself are the only available information to get an orientation of the endoscope.

If we can memorize the endoscopic scenes in association with the positional data given by the surgical navigation system equipped with the rigid-type endoscope, we may be able to identify the tip position of the flexible fiber scopes, from the images by using some image association systems. We can call this system as "Dejavu Scope" which resembles the "dejavy" we sometimes experience while walking around the old home-town of the childhood. For this purpose, the image association should be rotation-invariant, because the endoscope or fiber scopes may be rotated at an arbitrary angle during the inserting manipulation. At the same time, we have to memorize and recall quite a large number of images superimposed on a memory mechanism of a reasonable memory size.

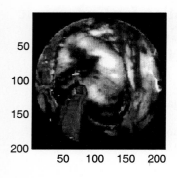

Fig. 1. An example of a surface image observed through endoscope introduced into the intra cranial ventricle.

The complex-valued associative memory (CAM) model proposed in this paper can handle gray-scale medical images and capable of rotation-invariant association.

2 Preliminaries

Fully complex-valued neural networks have previously been documented [3], [4], [5], [6]. We have already proposed a phase matrix image representation to apply CAM for image processing [7]. In this section, we outline CAM and the phase matrix image representation.

2.1 Construction of CAM

The state of k th complex-valued neuron of CAM, denoted by $x(k)$, is given by

$$x(k) = \exp(i\theta_K s(k)), \quad \theta_K = 2\pi/K, \quad s(k) = 0, 1, \cdots, K-1. \tag{1}$$

Let $\mathbf{W} = (w_{kj})$ be the complex-valued weight matrix among neurons, where w_{kj} denotes the connecting weight from the j th to the k th neuron. Let M be the number of neurons, $\mathbf{x}$ be an M-dimensional complex-valued column state vector of CAM and P vectors $\mathbf{x}^\alpha$ ($\alpha = 1, 2, \cdots, P$) be memorized into CAM. The weight matrix $\mathbf{W}$ can be calculated by the following:

$$\mathbf{W} = \mathbf{SS}^+, \quad \mathbf{S}^+ = \left(\mathbf{S}^{*t}\mathbf{S}\right)^{-1}\mathbf{S}^{*t}, \quad \mathbf{S} = \left(\mathbf{x}^1, \mathbf{x}^2, \cdots, \mathbf{x}^P\right) \tag{2}$$

where $\mathbf{S}^{*t}$ denotes conjugate transpose of $\mathbf{S}$. The fully connected CAM state transition equation can be given by

$$x_{next}(k) = \text{csgn}(y(k)), \quad y(k) = \sum_{j \neq k} w_{kj} x(j) \tag{3}$$

where

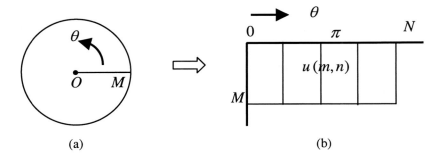

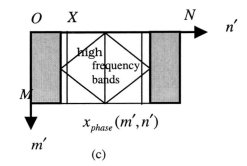

Fig. 2. (a) a circle image having a radius M. (b) a converted $M \times N$ rectangle image from the image (a). (c) the phase matrix of the image (b).

$$\operatorname{csgn}(y(k)) = \exp(i\theta_K\, s_{next}(k)),$$
$$s_{next}(k) = \left[\{\arg(y(k)) + \tfrac{1}{2}\theta_K\}/\theta_K\right] \pmod{K} \tag{4}$$

and the symbol $[c]$ presents the maximum integer not exceeding c.

2.2 Phase Matrix Image Representation

In order to deal with images in CAM, a phase matrix image representation was proposed [7]. Through the representation, images can be transformed into a phase matrix, which is suitable for CAM to use it. First, let the 2-dimensional discrete Fourier transform (2-D DFT) and inverse DFT (IDFT) of an $M \times N$ image $\mathbf{u} = \{u(m,n)\}$ be denoted by

$$\mathbf{v} = \{v(m',n')\} = DFT(\mathbf{u})$$
$$\mathbf{u} = \{u(m,n)\} = IDFT(\mathbf{v}). \tag{5}$$

Then, let
$$|v_{\max}| = \max\{|v(m',n')|;\ 0 \le m' \le M-1,\ 0 \le n' \le N-1\}, \tag{6}$$
$$\alpha(m',n') = \arg(v(m',n')),\quad \gamma(m',n') = \cos^{-1}(|v(m',n')|/|v_{\max}|). \tag{7}$$

The phase matrix $\mathbf{x}_{phase} = x_{phase}(m', n')$ of an image $\mathbf{u}$ is defined as

$$x_{phase}(m', n') = \exp(i(\alpha(m', n') + \gamma(m', n'))). \tag{8}$$

In this paper, let us denote this transform by

$$\mathbf{x}_{phase} = \{x_{phase}(m, n)\} = PMT(\mathbf{u}). \tag{9}$$

On the other hand, the inverse transform from $\mathbf{x}_{phase}$ to the original image $\mathbf{u}$ is given by the following equation.

$$\mathbf{u} = |v_{max}| \operatorname{Re}\{IDFT(\mathbf{x}_{phase})\}. \tag{10}$$

In the above equation, without using $|v_{max}|$, restoring the image $\mathbf{u}$ from $\mathbf{x}_{phase}$ is possible since it is just a constant value.

3 Rotation-Invariant Image Association System

3.1 Basic Idea

Arranging laterally radial pixels at every small angle around the center in a circle image shown in **Fig.2** (a), we can convert it into a rectangle image shown in **Fig.2** (b). Let the rectangle image be denoted by $u(m, n)$ and let P images $\mathbf{u}^1, \mathbf{u}^2, \cdots, \mathbf{u}^P$ be memorized into CAM. Note that rotational operation in the circle image corresponds to the lateral shift operation in the rectangle image. Suppose that some rotation in the circle image gives rise to q circular shift in the n-direction in the rectangle image, denoted by $u^\alpha(m, n-q)_c$. The phase matrix of $u^\alpha(m, n-q)_c$ is given as follows:

$$PMT\{u^\alpha(m, n-q)_c\} = \{x^\alpha_{phase}(m', n')\exp(-i\theta_N n'q)\}. \tag{11}$$

From this, we can see that the phase matrix of a rotated image can be obtained by shifting phase of the phase matrix $\mathbf{x}_{phase} = \{x^\alpha_{phase}(m', n')\}$. Furthermore, the quantity of the phase shift is equal in each column. So utilizing this property, we build a CAM from each column of $\mathbf{x}_{phase}$, denoted by $CAM(n')$ for a column number n'. Let $\mathbf{x}^\alpha_{phase}(n')$ denote an M-dimensional column vector for a number n' of the phase matrix $\mathbf{x}^\alpha_{phase}$. In order to reduce the time and memory space required for obtaining the weights, it is possible to remove some column vectors in the high frequency bands of $\mathbf{x}^\alpha_{phase}$ at the degree in which the image degradation is not conspicuous[8]. So we employ the low frequency column vectors

$\mathbf{x}_{phase}^{\alpha}(n')$, $n' = 0, 1, \cdots, X, N-X, \cdots, N-1$. The value of X determines the number of $CAM(n')$. Let us here define a new vector $\mathbf{x}_{in}^{\alpha}(n')$ as

$$\mathbf{x}_{in}^{\alpha}(n') = (\mathbf{x}_{phase}^{\alpha}(n'), 1). \tag{12}$$

This definition means we add one neuron, whose state is 0 phase, to $\mathbf{x}_{phase}^{\alpha}(n')$. Why this new vector is defined, will be described later. We then calculate the weight matrix $\mathbf{W}(n')$ of $CAM(n')$ using this new vector as follows:

$$\mathbf{W}(n') = \mathbf{S}(n) \, \mathbf{S}^{+}(n), \tag{13}$$

where

$$\mathbf{S}(n') = \left(\mathbf{x}_{in}^{1}(n'), \mathbf{x}_{in}^{2}(n'), \cdots, \mathbf{x}_{in}^{P}(n')\right). \tag{14}$$

By the way, CAM has the following property. When memorizing a state vector $\mathbf{x}_{in}^{\alpha}(n')$ into CAM by using Eq.(13), we can see that the phase-shifted state vectors of $\mathbf{x}_{in}^{\alpha}(n')$, expressed by $\mathbf{x}_{in}^{\alpha}(n')\exp(i\theta_K s(n'))$ ($s = 1, 2, \cdots, K-1$), are also generated as the fixed points of CAM. Therefore, if we set $K = N$ and simply memorize $\mathbf{x}_{in}^{\alpha}(n')$ into $CAM(n')$, the phase-shifted state vectors

$$\mathbf{x}_{in}^{\alpha}(n')\exp(i\theta_N s(n')) = \left(\mathbf{x}_{phase}^{\alpha}(n')\exp(i\theta_N s(n')), \exp(i\theta_N s(n'))\right) \tag{15}$$

are also generated as the fixed points of $CAM(n')$. This means they are memorized automatically. In this way, using Eq.(13), we can memorize both P circle images and their rotated images simultaneously. It is expected that $CAM(n')$ has an ability to retrieve a phase-shifted pattern given by $\mathbf{x}_{in}^{\alpha}(n')\exp(i\theta_K s(n'))$, in response to the presentation of an incomplete, noisy or rotated version of that pattern. However, what we need is the vector $\mathbf{x}_{phase}^{\alpha}(n')$, not $\mathbf{x}_{in}^{\alpha}(n')\exp(i\theta_K s(n'))$. This is why we added a neuron in Eq.(12). The state of the added neuron in a recalled pattern is expected to indicate the value of $s(n')$, if the recalling process is successful. Once the value of $s(n')$ is obtained, it is easy to recover the vector $\mathbf{x}_{phase}^{\alpha}(n')$ by shifting the phase of the recalled pattern by $-\theta_K s(n')$.

3.2 Proposed System

Fig.3 shows the block diagram of the proposed system. What follows is an explanation of the diagram.

1) Through "PMT" and "csgn_K" block, an $M \times N$ input image $\mathbf{u}$ is transform to a quantized phase matrix: $\mathbf{X}_{phase\_q} = \text{csgn}_K(PMT(\mathbf{u}))$.

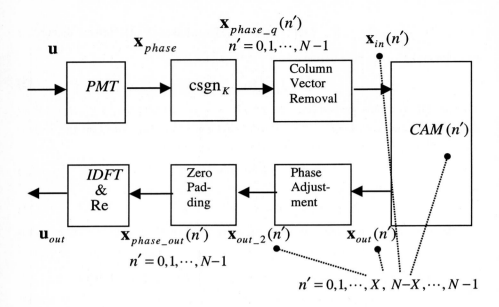

Fig. 3. Block diagram of the proposed rotation-invariant image association system.

2) In "Column Vector Removal" block, the high frequency column vectors of the matrix $\mathbf{X}_{phase\_q}$ are removed at the degree in which the image degradation is not conspicuous. The low frequency column vectors $\mathbf{x}_{in}(n')$, $n' = 0, 1, \cdots, X$, $N-X, \cdots, N-1$ are input to CAM.
3) CAM performs image association making use of the phase data.
4) In "Phase Adjustment" block, checking the value of $s(n')$, the phase shifting by $-s(n')\theta_K$ is performed for a recalled vector.
5) In "zero padding" block, zero column vectors are added in $\mathbf{x}_{out\_2}(n')$, $n' = X+1, \cdots, N-X-1$ and the preparation for $IDFT$ is completed.
6) In "$IDFT$ & Re" block, the output image $\mathbf{u}_{out}$ can be reconstructed by the following equation:

$$\mathbf{u}_{out} = \mathrm{Re}\left\{ IDFT\left(\mathbf{x}_{phase\_out}\right) \right\}.$$

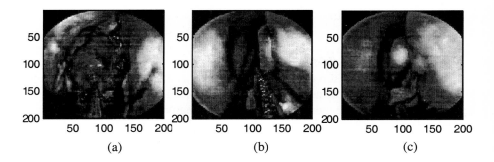

Fig. 4. Examples of memorized original surface images observed through endoscope.

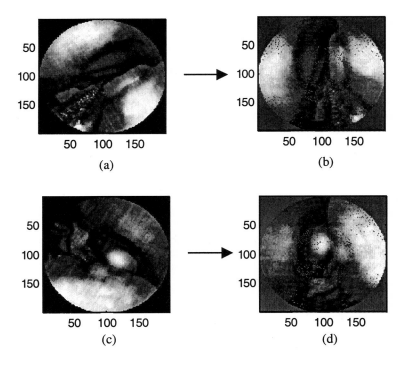

Fig. 5. (a), (c) : Examples of surface images input to the image association system. Images (a) and (c) are 60 and 120 degrees rotation of **Fig.4.** (b) and (c), respectively. (b), (d) : Recalled images from the image association system, when input images shown in (a) and (c) are given as key images.

4 Simulation Results

Let us here build an image association system that memorizes twenty gray scale circle images including the images shown in **Fig.4.**, which are observed through the endoscope. These images are with [0, 255] integer-valued gray levels and their radius $M = 100$, so the number of neurons in a CAM is $M + 1 = 101$. We took the other parameter as follows: the number of states in a neuron $K = 600$, the parameter $X = 60$ that controls the number of low frequency column vectors to be used to build $CAMs$ (so the number of $CAMs$ is $2X + 1 = 121$). **Fig.5** (b) and (d) show the recalled images when the rotated images shown in **Fig.5** (a) and (c) (60 and 120 degrees rotation of **Fig.4** (b) and (c), respectively) are given as input images to the system, respectively. One can see that non-rotated images were successfully recalled.

5 Conclusions

We proposed a new rotation-invariant gray scale image association system applying CAM combined with 2-D DFT process, and obtained some preliminary simulation results with respect to an image set including endoscopic images. It is possible that, by memorizing one image, CAM automatically memorizes its rotated images of all angles. This is the result from making good use of CAM's phase invariant property. Consequently, rotation-invariant image association system was realized.

References

1. Eiju Watanabe: Endoscopic surgery guided by neuronavigator, Clinical Neuroscience (in Japanese), Vol.16, No.12, pp.1419-1422 (1988)
2. E Watanabe, Y Mayanagi, Y Kosugi et al: Open surgery assisted by the neuronavigator, a stereotactic articulated, sensitive arm. Neurosurgery 28, pp.792-800 (1991)
3. Igor Aizenberg, Naum. Aizenberg, Joos Vandewalle: Multi-Valued and Universal Binary Neurons. Kluwer Academic Publishers (2000)
4 A. Hirose: Proposal of fully complex-valued neural networks, Proc. of IJCNN'92 Baltimore, U.S.A, vol.IV, pp.152-157, (1992).
5. H.Aoki: Analysis of equilibrium states of hopfield associative memory extended to complex number field. IEICE Trans., vol.J78-A, no.9, pp.1238-1241, (Sept. 1995)
6 S. Jakkowski, A. Lozowskik and J.M.Zurada: Complex-valued multistate neural associative memory, IEEE Trans. on Neural Networks, vol.7, no.6, pp.1491-1496, (Nov.1996)
7. H. Aoki, MR. Azimi-Sadjadi, Y. Kosugi: Image Association Using a Complex-Valued Associative Memory Model. IEICE Trans. Fundamentals, Vol.E83-A, No.9, pp.1824-1832, (Sept. 2000).
8. H. Aoki, Y. Kosugi: An Image Storage System Using Complex-Valued Associative Memories. 15th International Conference on Pattern Recognition Barcelona 2000 Vol. 2 pp.626-629 (Sept. 2000)

A Multi Layer Perceptron Approach for Predicting and Modeling the Dynamical Behavior of Cardiac Ventricular Repolarisation

Rajai El Dajani[1,2], Maryvonne Miquel[1,2] and Paul Rubel[1,2]

[1] Institut National des Sciences Appliquées de Lyon (INSA), Laboratoire d'Ingénierie des Systèmes d'Information (LISI), Bat 501, 20 avenue Albert Einstein, 69621 Villeurbanne, France
rmourid@lisi.insa-lyon.fr
miquel@if.insa-lyon.fr
rubel@insa.insa-lyon.fr
http://lisi.insa-lyon.fr

[2] INSERM XR121, 28 avenue Doyen Lépine, 69500 Bron, France
http://www.inserm.fr

Abstract. The QT interval measured on the body-surface Electrocardiogram (ECG), corresponds to the time elapsed between the depolarization of the first myocardial ventricular cell (beginning of the Q wave) and the end of the repolarisation of the last ventricular cell (end of the T wave). Abnormalities in the adaptation of the QT interval to changes in heart rate may facilitate the development of ventricular arrhythmia. None of the formulas previously proposed for the adjustment of QT for changes in heart rate provide satisfactory correction. This is due to the "memory phenomenon" (i.e. time delay) ranging up to 3-4 minutes, between a change in heart rate and the subsequent change in the QT interval. In this paper, patient specific predictive models based on a Multi Layer Perceptron are presented and their predictive performance is tested on real and artificial data.

1 Introduction

In Europe, 40% of all deaths of individuals who are 25-74 years of age are caused by cardiovascular disease [1]. Despite expanding insight into the mechanisms causing sudden cardiac death (SCD), the population at high risk is not yet going to be effectively identified, and mechanisms of SCD in subjects with apparently normal hearts are poorly understood. The development of a non-invasive method for the early detection of the heart pathologies underlying SCD becomes an important challenge for researchers in the field of biomedical computing.

One of the most studied measurements that are performed on the Electrocardiogram (ECG), is the QT interval one. It measures the time after which the ventricles are again repolarized. The study of the dynamicity of ventricular repolarisation in ambulatory patients is of major interest to assess the risk of sudden death [2].

The QT duration is influenced principally by the inverse of the heart rate, measured by the RR duration between two successive heart beats. QT duration is also influenced by gender, age, the central nervous system and circadian cycles. Bazzet already proposed a variation model of the QT-RR relationship in 1920 [3]. Since then many other relationships were proposed. These non-linear models are valid only for heart beats corresponding to steady rhythm periods lasting almost one minute. In the absence of the steady state situation, the study of the QT-RR relationship becomes more complex and the presently used methods are not adapted.

It has been shown by invasive studies that the QT interval has a delayed adaptation to sudden changes in heart rate in normal subjects. The QT-RR relationship seems to behave like a first order system with a time constant of about one minute [4]. The conception of predictive models would allow detecting life-threatening variations of the QT duration by comparing the predicted value to the real measured one. But up to now invasive techniques are the only ways for studying the ventricle response to sudden heart rate changes [4].

Preliminary studies [5] have suggested that global techniques such as the Multi Layer Perceptron (MLP) might be the preferred choice for modeling such a type of time series as they employ a limited set of kernel functions (nodes in the hidden layer) to represent the input space, and in theory can approximate any nonlinear function with arbitrary accuracy [6].

In this paper we propose to model the QT dynamic behaviour in function of the history of RR intervals by means of Multi Layer Perceptron. The networks will learn the following non-linear patient specific relationship:

$$QT_i = f(RR_i, RR_{i-1}, ..., RR_{i-M+1}, RR_{i-M}) \qquad (1)$$

where M is a time delay.

2 Methods

The complexity of physiological signals makes difficult to finalize and to validate the predictive models. Therefore we will also use, in addition to real data, simulated data for the design of the MLP and the assessment of their performances. Artificial data are used to simulate data that could have been recorded in invasive electrophysiology settings and to test the capacity of the MLPs to learn a first order step impulse response model.

2.1 Pre-Processing on RR and QT Signals

The ambulatory ECGs are recorded by means of a 3-channel analog Holter recorder. RR and QT sequences were chosen over the 24H recording using as a selection criterion the richness in variations of the RR interval.

The RR intervals and the QT duration are calculated using the "Lyon System" and the "Caviar" methods [7]. To pass from an unequally sampled to an equally sampled time series sequence, an over-sampling of 4Hz with linear interpolation is performed. The

RR and QT time series are low-pass filtered at 0.05Hz to eliminate the high frequencies (HF) and low frequencies (LF) components due to the parasympathic (HF) and the sympathic (LF) activities. Finally the filtered RR and QT sequences are down sampled to 0.5 sample per second by keeping one point over eight.

2.2 Artificial Neural Network

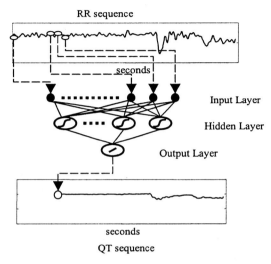

Fig. 1. Multi Layer Perceptron architecture used for the prediction of the QT time interval in function of the RR interval (inverse of the heart rate)

2.2.1 Neural Network Architecture

For the modeling of the QT dynamics we have chosen an artificial neural network implementing the MLP architecture with *NbE* entries, *k* sigmoidal hidden neurons and one linear output neuron as shown in Figure 1.

The number of entries *NbE* in the input layer will depend on the time delay M to be taken into account for modelling the relationship described in Equation (1). The number of entries *NbE* is determined by equation:

$$NbE = M \times 30 \qquad (2)$$

where M is the time delay expressed in minutes.

2.2.2 Architecture Dimensioning

To determine the appropriated MLP architecture, two ECG sequences belonging to "Clav" were selected. These sequences, recorded at 4:25 am and at 6:05 am, were used respectively as training and test sets. The mean square error (MSE) is taken as a measure for evaluating the performance:

$$MSE = \sum_{i}^{N} \frac{(QTp_i - QTm_i)^2}{N} \quad (3)$$

where QTp_i is the i-th network output, QTm_i the i-th target output and N the dimension of the input vector.

The delay M was set to vary from one minute up to eight minutes, and the number of hidden neurones from 2 to 22. The MLP with 120 entry neurones (i.e. M=4 minutes) and 10 hidden neurones gave the best results in learning the training set and in predicting the test set.

2.3 Artificial Data Description

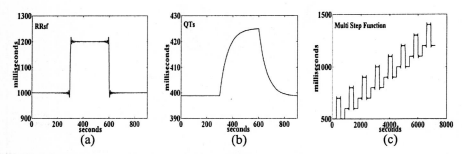

Fig. 2. Artificial data: a- The filtered step function. b- The first order response with a time constant of one minute. c- Eight step functions centred on several RR values.

To simulate a first order one-minute time constant behaviour of the QT-RR relationship, the following steps are performed:

- Generation of RRs, a RR sequence simulating a step function. RRs is composed of a 5 minute baseline, an abrupt change in amplitude of 200 milliseconds that is maintained for 5 minutes, a return to the baseline that is kept for 5 minutes. The total duration of the step function is 15 minutes.
- Computation of RRrep, the first order one minute time constant response to RRs:

$$RRrep_i = RRs_1 + [RRs_i - RRs_1] \times \left[1 - \exp\frac{1-i}{td}\right] \quad (4)$$

where td is the one minute time constant and RRs_1 is the first sample of RRs
- Filtering RRs using the same low-pass filter used for the real RR and QT data. An example of the filtered signal RRsf is given in Figure 2-a.
- Computation of QTs (Figure 2-b), the simulated QT, according to the following function where all values are in milliseconds [9]:

$$QTs_i = 8.7 * \sqrt{RRrep_{i-1}} + 123.7 \quad (5)$$

The data used for the training and the testing of the MLP are the RRsf and QTs.

2.4 Prediction Quality Evaluation

Beside the visual criterion, another criterion is needed to compare the dynamical behavior of real and predicted signals. Therefore, the standard deviation over the prediction error is used as a complementary criterion.

3 Results

3.1 Assessment of the Prediction Capacity of the MLPs for the Artificial Data Set

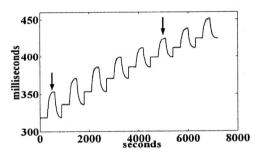

Fig. 3. The MLP's response for the multi steps function (in solid line) matches very closely the expected QT (QTs in dotted line). The two arrows indicate the step functions used for the training.

Two sets of artificial data are used to test the capacity of the neural network to learn the QT-RR relationship. Each set corresponds to a 15 minutes, RRsf and QTs sequence. The learning is stopped when the training error in Equation (3) is less than a given value. The generalization capabilities of the MLP is studied through predicting eight step functions, equally spaced by 100 msec and centered on different values between 600 msec and 1500 msec as shown in Figure 2-c.

The training set is composed of two step functions. The first set has a RRsf centered around 1100 msec, and the second set has a RRsf centered around 600 msec. They correspond respectively to the first and the sixth step functions in Figure 2-c. Training was stopped after 22.000 iterations when the learning error had reached a steady state and did not improve significantly.

Figure 3 shows the close dynamic behavior between the predicted signal and the expected one. The Multi Layer Perceptron succeeded in learning the transient dynamic behavior of the first order, one-minute response of the step impulse. It succeeded also in learning the steady state, non-linear QT-RR relationship defined by Equation (5).

3.2 Training the MLPs with Real RR and QT Time Series

To train the MLP, we have chosen a "Clav" sequence recorded during the night at 4:25 am. The training was stopped after 4000 iterations using the sequence recorded at 6:05 am as a test set.

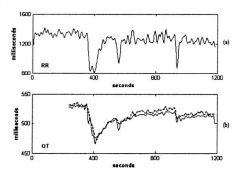

Fig. 4. Real RR and QT sequences (in full line) and the predicted QT (in dotted line) produced by the MLP.

Figure 4 shows the result of the network prediction for the RR sequence recorded at 7:35 am. The lower tracing of Figure 4-b corresponds to the measured QT interval (QTm) and the upper to the QT predicted by the MLP (QTp). The dynamic behaviour of QTp follows closely the real QT variations, with a standard deviation on the prediction error of 3.5 msec. The vertical shift between the two signals is due to a different time recording and to different activities of the sympathic and parasympathic systems.

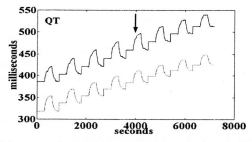

Fig. 5. Expected QT (lower tracing) and the MLP's response for the multi steps function (upper tracing). The arrow indicates the QT variation zone used for the training.

From figure 5, it can be noticed that the step response predicted by the MLP is very close to a first order step response for QT varying from 380 msec to 520 msec. The quality of the predicted signal is worsening when getting closer to the upper limit of the QT values.

In addition to sequences belonging to the patient "Clav" used for training, the neural model was also able to predict the night sequences of the other patients. Table 1 shows the prediction results for a subset of the tested night sequences, including the ECG sequences used in the training process. All standard deviations of the prediction error are less than 5 milliseconds, that is considered as an acceptable result. This value was chosen experimentally by watching sequences one by one.

Table 1. Prediction results on night sequences belonging to non-pathological patients. Clav sequences recorded at 04:25 and 06:05 were used respectively as training and testing set.

Patient	Sex	Age	Time of Recording	Standard Deviation (msec)
Clav	F	22	03:45	3,40
			04:25	2,25
			06:05	2,86
			07:35	3,51
Ardi	F	29	00:28	3,70
			05:45	3,70
Thou	M	20	02:50	4,22
			07:00	4,17
Dure	M	21	03:25	2,30
			05:35	1,80

4 Discussion

The capability of Multi Layer Perceptrons to learn and to correctly predict non-linear functions depends essentially on the architecture dimensioning (i.e. the number of entries and hidden neurons) and on the number of iterations. To minimise the training cost and the risk of over-design, the number of entries must be the lowest possible. But in the same time, the time delay M must be long enough to approach the non-linearity (1 minute time constant) of the studied signal. The number of entries can be reduced by changing the time delay value M or reducing the sampling rate. The time delay of 4 minutes found in this study is coherent with the findings that can be derived from invasive physiological tests: the step response of a one minute time constant first order system will reach 98% of the expected amplitude after 4 minutes. On the other hand, given the 0.05 Hz low-pass filter cut off frequency, it would be difficult to reduce the sampling rate below 0.5 Hz if we want to have a good quality of signal representation.

Predicting data outside the learning window is a big challenge for MLPs. As noticed in Figure 5, the amplitude variation doesn't closely follow the variations of the expected theoretical signal all over the range of the QT variations. It's hard to imagine that the MLP could "invent" a QT-RR relationship as in equation 5, when only a small part of RR and QT variations is used for the training. Using only two step functions but far away at the borders of the range of variation, we could, as shown in Figure 3, approximate both the static and the dynamic behaviour of the expected signal. The ECG sequences used for the training should display a large amount of RR and QT variations, and using a 24H signal for the training won't solve this problem.

Automatic methods need to be developed to select RR and QT sequences covering a wide range of variation.

Another feature is the vertical shift between the predicted and the measured QT. This shift depends on the time of the day and the activity of the central nervous system, and is in itself a valuable measurement that is expected to have a diagnostic importance. The same yields for several additional clinically usefully features such as the amplitude of the step impulse response, the time constant of the response and the symmetry of the rising and the falling edges, which might easily be derived for the step impulse response of the MLPs.

5 Conclusions

Although preliminary, our results indicate that Multi Layer Perceptrons are able to approach the non-linear aspects of the QT-RR relationship, and can model both the dynamic behaviour (response to a step impulse) and the steady state dynamic behaviour QT=f(RR) (response to different, fixed RR intervals).

The difference between the measured QT interval and the predicted one could be used to trigger an alarm each time a given threshold is passed. Further studies however are needed to determine such thresholds and to assess their predictive value as well as the value of the feature derived of the step impulse responses.

References

1. Holmberg, M., Holmberg, S., Herlitz, J. (ed): American Journal of Cardiol. The problem of out-of-hospital cardiac-arrest prevalence of sudden death in Europe today. (1999) 83-88
2. Fagundes, ML., Maia, IG., Cruz, F.E., Alves, P.A., Boghossian, S.H., Ribeiro, J.C., Sa, R.: Arq Bras Cardiol. Arrythmogenic cardiomyopathy of the right ventricle. Predictive value of QT interval dispersion to assess arrythmogenic risk and sudden death. (2000) 115-124
3. Bazzet, H.C.: Heart. An analysis of the time-relation of electrocardiograms. (1920) 353-370
4. Franz, M.R., Swerdlow, C.D., Liem, L.: Cycle-Length dependence of human ventricular action potential duration in the steady and non-steady state. In: Butrous, GS., Schwartz, PJ., (eds.): Clinical aspects of ventricular repolarisation. Farrand Press London (1989) 163-174
5. Rubel, P., Hamidi, S., Behlouli, H.: Journal of Electrocardiology. Are serial Holter QT, late potential, and wavelet measurment clinically useful? (1996) 52-61
6. Lapedes, A., Farber, R. : Nonlinear signal processing using neural networks : Prediction and system modeling. Los Alamos Nat. Lab., Los Alamos, CA, Tech. Rep. (1987) LA-UR-87-2662
7. Arnaud, P., Rubel, P., Morlet, D., Fayn, J., Forlini, M.C.: Method Inform Med. Methodology of ECG interpretation in the Lyon program. (1990) 393-402

Detection of Microcalcifications in Mammograms by the Combination of a Neural Detector and Multiscale Feature Enhancement

Diego Andina[1] and Antonio Vega-Corona[2]

[1] Universidad Politécnica de Madrid
Departamento de Señales, Sistemas y Radiocomunicaciones, E.T.S.I.
Telecomunicación, Madrid, Spain
andina@gc.ssr.upm.es
[2] Universidad de Guanajuato
F.I.M.E.E., Guanajuato, Mexico
tono@salamanca.uto.mx

Abstract. We propose a two steps method for the automatic classification of microcalcifications in Mammograms. The first step performs the improvement of the visualization of any abnormal lesion through feature enhancement based in multiscale wavelet representations of the mammographic images. In a second step the automatic recognition of microcalcifications is achieved by the application of a Neural Network optimized in the Neyman-Pearson sense. That means that the Neural Network presents a controlled and very low probability of classifying abnormal images as normal.

1 Introduction

Breast cancer is one of the main causes of mortality in middle-aged women, and the rate is increasing, especially in developed countries. Early detection constitutes the key factor for an effective treatment. Screening mammography is known as the most effective method for early detection of small, non-palpable breast cancers [1,2]. Screen/film mammography has been widely recognized as being the only effective imaging modality for the early detection of breast cancer in asymptomatic women. An early sign of disease in mammograms is the appearance of clusters of fine, granular microcalcifications whose individual grains typically range in size from 0.05-1 mm in diameter [3]. Microcalcifications are difficult to detect because of variations in their shape and size and because they are embedded in and camouflaged by varying densities of parenchymal tissue structures. An important branch of computer-aided diagnosis (CAD) methods for feature enhancement in mammography employs wavelet transforms [4,5] and statistical detection techniques. Our proposal for detecting microcalcifications in mammograms is to perform a multiscale edge detection [6] and analyse the properties of the edges that have been detected. The information obtained from the above process can be packed in a set of parameters, which constitutes the input to a Neural Detector optimized in the Neyman-Pearson sense. These detectors present an arbitrary low probability of classifying binary symbol "1" when symbol "0" is the correct decision. This kind of error, referred in the scientific literature as "false-positive" or "false alarm probability" has a high cost in this application and the possibility of controlling its maximum value is crucial.

2 Procedure for Detection of Microcalcifications

The method is divided in two steps. The first step accomplishes a feature enhancement to improve the visualization of any abnormal lesion. Multiscale wavelet representations of the mammographic images are used for this task [7]. A multiscale representation of a mammogram provides a set of images, each of them giving a different scale view of the initial mammogram. Abnormal mammographic lesions appear enhanced in one of more than one of the images, that means, their intensity profiles appear emphasized. These lesions appear as singularities easier to discriminate than in the original image. The second step obtains a set of parameters that compiles the information from the first step. Parameter examples are mean pixel intensity values and standard deviation of the pixel intensity values in the regions of interest. This set constitutes the input to a Neural Detector. For training the neural network, a data base with normal and abnormal mammograms (with biopsy proven abnormal lesions) is used.

This Neural Detector must have been previously trained to achieve a very low number of false positive errors, that is, a very low probability of classifying normal tissue as having microcalcificactions. This probability is also known as "false alarm probability" and the Neural Detector optimized minimizing the false alarm probability to an arbitrary value is known as Neyman-Pearson Neural Detector [8].

2.1 Characteristics of Mammographic Images

Although mammography currently is the best method for the detection of breast cancer, between 10% to 30% of women who have breast cancer and undergo mammography have negative mammograms. In approximately two thirds of these false-negative mammograms, the radiologist failed to detect the cancer that was evident retrospectively. The missed detections may be owing to poor image quality, eye fatigue or oversight by the radiologist. Double reading has been suggested, with the number of lesions found increasing by 15%. Thus, the goal of CAD research is to develop computer methods as aids to the radiologists, in order to increase diagnostic accuracy in mammography screening programs.

Microcalcifications are often a presenting sign among early breast cancers. On screening studies, 90% of all cases of nonpalpable ductal carcinoma in situ (DCIS) [9] and 70% of all cases of minimal carcinoma (infiltrating cancer smaller than 0.5 cm and all DCIS) were seen on the basis of microcalcifications alone.

The search of microcalcifications lends itself to computer detection methods because of their high clinical relevance and the lack of normal structures that have the same appearance. Individual microcalcifications appear as small (typically 0.05-1 mm) particulate objects of variable shape (from glandular to rod-shaped) and fairly uniform optical density. A typical example of microcalcification is presented in Figure 1. Although microcalcifications vary in outline and degree of elongation, the average form is roughly circular, with a tapered cross-sectional profile. Microcalcifications often appear in clusters.

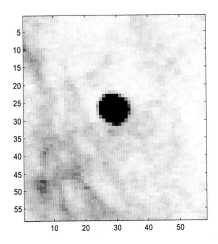

Fig. 1. Example of a circular microcalcification in a region of 1cm x 1cm.

The visibility of microcalcifications is often degraded by the high frequency texture of the breast tissue. Special attention must be paid in not to confuse singularities owing to breast tissue with those owing to microcalcifications.

A different type of lesions in mammograms are mass lesions. The computerized detection of mass lesions is different from that of microcalcifications because of some mass lesions and some normal parenchymal tissues mimic each other, thus making the interpretation difficult for both the human and the computer. In this work we are not involved with this type of lesions.

Once a lesion is detected, benignancy or malignancy must be determined. We are concerned only in detection of microcalcifications, without further analysis.

2.2 Wavelet Representations of Images and Multiscale Edge Detection

Let $\psi^1(x, y)$ and $\psi^2(x, y) \in L^2(R^2)$ two bidimensional wavelet functions and a integer j. We denote that:

$$\psi^1_{2^j}(x, y) = \frac{1}{2^{2j}} \psi^1(\frac{x}{2^j}, \frac{y}{2^j}) \tag{1}$$

$$\psi_{2^j}^2(x,y) = \frac{1}{2^{2j}} \psi^2(\frac{x}{2^j}, \frac{y}{2^j}) \qquad (2)$$

The wavelet transform of a function $f(x,y) \in L^2(R^2)$ at the scale 2^j has two components defined by the following convolutions:

$$W_{2^j}^1 f(x,y) = f * \psi_{2^j}^1(x,y) \qquad (3)$$
$$W_{2^j}^2 f(x,y) = f * \psi_{2^j}^2(x,y) \qquad (4)$$

We refer to the 2-D dyadic wavelet transform of $f(x,y)$ as the set of functions:

$$Wf = \left(W_{2^j}^1 f(x,y), W_{2^j}^2 f(x,y)\right)_{j \in Z} \qquad (5)$$

We use the term 2-D smoothing function to describe any function $\theta(x,y)$ whose integral over x and y is equal to 1 and converges to 0 at infinity. The image $f(x,y)$ is smoothed at different scales s by a convolution with $\theta_s(x,y) = (1/s2)\theta(x/s, y/s)$. If we define:

$$\psi^1(x,y) = \frac{\partial \theta(x,y)}{\partial x} \qquad (6)$$

$$\psi^2(x,y) = \frac{\partial \theta(x,y)}{\partial y} \qquad (7)$$

it can be easily proved [7, 10] that the wavelet transform can be rewritten

$$\begin{bmatrix} W_{2^j}^1 f(x,y) \\ W_{2^j}^2 f(x,y) \end{bmatrix} = 2^j \begin{bmatrix} \frac{\partial}{\partial x}(f * \theta_{2^j})(x,y) \\ \frac{\partial}{\partial y}(f * \theta_{2^j})(x,y) \end{bmatrix} \qquad (8)$$

$$\begin{bmatrix} W_{2^j}^1 f(x,y) \\ W_{2^j}^2 f(x,y) \end{bmatrix} = 2^j \nabla (f * \theta_{2^j})(x,y) \qquad (9)$$

The two components of the wavelet transform are proportional to the two components of the gradient vector of the function f smoothed at the scale 2^j. The first component measures how sharp $f(x,y)$ smoothed at a scale 2^j varies along horizontal directions, while the second component measures the variation along vertical directions. At each scale 2^j, the modulus of the gradient vector is proportional to:

$$M_{2^j} f(x,y) = \sqrt{\left|W_{2^j}^1 f(x,y)\right|^2 + \left|W_{2^j}^2 f(x,y)\right|^2} \qquad (10)$$

The angle of the gradient vector with the horizontal direction is given by:

$$A_{2^j} f(x,y) = \tan^{-1}\left(\frac{W_{2^j}^2 f(x,y)}{W_{2^j}^1 f(x,y)}\right) \qquad (11)$$

The sharp variation points of $f * \theta_s(x,y)$ are the points (x,y) where the modulus $M_s f(x,y)$ has a local maximum in the direction of the gradient given by $A_s f(x,y)$. We record the position of each of this modulus maxima as well as the values of the modulus $M_s f(x,y)$ and the angle $A_s f(x,y)$ at the corresponding locations. At fine scales, there are many local maxima created by the image noise, but at this locations, the modulus value has a small amplitude. We are interested in edge points whose modulus is larger than a given threshold at all scales. At coarse scales, the modulus maxima have different positions than at fine scales. This is due to the smoothing of the image by the function $\theta_s(x,y)$.

Sharp variations of 2-D signals are often not isolated but belong to curves in the image plane. Along these curves, the image intensity can be singular in one direction while varying smoothly in the perpendicular direction. It is well known that such curves are more meaningful than edge points by themselves beacuse they generally are the boundaries of the image structures. For discrete images, we reorganize the maxima representation into chains of local maxima to recover these edge curves. Then, we can characterize the properties of edges from the modulus maxima evolution across scales.

At the scale s, the wavelet modulus maxima detect the sharp variation points of $f * \theta_s(x,y)$. Some of these modulus maxima define smooth curves in the image plane along which the profile of the image intensity varies smoothly. At any point along the maxima curve, the gradient of $f * \theta_s(x,y)$ is perpendicular to the tangent of the edge curve. We thus chain two adjacent local maxima if their respective position is perpendicular to the direction indicated by the angle $A_s f(x,y)$. Since we want to recover edge curves along which the image profile varies smoothly, we only chain together maxima points where the modulus $M_s f(x,y)$ has close values. This chaining procedure defines an image representation that is a set of maxima chains.

3 Results

In this section we present preliminary results to support the proposed CAD method. First a multiscale feature enhancement is performed and secondly Neural Detector adapted to the preprocessed images is proposed.

3.1 Feature Enhancement

We perform a feature enhancement with a multiscale analysis. The objective in this step is to remove as much as possible information which is not relevant. We refer to intensity variations owing to parenchymal tissue structures and film noise. The first step is to obtain a dyadic wavelet transform of the full breast area with mother wavelets defined as in (6) and (7). Once obtained we can compute the modulus and the angle of the gradient vector (9) for each scale 2^j, as well as the sharp variation points inside the breast area.

The majority of these sharp variation points do not correspond to lesions in the mammogram, thus we can eliminate them. The removing criterion is to establish a threshold value, we keep the singularities whose gradient modulus values overcome this threshold. As a first threshold value we select a half of all maxima values.

At this point we have reduced the information content of the mammogram to a set of singularity points. The characteristics of this set determine the normality or abnormality of the mammogram.

Figures 2 and 3 show a region of a mammogram before and after the multi-scale feature enhancement. As it is shown, most of the noise and tissue variations are removed in the processed image, remaining the three present microcalcifications.

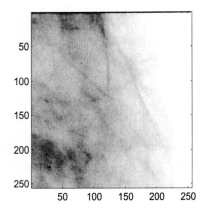

Fig. 2. Mammographic area with three proven microcalcifications

3.2 Neural Detection

We have selected an Artificial Neural Network to perform the detection task [11]. Artificial Neural Networks constitute a nonalgorithmic approach to information processing. These Neural Networks, which are capable of processing a large amount of information simultaneously, address problems not by means of pre-specified algorithms but rather by learning from examples that are presented repeatedly. The popularity of neural networks is due primarily to their apparent ability to make decisions and draw conclusions when presented with complex, noisy or partial information and to adapt their behaviour to the nature of the training data. In medical imaging, artificial Neural Networks have been applied to a variety of data-classification and pattern recognition tasks, such as the differential diagnosis of interstitial diseases, and have been shown to provide a potentially powerful classification tool [12].

We propose a three-layered feedforward Neural Network of the Multilayer Perceptron type with a backpropagation algorithm for the interpretation of the mammographic features. The enhanced image is divided in regions of size 8 x 8 pixels, corresponding to regions of 1.5mm x 1.5mm. Both the mean intensity

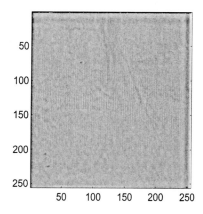

Fig. 3. The same mammographic area after the multiscale enhancement process

value and the standard desviation will be higher for regions containing a microcalcification, as can be observed in Figure 4. This figure shows the microcalcification in Figure 1 after enhancement. In consequence, these two parameters are selected as input to the Neural Network.

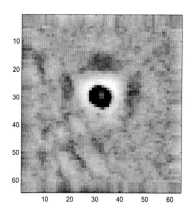

Fig. 4. Microcalcification in Figure 1 after enhancement

We have trained the three layer Neural Network with two inputs and one output, achieving a false alarm probability, P_{fa}, value of 0.001, with a detection probability, P_d, of normal mammograms of 50%. This early result predicts the success of the combination of wavelet-based an Neural Tools and motivates further investigations.

4 Conclusions

The proposed method emphasizes in one hand the ability of wavelets to perform feature extraction. In applications where the amount of information to manage is large they offer a tool of great potential that have been proved in several fields. On the other hand, the advantages typical of Neural Networks are applied to the problem of the automatic detection and characterization of singularities in mammograms. The Proposed Neural Network must be optimized in the Neyman-Pearson sense, leading to an automatic method of classification with a controlled and very low probability of classify abnormal mammograms as normal.

References

1. Feig, S.A., Hendrick, R.E.: Risk, Benefit and Controversy in Mammographic Screening. In A Categorical Course in Physics. A. G. Haus and M. J. Yaffe. (eds). Radiological Society of North America. (1993) 119–135
2. Feig, S.A.: Mammogramphic Evaluation of Calcifications. In A Categorical Course in Breast Imaging. (eds). Radiological Society of North America. (1995) 93–105
3. Bassett, L.W.: Mammographic Analysis of Calcifications. Radiologic Clinics of North America. **30** (1992) 93–105
4. Strickland, R.N., Hahn, H.I.: Wavelet Transform for Detecting Microcalcifications in Mammograms. IEEE Trans. on Medical Imaging, Vol. 15. **2** (1996) 218–229
5. Jeine, J., Deans, S.R.: Multirresolution Statistical Analysis of High-Resolution Digital Mammograms. IEEE Trans. on Medical Imaging, Vol. 16. **5** (1997) 503–515
6. Redner, R., Walker, H.: Mixture Densities, Maximum Likelihood and the EM Algorithm. SIAM Review. **26** (1994) 195–239
7. Mallat, S., Zhong, S.: Characterization of Signals from Multiscale Edge. IEEE Trans. on Pattern Analysis and Machine Intelligence, Vol. 14. **7** (1992) 710–732
8. Sanz-González, J.L., Andina, D.: Performance Analysis of Neural Network Detectors by Importance Sampling Techniques. Neural Processing Letters. Kluwer Academic Publishers. Netherlands. ISSN 1370-4621, Vol. 9. **3** (1999) 257–269
9. Feig, S.A., Shaber, G.S., Patchefsky, A.: Analysis of Clinically Occult and Mammographically Occult Breast Tumors. American Journal of Radiology, Vol. 128. (1997) 403–408
10. Meyer, Y.: Onddelettes et Operateurs. Herman, New York. (1990)
11. Mu, Y., Giger M.L., Doi, K.: Artificial Network in Mammography. Application to Decision Making in the Diagnosis of Breast Cancer. Radiology, Vol. 187. (1993) 81–87
12. Qian, W., Clarke, L.P., Kallergi, M., Clark, R.: Tree-Structured Nonlinear Filters in Digital Mammography. IEEE Trans. on Medical Imaging, Vol. 13. **1** (1994) 25–36

An Auto-learning System for the Classification of Fetal Heart Rate Decelerative Patterns

Bertha Guijarro-Berdiñas[1], Amparo Alonso-Betanzos[1],
Oscar-Fontenla-Romero[1], Olga Garcia-Dans[1], and Noelia Sánchez-Maroño[1]

Laboratory for Research and Development in Artificial Intelligence (LIDIA)
Department of Computer Science, University of A Coruña, Campus de Elviña s/n,
15071, A Coruña, Spain
{cibertha, ciamparo}@udc.es, {oscarfon, noelia}@mail2.udc.es,
OlguiGD@eresmas.com
http://www.dc.fi.udc.es/lidia/

Abstract. The classification of decelerations of the Fetal Heart Rate signal is a difficult and crucial task in order to diagnose the fetal state. For this reason the development of an automatic classifier would be desirable. However, the low incidence of these patterns makes it difficult. In this work, we present a solution to this problem: an auto-learning system, that combines self-organizing artificial neural networks and a rule-based approach, able to incorporate automatically to its knowledge each new pattern detected during its clinical daily use.

1 Introduction

Simultaneous monitoring of Fetal Heart Rate (FHR) and Uterine Pressure (UP) signals using a cardiotocograph allows to obtain information about fetal state by the relation existing between FHR patterns and certain fetal conditions. One of the most important patterns that can be observed in a cardiotocographic (CTG) record are decelerations. They are transitory decreases of the FHR with, at least, an amplitude of 15 beats per minute (bpm) and a duration of 15 s. They are usually associated with uterine contractions -see Fig. 1-, and constitute a worrying signal for fetal health. They can be classified in three types -early, late, and variable- according to their shape and temporal relationship with uterine contractions. Each type shows a fetal problem of different nature and significance. In a previous work we presented an automate solution for the classification of decelerative patterns [7], at present integrated in the intelligent CAFE (Computer Aided Fetal Evaluator) system, developed for the automation of the NST. Despite of its satisfactory results, it would be desirable for the classifier to be able to incorporate automatically to its knowledge each new pattern detected in its daily clinical functioning. However, the neural network model employed (an MLP) makes this difficult. In this work, we present a solution -substituting that integrated in CAFE- to this problem: a hybrid auto-learning system that combines self-organizing artificial neural networks (ANN) and a rule-based approach for the recognition of decelerations of the FHR.

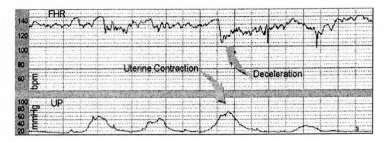

Fig. 1. CTG record showing the Fetal heart rate (FHR) and Uterine Pressure (UP)signals. Also a decelerative pattern and its associated contraction can be observed.

2 Background

There can be found in the literature several important contributions [1][6][10][11] that try to automate the analysis of CTG records. Nevertheless, the classification of decelerations in the early, late and variables types is practically an unresolved matter. As far as the authors know, only two works tackle this problem. In the first one [8], the classification is based on the estimation of the probability of each deceleration type -variable or non variable- by means of a linear equation in which the coefficients measure the correlation among several variables that represent deceleration's features. The accuracy levels reached were 92% for non-variable decelerations and 85% for the variable type. The second approach was proposed by us [7]. It is based on an MLP and tries to solve three main problems that have to be faced in this field. First, the feature set described in the literature to determine which type a deceleration belongs to is incomplete and imprecise. Second, a high inter and intra observer variability exists when experts face the classification of decelerations [9]. A third problem is the low incidence of this kind of patterns due to their ominous nature. We solved the first problem by feeding the network directly with the samples that composed the decelerative pattern, instead of extracted features. The second and third problems were overcome using patterns synthesized for printed CTG signals atlases where their classification is widely accepted as valid. An accuracy of 97.5% for the early type, 94.4% for the late type, and 96.9% for the variable type was obtained over 161 patterns. Nevertheless, the subtle differences among classes and the different prototypes that can be found inside each one suggest that the classifier's performance could degrade when it faces new patterns during its application in the clinical practice. For this reason, and taking into account the difficulty in obtaining patterns, it would be desirable for the classifier to be able to incorporate automatically to its knowledge every new pattern detected so as to increase its performance with time. However, none of the described approaches allow this possibility in a fast and simple fashion. With this objective in mind we proceed to the development of a new classifier.

3 Classifier Model Selection

Our classifier must meet some characteristics. First, it must be able to do incremental learning, incorporating new training data without having to retrain over the whole data set. Second, it must be self-organizing, adapting its structure to the knowledge progressively acquired. Third, the training process must not require the intervention of a human expert. Finally, there exists the dilemma about whether to use a supervised or a non supervised approach. The first one would eliminate the problem of the difficulty in obtaining an external reliable classification criterion. However, the low number of patterns available and the complexity of the application domain suggest that this approach could not reach satisfactory results -as it was later confirmed-, as it, normally, requires higher number of training data than the supervised approach. As the class of the patterns extracted from signal atlases are known, a supervised approach is also feasible. For these reasons, initially both types of learning will be considered. Following an analysis of the most interesting models we adopt that from the Adaptive Resonance Theory (ART). Specifically, the selected models were the Fuzzy ART for non supervised learning and the Fuzzy ARTMAP for a supervised one.

3.1 The Fuzzy ART Model

Fuzzy ART systems [3] are composed of three layer of neurons. Layer F_0 receives the input vector I= $(I_1,...,I_M)$, and its complement to 1. The output layer F_2 represents the number of categories in which the network has classified the input patterns. Every j node (j=1,...,N) in F_2 is connected to the input data through a weight vector $w_j = (w_{j1},...w_{jM})$. The middle layer F_1 receives both the inputs from F_0 and the information representing the active category for that input from F_2. For each input I the activation level T_j of each j node in F_2 is calculated as

$$T_j(I) = \frac{|I \wedge w_j|}{\alpha + |w_j|}. \tag{1}$$

where $\alpha \geq 0$ is the *choice parameter*, and $\wedge$ is the fuzzy AND operator (minimum). The output category is selected by the winner-takes-all criteria. *Resonance* occurs if the *match function*, that measures the similarity between the selected category and the input pattern, satisfies a *vigilance parameter* ρ

$$\frac{|I \wedge w_j|}{|I|} \geq \rho. \tag{2}$$

In this case, learning takes place by modifying the weights as

$$w_j(t+1) = \beta(I \wedge w_j(t)) + (1-\beta)w_j(t). \tag{3}$$

where $\beta \in [0, 1]$ is the *learning rate*. If the vigilance criterion is not met a *reset* occurs and a new winning output category is selected. This process continues until the searching criteria is met or no resonance occurs, in which case a new output category will be created.

3.2 The Fuzzy ARTMAP Model

The Fuzzy ARTMAP system [4] incorporates two Fuzzy ART modules -ART_a and ART_b- and constitute a supervised version of the original Fuzzy ART, so their functioning are quite similar. ART_a and ART_b interact trough another module F_{ab} or *map field*. During the supervised learning ART_a receives the inputs and ART_b the corresponding desired outputs, so both modules will make a categorization of their input vectors. Every j node of the ART_a's F_2 output layer, F_2^a, is connected to every node at the map field through a weight vector $w_{j_{ab}}$. The map field will establish associations among the categories in ART_a, and ART_b. If a mismatch occurs between the categories predicted by ART_a and ART_b, the vigilance parameter in ART_a will be increased in order to allow a searching for a new category in this module. Also, when a resonance in ART_a occurs, the learning rules will change the weights $w_{j_{ab}}$ for the ART_a active category J to learn to predict the ART_b active category K.

4 Materials and Methods

4.1 Description of the Training Data

As mentioned, the train/test data sets were synthesized from printed CTG signal atlases. These data has been proved to be equivalent to the *real* ones [7]. For each contraction-deceleration pair 66 data were extracted: 60 deceleration samples - patterns were normalized in time-, the amplitude of the contraction and the moments of beginning, maximum and ending points of the contraction and the minimum and ending points of the deceleration taking its beginning as the origin of the temporal scale. A data set composed by 53 early, 52 late and 54 variable decelerations was obtained.

4.2 Training and Testing Methodology

The requirement of external actions for the selected models in order to guide the training process are minimum. Once the values for the learning parameters are fixed, training finishes automatically when resonance occurs for all inputs.

To extrapolate the true error rate from the empirical error, according to our sample size, 10-fold Cross-validation was employed. Also, due to the different organization of classes when the training data are presented several times in a different order, different classification results can be obtained for the same test set -although the overall accuracy will be similar. To solve this lack of agreement, a *n voting strategy* is proposed [2] that for each test pattern selects the classification that obtained the highest number of votes among n simulations.

4.3 Software

ART Gallery was used for the implementation of the networks. It can be reached at http://cns-web.bu.edu/pub/laliden/WWW/nnet.frame.html.

5 Results

Being the 10-fold cross-validation very time consuming, we began analyzing the non supervised approach using the 75% of the data for training and the remaining cases for testing. Experiments were performed with different values for ρ -that determines the number and organization of the network's categories-, and for the learning rate β -that determines how the features of a new input are included in the assigned category. Table 1 shows some of the obtained results. In any

Table 1. Results obtained by the Fuzzy ART approach for several ρ and β values

ρ	β	Epochs	Categories	%Correct classifications
0.4	0.8	1	1	0.0%
0.5	0.2	28	3	52.5%
0.5	0.6	18	6	20.0%
0.5	0.8	10	6	17.5%
0.55	0.2	46	4	50.0%
0.55	0.6	10	4	45.0%
0.55	0.8	6	5	57.5%
0.6	0.6	7	4	45.0%
0.7	0.8	13	14	17.5%
0.8	1	1	17	12.5%
0.95	0.8	7	69	7.5%

case, the percentage of correct classifications were above 57.5%, neither it was possible to interpret the network's criteria for classifications, which were found to be random. This confirmed our suspects about the non-supervised approach. Therefore experiments with the fuzzy ARTMAP were carried out.

Before fine-tuning the network, the appropriate values for the parameters involved in the training were determined reserving again 25% of data for testing. Established ρ_{ab}, ρ_a, and β as 0.9, 0.7, and 0.75, respectively, the 10-fold cross-validation was performed. For every test set, a 7 voting strategy was applied in order to obtain the results shown in Table 2. For each test set it can be observed the percentage of cases correctly classified by the simulations that obtained the minimum and maximum number of correct answers, so as the mean over the 7 simulations and the result finally obtained after the application of the voting strategy. The estimated error rate E was 0.08 with a deviation from the true error rate E_{SD} equal to 0.02. Some performance indexes like the percentages of true positives (TP) and true negatives (TN), the specificity (Sp), sensitivity (S), positive predicted value (PPV), negative predicted value (NPV) and the accuracy (A) were also calculated and are shown in Table 3.

Table 2. Percentage of correct classifications obtained by 10-fold cross-validation and a 7 voting strategy

Testing subset	Minimum	Maximum	Mean	Voting
1	62.5%	87.5%	75.9%	87.5%
2	50.0%	87.5%	67.0%	81.3%
3	56.3%	100.0%	71.4%	87.5%
4	75.0%	93.8%	87.5%	100%
5	81.3%	100%	88.4%	100%
6	68.8%	100%	87.5%	100%
7	62.5%	81.3%	74.1%	81.3%
8	62.5%	93.8%	81.3%	93.8%
9	81.3%	100%	90.2%	100%
10	68.8%	87.5%	81.3%	87.5%

Table 3. Performance indexes calculated for the Fuzzy ARTMAP

Type	%TP	%TN	%S	%Sp	%PPV	%NPV	%A
Early	92.5	95.3	92.5	95.3	90.7	96.2	94.3
Late	94.2	93.5	94.2	93.5	87.5	97.1	93.7
Variable	88.9	99.1	88.9	99.1	98.0	94.5	95.6
Global	91.8	96.0	91.8	96.0	92.1	95.9	94.5

6 Discussion of the Fuzzy ARTMAP Results

The estimated error rate, $E = 0.082$, and its standard deviation for the true error, $E_{SD} = 0.022$, predict promising results for the classification system. Also, from Table 3 it can be deduced that the number of correct classifications is practically the same among classes and no systematic error is observed. Specifically, 94.3% of early type, 93.7% of late type, and 95.6% of variable type decelerations were correctly classified, thus surpassing the results obtained by Hamilton et al. [8]. In particular, for variable decelerations, being this type the most difficult to classify, the number of cases correctly assigned to this class was 98.0%, so it can be deduced that the network founded a adequate classification criteria. Regarding the late decelerations almost no pattern of this type was assigned to any other class (NPV = 97.1%), an important aspect if we take into account that this type is the most ominous one. Thus, the work performed confirms that the fuzzy ARTMAP network is an adequate approach to the deceleration classification problem. Although our previous approach based on a MLP reached slightly better results, the advantages of the self-organizing and the incremental learning capacities of the fuzzy ARTMAP, allow to conclude that this new proposed approach improves that actually integrated in the CAFE system, specially if we take into account another advantage of the fuzzy ARTMAP model: its fast evolution towards a better performance with only a few more training examples. This is demonstrated in Table 4 which shows how for a difference of only 39 cases the percentage of correct classifications increase from 75.0% to 82.5%, with almost no increment neither in the number of needed categories in ART_a to organize the input information nor the number of training epochs.

Table 4. Influence of the number of training cases in the Fuzzy ARTMAP's performance

Examples	Epochs	ART_a categories	%Correct classifications
80	3	12	75.0%
100	4	13	80.0%
119	5	15	82.5%

7 The Auto-learning Hybrid System

Our initial aim was to have a system able to retrain itself every time a new decelerative pattern is detected during the clinical use of CAFE. The selected Fuzzy ARTMAP model will make this easy. However, this is a supervised approach and a external classification criterion is needed, which is not easily obtainable due to the high inter and intra expert variability. As a solution we built a training shell, a small rule-based system, in charge of controlling the learning process. For every new pattern, it will try to obtain a reliable classification criterion and will activate the learning process whenever this criterion is available. In order to do it, clinicians will be asked to classify these patterns at different moments -at least, three times. The kappa measure [5] will be used to determine when the agreement between an expert and him/herself is good enough to establish his/her final classification for a pattern. This measure will be also calculated among clinicians. The retraining will take place whenever a good agreement for a given pattern is obtained among the criteria provided by, at least, three clinicians. After that, a significance test will be applied so as to know when the retrained network gives worst results than the original one and therefore to prevent a degradation of the system. If this occurs, a alarm e-mail will be sent to the system's maintainer. If on the contrary, a significant better result is obtained this will become the standard of comparison for future significance tests.

8 Conclusions

The classification of FHR decelerations is a difficult task due in part to the subtle differences among classes and the existence of different prototypes for each one. Therefore, it would be desirable for a classifier to be developed using a high number of patterns. However, these are difficult to obtain. As a solution to this problem we developed a hybrid system. Its core is a Fuzzy ARTMAP network for which results have demonstrated its ability to classify this kind of patterns. This network has been provided with a rule-based training shell whose knowledge has been designed to fire the retraining of the network with new patterns detected during its clinical use and to determine a reliable classification criterion to be provided to the network. In this way, a system for the classification of decelerative pattern was built that works as autonomously as possible.

Acknowledgements

This research has been funded in part by the pre-doctoral programme 2000/2001 and project PGIDT99COM10501 of the Xunta de Galicia.

References

1. Bernardes, J., Ayres-de-Campo,D., Costa-Pereira, A., Pereira-Leite, L., Garrido A.: Objective computerized fetal heart rate analysis. Int. J. Gynaecol. Obstet. **62(2)** (1998) 141-147
2. Carpenter, G., Grossberg, S. , Reynolds, J.H.: ARTMAP: supervised real-time learning and classification of nonstationary data by a self-organizing neural network. Neural Networks **4** (1991) 565-588
3. Carpenter, G., Grossberg, S., Rosen, D.B.: Fuzzy ART: fast stable learning and categorization of analog patterns by an adaptive resonance system. Neural Networks **4** (1991) 759-771
4. Carpenter, G., Grossberg, S., Markuzon, N., Reynolds, J.H.: Fuzzy ARTMAP: a neural network architecture for incremental supervised learning of analog multidimensional maps. IEEE T. Neural Network. **3** (1992) 698-712
5. Fleiss, J.L.: Statistical Methods for Rates and Proportions. John Wiley, New York (1981)
6. Greene, K.R.: Intelligent fetal heart rate computer systems in intrapartum surveillance. Curr. Opin. Obstet. Gynecol. **8** (1996) 123-127
7. Guijarro-Berdinas, B., Alonso-Betanzos, A., Fernandez-Chaves, O., Alvarez-Seoane, M., Ucieda-Pardinas, F.: A neural network approach to the classification of decelerative cardiotocographic patterns in the CAFE project. In: Proc. Int. ICSC/IFAC Symp. on Neural Computation/NC'98. Vienna, Austria (1998) 827-834
8. Hamilton, E., Kimanani, E.K.: Intrapartum prediction of foetal status. In: Rogers, M.S., Chang, A.M.Z. (eds.): Clinical Obstetrics and Gynaecology. International Practice and Research, Analysis of Complex Data in Obstetrics, Vol. 8(3). Balliere's Clinical Obstetrics and Gynaecology, London (1994) 567-581
9. Lotgering, F.K., Wallenburg, H.C.S., Schouten, H.J.A.: Inter-observer and intra-observer variation in the assessment of antepartum cardiotocograms. Am. J. Obstet. Gynecol. **125** (1982) 701-705
10. Rosen, B., Soriano, D., Bylander, T., Ortiz-Zuazaga, H., Schiffrin, B.: Training a neural network to recognize artefacts and decelerations in cardiotocograms. In: Proc. 1996 AAAI Spring Symposium on Artif. Intell. Med. Stanford, U.S.A. (1996)
11. Ulbricht, C., Dorffner, G., Lee, A.: Neural networks for recognizing patterns in cardiotocograms. Artif. Intell. Med. **12** (1998) 271-284

Neuro-Fuzzy Nets in Medical Diagnosis: The DIAGEN Case Study of Glaucoma

Enrique Carmona[1], José Mira[1], Julián G. Feijoo[2], Manuel G. de la Rosa[3]

[1] Dpto. de Inteligencia Artificial, Facultad de Ciencias, UNED, Paseo Senda del Rey 9
28040, Madrid, Spain
{ecarmona, jmira}@dia.uned.es

[2] Dpto. de Oftalmología Hospital Clínico San Carlos,Instituto de Investigaciones Oftalmológicas Ramón Castroviejo, Universidad Complutense, Madrid, Spain
mherrerad@sego.es

[3] Facultad de Medicina, Universidad de La Laguna, Tenerife, Spain
mgdelarosa@jet.es

Abstract. This work presents an approach to the automatic interpretation of the visual field to enable ophthalmology patients to be classified as glaucomatous and normal. The approach is based on a neuro-fuzzy system (NEFCLASS) that enables a set of rules to be learnt, with no a priori knowledge, and the fuzzy sets that form the rule antecedents to be tuned, on the basis of a set of training data. Another alternative is to insert knowledge (fuzzy rules) and let the system tune its antecedents, as in the previous case. Three trials are shown which demonstrate the useful application of this approach in this medical discipline, enabling a set of rule bases to be obtained which produce high sensitivity and specificity values in the classification process.

1. Introduction

Glaucoma is a disease characterized by the existence of a chronic and progressive lesion of the optic nerve caused by damage to the ganglion cells in the retina. This disease causes anatomic damage to the optic nerve and impairment of the visual function. The importance of this disease lies in its serious implications, since it causes progressive and irreversible loss of vision and is one of the most common causes of blindness in developed countries [10,12].

Consequently, measurement of the visual field (perimetry) is one of the most important ophthalmologic tests for the diagnosis of glaucoma. However no satisfactory solution has been found yet for addressing the problem of making an objective interpretation of the visual field (VF) results, and there are no standard criteria that, taking into account the distribution of the incipient glaucomatous lesions, enable them to be distinguished from lesions caused by other factors. Most of the research carried out into analyzing the perimetry results is based on clinical experience, and is inevitably influenced by pre-existent criteria on glaucomatous defects and their evolution. As a result, it may prove useful to carry out initial research, on an objective basis, into the characteristics of the glaucomatous defects, using mathematical database analysis methods which provide automatic VF analysis systems and in which the medical expert's involvement is minimal.

2. Patient Characteristics

A group of 218 glaucomatous patients and 62 individuals with no ocular pathology (normal) were used. In order to proceed to train the neuro-fuzzy network, 2/3 of each class were separated (145 pathological and 42 normal), and the remaining 1/3 (73 pathological and 20 normal) were used to check the resulting network. In both cases, training and test, the VF analysis was performed using the Octopus G1X program (Interzeag AG, Schilieren, Switzerland) and the first VF was ignored to rule out errors caused by the learning effect [4].

3. Data Processing

The information obtained when performing a VF analysis depends on the type of perimeter, the type of program, the matrix of points analyzed and the perimetric strategy employed. Generally, threshold determination is performed on a set of points of the VF for a particular stimulus of a determined size, display time and fundus illumination. Basically the stimuli are projected onto a spatial localization (which may vary depending on the matrix used) and their intensity is modified to determine the sensitivity threshold of the retina for each point. Eventually a set of threshold sensitivity measurements is obtained which the expert analyzes in order to assess the patient's condition. In an effort to automate and standardize the process of diagnosis, a series of numerical indices is used which manage to summarize the vast amount of information furnished by an ophthalmologic test of this kind.

Mean sensitivity (MS) is one of the most frequently used indices which, as its name implies, evaluates the mean value of all the sensitivity measurements (one for each point), each of which represents the minimum luminous intensity that the patient can perceive in each point of the VF analyzed. If a defect is considered to be the difference between the sensitivity value measured and that of a person of the same age considered to be statistically normal, the *mean defect* (MD) is defined as the mean of all these differences. Finally, another frequently used index corresponds to the calculation of the so-called *loss variance* (LV), which is simply the statistic variance applied to the previous defect values.

A database was created with the VF information furnished by Octopus. Each point tested (59 points altogether) was represented by the threshold sensitivity value in decibels (dB). The central 30° visual field area was divided into seven zones as shown in Figure 1(a). The choice of these areas was made by analyzing the correlation between each of the points with all the others and by choosing groups of points with a maximum degree of correlation between one another [5]. As can be seen, the zones eventually chosen are closely connected with the routes taken by the bundle of nerve fibers on their way from the different points of the retina to the papilla (emergence zone of the bundle of nerve fibers of the optic nerve), Figure 1(b). In [1, 3] VF valuation studies were performed which were also based on zones but with a different distribution to the one chosen in this instance.

Initially, 17 input variables were chosen: the mean defect and the loss variance in each of the seven zones chosen (MD1, MD2, MD3, MD4, MD5, MD6 and MD7, LV1, LV2, LV3, LV4, LV5, LV6 and LV7, respectively), the mean defect (MD) and the total loss variance (LV) and, finally, the number of points with a mean defect of over 5 dB ("Points>5dB").

If each variable is displayed on a coordinate axis opposite the others and when, in each case, a greater or lesser degree of overlapping between the normal and the pathological population is observed, the discriminating power of each of the input variables chosen can be established. This is relatively easy to perform given the limited number of classes involved in this particular classification: normal and glaucomatous patients. After performing this approximate study, the mean defect of zones 1, 2, 3, 4 and 5 and the number of points with a mean defect of over 5 dB proved to be the variables with most power of discrimination.

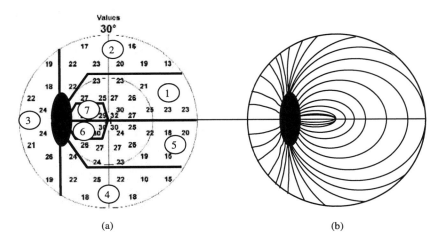

Figure 1. Seven zones into which the VF was divided (a) according to the evolution of the bundle of nerve fibers in the retina (a) An approximate diagram (b).

4. Method: Neuro-fuzzy Approach

The decade from 1965-1975 witnessed the first applications of fuzzy logic in the field of medicine. Specifically, in 1968 L.A. Zadeh presented the first article on the possibility of developing applications based on fuzzy sets for use in the field of biology [13]. Since then, fuzzy logic and other disciplines related to it (neuro-fuzzy systems, fuzzy clustering) have undergone increasing application in the field of medicine and biology [11].

Within the framework of taxonomy, as proposed in [7], for combining neural networks and fuzzy systems (fuzzy neural networks, concurrent neuro-fuzzy models, cooperative neuro-fuzzy models and hybrid neuro-fuzzy models), we have focused on the last approach, which consists in combining a fuzzy system and the work of a neuronal network in a homogeneous architecture. This system can be interpreted either as a special neuronal network with fuzzy parameters or as a fuzzy system implemented in distributed and parallel form.

The case studies of Nauck, Kruse and their collaborators should be encompassed here. This research group from the University of Magdeburg focuses its work on a fuzzy perceptron as a generic model for neuro-fuzzy approaches [6]. The idea is to use this fuzzy perceptron to provide a generic architecture which can be initialized with a priori knowledge and can be trained using neuronal learning methods. The training is performed in such a way that the learning result can be interpreted in the form of fuzzy linguistic if-then rules. Finally, in addition to the advantage of having a generic model for comparing neuro-fuzzy models, the fuzzy perceptron can be adapted to specific situations. In this case study, the NEFCLASS tool [8] is used to classify the VF data. Basically, NEFCLASS enables fuzzy rules to be obtained from data for classifying patterns in a specific number of classes. It uses a supervised learning algorithm based on fuzzy error backpropagation.

5. Learning

Learning can be performed in two facets: on the one hand, if a priori knowledge is not introduced into the network, the system tries to find the set of fuzzy rules that adapt best to the training patterns; on the other hand, once the best rules have been selected, the learning focuses on adapting the parameters that define each of the membership functions associated with the fuzzy sets that feature in the antecedent of each of the rules chosen.

When a priori knowledge exists, it is fed into the system in the form of fuzzy rules. The rule-learning stage is no longer necessary and the system focuses only on tuning the membership functions. There is a third option which involves the combination of the previous two options, that is to say, introducing a priori knowledge and letting the system add new rules to the rules that already exist so that, eventually, it tunes the precedents of them all.

Whichever option is chosen, NEFCLAS also obliges the user to choose the number of fuzzy sets with which each input variable will be partitioned, i.e. this magnitude is not learnt. Also, when the no a priori knowledge learning option is chosen, the user has the option of leaving the system with only the best n rules, before going on to the membership function adaptation stage, in which case there is one more parameter to initialize before launching the learning.

6. Results

Since it is possible to train the network with and without a priori knowledge, both cases were studied and produced the following results.

6.1 Without a Priori Knowledge

Various different trials were performed in this category, juggling with different input variables, using different fuzzy partitionings of each of the variables chosen and with the maximum number of rules. The best results were obtained for the configuration of parameters shown in Table 1, labeled as trial 1 and 2. Despite the fact that four rules less were used in trial 2 than in trial 1, the classification results are very similar, at the expense of increasing the partition of variable MD3 by just one more fuzzy set. The rule bases obtained after the network had been trained are shown in Figure 2 and Figure 3 for trial 1 and 2 respectively.

6.2 With a Priori Knowledge

In this case, the system starts with a set of rules (see Figure 4) which are input beforehand and are based on the expert's knowledge. That is to say, the training stage now only involves the tuning of the parameters associated with the membership functions of each of the precedents of the rules considered initially. Several tests were also carried out as regards the choice of the number of partitions of each of the variables selected. At the same time, the variable "Points>5dB" was included, which produced better results than in the previous case. Of all the trials carried out, the best is shown in Table 1, in the line labeled trial 3, where the classification results obtained are indicated. It can be seen how now, with a priori knowledge, there is a considerable improvement on those obtained in the previous two trials.

Table 1. Results of classifications for each trial

Trial	Input/MF	Knowledge	N° Rules	Errors		
				Training	Test	Total
1	MD1, MD2, MD4 = 3MF MD3, MD5 = 2MF	No	10	12 (6N+6P)	6 (1N+5P)	18 (7N+11P)
2	MD1, MD2, MD3, MD4 = 3MF MD5 = 2MF	No	6	12 (5N+7P)	5 (1N+4P)	17 (6N+11P)
3	MD1, MD2, MD4 = 2MF MD5, Ptos5dB = 2MF	Yes	6	6 (4N+2P)	2 (0N+2P)	8 (4N+4P)

MF = Membership function, N = Normal Patients, P = Pathological Patients
Total pattern = 280 (62N/218P), Training Pattern = 187 (42N/145P), Test pattern = 93(20N/73P)

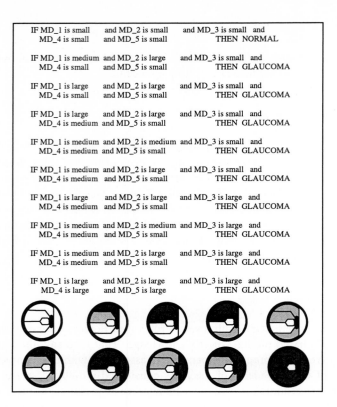

Figure 2. Rule base for trial 1 (no a priori knowledge)

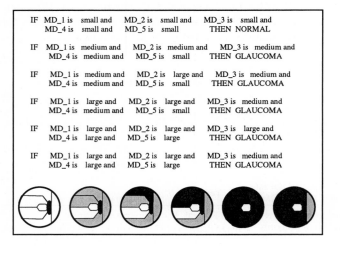

Figure 3. Rule base for trial 2 (no a priori knowledge)

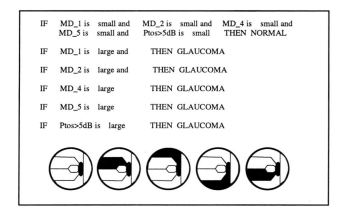

Figure 4. Selected rule base (a priori knowledge) for performing trial 3

7. Conclusions

Interpretation of the VF by the medical community is not yet based on a series of standards that allow a clear and precise classification of each of the pathologies that influence the deterioration of the VF to be made. Fortunately, different heuristics make it possible to orientate the decision-making process in one direction or another, so that a diagnosis can be made that is as reliable as possible. Nevertheless, the VF values obtained for a particular patient may not discriminate well enough to enable a reliable diagnosis to be made, meaning that the doctor has to resort to information furnished by other ophthalmologic tests. In this context, the use of systems that learn from data seem appropriate, as on the one hand they enable the diagnosis heuristics to be used and refined and on the other hand they enable new heuristics to be established.

The apparently large difference between the rule antecedents obtained from the trials performed with a priori knowledge training, or without it, is due mainly to a limitation of the NEFCLASS tool. When the system learns rules without a priori knowledge, all their antecedents have to participate on the basis of all the input variables, and it is only when a set of a priori rules is inserted that this possibility can be altered. In any case, the rule bases learnt in trial 1 and 2 express in a very similar manner what was inserted as a priori knowledge in the rule base of trial 3.

The VF indices referred to at the beginning, and which are calculated globally from 59 points on, could have been used directly as input variables. However the drawback in this case is that it is not possible to discriminate between a few abnormal points scattered over the VF and those grouped in localizations typical of glaucoma. As a result, we chose to work with those indices but calculated on the basis of the points that belonged to each of the zones into which the VF had been divided and which are highly related to the typical zones where scotomas appear in glaucoma.

As can be seen, on the basis of the results obtained, the network learning produces a set of fuzzy rules whose membership functions have been tuned on the basis of the set of training data. The great advantage of fuzzy rules compared with the possibility of using neural networks to solve the problem of classification is the high degree of linguistic interpretability of the former compared with the low interpretability of the learning result of the latter (black boxes). This high degree of interpretability has been maintained due to the NEFCLASS option of not permitting the learning of weights of the different rules obtained [9]. From the medical viewpoint, this is all crucial. In effect, when automatic diagnosis systems are used, doctors not only need to know what the pathology associated with the patient is, but also the reason for the classification performed by the system.

Table 2 shows the sensitivity and specificity associated with each of the tuned fuzzy rule bases and obtained in each of the three experiments. In all three cases it shows that the ability to detect glaucomatous pathology is greater than the ability to detect normal persons (higher sensitivity than specificity). But very good classification percentages are obtained in all cases (almost all of them above 90%). Finally, as might be expected, it should be stressed that the best sensitivity and specificity ratios are obtained when a priori knowledge (experiment 3) was used.

Table 2. Sensitivity and specificity for each rule base

Trial	Sensibility	Specificity
1	94.9%	88.7%
2	94.9%	90.3%
3	98.2%	93.5%

Although it is not always easy to compare case studies with different designs, the results obtained here can be compared with those contained in [2] based on expert systems, discriminating analysis and on neuronal networks. As can be seen, the results obtained in this instance are an improvement or are equal to the results shown there.

To summarize, this case study shows once again the important contribution of neuro-fuzzy systems to the field of medicine and, specifically, in this case, to the field of ophthalmology, by obtaining a set of rules that enable good classification results to be obtained among patients with glaucomatous pathology and normal individuals. Trials underway focus on the discrimination capacity of the neuro-fuzzy approach in terms of patients with VF disorders and who suffer from a different pathology to glaucoma.

Acknowledgments

We would like to thank the CICYT for its support, through the DIAGEN project (TIC-97-0604), in the context of which this case study has been carried out.

References

1. A. Antón, J.A. Maquet, A. Mayo, J. Tapia, J.C. Pastor, "*Value of Logistic Discriminant Analysis for Interpreting Initial of Visual Field Defects*", Ophthalmology Vol. 104, No. 3, pp. 525-531, 1997.
2. A. Antón, JC. Pastor, , "*La Inteligencia Artificial en el Diagnóstico Precoz del Glaucoma: Interpretación del Campo Visual*", aut: F.M. Honrubia, J. García-Sánchez, JC Pastor, Diagnóstico Precoz del Glaucoma, LXXIII Ponencia de la Sociedad Española de Oftalmología, ISBN: 84-8498-482-6, 1997.
3. P. Asman, A. Heijl, "*Glaucoma Hemifield Test. Automated Visual Field Evaluation*"", Arch Ophthalmol Vol. 110, pp. 812-819, 1992.
4. M.G. de la Rosa, A. Arias, J. Morales, J. García, "*Diagnóstico Precoz del Glaucoma: El Campo Visual*", aut: F.M. Honrubia, J. García-Sánchez, JC Pastor, Diagnóstico Precoz del Glaucoma, LXXIII Ponencia de la Sociedad Española de Oftalmología, ISBN: 84-8498-482-6, 1997.
5. M.G. de la Rosa, M. Gonzalez-Hernandez, M. Abraldes, A. Azuara-Blanco, "*A Quantification of Topographic Correlations of Threshold Values in Glaucomatous Visual Field*". Invest Ophthalmol Vis Sci. (ARVO Abstract), 2000.
6. D. Nauck, "*A Fuzzy Perceptron as a Generic Model for Neuro-fuzzy Approaches*", In Proc. Fuzzy-System'94, Munich, 1994.
7. D. Nauck, "*Beyond Neuro-Fuzzy: Perspectives and Directions*", Paper of Third European Congress on Intelligent Techniques and Soft Computing (EUFIT'95) in Aachen, 1995.
8. D. Nauck, R. Kruse, "*NEFCLASS-A Neuro-Fuzzy Approach for the Classification of Data*", In K. George, J.H. Carrol, E. Deaton, D. Oppenheim and J. Hightower, eds: Applied Computing 1995. Proc. of the 1995 ACM Symposium on Applied Computing, Nashville, pp. 461-465. ACM Press, New York.
9. D. Nauck, R. Kruse, "*How the Learning of Rule Weights Affects the Interpretability of Fuzzy Systems*", En Proc. IEEE International Conference on Fuzzy Systems (FUZZ-IEEE'98), pp. 1235-1240, Anchorage, AK, May 4-9, 1988.
10. HA. Quigley, "*Number of People with Glaucoma Worldwide*", Br J Ophthalmol. 80, pp. 389-393, 1996
11. H-N.L. Teodorescu, A. Kandel, L.C. Jain, "*Fuzzy Logic and Neuro-fuzzy Systems in Medicine and Bio-Medical Engineering: A Historical Perspective*", En: H-N. L. Teodorescu, A. Kandel, L.C. Jain (ed.), Fuzzy Logic and Neuro-fuzzy Systems in Medicine, pp. 3-16., CRC Press, USA, 1999.
12. B.Thylefors, AD. Negrel, "*The Global Impact of Glaucoma*". Bull Word Health Organization 72, pp.323-326, 1994.
13. LA. Zadeh, Biological application of the theory of fuzzy sets and systems, In: Proc. Int. Symp. Biocybernetics of the Central Nervous System, Little, Brown & Co., Boston, pp.199-212, 1969.

Evolving Brain Structures for Robot Control

Frank Pasemann[1,2], Uli Steinmetz[1], Martin Hülse[2], and Bruno Lara[2]

[1] Max Planck Institute for Mathematics in the Sciences, D-04103 Leipzig
[2] TheorieLabor, Friedrich Schiller University, D-07740 Jena

Abstract. To study the relevance of recurrent neural network structures for the behavior of autonomous agents a series of experiments with miniature robots is performed. A special evolutionary algorithm is used to generate networks of different sizes and architectures. Solutions for obstacle avoidance and phototropic behavior are presented. Networks are evolved with the help of simulated robots, and the results are validated with the use of physical robots.

1 Introduction

Starting from the hypothesis that higher information processing or cognitive abilities of biological and artificial systems are founded on the complex dynamical properties of neural networks, an artificial life approach to evolutionary robotics [4] is investigated. Complex dynamical properties of small neural networks, called neuromodules for obvious reasons, are in general caused by a recurrent coupling structure involving excitation and inhibition. Thinking about these neuromodules as functionally distinguished building blocks for larger brain-like systems, the (recurrent) coupling of these non-linear subsystems may lead to an emergent behavior of the composed system.

Because the structure of neuromodules, with respect to adaptive behavior of autonomous agents, is hard to construct [1], an evolutionary algorithm to develop neuromodules as well as the couplings between these subsystems is used. This algorithm is called ENS^3-algorithm for *evolution of neural systems by stochastic synthesis*. The ENS^3 has already been tested successfully on nonlinear control problems [6]. Furthermore, autonomous systems acting in a sensori-motor loop, like simulated or real robots, are an appropriate tool for studying the development of embodied cognition [7].

For the experiments reported in this paper the miniature Khepera robots [3] as well as the Khepera simulator [2] are used. With respect to a desired robot behavior, the neural networks are evolved with the help of the Khepera simulator. Those generating an excellent performance in the simulator are then implemented in the physical robot and tested in various physical environments. Finally, the structure of successful networks can be analysed. It turned out, that evolved networks with an outstanding performance can be comparatively small in size, making use of recurrent connections.

The ENS^3-algorithm is applied to networks of standard additive neurons with sigmoidal transfer functions. The algorithm sets no constraints, neither on the number of neurons nor on the connectivity structure of a network, developing network size, architecture, and parameters like weights and bias terms simultaneously. Thus, in contrast to genetic algorithms, it does not quantize network parameters and it is not primarily used for optimizing a given architecture. The algorithm will be outlined in section 2. For the solution of extended problems (more complex environments, more complex sensori-motor systems, more complex survival conditions, etc.) the synthesis of evolved neuromodules forming larger neural systems can be achieved by evolving the coupling structure between functionally distinguished modules. This may be done in the spirit of co-evolution of interacting species. We suggest that this kind of evolutionary computation is well suited for evolving neural networks, especially those with recurrent connectivity and dynamical features.

2 The ENS^3 evolutionary algorithm

For the following experiments we use networks with standard additive neurons with sigmoidal transfer functions for output and internal units, and simply buffers as input units. The number of inputs and outputs is chosen according to the problem; that is, it depends on the number of involved sensors and the necessary motor signals. Here networks will have to drive only the two motors of the Khepera. Thus, we use *tanh* as neural transfer function, setting bias terms to zero, to provide positive and negative signals for forward/backward operations of the motors. Nothing else is determined, neither the number of internal units nor their connectivity, i.e. self-connections and every kind of recurrences are allowed, as well as excitatory and inhibitory connections. Because input units are only buffering data, no backward connections to these units are allowed.

To evolve appropriate networks we consider a population $p(t)$ of $n(t)$ neuromodules undergoing a variation-evaluation-selection loop, i.e.

$$p(t+1) = S\,E\,V\,p(t)\ .$$

The *variation operator* V is realized as a stochastic operator, and allows for the insertion and deletion of neurons and connections as well as for alterations of bias and weight terms. Its action is determined by fixed per-neuron and per-connection probabilities. The *evaluation operator* E is defined problem-specific, and it is given in terms of a fitness function. This function usually has two types of terms: those defining the performance with respect to a given behavior task (system performance), and terms related to internal network properties like network size and the number of connections. After evaluating the fitness of each individual network in the population the number of network copies passed from the old to the new population depends on the *selection operator* S. It performs the differential survival of the varied members of the population according to evaluation results. During repeated passes through the variation-evaluation-selection loop the average performance of the populations will increase. If a satisfactory system performance is achieved, the evolution process will generate smaller networks of equal system performance, if terms in the fitness function are set appropriately.

3 Evolved networks for Khepera control

The goal of the following experiments is to generate a specific robot behavior like obstacle avoidance or phototropism. We use an average population size of 30 individuals. The time interval for evaluating these individuals was set to 2000 simulator time steps. There was also a stopping criterion for the evaluation of an individual: It was terminated when bumping into an obstacle. Furthermore, we influence the size of resulting networks by adding cost terms for neurons and connections to the given fitness functions.

3.1 The first experiment: Obstacle avoidance

In a first experiment a network which allows the simulated Khepera robot to move in a given environment without hitting any obstacle present in this environment is evolved. The final goal is to obtain networks, evolved in a simulated world, which produce a comparably successful behavior, when controlling the physical robot in its very different environment.

For solving this task, the six infra-red proximity sensors in the front, and the two in the rear of the robot provide the input to the network. To control the two motors of the Khepera, the signals of the two output neurons are used. Thus, initially the individual neuromodules have only eight linear input neurons as buffers and two nonlinear output neurons.

To reach the goal fitness function F is introduced which simply states: For a given time T go straight ahead as long and as fast as possible. This is coded in terms of the network output signals out_1 and out_2 as follows. First the quantities m_1 and m_2 are defined as follows:

if $out_i \leq 0$ then $m_i = 0$, else $m_i = out_i$, $i = 1, 2$.

Then, the fitness function is given by

$$F := \sum_{t=1}^{2000} \alpha \cdot (m_1(t) + m_2(t)) - \beta \cdot |m_1(t) - m_2(t)|, \tag{1}$$

with appropriate parameters α and β.

Starting with a simple environment (with less walls and obstacles than the environment shown in Fig. 2) it takes around 100 generations (depending on the parameter settings of the evolutionary program) to get individuals having satisfactory fitness values. Although generating a comparably good robot behavior, the corresponding neural networks can differ in size as well as in their connectivity structures.

The development of brain size (number of neurons and number of connections) of a typical evolutionary process can be followed in Fig. 1. At the beginning robots do not move or are spinning around having only one wheel active, but already after ten generations the first robot is moving slowly on a straight line. After thirty generations the fittest robots are exploring the accessible space,

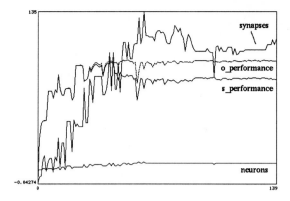

Fig. 1. Performance of simulated robots and the development of corresponding network size during the first 139 generations of an evolutionary process.

moving faster on straight lines and turning when near a wall. During this phase networks are still growing in general. The irregular development of network size corresponds to the different initial conditions from which robots in every generation have to start; they are more or less difficult to cope with. Then, around generation 45 the costs for neurons and synapses are increased to keep the size of the networks small. Therefore the curves for the output performance and the system performance are now splitting. At around generation 60 the number of hidden neurons of best networks stabilizes around 6, with networks having around 110 synapses. A further increase in costs for the network size is able to reduce the number of neurons and synapses without decrease in the performance of the controllers.

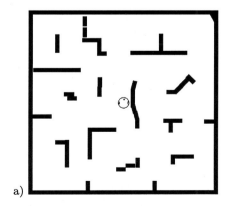 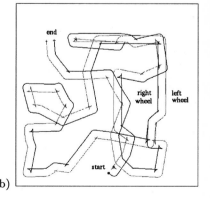

Fig. 2. a.) A simulator environment and b.) a robot path in this environment for 5000 time steps.

To make the simulator solutions more robust, in the sense that they can control the real robot in its very different physical environment equally efficient, the simulator environments gradually changed having more obstacles or walls, including for instance walls at 45 angles. An example environment is shown in Fig. 2a, together with a typical path of a simulated robot in this environment. The paths for the left and the right wheel start at the filled circles and end at the cross marks. It can be seen, that the robot turns left as well as right in different situations, achieving this by turning the corresponding wheel backwards for a short while.

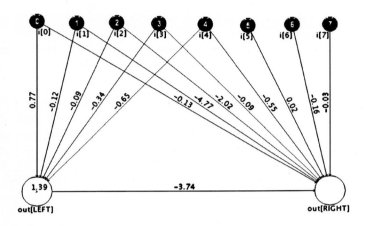

Fig. 3. The simplest evolved network which was able to generate an effective obstacle avoidance behavior for the real robot.

The most simple solution, which also generates a very good behavior of the real robot, is shown in Fig. 3. It uses no internal neurons, but only direct connections from inputs to the output neurons. What is remarkable about this solution, is its output neuron configuration. The neuron driving the left motor has a self-inhibitory neuron which also inhibits the second output neuron. Furthermore, it should be noted, that not all sensors are connected to the left or to the right output neuron, that the rear proximity sensors are connected only to right output neuron $out[0]$, and that the whole connectivity structure is highly asymmetric with respect to the left/right symmetry of the sensor and motor configuration. Remarkably, most of the connections are inhibitory. It is well known, that for a single neuron with positive self-connection larger than 1 there is an input interval over which a hysteresis phenomena [5] (flipping between two stable states) can be observed. It seems that in the shown network this effect is effectively used for instance to get out of situations like 45 angles between walls.

A second solution shown in Fig. 4 uses two hidden neurons and several closed signal loops. All neurons (hidden and output) use self-connections, and loops are of different length - like the 2-loop between the output neurons, the 3-loop

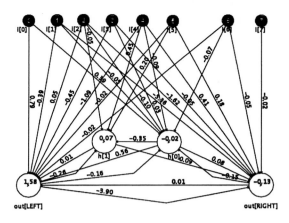

Fig. 4. A more complex network generating effective obstacle avoidance for the real robot.

involving neurons $out[0]$, $h[0]$ and $h[1]$, and the 4-loop involving $out[0]$, $h[0]$, $h[1]$ and $out[1]$. Furthermore, loops are even or odd, that is, the number of inhibitory connections in a loop is even or odd. The generated behavior in the real robot is slightly smoother than the one generated by the first network. The robot not only finds its way out of 45 angle walls but also maneuvers out of a narrow blind alley.

The general impression of the robots behavior is of course not only obstacle avoidance but that of an exploratory behavior: letting the robot move in its physical environment after a while it will visit almost all areas within reach, moving through small openings in the walls, wandering through narrow corridors and even coming out of dead ends. This of course cannot be achieved by pure wall following behaviors, as it is usually learned by robots.

3.2 The second experiment: Light seeking

For the second experiment in addition to the 8 proximity sensors, the 8 ambient light sensors of the Khepera are used; i.e., we now have 16 sensor inputs. The goal of this experiment is to find a light source as fast as possible and to stay there (eating). Because the first experiment already provided networks generating an exploratory behavior, for the initial population it is reasonable to use one of these network solutions. The additional light sensors are of course not yet connected. During the evolution process additional neurons will connect to these new inputs while the initial network configuration may change.

To find an appropriate solution the following fitness function is applied stating: For a given time and a given environment "eat" as much light as possible. This is coded as follows

$$F := \sum_{t=3D1}^{2000} \alpha \cdot in_f(t) + \beta \cdot | in_f(t) - in_r(t) | , \quad (2)$$

where α and β are appropriate parameters, and in_f and in_r are given in terms of the inputs $i[0]$, ..., $i[7]$ to the additional light sensors by

$$in_f := i[0] + i[2] + i[3] + i[5], \quad in_r := i[6] + i[7] \ .$$

Thus, for determining the fitness only four of the six front light sensors are used, and the proximity sensors are not evaluated at all.

Again, the ENS^3-algorithm is applied to the simulated robots with a few light sources now spread over the environments used also in the first experiment. As before, we gradually make the boundary conditions more complex; i.e., more walls and obstacles are introduced while reducing the number of light sources.

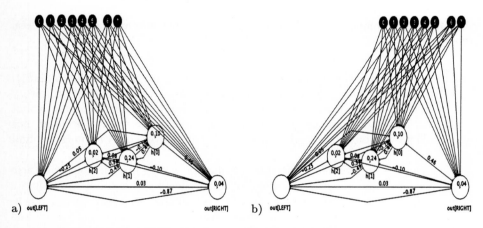

Fig. 5. a.) Connections of the proximity sensors and b.) of the ambient light sensors to the hidden and output neurons of an evolved network for phototropic behavior.

After having evolved networks for the simulated robots which seemed to generate the desired behavior in several different environments these networks are tested for the real robot. In fact, also the physical robot "looked" for a light source and got there straight when one was found. It even followed the light source, a small lamp, when this was moved by hand. The generated behavior seemed to be robust also in the sense that it was not much influenced by changing light conditions. Again there were reasonable solutions of different network size and structure. One interesting solution is shown in Fig. 5, where, for greater clarity, Fig. 5a shows the connections coming from proximity sensors and Fig. 5b those coming from the light sensors. It can be seen, that both types of input signals address all hidden and output neurons; i.e., both types of information are processed in one and the same network structure simultaneously. Furthermore the massively recurrent structure of this network is obvious, including also self-connections. Especially, the output neurons are again recurrently connected.

4 Final Remarks

The presented method to evolve neural networks for robot control turned out to be effective in two different directions. On one hand, the developed networks produced a robust robot behavior in the sense that one and the same network was not only able to cope equally good with different environments in the simulator, but turned out to have an excellent performance also when implemented in the physical Khepera and tested under different environmental boundary conditions. On the other hand, the ENS^3-algorithm produced a variety of recurrent neural networks, which can be analyzed with respect to their internal dynamical properties. This can reveal relations between these properties and the generated robot behavior. The mentioned hysteresis effect, which seemed to be responsible for effective turning in dead ends, may serve as a first example.

Our strategy is to progress from simple to more complex robot behaviors by evolving larger networks starting from initial populations of smaller functionally distinguished neuromodules. In principle there are two different techniques to achieve this. One method, called *module expansion*, starts from already evolved networks which can solve a subtask. Evolution will then add new neurons and connections to these, probably fixed, structures. It was applied in the experiment described above, where phototropic behavior was created from an existing exploration ability.

The second method may be called *module fusion*: two evolved functionally distinguished modules, with fixed architectures, are chosen for the initial population and the evolution process generates only an appropriate coupling structure to accomplish a more extensive behavior task. Here, of course, emergent properties have to be expected and are desired: The resultant behavior may not be due just to the "sum" of the basic abilities, but may be of a very different quality.

References

1. Husbands, P., and Harwey, I. (1992) Evolution versus design: Controlling autonomous robots, in: *Integrating perception, planning and action: Proceedings of the Third Annual Conferences on Artificial Intelligence*, IEEE Press, Los Alamitos.
2. Michel, O., *Khepera Simulator* Package version 2.0: Freeware mobile robot simulator written at the University of Nice Sophia-Antipolis by Oliver Michel. Downloadable from the World Wide Web at http://wwwi3s.unice.fr/~om/khep-sim.html
3. Mondala, F., Franzi, E., and Ienne, P. (1993), Mobile robots miniturization: a tool for investigation in control algorithms, in: *Proceedings of ISER' 93*, Kyoto, October 1993.
4. Nolfi, S., and Floreano, D. (2000) *Evolutionary Robotics: The Biology, Intelligence, and Technology of Self-Organizing Machines* MIT Press, Cambridge.
5. Pasemann, F. (1993), Dynamics of a single model neuron, *International Journal of Bifurcation and Chaos*, **2**, 271-278.
6. Pasemann, F. (1998), Evolving neurocontrollers for balancing an inverted pendulum, *Network: Computation in Neural Systems*, **9**, 495-511.
7. Pfeifer, R., and Scheier, C. (2000), *Understanding Intelligence*, MIT Press, Cambridge.

A Cuneate-Based Network and Its Application as a Spatio-Temporal Filter in Mobile Robotics

Eduardo Sánchez, Manuel Mucientes, and Senén Barro

Grupo de Sistemas Intelixentes (GSI)
Departamento de Electrónica e Computación, Facultade de Física,
Universidade de Santiago de Compostela,
15706 Santiago de Compostela, Spain
{teddy, manuel, senen}@dec.usc.es
http://www-gsi.dec.usc.es

Abstract. This paper focuses on a cuneate-based network (CBN), a connectionist model of the cuneate nucleus that shows spatial and temporal filtering mechanisms. The circuitry underlying these mechanisms were analyzed in a previous study by means of a realistic computational model [9, 10] of the cuneate. In that study we have used experimental data (intracellular and extracellular recordings) obtained in cat in *vivo* [2, 3] to guide and test the model. The CBN is a high-level description of the realistic model that allows to focus on the functional features and hide biological details. To demonstrate the CBN capabilities we have applied it to solve a filtering problem in mobile robotics.

1 Introduction

The cuneate nucleus is a part of the somato-sensory system and constitutes, in conjunction with the gracile nucleus, the dorsal column nuclei (DCN). Its afferent inputs, called Primary Afferent Fibers (PAF), are originated in both cutaneous and proprioceptive receptors of the upper body. The input signals are processed by a circuitry composed mainly by two different types of cells: projection neurons, also called cuneothalamic or relay neurons, and local neurons, also called interneurons.

Intracellular recordings obtained under cutaneous and lemniscal stimulation show that the afferent fibers can establish excitatory and inhibitory synaptic connections with the cuneothalamic neurons [2]. In addition, distinct types of recurrent collaterals with the capability of either exciting or inhibiting both cuneothalamic neurons and interneurons were also discovered [3]. With these data we can generate hypothesis about which circuits are implicated and also elaborate computational models to study their processing capabilities [9, 10]. The Cuneate-Based Network (CBN) is a connectionist model that describes the local circuitry of the cuneate, that means the circuitry without considering cortico-cuneate inputs. Our studies show that such circuit can detect dynamic patterns [10]. The CBN will be introduced in section 2.

To test the CBN capabilities we have applied it in a filtering problem in mobile robotics. The CBN performs a spatio-temporal filtering which goal is to improve the perceived trajectory made by a mobile obstacle. Section 3 explains that problem and how the network is integrated in a system that performs the task of collision avoidance of mobile obstacles [8]. The results, obtained in an experiment with a real robot, are shown in section 4.

2 Cuneate-Based Network

The Cuneate-Based Network focuses on the functional features of a realistic model previously developed. In that model we have studied: (1) the spatial filtering capabilities of the center-surround receptive fields and the recurrent lateral connections, and (2) the temporal filtering capabilities provided by presumed autoinhibitory connections. Figure 1 shows how the CBN architecture integrates these different connectivity in a single circuitry.

The output $y_j(t)$ of j-th neuron is calculated after applying a threshold function Ψ to the total input. This value is the result of adding, at a given time t, the contribution of: (1) afferent input $v_j(t)$, inhibitory lateral connections $u_j(t-\triangle t)$, and the output from the corresponding inhibitory neuron $o_j(t - \triangle t)$. u_j and o_j

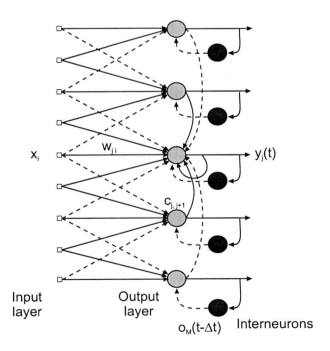

Fig. 1. CBN architecture.

are weighted by constant parameters α and β, respectively, that determine the strength of both types of connections:

$$y_j(t) = \Psi(z_j(t)) = \Psi(v_j(t) + \alpha u_j(t - \Delta t) + \beta o_j(t - \Delta t))\ j = 1, 2, ..., M \quad (1)$$

where Ψ is defined as follows,

$$\Psi(z_j(t)) = \begin{cases} z_j(t) & \text{si } z_j \geq U \\ 0 & \text{si } z_j < U \end{cases} \quad (2)$$

The contribution of afferent input to the j-th neuron is computed considering that middle inputs are excitatory and lateral inputs are inhibitory.

$$v_j = \sum_{i=n_j-C}^{n_j+C} w_{ji} x_i,\ j = 1, 2, ..., M \quad (3)$$

where v_j denotes the total afferent input to the j-th neuron, 2C+1 denotes the width of the receptive field, w_{ji} denotes the strength of the connection between the i-th input and the j-th output neuron (such that $w_{ji}=0$, if $i \leq 0$ or $i > N$), and x_i is the value of i-th input.

The lateral inhibition mechanism can be implemented with the popular Mexican-hat function. The contribution of these connections is the following:

$$u_j = \sum_{k=-K}^{K} c_{j,j+k} y_{j+k},\ j = 1, 2, ..., M \quad (4)$$

with K denoting the semi-width of the Mexican-hat and $c_{j,j+k}$ the value of the lateral connections between j-th and i-th neuron. Following the Mexican-hat distribution, $c_{j,j+k}$ is positive, i.e excitatory connection, for first-order neighbours, and negative, i.e inhibitory connection, otherwise (again, $c_{ji} = 0$, if $i \leq 0$ or $i > M$).

The temporal filtering mechanism relies in the inhibition of neurons of the output layer if the inputs persist in time. To implement this mechanism we have used the results from Koch et. al [7], that explain how the time constant of a biological neuron can vary as a function of the input activity. In our approach we have computed the difference ΔY between the current and the last input vector:

$$\Delta Y = y_j^{output}(t) - y_j^{output}(t-1) \quad (5)$$

The time constant τ will be a function of the variable ΔY:

$$\tau(\Delta Y) = \begin{cases} \tau + \Delta \tau & \text{if } \Delta Y \simeq 0 \\ \tau - K\Delta \tau & \text{if } \Delta Y \not\simeq 0 \end{cases} \quad (6)$$

The time constant τ can take any value from a range between $\tau_{min}=1$ and some maximum τ_{max}. The time constant defines the temporal window to integrate the inputs, which in turn, determines the total input sum to the interneuron $q_j(t)$:

$$q_j(t) = \sum_t^{t-\tau(\triangle Y)} y_j^{output}(t), j = 1, 2, ..., M \qquad (7)$$

Finally, the output of interneurons will depend on the result of a threshold function similar to the one that was introduced for the output layer.

3 The problem of collision avoidance of mobile obstacles

As it was explained before, we want to demonstrate the functional capabilities of the CBN by applying it on a real problem. We found a suitable one in the domain of mobile robotics. The task consists on filtering sonar data to improve the trajectory estimation of a mobile obstacle detected in the environment. This task is one of the subproblems encountered in the design of behaviours for collision avoidance of mobile obstacles [8]. This task can be decomposed in three different phases (figure 2): the perception of mobile objects, the appropriate decision-making to avoid them, and the execution of a certain motor action. Because the robot has to react dynamically to changes in the environment, the main challenge of this task is to achieve real-time performance.

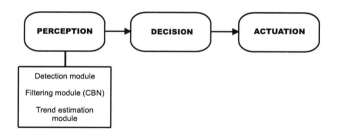

Fig. 2. Modules of the mobile obstacle avoidance task.

In the perceptual phase, the detection operation takes the raw data provided by a ring of sonar sensors and indicates a set of locations with high probability of being occupied by a mobile object. Unfortunately, this processing is not enough to estimate accurately both the direction and the speed of the obstacle. Some problems arose after the detection process:

1. Estimation of motion direction. Because of the motion of both robot and obstacle, the sonar shows a irregular trajectory of mobile obstacle points. Even if the trajectory tendency is clearly linear, in a local region the points seem to follow a random pattern (figure 3).

2. Estimation of speed. If the mobile obstacle is detected by a number of different ultrasound sensors, the data entries associated to that obstacle show space discontinuities, also called "jumps". These "jumps" appear when the sonar that tracks the motion changed. These discontinuities, abrupt in most cases, can be understood as important changes in the mobile obstacle speed (figure 3).

To solve these problems we have introduced the CBN after the detection operation (see figure 2). The overall system can detect obstacles, perceive obstacles trends and act accordingly to avoid collisions. With these features, the system is robust and can operate in real time.

4 Results

Before integrating the CBN with the architecture shown in figure 2, we have tested its individual performance with off-line simulations. For testing we have

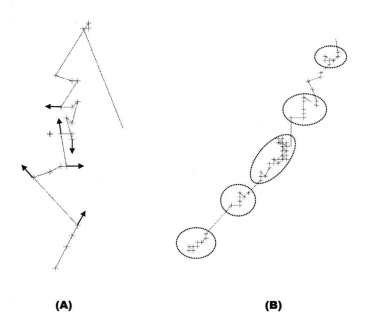

Fig. 3. Issues in the detection of mobile obstacles. (A) Set of points associated with a mobile obstacle. Arrows indicating the perceived direction at any given time is shown. Although the global trend is linear, the local trend is irregular and shows continuous changes in direction. (B) An example with data groupings is shown. These groupings are originated when there is a change in the ultrasound sensor that is performing the detection. These groupings impede an accurate speed estimation.

used real data obtained from the sensors and preprocessed by the detection stage. The goal of the simulations was to test if the CBN would be able to remove noise and improve the estimation of the trajectory of the mobile obstacles. The CBN was implemented in NSL (*Neural Simulator Language*) and the simulations were run on the NSL environment for Unix platforms.

The real experiments were performed with a *NOMAD 200* robot in the Department of Electronics and Computer Science, of our university. Initially we located the robot in one of the sides of the entrance. The robot begins its movement following a linear trajectory with the goal of reaching the other side avoiding mobile obstacles that would imply collision or impede the goal achievement. Figure 4 shows how a mobile obstacle is detected from the left side of the robot during three cycles. In the two following cycles the ultrasound sensors do not

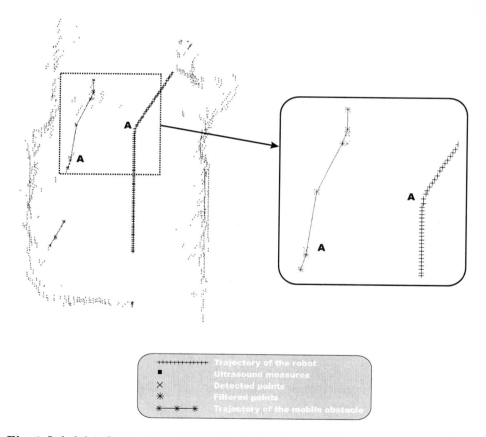

Fig. 4. Label A indicates the position of both the robot and the mobile obstacle at the time when robot begins to avoid the obstacle. The cross density in the robot trajectory indicates its speed at any given time. The less the speed, the more number of crosses, and vice versa.

provide new object information and the robot temporarily loose the trajectory. When the detection is retaken, the trend module perceives a collision situation. The object trend is, initially, *indifferent*, because no change in motion or angle is detected. As a result, the robot decides to implement the *observe* behavior and it will wait to detect either changes in the object behavior or a decrease of the collision time. In the next cycles, the mobile obstacle reduces its speed, from 41 to 34 cm/s, and so the parameter **collision time** increases. As a consequence, the control system selects the *cross first* behavior even though the mobile obstacle trend is still *indifferent*. The robot executes this behavior by turning right 30 degree and accelerating from 17.5 to 22.5 cm/s. Both actions are enough to avoid the possible collision and to continue the trajectory to achieve the initial goal.

5 Discussion

The three local mechanisms of the cuneate nucleus discussed in this paper (center-surround receptive fields, lateral recurrent connections and autoinhibitory connections) can be found in many other nucleus of the brain and have been used independently to develop other bio-inspired connectionist models. At this stage, the purpose of the CBN is not to provide a novelty general-purpose artificial neural network, but to integrate all mentioned mechanisms in a single architecture and analyze the functional capabilities of it.

The lateral inhibition, for example, is a well-known mechanisms to accomplish competitive computation. A reference model that captures this computational feature is the Didday model [4]. The goal here was to analyze competitive mechanism to understand the circuitry underlying the capturing process in frogs. The CBN architecture is though different in two aspects: (1) the number of required interneurons, and (2) the degree of inhibition exerted over each neuron in the output layer. The Didday model shows only one interneuron that, in turn, determines the same level of inhibition over the neurons in the output later. On the other side, the CBN shows an interneuron per neuron in the output layer and the degree of inhibition on each neuron is different.

Similarly, the center-surround receptor fields have also been a source of inspiration for other connectionist models. Buonomano and Merzenich [1] have developed a hierarchical network with this type of receptive field. This arrangement induces a response in the neurons of the output layer, whose temporal codification allows input pattern recognition. Other classical example can be found in the *Neocognitron* network [5]. This network was inspired by Hubel and Wiesel ideas about the hierarchical organization of the visual receptive fields.

Recently, autoinhibitory mechanisms have been used in attentional models in the visual system [6]. In this study, the autoinhibition is called *inhibition-of-return* and allows the system to change the focus of attention if new remarkable features appears in the environment. The functionality, very similar to the one presented in the CBN, consists on inhibiting, after some period of time, those winner neurons in the previous iteration with the aim to allow new incoming events to be detected.

From a functional point of view, the cuneate-based network performs a kind of spatial and temporal filtering, which was clearly showed by applying it on a collision avoidance of mobile obstacles problem. Center-surround receptive fields and lateral inhibition select salient maximal locations in local regions of the trajectory of the mobile obstacle. With this mechanism, CBN sends the most salient group of points of the mobile obstacle trajectory to subsequent processing modules. The temporal filtering removes the persistent objects, so it permits that other processing modules would consider those points relevant to the trajectory of the mobile obstacle. As a conclusion, the CBN has demonstrated the cuneate capabilities to perform an spatio-temporal filtering over incoming sensory information. As far as we know, this is the first connectionist model about the cuneate nucleus.

Acknowledgments

Authors would like to thank to **Laboratorios de Neurociencia y Computación neuronal (LANCON)**, and also, to acknowledge the support from grants TIC2000-0873 and PGIDT99PXI20601B.

References

1. Buonomano D. V. and Merzenich M. A neural network model of temporal code generation and position-invariant pattern recognition. Neural computation. Vol. 11 (1999) 103–116.
2. Canedo A., Aguilar J.: Spatial and cortical influences exerted on cuneothalamic and thalamocortical neurons of the cat. European Journal of Neuroscience. Vol. 12 **2** (2000) 2513–2533.
3. Canedo A., Aguilar J., Mariño J.: Lemniscal recurrent and transcortical influences on cuneate neurons. Neuroscience. Vol. 97 **2** (2000) 317–334.
4. Didday R. L. The simulation and modelling of distributed information processing in the frog visual system. Ph. D. Thesis. Stanford University. (1970).
5. Fukushima, K. Neocognitron: A self-organizing neural network model for a mechnsim of pattern recognition unaffected by shift in position. Biol. Cybern. Vol. 55 (1980) 5–15.
6. Itti L. and Koch C. Computational modelling of visual attention. Nature Reviews Neuroscience. Vol. 2 (2001) 1–9.
7. Koch C., Rapp M., Segev I.: A Brief History of Time Constants. Cerebral cortex. Num. 6 (1996) 93–101.
8. Mucientes M., Iglesias R., Regueiro C. V., Bugarín A., Cariñena P., Barro S.: Use of Fuzzy Temporal Rules for avoidance of moving obstacles in mobile robotics. Proceedings of the 1999 Eusflat-Estylf Joint Conference. (1999) 167–170.
9. Sánchez E., Barro S., Mariño J, Canedo A., Vázquez P.: Modelling the circuitry of the cuneate nucleus. In Lecture Notes in Computer Science. Volume I. Springer Verlag. Mira J. and Sánchez Andrés J. V. (Eds). (1999) 73-85.
10. Sánchez E., Barro S., Mariño J., Canedo A.: A realistic computational model of the local circuitry of the cuneate nucleus. Also included in this volume. (2001).

An Application of Fuzzy State Automata: Motion Control of an Hexapod Walking Machine

D. Morano[1] and L. M. Reyneri[2]

[1] Dipartimento di Elettronica – Politecnico di Torino
C.so Duca degli Abruzzi, 24 – 10129 Torino – ITALY
phone ++39 011 564 4170; fax ++39 011 564 4099
morano@athena.polito.it

[2] Dipartimento di Elettronica – Politecnico di Torino
C.so Duca degli Abruzzi, 24 – 10129 Torino – ITALY
phone ++39 011 564 4038; fax ++39 011 564 4099
reyneri@polito.it

Abstract. This paper describes an application of Fuzzy State Automata to the motion control of an hexapod walking machine. Fuzzy State Automata are used as reference generators for each leg of the robot managing each phase of the step; moreover interaction between the automata allows leg coordination without a higher level controller. This example shows some interesting points of Fuzzy State Automata like parameter tuning, emptying abandoned states and coordination between automata.

1 Introduction

1.1 Fuzzy State Automata

Fuzzy State Automata (FFSA) [2] are fuzzy systems [1] with memory. A Fuzzy State Automaton is similar to traditional Finite State Automaton (FSA), but transitions in the automaton are not triggered by crisp events but by fuzzy variables, and state transitions are fuzzy as well. It immediately results that, at any time, the whole system is not necessarily in one and only one well-defined state, but it may well be in more states at the same time, each one associated with its own *activity* value; calling μ_{S_j} the activity of the state S_j the only constraint is that total activity of the FFSA must be:

$$\sum_j \mu_{S_j} = 1 \qquad (1)$$

In control applications every state is usually associated to a controller; since the system can be in more states at the same time the controller will be given by a defuzzification function. State transitions are smoother and possibly slower, even if the controllers activated by each state are not designed to smooth the transitions. As a consequence, all the individual controllers may become as simple as a traditional PID, each one designed for a different target specific of that state.

FFSA can be of different types: time-independent and time-dependent; the former can either be synchronous or asynchronous, which somewhat reflect synchronous and asynchronous FSA, respectively. They differ only in the way state activity moves from one state to another, according to the degrees of membership of the fuzzy transitions. Roughly we could say that:

- in time-independent asynchronous FFSA, state transitions may take place as soon as inputs vary (yet transitions are not so abrupt as in traditional FSA);
- in time-independent synchronous FFSA, state transitions are computed in a way similar to asynchronous FFSA, but they are applied only at next clock cycle;
- in time-dependent FFSA, there is always a predefined intrinsic delay (usually larger than the clock period, if any) between input variations and the corresponding state transitions.

1.2 The hexapod walking machine

The Hexapod Walking Machine [3] is a mobile robot designed at Politecnico di Torino. It is composed of a main body of about 1m × 50cm × 10cm and six legs. Each leg is made of a thighbone and a shinbone both 50cm long and it has three degrees of freedom. At present mechanical parts are not definitively chosen yet and batteries and control system are not mounted yet; the final mass of the machine will be about 30kg.

Actually the purpose of this robot is the study of different control strategies applied to this class of vehicles.

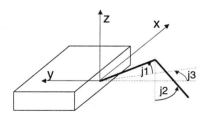

Fig. 1. The walking machine and the name convention of the joints. x is the walking direction

2 Control architecture

The control architecture is shown in figure 2; for simplicity only one leg is represented. The blocks outside the dashed box are the general parameters of the

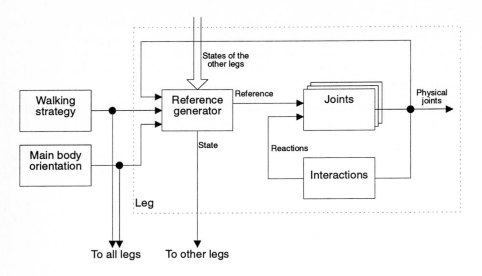

Fig. 2. Block diagram of the control

walk. The `walking strategy` represents how the hexapod must walk, for instance speed, height from ground, radius if the trajectory and vertical lift of the feet.

The `main body orientation` is the desired orientation of the main body during the walk.

The blocks inside the dashed box are the control of a single leg. The `joints` block represents the three joints of a leg with the controllers of their motors. Such controllers are as simple as a PID, which allows us to obtain good performance. `Interactions` represents the noise acting on the joints from the rest of the plant and the environment; this includes: contact with ground, gravity, inertia, etc. Most important ones are contact with ground and main body inertia. `Reference generator` must plan the leg's path. The trajectory of the leg can be roughly split in segments, so a possible solution is to build a simple reference generator that plans a polyline–shaped trajectory and the controllers of the joint motors must smooth the transitions between strokes to avoid mechanical stress. The used solution was to generate the leg trajectory with a FFSA: the fuzzy transitions smooth the path and simpler controllers can be used. The main advantage is that the project of the system is simple.

The inputs of such FFSA are:

– The current joints position and torque; this allows to know the leg position and whether the leg is on ground or it is hitting an hurdle
– The state of the other legs; this allows to coordinate the legs. The state of this FFSA is obviously available to other legs.

We plan trajectory in cartesian coordinates as they are easier to work with, but we need two additional blocks to convert cartesian to joint references and to convert joint to cartesian positions. Another block inside the reference generator allows the FFSA to change its parameters (speed, lift, range of expected irregularities, etc.) modifying the edges of the membership functions and the defuzzification function. The resulting block diagram is shown in figure 3.

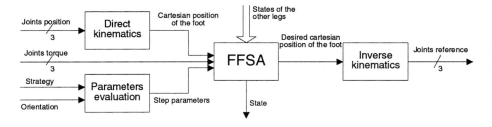

Fig. 3. Block diagram of the reference generator

3 Reference generators

In this section we describe two versions of an FFSA used as reference generators for the legs of the walking machine. The first one describes the trajectory of the step of a single leg when the correct height of the ground is unknown (but it is within a certain range); the second one is able to coordinate the legs: when a leg is finishing its thrust function and the next leg that must thrust the robot is not on ground yet the robot decelerates and waits for the flying leg hits ground.

These FFSA have been simulated inside the MATLAB/SIMULINK environment; the FFSA has been written as C S-function to obtain fast compiled code; the block were imported in SIMULINK for the simulation with the rest of the system: blocks for direct/inverse kinematics, ground hitting simulation and initial conditions.

The FFSA were implemented as *synchronous time-independent* with a sampling frequency of 100Hz since the control will be implemented as time–discrete with the same sampling frequency. The chosen membership function is the sigmoid function, that is defined by:

$$\mu(x) = \frac{1}{1 + \exp(\sigma(\alpha - x))} \qquad (2)$$

where σ is the "smoothness" of the function and α is its center.

Walking strategy and orientation of the main body were:

- Main body parallel to the ground
- Main body at a fixed height
- Rectilinear trajectory
- Constant walking speed (at least until this does not cause problems)

– Three thrusting legs at the same time; during the change of the thrusting legs group all legs shall be on the ground. Note that in this case every leg must be on the ground for at least half of the step time.

3.1 Single leg control

We first developed [4] [5] a FFSA for the step of a single leg, shown in figure 4. The step is split in six phases where each phase corresponds to a single state of the FFSA; since transitions between states are fuzzy, at the same time more than one phase could be active: this allows to obtain a smooth trajectory and low accelerations during transitions.

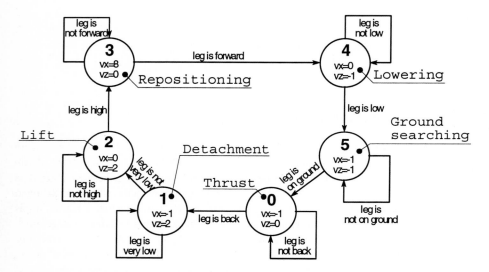

Fig. 4. FFSA of the single leg control. Note that speeds along x and z are normalized respect thrust and lowering speed

Transitions between states are a set of fuzzy rules; for instance the transitions outgoing from the state 0 correspond to the following rules:
   IF (STATE IS 0) AND (LEG IS BACK) THEN (NEXT STATE IS 1)
   IF (STATE IS 0) AND (LEG IS NOT BACK) THEN (NEXT STATE IS 0)
For each state the desired speed v of the foot is associated; the resulting speed is given by this defuzzification function:

$$v_x = \sum_j \mu_{S_j} v_{x,j} \qquad (3)$$

idem for v_y and v_z

where $v_{x,j}$ is the component of the speed along the x direction relative to the j-th state. In this example the speed along y will be always 0; the desired position

p of the foot at next sample time is given by

$$p(t+T) = p(t) + Tv(t) \tag{4}$$

where T is the sampling period.

State 0: thrust During this phase the foot is on ground and it thrusts the machine forward. Its speed is that of the machine but it is directed towards the opposite verse; there is no vertical speed. This state must be left when the foot position is back. The back position is relative to the main body and it is reveled by the position of the joint 3.

State 1: detachment When the foot is back it can be lifted from ground. During the detachment phase the foot is lift with a vertical speed, but at the same time the horizontal speed (the walking speed) is maintained to avoid foot slipping before complete detachment. This state must be left when the foot position is not very low; this position is not absolute because it is relative to the ground and the ground can present holes and small hurdles, so the not very low condition depends from the vertical position of the foot during the thrust phase (about a few centimeters from ground).

State 2: lift When the foot is detached from ground it is useless to maintain the horizontal speed; during this phase the horizontal speed is stopped and the foot is lifted vertically. This state must be left when the foot position is high; this position is relative to the main body and it depends from the maximum expected height of the small hurdles on the ground.

State 3: repositioning During this phase the foot is moved forward to prepare it to the next thrust phase. The vertical speed is stopped but the horizontal speed is directed towards walking direction and it must be faster than the walking speed, this because the thrust time must be *at least* the half of the total step time and:

- the horizontal path during the repositioning phase is longer than the path during the thrust phase
- the lift and the lowering of the foot need more time; higher hurdles and deeper holes need longer lift and lowering phases

The walking speed limit of the machine depends from the maximum speed of the joint motors and the maximum horizontal speed is reached during the repositioning phase; when the hurdles are higher the repositioning speed is faster and if the limit of the motor is reached the machine must decelerate its walking speed. This state must be left when the foot position is forward; this position is relative to the main body.

State 4: lowering During this state the leg is lowered vertically (without horizontal speed). This state must be left when the foot position is `low`; this position is relative to the main body. Note that the highest expected hurdle must be smaller than the `low` position because the foot can't thrust the machine without slipping yet.

State 5: ground searching During this phase the foot is lowered but the horizontal speed is the walking speed: in this case the foot is ready to touch the ground and to thrust the machine. Note that slipping is avoided only when this state is fully activated. This state must be left when the foot position is `on ground`; this condition is revealed by the current that goes through the motor of joint 1: when the foot hits the ground the motor absorbs more current. Note that the transition between states 5 and 0 is almost abrupt.

Considerations So far we don't care about coordination between legs, so the states of the other FFSA are not considered. The edges and the smoothness of the fuzzy sets were found by trial–and–error, trying to reach a body speed with smallest speeds and accelerations of the joints given an expected terrain.

A problem occurred during the simulation is that a residual activation of abandoned states remained and with the chosen defuzzification function the residual activation of a left state alters the desired speed.

There are two possible solutions:

- Changing the defuzzification function. A defuzzification function that "punishes" low activation states ignores abandoned states. In this case the rules of FFSA are respected, but an activation of the `thrust` state during the repositioning of the leg has not a much of sense.
- After calculating the activations, activations are recalculated "punishing" states with low activation. In this case abandoned states decrease their activation to 0, but the rules of FFSA are not fully respected.

The second solution has been chosen in this implementation: at every clock cycle, activations less than a threshold are multiplied for a number less than 1; activations are normalized again. The threshold and the reduction factor were chosen by trial–and–error; an acceptable threshold is 0.2 and a good reduction factor is 0.9. Note that this implementation changes its characteristics with the clock frequency: with a higher clock frequency states are emptied faster because the recalculation function is called more frequently.

During the states $2 \div 4$ it is not necessary to hold the y position of the foot, this FFSA does not care about this and the unnecessary motion causes a little consumption of power and a little noise; but to avoid this the reference generator must be heavily modified and more computation power is needed.

The desired trajectory of the foot is shown in figure 5; in this example during three different steps the foot hits the ground at three different heights.

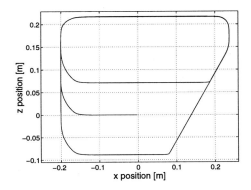

Fig. 5. Trajectory of a single leg; the y position is constant

3.2 Coordination control

The previous version of the FFSA allows coordination between legs after some modifications. Now the knowledge of the states of the others FFSA is fundamental, but the knowledge of *all* states of *all* the FFSA is redundant. The modifications involve the states **thrust** and **detachment**; moreover another state **wait** has been added (see figure 6). Note that the new state does not correspond to a phase of the step and its purpose is to coordinate the legs.

Let's introduce two new informations known by all legs:

- **Ready** is a fuzzy variable that establishes if the legs that are back can be detached from ground; this is given by:

$$\text{Ready} = (\text{LEG 1 IS on ground}) \text{AND} \ldots \text{AND}(\text{LEG 6 IS on ground}) \quad (5)$$

- **Speed** is a new parameter that is involved in the defuzzification function of the horizontal speed; it is a real number between 0 and 1 and the new defuzzification function for v_x is:

$$v_x = \text{Speed} \sum_j \mu_{S_j} v_{x,j} \quad (6)$$

Speed depends on the leg that is most in the wait state, therefore it can be evaluated as:

$$\text{Speed} = \text{NOT}((\text{STATE OF LEG 1 IS wait}) \text{OR} \ldots \text{OR}(\text{STATE OF LEG 6 IS wait})) \quad (7)$$

The purpose of the **Speed** signal is to decelerate (or even arrest) the legs in the thrust state when there are still legs not on ground; legs not on ground are still lowered (this signal doesn't affect vertical speed) and slipping is avoided because their horizontal speed has been lowered too.

Now the transition between **thrust** and **detachment** is executed only if the leg is **back** and the global condition **ready** is true: in this case all the legs thrust the main body, therefore the legs that are back can be lifted.

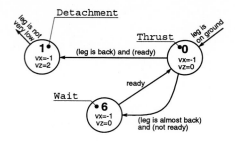

Fig. 6. Modifications of the FFSA to coordinate the legs; cyclic transitions are implicit

If a leg is **almost back** but the global status is **not ready** the condition is dangerous because the leg would reach the end of its range when not all the legs are on ground; in this case the **wait** state is activated. This state doesn't affect directly the speed of the leg, but it is involved in the evaluation of the parameter **speed**. When the machine is **ready** the transition to the thrust state is activated.

An example is shown in figure 7: this leg arrives "late" and the robot must decelerate; the lowering speed remains constant but the horizontal speed decreases, so the trajectory becomes more vertical.

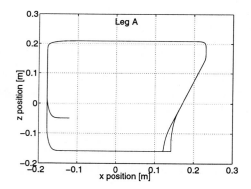

Fig. 7. Trajectory of a leg waited from the others

4 Conclusion

A parameter to evaluate the performance of the control is the ratio between the walking speed and the joint speed and acceleration. Since mechanical parts are currently in a study phase we don't know yet the maximum reachable joint speed and acceleration (due to the electromechanical parts characteristics and the performance of the joint controller).

We tried to optimize the parameters of the FFSA with these conditions:

- Maximum height of the hurdles: 10 cm
- Maximum hole deep without decelerating the robot: 5 cm
- Walking speed: 2 cm/s

We obtained the following results:

- Lift of the foot: 21 cm
- Maximum speed of the joint 3: 0.32 rad/s
- Maximum acceleration of the joint 3: 0.4 rad/s^2
- Maximum speed of the joint 1: 0.23 rad/s
- Maximum acceleration of the joint 1: 0.27 rad/s^2

It is possible to reduce the lift of the foot (it is almost high since it is more than the double of the height of the max hurdle), but is needed to increase the acceleration of the joint 3. When a better foot trajectory is needed and the joints are already at their limits the only solution is to reduce the walking speed.

Future enhancements can be the use of other membership functions in order to improve performance, the evolution of the FFSA in order to improve dexterity and the study of the `Parameters evaluation` block (see figure 3).

References

1. L. Wang, "Adaptative Fuzzy Systems and Control", *Prentice Hall*, Englewood Cliffs, New Jersey, 1994.
2. L.M. Reyneri, "An introduction to Fuzzy State Automata", Proc. of IWANN 1997, pp. 273–283.
3. F. Berardi, M. Chiaberge, E. Miranda, L. M. Reyneri, "A Walking Hexapod Controlled by a Neuro–Fuzzy System", Proc. of MCPA '97, Pisa (I), February 1997, pp. 95-104.
4. A. Comino, M. Grasso, "Controllo di robot esapode tramite logica fuzzy e realizzazione su DSP", Master Thesis, Politecnico di Torino (I), December 1998.
5. D. Morano, "Controllo di un robot esapode per movimentazione su terreni impervi", Master Thesis, Politecnico di Torino (I), December 2000.

Neural Adaptive Force Control for Compliant Robots

N. Saadia[(*)], Y. Amirat[(**)], J. Pontnaut[(**)] and A. Ramdane-Cherif[(***)]

[(*)] Laboratoire de robotique, Institute d'éléctronique. Université, USTHB, BP 32 Babs Ezzouar ,Alger, Algeria.
[(**)] LIIA-IUT de Creteil, 122 Rue Paul Armangot 94400 Vitry Sur Seine Cedex France
[(***)] Lab. PRiSM, Université de Versailles St-Quentin en Yvelines, 45 av. des Etats Unis 78035 Versailles, France
Email: saadia_nadia@hotmail.com
Email: amar.ramdane-cherif@prism.uvsq.fr

Abstract. *This paper deals with the methodological approach for the development and implementation of force-position control for robotized systems. We propose a first approach based on neural network to treat globally the problem of the adaptation of robot behavior to various classes of tasks and to actual conditions of the task where its parameters may vary. The obtained results are presented and analyzed in order to prove the efficiency of the proposed approach*

1 Introduction

Recently, the robot control problem during the phase of contact with the environment has received considerable attention in the literature [8] due to its complexity. Assembly tasks constitute a typical example of such an operation. These tasks are especially complex to achieve considering the stern constraints that they imply as the intrinsic parameters type of the task: mechanical impedance of the environment is non-linear and submitted to temporal and spatial changes, variable contact geometry according to the shape of parts, variable nature of surfaces. These difficulties increase when the system has to manipulate a class of different objects. In this paper, we focalize on the problematic of the force control of assembly tasks which constitutes a class of compliance tasks.

To obtain a robot's behavior, satisfying different situations of utilization, it necessary to design non linear adaptive control laws. The elaboration of these controls allows to overcome the difficulty to synthesize suitable control lows that are satisfactory over a wide range of operating points. Thus, several approaches are possible such as conventional adaptive controls [10], alternate methods or modern adaptive control such as intelligent control which uses experimental knowledge. The different approach analysis like H_∞ [6] [5] are not suited to our context because most of these strategies are complex to implement in real time context due to the dynamics of the system which requires short time sampling (less than 10 milliseconds) [7].

Learning adaptive control based on the utilization of connectionist techniques allows to solve directly the problem of non linearity of unknown systems by functional approximation. The objective of this paper is to propose tools which allow us to increase the control performances of contact interactions established between a robot and its environment, in complex situations by means of adaptability.

In the following of this article, we first present in section 2 a structure of type external forces control, which use the technique of reinforcement learning. In section 3, we describe the implementation of the proposed approach for the realization of various piece insertion according to different tolerances and speeds. In the last section, the experimental results obtained are presented and analyzed.

2   Neural Adaptive Control Structure

A control structure based on an external force control scheme is proposed (Figure 1). Indeed, this structure of control consists in hierarchical force control loop with regard to the position control loop. The law of force control that intervenes in the diagram works out, from the mistake by force (ΔF) between the desired contact force torque F_d and the measured one F_{cm}, an increment of position ΔX that modifies the task vector Xd according to the following relationship:

$$\Delta X = C_f (\Delta F) \tag{1}$$

where C_f represents the forces control law.

Considering the existing coupling between the different types of forces to which the robot is submitted (contact forces, gravitational forces end forces owed to the system's dynamics), for extract the contact force values from the measured ones, a FFNNM is used to estimate free motion forces.

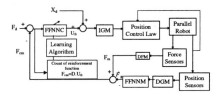

Fig. 1. Adaptive controller using associative reinforcement paradigm: DGM: the direct geometric model of parallel robot; IGM: the inverse geometric model of parallel robot; DFM: direct force model; X_d: the desired position trajectory; F_d: the desired forces trajectory; $\hat{F}_g$: the estimate free motion force torque; F_m: the measured forces torque; F_{cm} the estimate contact force torque.

2.1 Estimation of Contact Forces

In order to estimate the contact forces from the measured ones in a compliant motion, we propose an identification model based on a multi-layer neural network. In our approach, the task is released with constant velocity. This FFNNM is trained off-line using the data obtained from the robot's displacements in the free space. In such a situation, gravity and contact forces are de-coupled as contact forces are null. To build a suitable identification model (Fig 2), we proceed as follows: when the input $X_{m, k}$ is applied to the FFNNM, the corresponding output is $\hat{F}_{g,k}$ which approximates F_g such as:

$$\left\| \hat{F}_g - F_g \right\| = \left\| \hat{F}_g(X_{m,k}) - F_g(X_{m,k}) \right\| \leq \varepsilon \quad (2)$$
$$X_{m,k} \in P$$

where ε is the mistake of optimization, F_g the free motion force tensor in the operational space, $\hat{F}_g$ the estimated one and $X_{m,k}$ the position and orientation end-effector vector.

The input vector I_m of the neural network identifier model of free motion force is the position vector $X_m(k)$ of the end effector. The out put of the FFNNM can be computed using the input and the neural network's parameters W_m as:

$$\hat{F}_g = N_m(W_m, I_m) \quad (3)$$

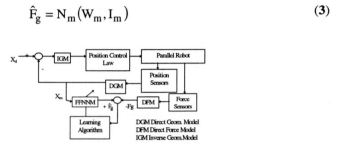

Fig. 2. FFNNM learning a free motion forces

2.2 Initialization of Neural Controller

The initialization of neural networks controller is done off line. The learning structure is given figure 3. While achieving a control of the system in the task space with a conventional controller of PID type, we constructed a basis of training (figure 3). This one is constituted of measured forces torque Fcm, and wanted ones F_d and control vector U_{fd} generated by the conventional controller. Values of mistakes to the previous sampling instants are taken in amount in the learning. Parameters of the controller PID have been adjusted experimentally for a speed of insertion of 5 mm/s of way to get a satisfactory behavior of the point of view of the minimization of contact forces. The period of sampling of the control is 5 ms. The input vector Ic of neural controller is constituted by force error vector e = F_d - F_{cm}. In this step the FFNNC is trained off-line using the data obtained from the robot's displacements in the constrained space under classical external force control. The learning structure is presented figure 3.

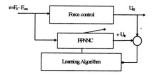

Fig. 3. Initialization of neural networks controller (FFNNC) scheme

The learning process involves the determination of a vector W_c^* which optimizes a performance function J_1:

$$J_1 = \frac{1}{2}(U_{fd} - U_{fr})^T R_c (U_{fd} - U_{fr}) \qquad (4)$$

where U_{fd} is the desired output, U_{fr} is the corresponding actual output and R_c is the weighting diagonal matrix. This last is function of the input of the neural network controller « Ic » and of its weight matrix W_c:

$$U_{fr} = N_c(W_c, I_c) \qquad (5)$$

2.3 Adaptive Neural Network Controller

A dynamic learning process may be formulated as follows:

$$W_{c,k+1} = W_{c,k} - \mu \frac{\partial J_{k+1}}{\partial W_{c,k}} \qquad (6)$$

where $W_{c,k}$ is the estimation of the weight vector et time t_k et μ is the learning parameter.

The output of the neural networks at time t_k can be obtained only by using the state and input of the network at past time. The learning process imply the determination of an $W_{c,k+1}$ vector which optimize the cost function J. This function is based on state criterion for process output and is as follow:

$$J = \frac{e^T * R * e}{2} \qquad (7)$$

where: $e = F_d - F_{cm}$; e^T is the transpose vector of e and R is the weighting diagonal matrix.

The learning algorithm consists in minimizing J [Fcm, k+1] with regard to weights of the network, its application permits to write :

$$\frac{\partial J}{\partial W_c} = -e^T * R * \frac{\partial F_{cm,k+1}}{\partial U_{fr}} * \frac{\partial U_{fr}}{\partial W_c} \qquad (8)$$

In the proposed structure (Figure 1), the back propagation method is based on the computation of the gradient of the criterion J using the learning paradigm called associative reinforcement learning [3].

In the proposed control approach, we chose a D matrix as reinforcement function. This choice permits to assimilate the system to control to a three-dimensional spring. The D matrix represents the model of behavior in contact with the environment. The dynamics of the interaction robot environment is described then below by the matrix relation:

$$F_{cm,k} = D * U_{fr,k} \qquad (9)$$

where:

$$\frac{\partial F_{cm,k}}{\partial U_{fr,k}} = D \qquad (10)$$

However, in the case of assembly tasks it appears discriminating to choose a linear type reinforcement function assimilating so the system to adaptable three-dimensional spring.

The learning algorithm consists in minimizing J with regard to weights of the network:

$$\frac{\partial J_k}{\partial W_{c,k}} = -e^T * R * D * \frac{\partial U_{f,k}}{\partial W_{c,k}} = -e^T * R * D * \frac{\partial N_{c,k}}{\partial W_{c,k}} \qquad (11)$$

This last equation shows that the implementation on-line of an adaptation algorithm of neural controller parameters makes that this last doesn't depend any more on the robot model but rather of the interaction robot-environment model. Thus, the proposed control structure present the advantage of not being subject to mistakes of identification of the system.

3 Implementation and Experimentation

In order to validate the proposed structure and to evaluate its performances, we have implemented it for the control of the parallel robot.

This first step consist to compute an optimal neural networks and to use it to control the system. The neural networks controller initialized off line using the structure presented figure 3 is implemented in order to control the parallel robot of assembling cell. This experimentation consist of inserting cylindrical part in a bore for a velocity of 5 mm/s, and a tolerance of 3/10 mm. The temporal evolution of the measured and desired contact force tensor are show figure 4a. The force component according to the axis of insertion (F_x) exhibits a peak because of the appearance of a contact between the part and the bore. Then, the components of contact force tensor decrease by oscillating around a corresponding desired value. Components F_y et M_z exhibits a peak because of the appearance of a contact between the part and the bore. Then, they decrease by oscillation around a residual value.

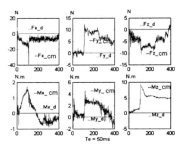

Fig. 4. a. Desired force torque (_d) and measured ones (_cm)

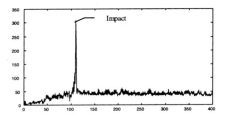

Fig. 4 b. Temporal evolution of criterion (off line learning)

The figure 4b shows the temporal evolution of criterion which toward to an asymptote of 45 when as its value at the impact time is 300. This asymptotic phase corresponds to a mistake on force torque components according to X of 5 NS, according to Y of 5 NS, according to Z of 2 NS, according to Mx of –0.5 Nm, according to My of –1 Nm and according to Mz of 5Nm. We can also notice that the decrease of the criteria is slow, since the final value is only reached after 300 samples. We have presented in this first step the off line learning of en external force control. The thus gotten neural controller has been tested on line to achieve the insertion of a cylindrical piece in a bore. Tests of insertion achieved show that the training out line of the controller permitted its generalization. The minimization of the criteria is not however not quite optimal since it exists a weak mistake in end of insertion.

To optimize the neural network controller (FFNNC), in real situation of utilization, an on line adjustment algorithm of FFNNC parameters is implemented. The on line training permits to refine the learning of the networks controller worked out off line. The adaptive controller performances have been evaluated on insertion tasks for different tolerances and velocities values. The comparison of temporal evolution of criterion obtained in the cases of an off-line and on-line learning (Figure 5) shows that the on-line adaptation algorithm leads to a better minimization of the criterion.

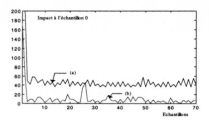

Fig. 5. Comparison of temporal evolution of criteria
    (a) off line learning (V=5mm/s, T=3/10mm, cylindrical section part)
    (b) on line learning (V=10mm/s, T=1/10mm, cylindrical section part)

In the case of an on-line adaptation, the criterion presents at the end of insertion an error weaker than in the case of an off-line learning. Indeed, in the case of the on line adaptation the criterion toward to an asymptote of 5, what corresponds to a mistake on force components according to X of 0.5 NS, according to Y of 2 NS, according to Z of 2 NS, according to Mx of 0.2 Nm, according to My of 2 Nm, according to Mz of 0.5Nm. Moreover, if velocity insertion increases, figure 6 shows that the criterion trace varies little.

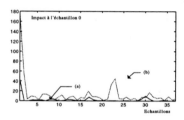

Fig. 6. Comparison of temporal evolution of criterion
    (a) on line learning (V=15mm/s, T=1/10mm, cylindrical section part)
    (b) on line learning (V=10mm/s, T=1/10mm, cylindrical section part)

Therefore, the on-line adaptation makes it possible to minimize the criterion with a residual error due to the sensors noise and perturbations that undergoes the system (for example frictions due to surface states). For the insertion of a parallelepiped part, figure 7 shows two phases: a phase where the criterion presents oscillations and a phase where it presents a closely identical behavior to the one corresponding to the cylindrical part insertion. Indeed, the learning base which has been constructed during the insertion of a cylindrical part, is not consistent in information on momentum. Thus, during the insertion of a parallelepiped part, oscillations of short duration and weak amplitude can appear and be considered as a normal phenomenon. In the case of parallelepiped section part the temporal evolution of the criteria toward an asymptote of 5, what corresponds to a mistake on force tensor components according to X of 2 NS, according to Y of -9N, according to Z of 0 NS, according to Mx of 0.002 Nm, according to My of 0.8 Nm, according to Mz of –0.5Nm,.

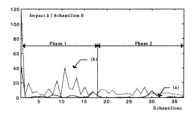

Fig. 7. Comparison of temporal evolution of criteria
  (a) on line learning (V=15mm/s, T=1/10mm, cylindrical section part)
  (b) on line learning (V=10mm/s, T=1/10mm, parallelepiped section part)

4 Conclusion

We presented in this paper a new approach of controller to treat the problem of the force control of active compliance robots witch are in interaction with their environment. The connectionist approach that is proposed rests on a methodology of control while considering the treatment of the task according to aspects off line and on line. Thus we proposed an external force control structure that uses the reinforcement paradigm for the on-line adaptation of the controller's parameters to variations of parameters of the task.

References

1. M.Y. Amirat. Contribution à la commande de haut niveau de processus robotisés et à l'utilisation des concepts de l'IA dans l'interaction robot-environnement. *PHD thesis* University Paris XII, Janvier 1996.
2. A. G. Barto and P. Anandan. Pattern recognizing stochastic learning automata. In *IEEE Transaction on Systems, Man, and Cybernetics*, 360-375, 1985.
3. N. Chatenet and H. Bersini. Economical reinforcement learning for non stationary problems. In Lecture notes in Computer Science 1327, Artificial Neural Network ICANN'97, 7th International Conference Lausanne, Switzerland Proceeding, pp. 283-288, October 1997.
4. G. Cybenco. Approximations by superposition of sigmoidal function. In *Advanced Robotics, Intelligent Automation and Active Systems*, pp 373-378, Bremen, September 15-17, 1997.
5 E. Dafaoui and Y. Amirat and J. Pontnau and C. François. Analysis and Design of a six DOF Parallel Manipulator. Modelling, Singular Configurations and Workspace. In *IEEE Transactions on Robotics And Automation*, vol. 14, pp. 78-92, Février 1998.
6. J.C. Doyle and K. Glover and P.P. Khargonecker and B.A. Francis. State space solutions to the standard H^2 and $H\infty$. In *IEEE Trans. On Automat. Contr.*, vol. 34, pp. 831-847, 1989.
7. B. Karan. Robust position/force control of robot manipulator in contact with linear dynamic environment, In *Proc. of the Third ECPD International Conference on Advanced Robotics, Intelligent Automation and Active Systems*, pp 373-378, Bremen, September 15-17, 1997.
8. O. Khatib. A unified approach to motion and force control of robot manipulators. In *IEEE J. Robot Automation*, 43—53, 1987.
9. S. Komada and al.. Robust force control based on estimation of environment. In *IEEE Int. Conf. on Robotics and Automation*, pp. 1362-1367, Nice, France, May,1992.
10. L. Laval and N.K. M'sirdi. Modeling, identification and robust force control of hydraulic actuator using $H\infty$ approach. In *Proceeding of IMACS*, Berlin, Germany, 1995.

Reactive Navigation Using Reinforment Learning in Situations of POMDPs

P. Puliti, G. Tascini, and A. Montesanto

Institute of Computer Science
University of Ancona
mailto: puliti@inform.unian.it

Abstract. The aim of this work is to individualize an architecture that allows the reactive navigation through an unsupervised learning based on the reinforcement learning. To reach the objective quoted, we used the Q-learning and one hierarchical structure of the architecture developed. To use these techniques in presence of Partially Observable Markov Decision Processes (POMDP) is necessary introduce some innovations: heuristic techniques for the generalization of the experience and for the treatment of the partial observability, a technique for the speed adjournment of the Q function and the definition of reinforcement policy adequate for the unsupervised learning of a complex assignment. The results show a satisfactory learning of the assignment of navigation in a simulated environment.

Introduction

In the reinforcement learning (RL) the information given to the agent is a numerical evaluation, the reinforcement, of the action done in a given state and of the state transition resulted. This approach was developed originally for markoviani decisional processes with a finished state space; in such particoular case varied techniques of RL are able to individualize the excellent behavior that the agent must hold. The application of the reinforcement learning to tasks in the real world presents two types of problems commonly. The space of state, for problems of this type, is usually endless. That makes impossible the exploration of all the possible actions in all the possible states in a finished time, so is necessary an opportune generalization of the experience also of states that are not explored already. The state is generally only partially observable. To adapt the RL to this condition could be used a specific approaches [1,2,3,4], or we could used one of the techniques of RL developed with the complete observability, replacing the sensorial state to the state [5,6,7]. The system robot/environment of this type of problem is characterized from a space of continuous state, and therefore also endless. The decisional process of the robot could be modeled like markovian and deterministic process. Using visual input, it is not possible build an observer of the state, that results therefore not observable. If we use the sensorial state instead of the state properly told, the process is neither markovian, neither deterministic, and this could make difficult the application of the techniques of reinforcement learning. The technique of the Q-learning is the most famous between those developed to implement the reinforcement learning. His principal advantages, comparing it with the other techniques developed, are constituted from: sturdiness as regards the not markovian characteristic and insensibility of the convergence as regards the exploration. Instead the principal disadvantages are due to the slowness of the learning and to the lack of formally justified solutions to the dilemma exploration-exploitation.

The architecture

The developed architecture acquires data from the world through a couple of television cameras, for effect the stereoscopic vision. The bump sensors must cover the whole perimeter of the robot. The data drawn from the vision is elaborate through two different runs: one to effect the raw localization, that is to individualize the area in which the robot is; the other to draw the vector of features that is used to realize the visual servoing. The control structure of the robot is composed by two levels gerarchically orderly. The upper level uses the information gotten through the raw localization and the bump sensors to activate or not the actions of other level that are implement by the lower level.

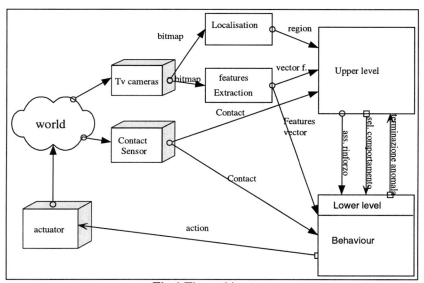

Fig.1 The architecture

The lower level implement the up level actions that are used from the upper level to effect the navigation. Each up level action is based on a visual servoing, and any of these actions integrates the data of the vision with those gotten from the bump sensors. The current sight is classified through a process that realize an opportune generalization on the visual input, departing from an image segmented in which is possible distinguish the walls from the floor.

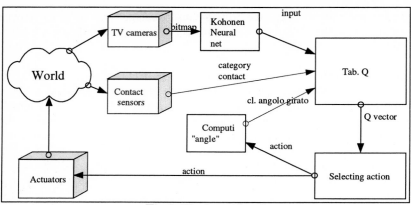

Fig.2: The lower level

The class of sight determined isused to individualize a vector of Q-values. This value is used to select an action, that it could be that to uper Q value, to exploit the learning previous effected, or a sub-optimal action according to the actual acquaintances, to the purpose of experiment the effect of this action. The action of low level choice is realized in the environment, closing the cycle of control. For any of the up level actions necessary for the navigation, only the class of sight could not be sufficiently meaningful for determine the action of low level that must be finished in a determined state (since different states could carry to the same class of sight: the cause are the not observability and the generalization). For the up level actions that manage the contacts with the walls, is used the "class of contact," that is represented bythe state of the bump sensors (of type on/ off), to specify the state in which is the robot.

Using the elaboration effected on the vision input, up level actions exist, that don't manage the contacts with the walls, for which an approach purely reactive is not sufficient, in as states that require actions of different low level could belong to the same class of sight. To resolve the ambiguity of the state in these cases, the robot is endowed with memory for the actions completed in the past, a number denominated "turned angle". This number notes of the actions of low level that carry to a change of direction of the robot completed beginning from the activation of the action of up level in course. The turned angle comes used for individualize a class, that could be used together to the class of sight in the actions that require it. The Q chart, above all when is used the classes of sight and of turned angle at the same time, could result of notable dimensions. In this case, the slowness of the learning based on the Q-learning could be excessive. For face this problem we used used a technique that exploits the rarity of the reinforcements diverged from zero characterizing this kind if approach. It allows to adjourn, at the same time, more values of the chart when the system obtain a not void reinforcement, without however comprimize the convergence of the chart. In this way we have a notable acceleration of the learning of the up level actions.

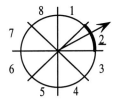

Fig.3: The angle representing the class of "change of direction".

Simulations

The developed architecture is implemented in a simulated environment, composed by curves and corridors' intersections. The only obstacles to the movement of the robot is constituted by the walls. The system has to face two kind of tasks. The first consists of the sail from a known point of the environment to an other known point, crossing an unknown run without beat against the walls. The second task requires of depart from an unknown zone and go in any known zone without beat. The two tasks have to be turns in the small possible time and crossing the shorter run.

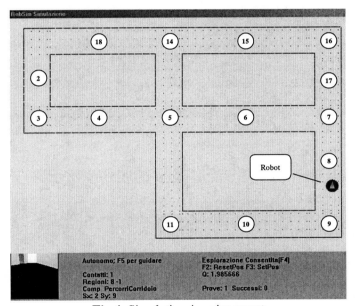

Fig.4: Simulations' environment

The robot vision and the first elaboration of the images is simulated through the rendering using the OpenGL graphic bookstore. The images coming rom the rendering are already segmented, so is therefore immediately possible distinguish the pixel belonging to the walls from those belonging to the floor. The operation of raw localization is simulated using a Kohonen SOM neural network to generalizes the class of sight. The execution of the task of navigation in the simulated environment is obtained defining 15 up level actions implemented from the lower level of the architecture. We define all the fifteen reinforcement policy starting only from three basic schemes.

Results

It in the first graph (Fig.5) we can see the behaviour of the robot during the learning of the up level action of "Right Curve"; particularly, we show the reinforcement gotten at the conclusion of the first 300 activations of the action. The Q chart, before the learning, was initialized with void values, therefore in the first activations the behavior of the robot was random. After a first phase of negative results, the action gets a initial phase of positive conclusions, follows from a phase of negative reinforcements.

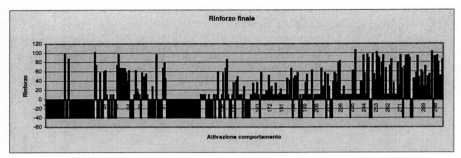

Fig.5: Trend of the learning of the behavior "curve to the right".

Beginning from the activation 120, the tendency toward the positive reinforcements is evident: the frequency of those negative ones is more and more low, while the mean of those positive is more and more elevated.

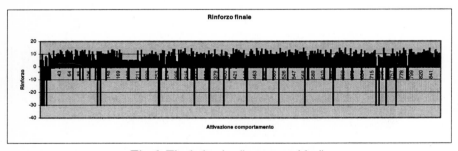

Fig.6: The behavior "cross corridor".

The action "Cross Corridor", that during the learning pahes is resulted the more activated in absolute, we can notice an initial negative phase clearly more short as regards "Bends Right", follows from a substantial stabilization of the performances for the following 800 activations. This action, different to what happens for "Right Curve", isn't disarmed to the first contact with the walls; that's why is more indicative the following graph (Fig.7) to examine the performances of the learning.

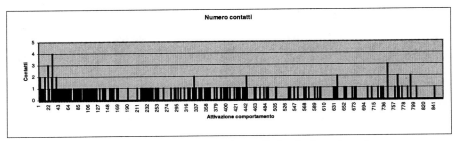

Fig.7: Number contacts with the walls in the learning phase

The graph (Fig.7) shows the number of contacts with the walls during the execution of the action in the phase of learning. From the brought data is evident that both the number and the frequency of the contacts with the walls decreases to the progress of the learning. In the next two graphics (Fig.8 and Fig.9) is possible to examine the performances of the two actions after a phase of very long learning, after a substantial stabilization of the relative Q charts. Is evident that the performances are very good: particularly "Right Curve", on 73 executions, has 72 positive terminations; for the action "Cross Corridor", we can see anly 4 contacts in 200 activations.

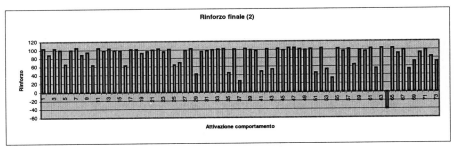

Fig.8: Behavior " right curve" after the convergence of the Q chart

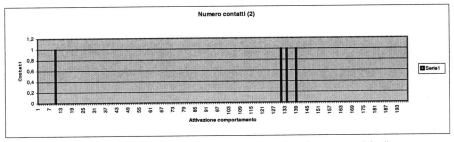

Fig.9: Number of contacts during the action "cross corridor" after the stabilization of the Q chart

Conclusions

In this work, also if only to the inside of simulations, we have gotten considerable results. Departing from images already segment according to the wanted criterion, and using the extraction of the features and the following elaboration through the Kohonen neural network, we obtained a sufficient generalization to get the learning of the tasks. The ambiguity on the state, due both to the intrinsic problems of the partial observabilit and to the effected generalization, has been stayed resolved, when necessary, using the memory for the actions completed in the past. The acceleration of the learning, particularly slow in the Q-learning, has obtained through the contemporary adjournment of more Q values with a procedure that maintains the coherence of the Q values learned. Finally, we have used an unsupervised learning for the necessary Q functions for the tasks execution through a complex strategy of reinforcements based on the available inputs. The individualization of the necessary strategy for each up level action has been simplified individualizing three basic strategy to use as prototypes for the different actions. In conclusion, we have show that it is possible to implement a system for the autonomous navigation with visual input without the need of a supervised learning.

References

[1] M. Kearns, Y. Mansour, A. Y. Ng, "Approximate Planning in large POMDPs via Reusable Trajectories", *NIPS*99*.
[2] E.A.Hansen, "Solving POMDPs by Searching in Policy space", *Proceedings of the 14th International Conference on Uncertainty in Artificial Intelligence*, Madison, Wisconsin, July 1998.
[3] P.L.Bartlett, J.Baxter, "Estimation and approximation Bounds for gradient-based reinforcement learning", *Thirteenth Annual Conference on Thirteenth Annual Conference on Computational Learning Theory*, 2000
[4] S.P. Singh, T. Jaakkola, M.I. Jordan, "Learning without state-estimation in partially observable markovian decision processes", *Machine Learning: Proceedings of the Eleventh International Conference*, 1994
[6] Leslie Pack Kaelbling, Anthony R. Cassandra, and James A. Kurien, ``Acting Under Uncertainty: Discrete Bayesian Models for Mobile-Robot Navigation," in *Proceedings of IEEE/RSJ International Conference on Intelligent Robots and Systems*, 1996.
[6] Leslie Pack Kaelbling, Michael L. Littman and Anthony R. Cassandra, ``Planning and Acting in Partially Observable Stochastic Domains," *Artificial Intelligence*, Vol 101, 1998.
[7] Michael L. Littman, Anthony R. Cassandra, and Leslie Pack Kaelbling. "Efficient dynamic-programming updates in partially observable Markov decision processes". Available as Brown University Technical Report CS-95-19.

Landmark Recognition for Autonomous Navigation Using Odometric Information and a Network of Perceptrons

Javier de Lope Asiaín[1] and Darío Maravall Gómez-Allende[2]

[1] Department of Applied Intelligent Systems
Technical University of Madrid
jdlope@eui.upm.es
http://www.sia.eui.upm.es/~jdlope/
[2] Department of Artificial Intelligence
Technical University of Madrid
dmaravall@fi.upm.es

Abstract In this paper two methods for the detection and recognition of landmarks to be used in topological modeling for autonomous mobile robots are presented. The first method is based on odometric information and the distance between the estimated position of the robot and the already existing landmarks. Due to significant errors arising in the robot's position measurements, the distance-based recognition method performs quite poorly. For such reason a much more robust method, which is based on a neural network formed by perceptrons as the basic neural unit is proposed. Apart from performing very satisfactorily in the detection and recognition of landmarks, the simplicity of the selected ANN architecture makes its implementation very attractive from the computational standpoint and guarantees its application to real-time autonomous navigation.

1 Introduction

Autonomous navigation is a hot topic in the field of advanced robotics, with a relative long existence since the first prototypes of mobile robots developed in the mid sixties. Through all these years two main approaches can be considered as consolidated paradigms for the design and implementation of autonomous navigation systems. The first one, chronologically speaking and undoubtedly the dominant approach, is based on the use of models of the environment in which the robot navigates. The second one has appeared as the field has matured and it has become an evidence the impossibility of building completely autonomous robots by relying exclusively on models or maps of the environments, mainly due to two reasons: (1) the lack of the required perceptual abilities from the part of the robots themselves and (2) the lack of appropiate formal and mathematical tools for environment's modelling. As a consequence, the so-called reactive paradigm [1,2] has emerged as a theoretical and practical alternative, or rather as a complement, to the conventional model-based approach. Confining our discussion to the model-based autonomous navigation, one key issue is to detect and recognize the so-called reference places or landmarks, which are basic elements for the map building endeavor [3,4,5]. In this paper we present a contribution to the landmark detection and recognition problem.

The detection of the reference places allows the creation of environment landmarks for use in the exploration, planning and navigation tasks. Besides the detection ability, in the exploration and navigation tasks the distinction between a new and an already created landmark —i.e. landmark recognition— is also required.

In [6] we propose a method for the detection of reference places that does not employ sensory information but the information provided by the robot's control subsystem. The changes in the behavior modes of the control subsystem generated by the presence of obstacles allow the creation of reference places.

In the sequel we describe two methods for the recognition of the reference places. In the first and more primitive method, the current distances of the robot to the already known landmarks are used for recognition purposes; in the second and much more powerful method a network of perceptrons recognizes the landmarks.

2 Distance-based Recognition of Reference Places

In this method the robot's position and those of the existing landmarks are utilized for recognition purposes. Then, let p_n be the current reference place to be recognized, with cartesian coordinates $\boldsymbol{p}_n = (x_n, y_n)$ and angular coordinate θ_n and let $P = \{p_0, ..., p_m\}$ be the set of existing landmarks with coordinates $\boldsymbol{p}_i = (x_i, y_i)$ and θ_i for each $p_i \in P$ as computed in the robot's reference system. If the euclidean distance between two landmarks p_i and p_n is:

$$d(\boldsymbol{p}_i, \boldsymbol{p}_n) = \sqrt{(x_i - x_n)^2 + (y_i - y_n)^2} \qquad (1)$$

then the nearest reference place p_c to the current landmark p_n is $p_i \in P$ such that minimizes (1); that is:

$$p_c \mid d_c = \min[d(\boldsymbol{p}_i, \boldsymbol{p}_n)] \ \forall p_i \in P \qquad (2)$$

in which d_c is the euclidean distance between p_c and p_n. The relationship between the orientation of both reference places is:

$$\delta_c = |\theta_c - \theta_n| \qquad (3)$$

The recognizer will classify the current landmark p_n as unknown if its distance to the nearest existing landmark is greater than a certain threshold t_d and if their orientation differs at least in some threshold t_δ. In a formal way:

$$p = \begin{cases} p_c & \text{iff } d_c < t_d \text{ and } \delta_c < t_\delta \\ p_n & \text{otherwise} \end{cases} \qquad (4)$$

The thresholds t_d and t_δ permit a posterior refinement in the recognition process in order to identify the landmarks with a greater exactness. Threshold t_d depends on the available precision in the estimation of the robot's position. Threshold t_δ determines the precision in the orientation of the robot's approach to a landmark. Obviously, $t_\delta = 2\pi$ means that the robot's orientation has no influence on the recognition process.

3 Landmark Recognition Using a Network of Perceptrons

The recognition of reference places by means of distance-based information works appropiately for highly structured environments and when the exact position of the robot is available, which unfortunately is not very common. Therefore, more robust methods for landmark recognition are necessary.

In the sequel we present an algorithm that employs a network of perceptrons for the landmark recognition task. The only information injected into the perceprons —i.e. the basic neural units— is the readings of the sonar sensors onboard the robot. This information is extremely simple making the method very attractive from the computational standpoint. A raw estimation of the robot's position can be used as well for the final recognition in case of a draw among several perceptrons.

Let $p_i \in P$ be a concrete, existing reference place. We define s_i as the sensory register of landmark p_i, with s_i being a vector of dimension M and formed by the M robot's sonar sensors measurements at place p_i.

The training set $S = \{s_1, ..., s_n\}$ is formed by the sensory registers taken during different visits to each individual landmark. Every reference place has at least one sensory register in the training set. We also define a function $f : P \times S \to [0, 1]$ that associates each landmark $p_i \in P$ with the sensory registers $s_i \in S$.

A perceptron with M inputs and a single output is associated with each landmark. Graphically, we can represent the network of perceptrons as in the Figure 1. Each landmark holds information related to its position in the environment, links to other connected landmarks and the perceptron.

Every perceptron is trained with the complete training set S in order to be properly activated whenever the input corresponds to a sensory register of the associated landmark. Formally stated, for each perceptron Π_i adscribed to landmark p_i, it holds:

$$\Pi_i = f(p_i, s_j) \ \forall s_j \in S \tag{5}$$

Therefore, when a new reference place p_n is detected during the recognition process, the perceptrons activated by the current sensory register are associated with the most similar existing landmarks:

$$P_c = \{p_i\} \ \forall \ \Pi_i = 1 \tag{6}$$

Afterwards, the landmark $p_c \in P_c$ nearest to the currently detected place p_n is selected:

$$p_c \mid d_c = min[d(\boldsymbol{p_i}, \boldsymbol{p_n})] \ \forall p_i \in P_c \tag{7}$$

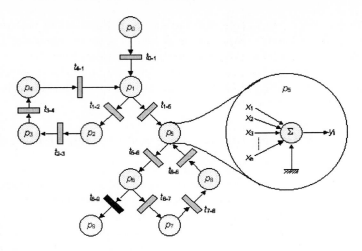

Figure 1. Net of landmaks with detailed information about a node

Finally, a test is carried out to decide whether or not both landmarks, the detected one and the existing one, are the same. This test is performed by applying a distance threshold t_d on d_c.

A positive test indicates that the current landmark is actually one of the already existing landmarks, i.e. p_c, and a negative test means that the current landmark p_n is a new reference place. In a formal way:

$$p = \begin{cases} p_c \text{ iff } d_c < t_d \\ p_n \text{ otherwise} \end{cases} \quad (8)$$

The t_d threshold can be made significantly greater than the threshold used in the distance-based recognition phase as the sensory information used by the networks of perceptrons has been able to dramatically refine the landmark recognition by a strong reduction of the search region.

4 Application to Autonomous Navigation of Mobile Robots

In our simulation and development environment the exact position of the robot is obviously available —unless we introduce simulated errors— and the distance-based recognition guarantees a perfect recognition of the reference places. Such accuracy permits an excellent validation of the model-building process and a reliable quality test of the neural network-based recognition method. The thresholds used in the distance-based recognition are $t_d = 15$ inches, that gives a slightly higher tolerance than the robot's radius of 10 inches and $t_\delta = \pi/4$ that guarantees a suitable rank of orientations for the type of landmarks occurring in our experiments —indoors environments—.

In Figure 2 are displayed several reference places that have been detected and recognized by just applying the distance-based recognizer. For this example, the threshold relative to the robot orientation has not been applied.

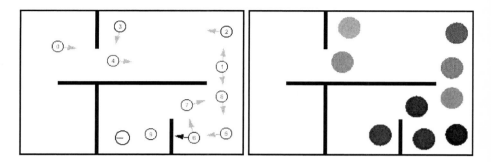

Figure 2. An example of landmarks detection and recognition employing only the distance-based method

In principle, the recognition of landmarks using the network of perceptrons has the important advantage of being almost independent on the errors of the estimated robot's position as it is based on the sensors readings. However, in our first experiments the results were unsatisfactory, even for simulated environments; the reason being quite obvious: the resemblance of the sensors readings for landmarks of similar shape: walls, corners, doors and so on. Concerning such problem, just observe landmarks p_2 and p_3 in Figure 3. Both reference places are of the same type —i.e. corners— although they are different physical landmarks. The sensory register for both landmarks are extremely similar: small magnitudes on the front side and on the right side and big magnitudes for the rest. In a similar fashion, landmarks p_1 and p_4 are of the same kind —i.e. bifurcations— and therefore present the same problem. In such situations, even with a sophisticated recognition algorithm, different physical landmarks of the same type cannot be correctly discriminated.

In order to overcome this serious problem and to increase the difference among the measurements produced by the set of landmarks, the coordinates of the robot's sensors are rotated to coincide with the robot's orientation θ_R. The formal expression of this base change is:

$$s_j = s'_i \mid j = \left(\frac{\theta_r}{2\pi/M} + i \right) \bmod M \qquad (9)$$

where s_j is the sensor readings after transforming the original readings s'_i; M is the number of sensors —which in our case have been placed in a totally simmetrical radial configuration— and $i, j \in [0, M)$ are indexes for each particular sensor.

In Figure 4 are displayed the landmarks detected and recognized by a network of perceptrons with thresholds $u_d = 30$ inches —i.e. three times the robot's

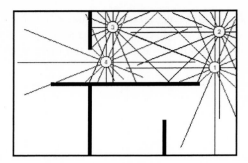

Figure 3. Average sensory register for two similar landmarks

radius— for the environment of Figure 2. During the exploration of this particular environment a 100% success ratio was obtained, for both new landmarks and already existing —i.e. revisited— landmarks: reference places p_1, p_5 and p_6.

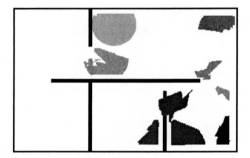

Figure 4. Regions corresponding to the landmarks detected and recognized using the perceptron network

Whenever a landmark cannot be recognized with absolute accuracy using the perceptron network, it is advisable to simultaneously employ the distance-based method in order to get an estimation of the robot's physical position and its distance to the nearest existing landmark. In spite of the position errors, the distance-based recognizer provides a valuable second vote for landmark detection and recognition. Eventually, in the worst case of the robot not being able to recognize a particular landmark, the only consequence is that the environment model will create a new landmark instead of employing the existing landmark associated with the corresponding region.

If a successful recognition has been performed when using the distance-based method as the background recognition process, then the corresponding sensory register has to be incorporated to the training set of the perceptron network. This new training element is incorporated inmediately after the distance-based landmark recognition has been accomplished in order (a) to avoid future errors in the landmark recognition by the perceptron network and (b) to update all the perceptrons with the new element. As the number of both the training instances and perceptrons is relatively small, the training duration is almost insignificant.

The formal description of the algorithm for the combined recognition process —i.e. the perceptron network and the distance-based recognizer— for both (1) recognition of a new landmark p_n and (2) the introduction of new instances in the training set S, is as follows.

1. If the reference place or landmark p_n can be recognized by means of expression (8), then the recognized landmark corresponds to landmark p_c, i.e. $p = p_c$.
2. Else, determine whether or not the reference place p_n can be recognized using expression (4).
   (a) If the recognition is performed, then the reference place is $p = p_c$ and the training set is consequently augmented $S = S \cup s_n$ with $f(p_c, s_n) = 1$ y $f(p_i, s_n) = 0 \ \forall p_i \neq p_c$ and the set of perceptrons Π_i are trained according to expression (5).
   (b) Else, the new landmark is $p = p_n$ and the training set is accordingly updated $S = S \cup s_n$ with $f(p_n, s_n) = 1$ and $f(p_i, s_n) = 0 \ \forall p_i \neq p_n$ and the set of perceptrons are trained according to expression (5).

5 Conclusions

Two methods have been proposed for the detection and recognition of landmarks in the creation of environment's models or maps, a fundamental issue in the autonomous navigation of mobile robots. The first method for landmark recognition employs the distance between the estimated position of the robot and the already existing landmarks. Due to the significant errors arising in the robot's position measurement, the distance-based recognition method performs quite poorly. For this reason we have introduced a much more robust and reliable second method, which is based on a neural network formed by perceptrons as the basic neuronal units. In order to optimize the success ratio in the detection and recognition of reference places we have combined both methods, with the distance-based recognizer acting as a background option.

The justification of a rather simple neuron prototype as the perceptron is twofold. In the first place, we have found through experimentation that more complex neural structures, for instance the multilayer perceptron trained with a well-tested learning algorithm, perform quite similarly to the network of perceptrons used in our work. The advantages in implementation, mainly storage requirements and computation time, for the simplest neural network are evident. In the second place, the computation burden in the training phase is almost insignificant. For environments with say 50 reference places and two samples in average for each landmark, the total computation time is always lower than 50 ms. Such fast performance is absolutely crucial for real-time navigation as the training phase can be executed under a low priority process, which for the architecture implemented in our mobile robot is almost compulsory [6,7]. Last but not least, the reconfiguration process in the number of neuron units due to the incorporation of a new landmark is immediate, as a new perceptron is added to the network whenever a new landmark is detected and the augmented neural network is easily re-trained. For any other neural network architecture different than the one used in our work, a maximum number of outputs should be defined at the beginning of an exploratory mission —i.e. the phase in which the robot creates the map of the environment by detecting landmarks— or it should be necessary to dynamically modify the outputs of the neural network. For multilayer perceptron networks the process could be even more involved because the hidden layers are much more difficult to update for each new neuron unit added to the output layer. An additional advantage of the network of perceptrons is the straightforward updating of the network parameters whenever a new landmark is introduced in the environment's model.

Acknowledgments

The work reported in this paper has been partly founded by the CICYT (Comisión Interministerial de Ciencia y Tecnología) under the project number TER96-1957-C03-02. All the experimental work has been carried out with a mobile robot platform Nomad-200.

References

1. R.A. Brooks. *The Cambrian Intelligence*. MIT Press, Cambridge, Massachusetts, 1999.
2. R.C. Arkin. *Behavior-based Robotics*. MIT Press, Cambridge, Massachusetts, 1998.
3. B.J. Kuipers. The spatial semantic hierarchy. *Artificial Intelligence*, 119:191–233, 2000.
4. A. Kurz. Constructing maps for mobile robot navigation based on ultrasonic range data. *IEEE Trans. on Systems, Man, and Cybernetics — Part B: Cybernetics*, 26(2):233–242, 1996.
5. F. Serradilla and D. Maravall. Cognitive modeling for navigation of mobile robots using the sensory gradient concept. In F. Pichler and R. Moreno-Díaz, editors, *Proc. of the Sixth Int. Workshop on Computer Aided Systems Theory (EUROCAST'97)*, pages 273–284. Springer-Verlag, New York, 1997.
6. J. de Lope, D. Maravall, and J.G. Zato. Topological modeling with Fuzzy Petri Nets for autonomous mobile robot. In A.P. del Pobil, J. Mira, and M. Ali, editors, *Proc. of the 11th Int'l. Conf. on Industrial and Engineering Applications of Artificial Intelligence and Expert Systems, IEA-98-AIE*, pages 290–299. Springer-Verlag, Berlin, 1998.
7. D. Maravall, J. de Lope, and F. Serradilla. Combination of model-based and reactive methods in autonomous navigation. In *Proc. on IEEE Int. Conf. on Robotics and Automation*, pages 2328–2333, 2000.

Topological Maps for Robot's Navigation: A Conceptual Approach

F. de la Paz López and J. R. Álvarez-Sánchez

Dpt. Artificial Intelligence - UNED (Spain). {delapaz,jras}@dia.uned.es

Abstract In this paper we present an example of incremental build up of a topological map to be used by a mobile robot which is programmed for navigation tasks. For this purpose we will use the concept of virtual expansion of the receptive field and invariant properties of the centre of areas in a region of open space around the robot. Then we propose, as basic elements in the building of this map, polygons of open space detected around the robot and referred to centres of area. The data structure that underlies our map is a graph where each node contains an open space polygon and where each arc represents the connectivity between these polygons.

1 Introduction

The study of different theories underlying programming of robots (Deliberative [5], Reactive, Behaviour-based [4] [3], Hybrid [6]) lead us to the conclusion that there are certain characteristics shared by all of them. These recurrences in all the architectures are more obvious if we try to obtain a unique generic model of description for all the autonomous robots. Usually we find the following conceptual components

Sensory Space: A set of functions that operate upon data coming from sensors and later processed to extract characteristics of the environment through algorithms of filtering, pattern matching etc.
Topological Space: A place where the different features obtained from the environment, with the purpose of building a map, and a set of functions, which give entity to those features, are stored. It is also considered as a memory space. Some times this last part of the calculus is not explicitly considered, although always underlies any causal relation between the sensory and decision spaces.
Decision Space: A set of functions in charge of selecting the proper motor actions.
Effector Space: A set of functions assigned to respond to orders from the decision space which control the movement of the robot.

In our previous works [1], [2] we already described the sensory space. In the present paper we will describe the topological space which is the place where the model of the environment will be build-up in a autonomous and incremental manner. Figure 1 shows the global conceptual scheme of the generic architecture in which the topological space is integrated.

We have also described elsewhere the virtual expansion of the receptive field of sensors and the concept of centre of areas that will be used later for feature extraction

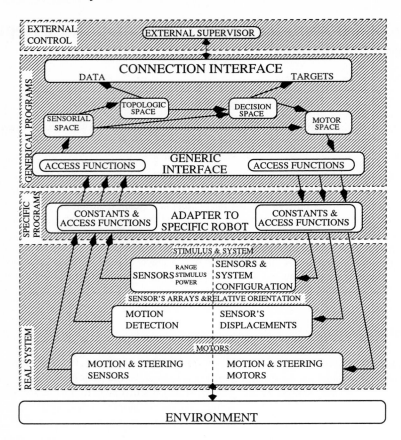

Figure 1. Conceptual scheme of the generic model for control of navigation

and classification of the different regions of the space [1]. The origin of the concept of "virtual sensors" came from the lateral inhibition process in biology. The results of this sensory expansion is the correction of punctual errors in the real sensors, as well as a significant increase of the precision in the representation process of the open space detected around the robot. The concept of "centre of areas" of a region of open space detected in the virtual sensors representation, is analogous to the physical concept of centre of mass. The position in the space corresponding to the centre of areas (CA) is invariant, no matter from which point was made the calculation. Provided that these points belongs to the same set of points we call this "region". If we transform the open space detected by the virtual sensors to obtain the representation referred to the centre of areas we get an invariant representation in front of translations between a set of positions in the same region. This property will enable us to identify all the points of this region with a canonical representative in the corresponding topological map.

2 Topological Space

Virtual sensors represents distances between the robot's centre and the different obstacles around it. Then we will take the linear envelope of the virtual sensors as a polygon of open space centred by the robot and reachable by it.

The purpose of the topological space is building an endogenous representation of the temporal sequence of polygons; these polygons represent the sensory space in each position of the robot all trough its trajectory. The sequence of this polygons constitute the input configuration to the topological space. It is in this space where polygons are first labelled and then associated to the nodes of a graph. These nodes contains a local memory where the representative features of polygons are stored. Complementary, the accessibility relations between those nodes are stored in the arcs of the graph.

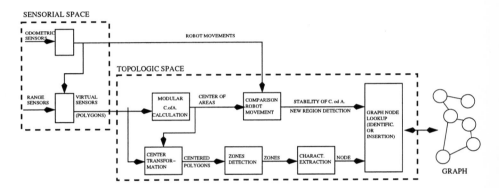

Figure 2. Block diagram of the topological space in relation to sensory space

The key problem at this level is the distinction between polygons. We need a set of invariants in front of a group of transformations of rotation and translation. To obtain these invariants we have used the cross-correlation of values of virtual sensors all through different time intervals. Consequently, subtasks needed for the global task in the topological representation of the space are:

T1. Calculus of the centre of areas and its stability relative to robot's movements in search of changes of region.
T2. Transformation of the open space polygons to equivalent ones centred at the CA.
T3. Detection of concavity zones in these polygons by means of differential interaction between neighbours.
T4. Feature extraction from concavity zones to build up measures of properties that specify a node
T5. Identification, labelling and interrelation between nodes corresponding to regions of the same centre of areas to check redundancies and update the topological map.

2.1 Centre of Areas and Its Stability

Starting from the polygon that represents the virtual sensors, we are able to calculate the position of the CA in a modular way[1].Each region can be represented by its own CA given that the position of this centre is invariant in front of translations of the point from which calculated. This theoretical construction allows us to simplify the spatial information used for navigation purposes since these invariant points are usually regions free of obstacles; this navigation will take place preferably there.

Figure 3 shows the representation of space points distributed over a squared area with a lateral aperture (fig. 3.a) and the correspondent CA's calculated from the sensors in those points (fig. 3.b). These calculations has been made placing a real robot in the positions marked in figure 3.a) and , henceforth, the results in figure 3.b) suffer from certain errors inherent to the measure process. Also figure 3.b) show us how the lateral aperture influence the position of the CA. In the event that we would represent the previously mentioned correspondence in a formal mathematical manner, not including the errors due to the measure process, we would have obtained the points symmetrically aligned only in the direction strictly orthonormal to the lateral aperture. The cause of it are increments of values of open space in front of aperture.

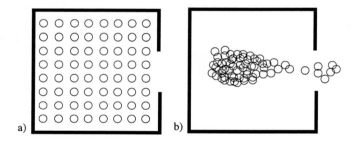

Figure 3. Correspondence of space points with the centre of areas using data from a real robot. a) Points from which the centre of areas has been calculated. b) Position of the corresponding centres calculated.

The centre of area is an invariant point for a given region of the open space around the robot. If the environment does not change, small movements of the robot will result in the same centre of areas. However, if the region is not homogeneous, the CA changes for small movements of the robot. This displacement of the CA that now we will name instability tell us two things. First, we are leaving a region and a new one will emerge with different properties. And second, in which direction the new region appears.

The relation between the CA displacement and that of the robot is a measure of how close or fare we are from a new stable centre, hence how far is a new stable region. If the robot is in a stable centre of areas, that relation will be very little. Per contra if the robot moves away from that point, the value of the relation will increase up to a maximum which is reached in the mid point between two stable regions.

2.2 Transformation of Centred Equivalent Polygons

The representation of open space polygons obtained in the sensory space is independent of sensory set rotations in relation to the centre of the system and adapts to the robot's displacements. Once we have this representation we must obtain another one now also independent of the robot position within a region defined by the range of the sensors (region of similitude or homogeneity)

Once the CA obtained we can translate any open space polygon to the equivalent polygon that would be seen if the robot be placed in the CA (fig. 4). The polygons can be clustered in groups (classes of equivalence) being these centred in the CA the representative of these classes (canonical polygons).

In this way, storing the polygons translated to the CA (centred), any other equivalent polygon can be compared and identified with the centred polygon used as canonical representative of the region.

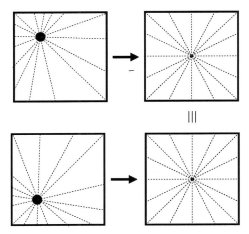

Figure 4. Polygons from two different positions which are equivalent in the representation of centre of area.

2.3 Modular Detection of Concavity Zones

Once transformed the sensory data into centred polygons we need to store some properties of these polygons which are invariants in relation to the CA and robust enough in front of small sensor changes, the measured position and environment. The main property that is always maintained is the general shape of the polygon. That is to say, the succession of concavities and convexities of the perimeter that correspond to obstacles or apertures in the real world. We call "zones" to this homogeneous and distinctive parts of the polygon.

Zones can be detected by means of differential operators that computes local changes of the polygon for each vertex. The second order spatial derivative give us information concerning the concavity of the zone (potential access to another region) or convexity (potential wall). Finally, is relevant the accentuation of differences in concavity which delimit different zones or parts of the polygon.

As in the previous calculus of the virtual sensors and CA [1], [2], the detection of concavity is also computed in a connectionist manner. A network with two layers is used as shown in figure 5 where the inputs to the first layer are the data defining the centred polygons. These data are in fact measures of the distances from the centre to the vertex that represent detected obstacles. A discrete set-down of non-recurrent lateral inhibition with receptive fields of a central element and two neighbours in the periphery, generate the corresponding zone marks.

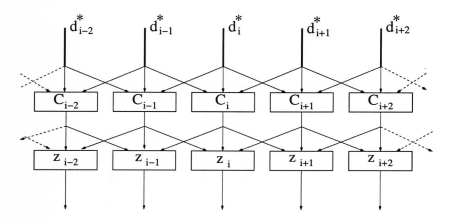

Figure 5. Detection of concavity zones by means of a lateral interaction network.

Input data d_i^* comes from the transformation of polygons centred on the centre of areas. Every one of the modules C_i in the first layer, calculate the first spatial derivative of second order taken as basis the distances d_i^*, d_{i-1}^* and d_{i+1}^*, and producing an output of type $\{+1, 0, -1\}$, which indicates the sort of local convexity of its receptive field. That is to say $C_i = \text{sign}\left(\Delta^2 d_i^*\right) = \text{sign}\left(\Delta d_i^* - \Delta d_{i+1}^*\right)$, with $\Delta d_i^* = d_i^* - d_{i+1}^*$. Those C_i values are then taken as inputs to the second layer where each module z_i performs a comparison between local concavity values in order to detect changes of zone. In this way each point is classified as belonging to: the inner part of a concave zone (IZ), an edge (SZ or EZ) or a transition zone (DC). This classification is attained by comparing C_{i-1}, C_i and C_{i+1} inputs using the look-up table shown in table 1. These outputs are later used to identify the border of the zones and, consequently, to detect specific characteristics of each zone. The figure 6 represents an example of results for those layers applied to a polygon part.

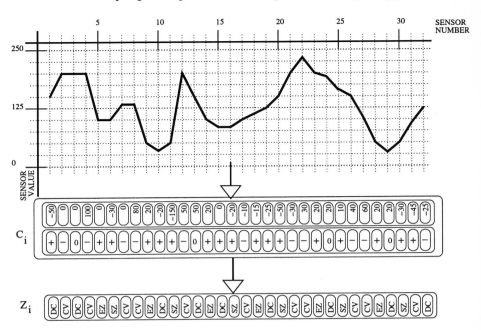

Figure 6. Example of calculations for zone detection of partial polygon.

Table 1. Look-up table for zone labelling.

C_{i-1}	C_i	C_{i+1}	zone label
+ or 0	+	−	SZ (start zone)
− or 0	+	+ or 0	EZ (end zone)
+ or 0 or −	−	+ or 0 or −	IZ (inside zone)
rest of combinations			DC (don't care)

2.4 Build-Up of the Graph

Once detected the zones within a centred polygon, we extract and store the following parameters: distance to the centre d_k, distance between both borders a_k, depth p_k and orientation, ψ_k, with respect to a fix direction in the space, as described in figure 7. To complete the topological information to be encoded in the graph, we add other fields of information concerning to: state of the connectivity of each zone with other regions (e_k, c_k), where e_k reports whether one region visited or not, and c_k identifies the node that zone is connected with. Finally, we add the total surface of the centred polygon s_i, and coordinates in an absolute system of reference, (x_c, y_c). In this way we obtain all potential nodes characterised by the numerical frame $N_i = (\{(d_k, a_k, p_k, \psi_k, e_k, c_k)\}, s_i, (x_c, y_c))$.

Once this information is compiled we proceed to compare it with every node included in the string that constitute the graph $\widetilde{G} = \{N_1, \ldots, N_i, \ldots, N_k\}$ (see fig. 8). This information is either added if there is not match, or identified if there is matching

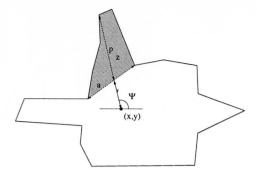

Figure 7. Illustration of the characterisation frame N_i for a centred polygon.

with any of pre-existing nodes. In the event of no coincidence, a new class of equivalence appears, that is a new label and consequently a new node with the characteristics derived from the new sort of zone. In case of coincidence, it can be asserted that it was a region already known where the robot is located, and we only need to update connectivity parameters between regions.

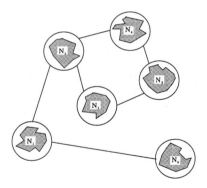

Figure 8. Construction of the graph into the topological space.

3 Conclusions

In this approach we have described the procedure that could be used by an autonomous robot to build up a topological map of its environment. This proposal is based on the virtual expansion of the receptive field and the invariance of the CA. Once the polygon of open space around the robot is obtained, we proceed to calculate the canonical representative of each class of equivalence by means of a transformation with respect to the position of the CA.

The invariant characterisation of different regions is finally obtained using a network of non-recurrent lateral interaction. From each one of these zones, the robot obtains a set of topologically relevant properties which are stored in a graph whose nodes represent zones and arcs represents the connectivity between regions. This graph is the robot's model of the environment and is embedded into a wider architecture. In the future, we intend to use this graph for planning and control the navigation based on objectives of an autonomous robot.

Acknowledgements

Authors wish to express their gratitude to Prof. Mira for his suggestions and comments about ideas and models described in this paper.

References

1. José Ramón Álvarez Sánchez, Félix de la Paz López, and José Mira Mira. On Virtual Sensory Coding: An Analytical Model of Endogenous Representation. In José Mira Mira and Juan V. Sánchez-Andrés, editors, *Engineering Aplications of Bio-Inspired Artificial Neural Networks*, volume 2 of *Lecture Notes in Computer Science*, pages 526–539. International Work-Conference on Artificial and Natural Neural Networks, Springer-Verlag, June 1999.
2. José Ramón Álvarez Sánchez, Félix de la Paz López, and José Mira Mira. Mode-based Robot Navigation using Virtual Expansion of the Receptive Field. *to appear in "Robotica" Special Issue on Sensorial Robotics*, 2001.
3. Rodney A. Brooks. A Robust Layered Control System for a Mobile Robot. *IEEE Transactions of Robotics and Automation*, RA 2(1):14–23, March 1986.
4. Maja J. Mataric. Behavior-Based Control: Main Properties and Implications. 1993.
5. N.J. Nilsson. Shakey the robot. Technical Note 323, AI Center, SRI International, 333 Ravenswood Ave., Menlo Park, CA 94025, April 1984.
6. Juan Romo Escansany, Félix de la Paz López, and José Mira Mira. Incremental Building of a Model of Environment in the Context of McCulloch-Craik's Functional Architecture for Mobile Robots. In Ángel Pasqual del Pobil, José Mira Mira, and Moonis Ali, editors, *Task and Methods in Applied Artificial Intelligence*, volume 2 of *Lecture Notes in Artificial intelligence*, pages 338–352. IEA-98-AIE, Springer-Verlag, June 1998.

Information Integration for Robot Learning Using Neural Fuzzy Systems

Changjiu Zhou[1], Yansheng Yang[2], and J. Kanniah[1]

[1] School of Electrical and Electronic Engineering, Singapore Polytechnic, 500 Dover Road, Singapore 139651, Republic of Singapore
zhoucj@sp.edu.sg
[2] Institute of Nautical Technology, Dalian Maritime University, Dalian, P.R. China
ysyang@mail.dlptt.ln.cn

Abstract. How to learn from both sensory data (numerical) and *a prior* knowledge (linguistic) for a robot to acquire perception and motor skills is a challenging problem in the field of autonomous robotic systems. To make the most use of the information available for robot learning, linguistic and numerical heterogeneous dada (LNHD) integration is firstly investigated in the frame of the fuzzy data fusion theory. With neural fuzzy systems' unique capabilities of dealing with both linguistic information and numerical data, the LNHD can be translated into an initial structure and parameters and then robots start from this configuration to further improve their behaviours. A neural-fuzzy-architecture-based reinforcement learning agent is finally constructed and verified using the simulation model of a physical biped robot. It shows that by incorporation of various kinds of LNHD on human gait synthesis and walking evaluation the biped learning rate for gait synthesis can be tremendously improved.

1 Introduction

Robots can only solve tasks after the tasks have been carefully analysed and added to the robot program by a human. To overcome the need for manual hard-coding of every behaviour, the ability to improve robot behaviour through learning is required. For a learning problem, there are many alternative learning methods can be chosen, either from the neural networks, the statistical or the machine learning literature. However, most of information obtained for robot learning is hybrid, that is, its components are not homogeneous but a blend of numerical data and linguistic information. We call the data of above types or vectors whose components are any of this, Linguistic-Numerical Heterogeneous Data (LNHD) [12]. There exists voluminous literature on the subject of making use of either sensory information or *a priori* knowledge for robotic learning and control [1, 4, 7, 11, 12]. However, very few papers are found in the field of LNHD integration.

Neural networks (NNs) and fuzzy logic (FL) are complementary technologies. NNs extract *numerical* information from systems to be learned or controlled, while FL techniques use *linguistic* information from experts. Neural Fuzzy Systems (NFSs) [6] being the product of both FL and NNs are computational machines with unique capabilities of dealing with both numerical data and linguistic information. Moreover, for such a hybrid system [3], to tune knowledge-derived models we can first translate domain knowledge into an initial structure and parameters and then use global or local optimisation approaches to tune parameters. Therefore, NFS is a naturally good choice to integrate LNHD for robot learning and control [1].

In this paper, to make the most use of the information available for robot learning, we investigate the LNHD integration using NFSs. Then a neural-fuzzy-architecture-based robot learning agent that can utilize LNHD by means of reinforcement learning [8, 13, 14] is constructed and verified using the simulation model of a physical biped robot.

2 Robot Learning

Robot learning refers to the process acquiring a sensory-motor control strategy for a particular movement task and movement system by trial and error. The goal of learning control is to acquire a task dependent control policy π that maps the continuous valued state vector **x** of a control system and its environment to a continuous valued control vector **u**:

$$\mathbf{u} = \pi(\mathbf{x}, \alpha, t) \tag{1}$$

Where the parameter vector α contains the problem specific parameters in the policy π that need to be adjusted by the learning system. Since the controlled system can be expressed as a nonlinear function $\dot{\mathbf{x}} = f(\mathbf{x}, \mathbf{u})$, the combined system and controller dynamics result in:

$$\dot{\mathbf{x}} = f(\mathbf{x}, \pi(\mathbf{x}, \alpha, t)) \tag{2}$$

From (2), we can see that the learning control means finding a function π that is adequate for a given desired behaviour and movement system. The general control diagram for a robot learning control is shown in Fig. 1.

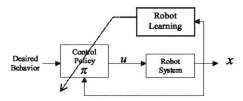

Fig. 1. General diagram for robot learning control

The following are some typical robot learning methods [7]: (1) Learning the control policy directly; (2) Learning the control policy in a modular way; (3) Indirect learning of control policies; (4) Imitation learning. Robot learning is a very complex problem. An important point should be noted is that most of the robot learning approaches only focus on using numerical data. Further more, Using the LNHD to train a robot is more powerful than other methods commonly used for robot learning. A trained robot is told more than whether it is wrong or correct for a particular action or sequence (reinforcement learning), the robot is also told what should have done (supervised learning). The robot can hence develop appropriate behaviour much more rapidly. In this paper, the LNHD will be integrated and incorporated in robot learning. Since a fuzzy reinforcement learning agent with neural fuzzy architecture can utilise both numerical and linguistic information [13, 14], it will be used to demonstrate how to use LNHD to improve robot learning.

3 Information Integration Using Neural Fuzzy Systems

Autonomous robotic systems depend on a large extent of the capability of the robotic system to acquire, to process, and to utilise both *sensory information* and *human knowledge*, to plan and execute actions in the presence of various changing or uncertain events in the robot-working environment [4]. As an example, let's consider the speed measurement of a mobile robot (see Fig. 2). The speed can be expressed as a real number if the sensors can precisely measure it. If the speed is evaluated non-numerically by a human observer, it may be interpreted as the linguistic term, such as the speed is "fast." It can also be represented as a single interval. Uncalibrated instruments could lead to this situation.

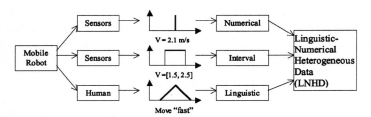

Fig. 2. Heterogeneous data fusion for a mobile robot.

3.1 Information Integration

When pieces of information issued from several sources have to be integrated, each of them represented as a degree of belief in a given event, these degrees are combined in the form $F(x_1, x_2, \cdots, x_n)$, where x_i denotes the representation of information issued from sensor i. A large variety of information combination operators F for data fusion have been proposed [2, 5].

In the framework of fuzzy sets and possibility theory, triangular norms (T-norms), triangular conorms (T-conorms), and mean operators are typical examples of conjunctive, disjunctive, and compromise fusion operators respectively. For example, the T-norm operators $\min(x, y)$, the T-conorm operators $\max(x, y)$, and many others [2].

3.2 LNHD Integration Using Neural Fuzzy Systems

Neural fuzzy systems (NFSs) retain the advantage of both NNs and FL and are capable of dealing with both numerically expressed and knowledge based information. Generally speaking, we may characterise NFSs in three categories [6]: (1) Neuro-fuzzy systems: the use of NNs as tools in fuzzy models. (2) Fuzzy neural networks (FNNs): fuzzification of conventional NN models. (3) Fuzzy-neural hybrid systems: incorporating FL and NNs into hybrid systems. The comparison of the above three LNHD integration approaches is illustrated in Fig. 3.

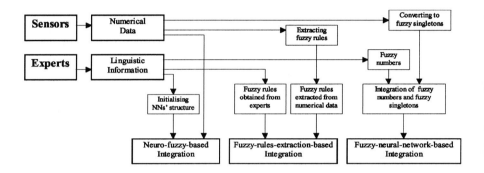

Fig. 3. The comparison of three LNHD integration methods using neural fuzzy networks

Neuro-fuzzy-based integration [6, 11]: The straightest way to integrate LNHD is to construct a neuro-fuzzy system using the fuzzy rules obtained from experts, and then fine-tune its parameters by the numerical data. One of the advantages of this approach is that the linguistic rules can initialise the model so that the robot does not learn from the scratch.

Fuzzy-neural-network-based integration [6]: There are no universally accepted models of FNNs so far. We may use a FNN with fuzzy supervised learning to process the integration of LNHD. The inputs, outputs, and weights of a neural network, which is the connectionist realisation of a fuzzy inference system, can be fuzzy numbers of any shape. The training data can be hybrid of fuzzy numbers and numerical numbers through the use of fuzzy singletons. In order to handle linguistic information in NNs, the fuzzy data is viewed as convex and normal fuzzy sets represented by α-level sets. A notable deficiency in this approach is that the computations become complex (e.g. multiplication of interval) and time-consuming.

Fuzzy-rules-extraction-based integration [11, 12]: In this approach, fuzzy rules are first extracted from input-output numerical data pairs. We can then construct the numerical data based fuzzy rule base (NFRB) with the degree of belief (DOB). The experts-based fuzzy rules (EFRB) with the DOB can be acquired from experts. The final uniform fuzzy rule base (UFRB) can be constructed through the integration of NFRB with EFRB by using fuzzy data fusion operators [2, 12].

4 Biped Robot Learning Using LNHD

In this section, a simulation is conducted to illustrate the proposed robot learning method using LNHD integration and its application to gait synthesis for a biped robot as shown in Fig. 4.

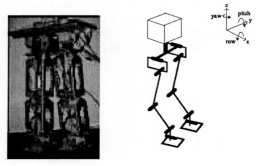

Fig. 4. A biped walking robot.

4.1 Biped Gait Synthesis

To synthesise the biped walking motion, it is required to take a workspace variable p from an initial position p_i at time t_i to a final position p_f at time t_f. The motion trajectory for $p(t)$ can be obtained by solving an optimisation problem. To achieve the dynamic walking, the zero moment point (ZMP) [9, 13] is usually used as a criterion, therefore, we can determine the biped motion to minimise the following performance index

$$\text{Minimize} \quad \int_{t_i}^{t_f} \left\| P_{zmp}(t) - P_{zmp}^d(t) \right\|^2 dt \qquad (3)$$

subject to the boundary conditions of both $p(t)$ and $\dot{p}(t)$ at time t_i and t_f, where P_{zmp} is the actual ZMP, and P_{zmp}^d is the desired ZMP position. The control objective of the gait synthesis for biped dynamic walking can be described as

$$P_{zmp} = (x_{zmp}, y_{zmp}, 0) \in S \qquad (4)$$

Where ($x_{zmp}, y_{zmp}, 0$) is the coordinate of the ZMP with respect to O-XYZ. S is the domain of the supporting area.

The proposed biped gait synthesising scheme is shown in Fig. 5. There are two independent fuzzy reinforcement learning (FRL) agents, namely, FRL Agent-x and FRL Agent-y for sagittal and frontal planes respectively. Each FRL agent only searches its relevant state-action space so as to speed up biped learning. Where $\Delta x_{zmp} = x_{zmp} - x_{zmp}^d$ is the displacement of the ZMP, $r_x(t)$ is the external reinforcement signal, and θ_x^d is the desired swing angle recommended by FRL Agent-x, while Δx_{zmp}, $r_x(t)$ and θ_y^d are for the frontal plane.

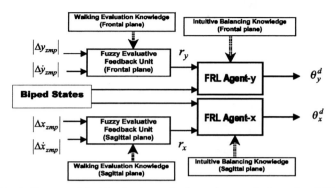

Fig. 5. The block diagram of the biped gait synthesiser using two independent FRL agents

4.2 The LNHD for Biped Learning

There are different kinds of information available for the gait synthesising [11, 13], such as the sampled data from the biped experiments (numerical); the intuitive walking knowledge (linguistic); and the walking knowledge which has been obtained from the biomechanics studies of human walking (linguistic + numerical).

Intuitive Biped Balancing Knowledge: There are many methods to derive fuzzy rules for the biped control. As an example, based on intuitive balancing knowledge, we can derive the following fuzzy rule to balance the biped robot in the sagittal plane.

  IF Δx_{zmp} is *Positive_Medium* AND $\Delta \dot{x}_{zmp}$ is *Negative_Small*
  THEN the body swings in *Positive_Small*

Walking Evaluation Knowledge: To evaluate biped dynamic walking in the sagittal plane, a penalty signal should be given if the biped robot falls down in the sagittal plane, that is, x_{zmp} is not within the supporting area in the sagittal plane.

$$r_x = \begin{cases} 0 & \text{if } x_{zmp} \in S_x \text{ and } |\Delta \dot{x}_{zmp}| \leq v_{xu} \\ -1 & \text{otherwise} \end{cases} \quad (5)$$

Where v_{xu} is used to limit the moving speed of the ZMP. Eq. (5) can only describe the simple "go – no go" walking states. It is less informative than human evaluation on walking. To provide more detailed information on evaluative feedback, for example, we can use following fuzzy rules to evaluate the biped dynamic balance in the sagittal plane.

  Rule: IF $|\Delta x_{zmp}(t)|$ *Zero* AND $|\Delta \dot{x}_{zmp}(t)|$ is *Small* THEN $r_x(t)$ is *Good*

Numerical data from measurements: For robot learning and control knowledge, the most common problem is that the human experts know how to control the robot but cannot express in words how to do it. Thus, transferring human empirical knowledge to a controller may turn out to be a difficult task. We refer this as *subconscious* knowledge. In this case, we may extract fuzzy rules from the action of an experienced manual operator [10].

4.3 Using LNHD to Improve the Biped Learning Rate

The LNHD used in biped gait synthesis is given in Table 1. From Fig. 6, we found that the learning speeds are comparable in term of different kinds of LNHD. For case A, the biped starts learning from a random setting. It achieves 120 seconds of walking at the 90th iteration (each iteration corresponds to each attempt of the biped to walk until a failure condition is encountered). For case B, the gait synthesiser is initialised by intuitive balancing knowledge. Starting from this configuration, the biped can walk a few seconds before the training. From case C and case D, it can be seen that the incorporation of different types of LNHD to the FRL agent can improve the biped learning rate tremendously.

Table 1. The LNHD for biped gait synthesis

Type	The description of the LNHD
Case A	No expert knowledge is available. Only numerical reinforcement signal r is used to train the gait synthesiser.
Case B	Only the intuitive biped balancing knowledge is used as the initial configuration of the gait synthesiser.
Case C	Both the intuitive biped balancing knowledge and walking evaluation knowledge are utilised.
Case D	Besides all the information used in case C, the fuzzy evaluative feedback, rather than numerical evaluative feedback, is included.

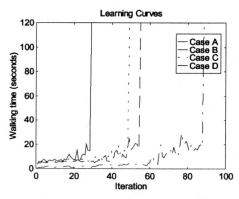

Fig. 6. The learning curves of the simulated biped robot using different types of LNHD.

5 Conclusions

We propose a NFS-based information integration approach for robot learning. It can utilize both numerical data from measurements and linguistic information from experts, *i.e.* LNHD, to acquire and improve robot behaviours. The simulation analysis demonstrates that it is possible for a biped robot to start with the initial configuration by the LNHD and to learn to refine it using the approach presented in this paper.

References

1. Akbarzadeh-T, M.R., Kumbla, K., Tunstel, E., Jashidi, M.: Soft computing for autonomous robotic systems. Computers and Electrical Engineering **26** (2000) 5-32
2. Bloch, I.: Information combination operators for data fusion: a comparative review with classification. IEEE Trans. on Systems, Man, and Cybernetics **26** (1996) 52-67
3. Bonissone, P.P., Chen, Y.-T., Goebel, K., Khedhar, P.S.: Hybrid soft computing systems: industrial and commercial applications. Proceedings of IEEE **87** (1999) 1641-1667
4. Ghosh, B.K., Xi, N., Tarn, T.J. (Eds.): Control in Robotics and Automation: Sensor-Based Integration. Academic Press (1999)
5. Hathaway, H.J., Bezdek, J.C., Pedrycz, W.: A parametric model for fusing heterogeneous fuzzy data. IEEE Trans. on Fuzzy Systems **4** (1996) 270-281
6. Lin, C.T., Lee, C.S.G.: Neural Fuzzy Systems: A Neuro-Fuzzy Synergism to Intelligent Systems. Prentice-Hall, Englewood Cliffs, NJ (1996)
7. Schaal, S.: Robot learning. Technical Report, University of Southern California (2000)
8. Sutton, R. S. and Barto, A. G.: Reinforcement learning: an introduction. MIT Press (1998)
9. Vokobratovic, M., Borovac, B., Surla, D. Stokic, D.: Biped Locomotion: Dynamics, Stability, Control and Application. Springer-Verlag (1990)
10. Zapata, G.O.A., Galvao, R.K.H., Yoneyama, T.: Extraction fuzzy control rules from experimental human operator data. IEEE Trans. Systems, Man, and Cybernetics **B29** (1999) 398-406
11. Zhou, C., Ruan, D.: Integration of linguistic and numerical information for biped control. Robotics and Autonomous Systems **28** (1999) 53-77
12. Zhou, C., Ruan, D: Fuzzy rules extraction-based linguistic and numerical data fusion for intelligent robotic control. In: Fuzzy If-Then Rules in Computational Intelligence: Theory and Applications, Ruan, D., Kerre, E.E. (eds.), Kluwer Academic Publishers (2000) 243-265
13. Zhou, C.: Neuro-fuzzy gait synthesis with reinforcement learning for a biped walking robot. Soft Computing **4** (2000) 238-250
14. Zhou, C., Meng, M., Kanniah, J.: Dynamic balance of a biped robot using fuzzy reinforcement learning agents," Fuzzy Sets and Systems (2001)

Incorporating Perception-Based Information in Reinforcement Learning Using Computing with Words

Changjiu Zhou[1], Yansheng Yang[2], and Xinle Jia[2]

[1] School of Electrical and Electronic Engineering, Singapore Polytechnic, 500 Dover Road,
Singapore 139651, Republic of Singapore
zhoucj@sp.edu.sg
[2] Institute of Nautical Technology, Dalian Maritime University, Dalian, P.R. China
ysyang@mail.dlptt.ln.cn

Abstract. In this paper, a general fuzzy reinforcement learning (FRL) agent that can utilise not only measurement-based information but also perception-based information by means of computing with words (CW) is proposed. By introducing fuzzy numbers and their arithmetic operations and fuzzy Lyapunov synthesis in the domain of CW, a set of stable fuzzy control rules can be derived from perception-based information. Moreover, based on a neuro-fuzzy network architecture, the fuzzy rules can be incorporated in the FRL agent to initialise its action network, critic network and evaluation feedback module so as to improve the learning. The performance and applicability of the proposed approach are illustrated through the practical implementation of learning control of an autonomous pole-balancing mobile robot.

1 Introduction

In neural learning methods, reinforcement learning (RL) agents [1, 4, 7] allow autonomous systems to learn from their experiences instead of exclusively from knowledgeable teachers. They are more appropriate than supervised learning (SL) for practical systems such as the robot learning and control when the input-output training data is not available. However, for real-time applications, it is difficult to search huge spaces for what constitutes a good action, and hence the learning is usually too slow. On other hand, human beings can always deal with perceptual and action uncertainty, and real-time constraints. Furthermore, human beings can utilise *perception-based information* to guide the reinforcement learning on those parts of the state-action space that are actually relevant for the task. For this reason, we conduct a research aimed at accelerating RL through the incorporation of both the perception-based information and the measurement-based information. To utilise the perception-based information in RL, a fuzzy RL agent [4, 9] with the neuro-fuzzy architecture is a naturally good choice.

A basic difference between perceptions and measurements is that, in general, measurements are crisp numbers whereas perceptions are fuzzy numbers or more generally fuzzy granules. Computing with words (CW), proposed and advocated by Zadeh [9], provides a foundation of a computing theory of perception (CTP) inspired by a remarkable human ability to perform a wide variety of tasks on the basis of fuzzy and imprecise information expressed in a natural language.

In this paper, by introducing fuzzy numbers and their arithmetic operations [2] and fuzzy Lyapunov synthesis [5] in the domain of CW, we demonstrate that the stable fuzzy control rules can be *systematically* rather than *heuristically* derived from perception-based information. Starting with this initial configuration, the fuzzy RL (FRL) agent can quickly refine its control knowledge and hence improve the robot behaviour through the experiment. This novel approach has been successfully applied to the learning control of an autonomous pole-balancing mobile robot.

2 Fuzzy Reinforcement Learning Agent

2.1 Structure of Fuzzy Reinforcement Learning Agent

The proposed FRL agent is shown in Fig.1. It is based on an actor-critic architecture [7]. An adaptive critic network provides an internal reinforcement signal to an actor, which learns a policy for controlling the process. It has three components [1]: the action selection network (ASN), the action evaluation network (AEN), and the stochastic action modifier (SAM). The ASN maps a state vector into a recommended action F using fuzzy inference. The AEN maps a state vector and a failure signal into a scalar score that indicates state goodness. With the multi-step prediction, the AEN can provide the ASN with a more informative internal reinforcement signal $\hat{r}$. The ASN can then determine a better action for the next time step according to the current environment state X and $\hat{r}$. The SAM uses both F and $\hat{r}$ to produce a desired control action. Based on the FRL architecture shown in Fig. 1 and the different kinds of available expert knowledge for reinforcement learning, we can construct different types of FRL agents.

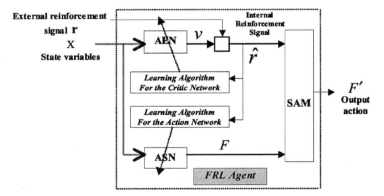

Fig. 1. The FRL agent and its learning system.

Table 1. Comparison of the different types of reinforcement learning agents

Types	Action network (ASN)	Critic network (AEN)	Evaluation feedback
RL agent	neural	neural	numerical
FRL agent (type A)	neuro-fuzzy	neural	numerical
FRL agent (type B)	neuro-fuzzy	neuro-fuzzy	numerical
FRL agent (type C)	neuro-fuzzy	neuro-fuzzy	fuzzy

Based on the available expert knowledge, the structural comparison of different types of FRL agents is given in Table 1.

RL Agent: It is a pure non-fuzzy RL agent [7]. Both action network and critic network can be constructed as a 3-layer neural network. Since the knowledge cannot be encoded into the RL agent, its learning starts from scratch. Thus, it usually needs a large number of trials for learning.

FRL agent (type A): It is similar to the GARIC model [1]. As the action network is expressed in a neuro-fuzzy framework, it can be constructed as a 5-layer neural network [1, 10]. Therefore, the control knowledge can be directly built into the action network as a starting configuration, while the critic network is initialised randomly as it remains as same as a RL agent (3-layer network).

FRL agent (type B): Both action network and critic network are constructed as neuro-fuzzy architectures. They can be implemented as 5-layer neural networks [4]. The learning rate of the FRL agent can be accelerated as both action and critic networks can house available knowledge as initial configuration.

FRL agent (type C): It has the same action network and critic network as type B. Moreover, it can accept fuzzy evaluative feedback is a kind of continuous signal and can provide a more informative external reinforcement signal, and hence improve the learning rate tremendously [9].

2.2 Learning in Fuzzy Reinforcement Learning Agent

The proposed learning system for the FRL agent is constructed by integrating two feedforward multiple-layer networks. One is a critic network (AEN) and another is an action network (ASN). The internal reinforcement signal from the critic network enables both ASN and AEN to learn without waiting for an external reinforcement, which may only be available at a time long after a sequence of actions has occurred.

Learning Algorithm for the Critic Network (AEN): The AEN evaluates the action recommended by the ASN and generates the internal reinforcement signal.

$$\hat{r}[t+1] = \begin{cases} 0 & start\ state \\ r[t+1] - v[t,t] & failure\ state \\ r[t+1] + \gamma v[t,t+1] - v[t,t] & otherwise \end{cases} \quad (1)$$

Where $\hat{r}[t+1]$ is the internal reinforcement at time $t+1$. $\gamma \in [0,1]$ is the discount rate. $v[t,t]$ is a prediction of the external reinforcement signal. $r[t+1]$ is an external reinforcement signal.

The temporal difference (TD) method [7] is employed to train the critic network. For the RL agent and the FRL agent (type A), their critical network architectures and training algorithms are the same as those in [7]. For the FRL agent (type B and C), its architecture and training algorithm are similar to those in [1, 10].

$$\Delta p_c = \eta_c \frac{\partial \hat{r}}{\partial p_c} = \eta_c \frac{\partial \hat{r}}{\partial v} \frac{\partial v}{\partial p_c} \tag{2}$$

Where η_c is the learning rate factor, p_c is the vector of all the weights in the critic network. $\partial v / \partial p_c$ can be computed using a back-propagation-like scheme.

Learning Algorithm for the Action Network (ASN): For the RL agent, the ASN structure is as same as that of a 3-layer neural controller. For the FRL agent (type A, B, and C), its architecture is the same as that of a 5-layer neuro-fuzzy controller. The goal of training the action network is to minimise the internal reinforcement signal $\hat{r}$ [1, 4]. In this paper, the input to output space mapping of the action network is denoted by $F_{p_a}(x)$. Where p_a is the vector of all the weights in the action network. The learning rule used to adjust the parameter value can be described as

$$\Delta p_a = \eta_a \frac{\partial v}{\partial p_a} = \eta_a \frac{\partial v}{\partial F} \frac{\partial F}{\partial p_a} \tag{3}$$

Where η_a is the learning rate factor. $\partial F / \partial p_a$ can be derived by applying the chain rule through the layers of the action network.

3 Initialising RL with Perception-based Information Using CW

3.1 Fuzzy Arithmetic Operations for CW

Perception-based information is usually expressed by a natural language. As a language, fuzzy arithmetic may be expressed in linguistic terms [2], making it possible to *compute with words* (CW) rather than numbers [8]. As an example, based on the physical intuition and the experience on balancing a pole (Fig. 3), we can get the perception-based information as shown in Table 2.

Table 2. Perception-based information for balancing a pole

	Perception–based information	Remarks
S1	$\dot{x}_1 = x_2$	From the state description.
S2	$\dot{x}_2 = \ddot{\theta}$ is proportional to the control u	The angular acceleration is proportional to the force applied to the cart.
S3	u is inversely proportional to $x_1 = \theta$ if $x_2 = \dot{\theta}$ is *small*	As the pole is falling over to the right (left) hand side, one must move his finger to the right (left) hand side at once.

Definition 1. Let A and B denote linguistic variables (fuzzy numbers) and let $* \in \{+,-,\cdot,/\}$, which denotes any of the four basic arithmetic operations. Then, we define a fuzzy set on $\Re$, $A*B$ by the following equation

$$\mu_{A*B}(z) = \sup_{z=x*y} \min(\mu_A(x), \mu_B(y)) \tag{4}$$

In this paper, we only consider triangular fuzzy numbers (TFN) with seven terms: PB (Positive Big), PM (Positive Medium), PS (Positive Small), Z (Zero), NS (Negative Small), NM (Negative Medium), and NB (Negative Big).

Theorem 1. Let $* \in \{+,-,\cdot,/\}$, and let A and B denote continuous fuzzy numbers. Then for the fuzzy set $A*B$ defined by Definition 1 is also a continuous fuzzy number. (The proof of this theorem is given in [2].)

Applying Definition 1, Theorem 1 and concepts on fuzzy linguistic approximation [11, 12], we can get the more general fuzzy-arithmetic-operating results of $C \cong A + B$ as given in Table 3. It will be used to extract fuzzy rules from the perception-based information (Table 2) to initialise RL so as to speed up its learning.

Table 3. Fuzzy-arithmetic-operation result of $C \cong A + B$

$C \cong A+B$		A				
		NM	NS	ZE	PS	PM
B	NM	NB	NB	NM	NS	ZE
	NS	NB	NM	NS	ZE	PS
	ZE	NM	NS	ZE	PS	PM
	PS	NS	ZE	PS	PM	PB
	PM	ZE	PS	PM	PB	PB

3.2 Extracting Fuzzy Rules for RL Using Fuzzy Lyapunov Synthesis and CW

Consider a Lyapunov function candidate $V(x_1, x_2) = \frac{1}{2}(x_1^2 + x_2^2)$ which can be used to represent a measure of the distance of the pendulum's actual state (x_1, x_2) and the desired state $(x_1, x_2) = (0,0)$. Differentiating V yields $\dot{V} = x_1 \dot{x}_1 + x_2 \dot{x}_2$. From S2 in Table 3, it can be rewritten as

$$\dot{V} \approx x_1 \dot{x}_1 + x_2 u = x_1 x_2 + x_2 u = x_2(x_1 + u) \tag{5}$$

Classical Lyapunov synthesis suggests that to guarantee $\dot{V} < 0$ by the design of u. Fuzzy Lyapunov synthesis follows the same idea but it is performed in the context of CW. The following is the linguistic description of Eq.(5).

$$LV(\dot{V}(x)) = LVx_2(LVx_1 + LVu) = Negative \tag{6}$$

Where $LV(\dot{V}(x))$, LVx_1, LVx_2, and LVu are linguistic values of $\dot{V}(x)$, x_1, x_2, and u respectively.

Theorem 2. If $V(x)$ is a Lyapunov function and the linguistic value $LV(\dot{V}(x)) = \textit{Negative}$ and $\textit{Supp}(\textit{Negative}) \subset (-\infty, 0]$, then the fuzzy controller designed by fuzzy Lyapunov synthesis is locally stable.

The detail description of Theorem 2 is given in [11, 12]. It provides a guidance to design a *stable* fuzzy controller *systematically*, rather than *heuristically*, using the perception-based information (Table 2). For example, if $x_2 = PM$, and $x_1 + u = NM$ is chosen, then from Eq.(6), $LV(\dot{V}(x)) = PM \cdot NM = \textit{Negative}$. Using Table 2, we can derive a set of fuzzy rules as shown in Fig. 2. From Theorem 2, the fuzzy controller is stable. Repeating the similar procedure, we can get a complete stable fuzzy control rules given in Fig. 2.

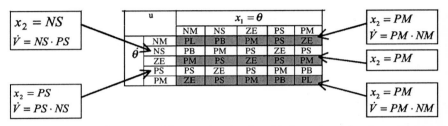

Fig. 2. Fuzzy control rules extracted from perception-based information using fuzzy Lyapunov synthesis and CW.

3.3 Incorporation of Perception-based Information in the FRL Agent

From Subsection 3.2, the perception-based information can be converted to a set of fuzzy rules by means of fuzzy numbers and their arithmetic operations in the frame of CW [11, 12]. Note that different kinds of knowledge expressed in form of fuzzy rules can be incorporated in various FRL agents as shown in Table 1 [9, 10]. In this way, we can extract fuzzy rules from perception-based information on control action (ASN), action evaluation (AEN), and reinforcement feedback evaluation respectively, and then make use of the above fuzzy rules to initialise FRL agents so as to speed up their learning. In the following section, we will demonstrate how the perception-based information on pole balancing can be incorporated in the FRL agent to control a pole-balancing mobile robot.

4 Learning Control of Autonomous Pole-Balancing Mobile Robot Using FRL Agent

To demonstrate the effectiveness of the proposed approach to incorporate perception-based information in FRL agents using CW, we conduct a real-time experiment of learning control of an autonomous pole-balancing robot using FRL agent. This project aims to design and fabricate an autonomous vehicle with an onboard TMS 320C32

DSP processor to participate in the Singapore Robotic Games (SRG). The vehicle is able to balance a free-falling pole by means of horizontal movement. The pole-balancing vehicle is shown in Fig. 3. While balancing the pole, it would also travel with a pre-designed slope profile (see Fig. 3). The vehicle with the highest number of successful cycles in a single untouched attempt within a predefined time slot will be considered the winning entry. The parameters of the physical robot are given as follows: the pole's length is $2l = 1m$, the mass of the pole is $m = 0.1kg$, and the mass of the cart is $m_c = 2.5kg$.

Fig. 4 (a) shows the trajectory of the pole angle using the initial fuzzy control rules (Fig. 2). It can be seen that the pole never falls down as the vehicle tracks the desired trajectory. It is demonstrated that the stable pole-balancing control can be achieved by the fuzzy control rules derived from perception-based information, though the pole swings very much. This may be because the perception-based information is too limited. To improve the pole-balancing performance, the FRL agent (type A) proposed in this paper is used to further tune the fuzzy control rules. The trajectory of the pole angle after the reinforcement learning is given in Fig. 4(b). The pole angle is always less than 0.1 *rad*. We also observed that the pole-balancing robot moves much smoother and faster after the reinforcement learning.

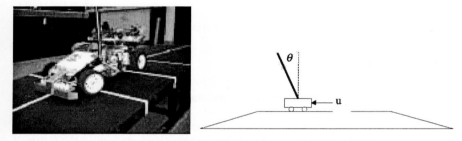

Fig. 3. The autonomous pole-balancing mobile robot and its pre-designed slope profile.

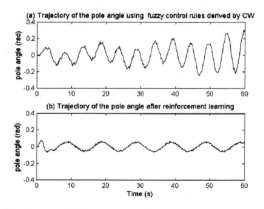

Fig. 4. Trajectory of the pole angle of the pole-balancing robot.

5 Concluding Remarks

A novel approach to incorporate perception-based information in FRL agents using fuzzy numbers and their arithmetic operations in the domain of CW is proposed in this paper. By introducing fuzzy Lyapunov synthesis, we demonstrate that the *stable* fuzzy control rules can be derived *systematically*, rather than *heuristically*. Start from this stable initial configuration, the FRL agent can learn to further tune its control behaviour quickly. The performance and applicability of the proposed method are illustrated through the practical implementation of the learning control of an autonomous pole-balancing mobile robot.

In this paper, only the FRL agent (type A) is demonstrated to use the perception-based information to control the pole-balancing robot. Since the perception-based information is very limited. How to utilise different kinds of FRL agents (Table 1) to integrate both measurement-based information and perception-based information to design an intelligent controller by means of CW will be a new challenge.

References

1. Berenji, H. R., Khedkar, P. S.: Learning and tuning fuzzy logic controllers through reinforcements. IEEE Trans. Neural Network **3** (1992) 724-740
2. Klir, G.J., Yuan, B.: Fuzzy Sets and Fuzzy Logic: Theory and Applications. Prentice Hall (1995)
3. Li, T.-H.S., Shieh, M.-Y.: Switching-type fuzzy sliding mode control of a cart-pole system. Mechatronics **10** (2000) 91-109
4. Lin, C.-T., Kam, M.-C.: Adaptive fuzzy command acquisition with reinforcement learning. IEEE Trans. on Fuzzy Systems 6 (1998) 102-121
5. Margaliot, M. and Langholz, G.: Fuzzy Lyapunov based approach to the design of fuzzy controllers. Fuzzy Sets and Systems **106** (1999) 49-59
6. Slotine, J.J.E., Li, W.: Applied Nonlinear Control, Prentice Hall (1991)
7. Sutton, R. S., Barto, A. G.: Reinforcement Learning: An Introduction. MIT Press (1998)
8. Zadeh, L.A.: From computing with numbers to computing with words – from manipulation of measurements to manipulation of perceptions. IEEE Trans. Circuits and Systems **I-45** (1999) 105-119
9. Zhou, C., Kanniah, J. Q. Meng: Intelligent robotic control using reinforcement learning agents with fuzzy evaluative feedback. In: Proc. of 4th FLINS. World Scientific Publisher (2000) 327-334
10. Zhou, C: Neuro-fuzzy gait synthesis with reinforcement learning for a biped walking robot. Soft Computing **4** (2000) 238-250
11. Zhou, C.: Fuzzy rules extraction for robot control using computing with words. In: Proc. 4th Asian Conference on Robotics and Its Applications, Singapore (2001)
12. Zhou, C., Ruan, D.: Fuzzy control rules extraction from perception-based information using computing with words. Int. J. Information Science (2001)

Cellular Neural Networks for Mobile Robot Vision

Marco Balsi[1], Alessandro Maraschini[1], Giada Apicella[1]
Sonia Luengo[2], Jordi Solsona[2], Xavier Vilasís-Cardona[2]

[1] Dipartimento di Ingegneria Elettronica, Università "La Sapienza",
via Eudossiana 18, 00184 Roma, Italy - balsi@uniroma1.it
[2] Universitat "Ramon Llull", Enginyeria "La Salle", Departament d'Electrònica,
Pg. Bonanova 8, 08022 Barcelona, Spain - xvilasis@salleURL.edu

Abstract. We show how Cellular Neural Networks are capable of providing the necessary signal processing to guide an autonomous mobile robot in a maze drawn on the floor. In this way, a non-trivial navigation task is obtained by very simple hardware, making real autonomous operation feasible. An autonomous line-following robot was first simulated and then implemented by simulating the CNN with a DSP.

1 Introduction

Navigation of a mobile autonomous robot in an unknown (or not completely known) environment can be achieved by use of visual information and feedback. When the environment is structured, the task required from the vision system is recognition of visual clues, and evaluation of their location relative to the robot position and orientation. Such information is then used to take suitable action by a control system.

Line following is one of the reference problems connected with navigation in a structured environment. It is a relatively simple, yet non-trivial problem that is relevant to real-life situations, such as road navigation, or motion through a maze (e.g. industrial environments). Line (or equivalent marking) following has been considered elsewhere. Solutions proposed in the literature may be classified into two main categories: simple solutions with very limited capability, or solutions for difficult, realistic problems obtained by use of rather complicated hardware.

An example of simple system is the ARGO partially autonomous vehicle [1]. It is a normal passenger car fitted with an automatic steering mechanism, controlled by a 486PC. The PC processes the images taken by two B/W cameras with the aim of finding the right line of the road, and keeping it in the appropriate position in the field of view. The processing is very simple, yet effective for the purpose.

The more complex systems are generally based on a computer which is either on board (for large vehicles, e.g. [2,3]), or remotely connected (e.g. [4]). In the quoted systems, images are processed by an additional unit, due to the necessity of high computing power. Of course, these vehicles can generally tackle fairly complex problems, such as unstructured road navigation, target following, obstacle avoidance, besides line following.

Cellular Neural Networks (CNN [5]) have been widely applied to sophisticated nonlinear real-time image processing tasks, so that they are natural candidates for silicon retina applications [6]. Their most appealing feature is the possibility of fully parallel analogue hardware implementation, which may include on-board image sensing capability [7], and the possibility of executing multiple consecutive and/or conditional operations without transferring images in and out of the network ("CNN Universal Machine" [8]). Such networks are the engine of our image processing system. CNNs have been applied to robot navigation tasks before, in particular to stereo vision [9, 10], and optical flow computation [11, 12]. A simple line-following problem was tackled by a hardware-implemented CNN by Szolgay et al. [13]. In this paper, we consider a simple, yet significant, robot navigation task as testbed. Our objective is to drive an autonomous robot equipped with a camera to follow a path in a maze of black straight lines on white background, crossing or joined at right angles. We have put an emphasis on using the simplest and least expensive hardware possible; real autonomy of the robot also called for compact and power-thrifty solutions. We decided to start by implementing a fully functional benchmark autonomous robot capable of navigating in a quite simple environment in order to thoroughly verify the validity of our approach. The results obtained will then be taken as a basis for addressing more complex navigation tasks.

2 Background: Cellular Neural Networks

Cellular Neural Networks are arrays of continuous-time dynamical artificial neurons (cells), that are only locally interconnected. Among the CNNs subclasses we shall resort to a Discrete-Time CNN model (DTCNN [14]) because, besides being also realisable as a fully parallel VLSI chip, is also more apt to our simulated implementation.

DTCNN operation is described by the following system of iterative equations:

$$x_{ij}(n+1) = \text{sgn}\left(\sum_{kl \in N(ij)} A_{k-i,l-j} x_{ij}(n) + \sum_{kl \in N(ij)} B_{k-i,l-j} u_{ij} u(n) + I\right) \quad (1)$$

where x_{ij} is the state of the cell (neuron) in position ij, that corresponds to the image pixel in the same position; u_{ij} is the input to the same cell, representing the luminosity of the corresponding image pixel, suitably normalised. Matrix A represents interaction between cells, which is local (since summations are taken over the set N of indices of cells that are nearest neighbours of the one considered) and space-invariant (as implied by the fact that weights depend on the difference between cell indices, rather than their absolute values). Matrix B represents forward connections issuing from a neighbourhood of inputs, and I is a threshold. The operation of the network is fully defined by the so-called cloning template $\{A, B, I\}$. Under suitable conditions, a time-invariant input u leads to a steady state $x(\infty)$ that depends in general on initial state values $x(0)$ and input u. Images to be processed will be fed to the network as initial state and/or input, and the result taken as steady state value, which realistically means a state value after some (order of 10 to 100 according to the task) time steps.

Many cloning templates have been designed for the most diverse tasks [15], so that some of the operations we need in this context were immediately obtained from the existing library.

3 Visual control system and image processing

The robot guidance problem is split into two processes, namely, the image processing to extract the mathematical features of the lines and the navigation of the robot to follow them. From the first, we expect to obtain the relative position of the mobile robot with respect to the line being followed, in order to correct possible deviations owing to misalignment of the wheels or mechanical or electrical fluctuations. This correction is to be performed by the navigation module. From the image processing we also need to extract information on the forthcoming crossings or bends, so as to decide which direction is to be taken and what is the correct moment to start turning. The mathematical information needed to control the robot is the angle between the robot and the path A_{th} and the distance from the centre from the line to the centre of the forward axis X_c. The result should be the steering angle of the front wheel A_{st}.

The image processing relies on the following stages: acquisition, preprocessing (cleaning and contrast enhancement), line recognition, extraction of line parameters (angle, position). Acquisition of course depends on the actual implementation, so it is treated below. For the pre-processing we used a so-called "small-object-killer" template (see table 1) to make a preliminary cleaning and binarisation of the acquired image (figure 1 (a,b)).

The calculation of the parameters of lines visible in an image is universally performed by the Hough transform [16]. Such approach is a very effective technique, yet computationally intensive, for obtaining direction and positions of all lines existing in an image. However, it becomes unnecessarily complicated when only few lines are present in the image. Our particular case demands of a reasonably fast response with our limited hardware. For this reason, toghether with the fact that the Hough transform is a global operation that does not lend itself to efficient implementation on the CNN, we chose to devise a different approach based on CNNs.

In order to perform direction and position evaluation, we need to get a thin line first. A "skeletonisation" algorithm, would be rather slow because it involves many iterations of an eight-step routine (*i.e.* cyclical application of eight cloning templates). For this reason, we resorted to the design [17] of a cloning template that extracts only one of the two edges of a stripe (it actually extracts only those parts of edges of black-and-white objects that lie on the left of the object itself). This template is given in table 1, and its effect is shown in figure 1 (c).

	A	B	I
small object killer	$\begin{pmatrix} 1 & 1 & 1 \\ 1 & 2 & 1 \\ 1 & 1 & 1 \end{pmatrix}$	0	0
left edge	2	$\begin{pmatrix} 0 & 0 & 0 \\ -1 & 0 & 0 \\ 0 & 0 & 0 \end{pmatrix}$	-1
vertically-tuned filter	2	$\begin{pmatrix} -1 & 0.5 & 1 & 0.5 & -1 \\ -1 & 1 & 1 & 1 & -1 \\ -1 & -1 & 5 & -1 & -1 \\ -1 & 1 & 1 & 1 & -1 \\ -1 & 0.5 & 1 & 0.5 & -1 \end{pmatrix}$	-13
small object killer	$\begin{pmatrix} 0 & 1 & 0 \\ 0 & 2 & 0 \\ 0 & -1 & 0 \end{pmatrix}$	0	0

Table 1. Templates used in the processing stage.

Of course, when dealing with approximately horizontal lines, we use a rotated version of this template, which extracts the upper edge.

The line position and orientation computation is based on two steps. We assume that a maximum of two, approximately orthogonal lines are within sight of the camera. This is not very restrictive, because we chose a setup in which the camera is oriented at angle that was chosen as a compromise between looking a reasonably long distance forward, and avoiding a large deformation due to perspective.

The first step is a directional filtering that extracts lines approximately oriented along two orthogonal directions. During normal operation these directions are the vertical and horizontal ones, corresponding to the line being followed and a possible orthogonal line following a bend or crossing. However, when the robot is turning or is largely displaced from all lines, it is necessary to switch to diagonal directions. As this situation is known to the controller, it will signal to the image processing stage which filters should be used.

Operation of a vertically tuned filter (table 1) is depicted in figures 2 and 3. Other direction-selective filters are obtaining by rotation of this cloning template. After the tuned filters have been applied, we get two images containing at most one line. Direction and position of the line is then extracted by performing an horizontal and vertical projection in order to read the positions of the first and last black pixels of the two projections. These four numbers, together with the information about which extreme of the line is closer to one of the borders of the image, are enough to compute direction and position of the line to the maximum precision allowed by image definition.

Extraction of the needed projections can be done by means of the so-called "connected-component-detector" cloning template (table 1). Operation of such template at an intermediate and final stage of processing is depicted in figure 1(d,e). It is apparent that besides obtaining the desired projections, also the information about which extreme is closer to the border can be obtained from examination of intermediate results.

Using the CNN operation described above, we have therefore obtained the parameters that are necessary for navigation. Such parameters are passed to the controller.

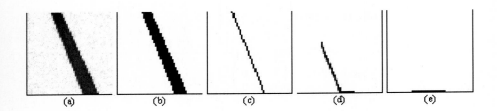

Fig. 1. Original image, taken from the actual camera (a), cleaning and binarisation (b), left edge (c), connected-component detector -intermediate result (d), final result (e).

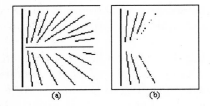

Fig. 2. Vertical line extracting filter: original image (a), result (b).

4 Fuzzy navigation module

For the sake of simplicity, we chose to implement a quite standard fuzzy controller for the line following task. The fuzzy rule base is built according to the Sugeno model [18] and consists of 15 rules, as listed in table 2

The global control strategy works as follows. With a single line in sight, the robot tries to align itself as fast as possible on it. When a second line gets into the view, the type of bend or crossing is assessed, by examining the relative positions of the two lines. The possible actions are, then, evaluated. If more than

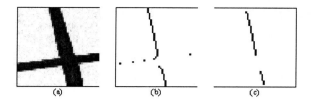

Fig. 3. Vertical line extracting filter applied to the image of a crossing (a) and after edge extraction (b): result (c).

		A_{th}				
		Large Positive	Small Positive	Almost Zero	Small Negative	Large Negative
X_c	Left	Medium Left	Small Right	Medium Right	Medium Right	Large Right
	Center	Medium Left	Small Left	Zero	Small Right	Medium Right
	Right	Large Left	Medium Left	Medium Left	Small Left	Medium Right

Table 2. Fuzzy control rule set for A_{st}.

one is possible (e.g. go straight on or turn right), than a command from a higher order level is called for (in our experiments we just implemented a pre-defined list of actions). If a turn is called for, we just switch the line to be followed. This involves also switching the directional filter that is selected for line following, and correctly interpreting angular displacement in order to choose the correct turn (left or right).

5 Hardware implementation

We realised a small autonomous robot that is guided by the algorithms proposed in this paper. It is a three-wheeled cart driven by a motor fitted on the single front wheel, that also steers by means of a second motor. The cart is approximately 27cm long, 18cm wide, 16cm tall, and weighs about 4Kg. A PAL camera is fitted on the front of the robot, oriented downwards, 30° from the horizontal. Images are grabbed and digitalised by dedicated circuitry implemented in a CPLD (ALTERA EPM7128STC100), which also performs image decimation to reduce unnecessary definition. Image processing (CNN simulation) is performed in a DSP (TMS320c32), and the control system (fuzzy rule base) is implemented in a 386-microprocessor-based microcontroller. A power board feeds the motors, and batteries are carried on board. This way the robot it is completely autonomous. The image processing stage can currently process one image per second. We estimated by simulation that this allows the robot to move smoothly at a speed of 2.5 cm/s. Of course if a CNN chip were used instead of the DSP, it would be possible to reach a much higher speed.

6 Simulation results

The robot layout and the algorithms have been thoroughly tested by employing a realistic simulation of the vehicle realised in Working Model, interfaced with a simulation of the fuzzy control and image processing system realised in Matlab. All mechanical and physical parameters of the actual robot were taken into account, as well as actual processing times of the mounted board, that has already been successfully tested. Mechanical mount and electrical testing of the robot is currently being completed, and we expect the robot to be fully functional soon. Meanwhile, figure 4 shows a simulation of the path followed by the robot, starting a little displaced ($A_{th} = 10°$, $X_c = 4$cm) and turning at a crossing.

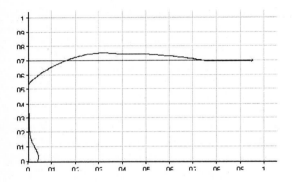

Fig. 4. Path followed by the front wheel of the robot. Axis marking in meters.

7 Conclusions

In this paper we describe the design of a fully autonomous visually guided robot, which is able to follow a line by means of CNN-based image processing and fuzzy control. The robot has been realised and is currently being tested. Accurate and realistic simulation demonstrate the effectiveness of the approach. We are currently working at a development of the image processing stage that will allow the robot to deal with broken/dashed and curved lines, and with vertical obstacles. The design presented allows straightforward integration of a CNN Universal Chip in place of the DSP-simulated CNN, that would allow for highly increased speed. We believe this work is a step in the direction of proving that CNNs are a good candidate for a silicon retina in robot guidance tasks.

Acknowledgements : We thank Prof. De Carli, of the Department of Computer Science of 'La Sapienza' University for his assistance in the simulation of the robot. We also thank Prof. Jordi Margalef of Enginyeria i Arquitectura La Salle for his assessment on the hardware development.

References

1. http://nannetta.ce.unipr.it/ARGO/english/index.html (as of Dec. 2000).
2. Chen K-H, Tsai W.-H. Vision-Based Autonomous Land Vehicle Guidance in Outdoor Road Environments using Combined Line and Road Following Techniques. Journal of Robotic Systems 1997; 14(10): 711-728.
3. Cheok KC, Smid G-E, Kobayashi K, Overholt JL, Lescoe P A Fuzzy Logic Intelligent Control System Paradigm for an In-Line-of-Sight Leader-Following HMMWV. Journal of Robotic Systems 1997; 14(6): 407-420.
4. Maeda M, Shimakawa M, Murakami S Predictive Fuzzy Control of an Autonomous Mobile Robot with Forecast Learning Function. Fuzzy Sets and Systems 1995; 72: 51-60.
5. Chua LO, Roska T The CNN Paradigm. IEEE Transactions on Circuits and Systems, part I 1993; 40(3): 147-156.
6. IEEE Transactions on Circuits and Systems, part I, Special Issue on Bio-Inspired Processors and Cellular Neural Networks for Vision. 1999; 46(2).
7. Liñán G, Espejo S, Domínguez-Castro R, Roca E, Rodríguez-Vázquez A CNNUC3: A Mixed-Signal 6464 CNN Universal Chip. Proc. of VII International Conference of Microelectronics for Neural, Fuzzy and Bio-inspired Systems, Granada, Spain, 1999: 61-68. http://www.imse.cnm.es/Chipcat/linan/chip-1.htm (as of Jan. 2001)
8. Roska T, Chua LO The CNN Universal Machine: an Analogic Array Computer. IEEE Transactions on Circuits and Systems, part II 1993; 40(3): 163-173.
9. Salerno M, Sargeni F, Bonaiuto V Design of a Dedicated CNN Chip for Autonomous Robot Navigation. Proc. of 6th IEEE International Workshop on Cellular Neural Networks and Their Applications (CNNA 2000), Catania, Italy, May 23-25, 2000: 225-228
10. Zanela A, Taraglio S A Cellular Neural Network Stereo Vision System for Autonomous Robot Navigation. Proc. of 6th IEEE International Workshop on Cellular Neural Networks and Their Applications (CNNA 2000), Catania, Italy, May 23-25, 2000: 117 -122
11. Shi BE A One-Dimensional CMOS Focal Plane Array for Gabor-Type Image Filtering. IEEE Transactions on Circuits and Systems, part I 1999; 46(2): 323-327.
12. Balsi M Focal-Plane Optical Flow Computation by Foveated CNNs. Proc. of 5th IEEE International Workshop on Cellular Neural Networks and their Applications (CNNA'98), London, UK, Apr. 14-16, 1998: 149 -154.
13. Szolgay P, Katona A, Eröss Gy, Kiss A An Experimental System for Path Tracking of a Robot using a 1616 Connected Component Detector CNN Chip with Direct Optical Input. Proc. of 3rd IEEE International Workshop on Cellular Neural Networks and Their Applications (CNNA'94), Rome, Italy, Dec. 18-21, 1994: 261 -266.
14. Harrer H, Nossek JA Discrete-time Cellular Neural Networks. International Journal of Circuit Theory and Applications 1992; 20: 453-467.
15. Roska T, Kék L, Nemes L, Zaràndy À, Brendel M CSL-CNN Software Library. Report of the Analogical and Neural Computing Laboratory, Computer and Automation Institute, Hungarian Academy of Sciences, Budapest, Hungary, 2000
16. Hough PVC, Method and means of recognizing complex patterns, U.S.Patent 3,069,654,1962. (See also : Jain AK, Fundamentals of digital Image Processing, Prentice Hall 1989.)
17. Zaràndy À The Art of CNN Template Design. International Journal of Circuit Theory and Applications 1999; 27: 5-23.
18. Terano T, Asai K, Sugeno M Fuzzy Systems Theory and its Applications. Academic Press, Boston, 1992

Learning to Predict Variable-Delay Rewards and Its Role in Autonomous Developmental Robotics

Andrés Pérez-Uribe* and Michèle Courant

Parallelism and Artificial Intelligence Group
Department of Informatics
University of Fribourg, Switzerland
Andres.PerezUribe@unifr.ch, http://www-iiuf.unifr.ch/~aperezu

Abstract. Researchers in the new field of "developmental robotics" propose to provide robots with so-called developmental programs. Similar to the development of human infants, robots might use those programs to interact with humans and their environment for extended periods of time, and become smarter autonomously. In this paper we show how a neural network model developed by neuroscientists can be used by an autonomous robot to learn by trial-and-error when considering rewards delivered at arbitrary times, as would be the case of developmental robots interacting with humans in the real world.

1 Introduction

A new trend in artificial intelligence research has been called "developmental robotics". In this new field, autonomous robots are supposed to be controlled by an intrinsic developmental program which *"develops mental capabilities through autonomous real-time interactions with its environment, by using its own sensors and effectors"* [14]. Similar to the development of human infants, robots might interact with humans and their environment for extended periods of time, and become smarter autonomously, under human (adult) supervision. This may be achieved by endowing the robots with epigenetic [9] mechanisms (i.e., a chain of developmental processes activated after the initial action of the genes - the description of the "developmental program") like adaptive categorization, trial-and-error learning, the construction of internal models, and imitation learning (See [5] and [7]).

A developmental robot similar to an adaptive organism has to deal with a continually changing environment. In order to "survive", it must be able to predict or anticipate future events by generalizing the consequences of its behavioral responses to similar situations experimented in the past. Predictions result after combining the mechanisms of *search* and *memory*, and give developmental robots the time to prepare their behavioral reactions, that is, to select one of a finite set of possible actions (innate reflexive behaviors [13]), and then, remember the actions that worked best and forget the inadequate actions [12].

---

* To whom correspondence should be addressed

Neuroscientists have identified a neural substrate of prediction and reward in experiments with primates. The so-called *dopamine neurons* have been shown to code an error in the temporal prediction of rewards [8]. Similarly, artificial systems can "learn to predict" by the so-called *temporal-difference* (TD) methods [12]. Temporal-difference methods are computational models that have been developed to solve a wide range of reinforcement learning problems including game learning and adaptive control tasks [12]. The system learns to predict rewards by trial-and-error, based on the idea that "learning is driven by changes in the expectations about future salient events such as rewards and punishments" [8].

In this paper, we use a biologically-plausible TD-learning architecture to learn to predict variable-delay rewards in a robot spatial choice task similar to the one used by neuroscientists with primates. Moreover, we analyze the robustness of the model by considering stimuli with variable amplitude. In Section 2, we describe learning by trial-and-error in artificial systems. In Section 3, we present the neural architecture developed by neuroscientists and its associated learning algorithm. Section 4 describes our experimental setup, Section 5 delineates the experimental results, and finally, Section 6 presents some concluding remarks.

2 Learning by Trial-and-Error

Reinforcement learning tasks are generally treated in discrete time steps: at each time step t, the learning system receives some representation of the environment's *state* $s(t)$, it *tries* an action a, and one step later it is *reinforced* by receiving a scalar evaluation $\lambda(t)$. Finally, it finds itself in a new state $s(t+1)$. This is what is called learning by trial-and-error in artificial systems. To solve a reinforcement learning task, the system attempts to maximize the total amount of reward it receives in the long run [12]. To achieve this, the system tries to minimize the so called *temporal-difference error* (also known as the effective reinforcement signal), computed as the difference between predictions at successive time steps.

In most machine learning applications, an automated trainer is programmed to deliver an immediate reward (or a punishment) when the learning system reaches a particular state. For example, in an obstacle avoidance learning task, a punishment signal may be delivered to the learning robot whenever it finds itself in a situation where one of its sensors is activated more than a certain threshold (which may correspond to the detection of an obstacle) [5].

In developmental robotic applications, there are no task-specific representations and the tasks may even be unknown [14]. Thus, the rewards (or the punishments) may be provided by external devices or humans at any moment (e.g., when the developmental robot achieves a goal or behaves incorrectly). In such cases, the reward signals may not be delivered (and received) at time $t+1$ (immediate reward) as in classical machine learning tasks, but after a variable lapse of time. One way to solve this problem is to use a time-delay neural network (TDNN) architecture [3], by implementing a kind of memory to represent the stimulus through time until the occurrence of a reward.

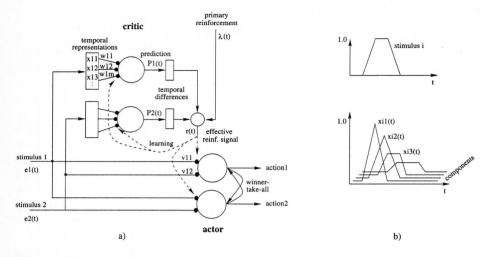

Fig. 1. a) Actor-critic architecture. The dots represent adaptive weight connections. The dashed lines indicate the influence of the effective reinforcement signal in those adaptive weights. b) Temporal stimulus representation.

3 Neural Architecture Implementing Temporal-Difference Learning

The neural model introduced by neuroscientists in [10] consists of an actor and a critic components. They receive the stimuli ($e_1(t)$ and $e_2(t)$ in the figure) coded as functions of time. The actor computes a weighted sum of the input stimuli to select an output action. A temporal stimulus representation allows the critic to associate a particular reward to the corresponding stimuli, even if the duration of the stimulus-reward interval is not null and variable. The critic generates the *effective reinforcement* signal, which is used to modify the adaptive synapses of the actor and critic components. The effective reinforcement signal takes a positive value if a desired event is better than predicted; a negative value, if the desired event is worse than predicted, and zero, if a desired event occurs as predicted [8]. Such signal improves the selection of behaviors by providing advanced reward information before the behavior occurs. Finally, a set of action and stimuli traces are used to record the occurrence of a particular stimulus or the taking of an action. Figure 2 presents the TD-algorithm in procedural form. The values of the most important parameters of the learning model were set following the work by Suri and Schultz [10]: $\gamma = 0.98$, $\eta_c = 0.08$, $\eta_a = 0.1$, $\delta = 0.96$, $\rho = 0.94$, and $\sigma = 0.5$ (notice that we use $\lambda(t)$ to refer to the external reinforcement signal, or 'primary reinforcement' and $r(t)$ to refer to the temporal difference error or 'effective reinforcement' to be consistent with Suri and Schultz [10], but $r(t)$ and $\delta(t)$ are more commonly used to refer to such signals, in the machine learning literature). In our implementation, we set the number of components of the temporal representation such that the model was

0. Initialize the actor and critic synaptic weights (w_{lm} and v_{nl} respectively), and the action and stimuli traces ($\bar{a}_n(t)$ and $\bar{e}_l(t)$ respectively).
1. Present a new input stimulus $e_l(t)$ to the actor-critic architecture.
2. Compute the temporal stimulus representation x_{lm} in the critic as follows: a) At the onset of the stimulus: $x_{l1} = 1, x_{l,m \neq 1} = 0$, b) Otherwise, k steps after the onset of the stimulus: $x_{l,m \leq k} = 0, x_{l,m>k} \rho^{m-1}$, where l is the stimulus, m is the component of its corresponding temporal representation, and ρ is a constant value in the range [0 1].
3. Select an action $a_n(t)$ based on the weighted sum of the stimuli traces and a uniformly distributed source of noise $\sigma_n(t) \in [0, \sigma]$: $a_n(t) = \sum_l v_{nl}\bar{e}_l(t) - \sigma_n(t)$. A winner-take-all rule prevents the system from performing two actions at the same time.
4. The critic component computes a weighted sum of a temporal representation of the stimuli to estimate the reward predictions ($P_l = \sum_m w_{lm} x_{lm}$), and generates the *effective reinforcement* signal as follows: $r(t) = \lambda(t) + \gamma P(t) - P(t-1)$, where, $\lambda(t)$ is the primary reinforcement, γ is a discount factor, $P(t)$ and $P(t-1)$ are the reward predictions at successive time steps, and $P(t) = \sum_l P_l(t)$ (i.e., the sum of the reward prediction for every stimulus l).
5. The change in the critic weights is proportional to prediction errors: $w_{lm}(t) = w_{lm}(t-1) + \eta_c r(t) max(0, x_{lm}(t-1) - \gamma x_{lm}(t))$, where, η_c is the learning rate of the critic.
6. The change in the actor weights is proportional to reward prediction errors, an action trace $\bar{a}_n(t)$, and the stimulus trace $\bar{e}_l(t)$, as follows: $v_{nl}(t) = v_{nl}(t-1) + \eta_a r(t) \bar{a}_n(t) \bar{e}_l(t)$, where, η_a is the learning rate of the actor.
7. The action and stimuli traces are updated as follows: $\bar{a}_n(t) = min(1, a_n(t) + \delta \bar{a}_n(t-1))$, and $\bar{e}_l(t) = min(1, \bar{e}_l(t-2) + \delta \bar{e}_l(t-1))$, where δ is a constant value in the range [0 1].
8. Goto step 1.

Fig. 2. The inner loop of the TD-learning algorithm associated to the time-delay actor-critic neural architecture with dopamine-like reinforcement signal (adapted from [10]).

able to associate a reward delivered up to 3 seconds after the selection of an action.

4 Spatial Choice Task

In a series of experiments with primates, Schultz and colleagues attached electrodes to the brains of monkeys to record electrical activity in dopamine-secreting neurons. The monkeys were trained to press a lever in response to a pattern of light to receive a reward (a squirt of juice). A trial was initiated when the animal kept its right hand on a resting key. Illumination of an instruction light above a left or right lever indicated the target of reaching. A trigger light determined the time when the resting key should be released and the lever be touched. In one of several experiments with monkeys, Schultz and his colleagues studied the dopamine neuron responses in a spatial choice task. In this task, the instruction and trigger lights appeared simultaneously, and both lights extinguished upon lever touch. The left and right instructions lights alternated semi-randomly. The monkey was rewarded for releasing the resting key and pressing the corresponding lever. It was not rewarded for incorrect actions like withholding of movement, premature resting key release or lever press, or incorrect lever choice [10].

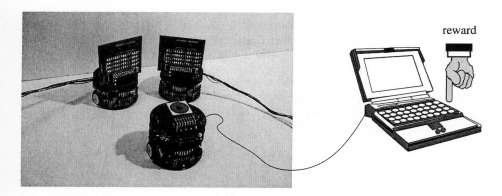

Fig. 3. Experimental setup. A Khepera robot with a linear vision system is placed in a workspace to replicate a primate learning task. Two other Khepera robots provided with MODEROK modules display a pattern of columns (only one robot at a time). The learning robot is intended to react to one such instruction pattern.

Even if a spatial choice task might be considered a "toy problem", it enables us to deal with the problem of learning by trial-and-error when considering rewards delivered at arbitrary times, as would be the case of developmental robots interacting with humans in the real world. In our experiments, an autonomous mobile robot is placed in a box with white walls (Figure 3). We have used the Khepera mobile robot [4] with a K213 vision turret developed by K-Team. It contains a linear image sensor array of 64 × 1 cells with 256 grey-levels each. The image rate is 2 Hz.

We programmed our robot to turn around itself covering an angle of 180 degrees. While the robot moves, it takes a snapshot of the environment and transmits the 64 pixel image to a host machine that implements the neural network model. A pre-processing phase is executed to take the linear image of the environment and compute the number of "drastic" changes of activation between neighboring sensor cells (i.e., to detect a pattern composed of several columns as in Figure 3). Such patterns are randomly displayed manually [6] or using two other Khepera robots provided with MODEROK modules (MOdule D'affichage pour le RObot Khepera), developed in collaboration with the Ecole d'ingenieurs de Fribourg. Those modules enable us to display a given pattern by turning-on some light-emitting diodes (LEDs) in a 16x4 array.

5 Experimental Results

In a first set of experiment, we used a thresholding mechanism to determine the onset of a stimulus: when more than four "drastic" changes of activation between neighboring sensor cells were found, we applied a stimulus of amplitude 1.0 to the right input of the model (stimulus 1), when the robot is on the right side of the environment, or to the left input, when the robot is on the left side of the environment (stimulus 2). The neural network model operates at steps of 100 ms to activate one of three possible actions, which correspond to the fact of *perceiving a pattern to the right* (action 1), *perceiving a pattern to the left* (action 2), and *perceiving no pattern at all* (action 3). In Figure 1, we show only two actions, for the sake of simplicity. When the model chooses the correct action, an experimenter provides a reward by pushing a button on the host machine (Figure 3). Otherwise, the system starts a new trial after 3 seconds of waiting for the reward. During a trial, a pattern of black and white columns is presented to the robot. The trial evolves as explained above. In average, the robot learned to respond correctly to 9 of the 10 last presentations of the pattern, after 37 trials. The average delay between the selected action and the corresponding reward (given by the experimenter) was 1250.2 ms. The maximum delay was 2876 ms, and the minimum delay was 510 ms (See [6] for more details).

In a second set of experiments, we let the robot learn without using the thresholding mechanism used in their inputs (we used MODEROK displays instead of the black and white patterns of the previous experiments). Indeed, when the robot moves and takes a snapshot of the environment, it does not always "perceive" the same number of columns in the instruction pattern. Therefore, the stimuli amplitude changes from trial to trial. Nevertheless, the system was able to learn the correct actions as in the previous experiment, but needed more trials in average.

To quantify the robustness of the learning system while using stimuli with variable amplitudes, we performed a set of simulations using the same model. After only 25 trials, the system was able to respond correctly 100% of the time, even if the stimuli amplitude randomly varied between 0.8 and 1.0. However, when we let the stimuli amplitude vary between 0.7 and 1.0, the learning system achieved a performance of 85% of correct answers after 75 trials. Finally, when we let the stimuli amplitude vary between 0.6 and 1.0, the system performed correctly only 60% of the time after 90 trials. Figure 4 represents a typical experimental run, where the stimuli amplitudes varied (randomly) between 0.6 and 1.0 from trial to trial. We found that such a neural network model efficiently deals with the problems of variable-delay rewards and variable stimuli amplitudes (a change of up to 20% in the stimuli amplitude did not change considerably its performance).

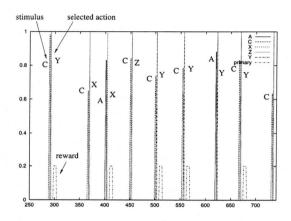

Fig. 4. Typical experimental run with variable stimuli amplitudes and variable-delay rewards. Stimuli 1 and 2 are represented by signals A and C. Actions 1, 2, and 3 are represented by signals X, Z, and Y respectively. Primary is the external reward provided by the experimenter.

6 Concluding Remarks

A developmental robotic system must be capable of adaptive categorization [13], learning by trial-and-error, imitation learning [1] and learning internal models of the world [11]. To achieve such a goal, neuroscience provides engineers with some knowledge about the computations in the cerebral cortex (unsupervised learning), the cerebellum (supervised learning of models), and the basal ganglia (reinforcement learning) [2]. In this work, we show that a time-delay actor-critic architecture (similar to the basal ganglia) implementing a temporal-difference learning algorithm enables an autonomous robot to learn a spatial choice task by a trial-and-error mechanism coupled with an externally-provided reward delivered at arbitrary times. Moreover, we described some experiments where we analyzed the robustness of the algorithm, by considering variable stimuli amplitudes. The approach appears as a promising mechanism for autonomous developmental robotic systems that learn from simple human teaching signals in the real world.

Acknowledgments

The authors wish to thank Prof. Wolfram Schultz for helpful comments and Dr. Thierry Dagaeff for his help with the first versions of the C-code of the neural network model. This work was supported by the Swiss National Science Foundation.

References

1. A. Billard. Learning motor skills by imitation: a biologically inspired robotic model. *Cybernetics and Systems*, 32(1-2), 2001 (in press).
2. K. Doya. What are the computations in the cerebellum, the basal ganglia, and the cerebral cortex. *Neural Networks*, 12:961–974, 1999.
3. K. J. Lang and G. E. Hinton. A time-delay neural network architecture for speech recognition. Technical Report CMU-DS-88-152, Dept. of Computer Science, Carnegie Mellon University, Pittsburgh, PA, December 1988.
4. F. Mondada, E. Franzi, and P. Ienne. Mobile robot miniaturization: A tool for investigating in control algorithms. In *Proceedings of the Third International Symposium on Experimental Robotics*, Kyoto, Japan, 1993.
5. A. Pérez-Uribe. *Structure-Adaptable Digital Neural Networks*, chapter 6. A Neurocontroller Architecture for Autonomous Robots, pages 95–116. Swiss Federal Institute of Technology-Lausanne, Ph.D Thesis 2052, 1999.
6. A. Pérez-Uribe. Using a time-delay actor-critic neural architecture with dopamine-like reinforcement signal for learning in autonomous robots. In S. Wermter, J. Austin, and D. Willshaw, editors, *Emergent Neural Computational Architectures based on Neuroscience*. Springer-Verlag, 2001 (to appear).
7. S. Schaal. Is imitation learning the route to humanoid robots? . *Trends in Cognitive Sciences*, 3(6):233–242, 1999.
8. W. Schultz, P. Dayan, and P. Read Montague. A Neural Substrate of Prediction and Reward. *Science*, 275:1593–1599, 14 March 1997.
9. M. Sipper, E. Sanchez, D. Mange, M. Tomassini, A. Pérez-Uribe, and A. Stauffer. A Phylogenetic, Ontogenetic, and Epigenetic View of Bio-Inspired Hardware Systems. *IEEE Transactions on Evolutionary Computation*, 1(1):83–97, April 1997.
10. R.E. Suri and W. Schultz. A Neural Network Model With Dopamine-Like Reinforcement Signal That Learns a Spatial Delayed Responde Task. *Neuroscience*, 91(3):871–890, 1999.
11. R.E. Suri and W. Schultz. Internal Model Reproduces Anticipatory Neural Activity. (available at http://www.snl.salk.edu/ suri), July 1999.
12. R.S. Sutton and A.G. Barto. *Reinforcement Learning: An Introduction*. The MIT Press, 1998.
13. J. Weng, W. S. Hwang, Y. Zhang, C. Yang, and R. Smith. Developmental Humanoids: Humanoids that Develop Skills Automatically. In *Proceedings the first IEEE-RAS International Conference on Humanoid Robots*, Cambridge MA, September 7-8 2000.
14. J. Weng, J. McClelland, A. Pentland, O. Sporns, I. Stockman, M. Sur, and E. Thelen. Autonomous Mental Development by Robots and Animals. *Science*, 291(5504):599–600, 2000.

Robust Chromatic Identification and Tracking

José Ramírez and Gianmichele Grittani

Grupo de Inteligencia Artificial
Universidad Simón Bolívar
Caracas, Venezuela
{jramire}@ldc.usb.ve

Abstract. The present work focuses on the implementation of a robust and fault tolerant global vision system for RoboCup Small League soccer teams. It is based on a vision control approach, in which vision processes are guided by necessity of information and knowledge about the environment. The object detection is based on a chromatic approach where chromatic patterns were modeled using a mixture of gaussian functions, trained with a stochastic gradient descent method. The implemented system meets, and in certain cases exceeds, the functionality required to participate in RoboCup and reported in related works.

1 Motivation

Though the vision system is just one of many pieces of a robotic soccer team, it is a very important one, since it provides the main sensor input to the control system. Many teams have found their vision systems to be an obstacle in the correct testing of their control algorithms, since they would occasionally fail and impede the control system to go on. This is the main motivation of the present work; its goal being, the construction of a robust and fault tolerant artificial vision system that will serve as a tool for latter research and for RoboCup Small League soccer teams.

Artificial vision systems rely on good image processing techniques to succeed. Generally, image processing algorithms are computationally intensive, so every bit of knowledge about invariant conditions of the ambient, counts towards faster processing [Klupsch et al.98]. Color markings and object shape knowledge is used to build a faster and more reliable vision system. Still, many problems need solution. The system must tolerate "salt and pepper" distortions in the image, changes in illumination conditions[1] caused by shadows of moving objects in the environment, occlusion of the ball caused by a robot, loss of a tracked object and marker color differences in the captured images.

Traditional vision systems did not use any information to guide the vision process. The present work uses all available information to focus the efforts of the vision system where needed. This takes advantage of the concept of a vision control [Arkin98], which analyzes the world and the information needs, to guide the vision process.

---
[1] RoboCup does its best effort to provide 700 to 1000 LUX of uniform light on the playing field.

2 Chromatic Pattern Processing

Since color markers are used for the identification of objects in RoboCup's small league, a way of defining these colors is needed. A naive approach would be to pick a color from one of the captured images and tell the system what object is identified by that color. The problem with this approach is that in the captured image, a color marker appears to have a richer variety of colors, due to camera noise and illumination differences over the object. Methods like template matching also have been tested with good results [Cheng et al.97]. Still, finding the best templates is not an easy task in certain cases.

An abstract way of defining colors is needed; one that treats this variety as an abstract unit related to an object. It is expected that color variations should be grouped together, forming clusters in a color space. These clusters appear in the image's histogram, like the one seen in figure 1, so all that needs to be done, is to find the center and radius of a cluster that will represent an abstract color. This could be done manually, by inspecting the histogram, but this is time consuming and difficult for a 3D space like color space. An alternative, is to model

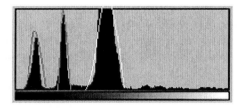

Fig. 1. Unidimensional image histogram showing a superimposed mixture of three gaussians modeling it.

the histogram using gaussian functions to represent each cluster, where the mean and variance define the abstract color's properties. In a gaussian mixture model like this, the histogram's density function, given by

$$p(x) = \sum_{j=1}^{M} P(j)p(x|j) \qquad (1)$$

is a linear combination of M gaussian basis functions of the form

$$p(x|j) = \frac{1}{(2\pi\sigma_j^2)^{\frac{d}{2}}} \exp\left\{-\frac{\|x - \mu_j\|^2}{2\sigma_j^2}\right\} \qquad (2)$$

where d is the space's dimension and $P(j)$ are the mixing parameters. Each basis function represents a class, and the mixture implements a classifier.

Given a data set $\{x^n\}_{n=1}^N$, the maximum likelihood solution can be found, for the means, variances and mixing parameters used. Minimizing the expression

$$E = -\log \mathcal{L} = -\sum_{n=1}^N \log p(x^n) \tag{3}$$

the following three equations are obtained.

$$\hat{\mu}_j = \frac{\sum_n p(j|x^n) x^n}{\sum_n p(j|x^n)} \tag{4}$$

$$\hat{\sigma}_j^2 = \frac{1}{d} \frac{\sum_n p(j|x^n) \|x^n - \hat{\mu}_j\|^2}{\sum_n p(j|x^n)} \tag{5}$$

$$\hat{P}_j = \frac{1}{N} \sum_{n=1}^N p(j|x^n) \tag{6}$$

These are highly coupled equations, and some method, like Expectation Maximization (EM) [Dempster77], must be used to obtain a solution from them. The problem with EM is that the whole data set must be available. If on-line sampling is to be done, this method can not be used. A stochastic gradient descent can be used instead [Tråvén91]: with a little algebraic work, equations (4), (5) and (6) can be expressed as recursive gradient descent equations over the data set dimension N, as follows

$$\mu_j^{N+1} = \mu_j^N + \eta_j^{N+1} \left(x^{N+1} - \mu_j^N \right) \tag{7}$$

$$(\sigma^2)_j^{N+1} = (\sigma^2)_j^N + \eta_j^{N+1} (\frac{1}{d} \|x^{N+1} - \mu_j^{N+1}\|^2 - (\sigma^2)_j^N) \tag{8}$$

$$P(j)^{N+1} = \frac{N}{N+1} P(j)^N + \frac{1}{N+1} p(j|x^{N+1}) \tag{9}$$

$$\eta_j^{N+1} = \frac{p(j|x^{N+1})}{(N+1) p(x^n)} \tag{10}$$

Using these equations, a mixture can be trained to model an image histogram, updating with each new pixel. This method has the property of convergence with small data sets.

3 Vision System Algorithms

Efficiency is the prerogative in almost all real time image processing applications. Initially, the system has no knowledge about object positions, so it must look for them in the whole image. To speed up the process, this is done as a coarse–fine search. In the coarse phase, the image is scanned every n pixels horizontally and vertically, looking for pixels that are different from the background color.

The pixel locations that are different are saved to be processed in the fine phase, where the size and type (color) of the object marker is determined. The pixel jump n is a fraction of the diameter of the smallest marker on the field. This guarantees that at least one pixel of each marker is to be found in ideal conditions. The tradeoff is between taking a small fraction, thus incrementing robustness and degrading efficiency, and taking a broad fraction, thus incrementing efficiency and risking robustness. Usually one fourth of the diameter is enough to find all objects.

The coarse phase discards from 70% to 90% of the image depending on the number of robots on the playing field, leaving just the important locations to make deep analysis. The fine phase visits each of the saved pixels, growing from them the shape of the marker, using an algorithm similar to the one used to fill flat images (see figure 2), if the pixel is not part of an already grown marker. This phase also fuses together broken markers, comparing their distance to the

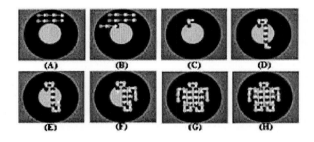

Fig. 2. Example of the Fine phase in the coarse–fine search, featuring the region growing ((c) to (h)) after finding a marker pixel ((a) to (c)).

specified diameter of a marker, to make sure that the pieces belong to the same marker. If after all the process, markers do not reach a reasonable fraction of the specified marker size, they are discarded as false positives.

At the end of all this sequence, the system has built a list with the position of all detected objects. Now it must detect the heading markers of the team's robots. To do so, it searches in the area of a square frame of fixed side, around the robot's marker, taking advantage of the fact that the heading marker is within a fixed distance from the central robot marker. Now that the system has all the available information, it must capture the next frame and track down the new positions of the objects. To do this efficiently, it uses the information gathered in the previous frame, searching for the object in a square around its old position, in the new image (see figure 3). The rest of the detection process is practically the same, with the exception that when looking for the object, the precise color of the marker is known (it must be the same as in the old image), so the search is exclusively for that color and not for any color that may represent an object, like it was done in the initial coarse–fine search.

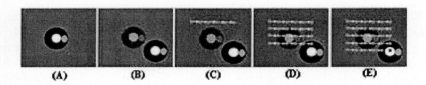

Fig. 3. Example of the tracking process. In the new image (b), a region around the old position of the object is searched, ((c) to (e)) until it is found again (e).

If after the tracking process some object is missing, the system initiates another coarse–fine search to locate it.

4 Results

4.1 Chromatic Pattern Detection

Figure 4(b) shows the chromatic patterns detected by the system, based on the initial figure 4(a). The system detected satisfactorily, all objects, showing a white circle surrounding the color markers. Notice that these circles almost fit exactly into the shape of the markers.

Fig. 4. Example of object detection. (a) Image captured by the frame–grabber. (b) White circles mark the detections done by the system.

4.2 Illumination Variations

Figure 5(a) shows the same configuration as in figure 4(a), with two thirds of the original illumination[2]. The system did not loose track of any objects, and the experiment had no influence on the processing speed of the system.

---
[2] The change is similar to the perturbations caused by moving objects in the vicinity of the field. It is not sharply noticeable since the camera used does auto gain adjustments and gamma corrections.

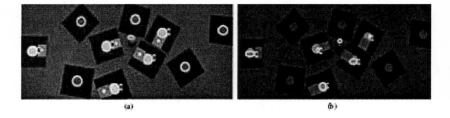

Fig. 5. Detection of objects under various lighting conditions.

To cause a dramatic effect, lights were turned off until one third of the illumination of figure 4(a) was reached (see figure 5(b)). In this case, the system was able to detect the yellow markers and the ball, since light deficiency causes the system to confuse the blue markers with the field's green color. Frame rate decreased to 8 fps since the system was continuously attempting to find the missing objects.

Tests were done within a range of illumination below the one given in RoboCup competitions. This does not limit the use of the system, since excess illumination can be controlled through the lens' iris and by preprocessing the captured image, in the frame-grabber. On the contrary, deficient illumination will always represent a problem, since colors tend to look similar.

4.3 Ball Visibility

To demonstrate how the system reacts to the ball occlusion problem, a test was conducted, partially covering the ball with the cardboard robots. Figures 6(a) and 6(b) show how the system still tracks the ball when only 50% and 25% respectively, is showing. In the latter case, the system lost track of the ball at times, recovering afterwards, for a total between 1 and 1.5 detections per second. Figure 6(c) shows the same experiment for a different zoom value, and 50% of

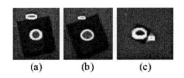

Fig. 6. Detection of a partially occluded ball. (a) 50% visible and (b) 25% visible. (c): Detection of a 50% occluded ball at small scale.

the ball covered.

A conclusion drawn from the tests is that the minimum visible surface required for detection of the ball is approximately 50 pixels (equivalent to a circle with a diameter of 8 pixels); this is considerably smaller than the expected occlusion situations found in RoboCup games.

4.4 Ball Tracking

In the last test, the ball was shot at various speeds between 1 and 6 meters per second, to observe how the system tracked it. The distance traversed by the ball and the frame rate were used to estimate the ball's speed.

Figure 7 shows how the system tracks the ball at various speeds. Notice that in figure 7(b) the system apparently placed four detections. This happens since screen refresh is slower than the frame processing rate. Also notice the gap between the pair of detections at the left and right sides. This was caused by a momentary loss of track of the ball, since it entered abruptly into the scene. Afterwards, the system recovered, marking the last two detections. The same

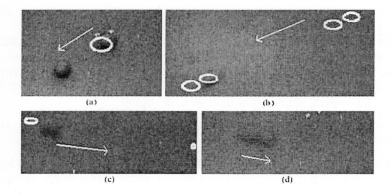

Fig. 7. Ball tracking at (a) 1.5 m/sec, (b) 3 m/sec, (c) 4 m/sec and (d) 5 m/sec.

peculiarity is observed in figure 7(c), in which the ball had a speed of 4 m/sec. This was the limit velocity found for the system, since the system did not detect a ball going at 5 m/sec. as shown in figure 7(d). It must be considered that the ball looks rather fuzzy, even for a human.

In past RoboCup events, the speeds reached by objects were around 3 m/sec, which means that the system is capable of tracking satisfactorily these objects.

5 Conclusions

The system meets the necessary functionality to be used in RoboCup competitions. It is robust and capable of recovery after the loss of a tracked object. It tracks objects at speeds higher than those seen in actual games, tolerates normal noise in illumination conditions and locates markers of sizes smaller than those used in RoboCup.

Using knowledge about invariant features of the environment in which the system will work, helps increase its efficiency and robustness.

Uninformed training of a mixture used to model image histograms leads in most cases, to local minima of little use for the application. This problem can be solved training the mixture based on an initial heuristic that corresponds to a state "close" to the trained state. The mixture training is then used for fine–tunning purposes. Under conditions of low illumination (near and under 150 LUX), the mixture tends to overlap classes, to compensate color diversity caused by the low stimulation of the camera's sensors. This does not represent a problem in RoboCup competitions, since the illumination given is in the range of 700–1000 LUX.

References

[Arkin98] Ronald C. Arkin, *Behavior–Based Robotics.* MIT Press, 1998, ISBN 0-262-01165-4

[Asada et al.97] Minoru Asada, Peter Stone, Hiroaki kitano, Alexis Drogoul, Dominique Duhaut, Manuela Veloso, Haijme Asama, and Sho'ji Suzuki, *The RoboCup Physical Agent Challenge: Goals and Protocols for Phase I.* Robot Soccer World Cup I / RoboCup–97, Springer-Verlag, 1997, pag. 42–61, ISBN 3–540–64473–3

[Cheng et al.97] Gordon Cheng and Alexander Zelinsky, *Real-Time Vision Processing for a Soccer Playing Mobile Robot.* Robot Soccer World Cup I / RoboCup–97, Springer-Verlag, 1997, pag. 144–155, ISBN 3–540–64473–3

[Dempster77] A. P. Dempster, N. M. Laird, D. B. Rubin, *Maximum Likelihood from incomplete data via the EM algorithm.* Journal of the Royal Statistical Society, B 39 (1), 1977, pag. 1-38

[Kitano et al.97] Hiroaki Kitano, Minoru Asada, Yasuo Kuniyoshi, Itsuki Noda, Eiichi Osawa and Hitoshi Matsubara, *RoboCup: A Challenge Problem for AI and Robotics.* Robot Soccer World Cup I / RoboCup–97, Springer–Verlag, 1997, pag. 1–19, ISBN 3–540–64473–3

[Klupsch et al.98] Michael Klupsch, Thorsten Bandlow, Marc Grimme, Ignaz Kellerer, Maximilian Lückenhaus, Fabian Schwarzer and Christoph Zierl, *Agilo RoboCuppers: RoboCup Team Desciption.* RoboCup-98: Robot Soccer World Cup II, 1998, pag. 431–438

[RoboCup99] RoboCup Home Page, *http://www.robocup.org.* Visited September 1999.

[Tråvén91] Hans G. C. Tråvén, *A Neural Network Approach to Statistical Pattern Classification by "Semiparametric" Estimation of Probability Density Functions.* IEEE Transactions on Neural Networks, 2 (3), 1991, pag. 366–377

Sequence Learning in Mobile Robots Using Avalanche Neural Networks

Gerardo Quero and Carolina Chang

Grupo de Inteligencia Artificial, Departamento de Computación y TI
Universidad Simón Bolívar
Apartado Postal 89000, Caracas 1080, Venezuela
{96-28854,cchang}@ldc.usb.ve
http://www.ai.usb.ve

Abstract. This paper describes the implementation of a neural network for sequence learning that is based on a neurocomputational theory of learning. The network is implemented on a physical mobile robot in order to learn to reproduce sequences of motor actions. At the onset of a conditioned stimulus the robot is presented with a sequence of visual stimuli that produce reactive motor actions of different duration. Initial results show that after learning the robot can approximate the motor sequence with no visual stimulation.

1 Introduction

Animal behavior has been classified in three main types [12,2]: reflexes, taxes, and fixed-action patterns. A reflex behavior is a rapid and involuntary response to a stimulus. Taxes are orienting movements towards or away from a stimulus. Finally, fixed-action patterns are response patterns of extended time duration. Contrary to a reflex, a fixed-action pattern can be motivated, and persists longer than the duration of the stimulus that triggered it.

Many reflex behaviors and taxes have been implemented successfully in physical mobile robots [17,4,1]. For example, Webb and colleagues have imitated the phonotaxis behavior of female crickets *G. bimaculatus*, who can locate and approach the calling song of the male [18,11]. Biologically inspired artificial neural networks have been implemented on robots, showing how adaptive responses can be learned by a robot through classical conditioning or operant conditioning [3, 15,5].

Fixed-action patterns are more complex behaviors than reflexes and taxes. However, some complex behaviors may be generated from simpler ones, *e.g.*, through hierarchies or sequences [2]. In particular, sequential behavior is fundamental to intelligence [14]. For instance, it is part of many human activities, such as reasoning, language, and problem solving. Sequence learning is not an easy task. It has been addressed using a variety of techniques, such as search based models, hidden Markov models, dynamic programming, neural networks, reinforcement learning, evolutionary computation, and so on.

In recent years we have developed a neural network model that learns simultaneously to produce obstacle avoidance behaviors and light approach behaviors in a mobile robot [6, 5]. Learning is achieved through classical conditioning and operant conditioning. The neural network is based on a detailed theory of learning proposed by Grossberg, which was designed to account for a variety of behavioral data on learning in vertebrates [9, 10].

This article describes further applications to robotics of ideas drawn from biological learning theories. We explore the problem of learning to generate sequences of motor actions that are initially triggered by unconditioned stimuli (UCSs). A sequence of UCSs follow the onset of a conditioned stimulus (CS). Preliminary results show that after sufficient training the CS alone triggers the sequence of motor responses learned by the robot. Related work has been reported by Verschure and Voegtlin [16].

Section 2 provides a brief overview of the outstar and the avalanche neural networks dynamics [7, 8]. The robotics platform and the experimental setup are described in section 3. Section 4 presents the results obtained after training the neural networks. Finally, section 5 presents the conclusions of this work.

2 Outstar and Avalanche Neural Networks

The outstar is an architecture for pattern learning. It can learn arbitrary patterns through repeated practice. In order to learn patterns the weights must be able to increase as well as decrease, thus a non-hebbian learning law is required.

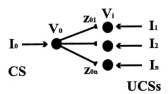

Fig. 1. The Outstar Network. Weights z_{0i} from the source cell V_0 to the border cells V_i store the long term memory of the neural network.

Figure 1 depicts an outstar network. The source cell V_0 learns the input pattern that activate the border cells V_i. The activation or the short term memory (STM) of the source cell V_0 and border cells V_i are given by:

$$\frac{d}{dt}x_0 = -a_0 x_0 + I_0(t) \qquad (1)$$

$$\frac{d}{dt}x_i = -a x_i + b[x_0(t - k\xi) - \Gamma]^+ z_{ki} + I_i(t) \qquad (2)$$

and the long term memory (LTM) traces z_{0i} obey the equation:

$$\frac{d}{dt}z_{ki} = -cz_{0i} + d[x_0(t-\xi) - \Gamma]^+ x_i \qquad (3)$$

the constants a_0, a, b, c, d, τ, and Γ are network parameters restricted to nonnegative values. The signal I_0 represents the conditioned stimulus while the signals I_i represent the unconditioned stimuli.

The outstar guarantees that under an appropriate sampling regime the network is capable of perfect learning, regardless of the initial STM and LTM patterns:

$$\lim_{t \to \infty} \vec{Z} = \lim_{t \to \infty} \vec{X} = \vec{\Theta} \qquad (4)$$

where $\vec{\Theta}, \vec{X}$ and $\vec{Z}$ are the normalized vectors formed from pattern variables:

$$\Theta_i = \frac{I_i}{\sum_{i=1}^n I_i} \qquad X_i = \frac{x_i}{\sum_{i=1}^n x_i} \qquad Z_i = \frac{z_{0i}}{\sum_{i=1}^n z_{0i}}$$

The pattern variables $\vec{\Theta_i}$, $\vec{X_i}$ and $\vec{Z_i}$ give a measure of how big the input, the STM activity and the LTM trace at site i are relative to the total.

A continuous input pattern that fluctuates in time can be learned by a group of outstars to an arbitrary degree of accuracy. The avalanche network (see figure 2 is a collection of outstars that are activated sequentially, in such a way that each outstar samples the pattern only during an interval of time. Every outstar sends an axon collateral to each cell V_i. The avalanche can be thought of as a cell whose axon has sequential clusters of collaterals that reach the common cells V_i.

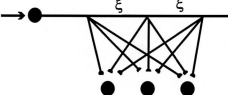

Fig. 2. The Avalanche Neural Network. Every ξ time units a signal traveling from the source cell body sequentially activates a synaptic knob of the axon. Each knob is an outstar that approximates the patterns at the border cells.

In the avalanche network, equations 2 and 3 become:

$$\frac{d}{dt}x_i = -ax_i + \sum_{k=1}^n b[x_0(t - k\xi) - \Gamma]^+ z_{ki} + I_i(t) \qquad (5)$$

$$\frac{d}{dt}z_{ki} = -cz_{ki} + d[x_0^{(2)}(t - k\xi) - \Gamma]^+ x_i \qquad (6)$$

where z_{ki} is the LTM trace from the k-th outstar to the cell V_i, n is the number of knobs or outstars in the cell axon, and ξ is the temporal resolution of the network.

3 The Robot and its Environment

We have implemented the avalanche neural network of figure 2 on a Khepera miniature mobile robot (K-Team SA, Préverenges, Switzerland). Khepera is a 55mm diameter differential drive robot (see figure 3(a)). It has eight infrared proximity sensors, and a color video turret. The robot is situated in a closed environment made out of LEGO bricks, as shown in figure 3(b). Panel (c) shows an image of the environment captured by the robot's camera. White LEGO bricks are used to build the walls. Occasionally the walls contain colored bricks: red, yellow, blue, or green.

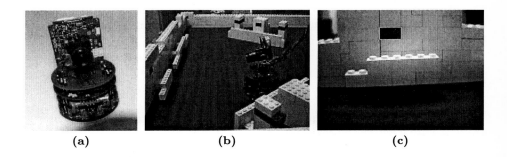

(a)   (b)   (c)

Fig. 3. The Khepera miniature robot **(a)** Khepera has 8 infrared sensors and a CCD color camera **(b)** The robot is situated in an environment made out of LEGO bricks. **(c)** The environment as seen from the robot's camera.

We use the *Khepera Integrated Testing Environment* (KITE) [13] to process the visual input. KITE is a tool for the evaluation of navigation algorithms for mobile robots. In particular, it features a segmentation module that detects and segments colored objects, automatically and in real time. For instance, in figure 3(c) KITE segments the red brick from the background. The figure shows the bounding box of the segmented object. KITE also provides a localization module that estimates the distance and bearing of the segmented objects with respect to the robot, based on visual looming. However, in this initial phase of our project we are taking advantage only of the segmentation module of KITE. Hence, the robot relies on its infrared readings to detect the proximity of the wall. We have added some sort of 'fences' (see panels (b) and (c) of figure 3) to prevent the size of the colored bricks from getting too big in the visual image. This is done because KITE does not segment large objects; they are seen as part of the background.

By default the robot moves forward. Detected colors elicit reactive motor responses. Red causes a right turn of 452°. Blue elicits a 463° left turn, while yellow elicits a 458° left turn. Green triggers a 20cm fast backward movement. Note that these reactive motor responses are arbitrarily. They were chosen this way only to facilitate our experimental setup, as discussed in section 4.

The goal of this work is to let the robot learn that some unconditioned visual stimuli (colored bricks) will arrive in a sequence after the onset of a unconditioned stimulus (a go signal). After sufficient training the onset of the unconditioned stimulus alone should elicit a sequence of motor responses that correspond to the sequence of unconditioned stimuli, even if these stimuli are not seen by the robot.

We use the avalanche neural network of section 2 to achieve sequential learning in the robot. The time resolution of the avalanche can be set as fine as desired but this may require a large number of outstars. We chose to set $\xi = 3.0$, and to perform a Fourth-order Runge-Kutta numerical integration of the avalanche equations 1, 5, and 6 using a fixed step size $h = 0.05$. To simplify our problem, for the time being we let the outstars encode the instantaneous motor actions (forward, left turn, right turn, backwards) instead of the unconditioned stimuli themselves.

4 Experiments

We train our robot with a sequence of four visual stimuli: red, blue, yellow and green. Figure 4 shows the trajectory that the robot produces in reaction to the visual stimuli. Training begins with the robot moving towards a wall that has a red brick (panels (a) and (b)). Detection of the red brick elicits a right turn, and then the robot moves towards a wall containing a blue brick (c). After a left turn the robot approaches a wall that has a yellow brick, which triggers another turn to the left (d). In panel (e) the robot has approached the wall that contains a green brick, so it moves backwards, stopping at the position shown in (f).

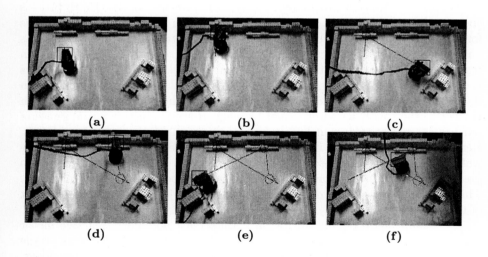

Fig. 4. Trajectory produced by the robot in response to visual stimuli. See text for details.

A top mounted camera tracks the robot's movements. The dotted lines of figures 4 and 5 represent the robot's trajectories. We put a color tag on the robot's top to help the tracking process. A white floor is used to enhance the contrast of the figures. Trajectories are displayed for visualization purposes only.

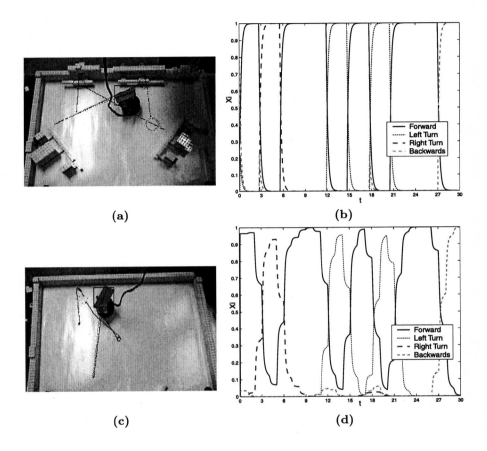

Fig. 5. Sequence learning of the Khepera robot. (a) In a training trial the robot encounters the sequence of visual stimuli red-blue-yellow-green. (b) Short Term Memory with visible stimuli, after 15 trial. (c) Performance of the robot when tested without the visual stimuli, after 15 training trials. (c) Short Term Memory with no visual stimuli, activated by the Long Term Memory of the network.

We train the robot with the same sequence of figure 5(a) for 15 trials. The starting position and heading of the robot are not controlled too carefully. Moreover, there is noise in the robot's sensors and motors. Thus, although the trials are similar, usually the stimuli are detected at different times, and the motor responses have a variable duration. Since the avalanche network learns space-time patterns of stimuli, we expect it to learn an average of all the training trials.

Figure 5(b) show a stable activation of the border cells V_i after 15 trials. The knobs or outstars of the avalanche encode accurately the robot's trajectory: forward, right turn, forward, left turn, forward, left turn, forward, and backwards. Note that the experiment was designed to allow the time resolution of the network be adequate for encoding the changes in motor actions.

The trajectory generated by the robot during testing is shown in figure 5(c). The robot performs the sequence of motor actions with no visual input, *i.e*, the color bricks and some walls are removed from the environment. The resulting trajectory is quite different from the trajectories produced during learning (panel (a)). However, the robot performs the motor actions in the correct order and with very good timing, as shown in panel (d). We are aware of the fact that small heading errors due to noisy motors or inexact duration of the turns can cause large trajectory errors.

5 Discussion

We have described our initial results on the implementation of avalanche neural networks for real time sequence learning in mobile robots. This type of network encodes the memory of a space-time pattern that can be arbitrarily complicated [8], for example, a piano sonata. Pure avalanche networks perform with no feedback. In humans and in animals some sequences of actions must be performed so rapidly that there is no time for feedback, for example, a cadenza played by a pianist or a escape response performed by lower animals.

The performance of the avalanche is ritualistic because it has no feedback. Once the sequence starts, it cannot be stopped. Clearly, organisms need some sort of feedback in order to adapt to the changes that take place in the environment. When they are combined with arousal cells, avalanches can become sensitive to feedback.

We are working on improving the accuracy in the sequences learned by the robot. Proprioceptive feedback could improve the trajectories generated by the robot. We could benefit as well from a fine time resolution of the network and a better encoding of the visual stimuli and the robot's fixed-action patterns.

Acknowledgments

The authors would like to thank Dr. Paolo Gaudiano, Director of the Boston University Neurobotics Lab and CEO of Aliseo Inc., who donated the Khepera robot and other pieces of equipment used in this project.

References

1. Ali, K. S., and Arkin, R. C.: Implementing schema-theoretic models of animal behavior in robotic systems. In *Proceedings of the fifth international workshop on advanced motion control* (1998).

2. Arkin, R. C.: *Behavior-based robotics.* Intelligent robotics and autonomous agents series, (1998), MIT Press.
3. Baloch, A., Waxman, A. M.: Visual learning, adaptive expectations, and behavioral conditioning of the mobile robot. *Neural Networks, 4,* (1991) 271–302.
4. Brooks, R. A.: A robust layered control system for a mobile robot. *IEEE Journal of Robotics and Automation,* (1986) *2*.
5. Chang, C., Gaudiano, P.: Application of biological learning theories to mobile robot avoidance and approach behaviors. *Journal of Complex Systems, 1*(1), (1998) 79–114.
6. Gaudiano, P., Chang, C.: Adaptive obstacle avoidance with a neural network for operant conditioning: experiments with real robots. In *Proceedings of the 1997 IEEE International Symposium on Computational Intelligence in Robotics and Automation (CIRA),*(1997) 13–18 Monterey, California.
7. Grossberg, S.: Some networks that can learn, remember and reproduce any number of complicated space-time patterns, I. *Journal of Mathematics and Mechanics,* (1969) *19*(1), 53–91.
8. Grossberg, S.: *Studies of Mind and Brain: neural principles of learning, perception, development, cognition and motor control.* (1982) Reidel, Boston, Massachusetts.
9. Grossberg, S. On the dynamics of operant conditioning. *Journal of Theoretical Biology, 33,* (1971) 225–255.
10. Grossberg, S., Levine, D. Neural dynamics of attentionally modulated Pavlovian conditioning: blocking, interstimulus interval, and secondary reinforcement. *Applied Optics, 26,* (1987) 5015–5030.
11. Lund, H. H., Webb, B., and Hallam, J.: A Robot Attracted to the Cricket Species Gryllus bimaculatus. In *Proceedings of the Fourth European Conference on Artificial Life,* (1997) 246–255. MIT Press.
12. McFarland, D.: *The Oxford Companion to Animal Behavior.* Oxford University Press (1981).
13. Şahin E., and Gaudiano P. : KITE: The Khepera Integrated Testing Environment. Proceedings of the First International Khepera Workshop. PAderborn, Germany. (1999) 199–208.
14. Sun, R., and Giles, C. L. : Sequence Learning: Paradigms, Algorithms, and Applications. Springer Verlag, LNAI 1828 (2001).
15. Verschure, P. F. M. J., Kröse, Ben J. A., Pfeifer, R.: Distributed adaptive control: The self-organization of structured behavior *Robotics and Autonomous Systems,* 9, (1992) 181–196.
16. Verschure, P. F. M. J., Voegtlin, T.: A bottom up approach towards the acquisition and expression of sequential representations applied to a behaving real-world device: distributed adaptive control III. *Neural Networks,* (1998) *11*(7-8), 1531–1549. 1998 Special Issue on Neural Control and Robotics: Biology and Technology.
17. Walter, W. G.: An Imitation of Life. *Scientific American, 182*(5), (1950) 42–45.
18. Webb, B.: Using robots to model animals: a cricket test *Robotics and Autonomous Systems,* (1995) *16,* 117–135.

Investigating Active Pattern Recognition in an Imitative Game

Sorin Moga, Philippe Gaussier, and Mathias Quoy

ETIS / CNRS 8051A, Groupe Neurocybernetique
ENSEA, 6, avenue du Ponceau,
F-95014, Cergy-Pontoise cedex, France
{moga,gaussier,quoy}@ensea.fr
http://www.etis.ensea.fr

Abstract. In imitation learning processes, the "student" robot must be able to perceive the environment and to detect one "teacher". In our approach of learning by imitation, we consider that the student tries to learn the teacher trajectory (temporal pattern). In this context, we propose a neural architecture for a mobile robot which detects its teacher using the optical flow information. The detected flow is used to initiate the imitative game. The main idea consists in using a pattern recognition system in order to allow the student to continue its imitative game even if the teacher is stopped. Since the movement detection and the pattern recognition systems work in parallel, they can provide different answers with different time constant. Neural fields equations are used to merge these information and to allow a stable dynamical behavior of the robot. Moreover, the stability of the decision making allows the robot to online learn to recognize the teacher from one image to the next.

1 Introduction

Our main goal is to design a neural network architecture which could permit the learning by imitation in an autonomous robot context. The *imitation* is generally mentioned in relation with other notions such as : emotions, consciousness of self, intentions or the existence of a supra modal-level. Unfortunately, there is not an universal accepted definition of imitation [6] and imitation is very hard to model in the perspective of designing autonomous learning architectures. Our works in learning by imitation [4] has begun with this question : what is the simplest architecture which could enable basics imitation capabilities ? In [5], we showed that proto-imitation (a simple, not intentional imitation) can be triggered by perception ambiguities.

We used this principle in a kind of "dance" task (the teacher robot must learn a sequence of movements with a particular timing). The interaction between the student and the teacher enables this imitative game (see [2] for more details). The Fig. 1 presents an overview of the general Perception-Action (PerAc) architecture used. The reflex path of the PerAc works as a movement tracking mechanism which consists in going towards to any perceived movement. The movement is detected using the optical flow information and we assume that only the teacher can move in the optical field of the student.

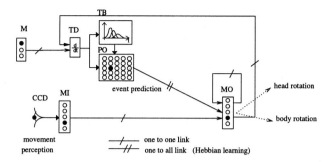

Fig. 1. A general diagram of the PerAc architecture use for learning the temporal aspects of a trajectory. CCD - CCD camera, M - Motivations, MI - Movement Input, MO - Motor Output, TD - Time Derivator, TB - time battery, PO - Prediction Output.

The second level of the architecture learns the temporal intervals between the successive robot orientations (i.e. a sequence of movements) and associates it to a particular motivation. The timing is learned by using a spectral decomposition of the time on a set of neurons (see [5] for details).

Using the architecture showed in Fig. 1, the *student* robot is able to learn trajectories. But the we have imposed a strong assumption : only the teacher is present in the student field of view and the teacher is perceived only if it generates movements. This assumption is enough to validate our proto-imitation principle. But, we can ask : what happened if the teacher stops, or if it is occulted by another object ? The answer is immediate : the student stops to follows the teacher and the imitative game is broken. One solution is to integrate in our architecture an active pattern recognition mechanism which enables the student to perceive the environment and to recognize the teacher. The idea of our solution is quite simple : using the perceived images (required for the movement detection), the student extracts the teacher position using a pattern recognition mechanism and merges this position with the perceived movement position (if any). An overview of the global architecture is presented on Fig. 2.

In section 2, the object recognition mechanism is presented. This mechanism is based on the detection of the interesting local views and the focusing mechanism on this views. In section 3, the integration of movement and object recognition mechanisms is introduced. At least, in order to reduce the processing time to acceptable values for a real time robot experiment, we show how neural field equations can help to parallelize this architecture.

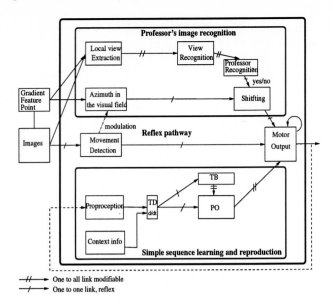

Fig. 2. A simplified figure of the model, allowing immediate and deferred proto-imitation. The system is designed as two parallel PerAc blocks, the "simple sequence learning" (bottom part of the figure) and the "teacher's image recognition" (top part) are the two perception level modulating the reflex pathway in the middle of the figure.

2 Object Recognition

In this section, we present the implementation of the mechanism for the object recognition. There are a lot of models for object recognition (see [3] for a review). In order to identify the objects, we must define the pertinent attributes and primitives. In our system, we use the output of a corner feature detector as attribute to categorize views.

The architecture of the focusing mechanism is shown in Fig. 3. The first step is to compute the gradient of the input image. We used a recursive, unidimensional realization of the Deriche filter. We preferred this technique to a pure neural technique for a speed processing purpose (the results are equivalents). The gradient image is convoluted with a Difference of Gaussian (DOG) filter. This convolution allows the existence of only one maximum in a region fixed by the DOG size. The output of the DOG filter is mapped into a map of interesting points. We define that a point is more interesting than another one if the corresponding DOG output is higher. In this manner, we introduce an order relation. It allows a classification of interesting points. By using a competition mechanism, the system successively focuses on each of these points enabling to find the position of local views of interest. This sequential strategy is directly inspired from monkeys and human visual system strategy.

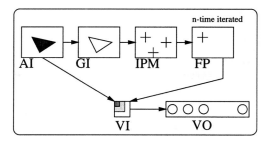

Fig. 3. The neural architecture of the recognition system. AI–sample image, GI–gradient image, IPM–interesting points map, FP–focus points, VI–visual input, VO–visual output. For all interesting points in the movement zone, the corresponding visual input is learned.

Till now, we have the position of the possible interesting points in the image. The next step is to find if an interest point corresponds to a teacher local view. Our system has no a priori knowledge of the visual shape of the teacher and it does not know if the teacher is present in its visual field. The only accessible information is that it must scan the interesting points, and probably, the teacher is near one or more focus points. We reject the idea of using color information (the teachers color is yellow) or an a priori shape (the teacher is a box) or any supervised learning algorithm (an operator indicate which focus point corresponds to the teacher or not) because all of these techniques are in contradiction with our goal : an autonomous learning architecture !

Our neural network only uses the single accessible information : the position of the movement area. The movement area is already computed by the proto-imitation mechanism and the position of this area is accessible. If the focus points are in this area then it is possible that these points belong to the teacher. And now, our algorithm can begin to discriminate the focus points.

In the learning stage, the student robot learns the *local views* centered on all the focus points situated inside of movement area. It local views are extracted after a log-polar transform and projected on the VI (Visual Input, see Fig. 3) map of neurons. This simulates the retinotopic projection of the retina on the primary cortical areas in the primate visual system. This preprocessing allows a weak scale and rotation invariance. The VIs are stored and recognized by the Visual Output (VO–a simple competitive structure). During the learning phase, the NN is assigned with a high global vigilance parameter (the same as describe in ART1). In the testing phase, the vigilance is set to a relative smaller value. The vigilance parameter, i.e. the switch between the learning and test phase, is modulated by the movement : the existence of the movement set the vigilance to a higher value and the absence of the movement to a smaller one.

If there is no detected movement then the system scans all the interesting points. The VO map will recognize the best matching between the learned local views and the scanned local views. The focus system scans all interesting points. For the current point, we compare the output of the AO map (see Fig. 5) with

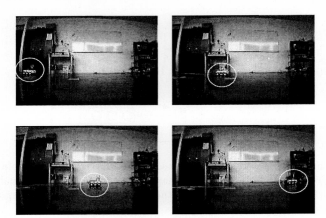

Fig. 4. Demonstration of recognition mechanism on still images of the teacher. The circle indicates where the teacher is recognized in the scene.

the the value of the best recognized local view. The value corresponding to the best recognized local view is stored in a buffer (B1 in Fig. 5). This value is set to zero for all new CCD sampled image. The output of the AO map is a confidence measure of the recognized local view correlated with an action. If the output of the AO map is higher that buffer value, i.e. the current local view is better recognized, then the position of the current interesting point is buffered in B2 (see Fig. 5) and the output of the AO is buffered in B1. After the scanning of all the points, the recognition system suggests the azimuth of the new movement according with the spatial location of the best recognized local view.

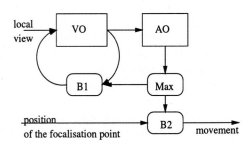

Fig. 5. The architecture of the action system. VO–visual output, AO–action output, B1 and B2–buffers, Max–maximum operator.

The association between a local view and an action is learned on the VO–AO links. The association uses a positive or negative reinforcement, which could affect further behaviors (is it necessary to follow this shape or not ?). Thus, although the movement detection cannot initiate any motor output of the student, recognizing a still teacher can permit to move towards it. The recognition of a still teacher induces the first movement in its direction. Fig. 4 shows an experiment where the teacher robot is still and the student robot is able to localize the teacher shape at different positions in a room.

3 Response integration

The global architecture (see Fig. 2) uses 2 ways in order to obtain the position of the teacher : the movement path and the local view path. The time to recognize the teacher robot with a sequential exploration of the steady image is longer that the time to detect moving areas. In practice, it is impossible to execute both tasks on the same computer in order to have a realistic execution time for the perception/action loop. Our choice was to run in parallel the two processes using the Parallel Virtual Machine (PVM) developed by Carnegie Mellon University.

A sketch of the parallel architecture is given in Fig. 6. Task2 performs the feature points extraction and shape recognition. Task1 performs data acquisition, movement detection and robot movement control. These two tasks are not fully synchronized in the sense that task1 is not waiting for the result of task2 for continuing its computation (non blocking reception). However, task2 needs the information from task1 for its computation, because it looks for the feature points only where movement is detected in the case there is a moving object. In the other case, the whole image must be processed. As expected, the parallel computation runs faster than the sequential one (see [7] for details). In average task2 runs faster than task1, mainly because of the time spent in the communications between the robot and the workstation.

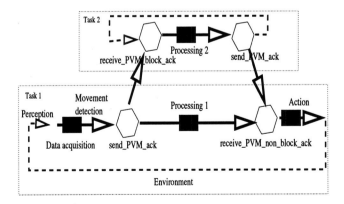

Fig. 6. Sketch of the computation done by the two tasks. Note that because the receive in task1 is non blocking, the computation effectively executed in parallel is not always what is marked as Processing1 and Processing2. Processing2 may be executed in task2 whereas task1 executes data acquisition and movement detection.

The main problem of two parallel processes was the integration of results of task1 and task2. How is it possible to integrate the results of non synchronized processes having different time constants? Neural field maps are a good solution to this problem. In a first time, we use them to perform the motor control as proposed by G. Schöner [8]. The motor group has to be a topological map of neurons using a dynamical integration of the input information to avoid forgetting the previously tracked target. A dynamical competition has also to be used to avoid intermittent switchings from a given target to another. The neural fields equations intensively studied in 70s can be a good solution to these problems. We will use the simplified formulation proposed and studied by Amari [1].

$$\tau \cdot \frac{f(x,t)}{dt} = -f(x,t) + I(x,t) + h + \int_{z \in V_x} w(z) \cdot g\left(f(x-z,t)\right) dz \quad (1)$$

Without inputs, the homogeneous pattern of the neural field, $f(x,t) = h$, is stable. The inputs of the system, $I(x,t)$, represent the stimuli information which excite the different regions of the neural field and τ is the relaxation rate of the system. $w(z)$ is the interaction kernel in the neural field activation. These lateral interactions ("excitatory" and "inhibitory") are modeled by a DOG function. V_x is the lateral interaction interval. $g(f(x,t))$ is the activity of the neuron x according to its potential $f(x,t)$.

One fundamental characteristic of the dynamic neuron models is their capability of memory and stable decision making. If the time constant are high enough events separated in time can be merged together and the decision can used both information without the draw back of classical linear filtering (no bifurcation capability when the external condition changes).

We tested our architecture in different conditions. The Fig. 7 shows the results obtain in a $38m^2$ room, using 2 Ultra Sparc 10 workstations. The student and the teacher robots are the Koala mobile robots using a CCD camera.

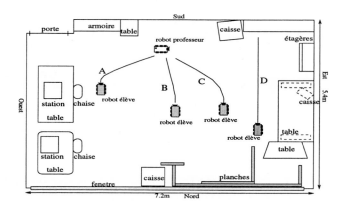

Fig. 7. Overview of one of the experiments.

In the learning phase, the student learns few feature points of the teacher. Then, the student robot is put at differents locations in the room. In order to test if the recognition system can initiate the imitative game, the teacher is stopped at the beginning (no movement is generated). In most of the cases (A,B and C trajectories in Fig. 7), the student recognizes the learned teacher and goes in its direction. But our recognition system is not perfect : the teacher can be not recognized (D trajectory in the Fig. 7). This happened when the local view of the the teacher is too different from the learned local view.

4 Conclusions

We showed, how a mobile robot can initiate an imitative game, based on the teacher visual shape learning. The recognition of the teacher is necessary for an autonomous learning mechanism. The complementarity between the movement information and the shape information is evident. The neural field is an amazing tool for information integration. We can imagine that we can integrate in our architecture anothers systems (for navigation, ...) by using the neural fields to integrate all the motor responses. In the same time, the PVM may be used in order to test more processes.

References

1. S. Amari. Dynamics of pattern formation in lateral-inhibition type neural fields. *Biological Cybernetics*, 27:77–87, 1977.
2. P. Andry, S. Moga, P. Gaussier, A. Revel, and J. Nadel. Imitation: learning and communication. In *The Society for Adaptive Behavior SAB'2000*, pages 353–362, Paris, 2000.
3. M. Boucart. *La reconnaissance des objets*. Presses universitaires de Grenoble, Grenoble, 1996.
4. P. Gaussier, S. Moga, J.P. Banquet, and M. Quoy. From perception-action loops to imitation processes: A bottom-up approach of learning by imitation. In *Socially Intelligent Agents*, pages 49–54, Boston, 1997. AAAI fall symposium.
5. P. Gaussier, S. Moga, M. Quoy, and J.P. Banquet. From perception-action loops to imitation processes: a bottom-up approach of learning by imitation. *Applied Artificial Intelligence*, 12(7-8):701–727, Oct-Dec 1998.
6. J. Nadel. *The Functionnal Use of Imitation in Preverbal Infants and Nonverbal Children With Autism*. Cambridge University Press, 2000.
7. M. Quoy, S. Moga, P. Gaussier, and A. Revel. Parallelization of neural networks using pvm. In J. Dongarra, P. Kacsuk, and N. Podhorszki, editors, *Recent Advances in Parallel Virtual Machine and Message Passing Interface*, pages 289–296, Berlin, 2000. Lecture Notes in Computer Science, Springer, no 1908.
8. G. Schöner, M. Dose, and C. Engels. Dynamics of behavior: theory and applications for autonomous robot architectures. *Robotics and Autonomous System*, 16(2-4):213–245, December 1995.

Towards an On-Line Neural Conditioning Model for Mobile Robots

Erol Şahin

Starlab Research Laboratories
Latour De Freins, Rue Engelandstraat 555, 1180 Brussels, Belgium
erol@starlab.net
http://www.starlab.org/

Abstract. This paper presents a neural conditioning model for on-line learning of behaviors on mobile robots. The model is based on Grossberg's neural model of conditioning as recently implemented by Chang and Gaudiano. It attempts to tackle some of the limitations of the original model by (1) using a temporal difference of the reinforcement to drive learning, (2) adding eligibility trace mechanisms to dissociate behavior generation from learning, (3) automatically categorizing sensor readings and (4) bootstrapping the learning process through the use of unconditioned responses. Preliminary results of the model that learn simple behaviors on a mobile robot simulator are presented.

1 Introduction

Mobile robots provide a challenging testbed for the development of neural models of learning. Neural models that are designed to learn using pre-recorded data sets or in unrealistic simulations, usually fail on mobile robots. Most of these models are plagued with implicit assumptions that are revealed when they are faced with the constraints of the real world. On-line learning is a fundamental constraint for learning algorithms on mobile robots. Once implemented on a robot, the model has to process a continuous stream of noisy data both for learning and testing in real time.

Animal learning research have unrevealed two basic forms of learning that allows animals to recognize the informative cues in their environment and take actions that increase their survival chance: Classical and operant conditioning. The classical conditioning can be illustrated by the following experiment: A hungry dog presented with food salivates. However, it does not salivate when a bell rings. Then the bell is rung prior to the presentation of food during several learning trials. After this, ringing of the bell alone yields salivation. Hence, classical conditioning enables the animal to recognize relevant stimuli in the environment. Here, food is called the *unconditioned stimulus* (UCS), salivation is called the *unconditioned response* (UCR), and the bell is called the *conditioned stimulus*. In operand conditioning, an animal learns to suppress behaviors leading to punishment, and exhibit behaviors leading to rewards.

2 Grossberg's neural conditioning model

In 1971, Grossberg [2] designed a detailed neural model to account for classical and operand conditioning in animals. By going through a thought experiment, Grossberg first set out the constraints, which he called psychological postulates, to be satisfied and then designed a minimal neural model to satisfy these. This model was then refined and studied in detail in later studies [3][4].

Recently Chang and Gaudiano [1] successfully implemented Grossberg's neural conditioning model to learn to generate avoidance/approach behaviors on two different robot platforms. This work follows up their work. Now we will first point out and discuss the limitations of their model and then propose a new model to tackle some of them. Note that some of these limitations are specific to Chang and Gaudiano's implementation whereas others are rooted in Grossberg's model.

– The model does not have a predictive nature. At a first glance, their results suggest that the model has a predictive nature. In the obstacle avoidance experiment, the model seemed to predict oncoming collisions with obstacles and avoid them by suppressing harmful behaviors in advance. However the model has no mechanism of making predictions since it could only learn the sensory cues that occur at the time of the collision *not* prior to the collision. This pseudo-predictive ability relies on the implicit assumption that the activation of sensory cues will get larger as the robot gets closer to a collision.
Note that, this limitation is specific to this implementation. Grossberg [3] had argued that the short term memory mechanism for the sensory cues would be able to keep the activity that occurred prior to the collision for learning. However, the mechanism that was proposed by Grossberg relies heavily on the dynamics of short term memory, and was omitted by Chang and Gaudiano.
– The model did not have a mechanism to create sensory cues (i.e. CS's) from raw sensor readings. Although raw sensor readings had sufficed for their experiments, the model needs to be extended to create sensory cues. Discovery of sensory cues should be done with feedback from the conditioning system.
– The model did not use a bootstrapping mechanism for learning. Motor commands were generated at random. Unconditioned response mechanism could not only not only naturally replace the random activation mechanism, but can also have the model learn on-line by ensuring that unconditioned responses would get take the robot out of bad states.

The following section describes the proposed model which tackles some of the problems mentioned above.

3 An on-line neural model of conditioning

In this section we outline the neural model that we propose to tackle the limitations of the original model.

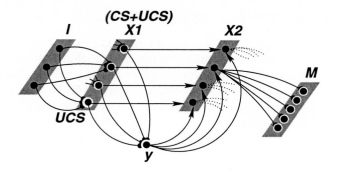

Fig. 1. Schematic description of the model.

The signals that causes activation and learning needs to be dissociated. Although this problem is not explicitly addressed in Grossberg's model, it is a well-known one and has been addressed by other conditioning models [6,5]. A different signalling mechanism, often called *eligibility traces*, needs to be used for enabling learning. Such a mechanism can sustain prior information about the activity of a neuron, without interfering with ongoing behavior generation. The cellular mechanisms that may implement such a mechanism in biological neurons through long lasting pre-synaptic changes are discussed in [6]

3.1 Behavior Generation

Figure 1 illustrates the structure of the neural model. The leftmost layer, I, indicates the sensory neurons. This layer is connected to the X_1 layer through weighted connections, Z_I. The neurons at the X_1 layer are divided into two groups: CS and UCS. The neurons in the CS group have adaptive weights and learn the prototypes of sensory patterns creating *learned sensory cues*. The UCS neurons represent the *innate* collision detectors. The activation of the neurons at this layer are given by

$$x_{1j}(t) = f_{CS/UCS}(\sum_i I_i(t) z_{Xj}(t)). \tag{1}$$

Here f_{CS} is a sigmoid function that bounds the activation of the CS neurons between 0 and 1. f_{UCS} is a hard threshold function that turns on when the activity of the corresponding input rises above the threshold signalling a collision.

The winner of X_1 layer, J, is then determined by $max(\{x_{1j}\}, v)$, where v is the vigilance parameter that sets the minimum level of neuron activity to win. Such a winner take all mechanism creates a mutually exclusive generation of behaviors.

The reinforcement neuron (named as the drive neuron by Grossberg), Y represents the reinforcement signal. The neuron is activated by the winner of the X_1 layer weighted by Z_y as

$$Y(t) = x_{1J}(t) z_Y(t) \tag{2}$$

If there are no winners at X_1 then Y is set to 0. Y energizes the learning through the rest of the model but does not affect the behavior generation.

At layer X_2 are the inputs from the X_1 layer and the Y converge. However this conversion is for learning only. During behavior generation, the winner take all activity of X_1 is merely copied into X_2.

Finally, the layer at the far right, M, represent the motor layer that control the robot. A one dimensional layer of neurons represent different angular velocities for the robot. The most active node generates a turn that it corresponds to. For instance, activating the leftmost node would generate a full left turn, whereas activation of the center node would move the robot straight ahead. A positive Gaussian shaped activation is placed at the center of this of the layer drives the robot straight ahead in the absence of any obstacles. The activation of neurons at this layer are as

$$m_l(t) = \exp{-(l-n)^2/\sigma^2} + \sum_k x_{2k}(t) z_{ml}(t). \qquad (3)$$

Here the first exponential term represents the Gaussian placed at the center of the layer (i.e. $n = 0$). The position of this Gaussian can be set to drive the robot to certain target. The second term is the summed inhibition caused through X_2 activation. When an obstacle is detected by the UCS or the CS neurons, the corresponding neuron at X_2 will become active and suppress the angular velocity nodes that may cause a collision.

3.2 Learning

Learning occurs when the change in the reinforcement Y is above a certain threshold. The reinforcement node releases an eligibility signal

$$Y^e(t) = Y(t) - Y(t-1) \qquad (4)$$

that enables learning throughout the model. The idea of using the time derivative of the reinforcement signal is central to whole class of conditioning models, for a review see [7])

The eligibility signal generated by the reinforcement node causes learning in three places. First, it is used for automatic categorization of sensory inputs into sensory cues at X_1. Initially all the CS nodes are *uncommitted* with all their incoming weights set to zero. If the eligibility signal received from Y is larger than a threshold (0.5), the eligibility of the X_1 neurons, $x_j^e(t) = x_j(t-1)$. are checked. If the the activation of the maximally eligible neuron (a CS node) is below the vigilance, v, then the next uncommitted neuron is employed. Incoming weights of the neuron are set to the normalized copy of the $I^e(t) = I(t-1)$. Here the eligibility mechanism is used for storing prior activations of the neurons, and learning them forms the basis for the predictive nature of the proposed mode. The normalization of weights during learning ensures that the closest CS node be maximally activated when a similar input comes. If the activation of the maximally eligible neuron is above the vigilance, no learning takes place.

Second, the weight of the maximally eligible X_1 neuron to Y is updated as

$$z_J(t) = z_J(t-1) + \gamma_1(\lambda Y - z_J(t-1)) \quad (5)$$

where γ_1 is the learning rate and $\lambda = 0.9$ is the discount rate of reinforcement in time.

Finally, the weights from X_2 to M are updated as follows to suppress the maximally active motor neuron:

$$z_{mkl}(t) = z_{mkl}(t-1) - \gamma_2 x_2^e(t) f_m(L) \quad (6)$$

where γ_2 is the learning rate, $f_m()$ is the suppression function that linearly suppresses the nodes from the L, the index of the maximally eligible node, to the center node.

4 Experimental Results

The model described above is implemented on the Webots mobile robot simulator (Cyberbotics SA, Switzerland) to control a simulated Khepera robot (K-team SA, Switzerland). The robot, shown in Figure 2-(a) is a miniature differential-drive robot with eight infrared proximity sensors. Only the six frontal sensors are used for this experiment. The two rear-facing infrareds are ignored.

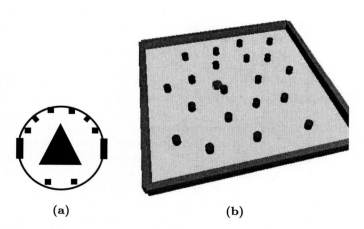

(a)   (b)

Fig. 2. (a) Top view of the Khepera, showing the placement of the infrared proximity sensors. (b) The environment of the simulated Khepera to learn obstacle avoidance.

Figure 2-(b) shows the environment of the simulated Khepera robot. The robot is left freely in this environment and no forms of supervision is done.

A gaussian activation centered at the motor map causes the robot to move in straight line unless an obstacle gets in its way. In this experiment six UCS neurons are used: each one detecting the collision on one of the sensors. The connection weights from the sensor layer are shown as bar graphs on the left side of Figure 3. The first three of the UCS nodes causes a full left turn as their UCR, whereas the remaining three causes a full right turn. The $UCS - UCR$ connections are the only innate connections in the model. Yet, the innate behaviors they produce are sufficient for bootstrapping the learning process. The weights of the first six sensory cues that were learned during the experiment are shown on the right side of Figure 3. It can be seen that they tend to be separate well in the sensory space and correspond to the sensory patterns that are prior to collisions.

Figure 4 plots the distance of the robot to the closest object with respect to time during learning. It can easily be seen that at the beginning, the robot tend to get closer to the objects, avoiding them through its innate UCR behaviors. As the learning process creates more predictive sensory cues and learns to activate the right avoidance behaviors, the robot is able to predict oncoming collisions and avoid the obstacles without getting too close.

Fig. 3. The connections weights from I layer to the first twelve nodes of X_1 layer is shown. On the left, six weight vectors show the innate connection weights to the six UCS nodes. On the right, first six weight vectors of the CS nodes are shown.

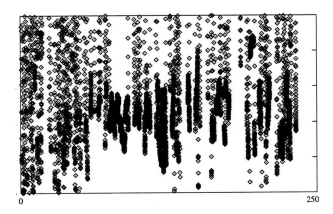

Fig. 4. The distance to the closest object as a function of time throughout learning.

5 Conclusions

We believe that neural models of learning have to run on-line on mobile robots. This requires that not only the core learning problem, but also seemingly peripheral problems like the sensory categorization, bootstrapping needs to be addressed. Grossberg's neural conditioning model creates a nice framework to study this. Although reinforcement learning, which was initiated through animal learning studies, provides a better computational analysis, we believe that a neural implementation of the same ideas can provide more insight to the problem.

References

1. Chang, C., Gaudiano, P.: Application of biological learning theories to mobile robot avoidance and approach behaviors. Journal of Complex Systems, **1**(1) (1998), 79–114.
2. Grossberg, S.: On the dynamics of operant conditioning. Journal of Theoretical Biology, **33**, (1971) 225–255.
3. Grossberg, S., Levine, D.: Neural dynamics of attentionally modulated Pavlovian conditioning: blocking, interstimulus interval, and secondary reinforcement. Applied Optics, **26**, (1987) 5015–5030.
4. Grossberg, S., Schmajuk, N. A.: A neural network architecture for attentionally-modulated Pavlovian conditioning: Conditioned reinforcement, inhibition and opponent processing. Psychobiology, **15**, (1987) 195–240.
5. Klopf, A. H.: The Hedonistic Neuron: A theory of memory, learning and intelligence. Hemisphere, Washington, D. C., (1982).
6. Sutton, R. S., Barto, A. G.: Toward a modern theory of adaptive networks: Expectation and prediction Psychological Review, **88**, (1981) 135–170.
7. Sutton, R. S., Barto, A. G.: A temporal-difference model of classical conditioning In Proceedings of the Ninth Conference of the Cognitive Science Society Hillsdale, NJ. Lawrence Erlbaum Associates, (1987).

A Thermocouple Model Based on Neural Networks

Nicolás Medrano-Marqués[1], Rafael del-Hoyo-Alonso[2], Bonifacio Martín-del-Brío[1]

[1] Departamento de Ingeniería Electrónica y Comunicaciones,
University of Zaragoza, Spain
{nmedrano, nenet}@posta.unizar.es
[2] Instituto Tecnológico de Aragón, María de Luna 7
50015 Zaragoza, Spain
rdelhoyo@ita.es

Abstract. Thermocouples are temperature sensors of common use in industrial applications. Classical mathematical models for these sensors consist of a set of two to four polynomial expressions reproducing their behaviour in different temperature ranges. In this work we propose a new 'one stage' model for these sensors covering the whole sensing range. The modelization has been carried out with an artificial neural network, which reproduces the sensor behaviour in the whole span, providing even better results than the standard piecewise model.

1 Introduction

Sensors represent the interface between real-world (physical) variables and processing systems. In order to design the electronics and programming of the processing system, an accurate model of the sensor behaviour is worthwhile [1, 2, 3].

A sensor (mathematical) model establish the physical_input/electrical_output characteristic of the sensor with some accuracy on a certain range. For instance, in the case of thermocouples (where a voltage is generated from the temperature difference of its two metal junction), the input-output function has been traditionally modelled by means of a set of n-th order polynomial expressions (their number and order depending on the thermocouple type [4, 5]).

In this paper a 'continuous' model for thermocouple sensors is presented. This model has been developed by means of a multilayer perceptron, which implements the mathematical relationship between volts and the corresponding temperature. This approach has three main features: i) only one expression defines the sensor behaviour in all the temperature range; ii) the neural model structure is the same for all kind of thermocouples, only changing the parameter values; and iii) the network parameters (weights) range in a far smaller span than that of the polynomial coefficients in the classical model.

This paper is structured as follows: In Section 2 a brief introduction to the applied neural model (the multilayer perceptron) is presented; Section 3 shows the classical polynomial-based model for thermocouples; Section 4 shows the proposed neural-based model and their results; in Section 5 a comparison of both modelling approaches is shown; finally, some conclusions of our work are presented.

2 Multilayer Perceptrons

Artificial neural networks (ANN) are computing systems based on the way the human brain processes the information [6, 7]. One of the main features of ANN is that these systems are able to learn input-output relationships from experimental data sets. ANN are said to be model-free, i.e., it is not necessary to assume a starting model function (linear, polynomial, exponential...) that relates the input-output data pairs.

In multilayer perceptron network (MLP), every connection has a weight that modifies the neuron input values; hence, an artificial neuron executes a weighted sum of the input data, followed by a non-linear mathematical operation, usually

$$y_i = \tanh(x_i) \quad or \quad y_i = \frac{1}{1+e^{-x_i}} \tag{1}$$

Where y_i is the output of the i-th neuron, and x_i the weighted sum of the neuron inputs:

$$x_i = \sum_j w_{ij} y'_j \tag{2}$$

In this expression, w_{ij} represents the weight connecting neuron j and neuron i, and y'_j is the output of the neuron j (and input of neurons in the next layer). Note that due to its non-linear characteristics, an ANN can approximate from slightly non-linear to highly non-linear functions.

The multilayer perceptron (MLP) [7] is the feed-forward neural network (data paths go from the network inputs to the outputs without feedback, Fig. 1) with a most number of applications, such as forecasting, optical character recognition, or function estimation. Sensor modelling can be considered a function estimation task, where the input data is the output of the sensor (voltage, current or resistance changes) and the output is the value of the measured physical magnitude. Several applications of neural networks to sensor modelling, linearization, etc., can be found in the literature [8, 9, 10].

3 Thermocouple Classical Models

Thermocouples are temperature sensors based on the Seebeck effect. In short, when two different metal or alloy wires are joined in a point a voltage is generated, which magnitude depends on the temperature of the junction. According to the metals or alloys used, there are several thermocouple types with different properties, such as

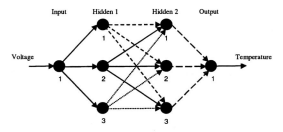

Fig. 1. Multilayer perceptron network with 1-3-3-1 architecture

temperature working range or behaviour. Every thermocouple type is identified by means of an uppercase letter, S, R, J, K, N, T, E and B [4, 5].

Classical thermocouple models consist of several 8th to 10th order polynomials that reproduce their behaviour (Figure 2 shows the polynomial V-T characteristic for a thermocouple of J type). Every polynomial represents the sensor response in a part of the whole temperature working range. For instance, Table 1 shows the polynomial coefficients for type J. As it is shown in that table, coefficient values varies from 10^3 to 10^{-31}, a very broad range that can difficult their use in some situations.

Differences between model output and real (physical) behaviour for the height thermocouple types are shown in Table 2.

4 Thermocouple Neural Model

In order to make the new model, and after having carried out several computer simulations, we have selected a 1-3-3-1 multilayer perceptron (one input neuron, two layers with three neurons every one, one output neuron, Fig. 1). Neurons (or processing elements) make use of the *tanh(x)* as non-linear output function (eqn. 1). The 22 weights of the 1-3-3-1 network are adjusted in the training process by means of voltage-temperature data presentations by using the Levenberg-Marquardt algorithm [11, 12]. Learning has been carried out by means of a set of 600 to 1400 points (depending on the sensor type) taken from the National Institute of Standards and Technology (NIST) tables [5]. Table 3 shows the obtained weights that model the J thermocouple behaviour. This new (neural) thermocouple model covers the whole sensor span (for instance, from −200 to +1200°C in a J thermocouple), giving the global errors shown in Table 4 for the different thermocouples.

In order to verify the generalisation capability of the model (whether errors are similar for patterns used and not used in the ANN development), a 10% of the data set randomly selected has been separated, training the network with the rest of the data. Figure 3 shows the network error for both training and test data sets. As Fig. 3 shows, errors corresponding to test patterns are similar to those of the training set, indicating a good generalisation ability of the proposed model.

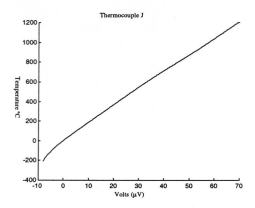

Fig. 2. V-T relation in a J type thermocouple

5 Models Comparison

We have compared the performance of the proposed ANN for thermocouple modelling to the classical polynomial approach. In Figure 4 are depicted errors due to classical polynomial piecewise model and those obtained in the proposed (continuous) neural model for every polynomial range. The figure shows that the proposed thermocouple model yields similar errors in all ranges, with only one mathematical model for all of them (it does not require dividing the temperature span in pieces for making a different modelling in every temperature interval, as is the case with the classical model).

Moreover, coefficients in the proposed model vary in a smaller range of numerical values. For instance, in the case of thermocouple type J (Table 3) weights range from 10^{-1} to 10^3, making easier their implementation in some cases (for instance, embedded systems with integer arithmetic). Similar weight ranges are found in the rest of thermocouple models.

Table 1. Polynomial coefficients for J thermocouple

	\multicolumn{9}{c}{Thermocouple J}								
	x^0	x^1	x^2	x^3	x^4	x^5	x^6	x^7	x^8
P1	0	1.95E-2	-1.22E-6	-1.07E-9	-5.90E-13	-1.72E-16	-2.81E-20	-2.39E-24	-8.38E-29
P2	0	1.97E-2	-2.00E-7	1.03E-11	-2.54E-16	3.58E-21	-5.34E-26	5.09E-31	0
P3	-3.11E3	3.00E-1	-9.94E-6	1.70E-10	-1.43E-15	4.73E-21	0	0	0

Table 2. Differences between thermocouple classical models and real behaviour (in degrees)

Type	S	R	J	K	N	T	E	B
Max. Error	1.18E-01	1.43E-1	6.47E-02	2.42E+01	2.44E+01	2.06E+01	2.04E+01	1.96E-1
Mean Error	2.55E-02	2.27E-2	1.48E-02	3.04E-01	3.09E-01	5.96E-01	2.97E-01	3.53E-2

Table 3. Weight values for neural network-based model for type J thermocouple

To	From	Input 1	Bias layer 1		
Hidden1 - PE1		-0.48571754	2.91642775		
Hidden1 - PE2		10.96377275	-14.63597918		
Hidden1 - PE3		2.12227842	-0.23630808		
		Hidden1 - PE1	Hidden1 - PE2	Hidden1 - PE3	Bias layer 2
Hidden2 - PE1		-7.33110103E+02	-1.79821331E+03	-1.33292341E+01	-1.05604092E+03
Hidden2 - PE2		1.98361282E+03	9.84495964E+02	3.78748841E+00	-9.87019334E+02
Hidden2 - PE3		3.24938477E+01	-1.24698233E+01	-8.95472278E-01	-4.01281737E+01
		Hidden2 - PE1	Hidden2 - PE2	Hidden2 - PE3	Bias layer 3
Output 1		-2.26484405E+02	-2.33476475E-01	-4.56929580E+02	6.83275727E+02

Table 4. Differences between the proposed neural model and real behaviour (in degrees)

Type	S	R	J	K	N	T	E	B
Max. Error	1.84E-01	2.44E-1	1.33E-01	1.81E+01	8.59E+00	4.20E-01	1.02E+00	2.13E-1
Mean Error	2.88E-02	2.43E-2	1.20E-02	1.65E-01	6.83E-02	3.08E-02	3.70E-02	3.47E-2

Comparing results presented in Tables 2 and 4, it is possible to check that the mean error is in general lower in the proposed model than in the classical piecewise one (although in both cases errors are very small). Table 5 shows the differences between both models and the real sensor behaviour at each polynomial range in degrees.

6 Conclusions

In this work a new continuous model for thermocouple sensors is presented. This model, based on the use of a multilayer perceptron, shows similar error values than the classical piecewise approximation. Otherwise, while in the classical approach the model structure (number and order of the polynomials) depends on the thermocouple (2, 3 or even 4 polynomials), the proposed model structure is the same for all types of them: the proposed 1-3-3-1 multilayer perceptron architecture modelizes every type of thermocouple. By changing the network weight values, one can achieve a very accurate model of the selected thermocouple, which covers the whole temperature working range. Thus, to change the model according to the temperature, as in the case of the classical piecewise model becomes unnecessary.

References

1. Webster, J.G.: The Measurement, Instrumentation and Sensors Handbook. CRC-Press, (1999)
2. Wisniewsky, M., Morawsky, R., Barwicz, A.: Modeling the Spectrometric Microtransducer. IEEE Transactions on Instrumentation and Measurement, vol. 48, no. 3, (1999) 747-752

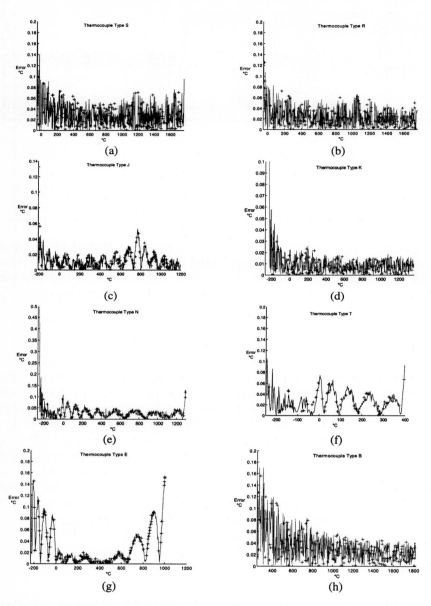

Fig. 3. Comparison between learning (continuous line) and test (crosses) data for thermocouples S (a), R (b), J (c), K (d), N (e), T (f), E (g) and B (h)

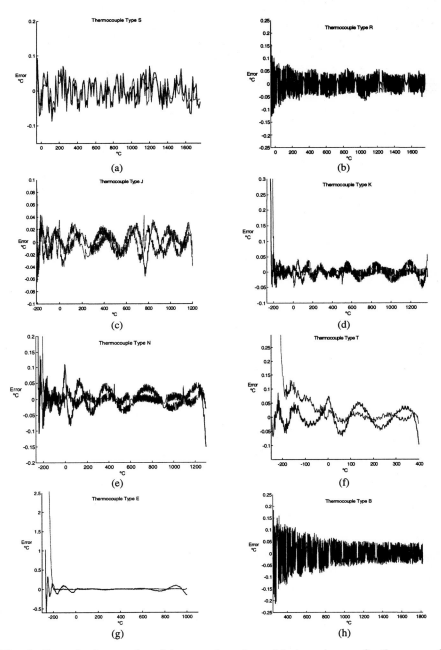

Fig. 4. Errors in the neural model versus the polynomial piecewise one for thermocouples S (a), R (b), J (c), K (d), N (e), T (f), E (g), B (h). Thick line corresponds to neural model; dotted thin line corresponds to polynomial model

Table 5. Differences between the neural and classical models per polynomial range compared to the real behaviour (in degrees)

Type	Section #1		Section #2		Section #3		Section #4		
	Pol. #1	Neural	Pol. #2	Neural	Pol. #3	Neural	Pol. #4	Neural	
S	0.1179	0.1840	0.0623	0.0743	0.0623	0.0744	0.0465	0.0952	Max. Error
	0.0394	0.0437	0.0245	0.0262	0.0233	0.0253	0.0233	0.0310	Mean Error
R	0.1434	0.2437	0.0533	0.0664	0.0359	0.0573	0.0374	0.0501	Max. Error
	0.0385	0.0393	0.0205	0.0231	0.0176	0.0180	0.0195	0.0191	Mean Error
J	0.0647	0.1327	0.0431	0.0562	0.0397	0.0205	-	-	Max. Error
	0.0155	0.0106	0.0141	0.0167	0.0140	0.0073	-	-	Mean Error
K	24.2	18.1	0.0506	0.0255	0.0574	0.0191	-	-	Max. Error
	1.76	0.9623	0.0152	0.0086	0.0178	0.0072	-	-	Mean Error
N	24.38	8.59	0.0451	0.0946	0.0394	0.1512	-	-	Max. Error
	1.7454	0.2707	0.0096	0.0297	0.0086	0.0231	-	-	Mean Error
T	20.63	0.42	0.0249	0.0968	-	-	-	-	Max. Error
	0.9944	0.0336	0.0073	0.0268	-	-	-	-	Mean Error
E	20.36	1.0242	0.0178	0.1533	-	-	-	-	Max. Error
	0.3759	0.0322	0.006	0.0549	-	-	-	-	Mean Error
B	0.1959	0.2125	0.0781	0.0828	-	-	-	-	Max. Error
	0.0578	0.0564	0.0263	0.0260	-	-	-	-	Mean Error

3. Dias Pereira, J., Postolache, O., Silva Girão, P.: A Temperature-Compensated System for Magnetic Field Measurements Based on Artificial Neural Networks. IEEE Transactions on Instrumentation and Measurement, vol. 47, no. 2, (1998), 494-498
4. The Omega Temperature Handbook. Omega Engineering Inc., (1998)
5. NIST web page: srdata.nist.gov/its90/main
6. Rojas, R.: Neural Networks, a Systematic Introduction. Springer-Verlag, (1996)
7. Haykin, S.: Neural Networks, a Comprehensive Foundation. Prentice Hall (2nd ed.), (1999)
8. Dempsey, G.L et al.: Control sensor linearization using artificial neural networks. Analog Integrated Circuits and Signal Processing, 13, (1997) 321-332
9. Massicotte, D., Legendre, S., Barwicz, A.: Neural-Network-Based Method of Calibration and Measure Reconstruction for a High-Pressure Measuring System. IEEE Transactions on Instrumentation and Measurement, vol. 47, no. 2, (1998) 362-370
10. Medrano-Marqués, N.J., Martín-del-Brío, B.: A General Method for Sensor Linearization Based on Neural Networks. Proc. of the 2000 IEEE International Symposium on Circuits and Systems ISCAS'2000. Geneva, Switzerland, (2000) 497-500
11. Hagan, M., Mohammad, B.: Training Feedforward Networks with the Marquardt Algorithm. IEEE Transactions on Neural Networks, vol. 5 no. 6, (1994) 989-993
12. Neural Networks Toolbox User's Guide, MatLAB Reference Guide, (1997)

Improving Biological Sequence Property Distances by Using a Genetic Algorithm

Olga M. Perez[1], F. J. Marin[2], and O. Trelles[1]

[1] Computer Architecture Department, University of Malaga,
29080 Malaga, Spain
{olga,ots}@ac.uma.es
[2] Electronic Department, University of Malaga,
29080 Malaga, Spain
fjmarin@uma.es

Abstract. In this work we present a genetic-algorithm-based approach to optimise weighted distance measurements from compositional and physical-chemical properties of biological sequences that allow a significant reduction of the computational cost associated to the distance evaluation, while maintaining a high accuracy when comparing with traditional methodologies. The strategy has a generic and parametric formulation and exhaustive tests have been performed to shown its adaptability to optimise the weights over different compositions of sequence characteristics. These fast-evaluation distances can be used to deal with large set of sequences as is nowadays imperative, and appear as an important alternative to the traditional and expensive pairwise sequence similarity criterions.

1 Introduction

The explosive growth of the number of biological sequences registered in data bases, stimulated by Genome Projects (such as the recently announced Human Genome) has modified the framework of several applications on the biological sequence analysis area. In most cases this new scenario is characterised by studies over large sets of sequences, which generates an enormous increase in the computational resource demand of the tools employed for their study, suggesting the need for new fast approaches, while maintaining the effectiveness of the methods.

A good example may be found in those applications that deal with large data sets, such as clustering and classification of sequences. Normally, these type of procedures include the repetitive task of computing the *distance* between each pair of sequences, followed by several data grouping stages. For most exhaustive algorithms [8][11] the computational demand in the calculation of *distance* grows with the square of the sequence lengths. Even though it could be thought that using heuristic approaches such as FASTA [9], or BLAST [1] would sufficiently expand the size of the problem that can be tackled, the increase in the computational demand, is, however, so high that we must complement this approach with other faster solutions.

Several authors have developed methods to compare sequences based on its amino-acid content, represented as different n-grams frequencies [14][15] in the line to reduce the computational cost. Other interesting approaches express the protein description in terms of general properties [4], aiming to exploit maximally the information not only present in the sequence but also in structure and other sources of biological related information (i.e., sequence length, aromatic rate or hydrophobic residues, etc.).

The underlying idea is that computing distances between sequences using those properties would significant reduce the computational cost of calculations. In this way routine clustering of large data sets, and even of whole data bases will be greatly facilitated, with a final outcome that reproduce closely already accepted biological results.

The general case when computing distance based on sequence properties is working over a multi-dimensional and heterogeneous property-space. The general weighted distance for two sequences i and j is defined as $D_{ij} = f(w, s_i, s_j)$ where w is the weight vector, s_i and s_j are the property *vector* of sequences *i*, and *j* respectively and *f* represents the distance function.

The key element in this type of approaches is the determination of the best combination of property weights (the w vector) to produce optimal distance measurements. We will describe a generic application that allows to explore the solution space through a genetic algorithm, and will illustrate the proposed approach with results from the optimisation of the weights for different property vectors over several data-sets of biological sequences. Along these tests, we will show how our proposal produce significant improvements on the distance measurements, that can be used to generate sequence groups that correspond to biologically meaningful protein subsets.

2 System and Methods

Genetic Algorithms (GA) are stochastic global search methods that mimic the natural biological evolution by means or working over a population of potential solutions applying the principle of survival of the fittest to produce each time better approximations to a solution [5].

At each generation, a new set of approximations is created by the process of selecting individuals according to their fitness in the application domain, and breading them together using operators borrowed from natural genetics. In our case, individuals represent putative solutions for the problem of optimise the property weights for a given distance function. We assume to have a collection of real property vectors (*pFile*) for which the clustering solution is known (*rFile*). Our goal can be stated in the following simple terms: search the property weights that optimise a given fitness function.

To assess how potential solutions perform in the application domain, we have implemented different indices that supply a measure of the clustering-quality: the *silhouette* coefficient [10] (implemented as default alternative), BD-index [6], Minkowski measure [12] and Jacard coefficient [2].

The power of the proposed algorithm lies in the fact that it has been developed as a generic and parametric genetic algorithm (*pGA*) that allows the optimisation of solutions for a broad diversity of property vectors combined by different distance functions. A pseudo-code outline of the *pGA* is shown in Table 1 with the relevant parameters (see Table 2).

Table 1. Pseudo-code that outline the parametric genetic algorithm described.

```
[1]    GetParameters ( pFile, rFile, I, N, G, M, C, S)
[2]    CreateInitialPopulation P_{g=0}(I,N);
[3]    for generation g=1 to G, do {
[4]        EvaluatePopulation(P_{g-1}, I);
[5]        Select (P_g from P_{g-1}, S)
[6]        Crossover(P_g, C);
[7]        Mutate(P_g, M);
[8]        if (ElitistStrategy) Elitist(P_g, E);
       }
```

Table 2. Application main parameters

Parameter	Description	Default value
N	Number of components in the property vector	None
I	Number of individuals in population	100
G	Number of generations	100
M	Mutation rate	1/100
S	Selection rate	1/2
C	Crossover rate	6/10
E	Elitism rate	5/100
pFile	Collection of real application property vectors	None
rFile	Clustering solution file	None

A population P_0 of *I* individuals representing property weight vectors (with *N* components), is initially set (step 2) with random real values in the range [0,1], satisfying:

$$\sum_{j=1}^{N} P_{i,j} = 1 ... \forall i = 1...I \quad (1)$$

In step 4 the objective function is used to evaluate each potential solution. We have implemented the notion of average *silhouette* value of a cluster [10] that reports the goodness of a clustering solution, that depends, in general, on how close the elements of a cluster are between them, and how far they are from the next closest cluster.

The family average silhouette value is used as a measure of the classification quality performed by each individual. This value lies between –1 and 1, where values near 1 represent a good classification while values that fall under 0 are accepted as badly classified (in fact, it is on average closer to members of some other cluster than the one to which it is assigned).

The need of an appropriated distance function is implicit in the evaluation procedure described in the previous paragraph. Our application implements two basic distance functions. The first one is the squared Euclidean distance which weighted formula for two property sequence vectors is:

$$D_{ij} = \sum_{k=1}^{n} w_k (s_{ik} - s_{jk})^2 \qquad (2)$$

where w_k is the weight for property k, s_{ik} is the property k of sequence i, and s_{jk} is the property k of sequence j. The second distance function we have implemented is one of the most used similarity metrics, the Pearson coefficient [9], that compute the distance for correlation metrics using the formula:

$$D_{ij} = 1 - \left(\frac{1}{\sum_{k=1}^{n} w_k} \sum_{k=1}^{n} \frac{w_k (s_{ik} - \overline{s_i})(s_{jk} - \overline{s_j})}{\sigma_{s_i} \sigma_{s_j}} \right) \qquad (3)$$

where w_k is the weight for property k, s_{ik} is the property k of sequence i, s_{jk} is the property k of sequence j, $\overline{s_i}$ is the average value of s_i , $\overline{s_j}$ is the average value of s_j and $\sigma_{s_i}, \sigma_{s_j}$ are the standard deviation of s_i and s_j respectively.

Once computed the fitness value for each individual in the population, the algorithm proceeds selecting individuals for reproduction (step 5) on the basis of their relative fitness. A maximal reproduction rate constrain for individual performance in one generation has been included to avoid highly fit individuals in early generations to dominate the reproduction causing rapid convergence to probably sub-optimal solutions.

The default selection strategy works by a *roulette wheel* mechanism to probabilistically select individuals based on the relative fitness value associated to each individual, in such a way that the number of offspring that an individual will produce is expected to be proportional to its performance.

An alternative selection procedure based on stochastic universal sampling, in which individuals are selected for reproduction with the same probability, has also been implemented. Both selection procedures can be switched along generations to allow a broader inspection of the solution space.

In step 6 selected *I/2* couple of individuals are mutually crossed to produce *I* new individuals that have part of both parent's genetic material. The number (to optionally allow multi-point crossover) and length of the fragments to be interchanged between two individuals is controlled by parameters, and because we are working with real-coded chromosomes that must satisfy Eq.1, both fragments must be normalised to maintain the constraint.

Finally, mutation is applied in step 7, to produce a new genetic structure. The role of mutation is providing a guarantee that the probability of searching any given solution will never be zero, and acting as a safety net to recover good genetic material that may be lost through the action of selection and crossover [3]. In our case, mutation is applied over two randomly selected gene positions (j and k) on the individual i (selected by using the mutation rate M) and its achieved by perturbing the P_{ij} and P_{ik} positions with a Δw value in the range $\max(-P_{ij}, P_{ik}-1) \leq \Delta w \leq \min(1 - P_{ij}, P_{ik})$ to satisfy Eq.1.

At this point a new population has been produced by selection, crossover and mutation of individuals from the old population, and a new generation can start with the fitness evaluation for this new population. In some cases it is desirable to deterministically maintain one o more of the best fitted individuals through the use of an *elitist strategy*, controlled by E parameter in step 8.

The algorithm works during G iterations. This ending alternative has been chosen because in most cases is difficult to formally specify a convergence criteria (the fitness may remain static for a number of generations before a new superior individual is found).

We conclude this section indicating that there is no algorithmic constraint with respect to the property vector employed. In particular, our approach may accommodate either homogeneous or heterogeneous compositions in the ranges the user wants to run depending on the type of analyses to be carried out.

3 Results

In this section we demonstrate the functionality for the proposed algorithm on two tests. Both are related with the clustering of biological sequences based on different sequence properties: di-peptide frequencies for complete and reduced amino-acid alphabets, compositional and physical-chemical properties.

To this end, several groups of sequences have been selected and clustered by using similarity distance [13], in such a way that biological well accepted results are available (clustering information forms *rFile* parameter). The idea is performing the clustering again, but now using fast-computing distances and evaluate how good the quality of the new outcome is. These fast-computing distances are based on weighted sequence properties, that will be optimised by using the proposed pGA.

We start with the set termed T*est-1* which is composed by 48 sequences belonging to five different protein families (see Table 3). The by-similarity distance (1.-similarity) has been computed using dynamic programming algorithm [11] with pam250 as score matrix, 12 and 4 as opening and extending gap penalties. The *silhouette* value gets a value of: 0.3659 that will be used as reference in this test.

Table 3. Protein sequence code for Test-1

Catalase	Citrate	Globin	Histone	G-proteins
CAT1_LYCES	CISW_BACSU	GB11_BOVIN	H2A1_HUMAN	HBA1_BOSMU
CATA_BOVIN	CISX_YEAST	GB13_MOUSE	H2A1_PEA	HBA1_IGUIG
CATA_CANTR	CISY_ACEAC	GB15_MOUSE	H2A1_TETPY	HBA1_PLEWA
CATA_DROME	CISY_ACIAN	GBA1_CAEEL	H2A1_YEAST	HBA1_XENBO
CATA_HAEIN	CISY_ECOLI	GBA1_COPCO	H2A2_XENLA	HBA3_PLEWA
CATA_MICLU	CISY_MYCSM	GBA1_DICDI	H2A3_VOLCA	HBAD_PASMO
CATA_PICAN	CISY_PIG	GBA2_DICDI	H2AM_RAT	HBAT_HORSE
CATA_PROMI	CISY_PSEAE	GBA2_NEUCR	H2AV_CHICK	HBAZ_CAPHI
CATA_YEAST	CISY_THEAC	GBQ1_DROME	H2A_EUGGR	HBA_CATCL
	CISZ_BACSU		H2A_PLAFA	HBA_LIOMI

A protein is a lineal sequence of amino-acids that are mathematically represented as a string of characters that belongs to the amino-acid alphabet (20 residues). The per-residue frequency of a sequence can be computed and combined in k-tuples or n-grams (n-consecutive residues) to encode the sequence information [14]. In this first test, the 2-gram frequency (a 400 properties vector) will be used to represent each sequence. When using an homogeneous weighting schema ($w_i=1/400$ being $i=1\ldots400$), the *silhouette* value is 0.0901 (Squared Euclidean distance) .

Now, we perform the optimisation of the weighting schema using pGA, with default parameters (100 individuals, and 100 generations). In Figure 1 the evolution of the fitness value is shown, with improvements in the *silhouette* value of more than two fold (up to 0,2102 using Eq.2 distance).

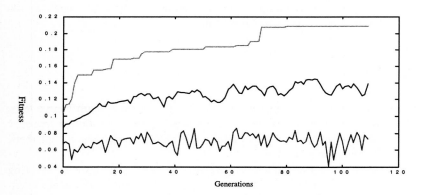

Fig. 1. Evolution of the *silhouette* value in Test-1, for the best fitted individual, average and worst. The observed variability in the value of the worst chromosome is a good indicator of the broad exploration of the solution space.

The 20-letter alphabet used can be regrouped or reduced to emphasize the diverse physicochemical properties of the molecular residues and maximise information extraction. For instance the original protein sequence can be represented by a six-letter alphabet based on the hydrophobicity scale [7] producing a new vector with 36 properties. Alternatives reductions with 11 and 8-letters alphabets have also been used (see Table 3). In all cases important improvements are observed comparing the *silhouette* value for clustering performed based on uniform weighted schemas and for the same clusters when using the optimised weights. In some cases, the goodness of the cluster-quality is even better than those obtained with exhaustive similarity distances.

In the next case, designated as *Test-2*, we will present the study over a set with 42 close-related sequences grouped in 5 clusters. It is interesting to study the behaviour of the algorithm under these conditions, since they represent one of the less favourable cases for our strategy to improve clustering quality.

In this case we have used a non-homogeneous property vector formed by the following sequence characteristics: sequence length, molecular weight, hydrophoby ratio, aromatic ratio, disulphide, low complexity region and trasmembrane region. Properties have been expressed in percent sequence length, except for molecular weight and sequence length (both as their logarithm). Each property is normalized by dividing by its standard deviation. Results obtained for the optimisation are summarised in Table 4, in which an important improvement is observed using either Euclidean and Pearson correlation distances.

DataSets	Properties	N	Dist. Fun.	Fitness value		
				Similarity	Uniform weights	pGA
Test-1	2-gram	400	SEW	0.3659	0.0901	0.2547
		400	PWC		0.2226	0.3629
		121	SEW		0.2012	0.3219
		121	PWC		0.2681	0.3777
		64	SEW		0.2605	0.3637
		64	PWC		0.2420	0.3415
		36	SEW		0.3368	0.4766
		36	PWC		0.3561	0.4731
Test-2	Features	7	SEW	0.5047	0.4019	0.4720
		7	PWC		0.4260	0.4939

Table 4. Results are shown for the data sets termed: Test-1 and Test-2, using 2-gram frequencies and physical-chemical sequence features. N column represent the length of the property vector used in each experiment. Two distance functions have been tested: squared Euclidean weighted distance (SEW) and Pearson weighted correlation (PWC). Similarity represent the reference value obtained by exhaustive distance methods, the next column represent uniform weighting schema and the pGA column contains the optimised values.

4 Discussion

In this work we present an application that can be applied to optimise the weighting schema used to compute property sequence distances. The application has been tested with diverse types of property vectors, achieving in all cases a substantial improvement in the quality of the measures. Our algorithm should be considered as a sort of automatic tool to determine the better combinations of weights when property vectors (homogeneous or heterogeneous) are expected to be used to compute clustering distances at an affordable computational cost.

The approach proposed in this work has also shown that, despite the reduction in the calculations that it achieves, it reproduces the results provided by other traditional biological sequence distance methods.

Acknowledgement

This work has been partially supported by grant 1FD97-0372 from the EU-FEDER Programme (Fondos Europeos de Desarrollo Regional).

References

1. Altschul, S.F., Madden T.L., Schaffer A.A., Zhang J., Zhang Z., Miller W. and Lipman D.J., (1997) "Gapped BLAST and PSI-BLAST: A new Generation of Protein DB search Programs", Nucleid Acids Research (1997) v.25, n.17 3389-3402
2. Everitt, B. (1993), "Cluster analysis", London: Edward Arnold, third edition.
3. Golberg, D.E., (1989), "Genetic Algorithms in Search, Optimisation and Machine Learning", Addison Wesley Publishing Company.
4. Hobohm, U. & Sander, C. (1995) "A sequence property approach to searching protein databases" J. Mol. Biol.251, 390-399.
5. Holland, J.H. (1975), Adaptation in natural and artificial systems.The University of Michigan Press.
6. Jain, A,K, and Dubes, R.L. (1998), "Algorithms for clustering data", Prentice-Hall
7. Nakata, K., (1995), "Prediction of zinc fingers DNA binding protein", Computer Application in the Biosciences, 11, 125-131
8. Needleman S.B. and Wunsch C.D. (1970) "A general method applicable to the search for similarities in the amino acid sequence of two proteins". J.Mol.Biol 48, 443-453
9. Pearson W.R. and Lipman D.J.; (1988), "Improved tools for biological sequence comparison", Proc.Natl,Acad.Sci. USA (85), 2444-2448
10. Rousseeuw, P.J. (1987) "Silhouettes: A graphical aid to the interpretations and validation of cluster analysis". J. of Computational and Applied mathematics,20:53-65.
11. Smith T.F. and Waterman M.S.(1981). "Comparison of Biosequences". Adv.in Aplied Maths, (2), 482-489
12. Sokal , R.R. (1977), "Clustering and classification: background and current directions", In Van Ryzin, J. ed., Classification and Clustering, 1-15, Acad. Press.
13. Trelles O., Andrade M.A., Valencia A., Zapata L., and Carazo J.M. (1998),"Computational Space Reduction and Parallelization of a new Clustering Approach for Large Groups of Sequences", BioInformatics v14 n5 439-451
14. Wu, C., Berry, M., Shivakumar, S. And Mclarty, J, (1995), "Neural Networks for Full-Scale Sequence Classification: Sequence Encoding with Singular Value Decomposition" machine Learning. 21, 177-193.
15. Wu, C. (1997) "Artificial neural networks for molecular sequence analysis" Computers Chem. Vol. 21, No. 4, 237-256

Data Mining Applied to Irrigation Water Management*

Juan A. Botía, Antonio F. Gómez-Skarmeta, Mercedes Valdés, and Antonio Padilla

Departamento de Ingeniería de la Información y las Comunicaciones,
Universidad de Murcia. Campus de Espinardo. E-30071 Murcia, Spain.
{juanbot, skarmeta, mvaldes, apadilla}@um.es

Abstract. This work addresses the application of data mining to obtain artificial neural network based models for the application in water management during crops irrigation. This problem is very important in the zone of the South-East of Spain, as there is an important lack of rainfall there. These intelligent analysis techniques are used in order to optimize the consumption of such an appreciated and limited resource.

1 Introduction

Data mining [4] in general and neural networks [6] in particular have been succesfully applied in the agricultural environment [5, 8]. The authors have previous experiences related with this issue [1], although using a rough and preliminary approach. The present work is dedicated to a concrete FAO (Food and Agricultural Organization) model which established the dependiencies between agroclimatical factors and expected amount of water consumption in a normal crop irrigation process. This is a mathematical and generic idealization of the world. As an alternative, we propposse the use of data mining techniques to induce a more concrete theory, for cultivations located in the Murcia region, in the South-East of Spain. In section 2 the problem of modeling irrigation water needs using one particular classic manner is introduced. Also, the necessity for a more precisse method is pointed out. Then in section 3, the available data is analysed and some initial conclusions that help in the design of modeling experiments are outlined. Section 4 is dedicated to the development of learning experiments performed, and to the analysis of results. Finally, in section 5, conclusions are given.

2 The Problem: Predicting Irrigation Water Needs

The prediction of irrigation water necessities is a key issue for a correct management of that resource. The problem is more important yet if it regards to crops

---

* Work supported by the European Comission through the FEDER 1FD97-0255-C03-01 project

in zones of the world that have a very low availability of water dedicated to agriculture. That is precissely the case of the south-east of Spain, the geographical zone in which this work is located.

The ET_0 is the reference crop evapotranspiration. This is the central parameter for the estimation of crop water need as it gives the amount of water that is lost from soil evaporation and plant transpiration. This is precissely the water that has to be put back to the cultivation. The ET_0 depends on the particular climate conditions and the kind of crop that is being used. Once the ET_0 is obtained, then a factor K_c is applied to it, in order to adjust the value to a concrete crop, in a concrete growth stage (i.e. fully grown crops need more water than crops just planted). So, the ET_{crop} factor is obtained, by means of $ET_{crop} = K_c \times ET_0$. Finally, irriagation water needed, Iw is obtained from the difference of ET_{crop} and effective rainfaill which is the amount of rain water useful for the crop, $P_{effective}$. The final expression is $Iw = ET_{crop} - P_{effective}$.

2.1 Class A Bucket Model

This work is focused in the estimation of the ET_0 parameter. This parameter has been traditionaly estimated from mathematical models. These models take into account agroclimatical factors that influence water comsumption by crops. In this work, the **Class A Bucket model** (CABM) will be studied. Its basic expression has a very simple formulation: $ET_0 = K_p \times E_0$ where K_p is obtained from the expression

$$K_p = a_0 + a_1 U + a_2 \overline{H}_r + a_3 d + a_4 \overline{H}_r^2 + a_5 d^2 + a_6 U \overline{H}_r^2 + a_7 d \overline{H}_r^2 \qquad (1)$$

where a_i, $1 \leq i \leq 7$ and d are ideally obtained coefficients. U is the mean wind speed, and $\overline{H}_r$ is the mean relative humidity. E_0 is the evaporation water, each day, of a calibrated bucket, located near the crop. K_p is a coefficient that depends on the bucket.

This method is, by its dependency of physical elements, very imprecisse. It has been used, in Murcia, inside a distributed information system, which is compound basically of 64 agroclimatical stations (i. e. hardware devices) that periodically take some meassures fron the environment, and a central server that gathers all that information, making it available through a relational database management system. Agroclimatical meassures are obtained, hourly, from the stations and sent to the server through a communication network. This data is aggregated into a DAY table. Also, data in the DAY table is aggregated and put into a WEEK relation. And this weekly data is aggregated also into a MONTH relation. This relation is also aggregated into a YEAR table. As it can be seen from equation 1, the ET_0 parameter can be estimated for a crop, using only $\overline{H}_r$, U and E_0 which are, precisely, parameters meassured from each station. A concrete view from the DAY table shows tuples compound by the attributes $ESTCOD$, $JULIAN$, $\overline{H}_r$, U, E_0, and ET_0. $ESTCOD$ is the code of the station that took the meassure and $JULIAN$ is the julian day since the first of january of 1970. The ET_0 value that appears in the tuple is precissely the estimated value, by means of the CABM. This method can become to be very inexact, at the point that many outliers could be produced (e. g. negative values for the ET_0). An expert supersees the weekly produced values, in order to appropiately correct misleading estimations. Obviously, there is a need here for a more accurate method for the estimation of the ET_0 parameter.

2.2 Backpropagation Learning Over a Sigmoidal One Hidden Layer Network

What is propossed in the context of this work, is the use of data mining. Particulary a simple sigmoidal neural network, with backpropagation learning, to obtain a model that is suited to this kind of problem.

A backpropagation learning multilayer perceptron is an adaptative network whose nodes perform the same function on incoming signals. This node function is usually a composite of the weighted sum and a differentiable nonlinear activation function, also known as the transfer function. Two of the most commonly used activation functions are the logistic function and the hyperbolic tangent function. Both functions approximate the signum and step functions respectively, and yet provide nonzero derivatives. In this case, it has been chosen the sigmoidal function. As output data is not limited to the $[0, 1]$ interval, it havs been normalized, so that it fits into $[0.1, 0.9]$.

The backpropagation learning rule is based on the derivatives of the error function with respect to the parameters to adjust. The adjust for the weight of the arc that comes from the node k and goes to the node i is $\Delta w_{ki} = -\eta \bar{\epsilon}_i x_i$ where $\bar{\epsilon}_i$ has the expresion

$$\bar{\epsilon}_i = o_i(1 - o_i)(t_i - o_i)$$

if the node is on the output layer and

$$\bar{\epsilon}_i = o_i(1 - o_i) \sum_{k \in O} w_{ik} \epsilon_k$$

if it is located in the hidden layer, where o_i is the output of the node i and t_i is the expected output. The O is the set of output node indexes connected to the hidden node i. A detailed study of this issue can be found in [7].

3 The Data

The information system addressed in section 2 was put into work in 1994. In that year, there were only a few stations working. Year by year, more stations have been added, disseminated through Murcia. There are 62 stations working nowadays. They meassure values form $\overline{H}_r$, U and E_0. The CABM can be represented as a function:

$$f_{cab} : \overline{H}_r \times U \times E_0 \to ET_0$$

Hence, the problem can be reduced to obtain a $\hat{f}_{cab}$ to approximate both, the correct behaviour of the CABM in some cases, and also the knowledge of the expert in others. The data from the daily table will be used for this task. It seems the most sutiable for that model. Next sections are dedicated to a preliminar study of the data in order to desing data mining experiments as best as possible.

3.1 Strong Relation between Attributes

If some care is taken in studying some statistics, like correlations, between attributes in the data set, it is soon discovered that the E_0 and the ET_0 show a very strong positive correlation coefficient, 0.99118. This shows that a mathematical expression modeling the relation between this two attributes would be very similar to a linear model. In fact, if we represent E_0 and ET_0 in a graph, obtainning the figure 1, we can see that relation between them is almost linear. A linear model could show a good modeling behaviour of f_{cab} due to this. It will be checked in section 4.1.

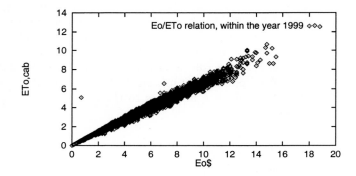

Fig. 1. Lineal relation between the E_0 and the $ET_{0,cab}$ parameter.

3.2 Null Values

It is necessary to study null values, in order to obtain correct data theories [3]. It is possible that values not present in a tuple, for a concrete attribute, could be correctly approximated and possible important facts hidden in the data would not be lost. In this manner, produced models would have more coverage of the problem.

In data coming from table DAY, there is a considerable proportion of non valid values. The absence of correct values is due, mainly, to temporary missfunction of stations. In table DAY there is a total of 53533 tuples. 30958 of that tuples are valid, and the rest have nulls. Both U and $\overline{H}_r$ attributes present less than 8% of null values. However, E_0 has 35% of its values with nulls. Hence, null values are a real problem in this context. Even, it could be possible that this data could not be useful in order to obtain $\hat{f}_{cab}$. But before pointing out wrong conclusions, some facts can be asserted. Both percentajes of null values for U and $\overline{H}_r$ are not so important. But, a 35% of null values for the E_0 parameter must be considered. Clearly, solutions ought to be found for the reconstruction of that values for E_0. A possible solution could be to use the mean of all values for E_0. However, the value of skewness for that parameter, 26.001, suggest that the poblation of values of E_0 as a random variable is not normally distributed.

3.3 Taking Values from Similar Dates, in the Same Station

However, another type of values reconstruction could be addressed here. It is reasonably to think that an E_0 value can be strongly related to another value, taken from the same station, the same date of another year (before or after). If the values of E_0 are represented through the days they were taken, from a concrete estation, then the (a) graph of figure 2 is obtained.

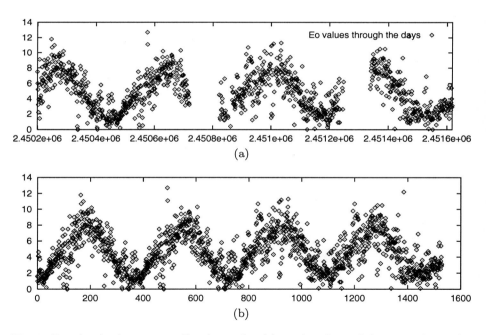

Fig. 2. Result of substituing null values, plot (a), with values of the same date, plot (b), the year before.

Substituting null values from other year, the same day, results in the (b) graph, in the same figure. Obviously, this methods benefits from the goodness of that data that shows an important regularity through the years. Using this method, only 608 tuples have been restored. There are still 12923 tuples from stations that have a percentaje of nulls bigger that it would be suitable for this method to be applied. Next section will address that problem.

3.4 Taking Values from Similar Dates, in Other Station

There are 38 stations that have an amount of nulls bigger than a 15%. The rest of them, that have correct or recovered values with the method explained in the former section, can be used as reference stations. If station E_a is the reference station of station E_b, that means that E_a is very similar, in terms of values for U, $\overline{H}_r$ and E_0 meassured in similar dates. Hence, E_a can be used to give values to E_b. The similarity meassure we have used has been the euclidean distance with the expression $d(E_a, E_b) = \sum_{i=1}^{n} \sqrt{(\hat{H}^i_{r,E_a} - \hat{H}^i_{r,E_b})^2 + (U^i_{E_a} - U^i_{E_b})^2}$ where i is the julian day in which the meassure was taken. In figure 3 the first six biggest clusters obtained with that criteria appear. Points labeled with + denote the surface of Murcia. The rest of points refer to distinct clusters. Notice that stations that are geographically near to a valid station are associated with it. The cluster of the main station AL41 is a good example of that fact.

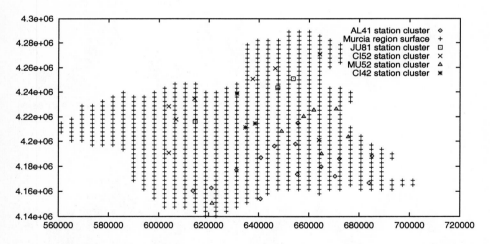

Fig. 3. Associations obtained, represented geographycally through the Murcia Region

4 Learning Experiments

In this section, learning experiments performed to obtain the best $\hat{f}_{cab}$, will be described. It has to be mentioned that, in order to compare all models, a cross-validation approach has been used, applying for all experiments, the same validation data set. The tuples of this set are all tuples, with valid values, from the 1998 year of the day table. This set is compound of 8095 tuples. Trainning proportion of the whole data set of each experiment must be adjusted, in order to match the fixed number of tuples for evaluation.

4.1 A Linear Regression Model

As it was remarked in section 3.1, relation between E_0 and ET_0 parameters is very strong. Hence, the possible suitability of a regression model must be checked. The model obtained with least squares method is

$$\hat{f}_{cab} = 0.66354 E_0 + 0.132150$$

Meassuring the goodnes of fit, for it, by means of the Rooted Mean Squared Error (RMSE), gives a trainning error of 0.27138. The generalization error is 0.25286, i. e. the one obtained in trying to generalize the evaluation set mentioned above. It seems to be a good model.

4.2 One Hidden Layer Sigmoidal Perceptron Models

Two groups of experiments have been arranged with a one hidden layer, sigmoidal perceptron with backpropagation learning. These experiments have been developed using a data mining software tool developed in our group [2]. The first group of experiments has been carried out with all tuples with no null values, from 1994 to 2001, except for the tuples of 1998 which are for evaluation. That gives a trainning proportion of 73.85% for a total of 30958 tuples. The number of epochs were 50.000 for all experiments, using a strong learning rate $\mu = 0.05$. Weights were randomly initialized and hidden nulls where in a range of 1 to 96, with hops of 5. That is the set of hidden nodes $\{1, 6, 11, \ldots, 96\}$. Errors are summarized in table 1 corresponding to columns labeled with TEVV and EEVV, for trainning error and evaluation error respectively. Best generalization error is obtained with 81 nodes in the hidden layer. It overcomes the error obtained with the linear model. The second experiments group has been arranged with the data set in which valid tuples and reconstructed ones are used for trainning. The same set as above has been used for evaluation, with a trainning proportion of 84.12%. That gives a total of 50285 tuples. The same range of hidden nodes and values for the rest of parameters has been applied. Corresponding results can be found in table 1, in columns TERV and EERV, for trainning error and evaluation error respectively.

4.3 Analysis of Results

Both data sets obtain similar models, regarding the evaluation error. The two different minimun errors obtained, 0.16419 and 0.16904 are very similar. However, there is a big difference between minimun trainning error, 0.12754 and 0.22116 of each data set learning. It must be noticed that the recovered tuples data set is a 60% bigger than the other data set and learning should be more complex (i. e. it would need more epochs) for it. The effectiveness of the null

Table 1. Trainning and valuation errors with both data sets.

Nodes	TEVV	EEVV	TERV	EERV	Nodes	TEVV	EEVV	TERV	EERV
1	1.19238	1.38206	0.411108	0.46175	51	0.13537	0.23017	**0.22116+**	0.18055
6	0.31979	0.38314	0.32427	0.37386	56	0.14567	0.23073	0.22889	0.18499
11	0.16291	0.24857	0.35851	0.38667	61	0.14720	0.23732	0.23481	0.19147
16	0.28416	0.51657	0.23810	0.23629	66	0.14330	0.22237	0.22160	**0.16904***
21	0.31988	0.39322	0.22488	0.17494	71	0.14090	0.20245	0.23480	0.21365
26	0.15170	0.23556	0.24529	0.22742	76	0.15718	0.22870	0.23517	0.23576
31	0.15402	0.23074	0.23800	0.24596	81	0.13654	**0.16419***	0.22765	0.19015
36	0.13432	0.20072	0.23934	0.22918	86	**0.12754+**	0.18809	0.22725	0.17385
41	0.13423	0.20619	0.23103	0.20122	91	0.14530	0.28109	0.22717	0.18448
46	0.15531	0.25566	0.24398	0.22562	96	0.14291	0.21121	0.22885	0.18115

values recovery method is demonstrated with the evaluation error means obtained for both data sets. Those are 0.31244 and 0.23441 respectively. A model obtained from recovered data should perform better with future data.

5 Conclusions

Data mining can be a powerful tool in the agricultural environment. In this work it has been demonstrated that a very well known and modeling tool, as it is a simple neural network with backpropagation learning, can be useful in this context.

References

1. J. A. Botía, A. F. Gomez-Skarmeta, M. Valdés, and Gracia Sánchez. Soft Computing Applied to Irrigation in Farming Environments. In *Conference Proceedings of the FUZZ-IEEE.*, volume 1, pages 505–512, San Antonio, Texas, May 2000.
2. Juan A. Botia, A. F. G. Skarmeta, Juan R. Velasco, and Mercedes Garijo. A proposal for Meta-learning through Multi-agent Systems. In Tom Wagner and Omer F. Rana, editors, *Lecture Notes in Artificial Inteligence (to appear)*. Springer, 2000.
3. Carla E. Brodley and Padhraic Smyth. The Process of Applying Machine Learning Algorithms. In *Workshop on Applying Machine Learning in Practice at IMLC-95.* 1995.
4. Usama Fayyad, Gregory Piatetsky-Shapiro, and Padhraic Smyth. Data Mining and Its Applications: A General Overview. In Jiawei Han Evangelos Simoudis and Usama Fayyad, editors, *The Second International Conference on Knowledge Discovery & Data Mining*. AAAI Press, August 1996.
5. HR. Ingleby and T.G. Crowe. Neural network models for predicting organic matter content in saskatchewan soils. *Canadian BioSystems Engineering*, 43(7), 2001.
6. J.S.R. Jang, C.T. Sun, and E. Mizutani, editors. *Neuro-Fuzzy and Soft Computing: A Computational Approach to Learning and Machine Intelligence*. Prentice-Hall, 1997.
7. Tom M. Mitchell. *Machine Learning*. McGraw-Hill, 1997.
8. M. L. Stone and G. A. Kranzler. Artificial Neural Networks in Agricultural Machinery Applications. In *ASAE Paper AETC 95052*, Chicago, Illinois, 1995. ASAE.

Classification of Specular Object Based on Statistical Learning Theory

Tae Soo Yun

Division of Internet Engineering, Dongseo University, San69-1,
Churye-Dong, Sasang-Gu, Pusan, South Korea
tsyun@ailab.knu.ac.kr

Abstract. This paper has presented an efficient solder joint inspection technique through the use of wavelet transform and Support Vector Machines. The proposed scheme consists of two stages: a feature extraction stage for extracting features with wavelet transform, and a classification stage for classifying solder joints with a support vector machines. Experimental results show that the proposed method produces a high classification rate in the nonlinearly separable problem of classifying solder joints.

1 Introduction

Solder joint inspection is one of the most challenging domains in which to apply automated visual inspection technology. This domain is more difficult than most other inspection areas due to the great variability in the appearance of acceptable solder joints and the highly specular nature of the flowed solder [1].

In recent years, many solder joint classification methods have been developed, which have been applied successfully in industrial fields. [1, 2, 3, 4,]. However, when input data space is too complex to determine decision boundaries for classification, the convergence rate of these methods may get worse and accordingly, a good performance of the classification may not be achieved [2, 5]. This is because their main concept is based on *empirical risk minimization* (ERM), which focuses on minimizing the training error without guaranteeing the minimization of the test error.

Therefore, in this paper, we proposed solder joint inspection technique through the use of wavelet transform and Support Vector Machines to solve these problems. The SVM has been recently introduced as a new method for pattern classification and non-linear regression [6]. Its appeal has origin in its strong connection to the underlying statistical learning theory. That is, SVM is an approximate implementation of the method of structural risk minimization (SRM), which shows that that generalization error is bounded by the sum of the training set error and a term depending on the Vapnik-Chervonekis (VC) dimension of the learning machine. For several pattern classification applications, SVM has been shown to produce better generalization performance than traditional techniques, such as NNs [7, 8].

The Wavelet transform enables to have an invariant interpretation of image at different resolutions in the form of coefficient matrices[9]. Since the details of solder joint image at different subband generally characterize different physical structures of the solder joint image, coefficients obtained from wavelet transform are very useful in classifying solder joints.

This paper uses Haar wavelet[10]-based features, directly extracted from solder joint images, as the input vectors to a support vector machine(SVMs) classifier to test our solder joint database and study the performance of the proposed method.

2 Design of Circular Illumination and Image Acquisition

For a specular surface, light is reflected such that the angle of incidence equals the angle of reflection. Therefore, we have to use a light source such as circular illumination to catch the whole solder joint surface.

Fig. 1. Experimental apparatus

As shown in Fig. 1, the experimental apparatus consisted of a CCD color camera and three color circular lamps(blue, red and green) with different illumination angles. Using this illumination, the surface patches of solder joints having the same slope show the same intensity patterns called iso-inclination contours in a camera image. The highlight pattern gives us good visual cues for inferring 3D shapes of the specular surface. Fig. 2 illustrates examples of the solder image obtained by the camera. The solder joints being inspected were divided into four classes according to their quality: Good, Excess, Insufficient, and No solder.

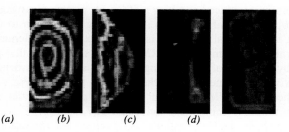

Fig.2. The types of solder joint: (a) excess (b) good (c) insufficient (d) no solder

3 Feature Extraction Using Wavelet Transform

The boundaries between the quality of solder joints are very ambiguous, therefore, it is essential to extract the features which is represents the characteristics of solder joints well for the efficient inspection. However, because the method which recognize the objects with geometrical features obtained by edge detection and thinning in spatial domain is difficult to extract specific features of objects relevant to the light, alignment, and rotation. Additionally, there are the problems of increasing the number of parameters according to the size of objects. Therefore, in this paper, we proposed a feature extraction method through the use of wavelet transform to solve these problems.

A wavelet transform with a set of Haar wavelets[10] is adequate for the local detection of line segments and global detection of line structures with fast computation. Recently, Wunsch[12] proposed a set of shape descriptors that represent a pattern in a concise way, which is particularly well-suited for the recognition of line objects. This descriptor set is derived from the wavelet transform of a pattern's contour. Therefore, a wavelet transform with a set of Haar wavelets can be used efficiently for extracting of features in solder joint images.

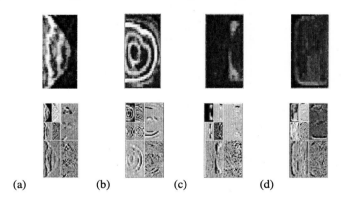

Fig. 3. Two-level wavelet transforms for different solder types: (a)good (b)excess (c) insufficient (d) no solder

By using Haar wavelets, the image of a solder joint is first passed through low-pass filters to generate low-low(LL), low-high(LH), high-low(HL), and high-high(HH) sub-images. These sub-images come under wavelet coefficients that capture any change in the intensity in the horizontal direction, vertical direction, and diagonal (or corners). The decompositions are repeated on the LL sub-image to obtain the next four sub-images. Fig.3 shows two-level wavelet transforms with a set of Haar wavelets for the different solder types.

After the 2-level wavelet transform is obtained from the solder images, 10 vertical projection features and 20 horizontal projection features are extracted from the LL sub-image. Also, 10 vertical projection features are extracted from the LH sub-image and 20 horizontal projection features from the HL sub-image. The features of the HH sub-image are obtained using the same method as used with the LL sub-image. This gives a total of 90 wavelet features corresponding to each 40 x 80 solder joint image.

Fig. 4 shows a graph of the projection features for each sub-image. In all cases, the values for good and excess solders are higher than those for insufficient and no solder. In the LL sub-image, while the values of good solders are high within a narrow range, the values for excess solders are widely distributed with higher values than those for good solders. In the case of an insufficient solder, because the solder fillet is made near to the lead part, the values of the vertical projection features in the LL sub-image are high relatively. The values for no solder cover a wide range as with the excess solders, however, the values are lower than those for the other types. This shows that the features obtained by the proposed method include sufficient information to classify these four solder types. Furthermore, the proposed method can lead to a dimensionality reduction of feature space without a matrix operation such as a principle components analysis(PCA). Once the feature vectors for our solder joint database are computed, they can be used to train an SVM classifier.

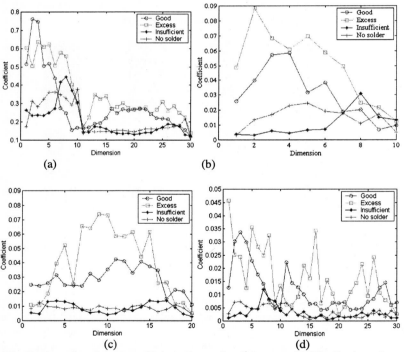

Fig. 4. Distribution of projected coefficients for each sub-image : (a) Distribution of wavelet feature for the LL sub-image (b) Distribution of wavelet feature for the LH sub-image (c) Distribution of wavelet feature for the HL sub-image (d) Distribution of wavelet feature for the HH sub-image

4 Classification with Support Vector Machines

Given a set of points $\{(\mathbf{x}_i, y_i)\}_{i=1}^{N}$ that belong to one of two classes, an SVM finds the hyperplane by leaving the largest possible fraction of points of the same class on the same side, while maximizing the distance of either class from the hyperplane. The hyperplane -called the Optimal Separating Hyperplane(OSH)- can be represented as:

$$o(x,w) = w^T z + b = \sum_i \alpha_i y_i K(x, x_i) + b \qquad (1)$$

where w is the weight vector, the vector z represents the image induced in the feature space due to the kernel $K(x, x_i)$, b is the bias and nonnegative variables α_i are *Lagrange multipliers* can be obtained by standard quadratic optimization technique[11]. A trained SVM determines the class of a test pattern according to the sign of eqn. 1. For linearly non-separable problems, such as the classification of a solder joint, the OSH can be identified through nonlinear kernel mapping into high dimensional space. The kernel used in our experiment is polynomial of degree 3 defined by:

$$K(\mathbf{x}, \mathbf{x}_i) = (\mathbf{x}^T \mathbf{x}_i + 1)^3 \qquad (2)$$

The approach taken to formulate this multi-class classification problem was to consider it as a series of binary classification problems. Fig.5 shows the architecture of solder joint classifier. The input to the classifier is 90-dimensional feature vector extracted from the solder joints image. Each element of feature vector was normalized to [0,1]. The number of kernel n_r for each rth SVM is the same as the number of support vectors obtained when training for classifying class r and the other classes.

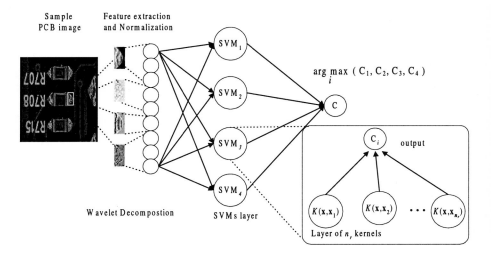

Fig. 5. The Architecture of the solder joints classifier.

5 Experimental Result

In order to verify the effectiveness and robustness of the proposed classifier, experiments were performed on various solder joints. Sample data on 402 solder joints was collected from commercially manufactured PCBs. 201 solder joint samples were used for training and 201 samples were used for testing. Each sample image was a window sub-image fitted to a solder joint with a size of 40 x 80 pixels.

To gain a better understanding of the relevance of the results obtained using SVMs, benchmark comparisons were carried out with other general classification methods, k-means and NNs, and the results are shown in Table 1.

Table 1. Comparison results with other classifier using wavelet features

Types	k-means (k=4)	NN_w	SVM_w	# of samples
Excess	88.23	92.16	100	51
Good	92	96	98	50
Insufficient	94	100	100	50
No Solder	100	100	100	50

Noise sensitivity experiments were performed to assess the robustness and flexibility of the proposed method. In this experiments, the classifier were trained with our solder joint database and tested with noisy images. New tested images were obtained by gradually adding zero mean random noise with different variance value from 10% to 50% to intensity of original solder joint images, and compared the accuracy with clean patterns.

Experiments were performed for a method using geometrical features (SVM_g) and wavelet features (SVM_w) respectively. SVM_g uses average RGB component value and highlight percentage as features obtained by image processing in spatial domain [4]. Fig. 6 shows the results. For comparison the results obtained from the NN-classifier are also presented. The better performance of SVM-based method suggests that in a noisy environment the decision boundary generated by the SVM was more effective in classifying the given solder joint image than that of the NNs.

How effective is to use any size of samples in classification is a matter of grave concerns. It is very difficult to determine the sample size in these classification problems in theoretical. Because, this is depends on complexity of problem or characteristics of training samples. In this experiments, varying the size of training samples, investigate the classification performance. As summarized in Table 2, SVM-based method (i.e. SVM_g, SVM_w) can produce a good result even with a small size of training data due to the nature of maximizing the distance of either class from the hyper-plane. Additionally, the proposed method can obtain the better generalization performance than traditional classification technique such as Neural Networks (i.e. NN_g, NN_w)

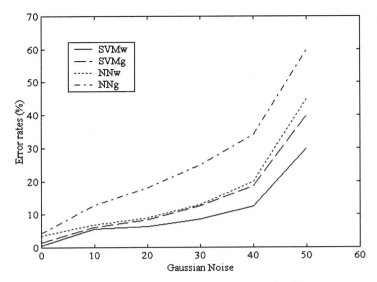

Fig. 6. Tolerance to Gaussian noise for each classifier.

Table 2. Classification performance for different size of training data.

# of samples	50	100	150	200
NN_g	88.24	90.6	92.54	94.05
NN_w	94.03	95.52	95.52	96.52
SVM_g	94.53	96.52	97.51	98.51
SVM_w	95.52	98.51	98.51	99.50

6 Conclusion

This paper has presented an efficient solder joint inspection technique through the use of wavelet transform and Support Vector Machines. The proposed scheme consists of two stages: a feature extraction stage for extracting multi-resolution features with wavelet transform, and a classification stage for classifying solder joints with a support vector machines. Using the wavelet transform, the proposed method was relatively simple compared with the traditional spatial domain method in extracting the feature parameters, and it could be used usefully for the feature extraction of objects like solder joints where the image is composed of contour patterns.

Also, our SVMs classifier based on statistical learning theory could classify effectively the solder joint with ambiguous boundary due to the nature of maximizing the distance of either class from the hyperplane.

In principle, the same techniques should apply to other object classification such as bin picking, SMD IC lead inspection, etc.

Reference

1. P. J. Besl, E. J. Delp, and R.C. Jain, "Automatic visual solder joint inspection, " *IEEE J. Robotics Automation*, vol. RA-1, pp. 42-56, Mar, 1985.
2. J. K. Kim and H. S. Cho, "Neural network-based inspection of solder joints using a circular illumination," *Image and Vision Computing*, vol. 13, pp. 479-490, 1995.
3. S. K. Nayar, A. C. Sandreson, L. E. Weiss, and D. D. Simon, "Specular surface inspecion using structured highlighting and Gaussian images, " *IEEE Trans. Robot. Automat.*, vol. 6, pp. 108-218, Mar. 1990.
4. T. S. Yun, K. J. Sim, H. J. Kim, "Support vector machine based inspection of solder joints using a circular illumination," *IEE Electronics Letters*, vol. 26, no. 11, pp. 949-951, 2000.
5. J. M. Zurada, *Introduction to Artificial Neural Systems*. St. Paul, MN:West, 1992.
6. V.Vapnik, *The Nature of Statistical Learning Theory*, New York:Springer-Verlag, 1995.
7. O. Chapelle, P. Haffner, and V. Vapnik, "Support vector machines for histogram-based image classification," *IEEE Trans. Neural Network*, vol. 10, no. 5, pp. 1055-1064, 1999.
8. B. Schölkopf, K. Sung, C. J. C. Burges, F. Girosi, P. Niyogi, T. Pogio, and V. Vapnik, "Comparing support vector machines with Gaussian kernels to radial basis function classifiers," *IEEE Trans. on Signal Processing*, vol. 45, no. 11, pp. 2758-2765, 1997.
9. S.G.Mallat, " A theory for multiresolution signal decomposition:The wavelet representation, *IEEE Trans.Pattern Anal.Mach.Intell*, 11(7), pp.674-693, 1989.
10. Haar, Zur Theorie der orthgonalen Funktionen System, *Math. Analysis 69*, 331-371, 1910.
11. HAYKIN, S.: Neural network: a comprehensive foundation, second edition, Prentice Hall, 1999.
12. WUNSCH, P. and LAINE, A.F.: 'Wavelet descriptors for multiresolution recognition of handprinted characters', *Patt. Recog.*, vol. 28, no. 8, pp. 1237-1249, 1995.

On the Application of Heteroassociative Morphological Memories to Face Localization

B. Raducanu, M. Graña

Universidad del Pais Vasco
Dept. CCIA, Apdo 649, 20080 San Sebastian
ccpgrrom@si.ehu.es

Abstract. Face localization is a previous step for face recognition and in some circumstances has proven to be a difficult task. The Heteroassociative Morphological Memories are a recently proposed neural network architecture based on the shift of the basic algebraic framework. They possess some robustness to specific noise models (erosive and dilative noise). Here we show how they can be applied to the task of face localization.

1 Introduction

Face detection can be defined as the problem of deciding the presence of a face in the image. Multiple face detection is the generalization to the presence of a set of several faces in the image. Multiple face detection is usually dealt with by solving many single face detection problems posed over a set of overlapping subimages extracted from the image by a sliding window. Face localization is the problem of giving the coordinates in the image of the detected faces. Face localization is answered giving the coordinates of the positive responses to the single face detection instances in the multiple face detection. From a statistical pattern recognition perspective, face detection can be considered as a two-class classification problem: given an image, it must be decided if it belongs either to the face class or to the non-face class. The main difficulty then is the appropriate characterization of the non-face class. Some works, based on neural networks, [3], [4] [8], [9], [10]. do it through a bootstrapping training strategy. Others, like the ones based on linear subspace transformation like the Principal Component Analysis (PCA) [5], [6] compute the likelihood of the face class on the basis of the distance to the face subspace characterized by the eigenfaces.The Local Feature Analysis [7] combines the PCA approach with structural approaches for face recognition and localization.An approach based on the Hausdorff distance between binary images is suggested in [11]. Geometrical approaches try to fit an ellipse to the face contour [13] or to detect some face elements and verify their relative distances [16], [14], [15]. Detection of face elements is difficult and a research subject by itself. Finally, approaches based on color processing [17], [18], [19] and [20] are very easy to realize, although prone to give high false positives rates. A sensible approach to more robust face localization is the combination of several methods into a multi-cue system. Recent works in this line, integrating geometrical and color-based approaches, are reported in [21], [22] and [23]. In this spirit we propose our work as another verification tool.

Morphological Associative Memories is a novel kind of neural network architectures recently proposed in [2], [1]. In these networks, the operations of multiplication and addition are replaced by addition and maximum (or minimum), respectively. In [2] and [1] the construction of the Heteroassociative and Autoassociative Morphological Memories (HMM and AMM) is done analogously to the construction of the Heteroassociative and Autoassocitive Hopfield Memories exchanging the matrix product for the min/max matrix product. The use of the minimum or maximum operator determines the erosive or dilative character of the morphological memory. The memory capacity of the AMM is not bounded by conditions of orthogonality of the input patterns. The memory capacity of the HMM, however, is conditioned to some kind of max/min orthogonality relations between the patterns. The sensitivity of both AMM and HMM to erosive and dilatative noise has been characterized. The construction of a robust HMM (insensitive to both erosive and dilative noise) is decomposed in the construction of an AMM on the input pattern kernels and an HMM that maps the input pattern kernels into the output patterns. The inconvenient of this approach, besides the difficulties in the definition of the pattern kernels, lies in the extremely high storage and computational demands imposed by the construction of an AMM of any practical utility. For this reason, in this paper the approach taken is that of defining an HMM based on eroded/dilated versions of the input patterns, and to apply simultaneously the dual minimum (erosive) and maximum (dilative) memories, obtaining a kind of hit-or-miss transform. The HMM is used as a classification tool for the face detection task. The HMM stores several face patterns, it is convolved witht the inspected images and its response is used to detect the faces in the image. The input patterns are eroded/dilated at several scales for the construction of the HMM in a scale space framework [24].

The paper is structured as follows. In section 2 we review the formal definition of HMM together with their properties. In section 3 we explain how the HMM is applied to the proposed task, and present some experimental results. And finally, in section 4, we present our conclusions and future goals.

2 Heteroassociative Morphological Neural Network

The work on Morphological Neural Networks stems from the consideration of an algebraic lattice structure $(\mathbb{R}, \vee, \wedge, +)$ as the alternative to the usual $(\mathbb{R}, +, \cdot)$ framework for the definition of Neural Networks computation [2], [1]. The operators $\vee$ and $\wedge$ denote, respectively, the discrete min and min operators (resp. sup and inf in a continuous setting). The approach is termed morphological neural networks because $\vee$ and $\wedge$ are the basic operators for the morphological erosion and dilation.

The activation of a neural unit becomes $y_i(t+1) = f\left(\left(\bigvee_j y_j(t) + w_{ij}\right) - \theta_i\right)$ or $y_i(t+1) = f\left(\left(\bigwedge_j y_j(t) + w_{ij}\right) - \theta_i\right)$ depending on the kind of operator chosen. Usually the activation function $f(.)$ is the identity and the threshold is ignored. The matrix formulation of the neural computation is based on the definition of the max/min matrix operators $C = A \vee B$ and $C = A \wedge B$ given, respectively, by $c_{ij} = \bigvee_k (a_{ik} + b_{kj})$ and $c_{ij} = \bigwedge_k (a_{ik} + b_{kj})$.

Following the analogy, given $(X, Y) = \{(\mathbf{x}^\xi, \mathbf{y}^\xi); \xi = 1, .., k\}$, a set of pairs of input/output patterns the heteroassociative neural network built up as $W = \sum_\xi \mathbf{y}^\xi \cdot (\mathbf{x}^\xi)'$ becomes in the setting of morphological neural networks:

$$W_{XY} = \bigwedge_{\xi=1}^{k} \left[\mathbf{y}^\xi \times (-\mathbf{x}^\xi)' \right] \quad M_{XY} = \bigvee_{\xi=1}^{k} \left[\mathbf{y}^\xi \times (-\mathbf{x}^\xi)' \right] \tag{1}$$

where $\times$ is any of $\vee$ or $\wedge$. It follows that the weight matrices are lower and upper bounds of the max/min products $\forall \xi; W_{XY} \leq \mathbf{y}^\xi \times (-\mathbf{x}^\xi)' \leq M_{XY}$ and therefore the following bounds on the output patterns hold $\forall \xi; W_{XY} \vee \mathbf{x}^\xi \leq \mathbf{y}^\xi \leq M_{XY} \wedge \mathbf{x}^\xi$, that can be rewritten.

$$W_{XY} \vee X \leq Y \leq M_{XY} \wedge X. \tag{2}$$

A matrix A is a $\vee$-perfect ($\wedge$-perfect) memory for (X, Y) if $A \vee X = Y$ ($A \wedge X = Y$). It can be proven that if A and B are $\vee$-perfect and $\vee$-perfect memories for (X, Y) then

$$A \leq W_{XY} \leq M_{XY} \leq B \text{ and } W_{XY} \vee X = Y = M_{XY} \wedge X. \tag{3}$$

Conditions for perfect recall on the memories are given by a theorem that states that W_{XY} is $\vee$-perfect if and only if $\forall \xi$ the matrix $\left[\mathbf{y}^\xi \times (-\mathbf{x}^\xi)' \right] - W_{XY}$ contains a zero at each row. Similarly M_{XY} is $\wedge$-perfect if and only if $\forall \xi$ the matrix $\left[\mathbf{y}^\xi \times (-\mathbf{x}^\xi)' \right] - M_{XY}$ contains a zero at each row. These conditions are rewritten for W_{XY} and M_{XY} respectively as follows:

$$\forall \gamma \forall i \exists j; x_j^\gamma = \bigvee_{\xi=1}^{k} \left(x_j^\xi - y_i^\xi \right) + y_i^\gamma \quad \forall \gamma \forall i \exists j; x_j^\gamma = \bigwedge_{\xi=1}^{k} \left(x_j^\xi - y_i^\xi \right) + y_i^\gamma. \tag{4}$$

Finally, let it be $\tilde{\mathbf{x}}^\gamma$ a noisy version of $\mathbf{x}^\gamma$. If $\tilde{\mathbf{x}}^\gamma \leq \mathbf{x}^\gamma$ then $\tilde{\mathbf{x}}^\gamma$ is an eroded version of $\mathbf{x}^\gamma$, or $\tilde{\mathbf{x}}^\gamma$ is subjected to erosive noise. If $\tilde{\mathbf{x}}^\gamma \geq \mathbf{x}^\gamma$ then $\tilde{\mathbf{x}}^\gamma$ is a dilated version of $\mathbf{x}^\gamma$, or $\tilde{\mathbf{x}}^\gamma$ is subjected to dilatative noise. Morphological memories are very sensitive to these kinds of noise. The conditions for the perfect recall of $\mathbf{x}^\gamma$ given a noisy copy $\tilde{\mathbf{x}}^\gamma$ for W_{XY}, that is, the conditions under which $W_{XY} \vee \tilde{\mathbf{x}}^\gamma = \mathbf{y}^\gamma$ are as follows:

$$\forall j; \tilde{x}_j^\gamma \leq x_j^\gamma \vee \bigwedge_i \left(\bigvee_{\xi \neq \gamma} \left(y_i^\gamma - y_i^\xi + x_j^\xi \right) \right) \text{ and} \tag{5}$$

$$\forall i \exists j_i; \tilde{x}_{j_i}^\gamma = x_{j_i}^\gamma \vee \left(\bigvee_{\xi \neq \gamma} \left(y_i^\gamma - y_i^\xi + x_{j_i}^\xi \right) \right).$$

Similarly for the perfect recall of $\mathbf{x}^\gamma$ given a noisy copy $\tilde{\mathbf{x}}^\gamma$ for M_{XY}, that is, the conditions under which $M_{XY} \wedge \tilde{\mathbf{x}}^\gamma = \mathbf{y}^\gamma$ are as follows:

$$\forall j; \tilde{x}_j^\gamma \leq x_j^\gamma \wedge \bigvee_i \left(\bigwedge_{\xi \neq \gamma} \left(y_i^\gamma - y_i^\xi + x_j^\xi \right) \right) \text{ and} \tag{6}$$

$$\forall i \exists j_i; \widetilde{x}_{j_i}^{\gamma} = x_{j_i}^{\gamma} \wedge \left(\bigwedge_{\xi \neq \gamma} \left(y_i^{\gamma} - y_i^{\xi} + x_{j_i}^{\xi} \right) \right).$$

These conditions (6), (5) and (2) are the basis for our approach. The conditions in (5) and (6) state that the matrix W_{XY} is robust against controlled erosions of the input patterns while the matrix M_{XY} is robust against controlled dilations of the input patterns. Therefore if we store in the W matrix a set of eroded patterns, the input could considered as a dilation of the stored pattern most of the times. The dual assertion holds for the M matrix. Also (2) determines that those cases when the output of both M and W memories are the same, then the output of both corresponds to the desired output. This holds in the case of interactions between the stored patterns. We will consider these matrices as approximations to the ideal memory of all the distorted versions of the input data, so that their output is an approximation to the response of this ideal memory. We applye a scale space approach to increase the robustness of the process.

Given a set of input patterns X and a set of output class enconding Y. We built a set of HMM $\{M_{XY}^{\sigma}, W_{XY}^{\sigma}; \sigma = 1, 2, ...s\}$ where each M_{XY}^{σ} is constructed from outputd and the input patterns eroded with an spherical structural object of scale σ, and each W_{XY}^{σ} is constructed from the outputs and input patterns dilated with an spherical structural object of scale σ. Given a test input pattern $\mathbf{x}$, the memories at the different scales are applied giving $\mathbf{y}^M = \bigcup_{\sigma=1}^{s} (M_{XY}^{\sigma} \bar{\wedge} \mathbf{x})$ and $\mathbf{y}^W = \bigcup_{\sigma=1}^{s} (W_{XY}^{\sigma} \veebar \mathbf{x})$. The final output is the intersection of these multi-scale responses:

$$\mathbf{y} = \mathbf{y}^M \bigcap \mathbf{y}^W. \tag{7}$$

In the case of face localization, the output is the classification of the image block as a face, which is given as a block of white pixels whenever the input image block is identified with any of the stored face patterns.

3 Experiments on Face Localization

As stated in the introduction, the target application is face localization. For this purpose a set face patterns is selected as the representatives of the face class. In the experiments reported here the set of face patterns is the one presented in figure 1. This small set shows several interesting features: faces are of different sizes, background has been manually removed, there is no precise registration of face features (some of the faces are rotated), and there is no intensity normalization (equalization or any other illumination compensation). Therefore, building this set of patterns corresponds to an almost casual browsing and picking of face images in the database.

Face localization is a two class classification problem, however, we formulate it as a response to a n-class classification problem. As described in the previous section, the M and W HMM's are built up to classify each image block as one of the face patterns. If it fails, the response is arbitrary and we consider

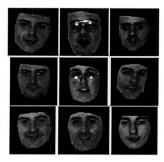

Fig. 1. Face patterns used in the experiments

the image block as a non-face block. The HMM's output are orthogonal binary vectors encoding the face pattern. Both memories are convolved with the image to search for faces. At each pixel the positive classification with one of the M^σ memories produces a square of face pixels of a size that is the half of the image block, and centered at this pixel position. The recognition at the different scales is added into an M-recognition binary image. The same process applies to the W^σ memories. The intersection of the face pixels recognized with each HMM is the final result, which is superimposed to the original image.

We have performed initial studies over a small database of 20 images with a varying range of scales. The average ROC curve over all the images relating the true and false positives obtained with scale ranges varying from $s = 1$ up to $s = 13$ is shown in figure 2. It can be appreciated that the approach obtains a high recognition rate (over 85%) with very small false recognition rates (less than 5%). As the scale range increases we reach the 100% of face recognition at the pixel level. Face pixels were labelled manually in a process which is independent of the selection of the face patterns. These results are very promising and we are planning the application of this approach to larger face databases, like the well-known CMU database. As a final result, we give in figure 3 some images with recognition results at scale 5.

4 Conclusions and Further Work

We propose the application of HMM for a realization of face localization that can be competitive with other graylevel based procedures. HMM give a relatively fast response because they only perform integer and max/min operations and its response does not imply the computation of an energy minimum. The main drawback of the HMM in general is their sensitivity to morphological noise: erosions and dilations of the image. However, inspired in the construction of the kernels in [2] and [1], we propose the utilization of the dual HMM memories as a kind of hit-or-miss transform. The eroded input patterns are used to built up the M memories and the dilated input patterns are used to built up the W

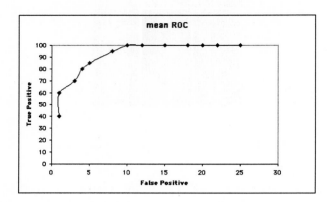

Fig. 2. Mean ROC of the face localization at pixel level across the experimental set of images

Fig. 3. Some results of face localization using patterns eroded/dilated to scales up to 5

memories. The intersection of their detections give a robust detection procedure. We explore the sensitivity of the approach to the erosion/dilation scale, in a multiscale space framework, for an instance of a set of training face patterns and a small database of images. The results are rahter encouraging. We plan to extend the experiments to larger databases, such as the CMU face localization database.

Acknowledgements The work has been developped under grants UE-1999-1 and PI-98-21 of the Gobierno Vasco (GV/EJ). B.Raducanu has a predoctoral grant from the University of The Basque Country (UPV/EHU).

References

1. G. X. Ritter, P. Sussner and J. L. Diaz-de-Leon, "Morphological Associative Memories", IEEE Trans. on Neural Networks, 9(2), (1998), pp. 281-292
2. G. X. Ritter, J. L. Diaz-de-Leon and P. Sussner, "Morphological Bidirectional Associative Memories", Neural Networks, Vol. 12, (1999), pp. 851-867
3. Rowley H. A., Baluja S., Kanade T., Neural Network-Based Face Detection, IEEE Trans. Patt. Anal. Mach. Int., vol. 20, no. 1, (1998), pp. 23-38
4. Sung K. K., Poggio T., Example-Based Learning for View-Based Human Face Detection, IEEE Trans. Patt. Anal. Mach. Int., vol. 20, no. 1, (1998), pp. 39-50
5. Moghaddam B. and Pentland A., Probabilistic Visual Learning for Object Detection, Proc. of International Conference on Computer Vision, IEEE Press, (1995), pp. 786-793
6. Turk M., Pentland A., Eigenfaces for Recognition. Journal of Cognitive Neuroscience, vol. 3, no. 1, (1991), pp. 71-8
7. Penev P. S., Atick J. J., Local Feature Analysis: A General Statistical Theory for Object Representation, Network:Computation in Neural Systems, vol. 7, (1996), pp. 477-500
8. Lin S. H., Kung S. Y., Lin L. J., Face Recognition/Detection by Probabilistic Decision-Based Neural Network, IEEE Trans. on Neural Networks, vol. 8, no. 1, (1997), pp. 114-132
9. Juell P., Marsh R., A Hierarchical Neural Network for Human Face Detection, Pattern Recognition, vol. 29, no. 5, (1996), pp. 781-787
10. Dai Y., Nakano Y., Recognition of Facial Images with Low Resolution Using a Hopfield Memory Model, Pattern Recognition, vol. 31, no. 2, (1998), pp. 159-167
11. Takács B., Comparing Face Images Using the Modified Hausdorff Distance, Pattern Recognition, vol. 31, no. 12, (1998), pp. 1873-1881
12. Marqués F., Vilaplana V., Buxes A., Human Face Segmentation and Tracking Using Connected Operators and Partition Projection, Proc. ICIP, 1999
13. Wang J., Tan T., A New Face Detection Method Based on Shape Information, Pattern Recognition Letters, vol. 21, no. 6-7, (2000), pp. 463-471
14. Leung T. K., Burl M. C., Perona P., Finding Faces in Cluttered Scenes Using Random Labeled Graph Matching, Proc. of The Fifth ICCV, http://HTTP.CS.Berkeley.EDU/~leungt/Research/ICCV95_final.ps.gz, 1995
15. Yow K. C., Cipolla R., Finding Initial Estimates of the Human Face Location, Technical Report TR-239, University of Cambridge, (1995)

16. Colombo C., Bimbo A. d., Real-Time Head Tracking from the Deformation of Eye Contours Using a Piecewise Affine Camera, Pattern Recognition Letters, vol. 20, no. 7, (1999), pp. 721-730
17. Lee C. H., Kim J. S., Park K. H., Automatic Human Face Location in a Complex Background Using Motion and Color Information, Pattern Recognition, vol. 29, no. 11, (1996), pp. 1877-1889
18. Yang G., Huang T. S., Human Face Detection in a Complex Background, Pattern Recognition, vol. 27, no. 1, (1994), pp. 53-63
19. Yoo T.-W., Oh I.-S., A Fast Algorithm for Tracking Human Faces Based on Chromatic Histograms, Pattern Recognition Letters, vol. 20, no. 10, (1999), pp. 967-978
20. McKenna S. J., Gong S., Raja Y., Modelling Facial Colour and Identity with Gaussian Mixtures, Pattern Recognition, vol. 31, no. 12, (1998), pp. 1883-1892
21. saber E., Tekalp A. M., Frontal-View Face Detection and Facial Feature Extraction Using Color, Shape and Symmetry Based Cost Functions, Pattern Recognition Letters, vol. 19, no. 8, (1998), pp. 669-680
22. Yin L., Basu A., Integrating Active Face Tracking with Model Based Coding, Pattern Recognition Letters, vol. 20, no. 6, (1999), pp. 651-657
23. Wang J.-G., Sung E., Frontal-View Face Detection and Facial Feature Extraction Using Color and Morphological Operators, Pattern Recognition Letters, vol. 20, no. 10, (1999), pp. 1053-1068
24. Jackway P. T., Deriche M., Scale-Space Properties of the Multiscale Morphological Dilation-Erosion, IEEE Trans. on Patt Anal. and Mach. Int., vol. 18, no. 1, (1996), pp. 38-516

Early Detection and Diagnosis of Faults in an AC Motor Using Neuro Fuzzy Techniques: FasArt + Fuzzy k Nearest Neighbors

J. Juez, G.I. Sainz*, E.J. Moya, and J.R. Perán

Department of Systems Engineering and Control. School of Industrial Engineering.
University of Valladolid. Paseo del Cauce s/n, 47011 Valladolid, Spain.
: gresai@eis.uva.es. Phone: +34 83 423000 Ext. 4401.

Abstract. An approach to detect and to classify incipient faults in an AC motor is introduced in this paper. This approach is based in an ART based neuro fuzzy system, (FasArt Fuzzy Adaptive System ART based), and in the fuzzy *k nearest neighbor algorithm* that is employing in an auxiliary way to complete the learning set.
A set of 15 non destructive faults has been tested and both a high degree of early detection and recognition has been reached. As well as, using the neuro-fuzzy nature of the FasArt model a database of fuzzy rules has been obtained permitting a fault description by linguistics terms.

1 Introduction

Several types of electrical motors are used extensively in a great variety of industrial environments. Safety, reliability, efficiency and performances are some of the most interesting aspects concerning to the motors, in order to reach high levels on these issues monitoring, on-line fault detection and diagnosis in automatical way are needed in the modern industry.

A review on the most usual motor problems, and the applied techniques are in (Finley and Burke, 1994), (Chow, 1997), (Benbouzid, 1999) and (Nandi and Tolyat, n.d.). Besides, in (Pouliezos and Stavrakakis, 1994) a detailed revision of techniques for real time fault monitoring in industrial processes is developed, some of them has been employed in order to make fault detection and diagnosis in electrical machine: radio frequency monitoring, particle analysis, parameter estimation, fuzzy logic, neural networks, etc. In this paper an approach based on fuzzy logic and neural networks has been selected[1].

Fuzzy logic and neural networks are two noninvasive techniques that has been employed successfully to problems in several areas, such as motors an other electrical engines (Arnanz et al., 2000), (Costa Branco and Dente, 1998), (Chow et al., 1999),etc. Theses neuro-fuzzy systems aim at combining the advantages of

---
* Author to whom all correspondence should be addressed.
[1] This work has been supported partially by the research national agency of Spain (CICYT) throughout the project 1FD97-0433.

the two paradigms: learning from examples and capacity for dealing with fuzzy information. In this paper a neuro-fuzzy ART based system is employed as kernel of the system for detecting and classifying of incipient faults in an early time before the machine eventually suffers a failure or a permanent damage.

The neuro-fuzzy proposed system use the neuro-fuzzy ART based model FasArt (Cano et al., 1996) and, auxiliary, the *fuzzy k nearest neighbor algorithm*. The FasArt model is a supervised model, has been applied successfully to several problems: pattern recognition (Sainz Palmero et al., December 2000), system identification (Cano et al., 1996), etc. In this case, it is combined to the well known *fuzzy k nearest neighbor algorithm* and applied to the detection and classification of faults in an AC motor. The integration between the two components of this approach is made in the learning stage, the *k mean fuzzy* algorithm permit to label the input values of the learning data set in an easier and effective way than if this process were made by hand.

Moreover, after the training and test stages of the neuro-fuzzy system, the knowledge stored into FasArt can be explained by a set of fuzzy rules that permits a better understanding of the knowledge involved in the process.

The paper is organized as follow, first a brief description of the neuro-fuzzy sistem FasArt, that is the kernel of the approach, is done. Then fuzzy k mean nearest neighbor algorithm and the way in which both are integrated are explained. Besides, a description of the experimental motor laboratory plant, the tests made and the results obtained are discussed in section 4.2. Finally the main conclusions achieved in this work are developed.

2 FasArt Model

FasArt (Fuzzy Adaptive System ART based) (Cano et al., 1996a) is a supervised neural network architecture based on the Adaptive Resonance Theory, in its model Fuzzy ARTMAP (Carpenter et al., 1992). The general structure of Fuzzy ARTMAP is maintained but FasArt replaces the boolean (ARTMAP) or "fuzzy" (Fuzzy ARTMAP) logic operators included in the neuron activation function by a true fuzzy operation, associating each category to a fuzzy set, for which each input pattern produces a membership degree. Due to these changes, FasArt represents several advantages over Fuzzy ARTMAP:

– It achieves a dual membership/activation function. A new activation function for each neuron k in F_2^a is defined (see Figure 1 (a)), as the product of fuzzy membership degrees on each input feature i.e.:

$$\eta_{R_k}(I) = \prod_{i=1}^{m} \eta_{ki}(I_i) \tag{1}$$

where $I = (I_1, \ldots, I_m)$ is the input pattern and $\eta_{ki}(I_i)$ is the membership degree of input I_i to unit k in layer F_2^a, given by:

$$\eta_{ki}(I_i) = \begin{cases} \max\left\{0, \frac{\gamma(I_i - w_{ki}) + 1}{\gamma(c_{ki} - w_{ki}) + 1}\right\} & \text{if } I_i \leq c_{ki} \\ \max\left\{0, \frac{\gamma(1 - I_i - w_{ki}^c) + 1}{\gamma(1 - c_{ki} - w_{ki}^c) + 1}\right\} & \text{if } I_i > c_{ki} \end{cases} \tag{2}$$

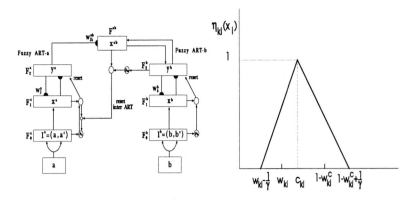

Fig. 1. FasArt network: (a) Architecture (b)Membership function.

where $W_k = (w_{k1}, w_{k1}^c, \ldots, w_{km}, w_{km}^c)$ and $C_k = (c_{k1}, \ldots, c_{km})$ are weight vectors associated to neuron k. Although triangular memberships functions have been selected here (see Figure 1 (b)), gaussian or bell-shaped functions could have also been used. This duality of the weights allows to represent each unit of F_2^a as a fuzzy set.

- Weight vector W_k has the same functionality as in Fuzzy ARTMAP, but a new weight vector, C_k, is defined for each neuron k of the F_2 levels, through a similar learning law as far W_k. C_k represents the central point of the triangular membership function, as shown in Figure 1(b).
- FasArt performance is affected by some user-tuned parameters inherited from Fuzzy ARTMAP: a vigilance parameter in each unsupervised module (ρ_a, ρ_b) controls the maximum allowed size of categories (in FasArt, the maximum support for fuzzy sets). During the test, ρ_a may be used to produce an *"unidentified"* answer if there is low confidence. Learning rates (β_a, β_b and β_a^c, β_b^c for the new weights vector C) determine how fast categories should include prototypes. Usually $\beta_a = \beta_b = 1$ (fast-learning).

In addition, the degree of fuzziness can be controlled through values of the parameter γ. Values of $\gamma \to \infty$ produce crisper fuzzy sets, while values of $\gamma \to 0$ increase the fuzzy nature of sets. This value determines the support of the fuzzy set.

- The relations stored in the inter-ART map, Figure 1(a), can be interpreted to construct a fuzzy rule set, with fuzzy sets in the input space defined by ART_a categories and in the output space by ART_b categories, obtaining rules of the following form:
  - IF I^a IS $R_j AND\ldots$ THEN I^b IS R_k

where I^a, I^b are linguistic variables and R_j, R_k are fuzzy sets in the input and output space respectively.

- The interpretation of FasArt as a fuzzy logic system permits the use of a defuzzification method in order to calculate a numeric output. In this case, the defuzzification method based on the average of fuzzy sets centers is employed. Therefore, given an input pattern $I = (I_1, ..., I_n)$ presented in the test phase, the output is obtained by:

$$y_m(I) = \frac{\sum_{l=1}^{N^b}\sum_{k=1}^{N^a} c_{lm}^b w_{kl}^{ab} \eta_{R_k}(I)}{\sum_{l=1}^{N^b}\sum_{k=1}^{N^a} w_{kl}^{ab} \eta_{R_k}(I)} \quad (3)$$

3 Fuzzy k-Nearest Neighbour Algorithm

This is the fuzzy version of the *fuzzy k nearest neighbor* decision rule (Keller et al., 1985). The crisp nearest-neighbor classification rule assigns an input sample vector to the class of its nearest neighbor. By another hand, the fuzzy k-nearest neighbor algorithm assigns class membership to a sample vector rather assigning the vector to a particular class. In addition, these membership values provides a level of confidence about the resultant classification.

The basis of the algorithm is to assign membership as a function of the vector's distance from its k nearest neighbors and those neighbors' memberships in the possible classes.

FasArt is a supervised system so the correct outputs to the inputs must provide in the learning stage. In the treated case, each data set has a fewer thousands of data and when several types of faults are considered and represented form overlapped clusters and trajectories so each input must be labeled with membership values to each fault sort. Using the fuzzy k nearest neighbor algorithm this task is done by an easier way than if the labels were provided by hand. But this is not a perfect method and it must be revised to avoid incorrect labellings, the mistakes are not too many.

4 Application to a Motor Laboratory Plant

The motor laboratory plant can be seen in Figure 2. This test motor set is composed by two motors of induction Leroy-Somer LS132ST with three phases, 4 poles, 28 rotor bars and 36 stator slots and a power of 5.5 Kw each one. The power supply frequency is 50 KHz and there is a delta connection. This motor plant has a sensor set to get a fine monitoring, sensors such as: voltage, current, temperature, magnetic flux, optical encoders, etc...

The experiment has involve only non destructive faults due to economic costs of the destructive tests. The types of faults tested were 15 which are listed next:

- Normal functioning.
- Several unbalanced power supply in the three phases.
- Unbalanced power supply in the three phases.
- Resistor stator variation ($\triangle$) in the three phases.
- Unbalanced mechanical load.
- Fault in the angular speed encoder.
- Fault in voltage sensor in the three phases.

Fig. 2. Motor plant. Leroy-Somer LS132ST, 5.5 Kw.

4.1 Data Acquisition and Processing

The data sets were acquired by a frequency of 25 MHz. For each fault type 2-4 data sets, containing each one 35000-40000 data vector with the variables that will be employed as inputs to the neuro-fuzzy system. These input variables are: 3 phase currents (I_1, I_2, I_3), 3 phase power (V_1, V_2, V_3) and angular speed (ω). One data set was employed in the learning stage and the rest of sets were employed for testing each considered fault.

In order to get a better performance in the classification process the electrical variables were processed using its effective values:

$$A_{ef}(t) = \left[\frac{1}{T} \sum_{k=0}^{\frac{T}{\Delta}T} A^2(t - k\Delta t)\Delta t \right]^{\frac{1}{2}} \qquad (4)$$

4.2 Experimental Results

The operative way in which these experiments were made reaching the stationary functioning mode (normal mode) of the motor then the failure were generated. Finally the motor come back to normal functioning when it were possible.

Using the data set generated as it was explained in the previous sections, the neuro-fuzzy system is able to learn to classify or to make diagnosis in each point of a fault trajectory throughout its temporal evolution, and in the test stage the system provide a set of confidential values of diagnosis to each one of the fault type considered for each time point of the trajectory.

Analyzing the output generated by the FasArt module, this is able to detect the learnt faults at an early time, when the fault is incipient, in this case the confidence output value of the normal functioning is not the best and/or there are some confidences values that have similar values to it. Also a diagnosis time is obtained when a new best confidence value is obtained and its value is bigger

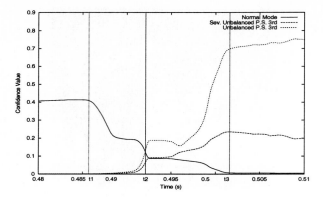

Fig. 3. Evolution of FasArt confidences values throughout of a fault evolution. (t_1), generation time, (t_2) detection time and (t_3) diagnosis time.

than the rest of fault alternatives. In Figure 3 can be observed these times and the evolution of the functioning of the motor, normal mode until t_1 in this time the fault is generated and at t_2 it is detected at t_3 the type of fault is identified. Both times, t_1 and t_2, are considered on a soft or not too restricted criterion so a hard time detection and diagnosis could be obtained by FasArt if it was necessary using a more restricted criterion.

Table 1. Detection and diagnosis time, and successful rate of diagnosis for each fault type ($\rho_a = 0.6$, $\gamma_a = 10$, $\rho_b = 0.9$, $\gamma_b = 10$)

Type of Fault	Detection Time (seg)	Diagnosis Time (seg)	Diagnosis
Several Unbalanced Power Supply 1st	0.0013	0.0232	100%
Several Unbalanced Power Supply 2nd	0.0013	0.0188	100%
Several Unbalanced Power Supply 3th	0.0008	0.0179	100%
Unbalanced Power Supply 1st	0.0013	0.0202	15%
Unbalanced Power Supply 2nd	0.004	0.0151	77%
Unbalanced Power Supply 3th	0.0052	0.0159	100%
Voltmeter 1st	0.0033	0.0077	100 %
Voltmeter 2nd	0.0059	0.0084	100%
Voltmeter 3th	0.0073	0.0089	100%
$\triangle$ Stator Resistor 1st	0.0094	0.0116	34%
$\triangle$ Stator Resistor 2nd	0.0075	0.0103	100%
$\triangle$ Stator Resistor 3th	0.0082	0.011	100%
Encoder	–	–	0%
Unbalanced Mechanical Load	–	–	0%

The experimental results are summarized in Figure 1 where in the secondth column the detection time for each type of fault is shown. Also a diagnosis time is obtained when a new best confidence value is obtained and its value is bigger than the rest of fault alternatives, thirdth column of the Table 1. Finally, a successful rate about the diagnosis made by the system can be seen in the fourth column of the table 1, this diagnosis is the classification generated by the system for each time value of the input variables throughout the fault evolution.

Observing these results, there are only two classes of failures that are not detected: encode failure and unbalanced mechanical load. In this case, using the real input variables it is not possible to detect them because it is confused with normal mode functioning, i.e., both cluster types are mixed.

The rest of faults have an early detection and diagnosis times. When the fault is classified the rate of successful is 100% but Several Unbalanced Power Supply 2nd has a 77% and other two faults have lower rate: Unbalanced Power Supply in the 1st, this fault is confused with Stator Resistor in the 1st phase and Several Unbalanced Power Supply in the 1st phase. The secondth is the Stator Resistor Variation in the 1st phase that has a rate about 34% it is confused with the by Normal functioning. Even thought these cases in that the system does not work so fine than the rest of faults, the diagnosis uncertainty is decreased to two possible fault alternatives.

Moreover, the knowledge stored by the fuzzy-neuro nature of the system can be expressed by fuzzy rules. The database obtained can be summarized in Table 2, in which is shown that the normal functioning mode can be explain using only three fuzzy rules and a similar way with the rest of modes or faults. One the most significant aspects is the rules number for a fault type is different depending of the electrical phase treated. It could be forced by the electrical unbalancing of the plant and the constructive aspects of the motor.

Table 2. Fuzzy database. Number of rules for each type of failure

Type of failure	Number of Fuzzy rules (in each phase)
Normal	3
Several Unbalanced Power Supply	5 (1st) 7(2nd) 6(3th)
Unbalanced Power Supply	3(1st) 5(2nd) 3(3th)
$\triangle$ Stator Resistor	3(1st) 4(2nd) 5(3th)
Unbalanced Mechanical Load	1
Encoder Failure	1
Voltmeter Failure	2(1st) 3(2nd) 3(3th)

5   Conclusions

In this paper the neurofuzzy model FasArt has been employed to make detection and diagnosis of faults in an AC motor. The use fuzzy k nearest neighbor algorithm as a base of prelabeling for the learning data set of the supervised neurofuzzy network has achieved to incorporate memberships values in adequated way as labels in the learning set that contains a few thousands of the overlapped fault data.

The results obtained by FasArt model, through its confidence values to each fault type, have permitted to detect and to make diagnosis of the most of the considered fault types. The successfully rate of classification/diagnosis (about 77% − 100%) and the detection and diagnosis times (about $10^{-4} - 10^{-3}$ seg) provided by this approach for 15 sort of faults, are able to make actions in order to avoid failures or permanent damage on the motor. Only two sort of fault are not detected and not classified, but in the cases in which the system does not work fine (about 15% − 34%) the possible alternatives of fault are reduced to two or three types of the considered.

Finally, a reduced fuzzy database is extracted from neurofuzzy system, in this way the knowledge stored in the system can be expressed by linguistics terms and a better understanding about the process can be effected and employed in other systems or tasks.

References

Arnanz, R., L. J. Miguel, E. J. Moya and J. R. Perán (2000). Model-based diagnostics of AC motors. In: *IFAC Symposium on Fault Detection, Supervision and Safety for Technical Processes (SAFEPROCESS'2000)*. Budapest, Hungary. pp. 1145–1150.

Benbouzid, M. E. H. (1999). Bibliography on induction motors faults detection and diagnosis. *IEEE Transactions on Energy Conversion* **14**(4), 1065–1074.

Cano, J.M., Y.A. Dimitriadis, M. Arauzo and J. Lopez (1996). Fasart: A new neurofuzzy architecture for incremental learning in system identification. In: *Proceedings of the 13th World Congress of IFAC, IFAC'96*. San Francisco, USA. pp. 133–138.

Carpenter, G., S. Grossberg, N. Markuzon and J. Reynolds (1992). Fuzzy ARTMAP: A neural network architecture for incremental supervised learning of analog multidimensional maps. *IEEE Transactions on Neural Networks* **3**(4), 698–713.

Chow, M. (1997). Motor fault detection and diagnosis. *IEEE Industrial electronics society News Letter* **42**, 4–7.

Chow, M., Altug S. and H. J. Trusell (1999). Heuristic constraints enforcement for trainig of and knowledge extraction from a fuzzy/neural architecture – part i: Foundation. *IEEEFuzzy* **7**(2), 143–150.

Costa Branco, P. J. and J. A. Dente (1998). An experiment in automatic modelling an electrical drive system using fuzzy logic. *IEEE Transactions on systemas, man and cybernetics- Part C* **28**(2), 254–261.

Finley, W. R. and R. R. Burke (1994). Troubleshooting motor problems. *IEEE Transactions on Industry Applications* **30**(5), 1383–1397.

Keller, J. M., M. R. Gray and Givens J. A. (1985). A fuzzy k-nearest neighbor algorithm. *SMC* **15**(4), 580–585.

Nandi, S. and H. A. Tolyat (n.d.). Fault diagnosis of electrical machines - a review. *Submitted to IEEE Transactions on Industry Applications.*

Pouliezos, A. D. and G. S. Stavrakakis (1994). *Real Time Fault Monitoring of Industrial Processes*. Kluwer academic publishers. The Netherlands.

Sainz Palmero, G.I., Y. Dimitriadis, J.M. Cano Izquierdo, E. Gómez Sánchez and E. Parrado Hernández (December 2000). ART based model set for pattern recognition: FasArt family. In: *Neuro-fuzzy pattern recognition* (H. Bunke and A. Kandel, Eds.). World Scientific Pub. Co.

Knowledge-Based Neural Networks for Modelling Time Series

Jacobus van Zyl[1]
Christian W. Omlin[2]

[1] Department of Computer Science
University of Stellenbosch
7600 Stellenbosch, South Africa
jvanzyl@cs.sun.ac.za

[2] Department of Computer Science
University of the Western Cape
7535 Bellville, South Africa
comlin@uwc.ac.za

Abstract. Various methods exist for extracting rules from data for classification purposes. We propose a new method for initializing a neural network used for time series modelling and prediction. We extract binary rules from a real valued time series and encode them into a neural network using an adaptation of KBANN. We test the method on the Lorenz system as well as on real world data in the form of a seismic time series. Results show that the method is successful in extracting and encoding prior knowledge. For the Lorenz system training time was halved and better generalization performance in the form of a lower mean squared error was obtained. Better training time and generalization performance was also obtained for the seismic time series.

1 Introduction

An important use for neural networks is time series analysis and time series prediction. The vast majority of real world time series are complex, often poorly understood or non-stationary. It is generally accepted that when expert domain knowledge is used in the training of neural networks, performance will be better compared to training without prior knowledge.

Obtaining rules from data with the use of neural networks can be divided into two broad fields: rule extraction and rule discovery [6]. Rule extraction assumes prior theories and then adapt them by learning on data. Rule discovery depends solely on learning based on data and assumes no prior theory.

Well known methods for knowledge acquisition are ID3 and C4.5. These methods can be used to obtain rules to make decisions based on attributes. The rules can then be encoded into a neural network. However, this is not as useful for time series modelling, since it is difficult to assign attributes to it.

Prior knowledge for time series often comes in the form of embedding dimension information, time window sizes or sampling frequencies. This information is extracted from the data by methods such as mutual information and false nearest neighbours. More information on the dynamical process itself may be obtained via methods such as Lyapunov exponent extraction, correlation dimension determination, power density spectrum analysis and non-stationarity detection [7]. Prior knowledge about the time series may also be applied to create additional "virtual" examples such as including examples $f(-x) = -f(x)$ for odd time series.

In this paper we propose a method of obtaining rules directly from the time series themselves. We then encode these rules into a neural network using the KBANN encoding method. We show performance results on the Lorenz system and on a real world time series: induced seismic events in mines.

2 Knowledge-Based Artificial Neural Networks

We use the method proposed in [2] to illustrate how Horn clauses can be encoded into feedforward networks. Other methods only differ in the way neuron inputs are combined (e.g. [1]). The construction of an initial network is based on the correspondence between entities of the knowledge base and neural networks, respectively. Supporting facts translate into input neurons, intermediate conclusions are modelled as hidden neurons, output neurons represent final conclusions; dependencies are expressed as weighted connections between neurons. The neuron outputs are computed by a sigmoidal function which takes as its argument a weighted sum of inputs.

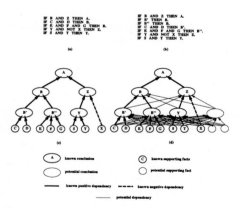

Fig. 1. Construction of KBANNs: (a) Original knowledge base (b) rewritten knowledge base (c) network constructed from rewritten knowledge base (d) network augmented with additional neurons and weights.

Given a set of if-then rules, disjunctive rules are rewritten as follows: The consequent of each rule becomes the consequent of a single antecedent which in turns becomes the consequent of the original rule. This rewriting step is necessary in order to prevent combinations of antecedents from activating a neuron when the corresponding conclusion cannot be drawn from such combinations. These rules are then mapped into a network topology; a neuron is connected via weight H to a neuron in a higher level if that neuron corresponds to an antecedent of the corresponding conclusion. The weight of that connection is $+H$ if the antecedent is positive; otherwise, the weight is programmed to $-H$. For conjunctive rules, the neuron bias [1] of the corresponding consequent is set to $(P - \frac{1}{2})H$ where P is the number of positive antecedents; for disjunctive rules, the neuron bias is set to $\frac{H}{2}$. This guarantees that neurons have a high output when all or any one of their antecedents have a high output for conjunctive and disjunctive rules, respectively. If the given initial domain theory is incomplete or incorrect, a network may be supplemented with additional neurons and weights which correspond to rules still to be learned from data.

If an initial domain theory is sparse, the network constructed from the prior knowledge may be too small for a given learning task. In particular, the number of hidden neurons which along with their weights correspond to intermediate conclusions may be insufficient. A heuristic search technique for dynamically creating hidden neurons during the learning process has been proposed [3]. After initial training, a set of tuning examples is used to identify poorly performing hidden units; new hidden units are added as long as a performance improvement can be observed. It has generally been observed that networks initialized with correct prior knowledge train faster and generalize better compared to networks trained without the benefits of an initial domain theory.

3 The Time Series

We test the networks on two chaotic time series: the Lorenz attractor and a real world seismic time series. It is well known that the Lorenz attractor is chaotic, and we will show that seismic time series has positive Lyapunov exponents.

We generate the Lorenz time series from the Lorenz differential equations given by:

$$\frac{dx}{dt} = \sigma(y - x) \; ; \; \frac{dz}{dt} = xy - bz \; ; \; \frac{dy}{dt} = rx - y - xz$$

The Lorenz time series is shown in Figure 2. We calculated the largest Lyapunov exponent (LLE) for our Lorenz time series and found it to be 0.235.

The seismic time series is obtained from actual measurements made in a South African gold mine. Most seismic monitoring systems are based on transducers. The transducers are usually used to measure ground motion relative to an inertial axis. A seismic monitoring system is obtained by placing transducers throughout the volume rock of interest. The signal is converted to an electrical signal, which can then be manipulated to extract useful information.

Seismic data passes through five stages: (1) monitoring each sensor continuously to decide when the signal becomes significant (called triggering); (2) ensuring that the signal represents a seismic event (validation); (3) deciding which

---

[1] The neuron bias offsets the sigmoidal discriminant function; it is not to be confused with the inductive bias of the learning process.

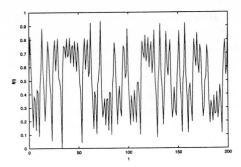

Fig. 2. The Lorenz Time Series

records from which sensors represent the same event (association); (4) extracting source and path parameters from the raw ground motion (seismological processing); (5) and inferring from past records what processes are acting within the monitored volume (interpretation) [4]. The time series we will model comes from stage four. One of the source parameters extracted, and the time series we will use, is the radiated seismic energy. The radiated seismic energy is extracted by using the following formula:

$$E = 4\pi \rho V_c S_{V2}$$

with ρ the rock density, V_C a body wave velocity at the source region, and S_{V2} the velocity spectrum given by

$$S_{V2} = 2 \int_0^\infty V^2(f) df$$

with V the ground velocity.

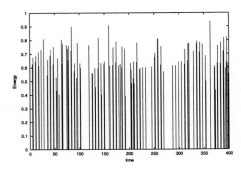

Fig. 3. The Seismic Time Series: Each point represents the energy of the biggest event within a four hour period, log-scaled between zero and one.

The LLE for the seismic time series was found to be 0.85. The time series is shown in Figure 3. The ultimate goal of seismic monitoring systems is to prevent accidents in mines by predicting big seismic events which can result in rock bursts.

4 KBANN Training

KBANN was designed to encode Boolean rules into a neural network. Since we will be encoding rules for a real valued time series, some adaptations must be made to the normal algorithm. We encode the rules in a Time Delay Neural Network (TDNN). It has been shown that TDNN's can model various types of systems, including chaotic systems. For the purposes of this paper, we will use a one-dimensional TDNN, but the principle can easily be extended to multidimensional time series.

4.1 Rule Extraction from a Time Series

We wish to speed up training of the neural network. We achieve this by encoding some form of prior knowledge into the network prior to training. The idea is to choose rules to bias the output of the TDNN high when it should be high. The extracted rules form the first layer of the TDNN. All these rules are then combined by a disjunctive rule at the output (see Figure 4).

We use a threshold α to determine which data points will be covered by rules. The time series is traversed from start to end, and whenever a real value $v > \alpha$ is found, a rule is generated. When the time series is exhausted, duplicate rules are removed. The time series can be traversed again to ensure that all the rules are valid. A rule is said to be valid when it causes no contradictions. A contradiction occurs when a rule causes the output to be high when the output should have been low. Note that each rule will eventually correspond to a neuron in the hidden layer. Thus, having too many rules will result in an excessively large network. Therefore, we choose this threshold fairly high.

To generate a rule, the n time series data points (with n the number of lags in the TDNN) leading to the rule data point are used as input to the rule. Since the time series is real valued, and KBANN encodes Boolean rules, the inputs must be transformed to binary values. Thus, threshold β must be chosen such that real values below β get assigned as zero and those equal or above as one. The choice of the threshold will influence the form of the extracted rules. We suggest using the mean of the time series as a threshold for binary encoding. Note that the size of the rule, i.e. the number of antecedents, will be equal to the number of lags of the TDNN.

4.2 Prior Knowledge Encoding

The extracted rules are encoded as conjunctive rules in the first layer of the TDNN using the normal KBANN method as described in Section 2. All the

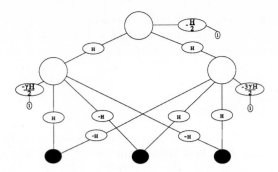

Fig. 4. The resultant network for the encoding of two rules: {100; 011}

conjunctive rules are then combined by one disjunctive rule in the second layer. This rule becomes the output.

Some adjustments to the normal KBANN method need to be made though, especially with respect to the bias values. Since KBANN was designed with binary input values in mind, it expects its inputs to be either one or zero. Real valued time series (scaled between zero and one) are seldom one. Thus, rules will seldom be activated since the neuron bias values for the conjunctive rules are too big to overcome. Therefore, we scale the bias values by a factor γ to adjust for this effect. Note, that the value of γ depends on the choice for β. During experiments, we found the output of the network quite sensitive to γ.

Since the network was designed with positive rules in mind, the output of the network prior to training will often be relatively high for most of the data points. If desired, the H value used to program the disjunctive rule, can be scaled by a factor δ. This will have the effect of adjusting the output to lower values. However, we expect error backpropagation training to quickly adjust this error automatically.

We design the network such that the input at the time delay must fit a rule exactly to activated the rule. Thus, we cannot randomly initialize weights for which the input should be zero; the collective effect of many small weights multiplied by large real valued inputs may be enough to active the rule. This effect is made worse by the downwards scaling of the bias as discussed above. Thus, we program the weights for zero inputs to be $-H$.

It is important to add a number of unprogrammed hidden neurons to allow the network to search for a better solution. During experiments, we often found that for larger H values, the rules are kept almost intact, with only small changes to the weights. Thus, if too little extra learning capability in the form of unprogrammed hidden neurons is available, the final result will not be as good.

5 Experiments

5.1 The Lorenz Time Series

We chose five time lags for the TDNN to model the Lorenz time series. Thus, all the conjunctive rules will have five inputs. We chose $\alpha = 0.9$ and $\beta = 0.5$. We found empirically that $\gamma = 0.3$ yields a good result. We chose not to scale the output and thus have $\delta = 1$. We extracted 48 rules, 5 of which remained after validation. The rules extracted were: $\{00111; 01111; 00001; 00011; 11111\}$. We chose the value of H as 2.0.

The training time for the initialized network is significantly less than that for an uninitialized network - often half the training time. The network constructed from the extracted rules was augmented with 15 additional hidden neurons. The training times for ten uninitialized networks, each with five time lags and 20 hidden neurons is compared to the training time of the constructed network. The results for one-step prediction on seen and unseen data is shown in Figure 5.

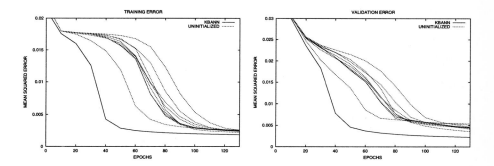

Fig. 5. Training and Validation Error for the Lorenz Time Series

5.2 The Seismic Time Series

The TDNN for the seismic time series has 40 input neurons. We chose $\alpha = 0.85$, $\beta = 0.65$, $\gamma = 0.3$, and $\delta = 1$. This resulted in 12 rules, with no invalid rules. We used the procedure of Snyders *et al* to determine the value of H [5]. The heuristic suggested using 1.0, and a few test cases indicated that this was indeed a good choice. We added 18 uninitialized hidden neurons to obtain the final network with 30 hidden neurons.

We compare the training results for the initialized network with those of no prior encoding. The result shown in Figure 6 suggests that the KBANN network reaches a good model of the training data much faster than training without prior knowledge.

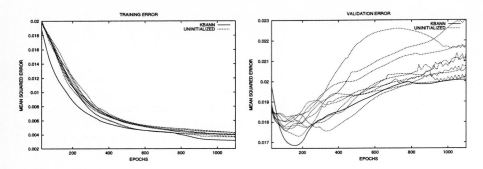

Fig. 6. Training and Validation Error for the Seismic Time Series

6 Conclusions

We proposed a method for extracting rules and encoding them into a neural network by using the KBANN method. We performed the experiment on two time series: the Lorenz system and a real world time series obtained from a seismic monitoring system in a gold mine. The results for the Lorenz system showed that for most runs, training time was halved. The generalization performance was also increased and the initialized network reached a lower MSE than any of the uninitialized networks.

Even though the effect was less pronounced for the seismic time series, the validation error reached by the initialized network reached a lower value than any of the uninitialized networks and remained lower than most of the uninitialized networks. Thus we conclude that rule extraction and encoding can prove beneficial to other time series as well.

References

1. Lacher, R.C., Hruska, S.I., Kuncicky, D.C. Clarke: Backpropagation learning in expert networks. IEEE Transactions on Neural Networks **3** (1) (1992) pp. 62–72
2. Towell, G.G., Shavlik, J.W.: Knowledge-based artificial neural networks. Artificial Intelligence **70** (1,2) (1994) pp. 119–160
3. Opitz, D., Shavlik, J.W.: Dynamically Adding Symbolically Meaningful Nodes to Knowledge-Based Neural Networks. Knowledge-Based Systems (1996) pp. 301–311
4. Mendecki, A.J.: Seismic Monitoring in Mines Chapman and Hall, London, UK (1997)
5. Snyders, S., Omlin, C.W.: What Inductive Bias Gives Good Neural Network Training Performance? International Joint Conference on Neural Networks (2000) pp. 445–450
6. Ishikawa, M.: Structural Learning and Rule Discovery, in Knowledge-Based Neurocomputing, Cloete, I., Zurada, J.M. (eds) MIT Press, London, England (2000) pp. 153–206
7. Drossu, R., Obradović, Z.: Datamining Techniques for Designing Neural Network Time Series Predictors, in Knowledge-Based Neurocomputing, Cloete, I., Zurada, J.M. (eds) MIT Press, London, England (2000) pp. 325–367

Using Artificial Neural Network to Define Fuzzy Comparators in FSQL with the Criterion of some Decision-Maker

R. Carrasco[1], J. Galindo[2], A. Vila[3]

[1] Caja General de Ahorros de Granada, Granada (Spain) Tél.: +34 958 244 783
rcarrasco@caja-granada.es
[2] Dpto. de Lenguajes y Ciencias de la Computación, Universidad de Málaga (Spain)
ppgg@lcc.uma.es
[3] Dpto. de Ciencias de la Computación e I.A., Universidad de Granada (Spain)
vila@decsai.ugr.es

Abstract. At present we have a FSQL server available for Oracle© Databases, programmed in PL/SQL. This server allows us to query a Fuzzy or Classical Database with the FSQL language (Fuzzy SQL). The FSQL language is an extension of the SQL language which permits us to write flexible (or fuzzy) conditions in our queries to a fuzzy or traditional database. In this paper we have incorporated a method of ranking fuzzy numbers using Neural Networks to compare fuzzy quantities in FSQL. The main advantage is that any user can to train his own fuzzy comparator for any specific problem We consider that this model satisfies the requirements of Data Mining systems (high-level language, efficiency, certainty, interactivity, etc) and this new level of personal configuration makes the system very useful and flexible.

1 Introduction

In the last years, the management applications have been popularized and widely extended. In these applications, the database management is very important. Thus, many times the DBMS (Database Management Systems) are used by the final users directly or through an interface program (front-end).

On the other hand, the fuzzy databases have been developed in the last years, rising up different models [14], among which they highlight the Prade-Testemale model, the Umano-Fukami model, the Buckles-Petry model, the Zemankova-Kaendel model and the GEFRED model by Medina-Pons-Vila [13]. This last model represents an eclectic synthesis of the different models which have appeared to deal with the problem of the representation and management of fuzzy information in relational databases. Besides, for the GEFRED model, the FSQL language [8,9,10], a fuzzy (or flexible) query language based on the SQL language, has been defined. This language allows us to express sentences taking into account the characteristics of imprecise information. Thus, as we will see, the FSQL queries allow to express fuzzy conditions, to calculate fulfillment degrees, to establish fulfillment thresholds... Some applications for FSQL Server can be found, for example, in [3,4,10,11].

First, we include a brief explanation of the main advantages of the FSQL SELECT sentence, in order to express fuzzy queries (a more detailed description of this and other FSQL sentences can be found in [8,10]). Then, we define a method of ranking fuzzy numbers using Artificial Neural Networks (ANN). After, we use the defined comparison method to compare fuzzy quantities in FSQL, allowing any user to train his own fuzzy ANN comparator. Thus, the trained ANN will be a special comparator for any specific problem.

We consider that this model satisfies the requirements of Data Mining (DM) systems [5,6,7] (high-level language, efficiency, certainty, interactivity, etc) and this new level of personal configuration makes the system very useful and flexible. Finally, we present an example used in our test, and we suggest some conclusions.

The example is in the context of a bank. This area needs a Data Mining system tailored to its needs, because this area manages very large databases and these data has a very concrete meaning. Thus, data must be treated according to this meaning.

Table 1. Some Fuzzy Comparators for FSQL.

fcomp	Significance	CDEG(A F_Comp B)
FEQ	Possibly Fuzzy Equal	$= \sup_{d \in U} \min(A(d), B(d))$, where U is the domain of A, B. A(d) is the degree of the possibility for $d \in U$ in the distribution A
FGT	Possibly Fuzzy Greater Than	$= 1$ if $\gamma_A \geq \delta_B$ $= \dfrac{\delta_A - \gamma_B}{(\delta_B - \gamma_B) - (\gamma_A - \delta_A)}$ if $\gamma_A < \delta_B \ \& \ \delta_A > \gamma_B$ $= 0$ otherwise
FGEQ	Possibly Fuzzy Greater or EQual	$= 1$ if $\gamma_A \geq \beta_B$ $= \dfrac{\delta_A - \alpha_B}{(\beta_B - \alpha_B) - (\gamma_A - \delta_A)}$ if $\gamma_A < \beta_B \ \& \ \delta_A > \alpha_B$ $= 0$ otherwise
MGT	Possibly Fuzzy Much Greater Than	$= 1$ if $\gamma_A \geq \delta_B + M$ $= \dfrac{\gamma_B + M - \delta_A}{(\beta_A - \alpha_A) - (\gamma_B - \delta_B)}$ if $\gamma_A < \delta_B + M \ \& \ \delta_A > \gamma_B + M$ $= 0$ otherwise M is the minimum distance to consider two attributes as very separate. M is defined in FMB for each attribute

2 FSQL: a Language for Flexible Queries

The FSQL language [8,9,10] extends the SQL language to allow flexible queries. FSQL has incorporate a Fuzzy Meta Objects so that make new definitions of operations and data are very flexible. We have extended the SELECT command to express flexible queries and, due to its complex format, we only show an abstract with the main extensions added to this command:

- **Fuzzy Comparators:** In addition to common comparators (=, >, etc), FSQL includes fuzzy comparators. There are some different kinds of fuzzy comparators. The most important is used to compare two trapezoidal possibility distributions A, B with A=\$[$\alpha A, \beta A, \gamma A, \delta A$] B=\$[$\alpha B, \beta B, \gamma B, \delta B$] (see Figure 1). Definitions of some of these fuzzy comparators are shown in Table 1. In the same way as in SQL, fuzzy comparators can compare one column with one constant or two columns of the same type. More information can be found in [9,10]. These definitions (Table 1) are based in typical possibility comparators in fuzzy set theory. However, this definitions are not quite satisfactory in all contexts. Thus, new definitions are interesting, like the definitions presented in this paper.
- **Fulfillment Thresholds γ:** For each simple condition a Fulfillment threshold may be established with the format <condition> THOLD γ, indicating that the condition must be satisfied with a minimum degree γ in [0,1] fulfilled.
- **CDEG(<attribute>) function**: This function shows a column with the Fulfillment degree of the condition of the query for a specific attribute, which is expressed in brackets as the argument.
- **Fuzzy Constants**: In FSQL we can use and store all of the fuzzy constants which appear in Table 2.

Table 2. Fuzzy constants of FSQL

F. Constant	Significance
UNKNOWN	Unknown value but the attribute is applicable
UNDEFINED	The attribute is not applicable or it is meaningless
NULL	Total ignorance: We know nothing about it
A=\$[$\alpha_A, \beta_A, \gamma_A, \delta_A$]	Fuzzy trapezoid ($\alpha_A \leq \beta_A \leq \gamma_A \leq \delta_A$): See Figure 1
\$label	Linguistic Label: It may be a trapezoid or a scalar (defined in FMB)
[n, m]	Interval "Between n and m" ($\alpha_A = \beta_A = n$ and $\gamma_A = \delta_A = m$)
#n	Fuzzy value "Approximately n" ($\beta_A = \gamma_A = n$ and n-$\alpha_A = \delta_A$=margin)

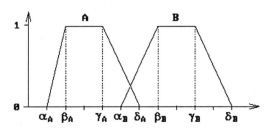

Figure 1. Trapezoidal possibility distributions: A, B.

3 Ranking of fuzzy numbers using ANN

Above we have explained the method to compare fuzzy attributes with FSQL. Of course, there are several methods to do this because the process of comparison of imprecise data is a diffuse process. Often this problem is not solved satisfactorily because it is difficult for a decision-maker to accept that some of the algorithmic methods in the literature [18] could adequately perform his own way to decide. Hence, the decision-maker would like to have a personal method, which implements his own criteria to compare fuzzy numbers especially in the face of conflictive cases like in Figure 2.

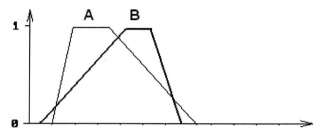

Figure 2. Conflictive case to compare fuzzy numbers.

Solving this problem seems a very appropriate field for ANN because for most people it is easy to see that the trained ANN has learnt with examples given by him. That is not so clear with the algorithmic methods. Requena et al. [15] propose a method of ranking fuzzy numbers using ANN. Next we describe such method shortly.

The problem is to compare two trapezoidal possibility distributions A, B with A=\$$[\alpha_A,\beta_A,\gamma_A,\delta_A]$ B=\$$[\alpha_B,\beta_B,\gamma_B,\delta_B]$. Therefore, the inputs to the network are 8 (4 for each fuzzy number) and 1 output with 3 possible values: A<B, A=B y A>B. The initial objective is to select an ANN structure capable to learn, and therefore to reproduce different comparison fuzzy numbers methods, considering for them the same topology and learning algorithm methods have coherent behavior which may not be found in the human decision-makers, and because it is a simpler and more direct way of obtaining learning examples. If the learning is good from different comparison methods [1,2,12,17] with these characteristics, then we can deduce that this model can be used to learn the behavior of any human decision-maker with a minimal level of coherence.

With the basic structures of 16 neurons with one hidden layer, 12+6 neurons for 2 hidden layers, initial values of 0.5 and 0.1 for the learning rate, backpropagation algorithm, and considering a basic learning set (with conflictive cases included) the authors obtain a sufficiently good learning of the behavior of any real decision-maker, when comparing fuzzy numbers.

4 Fuzzy Comparator in FSQL using ANN

Now, it is necesary to relate the FSQL environment with the ANN method explained. To do it, we will introduce a general definition of FSQL operators using ANN. We suppose that we want to comparate two trapezoidal possibility distributions A, B with A=\$[$\alpha_A,\beta_A,\gamma_A,\delta_A$] B=\$[$\alpha_B,\beta_B,\gamma_B,\delta_B$]. CDEG is used in FSQL to show a column with the fulfillment degree of the fuzzy condition. Therefore we will proceed to define CDEG for each fuzzy comparator using a procedure based on the ANN above explained.

Previously, we defined the function *fcomp_ann* (A, B, fcomp, dpc_fcomp$_i$) where:
- **fcomp** is a fuzzy comparator for FSQL (see Table 1)
- **dpc_fcomp$_i$** is the decisión personal criteria of the user *i* for the comparator *fcomp*. This parameter contains all information above the ANN trained by que user *i* (weights matrix) for the comparator *fcomp*.

Thus the user *i* trains a network (based on the ANN above explained) for the comparator *fcomp* thought a set of examples (fuzzy numbers with intersection with conflictive cases includes). For this examples he decide if A *fcomp* B is true except if the *fcomp* is FEQ then the user must to give the degree fulfillment in this set {0.2, 0.4, 0.6, 0.8}. Therefore the possible values of *fcomp_ann* are:

- If fcomp <> {FEQ}
  fcomp_ann (A, B, fcomp, dpc_fcomp$_i$) = 0 if *A fcomp B* with *dpc_fcomp$_i$*
  $\qquad\qquad\qquad\qquad\qquad\qquad\qquad$ = 1 otherwise
- If fcomp = {FEQ}
  fcomp_ann (A, B, fcomp, dpc_fcomp$_i$) = {0.2, 0.4, 0.6, 0.8}
  $\qquad\qquad\qquad\qquad\qquad\qquad\qquad$ degree fulfillment *A FEQ B* with
  $\qquad\qquad\qquad\qquad\qquad\qquad\qquad$ *dpc_fcomp$_i$*

How we have seen, CDEG returns the fulfillment degree in [0,1] of the fuzzy condition. But he function *fcomp_ann* defined (for the comparators of inequality) only allows us ranking the possibility distributions A and B. Therefore we needed a method in order to express such fulfillment degree. Requena et al. [16] to express the distance between two fuzzy numbers define the decision personal index DPI as an application from the fuzzy numbers set to raking in $\Re$ (real line). However when FSQL compare A and B do not have information on the rest of fuzzy numbers which it is a dynamic information (possible values of the fuzzy attributes). In order to solve this problem us we have developed the following method.

Firstly we define the function *df_ann (A, fcomp, dpc_fcomp$_i$)* which returns a value in [0,1] which is the crisp value that represents to *A* with the criteria of the user *i* learnt by the ANN for the *fcomp* comparator:

$$df\_ann\ (A, fcomp, fcomp, dpc\_fcomp_i) = (a_1 + a_2) / 2$$

where $a_1 = \text{Inf } \{a \in [\alpha_A, \delta_A] / fcomp\_ann\ (a, A, fcomp, dpc\_fcomp_i) = 1\}$
$a_2 = \text{Sup } \{a \in [\alpha_A, \delta_A] / fcomp\_ann\ (A, a, fcomp, dpc\_fcomp_i) = 1\}$

For the implementation of this function we have used a method of binary search in order to find a_1 and a_2. In practique to try to be more precise than the third decimal (for a_1 and a_2) place does not produce any improvement to represent A.

We have that $Ham\ (a,b)_{max,min}$ is a function witch returns the absolute difference between two real numbers a y b previously convert to $[0,1]$:

$$Ham\ (a,b)_{max,min} = |\ [(1 - max/(max - min)) + a * (1 / (max - min))]$$
$$- [(1 - max/(max - min)) + b * (1 / (max - min))]|$$

Now we proceed to define each one of the FSQL comparators considering that $max=\text{Max}(\delta_A, \delta_B)$ and $min=\text{Min}(\alpha_A, \alpha_B)$. The following definition can be incorporate without problems to the Fuzzy Meta Objects of FSQL:

$$\text{CDEG (A FGT B)} = \begin{cases} 1, & \text{if } \gamma_A >= \delta_B \\ 0, & \text{if } \gamma_B >= \delta_A \text{ or } fcomp\_ann\ (A, B, FGT, dpc\_FGT_i)=0 \\ Ham_{max,min}\ [(df\_ann\ (A, FGT, dpc\_FGT_i), \\ \qquad (df\_ann\ (B, FGT, dpc\_FGT_i)], \text{ otherwise} \end{cases}$$

$$\text{CDEG (A FGEQ B)} = \begin{cases} 1, & \text{if } \gamma_A >= \beta_B \\ 0, & \text{if } \beta_B >= \delta_A \text{ or } fcomp\_ann\ (A, B, FGEQ, dpc\_FGEQ_i)=0 \\ Ham_{max,min}\ [(df\_ann\ (A, FGEQ, dpc\_FGEQ_i), \\ \qquad (df\_ann\ (B, FGEQ, dpc\_FGEQ_i)], \text{ otherwise} \end{cases}$$

$$\text{CDEG (A FEQ B)} = \begin{cases} 1, & \text{if } \alpha_A=\alpha_B \text{ and } \beta_A= \beta_B \text{ and } \gamma_A=\gamma_B \text{ and } \delta_A=\delta_B \\ 0, & \text{if } \delta_A < \alpha_B \text{ or } \delta_B < \alpha_A \\ fcomp\_ann\ (A, B, FEQ, dpc\_FEQ_i), \text{ otherwise} \end{cases}$$

For the next comparator (MGT) we define $A_M=\$[\alpha_A-M, \beta_A-M, \gamma_A-M, \delta_A-M]$ with M is the minimum distance to consider two attributes as very separate. M is defined in Fuzzy Metaknowledge Base (FMB) for each attribute.

$$\text{CDEG (A MGT B)} = \begin{cases} 1, & \text{if } \gamma_A >= \delta_B+M \\ 0, & \text{if } \delta_A <= \delta_B+M \text{ or } fcomp\_ann\ (A,B,FGT, dpc\_FGT_i) = 0 \\ Ham_{max,min}\ [(df\_ann\ (A_M, FGT, dpc\_FGT_i), \\ \qquad (df\_ann\ (B, FGT, dpc\_FGT_i)], \text{ otherwise} \end{cases}$$

5 Example

This system (FSQL and ANN comparators) has been applied in a context of bank customers in a real life situation. Relevant attributes have been identified by the banking expert: payroll, credit card use level and average account balance. Payroll is a binary attribute that indicates if the client receives payroll through the financial company (value Y) or not (value N). The level of use of the credit card is represented by fuzzy values obtained through an analytic study in the company data warehouse system. The average account balance store fuzzy values based on the average of the last 12 months.

In this context, the following kind of queries are very useful and with a trained ANN for fuzzy comparators, the bank can obtain information from the database that it is not explicitly. Of course, the goodness of the resulting values depend on the training stage.

```
SELECT  AccountNumber, CDEG(*)
FROM    Customers
WHERE   Payroll = 'N'
   AND  Credit_Card_Use FEQ $Medium THOLD 0.7
   AND  Balance          FGT #300000 THOLD 0.7
ORDER BY 2 DESC;
```

6 Conclusions

The comparison of fuzzy quantities, a very important aspect in Data Mining problems [6], is not solved satisfactorily because it is difficult for a decision-maker to accept that some of the existent algorithms could adequately perform his own way to decide. Solving this problem seems a very appropriate field for ANN because for most people it is easy to see that the trained ANN has learnt with examples given by him. That is not so clear with the algorithmic methods. Requena et al. [14] propose a method of ranking fuzzy numbers using ANN.

In this paper we have incorporated a analogous method to compare fuzzy quantities in FSQL (Fuzzy SQL), an extension of SQL language in order to allow to store and to manage fuzzy values in a database. Thus any user can to train his own fuzzy ANN comparator for any specific problem. Besides, the computation speed is very good because the ANN topology.

FSQL includes a Fuzzy Meta Objects so that make new definitions of operations and data are very flexible. We have already applied FSQL to Data Mining processes [3,4] and others applications [10,11]. Therefore we incorporate the model of ANN to the Data Mining process.

References

[1] S.M. Bass, H. Kwakernaak, "Rating and ranking of multiple-aspect alternatives using fuzzy sets". Automatica 13, pp 47-58, 1997.
[2] L. Campos, A. González, "A subjective approach for ranking fuzzy numbers". Fuzzy Sets and Systems, 29, pp 145-153, 1989.
[3] R.A. Carrasco, J. Galindo, M.C. Aranda, J.M. Medina, M.A. Vila, "Classification in Databases using a Fuzzy Query Language". 9th International Conference on Management of Data, COMAD'98, Hyderabad (India), December 1998.
[4] R.A. Carrasco, J. Galindo, M.A. Vila, J.M. Medina, "Clustering and Fuzzy Classification in a Financial Data Mining Environment". 3rd International ICSC Symposium on Soft Computing, SOCO'99, pp. 713-720, Genova (Italy), June 1999.
[5] M. Chen, J. Han, P.S. Yu "Data Mining: An overview from a Data Base Perspective". IEEE Transac. On Knowledge and Data Engineering, Vol 8-6 pp. 866-883, 1996.
[6] M.A. Vila, J.C. Cubero, J.M. Medina, O. Pons "On the use of Soft Computing Techniques in Data Mining Problems". Technical report #DECSAI-96-01-12. Department of Computer Science and Artificial Intelligence. Universidad de Granada. Spain (1997).
[7] W.J. Frawley, G. Piatetsky-Shapiro, C.J. Matheus "Knowledge Discovery in Databases: An Overview" in G. Piatetsky-Shapiro, W.J. Frawley eds. "Knowledge Discovery in Databases" pp. 1-31, The AAAI Press, 1991.
[8] J. Galindo, J.M. Medina, O. Pons, J.C. Cubero, "A Server for Fuzzy SQL Queries", in "Flexible Query Answering Systems", eds. T. Andreasen, H. Christiansen and H.L. Larsen, Lecture Notes in Artificial Intelligence (LNAI) 1495, pp. 164-174. Ed. Springer, 1998.
[9] J. Galindo, J.M. Medina, A. Vila, O. Pons, "Fuzzy Comparators for Flexible Queries to Databases". Iberoamerican Conference on Artificial Intelligence, IBERAMIA'98, pp. 29-41, Lisbon (Portugal), October 1998.
[10] J. Galindo, "Tratamiento de la Imprecisión en Bases de Datos Relacionales: Extensión del Modelo y Adaptación de los SGBD Actuales". Ph. Doctoral Thesis, University of Granada (Spain), March 1999.
[11] J. Galindo, J.M. Medina, J.C. Cubero, O. Pons, "Management of an Estate Agency Allowing Fuzzy Data and Flexible Queries". EUSFLAT-ESTYLF Joint Conference, pp. 485-488, Palma de Mallorca (Spain), September 1999.
[12] R. Jain, "Tolerance Analisys using fuzzy sets". Internat. J. Systems Sci, 7, pp 1393-1401, 1976.
[13] J.M. Medina, O. Pons, M.A. Vila, "GEFRED. A Generalized Model of Fuzzy Relational Data Bases". Information Sciences, 76(1-2), pp. 87-109, 1994.
[14] F.E. Petry, ``Fuzzy Databases: Principles and Applications" (with chapter contribution by Patrick Bosc). International Series in Intelligent Technologies. Ed. H.-J. Zimmermann. Kluwer Academic Publishers (KAP), 1996.
[15] I. Requena, M. Delgado, J.L. Verdegay, "Artificial neural networks learn the criteria of a real decision-maker to compare fuzzy numbers". Fuzzy Sets and Systems 64, pp 1-19, 1994.
[16] I. Requena, M. Delgado, J.L. Verdegay, "A decision personal index of fuzzy numbers based on neural networks". Fuzzy Sets and Systems 73, pp 185-199, 1994.
[17] R.R. Yager, "Ranking fuzzy subsets over the unit interval". Proc.. 1978 CDC, pp 1435-1437, 1978.
[18] Q. Zhu, E.S. Lee, "Comparison and ranking of fuzzy numbers". Fuzzy Regression Analysis. J. Kacpryk and Fedrizzi, eds. Omnitech Press, Warsaw, Poland, pp 21-44, 1991.

Predictive Classification for Integrated Pest Management by Clustering in NN Output Space

M. Salmerón[1], D. Guidotti[2], R. Petacchi[2], and L.M. Reyneri[3]

[1] Corresponding author. E-mail: `moises@ugr.es`
Department of Computer Architecture and Technology,
Facultad de Ciencias / E.T.S.I.I., Universidad de Granada.
Campus Fuentenueva s/n. E-18071 GRANADA (SPAIN)
[2] Agricultural Sector, Agricultural Entomology Section,
Scuola Superiore SantAnna. Via Carducci 40. I-56100 PISA (ITALY)
[3] Department of Electronics, Politecnico di Torino.
c.so Duca degli Abruzzi 24. I-10129 TORINO (ITALY)

Abstract. In this paper we consider the successful hybridation of a two modern computational schemes, Clustering and Neural Networks, for the Predictive Classification of the future value of insect infestation levels for Integrated Pest Management (IPM) of olive groves. The predictive classification techniques employed allow managers to improve their work in two ways: first, by reducing sampling demands of the variables involved, which is a costly process; and second, by recognizing potential infestation problems a up to two weeks beforehand, in order to optimize the use of pesticide chemical products and thus reduce financial costs.

1 Introduction and Problem Posing

Neural network architectures and algorithms and clustering procedures have been developed over the past 10-20 years, and it has been experimentally found that they are methods of varying applicability to tasks of prediction, classification, system identification and parameter estimation in general. Several researchers claim these approaches are theoretically rather weak, but the range of successful applications expands every year.

One promising application for these techniques is the prediction of the future value of a time series or (what is equivalent), the estimation or prediction of the state of a dynamical system of more or less complexity at a future instant. Classical statistical time-series-analysis techniques are found to be somewhat inadequate when applied to systems with a medium to high degree of complexity. Neural and hybrid techniques then come into play as adaptive, fault-tolerant approximators and hence predictors. Depending however on the complexity of the problem dynamics, these procedures may fail when our goal is to get a numeric value of a prediction. Since normally the manager's objective will also be of a non-numeric, qualitative kind, a better approach in these cases may be to carry out *Predictive Classification* (a classification of the future value as belonging to one of several predefined groups or classes).

In this work we analyze the combination of Neural and Clustering techniques for Predictive Classification of olive fruit fly (*Bactrocera oleae Gmelin.*) infestation in the Liguria Region (north-east Italy). As the predicted variable we use an infestation index (attiva) used to determine a threshold for chemical treatment.

2 Predictive Classification and Neural Networks

Consider a single stochastic time series in discrete time represented by $\{x(t) : t = 1, 2, \ldots\}$ where the time variable t is an index within a discrete set (usually, the set $\mathbb{N}$). The range of series x is assumed to be the real number set $\mathbb{R}$. An *endogenous regular predictor* for x is defined as a 5-tuple $\mathcal{P}(x, k, N_\mathcal{P}, \delta_\mathcal{P}, \mathcal{R}_\mathcal{P})$ that indicates the association of a value $\hat{x}(t + k) \in \mathbb{R}$ for each real-valued *predictor vector* of the form

$$\mathbf{x}_\mathcal{P}(t) \equiv (x(t - \delta_\mathcal{P} \cdot (N_\mathcal{P} - 1)), x(t - \delta_\mathcal{P} \cdot (N_\mathcal{P} - 2)), \ldots, x(t - \delta_\mathcal{P}), x(t)) \in \mathbb{R}^{N_\mathcal{P}} . \quad (1)$$

Parameters $\delta_\mathcal{P}$, k and $N_\mathcal{P}$ are positive integers. k is usually called the *prediction horizon* and $N_\mathcal{P}$ indicates the number of predictors used, in this case lagged past values of the process $\{x(t)\}$. A general case would not require components of vector $\mathbf{x}_\mathcal{P}$ to be drawn from the same series we want to predict; they could also be values from other, "exogenous" processes (series or variables) considered useful or relevant to the prediction task. So, in the general case we have an *exogenous regular predictor*. In any case, prediction itself is performed by means of a rule or algorithm denoted by $\mathcal{R}_\mathcal{P}$ that implements the required mapping between the predictor vector $\mathbf{x}_\mathcal{P}(t)$ and the estimate $\hat{x}(t + k)$.

In many time-series applications, an accurate value for $\hat{x}(t + k)$ is required; this is the case, for example, in the study of chaotic processes such as the Mackey-Glass equation [10], because poor precision can lead to great divergence in the future predicted behaviour when estimations are iterated. Since the best predictor would require knowledge of the full probability structure of the process $\{x(t)\}$ to be incorporated into $\mathcal{R}_\mathcal{P}$, algorithms have been developed that construct more or less approximate predictors. The task of finding these algorithms is what constitutes *Time Series Analysis* [3]. Recently, Artificial Neural Networks (ANNs) techniques have been considered [11] as they claim to be universal function approximators [7] and thus candidates to implement the $\mathcal{R}_\mathcal{P}$ mapping to any desired degree of accuracy.

The problem is that the stochastic nature of $\{x(t)\}$ makes it impossible to exactly predict any future value $x(t + k)$. Even when the complete probability properties of x are known, the predictors must be compared to each other in terms of *expected values* of the prediction error. We no longer have a *function approximation* problem in the strict sense, and the universal approximation theorem is reduced to the fact that a neural network may find local minima of the prediction error surface and lead to suboptimal or biased forms of $\mathcal{R}_\mathcal{P}$.

2.1 Notion of Predictive Classifier

In other contexts, as indicated above, a numerical prediction may be neither possible nor even useful for the purposes of the application, especially when managerial decision support systems are involved. A qualitative *indicator response*, which may be simply a sign prediction (+1 or -1 whether the future value is expected to belong to a predefined class of interest or not) is of much more use for subsequent analysis and decisions. It may also be a much easier variable to predict and the difference in performance between optimal and sub-optimal predictors is usually less problematic with respect to "high-level" managerial uses. Therefore, in this work we restric ourselves to $\mathcal{R}_\mathcal{P}$ procedures implementing predictors $\mathcal{P} = (x, k, N_\mathcal{P}, \delta_\mathcal{P}, \mathcal{R}_\mathcal{P})$ that operate from $\mathbf{x}_\mathcal{P}$ over the binary-valued set $\{-1,+1\}$ (or equivalently, $\{0,1\}$), the mapping being defined in abstract form by

$$\mathcal{R}_\mathcal{P}(\mathbf{x}_\mathcal{P}) = \begin{cases} -1 & \text{if } \mathcal{R}_{\overline{\mathcal{P}}}(\mathbf{x}_\mathcal{P}) = \hat{x}(t+k) \in \Gamma_{-1} \\ +1 & \text{if } \mathcal{R}_{\overline{\mathcal{P}}}(\mathbf{x}_\mathcal{P}) = \hat{x}(t+k) \in \Gamma_{+1} \end{cases} \quad (2)$$

where we have a *raw predictor* system $\overline{\mathcal{P}} = (x, k, N_\mathcal{P}, \delta_\mathcal{P}, \mathcal{R}_{\overline{\mathcal{P}}})$ with a scalar range $\Gamma \subset \mathbb{R}$. Over set Γ a *binary partition* is established as $(\Gamma_{-1}, \Gamma_{+1})$ with the usual properties $\Gamma_{-1} \cap \Gamma_{+1} = \emptyset$ and $\Gamma = \Gamma_{-1} \cup \Gamma_{+1}$. The idea here is to use a standard regular predictor to get a gross numerical prediction, and then perform a discrimination between two classes —or, in general, any number of classes defined— in order to associate a final binary value $\hat{x}(t+k) \in \{-1,+1\}$ depending on the partition subset to which this raw predicted value belongs. Formally, this is a *classification procedure* over the raw prediction space Γ, and is also referred to in the statistics literature to *Discriminant Analysis* [8].

For some applications, a managerial user would find easier to interpret an indicator binary signal instead of a raw numerical value. This might also improve prediction accuracy since values that would be considered as bad responses with respect to a numerical prediction error can be counted as correct if the classifier has been constructed to correctly discriminate between ranges of the raw output. In the scalar case, classification is simply based on defining ranges for each class and giving as output the sign assigned to the range to which the raw prediction belongs. This is easily generalized to the case of Γ being a subset of the vector space $\mathbb{R}^r$. In this case $\mathcal{R}_{\overline{\mathcal{P}}}$ is no longer considered a scalar prediction $\hat{x}(t+k)$, but must be seen as a projection $\mathcal{R}_{\overline{\mathcal{P}}} : \mathbb{R}^{N_\mathcal{P}} \longrightarrow \Gamma \subset \mathbb{R}^r$ of the predictor vector $\mathbf{x}_\mathcal{P} \in \mathbb{R}^{N_\mathcal{P}}$ into a space with a different dimension r (normally $r < N_\mathcal{P}$). The transformation should be designed taking into account that classification within this new space must lead to a better class prediction than the "easy" one that we could perform over the original predictor space. In Figure 1 the cases of one-

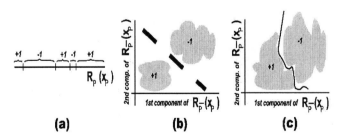

Fig. 1. Examples of low-dimensional Predictive Classification spaces: (a) Scalar classifier output (b) Two-dimensional linear predictive classifier (c) Two-dimensional non-linear predictive classifier.

and two-dimensional predictive classification spaces are schematically shown. General classification such as in case (c) must resort to non-linear separation rules, and NNs and other adaptive approaches are the best candidates to do this; we will examine this issue in subsection 2.3.

2.2 Classification in Neural Output Space

It has been argued that transformation from the predictor space in $N_\mathcal{P}$ dimensions into a new, reduced space in r dimensions followed by classification over the transformed space is a framework that allows us to perform Predictive Classification. To ensure that classification over the transformed space $\Gamma \in \mathbb{R}^r$ performs better for predictive purposes, we could study the conditions theoretically and try to derive the transformation, which would be an elegant but difficult process, or instead depend on techniques such as compaction of information or any adequate preprocessing of the predictor data and determine experimentally whether any improvement is gained.

We believe that techniques such as performing *Principal Component Analysis* (PCA) [9] on the raw predictor data, or preprocessing by non-linear, neural transformations supervisely trained by a classification target can enhance subsequent classification in predictive contexts. In the first case, dimensionality reduction is a by-product of the technique that encourages its use. In the second case, NN preprocessing serves as a *coarse-grain* transformation that hopefully makes posterior classification easier and efficient. This second approach has the clear advantage that available NN software with suboptimal prediction performance in numerical terms can be readily enhanced by the addition of a classification layer to turn it into a useful Predictive Classification tool. The two suggested options are schematized in Figure 2. Note that if the NN option is implemented "from scratch", an even better result will be obtained by training the NN directly with the classification desired as the target.

2.3 Alternatives for the Classification Stage

We now consider several alternatives for the classification step in a Predictive Classification setup. These are the usual choices encountered when we are faced with a standalone classification task, and so they are already widely known by the ANN community; nevertheless, we provide a brief review for the sake of convenience and completeness.

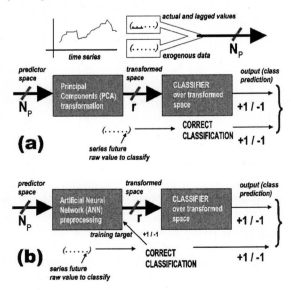

Fig. 2. Two examples of predictor space transformation for Predictive Classification: (a) Dimensionality reduction by a PCA procedure (b) Preprocessing by an ANN. The network acts as a coarse-grain transformation and is subject to supervised training by using the desired classification as the target.

1. ANN architectures such as perceptrons and MLPs have been applied to Discriminant Analysis or Classification tasks ([1] is a survey of application to classification and general pattern recognition problems). These classifiers are more flexible solutions that can handle difficult, non-linear separation regions, as opposed to *Linear Discriminant Analysis* (LDA) procedures. Linear separation is only a good choice when it can be assumed that the statistical distribution of predictor data within each class is multivariate Gaussian with a common covariance matrix for all classes (although class means need not be the same). In this case, we should predict the class for $\hat{x}(t+k)$ as the one maximizing the inner product $\beta_i^T \cdot \mathbf{x}_\mathcal{P}$ where vectors β_i are determined for each class based on Bayesian probability considerations, although it should be noted that the required assumptions are not very likely to be met in practice.
2. Classification may also be done by means of FDA (*Flexible Discriminant Analysis*), employing adaptive fitters such as MARS (Multivariate Adaptive Regression Splines) or BRUTO [6].

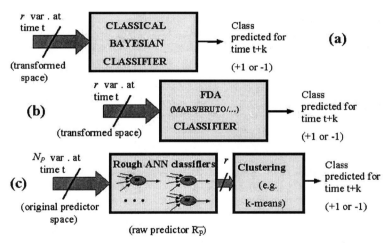

Fig. 3. Different methodologies for Predictive Classification with prediction horizon k: (a) Classical Bayesian classification (b) Improved classifier using Flexible Discriminant Analysis based on MARS / BRUTO / others (c) Classification by several different ANN coarse-grain classifiers followed by a Clustering step.

3. Some form of *Clustering* may be performed, such as the *k-means Clustering* algorithm [5] over an ANN-transformed space. In this setup, we could use different networks as raw scalar predictors (each one trained with the classification target values) and after this perform clustering. Experiments demonstrate that each resulting cluster normally encompasses target-related data, enabling easy classification of new predictor vectors. This is the methodology used for the experiments reported in Section 3.

The alternatives mentioned are illustrated in Figure 3. For the last option (set of ANNs + clustering), two key ideas may be noted: first, there is no special need to perform a very detailed training of the NNs. Rather, we want them to capture coarse-grain features of the training data, and therefore to be regarded as coarse-grain experts over a limited portion of the training data dynamics. Second, we would like to ensure that these NN approximate classifiers are representative enough of a diverse complexity, as represented by, say, the number of neurons or nodes in each layer and/or the training effort and error goal. Further extending this principle, we could also consider the choice of non-linearity, or even the number of layers itself. This might require considerable computational power. As far as this study is concerned, experimental results are reported based only on a range of layer sizes, but using a fixed non-linearity and a fixed two-layer neural architecture.

3 Experimental Setup and Results

We demonstrate the combination of slightly trained raw ANN predictors with k-means clustering to achieve efficient Predictive Classification of olive groves infestation; this prediction would then be integrated into a Pest Management System in the Italian region of Liguria.

3.1 Olive Fly Infestation Process and Measures

In the Liguria Region, known as one of the top world oil-producing areas, oil quantity and quality depend strongly on the infestation levels of the olive fruit fly [12]. The adult of *B.oleae* lays eggs inside olives, and preliminary stages develop and dig galleries inside the olives; adults then develop leaving the olive through an exit hole. Olives with exit holes usually fall to the ground leading to reduced production; olives with mature stages (pupae or 3rd stage larvae) decrease the olive oil quality. In the Liguria climate *B. oleae* develops about 3-4 generations between July and November. Several technicians were employed in 1999 to sample about 100 olives (one per plant) from each of 122 farms. They used a microscope to obtain a visual count of the number of fly eggs, first-, second- and third-stage larvae, pupae and damaged olives with exit holes. This sampling process is very costly and time-consuming.

The data are used to evaluate the necessity of applying chemical treatments (pesticides). A threshold of 15% of active infestation index is used to determine whether pesticide should be applied. The weekly *Attiva* infestation is defined for as the sum of the eggs and the first- and second-stage alive larvae count, that is: $\text{Attiva}(t) = U(t) + L_v(t,1) + L_v(t,2)$ and is a percentage of the sampled olives. In order to reduce sampling costs, a predictive system that warns in advance of any exceeding of treatment threshold could be used to support farm and technician decisions and to avoid sampling olives when it is not needed. The biological aim would be to make a predictor capable to forecasting two weeks in advance *whether the active infestation will exceed the 15% level* using data based on previous dynamics of the fruit fly, and the farm's geographical position and distance from the sea, known to influence *B. oleae* dynamics.

Entomologists have considered it useful to perform a Predictive Classification to try to assign $\text{Attiva}(t+2)$ either to a "+1" class, meaning the infestation is going to exceed the threshold, or "-1" when the future active infestation is predicted to lie below the threshold value. Other aggregate infestation indexes are also used as input information for our procedure. For simplicity, we will not enter into these definitions here. It might be expected that having a higher number of predictors would yield better predictions. This is only true up to a certain point, however, and thus after selecting variables for prediction of the 2 week-ahead Attiva infestation, $\text{Attiva}(t+2)$, some kind of *dimensionality reduction* that keeps only the most relevant variables would be desirable. We performed this with a PCA (*Principal Component Analysis*) procedure, as described in Section 2.2.

We have checked the procedure depicted in Fig. 3 (c) with infestation data from the 1999 season. We defined as the raw predictor components (vector $\mathbf{x}_\mathcal{P}$) the *Attiva*, *Target* (eggs, 1st and 2nd stage larvae, both alive and dead), *Dannosa* (3rd stage larvae, pupae, exit holes) and *Morta* (dead stage) infestation indexes, along with their 2 past lags. We also included the Latitude, Longitude and Distance from the sea, evaluated using spatial modelling. We also used as input the basic infestation components $U(t)$, $L_v(t,1)$, $L_v(t,2)$, $L_m(t,1)$, $L_m(t,2)$, $PA(t)$, $PI(t)$ and $FU(t)$ (the 5 latter being the dead larvae of second and third stages, the active and inactive pupae and the holed olive count, respectively).

Data was available for the 122 farms weekly from Aug 5, 1999 up to Oct 28, 1999. A dataset with predictor vectors was constructed using as target a binary variable representing whether the real Attiva infestation 2 weeks later was up/down the 15 % threshold, and time-evenly divided into a training set (691 predictor vectors with corresponding objectives) and a test set (617 vectors). Because of the differences in ranges, normalization and standardization of the training set was performed using the *MATLAB Neural Network Toolbox* (http://www.mathworks.com/products/neuralnet/). PCA for 98% of the variance was performed afterwards to reduce the predictor vectors from 23 to 11 components. Preprocessing parameters were saved for later application to the test portion. For the training data, 8 two-layer feedforward NNs were trained. Random numbers were uniformly chosen in the ranges [1,20] and [1,40] for the 1st and 2nd layer sizes, respectively; the training α parameter was also varied randomly between 0.01 and 0.50. This ensured some degree of variability/diversity among the NNs. The R-PROP enhanced backpropagation algorithm in the Toolbox was used for training. Trial NNs with a R-squared value less than 0.70 for the target variable were discarded, so when this step finished, we had a set of 8 good networks. Clustering based on the k-means algorithm was then performed over the 8 NN output space by using the maximum distance measure with the open-source R statistical package (http://www.stats.bris.ac.uk/R/).

It was found that the vast majority of points within cluster belonged to the same class; for example, only 6 of 325 points in Cluster 1 had a target value +1, whereas in Cluster 3 more than 95% of the vectors had +1 as a target. This was quite foreseeable from the preceding discussion and allows us to conjecture a rule for the new, real-time arriving vectors: we simply assign as output of the system the target value predominantly appearing in training at the vector's nearest cluster. For example, if we compare the incoming vector with the centroid matrix, and we determine (given the maximum distance measure) that the nearest cluster is Cluster 2, then we can look at the predominant target value obtained within this cluster for the training data, and simply assign this to the output of our predictive classification system. Using this rule, we obtained the classification results for the test dataset for this and another two methods: a Bayesian (LDA) classifier and a FDA/MARS adaptive classifier. The results are reported in Table 1. It is seen that the detection performance is optimal, and so this procedure can be quite useful for detecting in advance the farms most likely to be affected by infestation. This can enable the sampling effort to be concentrated on these farms, and thus chemical treatments can be directed at high-priority farms. The "false alarm" condition occurs in 20.7% of the farms, but the savings on sampling costs achieved by the 79.3% rate of correct results allow the managers to do a lot better than relying on more traditional approaches or on ad-hoc rules. It is especially important that the infestation-missed error rate has been minimized in order to prevent farms from losing production due to a prediction system fault.

4 Conclusions

We have illustrated the potential use of ANN and Clustering to improve the decision-making process in an agricultural context. These techniques allow managers and farms to improve their processes; integration of this new kind of knowledge into a decision-making support system would be an interesting way to enhance their daily work.

MEASURE	# CASES procedure	PCT	# CASES LDA	# CASES FDA-MARS
Detection (target=+1, predict=+1)	212 / 212	100.0	168/212	43/212
Savings (target=-1, predicted=-1)	321 / 405	79.3	354/405	353/405
False alarms (target=-1, predict=+1)	84 / 405	20.7	51/405	52/405
Misses (target=+1, predict=-1)	0 / 212	0.0	44/212	169/212

Table 1. Experimental results: comparison between the proposed Predictive Classifier and LDA or FDA-MARS discrimination as the last (classification) stage of infestation data preprocessed by 8 Neural Networks. "Detection" refers to prediction of future infestation over the 15% threshold when this does in fact occur. "Savings" refers to predicting low infestation when it is really low (thus saving pesticide costs). "False alarms" occur when the procedure predicts future infestation as being over the 15% threshold, but it is really under the threshold (this leads to unnecessary application of pesticide), and "Misses" means the procedure is unable to detect future dangerous (over 15%) olive infestation.

5 Acknowledgements

This work has been sponsored by Italian project COFIN-99 *Applicazioni di Tecniche di Previsione in Agricoltura*.

References

1. Bishop, C.M.: *Neural Networks for Pattern Recognition*. Oxford Univ. Press, 1995.
2. Bock, H.H., ed.: *Classification and Related Methods of Data Analysis*. North-Holland, 1998.
3. Box, G.E.P., Jenkins, G.M., and G.C. Riensel: *Time Series Analysis: Forecasting and Control*. P.Hall, 1994.
4. Dunteman, G.H.: *Introduction to Multivariate Analysis*. Sage Publications, 1984.
5. Hartigan, J.A., and M.A. Wong: *A K-means Clustering Algorithm*. Applied Statistics **28**, pp. 100–108, 1979.
6. Hastie, T., Tibshirani, R., and A. Buja: *Flexible Discriminant Analysis by Optimal Scoring*. Journal of the Americal Statistical Association **89**, pp. 1255–1270, 1994.
7. Haykin, S.: *Neural Networks: a Comprehensive Foundation*. P.Hall, 1999.
8. Huberty, C.J.: *Applied Discriminant Analysis*. Wiley, 1994.
9. Jackson, J.E.: *A User's Guide to Principal Components*. Wiley, 1991.
10. Mackey, M.C. and L. Glass: *Oscillation and Chaos in Physiological Control Systems*. Science **197**, pp. 287–289, 1977.
11. Weigend, A.S. and N.A. Gershenfeld, eds.: *Time Series Prediction: Forecasting the Future and Understanding the Past*. Addison-Wesley, 1994.
12. Zunin, P., Evangelisti, F., Tiscornia, E., and R. Petacchi: *Relation between Bactrocera oleae (Gmel.) Infestation and Oil Chemical Composition: Results of Two-Year Trials in an Eastern Ligurian Olive Grove*. Acta Horticolturae **356**, pp. 395–398, 1994.

Blind Source Separation in the Frequency Domain: A Novel Solution to the Amplitude and the Permutation Indeterminacies

Adriana Dapena and Luis Castedo

Departamento de Electrónica e Sistemas
Universidade da Coruña
Campus de Elviña s/n, 15.071. A Coruña. SPAIN
Tel: ++34-981-167000, Fax: ++34-981-167160, E-mail: adriana@des.fi.udc.es[*]

Abstract. This paper deals with the separation of convolutive mixtures of statistically independent signals (sources) in the frequency domain. The convolutive mixture is decomposed in several problems of separating instantaneous mixtures which are independently solved. In addition, we propose a method to remove the indeterminacies which occur when all the individual separating systems do not extract the sources in the same order and with the same amplitude. We will show that both the permutation and the amplitude indeterminacies can be solved using second-order statistics when the sources are temporally-white.

Keywords: Blind source separation, convolutive mixtures, blind deconvolution.

1 Introduction

The separation of convolutive mixtures of statistically independent signals (sources) is a fundamental problem in signal processing that arises in a large number of applications. The problem is called blind source separation (BSS) when the sources are recovered without resorting to any a priori knowledge of the sources or the mixing system. The BSS problem has been widely studied in the context of neural networks because BSS adaptive algorithms can be interpreted as unsupervised learning rules.

A way to recover the sources from convolutive mixtures is to work in the frequency domain by interpreting a time-domain convolution as several instantaneous mixtures [2, 4, 5, 7]. The main advantage of this interpretation is that the sources at each frequency bin can be recovered using many unsupervised learning algorithms proposed for separating instantaneous mixtures (see [3] and references therein). However, two additional indeterminacies must be solved because the sources can be recovered in a different order (permutation indeterminacy) and with different amplitude (amplitude indeterminacy) in some frequency bins.

---

[*] This work has been supported Xunta de Galicia (PGIDTOOPXI10504PR)

Although both indeterminacies must be removed in order to guarantee the correct reconstruction of the sources, previous work only addresses the permutation problem [2, 4, 5, 7].

The main contribution of this paper is to show that both the permutation and the amplitude indeterminacies can be solved using second-order statistics when the sources are temporally-white. In Section 2, we describe the problem of separating convolutive mixtures in the frequency domain. Section 3 introduces a compact representation of the observations in the frequency domain which will be used in the paper. In Section 4 and in Section 5 we will present a novel solution to the permutation indeterminacy and the amplitude indeterminacy, respectively. Section 6 presents some simulation results and Section 7 contains the conclusions.

2 Problem Statement

We will consider the following signal model. Let $\mathbf{s}(n) = [s_1(n), ..., s_N(n)]^T$ be the vector of N sources whose exact probability density functions are unknown. We assume that the sources are complex-valued, zero-mean, stationary, temporally-white, non-Gaussian distributed, statistically independent and have the same kurtosis sign. The sources arrive at an array of M sensors whose output, denoted by $\mathbf{x}(n) = [x_1(n), ..., x_M(n)]^T$, is a convolutive combination of the N sources

$$\mathbf{x}(n) = \sum_{k=-\infty}^{\infty} \mathbf{A}(k)\mathbf{s}(n-k) \qquad (1)$$

where $\mathbf{A}(k)$ is an unknown $M \times N$ matrix representing the mixing system. It is interesting to note that in the frequency domain, the convolutive mixture (1) takes the form

$$\mathbf{x}[k] = \mathbf{A}[k]\mathbf{s}[k] \qquad (2)$$

where $\mathbf{x}[k]$, $\mathbf{A}[k]$ and $\mathbf{s}[k]$ are formed by the observations, the mixing system and the sources in frequency ω_k, respectively. From (2), we deduce that the convolutive mixture can be interpreted as several instantaneous mixtures. Therefore, we can recover the sources at each frequency bin using a single layer neural network with output

$$\mathbf{y}[k] = \mathbf{W}^H[k]\mathbf{x}[k] \qquad (3)$$

where $\mathbf{W}[k]$ is the $M \times N$ coefficients matrix. Combining both (2) and (3) together, we can express the outputs as follows

$$\mathbf{y}[k] = \mathbf{G}[k]\mathbf{x}[k] \qquad (4)$$

where $\mathbf{G}[k] = \mathbf{W}^H[k]\mathbf{A}[k]$ is the overall mixing/separating matrix. The sources at the frequency ω_k are optimally recovered when each output extracts a single and different source. This means that the optimum matrix $\mathbf{G}[k]$ has the form

$$\mathbf{G}[k] = \mathbf{\Delta}[k]\mathbf{P}[k] \qquad (5)$$

where $\boldsymbol{\Delta}[k]$ is a diagonal matrix and $\mathbf{P}[k]$ is a permutation matrix. If the separating matrix $\mathbf{W}[k]$ is obtained independently at each frequency bin, the sources can be recovered in a different order (permutation indeterminacy) and with different amplitudes (amplitude indeterminacy). Removing both indeterminacies is not an important task when the separation of instantaneous mixtures is considered. However, this task is crucial in the separation of convolutive mixtures because the sources in the time domain are obtained from the outputs in all the frequency bins.

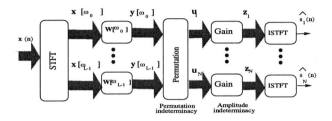

Fig. 1. Separating system in the frequency domain

According to the explanation above, the system shown in Figure 1 can be used to separate convolutive mixtures in the frequency domain. The first stage consists of applying the Short-Time Discrete Fourier Transform (STFT) to the observations. As a result, the convolutive mixture is transformed into several instantaneous mixtures. The sources at each frequency bin can be recovered using many existing algorithms proposed for separating instantaneous mixtures. In the next stages, the permutation and the amplitude indeterminacies are solved and, finally, the Inverse Short-Time Discrete Fourier Transform (ISTFT) is applied to the outputs.

3 Short-Time Discrete Fourier Transform of the Observations

The first stage of the separating system consists of applying the STFT to moving windows of observations. Towards this aim, we split each observation, $x_j(t)$, in R non-overlapped segments of K points, $\mathbf{x}_j(t_r) = [x_j(t_r), x_j(t_r+1), ..., x_j(t_r+K-1)]^T$ where $t_r = rK$, $r = 0 \cdots, R-1$ denotes the window position. Subsequently, we compute the L-points DFT of each window

$$x_j[k, t_r] = \sum_{m=0}^{K-1} x_j(t_r + m) e^{j2\pi km/L}, \quad k = 0, ..., L-1 \quad (6)$$

In a compact form, we can write the observation in the frequency domain as follows

$$x_j[k, t_r] = \mathbf{f}_k^T \mathbf{x}_j^e(t_r), \quad k = 0, ..., L-1 \quad (7)$$

where $\mathbf{f}_k = [1, e^{j2\pi k/L}, ..., e^{j2\pi k(K-1)/L}, ..., e^{j2\pi k(L-1)/L}]^T$ is the Fourier transform vector and the vector $\mathbf{x}_j^e(t_r) = [x_j(t_r), x_j(t_r+1), ..., x_j(t_r+K-1), 0, ..., 0]^T$ is an extended version of $\mathbf{x}_j(t_r)$ obtained by appending $L - K$ zeros. The relationship between observations and sources is approximated as

$$\mathbf{x}[k, t_r] = \mathbf{A}[k]\mathbf{s}[k, t_r] \tag{8}$$

In practice, we assume that the mixing system can be modeled as a FIR filter of order P. In this particular case, the mixing matrix $\mathbf{A}[k]$ are formed by coefficients

$$a_{ij}[k] = \mathbf{f}_k^T \mathbf{a}_i^e \tag{9}$$

where $\mathbf{a}_i^e = [a_i(0), a_i(1), ..., a_i(P-1), 0, ..., 0]^T$ is an extended version of the i-th column of the mixing matrix $\mathbf{A}$. The sources in the frequency domain have the form

$$s_j[k, t_r] = \mathbf{f}_k^T \mathbf{s}_j^e(t_r) \tag{10}$$

where $\mathbf{s}_j^e(t_r) = [s_j(t_r), s_j(t_r+1), ..., s_j(t_r+F-1), 0, ..., 0]^T$ is an extended version of the source vector.

4 Permutation Indeterminacy

In the previous section, we have transformed a convolutive mixture in several instantaneous mixtures. Then, the separating system at each frequency bin, $\mathbf{W}[k]$, can be found using many adaptive algorithms proposed to recover statistically independent sources from instantaneous mixtures of them (see [3] and references therein). If the separating systems are independently adapted, the sources may be extracted in a different order in some frequency bins. In this section we will propose a method to remove this indeterminacy using second-order statistics.

Let us assume that the output $y_i[0]$ extracts the source $s_p[0]$ and $y_l[k]$ extracts the source $s_t[k]$, i.e., $y_i[0] = g_{ip}[0]s_p[0]$ and $y_l[k] = g_{lt}[k]s_t[k]$. The cross-correlation between both outputs is given by

$$E[y_i[0]y_l^*[k]] = g_{ip}[0]\ g_{lt}^*[k]\ E[s_p[0]s_t^*[k]] \tag{11}$$

Using the equation (10) we obtain

$$E[y_i[0]y_l^*[k]] = g_{ip}[0]\ g_{lt}^*[k]\ \mathbf{f}_0^T\ E[\mathbf{s}_p^e(t_r)(\mathbf{s}_t^e(t_r))^H]\ \mathbf{f}_k^*$$

where $\mathbf{f}_k = [1, e^{j2\pi k/L}, ..., e^{j2\pi k(F-1)/L}, ..., e^{j2\pi k(L-1)/L}]^T$, $\mathbf{s}_j(t_r) = [s_j(t_r), s_j(t_r+1), ..., s_j(t_r+F-1), 0, ..., 0]^T$. Since the sources are temporally white and stationary, we obtain

$$E[y_i[0]y_l^*[k]] = g_{ip}[0]g_{lt}^*[k]\ E[s_p(n)s_t^*(n)]\ \hat{\mathbf{f}}_0^T\ \hat{\mathbf{f}}_k^*$$
$$= g_{ip}[0]g_{lt}^*[k]\ E[s_p(n)s_t^*(n)]\ \hat{\mathbf{f}}_k^H\ \hat{\mathbf{f}}_0 \tag{12}$$

where $\hat{\mathbf{f}}_k = [1, e^{j2\pi k/L}, ..., e^{j2\pi k(F-1)/L}]^T$. Let us assume that $\hat{\mathbf{f}}_k^H \hat{\mathbf{f}}_0 \neq 0$ and $g_{ip}[0]g_{lt}^*[k] \neq 0$, then the expression (12) will be non-zero when $y_i[0]$ and $y_l[k]$ extract the same source, i.e., when $y_i[0] = g_{ip}[0]s_p[0]$ and $y_j[k] = g_{jp}[k]s_p[k]$.

From the explanation above, we can devise a method to avoid the permutation indeterminacy. We compute the cross-correlation between the outputs at the first frequency bin and the outputs in other frequency bins, i.e., $E[y_i[0]y_l^*[k]]$. For each output at the first frequency bin, we select the outputs in the other frequency bins with the maximum cross-correlation. Then, we cluster the outputs corresponding to the same source as follows

$$\mathcal{U}_i = \{u_i[0] = y_i[0], u_i[1] = \max_{y_l[1]} |E[y_i[0]y_l^*[1]]|, ...,$$
$$u_i[L-1] = \max_{y_l[L-1]} |E[y_i[0]y_l^*[L-1]]|\} \quad i = 1, ..., N \quad (13)$$

Note that this criterion can be used only when $\hat{\mathbf{f}}_k^H \hat{\mathbf{f}}_0 \neq 0$. This occurs when kF/L is not an integer number.

5 Amplitude Indeterminacy

The amplitude indeterminacy occurs when the sources are recovered with different amplitude (or gain) in some frequency bins. This indeterminacy is difficult to remove because the matrix $\mathbf{\Delta}[k]$ in (5) is unknown for all the algorithms proposed to separate instantaneous mixtures. For a reduced number of separation algorithms, the magnitude of $\mathbf{\Delta}[k]$ at the separation convergence points has been analytically determined (see [1,6]). In this section we will show how to use this information to remove the amplitude indeterminacy.

We will consider the outputs into the set $\mathcal{U}_i$, i.e., $u_i[k] = \delta_{ii}[k]s_i[k]$, $k = 0, ..., L-1$. We can compute a new set of outputs $z_i[k]$, $k = 1, ..., L-1$ with the same amplitude as $z_i[0] = u_i[0] = \delta_{ii}[0]s_i[0]$ using the following expression

$$z_i[k] = \frac{\delta_{ii}[0]\delta_{ii}^*[k]}{|\delta_{ii}[k]|^2}u_i[k] = \frac{\delta_{ii}[0]|\delta_{ii}^*[k]|^2}{|\delta_{ii}[k]|^2}s_i[k] = \delta_{ii}[0]s_i[k] \quad (14)$$

Since we are assuming that $|\delta_{ii}[k]|^2$ is known, we only need to determine the term $\delta_{ii}[0]\delta_{ii}^*[k]$ in (14). Note that the cross-correlation between the outputs is given by

$$E[u_i[0]u_i^*[k]] = \delta_{ii}[0]\,\delta_{ii}^*[k]\,\hat{\mathbf{f}}_0^T\,E[\mathbf{s}_i^e(t_r)(\mathbf{s}_i^e(t_r))^H]\,\hat{\mathbf{f}}_k^* = \delta_{ii}[0]\,\delta_{ii}^*[k]\,\hat{\mathbf{f}}_k^H\,\hat{\mathbf{f}}_0 \quad (15)$$

As a consequence,

$$\delta_{ii}[0]\delta_{ii}^*[k] = \frac{E[u_i[0]u_i^*[k]]}{\hat{\mathbf{f}}_k^H \hat{\mathbf{f}}_0} \quad (16)$$

Finally, substituting (16) in (14), we deduce that the outputs $z_i[k]$, $k = 1, ..., L-1$ can be obtained as follows

$$z_i[k] = \frac{E[u_i[0]u_i^*[k]]}{\hat{\mathbf{f}}_k^H \hat{\mathbf{f}}_0\, |\delta_{ii}[k]|^2}u_i[k], \quad k = 1, ..., L-1 \quad (17)$$

where $\hat{\mathbf{f}}_k = [1, e^{j2\pi k/L}, ..., e^{j2\pi k(F-1)/L}]^T$, $\hat{\mathbf{f}}_0$ is an ones vector of dimension F, $E[u_i[0]u_i^*[k]]$ is the cross-correlation between two outputs into the set $\mathcal{U}_i$ and $|\delta_{ii}[k]|^2$ is the magnitude of the theoretical amplitudes that can be obtained from the convergence analysis of several separation algorithms (see [1, 6]).

6 Simulation Results

In this section we present several simulation results to validate the strategies proposed for solving the permutation and the amplitude indeterminacies. To recover the sources at each frequency bin, we will use the unsupervised algorithm proposed in [6] which adapts the coefficients of each $N \times N$ neural network, $\mathbf{W}_t[k]$, using the following recursion

$$\mathbf{W}_t[k] = \mathbf{W}_{t-1}[k] + \mu \mathbf{W}_t[k](4E[\mathbf{y}[k]\mathbf{y}^H[k]] - 4E[\mathbf{y}[k]\mathbf{y}^H[k]\mathbf{D}] + \mathbf{I}) \quad (18)$$

where $\mathbf{D} = Diag(|y_1[k]|^2, \cdots, |y_N[k]|^2)$ is a $N \times N$ diagonal matrix, $\mathbf{I}$ is the $N \times N$ identity matrix and the moments are estimated using the S samples of the outputs. The stability analysis of this recursion shows that in the separating point, the sources are recovered with an amplitude [6]

$$|\delta_{ii}[k]|^2 = \frac{F + \sqrt{F(E[|s_i(n)|^4] + 3F - 2)}}{2F(E[|s_i(n)|^4] + 2F - 2)} \quad (19)$$

Note that the equation (19) provides the magnitude of the resulting amplitude but not the phase.

We have generated 15,000 samples of two communication sources (a 4-QAM and a 16-QAM) which have been partitioned in 3,000 blocks of $F = 5$ samples. We have appended $P - 1 = 2$ zeros to each block. The sources have been passed through a mixing system modeled as FIR filters of order $P = 3$ obtained by truncating to three samples the following matrix:

$$\mathbf{H}(z) = \begin{bmatrix} -0.8\frac{0.1+z^{-1}}{1+0.1z^{-1}} & 0.5\frac{0.2+z^{-1}}{1+0.2z^{-1}} \\ 0.5\frac{0.5+z^{-1}}{1+0.5z^{-1}} & -0.8\frac{0.1+z^{-1}}{1+0.1z^{-1}} \end{bmatrix} \quad (20)$$

Table 1. Simulation results: Cross-correlations $|E[y_i[0]y_j^*[k]]|$

	$\|E[y_1[0]y_j^*[k]]\|$			$\|E[y_2[0]y_j^*[k]]\|$	
k	$E[y_1[0]y_1^*[k]]$	$E[y_1[0]y_2^*[k]]$	k	$E[y_2[0]y_1^*[k]]$	$E[y_2[0]y_2^*[k]]$
1	0.0134	0.2683	1	0.2641	0.0051
2	0.1856	0.0134	2	0.0101	0.1773
3	0.0714	0.0179	3	0.0078	0.0596
4	0.0682	0.0084	4	0.0127	0.0563
5	0.1795	0.0074	5	0.0299	0.1786
6	0.2767	0.0051	6	0.0046	0.2624

To recover the sources from the observations we have used $L = F+P-1 = 7$ frequency bins and windows of length $K = 5$. The coefficients of the separating systems have been computed using the algorithm (18) with $\mu = 5 \times 10^{-2}$ and $S = 3,000$. We have solved the permutation indeterminacy using the second-order moments shown in Table 1. From this table, we have clustered the outputs in two sets:

$$\mathcal{U}_1 = \{u_1[0] = y_1[0], u_1[1] = y_2[1], u_1[2] = y_1[2],$$
$$u_1[3] = y_1[3], u_1[4] = y_1[4], u_1[5] = y_1[5], u_1[6] = y_1[6]\}$$
$$\mathcal{U}_2 = \{u_2[0] = y_2[0], u_2[1] = y_1[1], u_2[2] = y_2[2],$$
$$u_2[3] = y_2[3], u_2[4] = y_2[4], u_2[5] = y_2[5], u_2[6] = y_2[6]\}$$

In the next stage, we have removed the amplitude indeterminacy using the theoretical values ($|\delta_{11}[k]|^2 = 0.1485$ and $|\delta_{22}[k]|^2 = 0.1444$) and the value of $\delta_{ii}[0]\delta_{ii}^*[k]$ shown in Table 2 which have been obtained using (16). For comparison, Table 2 also shows the true values obtained from the final separating/mixing matrix. We can see the similarity between the estimated and the true values. Finally, Figure 2 plots the recovered sources (part (a)) and the sources obtained without solving the amplitude nor the permutation problem (part (b)). It is apparent that both indeterminacies must be removed to recover the sources.

Table 2. Simulation results: values of $\delta_{ii}[0]\delta_{ii}^*[k]$

	$\delta_{11}[0]\delta_{11}^*[k]$	
k	Estimated value	True value
1	0.0868 + 0.1209i	0.0844 + 0.1202i
2	-0.0327 + 0.1452i	-0.0425 + 0.1442i
3	-0.1588 + 0.0228i	-0.1341 + 0.0590i
4	-0.1339 - 0.0744i	-0.1350 - 0.0628i
5	-0.0330 - 0.1401i	-0.0343 - 0.1423i
6	0.0951 - 0.1205i	0.0902 - 0.1182i

	$\delta_{22}[0]\delta_{22}^*[k]$	
k	Estimated value	True value
1	0.0860 + 0.1187i	0.0878 + 0.1124i
2	-0.0298 + 0.1390i	-0.0303 + 0.1373i
3	-0.1060 + 0.0819i	-0.1283 + 0.0641i
4	-0.1213 - 0.0359i	-0.1300 - 0.0616i
5	-0.0343 - 0.1391i	-0.0382 - 0.1365i
6	0.0816 - 0.1206i	0.0816 - 0.1139i

7 Conclusions

In this paper we have presented a new approach for the separation of convolutive mixtures of temporally-white signals. The problem is solved in the frequency domain by interpreting a convolutive mixture as several instantaneous mixtures which can be separated using many unsupervised learning rules.

Since the separating systems at each frequency bin are recovered independently, it may occur that the sources are recovered in a different order or with different amplitudes in some frequency bins. We have also proposed novel strategies that solve both the permutation and the amplitude indeterminacies. The permutation problem has been solved by grouping the outputs according to their

(a) Recovered sources solving both
the permutation and amplitude indeterminacies

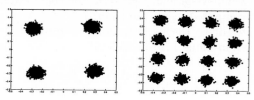

(b) Recovered sources without solving
the permutation nor the amplitude indeterminacies

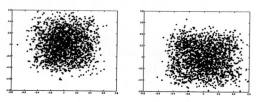

Fig. 2. Simulation results: recovered sources

cross-correlations. Then, the amplitude indeterminacy has been corrected taking into account the values predicted by the stability analysis.

References

1. S-I. Amari, T-P. Chen, A.Cichocki, "Stability Analysis of Learning Algorithms for Blind Source Separation", *Neural Networks*, vol. 10, no. 8, pp 1345–1351, August 1997
2. V. Capdevielle and C. Serviere and J.L. Lacoume, "Blind Separation of Wide-band Sources in the Frequency Domain", in *Proceedings of IEEE International Conference on Acoustics, Speech and Signal Processing (ICASSP 1994)*, pp. 2080-2083, 1994
3. J. F. Cardoso, "Blind Signal Separation: Statistical Principles", *Proceedings of IEEE*, vol. 86, no. 10, pp. 2009-2025, October 1998
4. G. Gelle, M. Colas and C. Serviere, "BSS for Fault Detection and Machine Monitoring Time or Frequency Domain Approach?", in *Proceedings of the Second International Workshop on Independent Component Analysis and Blind Signal Separation (ICA'2000)*, pp. 555-560, Helsinki, June 2000
5. S. Ikeda, N. Murata, "A method of blind separation based on temporal structure of signals", in *Proceedings of The Fifth International Conference on Neural Information Processing (ICONIP'98)*, pp.737-742, Kitakyushu, October 1998
6. C. Mejuto and L. Castedo, "A Neural Network Approach to Blind Source Separation", in *Proceedings of Neural Networks for Signal Processing (NNSP'97)*, pp. 486-495, USA, September 1997
7. P. Smaragdis, "Blind Separation of Convolved Mixtures in the Frequency Domain", in *Proceeding of International Workshop on Independence and Artificial Neural Networks*, Spain, February 1998

Evaluation, Classification and Clustering with Neuro-Fuzzy Techniques in Integrate Pest Management

E. Bellei[1], D. Guidotti[2], R. Petacchi[2], L.M. Reyneri[1], and I. Rizzi[2]

[1] Department of Electronics, Politecnico of Torino
C.so Duca degli Abruzzi 24, I-10129 Torino, Italy
phone ++39 011 564 4045; fax ++39 011 564 4099
phone ++39 011 564 4038
E-mail: reyneri@polito.it, bellei@snoopy.polito.it

[2] Agricultural Sector, Agricultural Entomology Section, Scuola Superiore S.Anna of Pisa,
Via Carducci 40, I-56100 Pisa, Italy
phone ++39 050 883442; fax ++39 050 883215
E-mail: ruggero@sssup.it

Abstract.

In the present article are described the results obtained by the application of neuro-fuzzy methodologies in the study of Bactrocera Oleae (olive fly) infestation in Liguria region olive grows.

The main aim of this project is create an informatic decisional support for experts in the applications of Integrated Pest Management strategies against the Bactrocera Oleae infestation. This system will suggest an appropriate treatments for each monitored farm to optimize the quality of the olive oil and the economic and environmental impact of these treatments.

Forecast and statistical analyses on agronomic data sets like the case in study (the growth of olive fly), are actually made using standard approaches like analytical ones; this kind of data are very variable and non-linear, characteristics which make them complex to be treated mathematically. Agronomic research needs to introduce new analysis techniques for taking data and information, for example neuro-fuzzy techniques that allow a large use of infestation data with a good flexibility degree.

1. Biological Introduction

The quality of an agrarian product deals with its geographic and chemical-physical characteristics but for a good classification of the oil quality, many other factors shall be considered like growing and transformation ones.

The olive fly infestation and the used defence techniques influenced strongly the olive production and the oil quality, therefore it is mandatory to carry out studies on the growing cycle of Bactrocera Oleae, annual behaviour of infestation, monitoring and control methodologies.

When the olive has overtaken the phase of hard stone, the female lays eggs and after few days the larva comes out. Larvae presents 3 different growing stages (L1,L2,L3) that grow up eating the olive; when the larva is mature comes out from olive, falls to the ground and becomes a pupa.

The olive fly develops 3 complete generations per year and sometimes can begin a fourth ones if the climatic conditions are favorable; the beginning of infestation is connected to the annual temperature's variation and to the micro-climate.

The infestation has been clustered in 5 different indexes expressed in percent on total olive sampled: Active (eggs, I and II stage alive larvae), Dead (I and II stage dead larvae), Damaging (III stages larvae and pupae), Total (the sum of the above) and Target (eggs, I and II stage larvae).

The infestation considered for this research is the target one; it is now in study the contribution of each other kind of infestation in the optimization of oil quality.

This project started on March 2000 with the monitoring and collection of data from a several number of oil farms in Liguria area.

The monitoring shall test the presence and growth of harmful bugs, following the dinamic infestation in the olive-grove to define the best treatment.

There are two kinds of monitoring:
- adult capture weekly made with plexiglass trap, the experience shows that is not possible to find a correlation between the level of capture and the infestation
- olive collection allows the estimation of infestation percentage in the olive-grove.

The actual defence methodology is to apply the treatments only if necessary, and infestation studies allow to carry out aimed treatment, with low quantity of pesticide to reduce the environment impact.

The work made until now is:
- characterization of curves with a WRBF neuro-fuzzy network (see par. 3)
- evaluation of reliability of each infestation measures with a fuzzy system (see par. 4)
- characterization of temperature annual cycle with a WRBF network (not described in this article)
- clustering of infestation (few results, see par. 5)

2. WRBF-n Neuron

To create a simulation model of 'fly-tree-environment' system are analysed biological data of the fly, the environmental data where the fly grows and the infestation development.

Since this system is not linear, very variable and difficult to be represented by traditional methods that make use of mathematical equations; the present work has used, as an alternative, WRBF-n neural network [2], which has a big flexibility and can learn from experience thanks to an appropriate training alghoritms.

A WRBF-n neuron (weighted radial basis function) is an unification of different paradigms like Perceptron, Wavelet neuron and Fuzzy system; the best advantage of this neuron is that unites a training alghoritm with the experience of experts as fuzzy rules.

Each WRBF neuron is characterized by an order n, that is the neuron's metric, a weight matrix w, a centre matrix c, an optional bias θ and an activation function F(z). The output of a WRBF-n neuron is :

$$y_j = \begin{cases} F_j\left(\sum_i (x_i - c_{ji}) \cdot w_{ji} + \vartheta_j \right) & \text{for} \quad n = 0 \\ \\ F_j\left(\sqrt[n]{\sum_i |x_i - c_{ji}|^n \cdot w_{ji}^n + \vartheta_j} \right) & \text{for} \quad n \neq 0 \end{cases} \quad (1)$$

3. Characterization of Infestation with WRBF Network

The study of fly infestation during one year, shows the growth of 3 (or at least 4) different generations of fly as shown in Fig.1; often the three generations are not so evident due to evolving of the climatic conditions.
To make the clustering for kinds of infestation (see par.5) is important to describe in a simple way a curve using known math functions that allow to extract a limited numbers of parameters which can describe very well the curve.
The tests made with real data, have shown that the behavior of generations can be satisfactorily described by 3 gaussian like curves; each curve is characterized with 3 parameters:
- the center of gaussian function that is the sample corrisponding to the maximum infestation
- the height of gaussian that is the value of maximum level of infestation
- the dispersion of gaussian, namely the width of curve.
The characterization of curve with the interpolation of 3 gaussian curves, has been made building a two layers feed-forward neural WRBF-n network with a training back-propagation algorithm. The network has only one input, the day of the year, and gives as output the characterized percentage of infestation.
In the hidden layer there are 3 WRBF-n neurons, one for each generation, each one has a gaussian activation function and an order equal to 2. The tests made using a WRBF-3 showed in some case a better error than the WRBF-2 one.
In the output layer there is one linear neuron WRBF-($+\infty$) to interpolate on the max of each curves; the tests have been made also using a WRBF-0 to interpolate on the sum of curves but the results are not so good like with infinite order.
The network makes the characterization considering initially each measure with the same reliability (degree=0.9) and then associating to each measure the relative reliability computed applying suitable entomological rules (the system for the reliability is described in par.4).
The results obtained are very encouraging in fact the standard error between the real infestation and the output of the trained network is quite low.

Here below the results obtained by 7 farms in Liguria, without the reliability of measures.

Farm	Standard error with WRBF-2	Standard error with WRBF-3
F1	7.47 %	7.29 %
F2	3.48 %	3.01%
F3	3.3 %	3.91 %
F4	7.78 %	7.07 %
F5	4.8 %	4.27 %
F6	2.86 %	1.43 %
F7	2.34 %	2.1 %
Average	4.57 %	4.15 %

Table 1.

In Figure 1 there is an example of the characterization of the infestation in the farm F6, using a WRBF-3 network. On the x axis there are the day of sample and on the y axis the value of infestation.

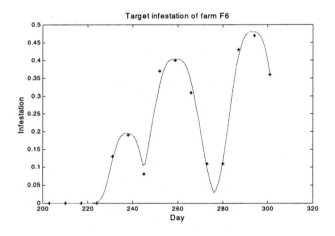

Fig. 1. Infestation of F6: continuos line is the output of WRBF-3, asterisks are measurements.

4. Reliability of Measures with a Neuro-Fuzzy System

In field condition, the infestation monitoring performed on the samples of olives could be often affected by further errors (high spatial heterogeneity, inexperience to detect mortality in younger larvae,...) therefore it is mandatory to establish a method to evaluate the error and to associate a reliability degree to each measure.
The knowledge on B.oleae development based on scientific studies and experts experience on analysis of data quality, allow formulate heuristic rule to detect data errors.

A way to understand the reliability level of an infestation measure, is to identify two kinds of errors based on comparison of fruit fly infestation data among monitoring at one week distance; the two kind of errors are unexpected increase and unexpected decrease of the infestation.

With the values of all active (L1a, L2a, L3a, Pa) and inactive (L1i, L2I, L3I, Pi) stages of fly growth together exit hole (Eh) and the sum of all the stages of fly growth (Tot), are calculated with 5 equations and the result of each one is evaluated trough 4 fuzzy rules to determine the relative reliability in terms of : high reliability, medium, low and very low reliability.

The minimum among these results represents the final reliability of the measure.

An example of one of the 5 equations is the following:

IF '$(Pa + Pi)_{(t)} - (Pa + Pi)_{(t-1)} + L3a_{(t-1)}$' is 'small' THEN 'reliability' is 'high'
IF '$(Pa + Pi)_{(t)} - (Pa + Pi)_{(t-1)} + L3a_{(t-1)}$' is 'medium' THEN 'reliability' is 'medium'
IF '$(Pa + Pi)_{(t)} - (Pa + Pi)_{(t-1)} + L3a_{(t-1)}$' is 'big' THEN 'reliability' is 'low'
IF '$(Pa + Pi)_{(t)} - (Pa + Pi)_{(t-1)} + L3a_{(t-1)}$' is 'very big' THEN 'reliability' is 'very low'

For each equation, it has been created a fuzzy system by a WRBF-5 neural network with three layers in which the first hidden layer contains so many neurons as the number of rules (4) the second one is a normalization layer for the first step of de-fuzzyfication and the output layer has one linear neuron corresponding to the value of reliability.

Each measure has associated five values of reliability (one for fuzzy system) and to find the final reliability of the measure is calculated the minimum among these 5 results with one simple linear WRBF-($-\infty$) neuron.

In the Fig.2 it is shown the system for reliability of measure.

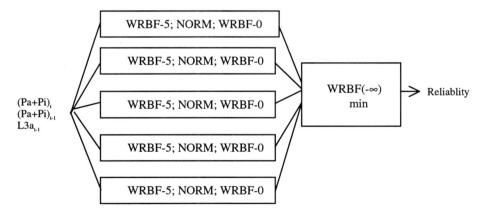

Fig. 2. Neuro-fuzzy system for reliability of measures

The value of 'small', 'medium', 'big' and 'very big' of rule and 'very low', 'low', 'medium' and 'good' reliability has been defined considering the experts knowledge; the choice is to consider the 3% of infestation as a minimum level for rules activation, these level will be compared with statistical analysis of data from field trials.

The reliability evaluation system had been tested on a large number of farm and the results compared with the crisp applications of the tests on the same data infestation performed by the agronomists.

The agreement among the results is very good and encourage for the future applications in other field.

	Human	Expert
Neuro-fuzzy system	Not reliable data	Reliable Data
Not reliable Data	87.5 %	12.5 %
Reliable Data	9.09 %	90.91 %

Table 2.

Below there are two examples of results obtained by application of the neuro-fuzzy system; low reliabilities (degree=0.1) noticed by the system have been confirmed by the agronomists.

Day	Egg	L1a	L1I	L2a	L2i	L3a	L3i	Pa	Pi	Eh	Reliability
1999-9-2	4	2	0	7	2	10	4	3	8	6	0.9
1999-9-9	9	8	2	0	0	0	0	0	0	0	0.1

Day	Egg	L1a	L1I	L2a	L2i	L3a	L3i	Pa	Pi	Eh	reliability
1999-8-5	4	2	0	7	2	10	4	3	8	6	0.89
1999-8-12	9	8	2	0	0	0	0	0	0	0	0.1

Table 3.

There is a good advantage in using this automatic system because the time that should spend the agronomists for estimate the reliability of all measures for each farm is very large.

The estimated reliability is associated at each measure date and than used in the training algorithm of the WRBF neural network (see par.2), as a product factor; this results in a better characterization of infestation curve.

The standard error between the target and the output of the trained network is generally worst than the one obtained by using the same reliability for each measure, but the parameters that describe the curve are more reliable than the previous.

A new neural network with 7 neurons that implements 2 other new rules (one for the decrease of dead infestation and one for the damage infestation) is now under testing and the results shows a better agreement with the evaluations of agronomists.

5. Clustering of Infestation Parameters

The annual infestation of Bactrocera Oleae is characterized by centers and by the level of each generation.

The infestation's development depends on the geographic position of farm and by meteo condition of the year, so in standard climate conditions, the infestation of each farm can be clustered on centers and entities.

From a first analysis of infestation development in different farms in Liguria, we evaluate the possibility to create a classification based on a number of clusters from a minimum of 7 to a maximum of 15.

For the clustering, a neural network WRBF has been created with 6 inputs (3 centers and 3 entities of infestation) and one output neuron that gives the memebership cluster of the input (see par.3).

The WRBF neural network created for the clustering has two hidden layers of neuron: the first one contains as many neuron as everyone as clusters, everyone with gaussian activation, while the second layer is a normalization layer (from which we obtain the target) see Fig. 3.

The ouptut of the normalization layer is the target for the back-propagation training [2].

Training network will give the membership infestation class for each farm and for each year, using a WTA neuron (infinite normalization).

Validation of this neural network, now under testing, requires to search for similar characteristics among infestations of same class to optimize the classification.

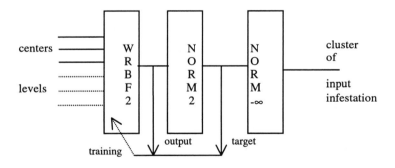

Fig. 3. WRBF network for the clustering of infestation.

6. Conclusion

Actually it is under testing the fuzzy system for the reliability of measurements and it is evaluating the possibility to use a neuro-fuzzy system with 7 equations to find more exactly an error.

Moreover, it is in study also a neuro-fuzzy system to evaluate the reliability of infestation at generation level.

The characterization of annual median temperature to normalized the parameters used for the clustering is made with a sinusoidal WRBF-2 network but is under analisys by the agronomist, to find the best correlation with infestation.

We are also evaluating the possibilty to use for the clustering other meteo data like day-degrees or percentage of umidity.

Now we are validating with experimental data the clustering results of the infestation curves using the parameters obtained by the characterization with the WRBF neural network and the neuro-fuzzy system for the reliability.

We are also searching a link among meteo data (temperature and umidity) and the infestations.

In future a WEB-based environment will be developed for acquisition, processing, classification, forecast and management of disinfestation treatments.

The possibility to apply the same methods to other agricultural processes and problems will also be analyzed.

The program will finish with the availability of a decision-support environment tested on the olive fruit fly infestation.

References

[1] Petacchi, R.: *Dacus oleae (Gmelin): primi risultati di uno studio poliennale sulla dinamica dell'infestazione in due biotopi della Liguria di levante*. Frustula Entomologica, N.S., **VXII** (XXV), 1989.

[2] Reyneri L.: *Unification of Neural and Wavelet Networks and Fuzzy system*, IEEE Transactions on Neural Networks, Vol.10, No.4, July 1999

[3] Rajesh N. Davè, Raghu Krishnapuram.: *Robust Clustering Methods: A Unified View*, IEEE Transactions on Fuzzy Systems, Vol. 5, No.2, May 1997

[4] Haykin, S.: *Neural Networks: a Comprehensive Foundation*. New York: McMillan, 1994

Acknowledgement

This work has been sponsored by Italian project COFIN – 99 "Applicazioni di tecniche di previsione in Agricoltura".

Inaccessible Parameters Monitoring in Industrial Environment: A Neural Based Approach

Kurosh Madani[1] and Ion Berechet[2]

[1] Intelligence in Instrumentation and Systems Lab. (I²S)
SENART Institute of Technology - University PARIS XII
Avenue Pierre Point, F-77127 Lieusaint, France
madani@univ-paris12.fr

[2] CREATA – R&D Intelligence
Z.I. – Avenue Victoire
F-13106 Rousset Cedex, France

Abstract: In the case of a large number of applications, especially complex industrial ones, the knowledge on system's (process, plant, etc.) parameters during the operation of the system is of major importance. However, in real cases, there are always parameters, which are not accessible. In the present work, we focus our interest around the extraction possibility of information relative to inaccessible parameters. Of course, such dilemma becomes a very complex and difficult problem in a general context. However, we will discuss some realistic (and especially, realizable) conditions for which a solution could be approached. In proposed approach, we use the neural network's learning and a synaptic weight based indicator to detect changes related to system's inaccessible parameters. Experimental results relative to a real industrial process have been reported validating our approach.

1. Introduction

In large case of applications, especially complex industrial ones, the knowledge on system's (process, plant, etc.) parameters during the operation of the system is of major importance. The knowledge of such parameters (especially internal parameters) is a chief condition to predict the system's evolution and related decisions. In theory, such parameters could be obtained by implementation of as many as needed appropriated sensors. However, in real cases (real systems), only a very limited number of sensors could be implemented. The limitation of sensors number is related to several reasons among which: cost constraints, implementation difficulty, monitoring or supervision difficulties, etc..

In the present work, we focus our interest around the following question: is it possible to extract information relative to inaccessible parameters ? Of course, the above posed question become a very complex and difficult problem in a general context and we don't pretend, here, to solve it in such context. However, we will discuss some realistic (and especially, realizable) conditions for which the answer to that question could be approached and a solution could be proposed. Our approach is based on use of Artificial Neural Networks (ANN).

The most attractive property of Artificial Neural Networks models is their learning and generalization capabilities. The mentioned property of ANN makes such techniques suitable for design of new solutions in a large number of areas where conventional techniques show their limitations and offer original perspectives to overcome the above mentioned difficulties [1], [2], [3], [4].

The problem on which we are interested here deals with several domains as Intelligent Measurement, Identification and Behavior Prediction. Concerning the use of ANN in the above-mentioned topics, a number of advances has been accomplished in each area (especially, the identification dilemma has been the center of interest of a large number of works). For example, in their paper (see [5]), Sachenko & al. discuss sensor errors prediction using neural networks showing how such models can be used to increase the accuracy of physical quantity measurements by prediction of sensor drift. They show that the best prediction quality is provided when the predicting neural network is trained on real data about sensor drift obtained by testing and calibration [5]. In another paper, Sachenko deals with Intelligent Measurement and Control System (IMCS) development on the basis of distributed sensor networks [6]. In [7], Zhang and Ma, propose a neural structure to extract Principle Component for direction-of-arrival (DOA) prediction. On the other hand, Yang and Jiao propose ANN for solving a class of generalized eigenvalue problems of matrix pair for signal processing and identification [8].

To approach a solution to the above posed problem, we propose a different slant combining neural based Intelligent Measurement and identification. We use a synaptic weight based indicator (which will be defined later) to detect changes related to system's inaccessible parameters. Moreover, we use the neural network's learning to sample the behavior of those inaccessible parameters. Conform to results of [5], we have used real data, relative to a real industrial process, to validate our approach.

The paper is organized as follow: in section 2, we formulate the problem, discussing supposed conditions, and we expose our solution to the posed problem. In section 3, we present and discuss experimental results validating the proposed approach. In the last section we give the conclusion, discuss some areas of possible applications and give future perspectives.

2. Problem Formulation and Proposed Approach to It's Solution

Let us consider a system with a given number of inputs, outputs and internal parameters. We suppose system inputs and outputs to be always accessible (trough related sensors). However, concerning the system's internal parameters (except those for which appropriated sensors have been implemented), if they may be accessible for a limited period (for example, during the system installation or during maintenance period) they become normally inaccessible during the operation phase. Thus, we suppose that these parameters may be accessible for a given period and then, they will be supposed to be inaccessible. The question (problem) is: *is it possible to extract information relative to inaccessible parameters?*

Before analyzing the above posed problem, let us review the consequence of a perturbation on input data in a neural network.

2.1 Variation Flow in a Neural Network

Let us consider a 3 layers (input, hidden and output layers) standard neural network with a Back-Propagation learning rule. The input layer includes M neurons, the hidden layer includes P neurons and the output layer includes N neurons. Figure 1 represents such neural network, where $\mathbf{X} = (X_1, \ldots, X_j, \ldots, X_M)^T$ represents the input vectors, with $j \in \{1, \cdots, M\}$, $\mathbf{h} = (h_1, \ldots, h_j, \ldots, h_P)^T$ represents the hidden layers output and $\mathbf{S} = (S_1, \ldots, S_i, \ldots, S_N)^T$ the output vector with $i \in \{1, \cdots, N\}$. We note W_{kj}^H and W_{ik}^S synaptic matrixes elements, corresponding to input-hidden layers and hidden-output layers respectively. The hidden layer's neurons are supposed with a non-linear activation function F(.) and the output layer's ones are supposed with a linear activation function. Finally, V_k^H and V_i^S, defined by relation (1), represent the synaptic potential (activity) vectors components of hidden and output neurons, respectively (e.g. vectors V^H and V^S components).

$$V_k^H = \sum_{j=1}^{j=M} W_{kj}^H \cdot X_j \quad \text{and} \quad V_i^S = \sum_{k=1}^{k=P} W_{ik}^H \cdot h_k \quad (1)$$

We suppose that mentioned synaptic weights of the neural network have been adjusted thank to a learning process. Taking into account such considerations, the hidden and output layers output vectors will be given by relations (2).

$$\mathbf{h} = \mathbf{F}(V^H) = (F(V_1^H) \quad \cdots \quad F(V_k^H) \quad \cdots \quad F(V_P^H))^T$$
$$\text{with } h_k = F(V_k^H) \text{ and,} \quad (2)$$
$$\mathbf{S} = (V_1^S \quad \cdots \quad V_i^S \quad \cdots \quad V_N^S)^T$$

Let us now consider some perturbation ΔX on j-th input: the j-th input component would thus become $X_j + \Delta X$. The new synaptic potential vector corresponding to hidden neurons will then be given by relation (3).

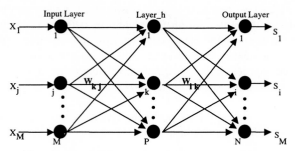

Fig. 1. Bloc-diagram of a 3-layers neural network.

$$\mathbf{V}^H = \begin{pmatrix} \sum_{q \neq j} W_{1q}^H \cdot x_q + W_{1j}^H (x_j + \Delta x) \\ \vdots \\ \sum_{q \neq j} W_{kq}^H \cdot x_q + W_{kj}^H (x_j + \Delta x) \\ \vdots \\ \sum_{q \neq j} W_{Pq}^H \cdot x_q + W_{Pj}^H (x_j + \Delta x) \end{pmatrix} = \begin{pmatrix} V_1^H + W_{1j}^H \Delta x \\ \vdots \\ V_k^H + W_{kj}^H \Delta x \\ \vdots \\ V_P^H + W_{Pj}^H \Delta x \end{pmatrix} \quad (3)$$

and then, the hidden layer's new output vector will be given by:

$$\mathbf{h} = \left(F(V_1^H + W_{1j}^H \Delta x) \quad \cdots \quad F(V_k^H + W_{kj}^H \Delta x) \quad \cdots \quad F(V_P^H + W_{Pj}^H \Delta x) \right)^T$$

Using Taylor development, a given k-th neuron's output will be subject to a modification expressed by relation (4), where F'(.) represents the derivative of the activation function F(.). In the same way, the neural network's output will generate an output error vector expressed by relation (5).

$$\Delta h = F(V_k^H + W_{kj}^H \Delta x) - F(V_k^H) = F'(V_k^H) W_{kj}^H \Delta x \quad (4)$$

$$\boldsymbol{\varepsilon} = (\varepsilon_1 \quad \cdots \quad \varepsilon_i \quad \cdots \quad \varepsilon_N)^T = \left(\sum_{k=1}^{P} W_{1k}^S \Delta h_k \quad \cdots \quad \sum_{k=1}^{P} W_{ik}^S \Delta h_k \quad \cdots \quad \sum_{k=1}^{P} W_{Nk}^S \Delta h_k \right)^T$$

$$\boldsymbol{\varepsilon} = \left(\sum_{k=1}^{P} W_{1k}^S F'(V_k^H) W_{kj}^H \Delta x \quad \cdots \quad \sum_{k=1}^{P} W_{ik}^S F'(V_k^H) W_{kj}^H \Delta x \quad \cdots \quad \sum_{k=1}^{P} W_{Nk}^S F'(V_k^H) W_{kj}^H \Delta x \right)^T \quad (5)$$

This relation shows the repercussion of an input variation on hidden and output neurons responses. A new learning of the neural network will modify synaptic weights of both input-hidden and hidden-output synaptic matrixes ($\mathbf{W}^H$ and $\mathbf{W}^S$ synaptic matrixes). Taking into account the relation (5) and with respect to the standard Back-Propagation learning rule (or a derivative of this rule) expressed by relation (6) giving the synaptic weight update connecting the j-th neuron to the i-th neuron of two adjacent layers of a given neural network, one can write relations (7) and (8).

$$\Delta W_{i,j} = -\eta \frac{\partial \varepsilon_i}{\partial W_{i,j}} \quad (6)$$

$$\Delta W_{i,k}^S \propto -\eta \, F'(V_k) W_{k,j}^H \Delta X \quad (7)$$

$$\Delta W_{k,j}^H \propto -\eta \, F'(V_k) \Delta X \quad (8)$$

These relations show that synaptic weights of the neural network will be modified proportionally to the input perturbation.

2.2 Definition and Interpretation of a Synaptic Indicator

In the previous sub-section, it has been shown that thank to the learning process, the synaptic matrix relative to a pair of layers will contain information relative to all changes on input data of the concerned pair of layers. In the present sub-section we define an indicator, that we call Synaptic Indicator (SI). Such indicator is defined according to the relation (9). The Figure 2 shows the principle of definition of SI relative to a pair of adjacent layers. As one can remark, if the input layer (of the pair of layers) includes M neurons, then M Synaptic Indicators could be available. On the other hand, one can remark that the j-th indicator will contain the changes corresponding to the j-th input neuron (j-th parameter).

$$I_j = \sum_{i=1}^{i=N} W_{i,j} \quad (9)$$

Let us now interpret the Synaptic Indicator, defined by previous relation and the figure2. According to that figure, and after a learning process, the considered pair of layers will act as some non-linear function identifier minimizing the output neurons errors. If G(.) is the learned transfer function, then the considered pair of layers identify this function realizing Y = G (X).

Let us now back to the relation (9). This relation could also be written as: $I_j = \sum_{i=1}^{i=N} W_{i,j} \cdot 1$. It could be remarked that this relation corresponds to the outputs computation in a two layers neural network with linear activation function, where the input vector is identity vector **1** and where the synaptic matrix corresponds to the transposed of the synaptic matrix of figure 2. So, the above-defined synaptic indicator could be interpreted as the response of a linear neural network, which approximates the function $G^{-1}(.)$ (which represents the G(.) inverse function), to some normalized input stimulus (some normalized test stimulus). Figure 3 shows the presented interpretation.

2.3 Inaccessible Parameters Monitoring Using Synaptic Indicator

Taking into account the previous considerations, let us now consider a system with conditions (hypothesis) mentioned at the beginning of the section 2: a system with a given number of inputs, outputs and internal parameters; inputs and outputs are supposed to be always accessible, while the system's internal parameters, they may be accessible for a given period then, they will be supposed to be inaccessible. The internal parameters accessibility during some determined period is a realistic condition, which could be realizable in a large case of real world industrial environments: parameters accessibility during maintenance, parameters characterization during the test phase of the plant's (or system) manufacturing, etc.. Let us also consider a 3 layers neural network (in conformity with figure 1) identifying such system. That means that the mentioned neural network pre-learns some appropriated database, including data relative to system (or plant) inputs, outputs and internal parameters behavior. Then, a reference set of synaptic indicators relative to system's internal parameters is generated.

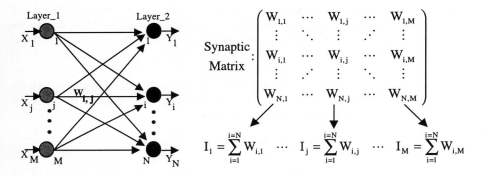

Fig. 2. Principle of Synaptic Indicator definition.

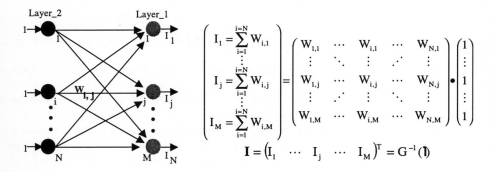

Fig. 3. Interpretation of Synaptic Indicator.

After the pre-learning phase, such neural network, acting as a virtual sensor, could be used to detect changes related to system's internal parameter, through the above-defined synaptic indicator. This operation concerns the monitoring phase during which internal parameters are supposed to become directly inaccessible. In monitoring operation mode, the neural network learns periodically, only accessible data (inputs and outputs) leading to a new synaptic matrix. This new synaptic matrix is then used to generate a new set of synaptic indicators, which is compared to the reference set of synaptic indicators (obtained from pre-learning phase). One can remark that in our technique, we use the ANN learning capability to sample changes related to inaccessible parameters. Figures 4 and 5 show the bloc diagrams of the pre-learning (during which the system's internal parameters are supposed to be accessible) phase and the monitoring phase (during which those internal parameters become inaccessible) respectively.

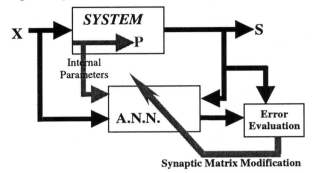

Fig. 4. Neural network pre-learning phase bloc-diagram.

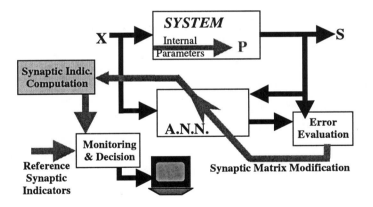

Fig. 5. System's parameters monitoring phase bloc-diagram.

3. Experimental Results

To validate the above exposed concept, we have used our technique in the case of real world industrial process internal parameters monitoring. The industrial process is a drilling rubber plant, which is used to manufacture a set of products with complex profiles (for car industry) needing several different kinds of plastics. Because of industrial confidentiality, we will not give the nature of each parameter. That is why these parameters will be called P_j with $j \in \{1,\cdots,8\}$. The input of the process corresponds to some rotation velocity and the output of the process is related to some lineal speed of produced matter. The output (final product) quality (profile, mechanical and acoustic properties) is correlated to the mentioned internal parameters, which are not directly accessible during process operation. A normalized database, including input, output and internal parameters, variations, has been constructed. This database has been divided on two databases: one used for pre-learning phase, and the other one for validation.

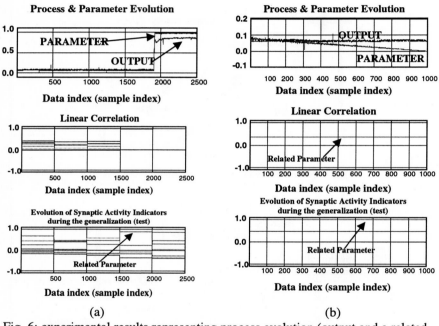

Fig. 6: experimental results representing process evolution (output and a related internal parameter), a linear correlation based estimation of internal parameters evolution and the evolution of the same internal parameters given by above defined synaptic indicator (SI) based monitoring approach.

The figure 6 shows experimental results comparing a conventional linear correlation based estimation of internal parameters evolution with the evolution of the same internal parameters given by above defined synaptic indicator (SI) based monitoring approach. In the figure 6-a, corresponding to a strong variation of internal parameters (leading to a strong variation of process) all 8 parameters (which are each other dependents) have been plotted. As one can remark, the conventional correlation based approach doesn't permit to detect the most related parameter to such output variation. SI based approach leads to determine the concerned internal parameter. It permits also to detect other related parameters. The figure 6-b shows the case of a soft variation. Same kind of improvement has been obtained.

4. Conclusion

In the present work, we focus our interest around the extraction possibility of information relative to inaccessible parameters. We discussed some realistic (and especially, realizable) conditions for which a solution has been approached. In proposed approach, we use the neural network's learning and a synaptic weight based indicator to detect changes related to system's inaccessible parameters. The obtained results show feasibility and efficiency of the resented approach comparing to conventional correlation based estimators. However, several problems are still open. Among problems to solve, the point related to fine quantification of relationships between synaptic indicator variation and internal parameters behavior. We are working now to understand those relationships and their dependence on process nature and other related problems.

References

[1] M. MAYOUBI, M. SCHAFER, S. SINSEL, "Dynamic Neural Units for Non-linear Dynamic Systems Identification", From Natural to Artificial Neural Computation, LNCS Vol. 930, Springer Verlag, 1995, pp.1045-1051.
[2] Faller W., Schreck S., "Real-Time Prediction Of Unsteady Aerodynamics : Application for Aircraft Control and Manoeuvrability Enhancement", IEEE Transactions on Neural Networks, Vol. 6, N¡ 6, Nov. 1995.
[3] Y. MAIDON, B. W. JERVIS, N. DUTTON, S. LESAGE, "Multifault Diagnosis of Analogue Circuits Using Multilayer Perceptrons," IEEE European Test Workshop 96, Montpellier, June 12-14,1996.
[4] I. Grabec, "Continuation of Chaotic Fields by RBFNN", Lecture Notes in Computer Science - Biological and Artificial Computation : From Neuroscience to Technology, Edited by : Jose Mira, Roberto M. Diaz and Joan Cabestany - Springer Verlag Berlin Heidelberg 1997, pp. 597-606.
[5] A. Sachenko, V. Kochan, V. Turchenko, V. Golovko, J. Savitsky, A. Dunets, T. Laopoulos, "Sensor errors prediction using neural networks", Proceedings of the International Joint Conference on Neural Networks, International Joint Conference on Neural Networks - IJCNN'2000, Jul 24-Jul 27 2000, 2000, Como, Italy, p 441-446.
[6] A. Sachenko, "IMCS development on the basis of distributed sensor networks", IEEE AFRICON Conference, v1, 1999 IEEE, Piscataway, NJ, USA, p 345-350.
[7] Y. Zhang, Y. MA, "CGHA for Principal Component extraction in the complex domain", IEEE Trans. On Neural Networks, Vol. 8, No. 5, September 1997, pp. 1031-1036.
[8] H. B. Yang, L. C. Jiao, "Neural network for solving generalized eigenvalues of matrix pair", 2000 IEEE Interntional Conference on Acoustics, Speech, and Signal Processing M1 2000 Jun 5-Jun 9, p 3478-3481.

Autoorganised Structures for Extraction of Perceptual Primitives

Marta Penas[1], María J. Carreira[2], and Manuel G. Penedo[1]

[1] Dpto. Computación. Fac. Informática, Universidade da Coruña. 15071 A Coruña. SPAIN. {infmpc00,cipenedo}@dc.fi.udc.es
[2] Dpto. Electrónica e Computación. Fac. Física, Universidade de Santiago de Compostela. 15782 Santiago de Compostela. SPAIN. mjose@dec.usc.es

Abstract In this work we have used directional features extracted from Gabor wavelet responses to compare different auto-organised networks in order to extract perceptual primitives without taking into account the kind of images to analyse. This is an adequate problem to prove the performance of these models because of the high dimensionality of the input space. Three different models have been analysed: self-organised maps, growing-cell structures and growing neural gas. Results have proved that growing-cell structures generalise better all directional perceptual primitives we are searching for, and they do not provide very noisy images.

1 Introduction

The phenomenon of perceptual organisation enables humans to detect such relationships in the word as collinearity, parallelism, connectivity, and repetition. In computer vision, perceptual grouping is the study of how features are clustered for object recognition. Inspired by biological studies, specially the Gestalt school, its purpose is grouping feature elements prior to recognition. Perceptual organisation has been studied by investigators in psychology [1, 2] in an attempt to classify the behaviour of grouping phenomena in the human visual system, with laws of symmetry, proximity, simplicity, closure, etc. proposed as the mechanism for grouping features such as edges, corners or regions.

Sarkar and Boyer [3] proposed a classification of perceptual organisation in computer vision dependent on the kind of images being analysed. They considered feature types to be organised stratified in layers of abstraction: signal level, primitive level, structural level, and assembly level. We will focus this classification on 2D organisation. Signal level is mainly concerned with organising pixels or interest points into extended regions, edge chains or dot clusters. Primitive level concerns organisation of edge chains and regions by means of a search for contiguous edge pixels similar in such attributes as curvature and contrast. Features produced at the primitive level are organised at the structural level, concerned with search for structures such as ribbons, corners, polygons, closed regions, and strands. These basic structures are then organised at the assembly level in order to identify arrangements on them.

Our work is embedded in the signal level described above. We will try to detect perceptual primitives using the directional properties that Gabor filters provide as an alternative to classical edge detectors. Each pixel will be classified using a self-organised structure in order to reduce the high dimensionality of the input space provided by the application of 24 Gabor filters (3 frequencies and 8 orientations). In this work, we have performed a study of neuronal models which, by means of no-lineal projections, allow a principal component extraction with a topological ordering of generated clusters based on input space data. By means of a later analysis, these clusters can be grouped in order to permit a correct grouping of the data to be analysed. The auto-organised structures which will be compared are self-organised maps, growing-cell structures and growing neural gas.

This paper is organised as follows. In Sec. 2 we introduce the methodology for computation of initial Gabor properties, which we have developed in order to reduce the high Gabor filters computation time. Sec. 3 is an introduction to the different auto-organised structures we will analyse and Sec. 4 shows the results obtained with each one, applying them to synthetic and real images. Sec. 5 concludes the work.

2 Efficient Spatial-Domain Implementation of Gabor Wavelets

Gabor wavelets are complex exponential signals modulated by gaussians employed in a great variety of techniques for image processing, like compression or edge detection. One of their main features is the maximisation of conjoint localisation in spatial and frequency domains.

In spatial domain and using polar coordinates, mathematical expression of two dimensional Gabor wavelets is:

$$g_{x_0,y_0,f_0,\theta_0} = \exp\left\{i\left[2\pi f_0 \left(x\cos\theta_0 + y\sin\theta_0\right) + \phi\right]\right\} gauss\left(x - x_0, y - y_0\right) \quad (1)$$

where

$$gauss(x,y) = a\exp\left\{-\pi\left[a^2 \left(x\cos\theta_0 + y\sin\theta_0\right)^2 + b^2 \left(x\sin\theta_0 - y\cos\theta_0\right)^2\right]\right\} \quad (2)$$

The shape of the wavelet depends on parameters a and b, while x_0, y_0, f_0 and θ_0 represent localisation in spatial and frequency domains, respectively. In this report we have worked with the even part of Gabor wavelets centred at the origin ($x_0 = 0, y_0 = 0$), symmetric ($a = b$) and null phase ($\phi = 0$).

There are three alternative implementations of Gabor wavelets: frequency domain, multi-resolution frequency domain and multi-resolution spatial domain implementation.

Spatial implementation [4] is based on the convolution of reduced versions of the image with unidimensional filters of 11 components obtained from the decomposition of two dimensional Gabor filters. The objective of this implementation is the reduction of computational requirements in frequency domain and multi-resolution frequency domain implementations.

3 Auto-organised Structures for Extraction of Perceptual Primitives

Reduction of input space dimensionality is necessary in order to fix an objective criterion to perform a right feature grouping that work well with any kind of images to process. Thus, the classified output space must maintain the topology of the input space, being faithful to existent characteristics. Lastly, this reduction should lead to a generalisation of results so that output space could perform a right feature grouping independently on the kind of images to process.

In the following subsections, we will give a brief introduction to the auto-organised structures employed in order to achieve this dimensionality reduction, that is, self-organised maps, growing-cell structures and growing neural gas.

3.1 Self-Organised Maps (SOMs)

Self-organised maps (SOMs) [5] are topological maps whose size and structure are determined prior to their training, so the resulting mappings are conditioned by this election. The weights of the processing elements represent clusters associated to groups of patterns in the training set. By other hand, in the network structure a neighbour relation is established between the processing elements that configure it in such a way that the topological ordering existent in the training set is projected to the output map generated by the network.

SOM's performance is divided into various processes. The initialisation process fixes the size of the structure, the weights and learning parameters, the size associated to the region of interest and the maximum number of cycles in the training process. During the learning process, the weights of the winning element, selected in base to Euclidean distance to input vector, and those of all elements in its region of interest are modified in the presence of a determined pattern. Weight modification of neuron i, (w_i) is as shown in Eq. 3:

$$w_i(t+1) = w_i(t) + \gamma * h(i,k) * (x(t) - w_i(t)) \forall i \in S \tag{3}$$

where γ is the learning rate, $h(i,k)$ is a function dependent on the distance between units and maximum over the winning element and S is the region of interest centred on the winning neuron.

At the end of the training process, the output map will represent, with the limitations imposed by the size and structure of the model, the input space where processing elements keep on the same topological relations existent in the training set, representing also their probabilistic distribution.

3.2 Growing-Cell Structures (GCS)

This model [6] is based on Kohonen self-organised maps, but it tries to eliminate the restrictions of the a priori definition of the network size incorporating a mechanism to add new processing elements when needed, maintaining the network topology. Learning begins with a K-dimensional structure of $K + 1$ nodes and $(K + 1) K /2$ connections where each processing element is associated with a parameter called the resource value that indicates where new processing elements must be inserted.

GCS performance is divided into various processes. In the initialisation process, the K-dimensional structure, weights, learning rate and resource value for each processing element are fixed. Training process has two different phases: In the first one, a number of P patterns is presented to the map. For each of them, the winning neuron is determined updating its weights and resource value and those of its direct neighbours. This process is repeated Q times and then growing phase starts. In this phase a new processing element is added between two processing elements determined by resource value. Weights and neighbourhood relations of new processing element are established and those of its neighbours are updated. The next Q patterns are presented until the stop criterion is satisfied.

3.3 Growing Neural Gas (GNG)

This auto-organised model [7] gives flexibility to the structure, both in size and in topology, as it allows insertion of new processing elements, modification of neighbour relations and even elimination of these relations. The idea of the structure topology modification was introduced to achieve that these structures faithfully represent the input space, eliminating those processing elements that because of constraints of vicinity must be placed over areas in the input space where the probability is almost null.

As in previous sections, GNG performance is divided into different processes. In the initialisation process, the structure has two disconnected processing elements. Weights, learning rate and resource value of these elements are fixed. Learning process has two phases: In the first one, P input patterns are presented to the network organised in groups of Q patterns ($P >> Q$). For each input pattern the two elements with closest weights are determined. If they do not have neighbour relation, it is created and the connection weight is initialised to zero. If they have neighbour relation, its weight is modified adding one to it and connection weights between the winner and all its direct neighbours are reduced. Next, the weights of the neighbourhood relations are examined. If any of them has a weight less than a certain threshold T, that relation is broken and if in consequence an element is isolated, it is also removed. In next stage, weights are modified as in SOMs, with the region of interest constrained to direct neighbours. Once the weights are modified, the resource values of those processing elements are updated. The second phase, insertion of new processing elements is as in previous section. Next group of Q patterns is presented to the network until the stop criterion is satisfied.

4 Results

Results obtained from Gabor decomposition have been analysed by means of different neural network models. Neural network inputs have been constructed assigning an 8 component vector to each pixel of original image and each frequency channel considered. This vector contained the results previously obtained from Gabor analysis in 8 different orientations. A ninth component, called response, was added to previous ones in order to determine the presence of an important component in one of the main orientations considered. The first eight components of the vector were scaled such that their modulus was equal to the response in order to reduce the inter-image variability.

Images in Fig. 1 have been used to show and compare the results obtained with each neural network model considered. We have used a concentric circles synthetic image in order to consider all possible orientations, a corridor and a FLIR aerial image representing a bridge.

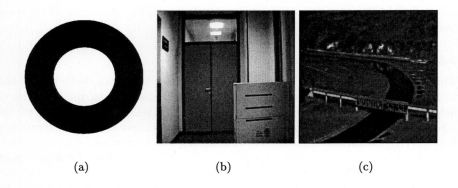

(a)   (b)   (c)

Figure 1. (a)Synthetic image, (b)Corridor and (c)Bridge.

The training set for the SOM was generated from a set of artificial images containing almost the same proportion of lines in each of the main orientations. Background was not included in this training set. The goal was to have a similar number of processing units for each main orientation.

The SOM-PAK package developed at Helsinki University was used to train the maps. Fig. 2(a) shows the structure of the 12x12 SOM after training. This training process was divided in two phases. First one is the ordering phase, with learning rate 0.05 and neighbourhood size 12. The second phase is longer than first, with learning rate 0.02 and neighbourhood size 4.

Training set for GCS was generated as in SOM, but this time including the background. Objective was generation of a set of clusters with a similar proportion of processing units, each cluster representing one of the main orientations considered or the background.

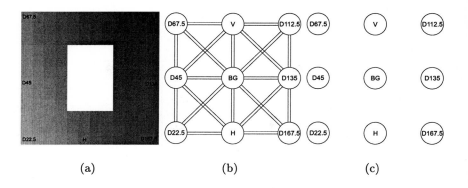

(a)   (b)   (c)

Figure 2. (a)12x12 SOM representation with neurons representing each main perceptual feature, (b)Symbolic Representation of the final GCS map and (c)Symbolic representation of the final GNG map.

Fig. 2(b) shows symbolic structure of a two dimensional GCS map with 200 processing units after training $P = 54374$ input patterns with $Q = 6000$ adaptation steps and taking 0.06 as the factor for winning processing unit weight's update, 0.002 as the neighbour weight's update and 0.05 for resource value's update. Dynamic structure generation (Fig. 2(b)) causes the formation of clusters around processing units representing each orientation or background. These clusters are represented as circles and connections between them as intermediate lines.

GNG is the last model considered to analyse the results of Gabor decomposition. This model was trained with two different kinds of training sets. The first was constructed as in SOM and GCS case, including and excluding the background. The second was based on real images containing approximately the same proportion of all directional features. With all the training sets employed, results were worse than those obtained with previous model.

Fig. 2(c) shows structure of a two dimensional GNG map with 100 processing units after training taking $Q = 3000$ adaptation steps between insertions, 0.2 as winning neuron's learning rate, 0.006 as direct neighbour's learning rate, $T = 50$ as age for deletion and 0.995 for resource value reduction. Dynamic structure generation yields, as in GCS case, to the formation of clusters around processing elements representing main orientations and background. Similarly, removal of edges and processing units motivates an excessive specificity, that is, an excessive separation of clusters.

Results obtained from test images shown in Fig. 1 are depicted in Fig. 3. First row shows results obtained from SOMs, where noise presence is relevant, and there are some directional primitives not detected by the map. Second row shows results obtained by the GCS map. These results, compared with those obtained from SOM analysis, show an important noise reduction and a better detection of perceptual directional primitives. Finally, third row shows results

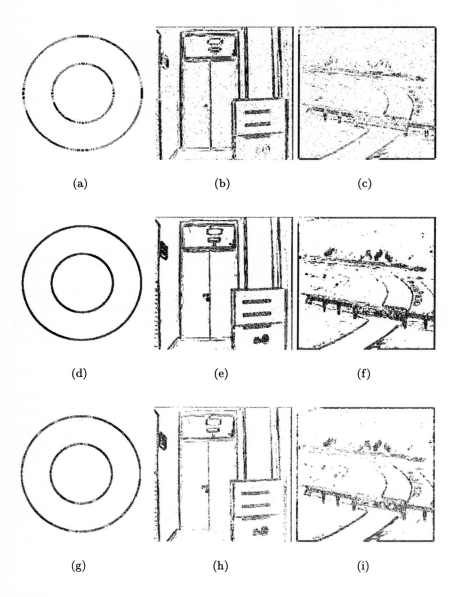

Figure 3. Results from Fig. 1. First line corresponds to SOM, second line to GCS and third line to GNG.

obtained by GNG map. As in GCS noise reduction is important but this time there is an important lost of directional primitives as in SOM.

5 Discussion and conclusion

In results obtained from growing cell structures, noise reduction is important with respect to those obtained from SOM maps. This noise reduction is motivated from dynamic structure generation, which yields to the formation of clusters around winning elements of different main orientations or background.

Results from SOM are noisy because the background neurons are not enough separated from the neurons belonging to other directions. The lost of some intermediate directions comes from the association of some of these directions to the background. Results from GCS and GNG are less noisy. This is because the vicinity between clusters is lower, associating almost all the noise to the background cluster.

Results obtained from GNG maps show an important lost of directional primitives with respect to those of GCS map. This is motivated by the removal of cells and transition elements in GNG learning algorithms, with the lost of intermediate directions. This leads to an excessive separation between clusters corresponding to different orientations, that is, a bad generalisation capacity.

As a conclusion of previously exposed arguments, GCS is auto-organised structure that best fit requirements necessary for posterior perceptual organisation of results obtained by Gabor wavelet analysis.

References

[1] Palmer, S. E.: The Psychology of Perceptual Organisation: A Transformational Approach. Human and Machine Vision, J. Beck, B. Hope and A. Rosenfeld eds., 269–339. Academic, New York (1983)
[2] Ross, W. D., Grossberg, S., Mingolla, E.: Visual Cortical Mechanisms of Perceptual Grouping: Interacting Layers, Networks, Columns and Maps. Neural Networks **13(6)** (2000) 571-588
[3] Sarkar, S., Grossberg, K. L., Mingolla, E.: Perceptual Organisation in Computer Vision: A Review and a Proposal for a Classificatory Structure. IEEE Transactions on Systems Man and Cybernetics **23(2)** (1993) 382–399
[4] Nestares, O., Navarro, R., Portilla, J., Tabernero, A.: Efficient Spatial-Domain Implementation of a Multiscale Image Representation Based on Gabor Functions. Journal of Electronic Imaging **7** (1998) 166–173
[5] Kohonen, T.: Self-Organised Formation of Topologically Correct Feature Maps. Biological Cybernetics **43** (1982) 267–273
[6] Frizke, B.: Growing Self-organising Networks - Why?. 4th European Symposium on Artificial Neural Networks, ESANN (1996) 61–72
[7] Martinetz, T., Schulten, K.: A 'Neural-Gas' Network Learns Topologies. Artificial Neural Networks. Proceedings of the 1991 International Conference, ICANN-91 **1** (1991) 397–402

Real-Time Wavelet Transform for Image Processing on the Cellular Neural Network Universal Machine

Victor M. Preciado[1]

[1] Dpto. de Electronica e Ingenieria Electromecanica, Area de Ingenieria de Sistemas y Automatica, Escuela de Ingenieros Industriales, Avda. de Elvas s/n, 06071 Badajoz. (Spain)
vpdiaz@unex.es

Abstract. A novel algorithm for achieving Wavelet transform on the Cellular Neural Network Universal Machine (CNN-UM) visual neuroprocessor is presented in this work. The CNN-UM is implemented on a VLSI programmable chip having real time and supercomputer power. This neurocomputer is a large scale nonlinear analog circuit made of a massive aggregate of regularly spaced neurons which communicate with each other only through their nearest neighbors. VLSI implementation of this circuit is feasible due to its locally connectivity and fixed output function of each cell consisting of a piece-wise linear saturation function imposed by the difficulty of realizing non-linearities in hardware. In the next, implementation of wavelet transforms by means of an analog algorithm is presented. Thus, we can use the CNN-UM in solving real-time applications where wavelet are an essential step like computer-vision algorithms for stereo vision, binocular vergence control, texture segmentation and face recognition. .

1 Introduction

Analog Cellular Neural Network (CNN) array computer arises as an alternative to traditional digital processors in many image processing applications. This novel computation paradigm is capable of making in a single chip Tera equivalent operations per second [2], [3]. A 4096 analog CNN processor array is able to perform complex space-time image analysis, being much faster than a camera-computer system in manifold applications. Both chips have been implemented in CMOS technology [5], [6] and they are managed by a 32-bit high-performance low-cost microcontroller.

In this work, CNN-Universal Machine (CNN-UM) actual capabilities and how they can be used to solve many image processing tasks are presented [9], [12]. Next, it will be shown a novel method for adjusting an arbitrary output function by means of piece-wise linear saturation output functions. This approximation process provide us a method for correlating images by using logarithmic and exponential approximations. Thus, we can perform any kind of wavelet transforms by means of this correlation step if we are able to generate any wavelet base.

2 Cellular Neural Network Universal Machine Paradigm

The mathematical model for the Cellular Neural Network is provided and, based on it, the CNN-UM computing framework is presented.

2.1 CNN Mathematical Model

Many computational problems can be reformulated as well-defined tasks where the signal values are placed on a regular 2-D grid, and direct interactions between signal values are limited within a finite local neighborhood. This is just the basis idea of Cellular Neural Network (CNN´s): An array of analog dynamic processors whose cells interact directly within a finite local neighborhood [2], [3], [4]. The local CNN connectivity allow its implementation as VLSI chips that can operate at a very high processing speed and complexity: 0.3TeraXPS performance for a 10x10mm^2 chip using a 2-µm technology in a robust implementation. The dynamic of the array can be described by the following differential equations:

$$C \frac{dV_{xij}(t)}{dt} = -R_x \cdot V_{xij}(t) + \sum_{Cel(kl) \in N_r(ij)} a_{kl} V_{ykl}(t) + \sum_{Cel(kl) \in N_r(ij)} b_{kl} V_{ukl} + I \quad (1)$$

$$V_{ykl} = \frac{1}{2} \cdot (|V_{xkl} + K| - |V_{xkl} - K|) \quad (2)$$

$$N_r(i,j) = \{C(k,l) | \max\{|k-i|, |l-j|\} \le r\} \quad (3)$$

where i, j refers to a point associated with a cell on the 2-D grid, and k, l is a point in the neighborhood within a radius r of the i,j cell. Equation (1) describes a nonlinear dynamical system due to the equation (2) that includes in our system a piecewise linear saturation function. In the equation (3) the concept of neighborhood is defined. C is an input capacitor, R an input resistance and I an input bias current. $V_{ykl}(t)$ represents the neural activity, whereas $V_{yij}(t)$ is the output of the network and V_{uij} is a fixed external input to the network. B and A are connection matrices that describe the input and feedback connectivity respectively.

2.2 Architecture Framework of the CNN-UM

The CNN is the foundation for a computation paradigm called the Cellular Neural Network Universal Machine [9]. The CNN supply the supercomputation power for accomplish Tera equivalent operations per second in a single chip. But this potentiality must be driven by means of an adequate framework, yielding the CNN-UM Architecture.

First we need a control unit provided by a simple low-cost microprocessor where we can load the analog programs to be performed into the CNN. Besides, the CNN is provided by four local analog and four local binary memories that allow us to save partial results of our analog algorithm. A local logical unit inside each cell permit us achieve logical operations involving logic memories and CNN output.

3 Adjusting an Arbitrary Saturation Output Function

The classical CNN cell output function is presented in equation (2). In a first step we must change the parameters that define the shape of the normalized saturation function. The four parameters that define the shape can be expressed in two ways:
- *Two points definition*, in such a way that we have the saturation shape fixed if we set the coordinates of the two extreme points. This coordinates are (l,d) and (r,u), like can be seen in Fig. 1 (left).
- *Template parameters definition* where the shape of the saturation is expressed with the parameters that appear explicitly in the templates that will be used in the change of shape process. These parameters are: central point y-coordinate (or average value) m, central point x-coordinate c, slope in the central zone a/b, and output swing $2/b$, like is shown in Fig. 1 (left).

Both definitions are easily related by the following set of equations:

$$a = \frac{2}{r-l} \qquad b = \frac{2}{u-d} \qquad c = \frac{r+l}{2} \qquad m = \frac{u+d}{2} \qquad (4)$$

The shape of the normalized saturation function can be transformed into an arbitrary shape saturation function parameterized by means of the templates expressed in eq. 5.

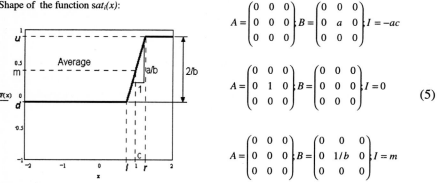

$$A = \begin{pmatrix} 0 & 0 & 0 \\ 0 & 0 & 0 \\ 0 & 0 & 0 \end{pmatrix}; B = \begin{pmatrix} 0 & 0 & 0 \\ 0 & a & 0 \\ 0 & 0 & 0 \end{pmatrix}; I = -ac$$

$$A = \begin{pmatrix} 0 & 0 & 0 \\ 0 & 1 & 0 \\ 0 & 0 & 0 \end{pmatrix}; B = \begin{pmatrix} 0 & 0 & 0 \\ 0 & 0 & 0 \\ 0 & 0 & 0 \end{pmatrix}; I = 0 \qquad (5)$$

$$A = \begin{pmatrix} 0 & 0 & 0 \\ 0 & 0 & 0 \\ 0 & 0 & 0 \end{pmatrix}; B = \begin{pmatrix} 0 & 0 & 0 \\ 0 & 1/b & 0 \\ 0 & 0 & 0 \end{pmatrix}; I = m$$

Fig. 1. Parameters modified in the adjusting process (m, a/b, $2/b$ and c). These values can be also expressed depending on l, r, u and d (bolded in the figure).

3.1 Modification of the Piece-Wise Linear Saturation Function

All these changes can be performed with the three templates expressed in eq. (5). The first and third templates provide non-dynamical networks, thus they can be solved in only one time constant of the circuit. The second one represents a dynamical network that realize the normalized saturation function

4 Approximating an Arbitrary Function

Among the several criteria that can be contemplated in an approximation process, we have selected an error function defined with the infinite norm. In this case, the error function we must minimize is $\varepsilon = \|y_i - (ax_i + b)\|_\infty = \max\{|y_i - (ax_i + b)|\}$.

Actual CNN-UM Chip is implemented in analog VLSI technology and can perform 2-D signal processing tasks into a digital equivalent accuracy of 7 bits. The normalized saturation function has an output swing of 2 units, therefore this accuracy represent an uncertainty in the signal operation equal to $2/2^7 = 2^{-6}$. Thus, this value is an suitable measure for the error represented by the infinite norm. In this way, the error introduced in the approximation process is diluted by the error of the chip operation.

The line that minimize the infinite norm in a interval $[a,b,]$ is called the Chebyshev line and it can be determined by means of the *Error Property Theorem* [10]: "A n-orden polynomial P_n minimize the infinite norm with respect to a function f(x) defined into a interval [a,b] if $|f- P_n|$ take the value of the infinite norm in at least n+2 points into the interval [a,b] with alternative sign".

In the case of a function with positive second derivative $f''(x)>0$, $\forall\ x\in[a,b]$ it can be demonstrated that the Chebyshev line (or first order polynomial) is defined by:

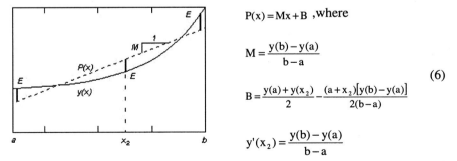

$$P(x) = Mx + B \text{ , where}$$

$$M = \frac{y(b) - y(a)}{b - a}$$

$$B = \frac{y(a) + y(x_2)}{2} - \frac{(a + x_2)[y(b) - y(a)]}{2(b - a)} \quad (6)$$

$$y'(x_2) = \frac{y(b) - y(a)}{b - a}$$

Fig.2. Chebyshev line conditions for a f(x) with $f''(x)>0$, $\forall\ x\in[a,b]$. In a first-order polynomial the maximum difference must be reached in three points.

In the case of a function with positive second derivative $f''(x)>0$, $\forall\ x\in[a,b]$ similar results can be reached.

This criterion gives us a direct method to perform a pieced Chebyshev approximation of logarithmic and exponential functions in a forward way by solving equation (7) in the case of approximating a logarithmic function and equation (8) in the exponential case.

$$2 \cdot E = \ln\left(\frac{x_i - x_{i-1}}{\ln(x_i) - \ln(x_{i-1})}\right) + \left(\frac{x_i - x_{i-1}}{\ln(x_i) - \ln(x_{i-1})}\right)^{-1} \cdot x_{i-1} - \ln(x_{i-1}) - 1 \quad (7)$$

$$2 \cdot E = \frac{\exp(x_i) - \exp(x_{i-1})}{x_i - x_{i-1}} \cdot \ln\left(\frac{\exp(x_i) - \exp(x_{i-1})}{x_i - x_{i-1}}\right) + \frac{\exp(x_i) - \exp(x_{i-1})}{x_i - x_{i-1}} \cdot x_{i-1}$$
$$- \exp(x_{i-1}) - \frac{\exp(x_i) - \exp(x_{i-1})}{x_i - x_{i-1}} \quad (8)$$

where E is the equivalent analog accuracy that the VLSI neuromorphic circuit performs (2×2^{-7}) and x_i is the right extreme of an approximation interval. We must chose a first x-coordinate point x_0 and solve iteratively eq. (7) and (8) until we reach the another extreme. Once we have calculated every x_i we can determine the y-coordinate of each point by adding E to the value of $exp(x_i)$ in the case of exponential approximation and subtracting the same value to $ln(x_i)$ in the logarithmic approximation case. In this way we obtain the following values for the extreme points of the Chebyshev lines that approximate the exponential (table 1) and the logarithmic (table 2) functions. Template coefficients are determined by means of eq. (3), taking into account that $l=x_{i-}$, $r=x_i$, $u=y_i$ and $d=y_{i-}$.

Table 1. Extreme points and template coefficients for each approximation stage in the exponential approximation.

Step	Points in the Chebyshev line	Template coefficients
1st	(-1,0.35) to (-0.31,-0.72)	a=2.90, b=5.47, c=-0.65, m=0.55
2nd	(-0.31,-0.72) to (0.20,1.22)	a=3.86, b=4.02, c=-0.05, m=0.25
3rd	(0.20,1.22) to (0.61,1.84)	a=4.87, b=3.21, c=0.41, m=0.31
4th	(0.61,1.84) to (1,2.70)	a=5.9, b=2.67, c=0.78, m=0.37

Table 2. Extreme points and template coefficients for each approximation stage in the logarithmic approximation.

Step	Points in the Chebyshev line	Template coefficients
1st	(0.36,-0.9) to (0.60,-0.48)	a=8.33, b=4.76, c=0.48, m=-0.69
2nd	(0.60,-0.48) to (1.00,0.01)	a=5, b=4.08, c=0.8, m=0.25
3rd	(1.00,0.01) to (1.67,0.52)	a=2.98, b=3.92, c=1.33, m=0.25
4th	(1.67,0.52) to (2.17,1.01)	a=4, b=3.92, c=1.92, m=0.25

5 CNN Structure to Approximate

After calculating the Chebyshev lines we must carry the approximation results into a CNN algorithms.

For this purpose we build the non-linear function overlapping several linear saturation. For this purpose we can take the Chebyshev line at the left of the approximation piece-wise linear function, and make a first approximation step with the extreme points of this piece. We take the following pieces and calculate the template coefficients taking into account that adding all the outputs must result in the approximating function. With this purpose we subtract y_i to each pair (y_i, y_{i-1}) that define the Chebyshev line for each piece in which i>0. These summands can be seen in Fig. 3. Proceeding in this way we demonstrate in the following that the accumulated output follows the non-linear curve taking into account the Chebyshev criterion.

Knowing that $\hat{f}(x) = y_{i-1} + M_i \cdot (x - x_{i-1})$, where M_i is the slope of the Chebyshev line into the interval and $x \in [x_{i-1}, x_i]$, if we add every saturation function we obtain the same:

$$\sum_{k=1}^{n} sat_k(x) = \sum_{k=1}^{i-1} sat_k(x) + sat_i(x) + \sum_{k=i+1}^{n} sat_k(x) = y_1 + (y_2 - y_1) + (y_3 - y_2) + \ldots$$
$$+ (y_{i-1} - y_{i-2}) + M_i \cdot (x - x_{i-1}) + 0 + 0 + \ldots + 0 = y_{i-1} + M_i \cdot (x - x_{i-1}) = \hat{f}(x)$$

Thus, performing logarithmic approximation of two images, parallel point by point addition and realizing the exponential function of the resulting image, we can obtain the multiplication of initial images.

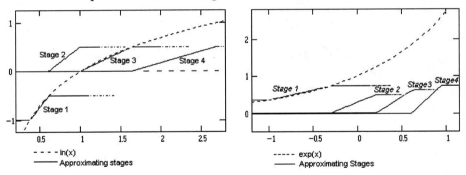

Fig. 3. Different piece-wise linear saturation stages whose addition performs the approximation of a logarithmic and exponential functions.

6 Performing Wavelet Transform

In this point we are going to present a well-known kind of Wavelet Transform called Gabor transform and its implementation on the CNN-UM. The responses of orientation selective cells in the visual cortex can be modeled using this kind of wavelet filters [11]. The impulse response of a two dimensional Gabor filter is a bidimensional complex exponential modulated by a gaussian

$$g(x, y) = \frac{1}{\sqrt{2\pi}\sigma} e^{-(x^2+y^2)/2\sigma^2} e^{j(w_{x0}x+w_{y0}y)} \qquad (9)$$

that is tuned to spatial frequency (w_{x0}, w_{y0}). 2-D Gabor filters have found applications in computer-vision algorithms for stereo vision, binocular vergence control, texture segmentation and face recognition.

We can use our correlation algorithm in making the Gabor (or a general Wavelet transform) in a parallel way [7], [8]. Thus, we can consider the impulse response in eq. (9) and truncate the Gaussian where it falls a 98% with respect to its central value. The Geometric configuration that results in this truncation is a circle that can be joined with identical truncated impulse response into a hexagonal grid in order to take advantage of the compactness of this topology, as can be seen in Fig. 5. By means of the correlation step we obtain the value of the Gabor transform in the central points of the gaussians. Making an iterative shift and performing the correlation step previously developed we can perform Gabor transform in real-time.

Fig. 4. Hexagonal distribution of truncated Gabor impulse response with different spatial frequencies. This image can be correlated with the input one in order to obtain the Gabor transform that corresponds to the central points of the gaussians by means of the method developed in this work.

7   Conclusions and Further Works

A novel method to perform any non-linear saturation output function on the CNN-UM neuromorphic circuit has been shown in this work. This universal machine is implemented on a single chip with a normalized saturation output function, and we combine this function in a proper manner to perform the adjustment with the Chebyshev criterion in order to minimize the infinite norm. This criterion allows us to take into account the 7 bits of equivalent digital accuracy that this chip performs in order to not introduce any error due to the approximation process.

Next we have adjusted exponential and logarithmic function to succeed in making real-time correlation between images that provide us a general method to fulfill any kind of integral transform on the CNN-UM.

Lastly we have commented how the Gabor Wavelet transform, that has been used in modeling the responses of orientation selective cells in the visual cortex, can be implemented on this chip. The applications of this transform can be found in several fields like computer-vision, algorithms for stereo vision, binocular vergence control, texture segmentation and face recognition.

References

1. Burden R.L., Douglas J., "Numerical Analysis", PWS Publishing Company, 1993.
2. L. O. Chua, L. Yang. "Cellular Neural Networks: Theory". IEEE Trans. on Circuits and Systems. Vol. 35, No. 10. pp. 1257-1272. October 1988.
3. L. O. Chua, L. Yang. "Cellular Neural Networks: Applications". IEEE Trans. on Circuits and Systems. Vol. 35, No. 10. pp. 1273-1290. October 1988.
4. L. O. Chua, T. Roska. "The CNN Paradigm". IEEE Trans. on Circuits and Systems. I: Fundamental Theory and Applications. Vol. 40, No. 3. pp. 147-156. March 1993.
5. J. M. Cruz and L. O. Chua, "Design of high speed high density CNN´s in CMOS technology", Int. J. Circuit Theory and Applications, pp. 555-572, vol. 20, 1992.
6. S. Espejo, R. Carmona, R. Domínguez-Castro, A. Rodriguez-Vazquez, "A VLSI oriented continuous-time CNN model", Int. Journ. Of Circuit Theory and Applications, vol. 24, pp. 341-356 (1996).
7. R.C. Gonzalez and R.E. Woods, "Digital Image Processing", Addison-Wesley Publishing Company, Inc., Reading, Massachusetts, 1992.
8. S. Mallat, "A Wavelet Tour of Signal Processing", Academic Press, 1997.
9. T. Roska and L.O. Chua, "The CNN Universal Machine: An Analogic Array Computer", IEEE Trans. on Circuits and Systems. Vol. 40, No. 3. pp. 163-173. March 1993.
10. F. Scheid and R.E. Di Constanzo, "Schaum's Outline Numerical Analysis", McGraw Hill, 1991.
11. B.E. Shi, "Gabor-Type Filtering in Space and Time with Cellular Neural Networks", IEEE Transactions on Circuits and Systems, Vol. 45, No. 2, 1998.
12. P.L.Venetianer, F.Werblin, T.Roska and L.O.Chua, "Analogic CNN Algorithms for some Image Compression and Restoration Tasks", IEEE Transactions on Circuits and Systems, Vol. 42, No. 5, 1995.

OBLIC: Classification System Using Evolutionary Algorithm

J.L. Alvarez[1], J. Mata[1], and J.C. Riquelme[2]

[1] Universidad de Huelva, Spain,
{alvarez,mata}@uhu.es
[2] Universidad de Sevilla, Spain,
riquelme@lsi.us.es

Abstract. We present a new classification system based on Evolutionary Algorithm (EA), OBLIC. This tool is an OBLIque Classification system whose function is to induce a set of classification rules no hierarchical from a database or training set. The core of the algorithm is a EA with real-coded and Pittsburgh approach. Each individual is composed by a no fixed classification rules set what split in regions the search space. The fitness of each classification is obtained by means of the exploration of these regions. The result of the tool is the best classification obtained in the evolutionary process.
This paper describe and analyze this new method by comparing with other classification systems on UCI Repository databases. We conclude this paper with some observations and future projects.

1 Introduction

Nowadays, the growing of data acquisition technologies produces a huge volume of information. Actually, people can't handle this volume of information by manual procedures. Thus, there has been an explosion of machine learning and Data Mining systems that attempts to knowledge discovery of this information. Between the most popular techniques are the classification systems. The classification systems have as objective to induce a rules set from a labeled database.

Decision trees are a particularly useful technique to hierarchical classification system [8][7]. These techniques generate a decision tree from which the classification is interpreted. On the other hand, EA's [2][5][9] have been applied in numerous machine learning and Data Mining problems with excellent results [1][3][4].

In this paper, we describe the theoretical base and development of a classification system based on EA whose focus is to induce a classification rules set from a labeled database with numerical attributes. Each potential solution (classification) consists of a rules set that splits into regions the search space. The fitness of each possible classification is evaluated in the evolution process attending to the correctly classified elements by each region.

The developed tool is framed inside the denominated oblique classification systems. This type of tools restricts its use domain to databases with exclusively numeric attributes. For it, the databases with exclusively symbolic or heterogeneous attributes should be adapted.

In the following sections the basic concepts of the tool are detailed. In section 2 we start with brief description of the EA framework used. We follow with a overview about OBLIC implementation in section 3. Then, in section 4 we compare the results obtained on UCI Repository database [6], with the traditional classification systems C4.5 [8] and OC1 [7]. Finally, in section 5, we conclude with some observation about the development tool and future projects.

2 Algorithm

The core of the algorithm is a EA with real-coded individual (potential solutions). Each individual consists of a vector set, where each vector represents a classification rule. A classification rule consists of a float numbers vector, that it describes the equation of hyperplane that splits into regions the search space. The evaluation function determines the fitness of these individuals according to the number of correctly classified elements of training set. The crossover and mutation genetics operators generate the next population in this evolutionary process.

In the following subsections we approach with more detail each one of the significant elements of the EA. The most significant parameters are shown in section 3.

2.1 Individual Data Structure

Each potential classification are composed a string set of float numbers vectors. In this words, a two-dimensional matrix where each ith string represent the ith classification rule. Thus, each jth column represent the jth float coefficient of the hyperplane equation for the ith rule.

$$a_{i1}X_1 + a_{i2}X_2 + \cdots + a_{id}X_d + a_{d+1} \tag{1}$$

Thus, for d dimensional search space, the ith string implies the classification rule shown in the equation 1.

a_{11}	a_{12}	$\cdots$	a_{1d+1}
a_{21}	a_{22}	$\cdots$	a_{2d+1}
$\cdots$			
a_{m1}	a_{m2}	$\cdots$	a_{md+1}

Fig. 1. Individual data structure

$$a_{11}X_1 + a_{12}X_2 + \cdots + a_{1d}X_d + a_{1d+1} = 0$$
$$a_{21}X_1 + a_{22}X_2 + \cdots + a_{2d}X_d + a_{2d+1} = 0$$
$$\vdots$$
$$a_{m1}X_1 + a_{m2}X_2 + \cdots + a_{md}X_d + a_{md+1} = 0 \qquad (2)$$

For the individual representation, we use the Pittsburgh approach, thus, each individual in the population consists of a variable-length rules set. Graphically, figure 1 show the individual structure for d dimensional search space $(X_1, X_2, \cdots, X_d$ attributes) and m-rules. Equation 2 show the rules set deduced from the individual of the figure 1.

2.2 Initial Population

The initial population of the individuals is choosing randomly. For each rule, d+1 float values are generated, where d is the number of attributes or dimension of the search space. These values represent the equation of the hyperplane of each classification rule according to 1. The number of rules of the individuals is selected between 1 and an maximum value and the coefficients of each rule are chosen about a range. This values are detailed in section 3.

2.3 Evaluation Function

The fitness of a potential classification is obtained by the exploration of search space. So, the rules of a classification split in regions the search space, this regions are analyzed and the accuracy of the classification is returned as fitness.

Thus, we establish an ordination of classification rules of the individual. This ordination allows a binary coding of the regions. Each region is coded by a bit series, where the ith element corresponds with the ith classification rule. The value of the ith bit is 0 if the region is to the left of the ith rules and it is 1 if the region is to the right.

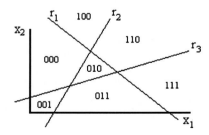

Fig. 2. Binary coded regions

Figure 2 shows an example of the binary coding of a individual about a two-dimensional space (two attributes: X_1, X_2) with three rules.

$$f(n) = Pos(n) - Neg(n) - Rules(n) * FIR \qquad (3)$$

The items of the database are examined about each region to obtain the fitness of each individual. Thus, a region will be labeled with majority class. The items with the same class that a region is positive cases, while the cases with a different class are negative cases for this region.

Let n be individuals, $Pos(n)$ be total positive cases, $Neg(n)$ be total negatives cases, $Rules(n)$ be the number of rules of the classification and FIR be the Factor Influence of Rules, between 0 a 1, the fitness of the individuals is evaluated by maximized 3.

The FIR factor allows to determine the influence of the number of rules in the classification. So, if this value is near to 0 then the number of rules has very low influence, and if this value is near to 1 then the number of rules has very high influence.

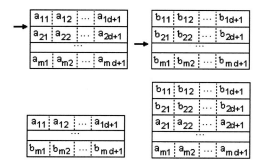

Fig. 3. Crossover operator

2.4 Genetic Operators

New individuals are gererated by the crossover operator. The crossover operator selects two individuals and produces a new individual of them. Thus, a crossing point is selected in each father individual and the new individual is created by the previous genes to the crossing point of one and the genes next to the point of crossing of the other one. Graphically, the figure 3 show this crossover operator.

The mutation operators alter the individuals of a population. There are three mutation operators. This operators are:

- Alters
- Add
- Rest

This mutation operators affect the individuals randomly. Thus, the crossing possibility is raffled and if this is bigger than the mutation rate then one of the mutation operators will be applied. The mutation operator is selected randomly.

The first mutation operator alters one o more genes (coefficients) of the individual, randomly. The genes are altered a randomly proportional quantity with equal percent mutation. The second, Add mutation operator, add new genes for individual. Thus, one new rule is added to classification. And the Rest operators eliminate genes for one existent rule. Thus, one existent rules is suppressed of the classification.

Table 1. EA parameters

Parameter	Value
individuals	100
generations	200
Mutation rate	80%
Mutation percent	70%
rules individuals	12
Range coefficients	[-10,10]
FIR	10%

3 Implementations

The previous algorithm has been developed as OBLIC. The values of the principal parameters are shown in table 1.

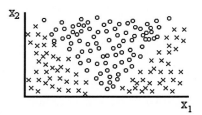

Fig. 4. Synthetic database

3.1 Practical Implementation

In this subsection, we show the results obtained on a synthetic dataset. The dataset is composed by two attributes (X_1, X_2) and two class (x, o). The classes have been distributed with 4 by random values of the X_1 attribute. An approximate representation is shown in the figure 4.

$$X_2 = (X_1 - 5)^2 \qquad (4)$$

The results obtained for the previous database are:

```
Training set: synthetic.dat
200 cases. 2 attributes. 2 class.

Classification
==============
Regions
-------
Region 0 -> Class x. Pos: 56. Neg: 0
Region 2 -> Class o. Pos: 89. Neg: 0
Region 3 -> Class x. Pos: 55. Neg: 0

Rules
-----
(-6.00)X1 + (-3.00)X2 + (30.00) = 0
(-6.00)X1 +  (3.00)X2 + (30.00) = 0

#Rules: 2     Errors: 0 ( 0 %)
```

The results show the regions and the rules for the classification. For each region, it shows its class and the positives and negatives cases. After, the classification rules are shown in the order to binary coding of the regions. Finally, the numbers of the rules, the numbers of the negatives cases and the percentage of errors are shown. The figure 5 represent, graphically, the result on the synthetic database.

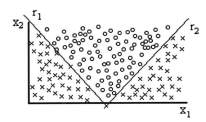

Fig. 5. Classification of synthetic database

4 Results

The performance of this method is compared with traditional classification system C4.5 and OC1 on real UCI Repository database. In table 2 we show the databases used. The results are summarized in the table 3.

In the table 3 we summarize the results obtained by the C4.5, OC1 and Oblic classifiers systems on five UCI Repository databases (bupa, iris, new-troid, pima, glass). For each one, we are shown the number of rules ($\#Ru.$) and the error percentage ($\%Er.$) obtained during the phase of tests.

The results have been generated starting from four tests for each database. In each test we have used 80% of the items for training set and 20% for test set. For each database, the results show the average value of tests.

Table 2. Databases

Databases	#Attr.	#Cases	#Cases-test	#Classes
BUPA	7	345	69	2
IRIS	4	150	27	3
TROID	5	215	43	3
PIMA	8	768	153	2
GLASS	10	214	42	6

Table 3. Comparative results

	C45		OC1		OBLIC	
	#Ru.	%Er.	#Ru.	%Er.	#Ru.	%Er.
BUPA	26	12.31	3	15.21	12	12.60
IRIS	4	2.77	3	1.85	3	0.92
TROID	8	1.74	5	1.16	6	0.58
PIMA	21	16.01	21	9.96	12	19.44
GLASS	5	1.19	5	1.19	5	1.19

5 Conclusion

In this paper, we present the theoretical aspects and we offer the implementation of a classification system based on an EA. This tool offers an alternative to the decision trees classification systems.

Obviously, this tool has the disadvantages characteristic of the AE, as the randomness and the high time of computation. But it presents other very significant advantages.

About the advantages we highlight: the classification rules is not hierarchical and the division criterion is not established a priori.

The non-hierarchical classification allows that the rules can be evaluated independently and no order. And the division criterion will be induced from the training set, itself.

Nowadays, we are working on new projects where a classification system is been developed. This new tools permits establish the number of rules and/or the maximum error rate of the final classification.

References

1. K.A. De Jong: Using Genetic Algorithms for Concepts Learning". Machine Learning, **13**, pp. 161-188, (1993).
2. D.E. Golberg: Genetic Algorithms in Search, Optimization and Machine Learning". Addison-Wesley Pub. Company, inc., 1989.
3. C.Z Janikov: A Knowleged-Intensive Genetic Algorithm for Supervised Learning". Machine Learning, **13**, pp. 189-228, (1993).
4. J.R. Koza: Concept Formation and Decision Tree Induction using the Genetic Programming Paradimg. Springer-Verlag, (1991).
5. Z. Michalewicz: Genetics Algorithms + Data Structures = Evolution Programs. Third Edition. Springer-Verlag, (1999).
6. P.M. Murphy, D.W. Aha: UCI Repository of Machine Learning Databases. http://www.ics.uci.edu". Department of information and Computer Science, University of Californnia, (1994).
7. S.K. Murthy, S. Kasif, S. Salzberg: A System for Induction of Obliques Decision Tress. Journal or Artificial Intelligence Research, vol. **2**, pp. 1-32, (1994).
8. J.R. Quinlan: C4.5: Programs for Machine Learning. Morgan Kaufmann Pub., (1993).
9. A.H. Wright: Genetic Algorithm for Real Parameter Optimization. Morgan Kaufmann Pub., (1991).

Design of a Pre-processing Stage for avoiding the Dependence on TSNR of a Neural Radar Detector

Pilar Jarabo Amores, Manuel Rosa Zurera, and Francisco López Ferreras

Departamento de Teoría de la Señal y Comunicaciones
Escuela Politécnica. Universidad de Alcalá
Ctra. Madrid-Barcelona, km. 33,600
28.871 Alcalá de Henares - Madrid, Spain
{mpilar.jarabo, manuel.rosa, francisco.lopez}@uah.es

Abstract. A new pre-processing stage for neural radar detectors is presented in order to reduce the detector performance dependence on the Training Signal-to-Noise Ratio (TSNR). The proposed scheme combines Time-frequency Analysis for transforming radar echoes into a feature space where the detection task is easier, and Principal Component Analysis for dimensionality reduction. The results are compared with those obtained when using a single MLP, demonstrating that the new detection scheme can match the best receiver operating characteristic of the single MLP radar detector, for any value of TSNR, avoiding the laborious trial-and-error process that is necessary to select the optimum TSNR for a single MLP radar detector.

1 Introduction

The classical solution for radar detection uses a matched filter and assumes that the interference is modeled as Additive-White-Gaussian Noise (AWGN). The actual case is not so simple, because the interference is not gaussian and may be non-stationary. A summary of the early efforts in automatic target recognition using neural networks can be found in [1] and [2]. Neural networks can implement the Bayesian optimum detector [3] and it has been demonstrated that the back-propagation algorithm applied to a feed-forward network approximates the optimum Bayesian classifier, when using the mean square error criterion. In [4], the performance of a Multilayer Perceptron (MLP) for non-fluctuating targets is analyzed, putting special emphasis on the training conditions that guarantee the optimum performance. Taking these results as starting point, we have analyzed the performance of the MLP using more elaborated target models, showing that the slow fluctuating ones are the most difficult to be detected [5]. An important conclusion extracted from the previous works is the dependence of the receiver operating characteristic on the selected Training-Signal-to-Noise-Ratio (TSNR). This optimum TSNR is different for each target model, making the optimum training too complex.

Motivated by the pour performance for slow fluctuating targets, and the critical dependence on the TSNR, we have designed a pre-processing stage that reduces the dependence on the TSNR, improving the generalization capabilities of the system for SNR values different from the TSNR. It is based on Time-frequency Analysis, followed by Principal Components Analysis (PCA).

2 Problem Formulation

A pulse radar detection environment has been simulated. The instantaneous Radar Cross Section (RCS) is proportional to the effective pulse power (the average power during the pulse). If we assume AWGN of zero mean and unity variance, the instantaneous signal-to-noise ratio expressed in natural units is equal to the effective pulse power, and so proportional to the instantaneous RCS.

As radar returns we have used the Swerling 1 model (Sw1) [6] [7], that applies to complex targets consisting of many independent scatters of approximately equal echoing areas. As examples of such targets we can consider a large aircraft, rain clutter, terrain clutter, etc. Sw1 targets are the most difficult to detect using a single MLP [5]. The effective pulse power can be modeled as a random variable with exponential p.d.f. and, as a normalization criterion, its mean value can be considered as the mean signal to noise ratio expressed in natural units (snr):

$$f(x) = \frac{1}{snr} e^{\frac{-x}{snr}} u(x) \tag{1}$$

The p.d.f. of the radar receiver output voltage envelopes, y, can be modeled as a Rayleigh random variable:

$$y = \sqrt{2x} \Rightarrow f(y) = \frac{y}{snr/4} e^{\frac{-y^2}{snr/2}} u(y) \tag{2}$$

For low snr, the voltage envelope p.d.f. and the noise p.d.f. overlap each other significantly and the neural network learns to distinguish low power signal patterns from noise patterns, giving rise to high probability of detection (P_d), at the expense of a high probability of false alarm (P_{fa}). For high snr, the two p.d.f.'s separate out, and the neural network learns to distinguish noise patterns from high power signal patterns. But in actual environments, the SNR can be very low making the P_d too low for practical P_{fa} values.

¿From this discussion, an important question arises: What is the optimum TSNR for maximizing P_d maintaining low P_{fa} and ensuring a good performance for different SNR values? Due to the heuristic nature of neural networks learning, the only way of determining this optimum TSNR value is a trial and error procedure, making the design of the neural detector too laborious. Figure 2 represents the receiver operating characteristic of a detector based on a single MLP with 16 input nodes, 8 processing units in the hidden layer and 1 output node, trained with $TSNR = 15dB$ (dotted line) and with $TSNR = 5dB$ (continuous line) for $SNR = 7dB$. The dependence on the TSNR is evident.

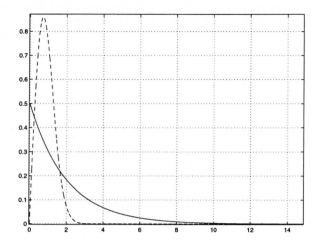

Fig. 1. P.d.f. of the effective pulse power (continuous) and the voltage envelope (dotted) for $SNR = 3dB$

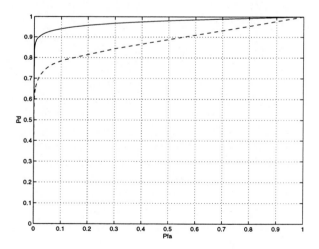

Fig. 2. Detector ROC curves for $SNR = 7dB$: $TSNR = 5dB$ (continuous) and $TSNR = 15dB$ (dotted)

3 Proposed Pre-processing Stage

In this paper we propose a pre-processing stage in order to reduce the dependence on the TSNR of the radar neural detector.

3.1 Time-frequency Analysis

Time-frequency analysis provides a tool for transforming the input vector and distributing the information throughout the time-frequency space, making the information extraction easier. The Wigner-Ville Distribution has been selected as in [8]. It is calculated using expression 3.

$$W_x(n,k) = \sum_{m=1}^{L} x[n+m]x^*[n-m]e^{-j\frac{2\pi k}{L}m}$$
$$n = 0, 1, 2, \ldots, L-1$$
$$k = 0, 1, 2, \ldots, L-1$$
(3)

where $x[n]$ denotes a sample of the L-length complex-valued discrete-time received signal. In our study, dealing with Sw1 targets, all the samples in a pattern have the same magnitude. So, the values of the WVD for noise-free patterns are close to zero except those concentrated around low frequencies (figure 3). On the other hand, WVD images of noise patterns have significant values throughout all the time-frequency plane as can be observed in figure 4.

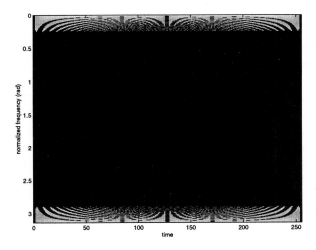

Fig. 3. WVD (magnitude) of a noise-free Sw1 pattern

3.2 Dimensionality reduction stage

The main disadvantage of the WVD is the significant increase in the problem dimension. We work with 8-length complex vectors, that give rise to 8 by 8

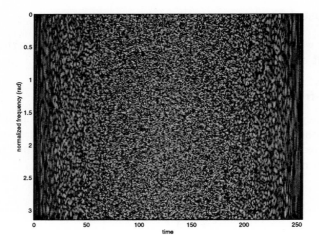

Fig. 4. WVD (magnitude) of a noise pattern

real matrices in the WVD domain. So, a dimensionality reduction is needed in order to reduce the number of inputs of the MLP, and so, to increase the network generalization capability. Using Principal Components Analysis (PCA), not only a dimensionality reduction is achieved, but the resulting coefficients are uncorrelated and, as a result of the well known heuristic rule, the learning process is accelerated [9] [10]. The Generalized Hebbian Algorithm (GHA) is used for training. The synaptic weight q_{ij}, that connects the source node j to the linear neuron i of the output layer of a feature extractor is adapted according to 4.

$$q_{ij}(n+1) = q_{ij}(n) + \eta \hat{x}(n)[x_j(n) - \sum_{k \leq i} q_{kj}(n)\hat{x}_k(n)] \qquad (4)$$

where $\mathbf{x}$ is a column of the WVD image, $\hat{\mathbf{x}}$ is the output vector, η is the learning parameter and n is the iteration step. Due to the simplicity of noise-free patterns, a column of the WVD of these patterns can be represented with the projection onto the first eigenvector without significant loss of information. On the other hand, noise patterns produce significant projections onto all the eigenvectors. This reasoning led us to use a set of eight eigenvectors (one for each column of the WVD), in order to obtain a good representation of signal patterns and reducing the contribution of noise patterns. Unfortunately, in actual environments, there are not noise-free patterns and the SNR values are usually very low. So the vector applied to the MLP is composed of a slow varying component and a fast varying one that is the contribution of the noise and the WVD cross-terms. So, the task is not very different from the classification of raw data.

As a solution, a second set of feature extractors (PCA) has been trained with images of only-noise patterns. Obviously, each column of these WVD images is not well represented with only one component, but thinking about maintaining the same MLP structure (16/8/1) we keep only the first eigenvector of each WVD column.

Once the 16 PCAs have been trained, if $\mathbf{P}_{Sw1}$ and $\mathbf{P}_{noise}$ are the real matrices whose columns are the eigenvectors associated to the WVD columns of the noise-free Sw1 targets and the noise patterns respectively, and $\mathbf{W}_x$ is the WVD image of an input pattern, the output of the first and second sets of PCAs can be calculated with 5 and 6. These are the inputs to the MLP classifier.

$$\mathbf{c}_1 = diag(\mathbf{W}_x^T \mathbf{P}_{Sw1}) \tag{5}$$

$$\mathbf{c}_2 = diag(\mathbf{W}_x^T \mathbf{P}_{noise}) \tag{6}$$

4 Results and Conclusions

In order to assess the performance of this detection scheme, the following experiment has been carried out:

1. Nine training sets have been created for different TSNR values, each one of 10,000 noise and signal-plus-noise patterns randomly distributed with equal a priori probability. The selected TSNR values are $0dB$, $3dB$, $5dB$, $8dB$, $11dB$, $18dB$, $21dB$ and $24dB$. For each training set, a validation one has been created with the same TSNR.
2. The PCAs at the pre-processing stage have been trained with 100,000 simulated noise-free patterns and 100,000 simulated noise patterns.
3. The training and validation sets described above have been pre-processed in order to obtain the training and validation data for training the MLP with the back-propagation algorithm.
4. The selected size of the MLP is 16/8/1 (16 input nodes, 8 neurons in the hidden layer and 1 output node) and is finished by a threshold detector.
5. To assess the performance of the radar detector, we have estimated the P_{fa} and P_d for different values of SNR, using Montecarlo simulations. Data sets of 80,000 noise patterns have been created in order to guarantee an accurateness of 0.05 (relative error of 5%) for a minimum P_{fa} value of $5 \cdot 10^{-3}$.

Figure 5 shows the P_{fa} versus the threshold for different TSNR values. The first conclusion that could be extracted is that there is an optimum TSNR value of 18dB, because lower probability of false alarm is obtained for the same threshold value, compared with other TSNR values. But we have to take into consideration that the performance of a detector is not determined by this fact, but the receiver operating characteristic (ROC). When ROC curves are represented for different TSNR values, an impressive result is obtained: all of them are more or less superposed, being difficult to distinguish one from the others, as can be observed in figures 6 and 7.

So, using the pre-processing stage proposed in this paper, not only the performance of the radar detector implemented with a neural network is nearly

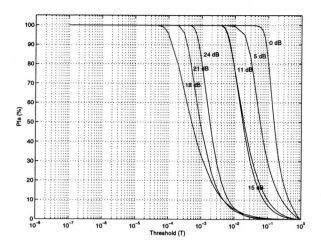

Fig. 5. P_{fa} vs. Threshold for different $TSNR$ values

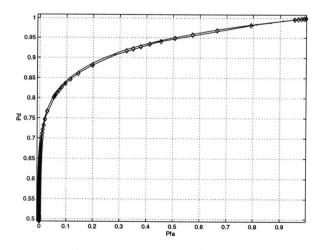

Fig. 6. ROC curves for $SNR = 3dB$: $TSNR = 3dB$ (continuous), $TSNR = 18dB$ (diamonds) and $TSNR = 24dB$ (dotted)

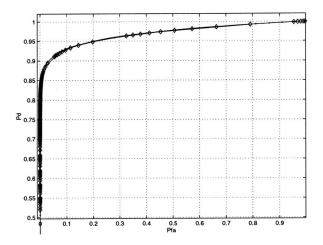

Fig. 7. ROC curves for $SNR = 7dB$: $TSNR = 3dB$ (continuous), $TSNR = 18dB$ (diamonds) and $TSNR = 24dB$ (dotted)

independent of the TSNR value, but the performance for every TSNR value is similar to the best one for a single neural network detector (without pre-processing stage). As a result of this, the neural network training is simplified.

References

1. Watterson, J.W.: An Optimum Multilayer Perceptron Neural Receiver for Signal Detection. IEEE Trans. on Neural Networks, Vol. 1, N 4 (1990)
2. Ruck, D.W., Rogers, S.K., Kabrisky, M., Oxley, M.E., Suter, B.W.: The Multilayer Perceptron as an Approximation to a Bayes Optimal Discrimination Function. IEEE Trans. on Neural Networks, Vol. 1, N 4 (1990) 296–298
3. Wan, W.A.: Neural Network Classification: a Bayesian Interpretation. IEEE Trans. on Neural Networks, Vol. 1, N 4 (1990) 303–305
4. Andina, D.: Optimization of Neural Detectors. Application to Radar and Sonar. Ph.D. Thesis (Universidad Politecnica de Madrid) (1995)
5. Jarabo, P., Rosa, M., et al.: Performance Analysis of a MLP Based Radar Detector for Swerling Targets in AWGN. Procedings of IASTED Int. Conference on Signal and Image Processing, Las Vegas, U.S.A. (2000) 474–479
6. Eaves, J.L., Reedy, E.K.: Principles of Modern Radar. Van Nostrand Reinhold (1987)
7. Skolnik, M.: Radar Handbook. Second edition. McGraw-Hill, Inc. (1990)
8. Haykin, S.: Modular Learning Strategy for Signal Detection in a Nonstationary Environment. IEEE Trans. on Signal Processing, Vol. 45, N 66 (1997) 1619–1637
9. Principe, Euliano, Lefebvre: Neural and Adaptive Systems. Fundamentals Through Simulations. John Wiley&Sons, Inc. (2000)
10. Haykin, S.: Neural Networks. A Comprehensive Foundation. Second Edition. Prentice-Hall, Inc. (1999)

Foetal Age and Weight Determination Using a Lateral Interaction Inspired Net

A. Fernández-Caballero[1], J. Mira[2], F.J. Gómez[1], and M.A. Fernández[1]

[1] Departamento de Informática, Universidad de Castilla-La Mancha, 02071- Albacete, Spain
{caballer, fgomez, miki}@info-ab.uclm.es
[2] Departamento de Inteligencia Artificial, UNED, c/ Senda del Rey 9, 28040- Madrid, Spain
jmira@dia.uned.es

Abstract. The clinical estimate of the foetal weight is probably one of the most difficult parameters to obtain in the prenatal control. Only very accurate foetal body measurements reflect the gestation age and the weight of the foetus. A model is presented that performs an automated foetal age and weight determination from ultrasound by means of biparietal diameter, femur lenght and abdominal circumference parameters. The model proposed in this paper exploits the data in the images in three general steps. The first step is image preprocessing, to highlight useful data in the image and suppress noise and unwanted data. The next step is image processing, which results in forming regions that can correspond to structures or structure parts. The last step is image understanding, where knowledge on the specific problem is injected.

1 Introduction

By means of the prenatal control it is possible to watch over the evolution of the pregnancy. The general objectives of the prenatal control are: (a) to identify factors of risk, (b) to diagnose the gestation age, (c) to diagnose the foetal state, and, (d) to diagnose the maternal state.

The elements that are used for the calculation of the gestation age are the amenorrhoea time, starting from the first day of the last menstruation, and the uterine size. The ignorance of the gestation age constitutes for itself a factor of risk. That's why it is so important to have ultrasound resources.

The clinical elements that allow to evaluate the foetal state are: (a) the foetal heartbeats, (b) the foetal movements, (c) the uterine size, (d) the clinical estimate of the foetal weight, and, (e) the clinical estimate of the amniotic liquid volume. The heartbeats can be identified with ultrasound from the tenth week of pregnancy. The uterine size may be obtained using a flexible ribbon and permits to estimate the foetal size in each prenatal control. The clinical estimate of the foetal weight is probably one of the most difficult parameters to obtain in the prenatal control, since it demands a good piece of experience for its determination. The error of estimate of the foetal weight in pregnancy during the third trimester is about a ten percent.

The need for a quick and easy method for estimating foetal weight in uterus has been clearly established by a great number of professionals. Estimates by abdominal palpation and foetal hormone production have proved to be of limited value [17]. Only very accurate measurements of the foetus allow dating of the pregnancy and serial assessment of foetal growth in comparison with previous measurements.

2 Foetus Age and Weight Determination

2.1 The Most Important Measures

In the 1980s the assessment of intrauterine growth retardation using ultrasonic parameters was the subject of many research papers. The aim to predict foetal weight from computer-generated equations produced normalised tables for every measurable parts of the foetal body [15] [18]. The arrival of the real time scanners have added further impetus to ultrasound techniques and have established ultrasound as the most important imaging modality on Obstetrics and Gynaecology.

The advent of ultrasound in Obstetrics has created the new speciality called prenatal diagnosis that has developed since its early conception. Foetal body measurements reflect the gestation age of the foetus. The following measurements are usually made:

1. The crown-rump length (CRL). This measurement can be made between 7 to 13 weeks and gives very accurate estimation of the gestation age.
2. The biparietal diameter (BPD) and the head circumference (HC). The biparietal diameter is the diameter between the two sides of the head. This is measured after 13 weeks. The HC is less often used than the BPD.
3. The femur length (FL). This is the measure of the longest bone in the body and reflects the longitudinal growth of the foetus. Its usefulness is similar to the BPD.
4. The abdominal circumference (AC). This is the single most important measurement to make in late pregnancy. It reflects more of foetal size and weight than age. Serial measurements are useful in monitoring growth of the foetus.

2.2 The Domain Knowledge

Expert or knowledge-based systems are the commonest type of Artificial Intelligence in Medicine system in routine clinical use. They contain medical knowledge, usually about a very specifically defined task [4]. This is precisely our aim in this paper. So we are going to explain the domain knowledge on the measures described before. But we have restricted to those measures characteristic of the second and third pregnancy trimester.

The BPD remains the standard against which other parameters of gestation age assessments are compared. The BPD should be measured as early as possible after 13 weeks of dating. The problem of head moulding as it relates to the accuracy BPD of the foetal head has long been recognised. The anatomical landmarks used to ensure the accuracy and reproducibility of the measurement include: (a) a midline falx, (b) the thalami symmetrically positioned on either side of the falx, (c) visualisation of the septum pellucidum on one third the front-occipital distance. The BPD increases from about 2.4 cm at 13 weeks to about 9.5 cm at term. A wrong measurement plane can produce errors up to 2 cm.

The FL is a mandatory measurement. By convention, measurement of the FL is considered accurate only when the image shows two blunted ends. The lateral surface of the femur is always straight and the medial surface is always curved. The use of FL in dating is similar to the BPD, and is not superior unless a good plane cannot be obtained. The FL increases from about 1.5 cm at 14 weeks to about 7.8 cm at term.

Measurement of the AC should be made as accurately as possible. The best plane is the one in which the portal vein is visualised in a tangential section. The plane in which the stomach is visualised is also acceptable. The outer edge of the circumference is measured. With a good AC, one will be able to arrive at a very accurate foetal weight. Indeed, the weight of the foetus at any gestation age can be estimated with great accuracy using polynomial equations containing the BPD, FL and AC. One such possible equation is [15]:

$$\log_{10} W = -1.7492 + 0.166 \cdot BPD + 0.046 \cdot AC - 0.002646 \cdot AC \cdot BPD \qquad (1)$$

3 The Model

The model proposed in this paper falls into the data-driven approaches and exploits the data in the images in three general steps. The first step is image pre-processing, to highlight useful data in the image and suppress noise and unwanted data. To fulfil this step some standard image filters have been employed. The next step is image processing, which results in forming regions that can correspond to structures or structure parts. Here, we have used the Lateral Interaction in Accumulative Computation Model [5]. This model formally splits into four stages, as depicted on figure 1. The last step is image understanding, where all knowledge on the specific problem is injected.

Precise segmentation of underlying structures in medical images is a top cue. The existing work on image segmentation can be typically categorised into two basic approaches [2]: region-based methods relying on the homogeneity of spatially and temporally features, and, gradient-based methods looking for some kinds of boundaries. Our approach integrates both concepts to limit the intrinsic problems of both methods.

Step 1. Image Pre-processing by Standard Filters

A major problem in medical image analysis is the potential variability in the image characteristics and object appearance. This first step aims to maximally improve the input images (ultrasonography images) in order to optimise the image processing step in time and quality. This problem has firstly been addressed by selecting a set of well-known standard filters used in image pre-processing.

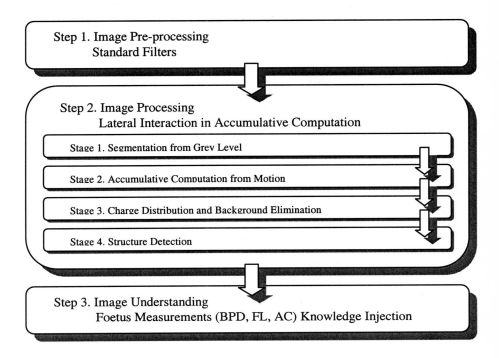

Fig. 1. The model

The final goal of step 1 is to reduce the textured aspect of the embryonic ultrasonography images and to concentrate on the desired regions or structures. In our particular cases, we have to highlight any aspects of the structure of the skull, the femur and the abdomen, whereas the rest of the image has to be considered as undesired background.

Step 2. Image Processing by Lateral Interaction in Accumulative Computation

By image processing in automatic analysis of medical images we include all those techniques that help in diagnostic. We could mention among others: (a) objective quantitative parameter extraction on shape and texture, (b) change detection among two images, (c) information fusion from several modalities, (d) comparison of images from two different patients, (e) probabilistic anatomical and functional atlas construction, (f) motion measures of dynamic organs, and, (g) dynamic visualisation of images [1].

Our model at this step 2 takes advantage of all information concerning motion analysis in foetal ultrasound. Motion analysis in dynamic image sequences is a really hard matter [9] and some approaches have been used in medicine [3] [10] [14] [8] [16]. This generic model is based on a neural architecture, with recurrent and parallel computation at each specialised layer, and sequential computation between consecutive layers. The model is based on an accumulative computation function [6] [7], followed by a set of co-operating lateral interaction processes performed on a functional receptive field organised as centre-periphery over linear expansions of their input spaces [11] [12] [13]. The model also incorporates the notion of double time scale at accumulative computation level present at sub-cellular micro-computation [7].

Any stage of step 2 is implemented as a neural layer as depicted on figure 2. This figure shows the intra- and interconnections of any element (i,j) of any one of the four layers. At each layer n, element (i,j) receives an input from element (i,j) of layer $n-1$ and sends an output to element (i,j) of layer $n+1$ at global time scale t. At local time scale T $(t = k \cdot T)$, intraconnectios take place in the sense that data present at element (i,j) is exchanged with its neigbours data. These neighbours are $(i-1,j)$, $(i+1,j)$, $(i,j-1)$ and $(i,j+1)$.

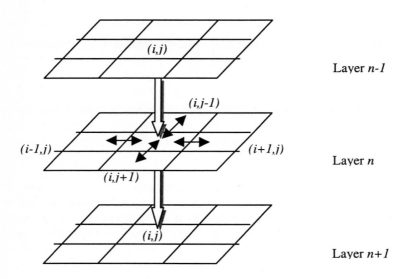

Fig. 2. A module's connections

Stage 2.1. Segmentation from Grey Level

The aim of this stage 2.1, corresponding to layer 1, is to determine in what grey level stripe *GLS* a given pixel (i,j) falls. We consider the pre-processed image to segment to be the input to this stage.

$$GLS_k(i,j,t) = \begin{cases} 1, & \text{if } GL(i,j,t) \in [(256/n)k, (256/n)(k+1)[\\ 0, & \text{otherwise} \end{cases}$$

where n = number of grey level stripes
 k = grey level stripes
 GL = grey level

Stage 2.2. Accumulative Computation from Motion

Supstep 2.2 (layer 2) makes use of n sub-layers, one for each of the chosen grey level stripes. This stage incorporates the described lateral interaction mechanisms. We are to explain how this stage works on each of the central elements (i,j) at any sub-layer k. This stage is capable of modelling the motion on the image, starting from the pixel's grey level stripe and the element's state or permanence value PM. At each time instant t, the permanence value is obtained in two steps. (1) At global time, a complete discharge (v_{dis}), a saturation (v_{sat}) or partial discharge (v_{dm}) due to the motion detection, that's to say, due to a change in the grey level stripe, and, (2) at local time, a partial recharge (v_{rv}) due to the lateral interaction on the partially charged elements that are directly or indirectly connected to maximally charged elements.

$$PM_k(i,j,t) = \begin{cases} v_{dis}, & \text{if } GLS_k(i,j,t) \equiv 0 \\ v_{sat}, & \text{if } GLS_k(i,j,t) \equiv 1 \cap GLS_k(i,j,t-\Delta t) \equiv 0 \\ PM_k(i,j,t-\Delta t) - v_{dm}, & \text{if } GLS_k(i,j,t) \equiv 1 \cap GLS_k(i,j,t-\Delta t) \equiv 1 \end{cases}$$

$$PM_k(i,j,T) = PM_k(i,j,T-\Delta T) + v_{rv}$$

Stage 2.3. Charge Distribution and Background Elimination

Starting from the values of the permanence memory PM in each pixel, it is possible to obtain the silhouette of all the parts of a moving object starting from the spots left by the different grey level stripes. That's why, at this point, the charge of the permanence values is homogeneously distributed among all the elements that have the same grey level value, provided that they are directly or indirectly connected to each other through the necessary lateral interaction mechanisms of recurrent type. This way, a double objective will be obtained at this layer 3. First, the one of diluting the charge due to the background false motion detected on the image, only keeping the movement of the desired structures of the scene. And secondly, the one of obtaining a common parameter to all the elements of a same part of a structure.

$$C_k(i,j,t) = PM_k(i,j,t)$$

$$C_k(i,j,T) = \frac{1}{5}(C_k(i,j,T-\Delta T) + C_k(i-1,j,T-\Delta T) + C_k(i+1,j,T-\Delta T) + \\ + C_k(i,j-1,T-\Delta T) + C_k(i,j+1,T-\Delta T))$$

Stage 2.4. Structure Detection

So far, by means of the necessary co-operative calculation mechanisms, attention has been captured on anything that has moved at any grey level stripe, and motion due to the background has been eliminated. Now, at this layer 4, it is necessary to fix as a new objective to distinguish the different objects that conform the different parts of the structures obtained on a grey level stripe basis (spots). The discrimination of these structures is also performed by lateral interaction of recurrent type. Now, we will no longer work with sub-layers, but rather all the information of the n sub-layers of stage 2.3 is integrated in a single layer. In this stage 2.4, the charge again will be homogeneously distributed among all the elements that have a charge value superior to a minimum threshold and that are physically connected to each other.

$$S(i,j,t) = \max(C_k(i,j,t))$$

$$S(i,j,T) = \tfrac{1}{5}(S(i,j,T-\Delta T) + S(i-1,j,T-\Delta T) + S(i+1,j,T-\Delta T) + \\ + S(i,j-1,T-\Delta T) + S(i,j+1,T-\Delta T))$$

Step 3. Image Understanding by Foetus Measurements Knowledge Injection

In this last step all general knowledge on foetus measurements to obtain the age and the weight of the embryo is injected. Four stages are needed to obtain both parameters. (1) BPD determination, (2) FL determination, (3) AC determination, and, (4) age and weight calculation. We next offer the rules as applied from domain knowledge.

```
BPD determination

Locate Skull;
Locate Biparietal extremities;
Obtain BPD;

FL determination

Locate Femur;
Locate Femur extremities;
Obtain FL;

AC determination

Locate Abdomen;
Locate Abdomen circumference;
Obtain AC;

Age and weight calculation

Find Age-BPD from BPD-Chart; Output Age-BPD;
Find Age-FL from FL-Chart;   Output Age-FL;
Calculate Weight from Equation 1;
```

4 Learning in Lateral Interaction in Accumulative Computation

Learning in lateral interaction in accumulative computation starts from the knowledge of the influence of the basic parameters of the model. Learning in lateral interaction in accumulative computation model consists in adjusting the parameters of the diverse layers to offer the best processing result of the image sequence when obtaining the silhouettes of moving elements present in the scene.

During the learning process, previous to the normal operation process, the architecture is offered an input image sequence, as well as the following reinforcement parameters (see figure 3):
- *Number of moving elements (S_m)* to be detected in the sequence
- *Maximum size of a silhouette (S_{max})* to be detected in the sequence
- *Minimum size of a silhouette (S_{min})* to be detected in the sequence

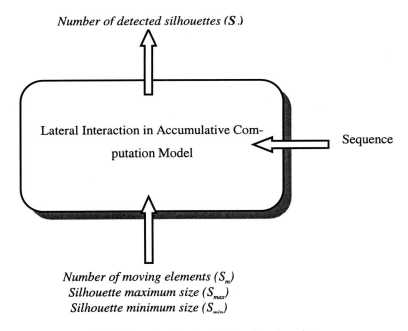

Fig. 3. Inputs and outputs during learning phase

Due to its simplicity, it doesn't seem necessary to explain the reinforcement parameter *Number of moving elements (S_m)*. The other two parameters arise from the domain knowledge of lateral interaction in accumulative computation model. It is indispensable to introduce parameters *Maximum size of a silhouette (S_{max})* and *Minimum size of a silhouette (S_{min})* to capture the attention on those objects whose silhouette falls between these two magnitudes.

Learning turns, in our case, into an iterative process where, for a given scene, the model is nurtured by a same image sequence, just modifying the basic parameters until the number of silhouettes obtained at layer 4 is close enough to *Number of moving elements* (S_m). The output obtained at layer 4 is *called Number of Detected Silhouettes* (S_d).

The basic parameters of lateral interaction in accumulative computation model have been classified into two groups:
- *Parameters with constant values that don't evolve during the learning phase.* These are v_{dis} (minimum permanency value) and v_{sat} (maximum permanency value) at layer 2.
- *Parameters with values that do evolve during the learning phase.* These are: n (number of gray level bands) at layer 1; v_{dm} (discharge value due to motion detection), v_{rv} (recharge value due to vicinity), and, θ_{per} (threshold) at layer 2; θ_{ch} (threshold) at layer 3; θ_{obj} (threshold) at layer 4.

So, we use an error minimization function in the sense that the problem is now to find a procedure of estimating a set of values that best leads to the desired solution. In other words, we have to look for a set of optimal values

$$C^* = \left(n^*, v^*_{dm}, v^*_{rv}, \theta^*_{per}, \theta^*_{ch}, \theta^*_{obj} \right)$$

that minimize error function

$$E = \left| S_m - \frac{1}{k} \sum_{t=0}^{k} S_d(t) \right|$$

where k is the number of images that form the learning sequence,

S_m is the number of moving elements to be detected (constant through the whole training sequence),

$S_d(t)$ is the number of detected silhouettes at time instant t.

5 Results

The model has been tested with a series of ultrasound images of Sara Gómez at a gestation age of aproximately 18 weeks. This important information has helped us to confirm the results of our model. You may appreciate the results of the image pre-processing and image processing steps of some of Sara's skull, femur and abdomen images on figures 4, 5 and 6, respectively.

You may appreciate on columns (a) the original input images. Column (b) presents the pre-processed images where only a few pixels of interest are taken. Lastly, on column (c) you may observe how our proposed model is capable of obtaining more significant information from the described region growing technique. Finally, table 1 shows the results of the image understanding step.

These results suggest that the proposed model is able to obtain accurate data for an automated foetal age and weight determination.

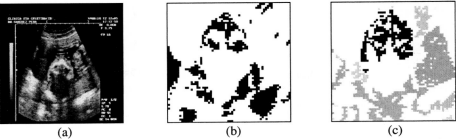

Fig. 4. Application of image processing model to Sara's skull. (a) Original image. (b) Result of pre-processing step. (c) Result of processing step.

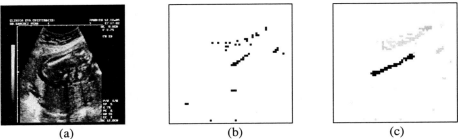

Fig. 5. Application of image processing model to Sara's femur. (a) Original image. (b) Result of pre-processing step. (c) Result of processing step.

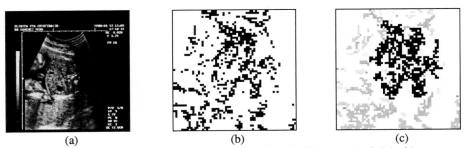

Fig. 6. Application of image processing model to Sara's abdomen. (a) Original image. (b) Result of pre-processing step. (c) Result of processing step.

Table 1. Results of image understanding model to Sara's ultrasound images

```
Computed Age-BPD   = 17.6   weeks
Computed Age-FL    = 18.2   weeks
Calculated Weight  =  0.231 kg
```

References

1. Ayache, N.: L'analyse automatique des images médicales. Etat de l'art et perspectives. Technical Report 3364, INRIA, France (1998)
2. Ballard, D.H., Brown, C.M.: Computer Vision. Prentice Hall (1982)
3. Benayoun, S., Ayache, N.: 3D motion analysis using differential geometry constraints. International Journal of Computer Vision 26:1 (1998) 25-40
4. Coiera, E.: Guide to Medical Informatics, the Internet and Telemedicine. Chapman & Hall Medical (1997)
5. Fernández-Caballero, A.: Modelos de interacción lateral en computación acumulativa para la obtención de siluetas. Unpublished Ph.D. dissertation (2001)
6. Fernández, M.A., Mira, J.: Permanence memory: A system for real time motion analysis in image sequences. IAPR Workshop on Machine Vision Applications, MVA'92 (1992) 249-252
7. Fernández, M.A., Mira, J., López, M.T., Alvarez, J.R., Manjarrés, A., Barro, S.: Local accumulation of persistent activity at synaptic level: Application to motion analysis. In: Mira, J., Sandoval, F. (eds.): From Natural to Artificial Neural Computation, IWANN'95, LNCS 930. Springer-Verlag (1995) 137-143
8. Kraitchman, D., Young, A., Chang, C.N., Axel, L: Semi-automatic tracking of myocardial motion in MR tagged images. IEEE Transactions on Medical Imaging 14:3 (1995) 422-433
9. McInerney, T., Terzopoulos, D.: Deformable models in medical image analysis: A survey. Medical Image Analysis 1:2 (1996) 91-108
10. McVeigh, E.R.: MRI of myocardial function: Motion tracking techniques. Magnetic Resonance Imaging 14:2 (1996) 137-150
11. Mira, J., Delgado, A.E., Alvarez, J.R., de Madrid, A.P., Santos, M.: Towards more realistic self contained models of neurons: High-order, recurrence and local learning. In: Mira, J., Cabestany, J., Prieto, A. (eds.): New Trends in Neural Computation, IWANN'93, LNCS 686. Springer-Verlag (1993) 55-62
12. Mira, J., Delgado, A.E., Manjarrés, A., Ros, S., Alvarez, J.R.: Cooperative processes at the symbolic level in cerebral dynamics: Reliability and fault tolerance. In: Moreno-Díaz, R., Mira, J. (eds.): Brain Processes, Theories and Models. The MIT Press (1996) 244-255
13. Moreno-Díaz, R., Rubio, F., Mira, J.: Aplicación de las transformaciones integrales al proceso de datos en la retina. Revista de Automática 5 (1969) 7-17
14. Rueckert, D., Sánchez-Ortiz, G.I., Burger, P.: Motion and deformation analysis of the myocardium using density and velocity encoded MR images. In: Proceedings of the 10th International Symposium and Exhibition on Computer Assisted Radiology (1996)
15. Shephard, M.J., Richards, V.A., Berkowitz, R.L., Hobbins, J.C.: An evaluation of two equations for predicting fetal weight by ultrasound. American Journal of Obstetrics and Gynecology (1982) 142:147.
16. Shi, P., Robinson, G., Chakraborty, A., Staib, L. Constable, R., Simusas A., Duncan, J.: A unified framework to assess myocardial function from 4D images. In: Computer Vision, Virtual Reality and Robotics in Medicine, LNCS 905. Springer-Verlag (1995) 327-337
17. Warsof, S.L., Gohari, P., Berkowitz, R.L., Hobbins, J.C.: The estimation of fetal weight by computer-assisted analysis. American Journal of Obtetrics and Gynecology 128:8 (1977) 881-892
18. Woo, J.S.K.: Obstetric ultrasound: A comprehensive guide to ultrasound scans in pregnancy. http://www.ultrasound.net/ (2000)

Inference of Stochastic Regular Languages through Simple Recurrent Networks with Time Delays*

Gustavo A. Casañ and M. Asunción Castaño

Dpto. Ingeniería y Ciencia de los Computadores
Universidad Jaume I. Castellón, Spain.
{ncasan,castano}@inf.uji.es

Abstract. Previous work in the literature has shown that, using a local representation of the alphabet, simple recurrent neural networks were able to estimate the probability distribution corresponding to strings which belong to a stochastic regular language. This paper carries on with the empirical works in the matter by including input time delays in simple recurrent networks. This technique could sometimes avoid the use of fully-recurrent architectures (with high computational requirements) to learn certain grammars. Therefore, we could avoid the problems of memory that arise using networks with simple recurrences.

1 Introduction

Grammatical inference has been encourageley employed in fields such as language processing, image processing, robotics or biology. Using this paradigm, formal models (grammars) are automatically learned from a finite set of input samples. Traditionally, grammatical inference has been undertaken with symbolic techniques; however, connectionist computation has recently proved to be an alternative (or complementary) approach, as it is reviewed in [2].

Works carried out on this matter in the literature trained neural architectures either to *accept or reject strings* belonging to an specific regular language, or to *predict the possible successor(s)* of each character in the string. The first paradigm consists of presenting strings to the net so that it should accept the strings which belong to the regular language to be inferred and reject the strings which does not belong to the language. The other paradigm consists of predicting the possible successor(s) that must follow each symbol of the strings generated by the grammar to be learned.

This paper focuses on the predictive approach and is based on the work carried out in [1] and [2], where the induction of stochastic regular grammars (SRGs) was studied. SRGs are interesting because they are appropriate tools to approach real world pattern recognition tasks which exhibit noise and distortions. Experiments presented

---

* Partially supported by the Spanish Fundación Bancaja, project P1A99-10.

in the previous papers showed that neural networks were not only able to predict the next symbol(s) of strings randomly-generated by a regular grammar, but they were also able to automatically estimate the probability distribution corresponding to the strings generation. Both Elman simple recurrent networks [3] and first-order fully-recurrent networks [7] were employed in those works. However, fully-recurrent nets are computationally expensive. On the other hand, the cost of Elman networks is lower but it seems that they are not able to infer certain (non-stochastic) grammars which require to remember events which have occurred a long time ago [6]. The experiments presented in this paper approach the prediction of SRGs through an Elman network in which input time delays are included in order to increase its memory.

This paper is organized as follows: Section 2 describes the connectionist architecture employed to infer SRGs, as well as the procedure used to train it. Section 3 presents the SRGs to be learned in the experimentation and reports the performances obtained. Finally, Section 4 discusses the conclusions of the experimental process.

2 Neural Networks for the Inference of Stochastic Regular Grammars

2.1 Network Architecture

In accordance with the nature of the task, a connectionist model with an explicit representation of time is required. Therefore, the basic neural architecture adopted in the experimentation of this paper is a simple recurrent network (SRN) introduced by Elman in [3]. This net has the typical connections from inputs to hidden units and from hidden to output units; in addition, the inputs and additional context units containing the state of hidden activations in the previous step, are fed into the hidden layer.

However, some translation experiments [2] have shown that translation rates could be improved by including time delays to this SRN. Therefore, in order to increase the performance of the inferred connectionist models of this paper, the preceding context (symbols) of the input string is also presented to the input of the Elman SRN. Figure 1 illustrates the resulting neural topology.

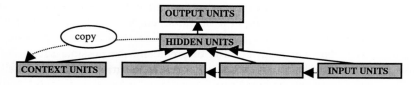

Fig. 1. Elman simple recurrent network with delayed inputs.

In both SRNs, with and without time delays, the input units and the output layer are designed according to a *local representation* of the alphabet. This means that every input and output is dedicated to the representation of one of the possible terminal symbols of the grammar. However, the local encoding would be only required for the output layer in order to approach the prediction task. An additional unit (and the corresponding symbol in the alphabet) is included to mark the beginning and the end of the strings. This unit allows to predict the first symbol of the string as well as the end of the string. Two different units to code the beginning and the end of the string could be also adopted. However, experimental results carried out by the authors have shown that the results achieved were similar to those obtained using only one unit.

2.2 Training Procedure

The neural architectures described above are trained using an on-line version of the backward-error propagation algorithm [4]; that is, a gradient-truncated version of a full-descendent procedure. Strings are presented character by character to the net so that the next character of the string may be predicted; that is, for each input symbol (character of the string) the target output consists of the character that followed this input symbol in such string. The output character is then applied to the input units and the next character is used as the target and son on.

For the SRNs with delays, strings are presented to the net in a similar way. However, if we assume n delayed inputs, the net has at the input layer both the current input character and the previous n characters in the string.

A sigmoid function (0,1) is used as the non-linear activation function. Consequently, context activations are initialized to 0.5 at the beginning of every string. In order to estimate appropriate values for the learning rate and momentum, the net is trained for a presentation of the complete learning corpus (epoch). Training continues using the learning rate and momentum with the lowest mean squared error. After every training epoch, the net is evaluated on a different (validation) set of strings randomly generated by the same SRG. Learning stops when some established criterion is verified.

2.3 Correct Prediction Criterion

Let the product $p_1 p_2 ... p_m$ be the probability of generating a given string $x_1 x_2 ... x_m$ of length m by the (unrestricted and deterministic) SRG to be learned, where each p_i corresponds to the probability of the grammar rule used to generate x_i; and let the product $q_1 q_2 ... q_m$ be the probability of predicting the same string with the inferred network, where q_i corresponds to the output activation associated to x_i. This *string* is considered as *correctly predicted* if the following condition is verified:

$$\left| \ln \left(\frac{(p_1 p_2 ... p_m)^{1/m}}{(q_1 q_2 ... q_m)^{1/m}} \right) \right| < Threshold \ . \tag{1}$$

Note that the expression between brackets corresponds to the *normalized likelihood quotient*. The leftmost term of (1) should be very close to zero for every sample in the validation and/or test set, since our aim is to achieve $p_i = q_i$, $i=1,...,m$.

3 Experimentation

3.1 Stochastic Regular Grammars to be Learned

The inference of stochastic regular grammars (SRGs) using the above proposed SRNs has been evaluated by applying them to two different (deterministic) SRGs, G_s and G_D, shown in Figure 2 and 3, respectively. These grammars correspond to the simple and double Reber grammar, respectively, with an arbitrary probability distribution. They were chosen because they have been employed in the literature as benchmarks for inference experiments [5-6]. One interesting feature of the double Reber grammar is the prediction of the second to the last character in the string, since it only depends on the second symbol of the input string; this means that learning a powerful context inside the net is required in order to learn the grammar.

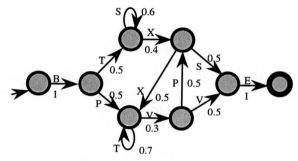

Fig. 2. Simple Reber grammar, G_s.

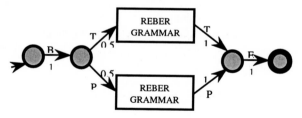

Fig. 3. Double Reber grammar, G_D

3.2 Training, Validation and Test Corpora

For each of the two different SRGs studied 30,000 strings were randomly generated for training the SRNs. These strings ranged from 5 to 57 symbols for G_s and from 9 to 54 for G_D. The average length of the strings was approximately 10 characters for G_s and 12 for G_D. Each experiment was repeatedly evaluated on 10,000 new samples corresponding to the validation set. Performance was finally assessed on a different corpora of 10,000 test samples.

3.3 Features of the Networks

The neural topology of the Elman SRNs adopted in the experimentation has a single hidden layer of first-order connections. Preliminary experiments revealed that 20 neurons were appropriated for the hidden layer of the SRNs considered to infer G_s, while 40 neurons were necessary for G_D.

The alphabet of both grammars to be learned coincides and has 6 symbols. Consequently, according to a local representation of the alphabet, 6 binary input units and 6 outputs were employed in all of the SRNs.

In both experiments, Elman SRNs with and without time delays were considered for comparison purposes. 1 and 2 input delays were adopted to learn G_s and 1, 2 and 4 input delays for G_D.

3.4 Results for the Simple Reber Grammar

The probabilities of the simple Reber grammar G_s were inferred by training the Elman SRNs previously described with 20 hidden neurons and with or without input delays. The phase of parameter estimation suggested 0.1 and 0.3 as adequate values for the learning rate and momentum, respectively, of the net without delays; 0.7 and 0.1 for the net with 1 input delay and 0.1 and 0.1 for the net with 2 input delays. The performance evolution of each trained network was repeatedly evaluated on the 10,000 validation strings, after every presentation of the complete training set. Learning stopped after 50 epochs and the resulting learned models were then assessed on the 10,000 test strings.

Table 1 shows the correct test string prediction rates, assuming a value of 0.08 for the prediction threshold. These results show that very good performances were achieved. On the other hand, we can observe that test prediction rates were greater than 99.5% without delays and with 1 input delay; however, this rate decreased to 94.8% with 2 input delays. That is because a SRN with 20 hidden units and without delays is enough to infer G_s and additional input delays only increase the complexity of the SRN to be trained.

Figure 4 depicts the evolution in successive validations of the normalized likelihood ratio expressed in terms of the average of the absolute logarithms (left side of (1)). These results show that the logarithms became very close to zero, which means that the network's outputs really approached the probabilities of the grammar.

Table 1. Test string prediction rates for G_s and G_D with and without input delays.

\# Delays	G_s Prediction rate	\# Delays	G_D Prediction rate
0	99.5%	0	87.0%
1	100%	1	85.9%
2	94.8%	2	89.9%
		4	100%

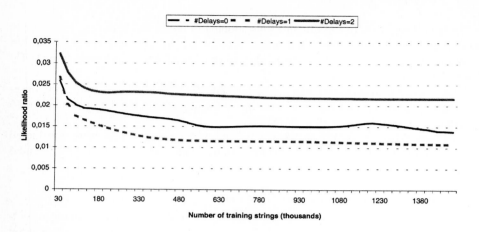

Fig. 4. Evolution of the average absolute logarithm of the normalized likelihood ratio during validation for G_s with and without input delays.

A more explicit display of the degree of approximation to the true stochastic language probabilities can be given by plotting the probability distribution supplied by the learned network compared with the corresponding probabilities associated with the real grammar. This is shown in Figure 5 for the SRN with 1 input delay. In this figure, the actual probability of generating any string of a given length is compared with the sum of the probabilities supplied by the net for all the strings of that length. The results also confirm the accuracy of the estimations.

Let us suppose a grammar string with length m in which $p_i=q_j$, $p_j=q_i$ and $p_k=q_k$, $k \neq i,j$ $i,j,k=1,...,m$; attending to criterion (1), these wrong probabilities would be compensated and the string could be considered as correctly predicted. In order to avoid such situations, this criterion was changed to the following expression:

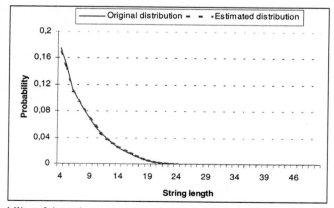

Fig. 5. Probability of the strings generated with G_s and probability estimated through the SRN with 1 input delay, according to the length of the strings.

$$\frac{1}{m}\sum_{i=1}^{m}\left|\ln p_i - \ln q_i\right| < Threshold \ . \tag{2}$$

However, other experiments showed that the results corresponding to the values of the leftmost term in (2) as well as the rates of strings correctly predicted were similar to those obtained adopting the expression (1).

3.5 Results for the Double Reber Grammar

The probabilities of the double Reber grammar G_D were later induced by training the Elman SRNs described in Section 3.3 with 40 hidden neurons and with or without input delays. The values of the learning rate and momentum adopted were, respectively, 0.1 and 0.3 for both the net without delays and the net with 2 input delays, 0.5 and 0.1 for the net with 1 delay and 0.1 and 0.1 for the net with 4 delays. The performance evolution of each trained network was repeatedly evaluated on the validation strings, after every presentation of the complete training set. Since this grammar was more difficult to learn than G_s, learning stopped after 250 epochs. The resulting learned models were then evaluated on the test strings.

Table 1 shows the test string prediction rates, assuming a value of 0.15 for the prediction threshold. These results show that good test prediction rates were also obtained in this case (100% for the net with 4 delays). On the other hand, the inclusion of input delays in the inferred networks considerably increased the prediction performance. Nevertheless, it should be noted that the results presented in Sections 3.4 and 3.5 for inferring G_s and G_D, respectively, cannot be exactly compared since different prediction thresholds were adopted.

Figure 6 depicts the probability distribution supplied by the network with 4 input delays compared with the corresponding probabilities associated to the real grammar. We compare both probability distributions for different strings lengths in a similar way that in Section 3.4. The results confirm again that the network's outputs approached the probability distribution of the grammar to be inferred.

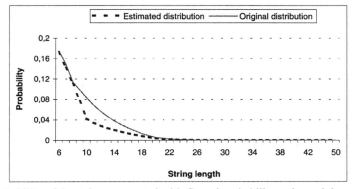

Fig. 6. Probability of the strings generated with G_D and probability estimated through the SRN with 4 input delays, according to the length of the strings.

4 Conclusions and Future Work

Previous works in the literature [1], [2] have shown that, using a local representation of the alphabet, SRNs trained to predict the next symbol in a substring generated by a stochastic regular grammar approached the probability distribution of this grammar. However, it seems that these networks are not able to infer certain grammars which require to remember events which have occurred a long time ago [6]. Our work shows empirical evidence supporting that prediction performance can be increased by including input time delays in the SRNs. The results are more outstanding when the net is not big enough to learn the grammar to be inferred correctly.

Experiments showed in this paper open several topics to be studied in the future. First, the influence of both the topology and the probability distribution of the source grammar on the behavior of the net. Second, by looking at the successive net outputs for every symbol of the alphabet, we can try to extract the whole structure and the probability distribution of the stochastic grammar using simpler techniques than those used so far in the literature. And finally, we plan to study the relationship between n-gramms for language modeling and the predictive connectionist approach which includes input time delays.

References

1. M. A. Castaño, F. Casacuberta, E. Vidal. *Simulation of Stochastic Regular Grammars through Simple Recurrent Networks*. In "New Trends in Neural Computation". Lecture Notes in Computer Science, vol. 686, pp. 210--215. Eds. J. Mira, J. Cabestany, A. Prieto. Springer-Verlag. 1993.
2. M. A. Castaño. *Redes Neuronales Recurrentes para Inferencia Gramatical y Traducción Automática*. Ph.D. dissertation. Universidad Politécnica de Valencia. 1998.
3. J. L. Elman. *Finding Structure in Time*. Cognitive Science, vol. 2, no. 4, pp. 279--311. 1990.
4. D.E. Rumelhart, G. Hinton, R. Williams. *Learning Sequential Structure in Simple Recurrent Networks*. In "Parallel distributed processing: Experiments in the microstructure of cognition", vol. 1. Eds. D.E. Rumelhart, J.L. McClelland and the PDP Research Group, MIT Press. 1986.
5. A.W. Smith, D. Zipser. *Learning Sequential Structure with the Real-Time Recurrent Learning Algorithm*. International Journal of Neural Systems, vol. 1, no. 2, pp. 125--131. 1991.
6. D. Servan-Schreiber, A. Cleeremans, J.L. McClelland. *Graded State Machines: The Representation of Temporal Contingencies in Simple Recurrent Networks*. Machine Learning, no. 7, pp. 161--193. 1991.
7. R.J. Williams, D. Zipser. *Experimental Analysis of the Real-Time Recurrent Learning Algorithm*. Connection Science, vol. 1, no. 1, pp. 87--111. 1989.

Is Neural Network a Reliable Forecaster on Earth? A MARS Query!

Ajith Abraham & Dan Steinberg[†]

School of Computing & Information Technology
Monash University, Churchill 3842, Australia
Email: ajith.abraham@infotech.monash.edu.au

[†] Salford Systems Inc
8880 Rio San Diego, CA 92108, USA
Email:dstein@salford-systems.com

Abstract. Long-term rainfall prediction is a challenging task especially in the modern world where we are facing the major environmental problem of global warming. In general, climate and rainfall are highly non-linear phenomena in nature exhibiting what is known as the "butterfly effect". While some regions of the world are noticing a systematic decrease in annual rainfall, others notice increases in flooding and severe storms. The global nature of this phenomenon is very complicated and requires sophisticated computer modeling and simulation to predict accurately. In this paper, we report a performance analysis for Multivariate Adaptive Regression Splines (MARS) [1] and artificial neural networks for one month ahead prediction of rainfall. To evaluate the prediction efficiency, we made use of 87 years of rainfall data in Kerala state, the southern part of the Indian peninsula situated at latitude-longitude pairs (8°29• N - 76°57• E). We used an artificial neural network trained using the scaled conjugate gradient algorithm. The neural network and MARS were trained with 40 years of rainfall data. For performance evaluation, network predicted outputs were compared with the actual rainfall data. Simulation results reveal that MARS is a good forecasting tool and performed better than the considered neural network.

1. Introduction

The parameters that are required to predict rainfall are enormously complex and subtle even for a short time period. The period over which a prediction may be made is generally termed the event horizon and in best results, this is not more than a week's time. It has been noted that the fluttering wings of a butterfly at one corner of the globe may ultimately cause a tornado at another geographically far away place. Edward Lorenz (a meteorologist at MIT) discovered this phenomenon in 1961, which is popularly known as the butterfly effect [9]. In our research, we aim to find out how well MARS and neural network models are able to capture the periodicity in these patterns so that long-term predictions can be made [7]. This would help one to anticipate the general pattern of rainfall in the coming years with some degree of confidence.

We used an artificial neural network using the scaled conjugate gradient algorithm and MARS for predicting the rainfall time series. Both models were trained on the on the rainfall data corresponding to a certain period in the past and predictions were made over some other period. In Section 2, we present some theoretical background on MARS followed by a discussion of artificial neural networks in Section3. In section 4, the experimental set up is explained followed by discussions and simulation results. Conclusions are provided at the end of the paper.

2. Multivariate Adaptive Regression Splines (MARS)

Splines can be considered an innovative mathematical process for complicated curve drawings and function approximation. Splines find ever-increasing application in the numerical methods, computer-aided design, and computer graphics areas. Mathematical formulae for circles, parabolas, or sine waves are easy to construct, but how does one develop a formula to trace the shape of share value fluctuations or any time series prediction problems? The answer is to break the complex shape into simpler pieces, and then use a stock formula for each piece [3]. To develop a spline the X-axis is broken into a convenient number of regions. The boundary between regions is known as a knot. With a sufficiently large number of knots virtually any shape can be well approximated. While it is easy to draw a spline in two dimensions by keying on knot locations (approximating using linear, quadratic or cubic polynomial regression etc.), manipulating the mathematics in higher dimensions is best accomplished using basis functions.

The MARS model is a spline regression model that uses a specific class of basis functions as predictors in place of the original data [1]. The MARS basis function transform makes it possible to selectively blank out certain regions of a variable by making them zero, allowing MARS to focus on specific sub-regions of the data. MARS excels at finding optimal variable transformations and interactions, as well as the complex data structure that often hides in high-dimensional data [2] [6].

Given the number of predictors in most data mining applications, it is infeasible to approximate a function $y=f(x)$ in a generalization of splines by summarizing y in each distinct region of x. Even if we could assume that each predictor x had only two distinct regions, a database with just 35 predictors would contain 2^{35} or more than 34 billion regions. This is known as the curse of dimensionality. For some variables, two regions may not be enough to track the specifics of the function. If the relationship of y to some x's is different in three or four regions, for example, with only 35 variables the number of regions requiring examination would be even larger than 34 billion. Given that neither the number of regions nor the knot locations can be specified a priori, a procedure is needed that accomplishes the following:

- judicious selection of which regions to look at and their boundaries, and
- judicious determination of how many intervals are needed for each variable.

A successful method of region selection will need to be adaptive to the characteristics of the data. Such a solution will probably reject quite a few variables (accomplishing variable selection) and will take into account only a few variables at a time (also reducing the number of regions). Even if the method selects 30 variables for the model, it will not look at all 30 simultaneously. Similar simplification is accomplished by a decision tree (e.g., at a single node, only ancestor splits are being considered; thus, at a depth of six levels in the tree, only six variables are being used to define the node).

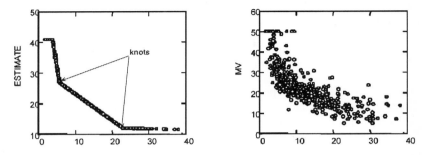

Figure 1. MARS data estimation using spines and knots (actual data on the right)

MARS Smoothing, Splines, Knots Selection and Basis Functions

A key concept underlying the spline is the knot, which marks the end of one region of data and the beginning of another. Thus, the knot is where the behavior of the function changes. Between knots, the model could be global (e.g., linear regression). In a classical spline, the knots are predetermined and evenly spaced, whereas in MARS, the knots are determined by a search procedure. Only as many knots as needed are included in a MARS model. If a straight line is a good fit, there will be no interior knots. In MARS, however, there is always at least one "pseudo" knot that corresponds to the smallest observed value of the predictor. Figure 1 depicts a MARS spline with three knots.

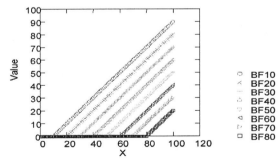

Figure 2. Variations of basis functions for c = 10 to 80

Finding the one best knot in a simple regression is a straightforward search problem: simply examine a large number of potential knots and choose the one with the best R^2. However, finding the best pair of knots requires far more computation, and finding the best set of knots when the actual number needed is unknown is an even more challenging task. MARS finds the location and number of needed knots in a forward/backward stepwise fashion. First a model which is clearly overfit with too many knots is generated, then, those knots that contribute least to the overall fit are removed. Thus, the forward knot selection will include many incorrect knot locations, but these erroneous knots should eventually, be deleted from the model in the backwards pruning step (although this is not guaranteed). Thinking in terms of knot selection works very well to illustrate splines in one dimension; however, this context is unwieldy for working with a large number of variables simultaneously. Both concise notation and easy to manipulate programming expressions are required. It is also not clear how to construct or represent interactions using knot locations. In MARS, Basis Functions (BFs) are the machinery used for generalizing the search for knots. BFs are a set of functions used to represent the information contained in one or more variables. Much like principal components, BFs essentially re-express the relationship of the predictor variables with the target variable. The hockey stick BF, the core building block of the MARS model is often applied to a single variable multiple times. The hockey stick function maps variable X to new variable X^*:

max $(0, X - c)$, or
max $(0, c - X)$.

In the first form, X^* is set to 0 for all values of X up to some threshold value c and X^* is equal to X for all values of X greater than c. (Actually X^* is equal to the amount by which X exceeds threshold c.) The second form generates a mirror image of the first. Figure 2 illustrates the variation in BFs for changes of c values (in steps of 10) for predictor variable X, ranging from 0 to 100. MARS generates basis functions by searching in a stepwise manner. It starts with a constant in the model and then begins the search for a variable-knot combination that improves the model the most (or, alternatively, worsens the model the least). The improvement is measured in part by the change in Mean Squared Error (MSE). Adding a basis function always reduces the MSE. MARS searches for a pair of hockey stick basis functions, the primary and mirror image, even though only one might be linearly independent of the other terms. This search is then repeated, with MARS searching for the best variable to add given the basis functions already in the model. The brute search process theoretically continues until every possible basis function has been added to the model. In practice, the user specifies an upper limit for the number of knots to be generated in the forward stage. The limit should be large enough to ensure that the true model can be captured. A good rule of thumb for determining the minimum number is three to four times the number of basis functions in the optimal model. This limit may have to be set by trial and error.

3. Artificial Neural Network (ANN)

ANN is an information-processing paradigm inspired by the way the densely interconnected, parallel structure of the mammalian brain processes information. Learning in biological systems involves adjustments to the synaptic connections that exist between the neurons [5]. Learning typically occurs by example through training, where the training algorithm iteratively adjusts the connection weights (synapses). These connection weights store the knowledge necessary to solve specific problems.

Backpropagation (BP) is one of the most famous training algorithms for multilayer perceptrons. BP is a gradient descent technique to minimize the error E for a particular training pattern. For adjusting the weight (w_{ij}) from the i-th input unit to the j-th output, in the batched mode variant the descent is based on the gradient ∇E ($\frac{\delta E}{\delta w_{ij}}$) for the total training set:

$$\Delta w_{ij}(n) = -\varepsilon * \frac{\delta E}{\delta w_{ij}} + \alpha * \Delta w_{ij}(n-1) \tag{1}$$

The gradient gives the direction of error E. The parameters ε and α are the learning rate and momentum respectively [4].

In the Conjugate Gradient Algorithm (CGA) a search is performed along conjugate directions, which produces generally faster convergence than steepest descent directions. A search is made along the conjugate gradient direction to determine the step size, which will minimize the performance function along that line. A line search is performed to determine the optimal distance to move along the current search direction. Then the next search direction is determined so that it is conjugate to previous search direction. The general procedure for determining the new search direction is to combine the new steepest descent direction with the previous search direction. An important feature of the CGA is that the minimization performed in one step is not partially undone by the next, as it is the case with gradient descent methods. The key steps of the CGA is summarized as follows:

- Choose an initial weight vector w_1.
- Evaluate the gradient vector g_1, and set the initial search direction $d_1 = -g_1$
- At step j, minimize $E(w_j + \bullet d_j)$ with respect to $\bullet$ to give $w_{j+1} = w_j + \bullet_{m} d_j$)
- Test to see if the stopping criterion is satisfied.
- Evaluate the new gradient vector g_{j+1}
- Evaluate the new search direction using $d_{j+1} = -g_{j+1} + \bullet_j d_j$. The various versions of conjugate gradient are distinguished by the manner in which the constant $\bullet_j$ is computed.

An important drawback of CGA is the requirement of a line search, which is computationally expensive. The Scaled Conjugate Gradient Algorithm (SCGA) is basically designed to avoid the time-consuming line search at each iteration. SCGA combine the model-trust region approach, which is used in the Levenberg-Marquardt algorithm with the CGA. Detailed step-by-step descriptions of the algorithm can be found in Moller[8].

4. Experimental Setup Using Neural Networks and MARS

The 87 years (1893-1980) rainfall data was standardized and the first 40 years was used for training the prediction models and the remaining for testing purposes. We used 12 inputs (previous 4 years rainfall data) to predict the amount of rain to be expected in each month of the fifth year. Experiments were carried out on a Pentium

II 450MHz machine and the codes were executed using MATLAB. Test data was presented to the network and the output from the network was compared with the actual data in the time series.

- **ANN – SCG Algorithm**

    We used a feedforward neural network with two hidden layers consisting of 12 neurons each. We used log-sigmoidal activation function for the hidden neurons. The training was terminated after 600 epochs.

- **MARS**

    We increased the number of basis functions in steps of five and selected 1 as the setting of minimum observation between knots. To obtain the best possible prediction results (lowest RMSE), we sacrificed speed (minimum completion time).

- **Performance and Results Achieved**

    Figure 3 illustrates the training performance of the proposed neural network. Table 2 summarizes the comparative performances of MARS and neural network. Figure 4 shows the change in RMSE values for change in number of basis functions.

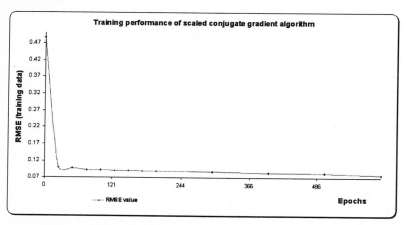

Figure 3. Training performance of SCGA for 600 epochs

Table1. Training and testing performance comparison

Model	RMSE		Epochs	Training time (seconds)
	Training set	Test set		
MARS	0.0990	0.0832	-	5
ANN-SCG	0.0780	0.0923	600	90

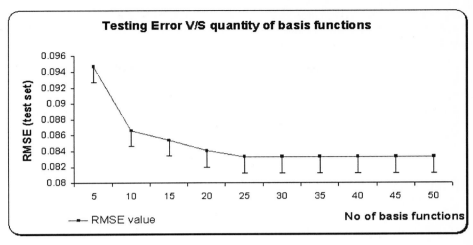

Figure 4. MARS: Test set RMSE convergence

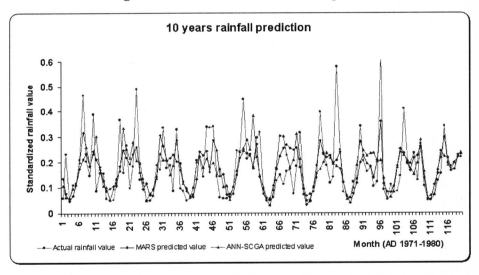

Figure 5. Test results showing one-month ahead prediction of rainfall for 10 years using MARS and ANN-SCGA.

5. Conclusion

In this paper, we attempted to forecast rainfall (one month ahead) using MARS and a neural network trained using SCGA. As the RMSE values on test data are comparatively less, the prediction models are reliable. As evident from the entire test results (for 47 years), there have been few deviations of the predicted rainfall value from the actual. In some cases it is due to delay in the actual commencement of monsoon, EI-Nino Southern Oscillations (ENSO) resulting from the pressure oscillations between the tropical Indian Ocean and the tropical Pacific Ocean and their quasi periodic oscillations [10]

Prediction results reveal that MARS is capable of outperforming neural networks in terms of performance time and lowest RMSE. In our experiments, we used only 40 years training data to evaluate the learning capability of the models. Network performance could have been further improved by providing more training data. It will be interesting to study further the robustness of MARS when compared to neural networks.

References

[1] Friedman, J. H, Multivariate Adaptive Regression Splines, Annals of Statistics, Vol 19, 1-141, 1991.

[2] Steinberg, D and Colla, P. L, MARS User Guide, San Diego, CA: Salford Systems, 1999.

[3] Shikin E V and Plis A I, Handbook on Splines for the User, CRC Press, 1995.

[4] Abraham A and Nath B, Optimal Design of Neural Nets Using Hybrid Algorithms, In proceedings of 6th Pacific Rim International Conference on Artificial Intelligence (PRICAI 2000), pp. 510-520, 2000.

[5] Hagan M T, Demuth H B and Beale M H, Neural Network Design, Boston, MA: PWS Publishing, 1996.

[6] MARS Information Center: http://www.salford-systems.com

[7] Abraham A, Philip N S and Joseph K B, Will we have a Wet Summer? Soft Computing Models for Long-term Rainfall Forecasting, In Proceedings of European Simulation Conference ESM 2001, Prague, June 2001. (Submitted)

[8] Moller A F, A Scaled Conjugate Gradient Algorithm for Fast Supervised Learning, Neural Networks, Volume (6), pp. 525-533, 1993.

[9] Butterfly Effect information center: http://www.staff.uiuc.edu/~gisabell/buttrfly.html

[10] Chowdhury A and Mhasawade S V, Variations in Meteorological Floods during Summer Monsoon Over India, Mausam, 42, 2, 167-170,1991.

Character Feature Extraction Using Polygonal Projection Sweep (Contour Detection)

Roberto J. Rodrigues[1], Gizelle K. Vianna[2], Antonio C. G. Thomé[3]

AEP/NCE- Núcleo de Computação Eletrônica/UFRJ, Caixa Postal 2324, Ilha do Fundão, Rio de Janeiro, RJ, Brasil

[1]rjr@nce.ufrj.br, [2]kupac@posgrad.nce.ufrj.br, [3]thome@nce.ufrj.br

Abstract. It is presented in this paper a new approach to the problem of feature extraction. The approach is based on the edge detection, where a set of feature vectors is taken from the source image. The images under this investigation are considered to be manuscript characters and the features are obtained by the distance from the contour of each character to several observation points placed around the image. Such observation points are arranged along different geometric polygons built in a way to surround the image. The approach is evaluated against the naïve bitmap matrix considering different types of polygon. The discrimination power of each method is computed using both statistic and neural network entries. The proposed approach provides also good response to the scale, rotation and translation problems in addition to discrimination.

Keywords: Feature Extraction, Manuscript Characters Recognition, Cursive Characters, Pattern Recognition, Image Segmentation, Neural Networks.

1 Introduction

Cursive character recognition is a hard and complex task to be achieved. Due to a substantial amount of parameters and individuals features of cursive written, there is not yet a known recognition procedure, which performance had been shown to be good enough under all kinds of situations.

The performance of an automatic recognition system depends deeply on the quality of the originals, before and after its digitalization. To minimize the effects due a poor quality of the data, it is very common the usage of some image compensation technique. These techniques include contour enhancement, line and underline removal, noise removal, and others. The most common problems related to quality and difficulties of a text-based image processing are noise, distortion, translation, rotation, style variation, scale, texture and trace.

The recognition process of manuscript text, which can be cursive, not cursive or any type of printed material includes several steps that goes from an ordinary analysis of the individual characters up to the utilization of lexical or contextual information. These last procedures are used to validate the text as a whole. In general, the very first steps include digitalization, capture and segmentation of the image into individual characters. Capture and segmentation are fundamental for the entire process and, in general, they are very complex and difficult to implement.

Cursive character segmentation deals with problems like slanted, underlined and connected characters. Feature extraction is another important step, some of the known techniques are based on the digits themselves, through some kind of image processing, such as thinning or skeleton algorithms. There are procedures that generate features vectors, based on the direction of the segment lines, ending points, intersection points and cycles. A validation procedure is commonly applied after recognition and it is based on contextual libraries.

One of the currently most used approach is based on the extraction of lines and points for a later contour analysis. However, this method shows itself too complex and requires a set of extremely interdependent parameters. The relevance of such technique remains on the fact that it is very tolerant to distortion and style variation. The investigation related in this paper focus on feature extraction and coding for the problem of cursive character recognition. Such step, takes part in the recognition system that is currently being built in the UFRJ's IC Lab.

2 Segmentation

Cursive character segmentation deals with problems that can generate a substantial amount of difficulties along the process. Some of these problems appear more often like connected or slanted characters, non-numeric elements and underlined characters. The segmentation method proposed for this investigation was described in [1] and is based on the use of profile projection histograms. Profile projection histogram represents a structure that stores the result of the image projection over each one of the existing dimensions. Each dimension assigns a vector for storing the number of pixels that present energy above a predefined threshold (usually established as the background color). After that, a set of successive refinements and decision algorithms are applied on the resultant vector, up to the point where a satisfactory segmentation is achieved.

This method can be separated in 3 stages: image quality compensation, initial segmentation and successive refinements. The image quality compensation step is used to improve the quality of the scanned image, reducing or enhancing details as noises and contrasts. Initial segmentation is the step where the image is initially segmented and it is based on the projection histogram. The algorithm is fast and it is able to segment all those disconnected characters. Dots and lines can be removed using the data retrieved from the histogram horizontal vector. The last product generated by the initial segmentation is the identification of those possible connected characters.

The refinement phase may includes several recursive attempts to segment those connected characters identified on the initial segmentation. The first attempt always replies the histogram strategy readjusting all refinement parameters. This procedure allows the segmentation of those weakly connected characters. Additional steps, when necessary, are based on other segmentation techniques. The result described in [1] over our database, provided a correct segmentation rate of 95.20% with the first level of refinement.

3 Features Extraction and Coding

The implementation of a character recognition procedure through neural networks requires the construction of input vectors taken from each source character image. Pattern discriminative ability and dimensionality reductions are two relevant problems to work with at this point. The feature extraction method described in this paper represents a novel derivative of the ordinary contour detection methods. It is based on the computation of a set of distances from the image contour up to a reference polygon. The reference polygon can be of any number of sides. It has to be placed surrounding the image, as shown in the figure below. Using the provided set of formulas one can build polygons of any number of sides. This method not only proved to provide good discrimination but also to mitigate the problems related to scale, rotation and translation of the image.

Fig. 1. sequence of square scanning procedure

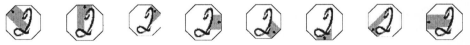

Fig. 2. Sequence of octagon scanning procedure

Fig. 3. Representation for the circle scanning procedure

Fig. 4. Geometry representation of the method.

Using the power of reconstruction of the original image as a visual and first benchmark, it was observed that odd sided polygons were not as good as the even ones. It was then decided to concentrate the investigation over the square (figure 1), the octagon (figure 2) and the circle (figure 3). This last one in two versions: one based on the radius and other on the diameter. As showed by the figures, the input vector is coded based on predefined dimensions and the features are extracted as the distances counted in number of pixels from each side of the polygon to the contour of the image.

4 Evaluation Environment

The database used to evaluate the proposed technique was taken from forms filled up by several persons inside campus with different scholarship level. In each form the custom was asked to provide five different zip codes in a free style cursive written. The forms were then scanned in a 8 bit gray scale scanner and the images were taken with a resolution of 200 dpi. After the segmentation process, each character image was converted to a binary scale [0, 1] and stored in a 16x16 matrix of pixels.

The database was formed exclusively with numerical and isolated digits, from 0 to 9, resulting from the segmentation process. Some of then were not well segmented and many others presented problems of rotation, scale translation, texture and trace. The total amount of available data is shown in the table bellow and as can be seen, the amount of each number is not homogeneous.

As mentioned before, the evaluation objective was to find out if the proposed approach was able to improve the discrimination performance. In order to tackle such goal it was chosen to use some statistical metrics such as Fischer and Distance Discriminating [1, 2, 3, 4] and also some nonlinear methods through neural network models. Both approaches are described in the following sections. The comparative analyses took in consideration the bitmap matrix coding and the ones obtained from the usage of three different polygons: square, octagon and circle. Where the last one was subdivided into two classes: one based on the radius distance and the other on the diameter distance.

Table 1. Individual digits amount distribution

Digits	0	1	2	3	4	5	6	7	8	9
Samples	317	145	217	91	84	173	51	39	43	36

5 Evaluation by Statistical Metrics

5.1 Distance Criterion

The first applied metrics was the Distance Based Discriminant Ratio. It is based on the relative distance from each pattern and the center of mass of the existing classes. In order to compute this ratio it is first necessary to arrange the database into different classes G_i, each one uniquely formed by digits representing the same value, that is, "i" varying from 0 to 9. The distance ratio is computed through the following steps:
1. Computation of the center of mass of each class:

$$cm_{i,j} = \frac{1}{n}\sum_{k=1}^{n} p_{k,j}, \quad 0 \le i \le 9, 1 \le j \le 256 \text{ and } k \in \{G_i\}$$

2. Computation of patterns distance to the center of mass of each class

$$d_{i,j} = \frac{1}{L}\sum_{L}\left(\sum(cm_{i,k} - a^L{}_{j,k})^2\right)^{\frac{1}{2}}, \quad 0 \le i, j \le 9, L \in \{G_j\}$$

Which means the cumulative distance from all patterns from the class "j" to the center of mass of the class "i". This step ends with a set of 100 $d_{i,j}$s, from $d_{0,0}$ to $d_{9,9}$.
Computation of the relative distance from the patterns belonging to one class to the center of mass of the other classes: $rd_{i,j} = \dfrac{d_{i,i}}{d_{i,j}}, \quad 0 \le i,j \le 9$

3. Finally, from the relative distance "rd" two other values are computed: the summation of rd_s (SRD) and the summation of the srd_s (SSRD). This last parameter was used as a measure of the discrimination power of the extracted features.

$$SRD_i = \sum_{j=0}^{9} rd_{i,j}, \quad 0 \le i \le 9 \quad \text{and} \quad SSRD = \sum_{i=0}^{9} SRD_i$$

The following tables present the results from the distance ratio method applied to the square and the bitmap-matrix feature extraction approaches. As can be seen, both tables show the diagonal equal 1, which represents the relative distance from one set to itself and all the others ideally less than 1, which means the relative distance between two different set of digits, G_i and G_j where $i \ne j$.

As can be noted through the values on both tables, square method presents better coding for the patterns once it provides classes more distant from each other than the ones provided by the bitmap. Looking at the table below, the SSRD measure shows that the square polygon is even better than the octagon and both approaches of the circle. Bitmap is the last in the rank.

Table 2. Square approach, relative distance map.

<table>
<tr><td colspan="12" align="center">SQUARE – RELATIVE DISTANCE MAP</td></tr>
<tr><td colspan="12" align="center">CLASS</td></tr>
<tr><td></td><td></td><td>0</td><td>1</td><td>2</td><td>3</td><td>4</td><td>5</td><td>6</td><td>7</td><td>8</td><td>9</td></tr>
<tr><td rowspan="10">CENTER OF MASS</td><td>0</td><td>1.00</td><td>0.52</td><td>0.91</td><td>0.73</td><td>0.69</td><td>0.78</td><td>0.73</td><td>0.63</td><td>0.73</td><td>0,69</td></tr>
<tr><td>1</td><td>0.42</td><td>1.00</td><td>0.60</td><td>0.64</td><td>0.67</td><td>0.66</td><td>0.62</td><td>0.77</td><td>0.70</td><td>0,66</td></tr>
<tr><td>2</td><td>0.70</td><td>0.64</td><td>1.00</td><td>0.90</td><td>0.81</td><td>0.90</td><td>0.86</td><td>0.78</td><td>0.87</td><td>0,81</td></tr>
<tr><td>3</td><td>0.60</td><td>0.67</td><td>0.85</td><td>1.00</td><td>0.84</td><td>0.94</td><td>0.84</td><td>0.84</td><td>0.87</td><td>0,89</td></tr>
<tr><td>4</td><td>0.60</td><td>0.70</td><td>0.82</td><td>0.85</td><td>1.00</td><td>0.85</td><td>0.82</td><td>0.88</td><td>0.85</td><td>0,90</td></tr>
<tr><td>5</td><td>0.61</td><td>0.66</td><td>0.84</td><td>0.93</td><td>0.82</td><td>1.00</td><td>0.90</td><td>0.81</td><td>0.91</td><td>0,88</td></tr>
<tr><td>6</td><td>0.64</td><td>0.66</td><td>0.86</td><td>0.87</td><td>0.82</td><td>0.92</td><td>1.00</td><td>0.76</td><td>0.90</td><td>0,79</td></tr>
<tr><td>7</td><td>0.53</td><td>0.76</td><td>0.75</td><td>0.83</td><td>0.87</td><td>0.81</td><td>0.75</td><td>1.00</td><td>0.85</td><td>0,89</td></tr>
<tr><td>8</td><td>0.61</td><td>0.70</td><td>0.85</td><td>0.88</td><td>0.85</td><td>0.91</td><td>0.89</td><td>0.85</td><td>1.00</td><td>0,88</td></tr>
<tr><td>9</td><td>0.60</td><td>0.67</td><td>0.81</td><td>0.90</td><td>0.90</td><td>0.90</td><td>0.78</td><td>0.89</td><td>0.88</td><td>1,00</td></tr>
<tr><td colspan="2">SRD</td><td>5.30</td><td>5.98</td><td>7.29</td><td>7.52</td><td>7.27</td><td>7.65</td><td>7.19</td><td>7.21</td><td>7.54</td><td>7.40</td></tr>
</table>

Table 3. Bitmap approach, relative distance map.

<table>
<tr><td colspan="12" align="center">BITMAP MATRIX – RELATIVE DISTANCE MAP</td></tr>
<tr><td colspan="12" align="center">CLASS</td></tr>
<tr><td></td><td></td><td>0</td><td>1</td><td>2</td><td>3</td><td>4</td><td>5</td><td>6</td><td>7</td><td>8</td><td>9</td></tr>
<tr><td rowspan="10">CENTER OF MASS</td><td>0</td><td>1.00</td><td>0.77</td><td>0.90</td><td>0.85</td><td>0.87</td><td>0.91</td><td>0.89</td><td>0.84</td><td>0.88</td><td>0,90</td></tr>
<tr><td>1</td><td>0.86</td><td>1.00</td><td>0.94</td><td>0.91</td><td>0.92</td><td>0.92</td><td>0.91</td><td>0.95</td><td>0.92</td><td>0,92</td></tr>
<tr><td>2</td><td>0.92</td><td>0.88</td><td>1.00</td><td>0.91</td><td>0.92</td><td>0.93</td><td>0.94</td><td>0.94</td><td>0.95</td><td>0,94</td></tr>
<tr><td>3</td><td>0.87</td><td>0.89</td><td>0.93</td><td>1.00</td><td>0.89</td><td>0.96</td><td>0.90</td><td>0.89</td><td>0.92</td><td>0,94</td></tr>
<tr><td>4</td><td>0.89</td><td>0.87</td><td>0.93</td><td>0.87</td><td>1.00</td><td>0.92</td><td>0.93</td><td>0.94</td><td>0.92</td><td>0,94</td></tr>
<tr><td>5</td><td>0.90</td><td>0.89</td><td>0.94</td><td>0.94</td><td>0.93</td><td>1.00</td><td>0.94</td><td>0.91</td><td>0.94</td><td>0,95</td></tr>
<tr><td>6</td><td>0.90</td><td>0.85</td><td>0.94</td><td>0.88</td><td>0.92</td><td>0.94</td><td>1.00</td><td>0.89</td><td>0.93</td><td>0,91</td></tr>
<tr><td>7</td><td>0.87</td><td>0.92</td><td>0.95</td><td>0.88</td><td>0.95</td><td>0.92</td><td>0.90</td><td>1.00</td><td>0.94</td><td>0,94</td></tr>
<tr><td>8</td><td>0.88</td><td>0.86</td><td>0.96</td><td>0.89</td><td>0.91</td><td>0.93</td><td>0.93</td><td>0.92</td><td>1.00</td><td>0,94</td></tr>
<tr><td>9</td><td>0.89</td><td>0.87</td><td>0.94</td><td>0.92</td><td>0.93</td><td>0.95</td><td>0.91</td><td>0.93</td><td>0.94</td><td>1,00</td></tr>
<tr><td colspan="2">SRD</td><td>7.98</td><td>7.80</td><td>8.44</td><td>8.06</td><td>8.24</td><td>8.38</td><td>8.24</td><td>8.22</td><td>8.34</td><td>8.37</td></tr>
</table>

Table 4. Performance measures using SSRD.

METHOD	Square	Circle - radius	Circle - diameter	Octagon	Bitmap matrix
SSRD	70.34	72.02	72.37	73.43	82.07

1.2 Fisher Criterion

To compute Fisher's discriminant ratio it is also necessary to separate the database into different classes G_i, with "i" from 0 to 9. Now it is also necessary to define a general class G with all patterns in it. The ratio is computed through the following steps:

1 Computation of the inner covariance matrices (inside each class)

$$\overline{m}_i = \frac{1}{N_i} \sum_{k \in G_i} \overline{a}_k, \qquad Sw_i = \sum_{k \in G_i} (\overline{a}_k - \overline{m}_i)(\overline{a}_k - \overline{m}_i)^T, 0 \leq i \leq 9$$

2 Computation of the cumulative inner matrix: $SW = \sum_{i=0}^{9} Sw_i$

3 Computation of the total covariance matrix: $\overline{m} = \dfrac{1}{N}\sum_{k=1}^{N}\overline{a}$

$$ST = \sum_{k=1}^{N}(\overline{a}_k - \overline{m})(\overline{a}_k - \overline{m})^T, \quad N = \text{total number of patterns}$$

1 Computation of the intra classes covariance: $ST = SW + SB$, The intra classes covariance (SB), can be approximated by

$$SB = \sum_{i=1}^{c} N_i(\overline{m}_i - \overline{m})(\overline{m}_i - \overline{m})^T, \quad c = \text{number of classes}$$

2 The discriminating ration (DR) is then obtained from the SW and SB matrices

$$DR = \dfrac{trace(SB)}{trace(SW)}$$

Using this discriminating ratio, it was found again that square method presents the best result (Table 5), octagon came to the second place and the relative order of the others did not change.

Table 5. Fisher Ratio

Method	Square	Octagon	Circle – radius	Circle - diameter	Bitmap matrix
DR	1.1654	1.1541	1.1451	1.1424	1.1248

6 Evaluation by Neural Networks

Two different types of neural network model were used to evaluate the nonlinear discrimination power of the proposed feature extraction technique. In the first experiment there was used a single MPL – Multi Layer Perceptron network, trained with the back propagation algorithm with 10 output neurons (Table 6). The network was trained to learn all 10 digits. One and two hidden layer was evaluated with different number of neurons in each one. Two distinct propagation functions were used: linear and sigmoid. The obtained results are shown in Table 7.

Table 6. Type I network output scheme

Character	0	1	...	9
Output	0000000001	0000000010	...	1000000000

Table 7. One hidden layer, linear propagation function network performance. The colored cells show the best results

1 Hidden Layer Linear Propagation Type I Network					
Neurons	Square	Octagon		Radius	Bitmap
2	0.2176	0.1472	0.2424	0.2301	0.2243
4	0.2399	0.1855	0.2957	0.2439	0.2210
8	0.3453	0.2185	0.3831	0.2934	0.2597
10	0.3905	0.1373	0.3442	0.3304	0.2519
16	0.4446	0.1512	0.4126	0.3875	0.2740

20	0.4655	0.1333	0.4076	0.3843	0.3249
30	0.4351	0.1386	0.4550	0.4401	0.3790
32	0.4378	0.1307	0.4511	0.4320	0.3591
40	0.3912	0.1294	0.4464	0.4345	0.4055

For the second experiment a separated network for each output was used. There was 10 networks, each one with a single output. Table 8 below summarizes the results obtained with the experiments.

Table 8. Best results with Type I - single network with 10 outputs and Type II - set of 10 networks with one output each.

Methods	TYPE I				TYPE II	
	1 layer	2 layers			2 layers	
	purelin	purelin/purelin	logsig/purelin	logsig/logsig	purelin/purelin	logsig/logsig
Square	0.4655	0.4473	0.6899	0.8554	0.4459	0.8716
Octagon	0.2185	0.3545	0.5254	0.6475	0.3696	0.6832
Diameter	0.4550	0.4550	0.6165	0.6683	0.4576	0.6576
Radius	0.4401	0.4464	0.6150	0.6865	0.4545	0.7147
Bitmap	0.4055	0.4022	0.6983	0.7812	0.4088	0.7680

7 Conclusions

As described, the proposed approach, manly the one based on the square polygon, proved to provide very good discrimination power to be use with neural network models. The on going research is now addressing the hexagon extraction, the use of additional information added to the polygon vector and also the comparison with other approaches found in the literature.

Bibliography

1. Rodrigues, R. J.; Thomé, A. C. G.; "Cursive character recognition – a character segmentation method using projection profile-based technique", The 4th World Multiconference on Systemics, Cybernetics and Informatics SCI 2000 and The 6th International Conference on Information Systems, Analysis and Synthesis ISAS 2000 – Orlando, USA – August 2000
2. Rodrigues, R. J.; Thomé, A. C. G.; "Reconhecimento de dígitos cursivos – um método de segmentação por histogramas", a ser publicado no SCI
3. Srikantan, G.; Lam; S. W.; Srihari, S, N; Gradient-Based Contour Encoding For Character Recognition, Pattern Recognition, Vol. 29, No. 7, pp. 1147-1160, 1996
4. Suen, C.Y.; Berthold, M.; Mori, S.; Automatic Recognition of Handprinted Characters – The State of The Art,Proceedings of the IEEE, Vol. 68, No. 4, April 1980
5. Trier, O. D.; Jain, A.K.; Taxt, T.; "Feature Extraction Methods for Character Recognition – A Survey"; Pattern Recognition, Vol. 29, No. 4, pp. 641-662, 1996
6. Verschueren, W.; Schaeken, B.; Cotret, Y. R.; Hermanne, A.; Structural Recognition of Handwritten Numerals, CH2046-1/84/0000/0760$01.00@1984 IEEE

7 Yang, L.; Prasad, R.; "Online Recognition of Handwritten Characters Using Differential Angles and Structural Descriptors", Pattern Recognition Letters, no 14, Dec-93, North-Holland, pp: 1019-1024.

Using Contextual Information to Selectively Adjust Preprocessing Parameters

Predrag Neskovic and Leon N Cooper

Physics Department and Institute for Brain and Neural Systems, Brown University
Providence, RI 02912, USA
email: Predrag_Neskovic@brown.edu and Leon_Cooper@Brown.edu

Abstract. It is generally accepted that some of the problems and ambiguities at the low level of recognition and processing can not be resolved without taking into account contextual expectations. However, in most recognition systems in use today, preprocessing is done using only bottom-up information. In this work we present a working system that is inspired by human perception and can use contextual information to modify preprocessing of local regions of the input pattern in order to improve recognition. This is especially useful, and often necessary, during the recognition of complex objects where changing the preprocessing of one section improves recognition of that section but has an adverse effect on the rest of the object. We present the results of our recognition system applied to on-line cursive script where in some cases the error rate is decreased by 20%.

1 Introduction

A significant amount of work over the past several years has been put into constructing sophisticated preprocessing techniques and complex systems for recognition of on-line handwriting. Most of the state-of-the-art systems are based either on hidden Markov models (HMMs) [8, 5] or a combination of HMMs and neural networks (NNs), so called hybrid HMM/NN models [12, 1, 2]. An important characteristic of these systems is that they are bottom-up approaches since higher level modules do not influence processing of low-level modules. Backpropagation NNs and HMMs use feedback, but only during the training phase and the recognition stage is purely feed-forward.

Although these sophisticated systems did improve handwriting recognition, there are some problems and ambiguities at the low level of processing that cannot be resolved without taking into account cognitive level expectations.

Construction of a recognition system that is able to use feedback information during the recognition phase is therefore of great importance.

Despite the fact that human visual system is one of the finest examples of successful use of contextual information, not many recognition systems are based on human perception. One of the few working systems for handwriting recognition that is inspired by human reading skills was constructed by Lecolinet et al. [3]. The system is based on the model suggested in [6] and can capture some of the characteristics of human perception. However, it is very limited in scope and has been tested on a very small dictionary consisting of only 32 words.

Inspired by some properties of human vision we have constructed a recognition system that has been successfully applied to recognition of on-line cursive script [10, 9]. Our solution is not restricted only to on-line cursive writing and the approach is easy to extend and apply to segmentation and recognition of 2D objects.

In our previous work we showed how selective attention, guided by contextual expectations, can be used during segmentation and recognition process [9]. In this work we present another application of context-guided selective attention and demonstrate how it can be used to improve recognition by adjusting the preprocessing parameters. If the correct classification of the unknown pattern is not possible, the system is able to locate an area where the error or ambiguity is and apply different preprocessing on that region.

2 Preprocessing and related problems

An integral part of most of the artificial recognition systems currently in use is the preprocessor. There are various reasons for using preprocessing. An input signal often contains noise, and preprocessing can sometimes eliminate what seems to be irrelevant information. Furthermore, preprocessing presents a useful data reduction procedure which has at least two advantages for NNs-based recognition systems: a) NNs have better generalization properties if the dimensionality of the input vector is small, and b) it is much simpler and faster to train the network on smaller data sets. However, each of these advantages has some drawbacks since the preprocessing often results in the loss of important information.

The input signal to our system is transformed through the preprocessor into a feature vector in the following way: A raw data file, representing a word from a dictionary, contains information about x and y pen positions recorded every 10 milliseconds. The input signal, X, is first transformed into *strokes*, which are defined as lines between the points with zero velocity in the direction of y-axis. Each stroke is characterized by a set of features [11]: the x and y size of the stroke, the net motion of the pen half way through the stroke, the velocity in the x direction, and the ratio of the frequency in the x direction ω_x, to the frequency in the y direction, ω_y. The preprocessor extracts these features from the strokes, and outputs the feature vector to the segmentation network.

Some of the difficulties associated with detection of strokes and calculation of features involve: smoothing, normalization, and reordering of strokes (see [4] for a review of preprocessing techniques for on-line cursive handwriting). For example, the purpose of normalization of strokes is to allow scale invariant recognition. One possibility is to normalize them with respect to the size of the small letters [11]. The problem with this choice is that it requires the recognition of what is to be recognized. Similarly, one can show problems with any other choice for the normalizing factor. Another problem is noise reduction. A small stroke, or a

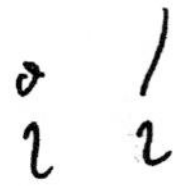

Fig. 1. The "dot" over the letter "i" can be made up of few strokes, and its size is not always small.

set of very short strokes, often represents noise. However, if the noise-like structure represents the dot above "j" or "i", then it can not be considered "noise" but an important part of the letter. One criterion that is used for detecting strokes that represent dots above "i" and "j" is: "a dot is the stroke of short length that is preceded by a pen-up stroke, and followed by a pen-up stroke" [7]. In most situations that is the case, but sometimes a "dot" consists of several strokes, as shown in Figure 1 on the left. The value of the threshold parameter that would control the filtering of "small" strokes is not always possible to determine in advance. What is noise, in most cases, depends on context. The size of a stroke depends on strokes (or any other "reference object") to which we compare it. In Figure 1 (right), the "dot" above "i" is larger than the size of the strokes that the letter consists of. In Section 4 we will show how to use contextual information to deal with this problem.

3 The segmentation and binding networks

The feature vector is supplied to the segmentation network, Figure 2, which has the task of detecting letters. The segmentation network is a multi-layer feed forward neural network based on a weight sharing technique. This particular architecture was proposed by Rumelhart [11]. The output layer of the network consists of units called *letter detectors*. The outputs of the letter detectors form a *detection matrix*. Elements of the detection matrix represent recognition probabilities of the letters from the alphabet. Each row of the detection matrix represents one letter and each column corresponds to the position of the letter within the pattern. The units that are in the same row have restricted and overlapping receptive fields and share the same weights. We will denote by $p(\alpha^c|\Delta x)$ the probability that a section, Δx, of the input pattern, over which the letter detector is positioned, represents a letter of the $c-th$ class, $c \in [1,...,26]$.

The next stage of information processing is performed by the *binding* network [9]. The input to the binding network consists of the elements of the detection matrix and the task of the network is to select, or bind, a subset of the elements from the detection matrix such that they represent a dictionary word. It is important to stress that the set of the selected elements from the detection matrix that represents a given dictionary word is equal to the number of the letters in the word and not to the number of columns of the detection matrix (in contrast to HMMs) [9].

The network has a hierarchical structure and consists of several layers of processing units (for a detailed description see [9] and references therein). During the recognition process, the system focuses its attention on different locations/letters from the input at different times. It probes the pattern in a way similar to saccadic eye movements and slowly integrates information, from different regions of the pattern, into a perception of a word. At this stage of recognition, contextual information is used to guide selective attention in choosing the fixation points. The system chooses the location on which to saccade if the region around that location has high support from surrounding letters (top-down support) and high support from the pattern itself (bottom-up support).

In addition to providing the global characterization of the pattern, the binding network produces a description of the local sections of the pattern. Let us denote by α_i^n the $i-th$ letter of the $n-th$ dictionary word. Then the probability that the section of the input pattern over which a letter detector is positioned, Δx, represents the $i-th$ letter of the $n-th$ dictionary word is given as $p(\alpha_i^n|\Delta x_i)$. The binding network makes an assignment $p(\alpha^c|\Delta x) \to p(\alpha_i^n|\Delta x_i)$, so that each

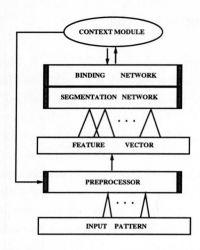

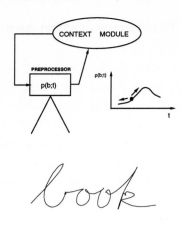

Fig. 2. Overview of the system.

Fig. 3. Preprocessing threshold is modified to maximize recognition by the letter detector.

of the selected elements from the detection matrix now represents one letter of the given dictionary word. Moreover, the binding network provides the estimates of finding the $i-th$ letter (of the $n-th$ dictionary word) at its location, $l(\alpha_i^n)$, given the location of any other $(j-th)$ letter of the word, $p(l(\alpha_i^n)|l(\alpha_j^n))$, and given the locations of all the letters of the word $p(l(\alpha_i^n)|l(\alpha_j^n),...,l(\alpha_{L_n}^n))$, where L_n is the number of letters in the $n-th$ dictionary word.

4 Preprocessing and feedback

As discussed in Section 2, it is not always possible to resolve some of the problems on the preprocessing level since at the time the preprocessing is done some information, necessary for choosing optimal preprocessing parameters, is missing. No matter how sophisticated the preprocessor is, or how well adjusted the parameters that control the preprocessor are, we know that in some cases they will fail. Therefore, the problem that we want to address is not how to improve the estimates of the parameters but how to detect when they are wrong and how to correct them in that case.

Our approach is to use a preprocessor whose parameters can be selectively modified in accordance to top level expectations. It is important to stress that the changes of the parameter values, using the feedback, are done *locally*, which means that the changed values are used for preprocessing only certain spatial regions of the input pattern. This is useful, and often *necessary*, during the recognition of complex objects where changing the preprocessing of one section improves recognition of that section but has an adverse effect on the rest of the object.

Let us now assume that the binding network classified the pattern as the $n-th$ dictionary word. The probability $p(\alpha_i^n|\Delta x_i)$, associates the section Δx_i of the pattern with the $i-th$ letter of the $n-th$ dictionary word. This probability implicitly assumes certain values of preprocessing parameters that have been used while processing the section Δx_i. If there are k different parameters controlling the preprocessor, the functional dependence on them can be written as

$$p(\alpha_i^n|\Delta x_i) = F(\Delta x_i; t_i^1, ..., t_i^k), \qquad (1)$$

where the subscript i means that all these parameters are used for preprocessing only the section Δx_i.

In the beginning of the recognition process, the default values for the threshold parameters are used for preprocessing the whole pattern. The system outputs the ranked set of dictionary words. If there is a confusion among top ranked words, or if the top ranked word is not recognized with acceptable confidence, the system initiates the process of adjusting the preprocessing parameters on localized regions. Only one word at a time is selected for re-preprocessing. Supplied by the two estimates for each of the letters of a given word, the detection estimate $p(\alpha_i^n|\Delta x_i)$ and the location estimate $p(l(\alpha_i^n)|l(\alpha_1^n),...,l(\alpha_{L_n}^n))$, the system locates sections with low value of either one, or both, of these estimates. This is in contrast with the use of selective attention during the first phase of "exploratory" recognition where the system was looking for salient regions (those with high location and recognition estimates) in the pattern in order to correctly segment/recognize the pattern. This time, during the "confirmatory" recognition, the system is looking for suspicious regions of the pattern, those with low value estimates.

As an example of the parameter used by the preprocessor is the "noise" threshold. The preprocessor removes all the strokes smaller than the selected threshold size, from the section of the pattern on which it operates. The pattern illustrated in Figure 3 was initially associated with the words "book" and "look". In order to resolve the confusion, the system located the section of the pattern representing the letter "b" for additional processing. The threshold value, t, was varied until the maximum recognition estimate, $p(b;t)$, for the letter "b" was reached. At the same time no value of the threshold parameter could have produced a better recognition estimate for the letter "l". It is interesting to mention that the optimal value of the threshold parameter used on the section representing the letter "b", resulted in a lower recognition rate when applied to the section of the image associated with the letter "k". Therefore, two different values of the threshold parameter were applied to two different regions of the image.

5 Discussion and results

The selection of the best set of parameter values can also be accomplished without using feedback information, in a straightforward manner by trying all possible parameter values, on every section of the pattern, and then selecting the best ones. However, this method is computationally prohibitive and it can be easily shown that the number of different parameters combinations is exponential with respect to the number of sections that are supplied to the segmentation network. With our approach the number of possible combinations is linear with respect to the number of letters. Moreover, the straightforward change of preprocessing parameters wouldn't necessarily guarantee better performance since we can imagine a situation where a specific combination of parameters "creates" a non-existing word. In our method, due to contextual constraints, the change of parameters has always improved the recognition.

We have tested our system on a dataset of on-line cursive words collected from 100 different writers, where each writer wrote 1000 words [11]. In our previous work [9] we showed that compared to the dynamic programming-based postprocessor, and results reported by Rumelhart [11], our recognition performance is significantly better, and for some writers the relative error rate is reduced by more than 32%. Similarly, we demonstrated [10] that HMM-based postprocessing can be successfully replaced by our model. In this work we report that the error rates can be further reduced by selectively changing the value of the threshold parameter that controls the amount of noise the preprocessor should remove. The system's performance depends on the writer on which it is being tested. For some writers, with clean handwriting, the error rate decreased by less than 5%, but on some "sloppy" writers the error rate decreased by almost 20%. We think that including the feedback in controlling other preprocessing parameters will further improve our results.

The re-preprocessing procedure can, in principle, be time consuming, if invoked too often and on many regions of the pattern. However, in practice this is not the case since only a few dictionary words are usually associated with a pattern and only local regions of the pattern are chosen for additional processing. In our experiments we found that recognition speed affected only "sloppy" writers and the algorithm was not called many times for "clean" writers. This is another important feature of our algorithm that is similar to human perception since reading sloppy writing is more time consuming than reading clean writing. In addition, the frequency of invoking the re-preprocessing algorithm can be easily controlled by setting the appropriate confidence threshold for word recognition, as explained in the previous section.

References

1. Y. Bengio, Y. LeCun, C. Nohl, and C. Burges. Lerec: A NN/HMM hybrid for on-line handwriting recognition. *Neural Computation*, 7:1289–1303, 1995.
2. H. Bourlard and C. Wellekens. Links between hidden Markov models and multi-layer perceptrons. *IEEE Transactions on Pattern Analysis and Machine Intelligence*, 12:1167–1178, 1990.
3. M. Cote, E. Lecolinet, M. Cheriet, and C. Suen. Automatic reading of cursive scripts using a reading model and perceptual concepts. *Internatinal Journal of Document Analysis and Recognition*, 1997.
4. W. Guerfali and R. Plamondon. Normalizing and restoring on-line handwriting. *Pattern Recognition*, 26(3):419–431, 1993.
5. J. Hu, M. Brown, and W. Turin. HMM based on-line handwriting recognition. *IEEE PAMI*, 18(10):1039–1045, 1996.
6. J. McClelland and D. Rumelhart. An interactive activation model of context effects in letter perception. *Psychological Reviews*, 88:375–407, 1981.
7. P. Morasso, L. Barberis, S. Pagliano, and D. Vergano. Recognition experiments of cursive dynamic handwriting with self–organizing networks. *Pattern Recognition*, 26(3):451–460, 1993.
8. K. Nathan, H. Beigi, J. Subrahmonia, G. Clary, and H. Maruyama. Real-time on-line unconstrained handwriting recognition using statistical methods. In *International Conference on Acoustics, Speech and Signal Processing*, volume 4, pages 2619–2613, 1995.
9. P. Neskovic and L. Cooper. Neural network-based context driven recognition of on-line cursive script. In *7th International Workshop on Frontiers in Handwriting Recognition*, pages 352–362, 2000.
10. P. Neskovic, P. Davis, and L. Cooper. Interactive parts model: an application to recognition of on-line cursive script. In *Advances in Neural Information Processing Systems*, 2000.
11. D. E. Rumelhart. Theory to practice: A case study – recognizing cursive handwriting. In E. B. Baum, editor, *Computational Learning and Cognition: Proceedings of the Third NEC Research Symposium*. SIAM, Philadelphia, 1993.
12. M. Schenkel, I. Guyon, and D. Henderson. On-line cursive script recognition using time delay neural networks and hidden markov models. *Machine Vision and Applications*, 8:215–223, 1995.

Electric Power System's Stability Assessment and Online-Provision of Control Actions Using Self-Organizing Maps

Carsten Leder[1] and Christian Rehtanz[2]

[1]Institute of Electric Power Systems, University of Dortmund, 44221 Dortmund, Germany
leder@ieee.org

[2]ABB Corporate Research Ltd., 5405 Baden-Dättwil, Switzerland
christian.rehtanz@ieee.org

Abstract. Power utilities are interested in operating their grid closer to technical limits. Moreover competition leads to system states which the operators in control centers are not familiar with. In order to operate the higher stressed power system secure, even in critical situations, an efficient security assessment must provide high-quality state information instead of thousands of single values. Furthermore, the energy management system (EMS) must give proposals for control actions. The Self-Organizing Map (SOM) supports both tasks efficiently. The paper presents a SOM-based solution for fast security assessment and the provision of control actions. The application to a real power system also shows the capability of the tool for expressive visualization.

1 Introduction

The increasing competition in electrical energy markets leads to changing requirements for power systems supervision and control. The grid companies aim to maximize their profit by making optimal use of their investment, the network. That means the operation of the power system closer to technical limits. Furthermore, the growing number of market activities, especially of wheeling contracts, results in new system's states, which the operators in the control center are not well familiar with. The two mentioned aspects produce a discrepancy, because the operation nearby the stability limit requires experiences and always an extensive knowledge about the system's state.

Therefore, an EMS must support the operators in the control center with high-quality information about the current operating state, the future development and effective stabilizing control actions to prevent insecure system's states. On the basis of the incoming measurements security assessment tools determine all important kinds of security limits. Conventional methods for security assessment are well known. The transient stability can be determined by dynamical simulations. Checking of thermal and voltage limits just as voltage stability bases on load flow equations. For the assessment of all different kinds of stability a lot of computational effort is needed. The fastest applied analytical tools need more than 30 minutes to determine the overall security of a large transmission system, which is not sufficient in highly-stressed situations. Because of this computational burden, the use of artificial neural networks (ANN) for solving the problem has been the subject of recent research projects [1,2]. Unsupervised learning methods as a sub-area of ANN-methods have the advantage, that they need less training time than the supervised ones. Therefore, especially SOM has been used for solving several tasks in power system's security assessment, e.g. for transient stability determination [3]. First results using SOM for voltage stability analysis are presented in [4,5]. Beside security assessment the SOM promises good results for the provision of control actions for the operators, which are presented here for the first time.

More details on power system's security limits are given in section 2. This will be followed by the description of an online security assessment tool using SOM in section 3. How to support control engineers with intelligent actions will be presented in section 4. Both sections 3 and 4 include examples with realistic data.

2 Security Limits of Power Systems

The area of operation of a power system is restricted by several limits. The voltage magnitude at all buses must stay within a given range, in order to guarantee supply quality. A permanent thermal rating of a network device above a maximum value would mean the loss of the device. Furthermore, voltage stability, short-circuit power limits, and static stability need not to be violated. During the transition from one operating state towards another the transient stability must be guaranteed. The following describes a conventional way of security assessment.

At first a load flow calculation determines, if an actual base case fulfills the thermal and voltage limits, and if the system is voltage stable. A dynamic simulation assesses the transient stability for several contingencies. Which contingencies have to be considered must be determined separately. Modifying an operation variable of the system stepwise, e.g. the load, this procedure can be executed until one of the limits has been reached. An optimization process for finding the limit can also be used. Fig. 1 shows this with the two left bars (arrow a) for voltage stability and b) for transient stability. Doing this, several load flow calculations and dynamical simulations are needed.

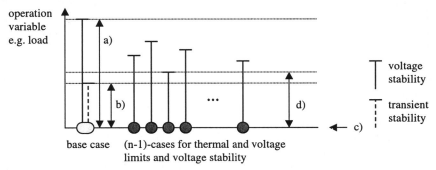

Fig. 1. Different kinds of security assessment

Additionally, selected (n-1)-cases have to be assessed (cases c). The cases which have to be considered can be selected according to the reliability. The situation with the lowest security defines the actual security value (arrow d). If the transient stability limit is lower, b) is the actual value. This value can be expressed by using the unit of the modified operation variable, e.g. MW for the load, and will be called *LI*.

3 Online Security Assessment Using Self-Organizing Maps

As mentioned before, conventional tools for security assessment are very time-consuming. Considering that for the use of ANN-tools the generation of a huge number of training scenarios is necessary, only unsupervised learning can reach the aim of acceptable training time, because supervised learning requires the calculation of a security indicator for each training vector using the described analytical tools. The SOM as an unsupervised learning method allows to find groups and structures within the input space, representing the potential power system's operating area. After the training, the security indicator can be calculated for a small number of feature vectors belonging to the neurons in an acceptable time. The following text presents the principles of SOM, an algorithm for the use of SOM in an EMS, and its application to a real power system.

3.1 Self-Organizing Map

A SOM maps a high dimensional input space to a low dimensional output space [6]. Fig. 2 shows the structure of the SOM and the example of a two-dimensional section of an input space x with two different clusters. The state information y of the clusters is colored pale and dark. The training of the SOM is done by feature vectors w_j in a way that their mean distances to the training vectors are minimized. The feature vectors are structured in a neighborhood grid. If the grid is two-dimensional, the SOM offers the possibility for the visualization of its mapping [4,5].

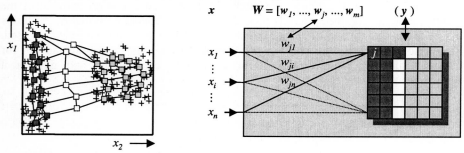

Fig. 2. Structure of a SOM with the example of a two-dimensional section of a state space

3.2 Application Strategy

The strategy for the application of SOM to the described power system's problem is shown in Fig. 3. The first step is the buildup of the training set. It consists of historical process data and simulated situations, in order to consider all possible system states, even those which occur very seldom in practice. The original dimension of the input space consists of hundreds of values. Using the well known statistical method of principal component analysis (PCA) [7] combined with the engineering process knowledge the dimension could be reduced significantly. After the training of SOM stability indicators are calculated for the resulting feature vectors with the analytical tools. The calculation of effective control actions for feature vectors, which represent critical states of the power system, is the last step of the preparation process (see Section 4).

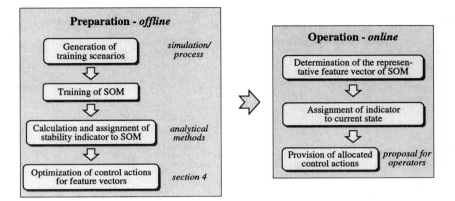

Fig. 3. Strategy for assessment of stability indicator and control actions with SOM

During the online operation of the stability tool the input vector consists of the actual measurements. After the determination of the winner neuron the security indicator which has been allocated during the training can be used to assess the actual operation state. If a critical system's state requires stabilizing control actions, the SOM provides the most efficient ones as choice for the operators.

3.3 Validation of Conformity

For the validation of conformity the described strategy has been applied to the voltage stability assessment for a real power system, consisting of 130 nodes and about 200 transmission devices. The training data set is made of 5000 different state vectors including contingency and non-contingency situations. In order to represent all practical relevant operation conditions the load level and the generation pattern are distributed randomly within their technical limits. After a correlation analysis the training vector has a dimension of 12 components consisting of the generator's reactive power reserve and active power injection. The selected inputs are normalized to the interval [0..1], because this guarantees equal weights for each component. The next step of the preparation process is the calculation of the stability value, maximum possible load increase LI, for each feature vector and its assignment to SOM.

Fig. 4 shows on the left the resulting 15x15 SOM, which bases on a grid topology. The colors are representing the stability measure. Dark colors indicate system's states close to the stability limit and bright colors indicate very stable ones. For the verification of the mapping quality, the analytically determined LI values are compared with the response of SOM, for 500 test vectors. Fig. 4 (right) shows the graph of the mapping error. The response of SOM LI_{SOM} and the error $E=LI_{SOM} -LI_{CAL}$ are given for each test vector. The graph illustrates the good accuracy of SOM, which can be confirmed by the small standard deviation of the error's frequency distribution of 8 %. Extreme values of the mapping error are very seldom and belong only to very few test vectors. A quit positive result is the fact, that the mapping error increases according to *LI*. A more effective creation of the training set, which means the limitation to only that states which occur in practice and not to all technically possible states, would increase the mapping quality.

Fig. 4. SOM colored according to the stability level (l) and the mapping error for test

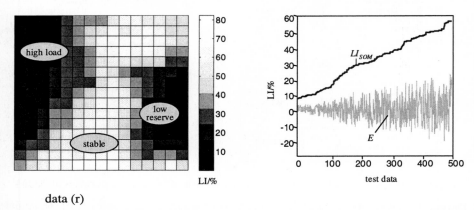

data (r)

Quite important for the application of SOM is the possibility of user friendly visualization. The color-coded information representation allows a fast and secure perception of high-quality information for the system operator.

4 Control Actions

The assessment of the system's security is of no use if information about efficient control actions is not available. Not only the experiences of operators are sufficient to find the best reaction, especially in the case of new situations and of an operation close to limits. Therefore, the EMS must support the staff with useful proposals for control actions. Table 1 lists the portfolio of available actions for power system's control.

Table 1. Possible control actions in power systems

system part	control action
load	load shedding and limited load control
network	switching of transformer tap-changer
	switching of reactive power compensation
	changing of topology
generation	control of active and reactive power of generators

The first step for the assignment of control actions to feature vectors of SOM is the ranking of the available actions according to their expense. After that the influence of a single action or a combination of actions to the systems security must be calculated. Beginning with the less expensive action this must be repeated until the security margin is sufficient for a secure operation. This most efficient action can be assigned to the feature vectors of the SOM and be used in critical situations during online operation.

Fig. 5 shows an example of the described strategy. The color of each field indicates the most efficient control action. An interactive access to the control action via a mouse-click guarantees the quick execution during online operation.

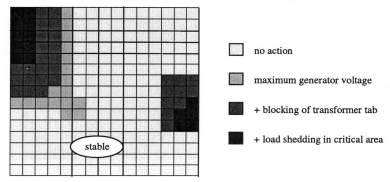

Fig. 5. SOM colored according to assigned control actions

5 Conclusions

The operation of the power system in a new market environment means higher stressed system's states and frequently changing situations. Only a high-quality security assessment and an adequate visualization of its results guarantee further on a secure and reliable electricity supply. The paper proposes the use of SOM for real-time security assessment. The unsupervised learning minimizes the computational effort, because the training of the data structure and the analytical assessment of feature vectors are separated. The application to a real power system illustrates that SOM is sufficient to solve the given problem. Furthermore it allows the pre-calculation of control actions and their fast provision during critical situations. The future research work must be focused on adapting the training set more adequately to real situations instead of distributing it randomly between technical limits in order to increase the mapping quality and to prevent great mapping errors.

References

[1] Dillon, T. S., Niebur, D.: "Neural Networks Applications in Power Systems", CRL Publishing, London, 1996
[2] Germond, A. J., "AI Techniques applied to power systems", Proc. of EPSOM98, Zürich, 1998
[3] Cho, H.-S., Park, J.-K., Kim, G-W.: "Power System transient stability analysis using Kohonen neural networks", Engineering Intelligent Systems, vol. 7, no. 4, Dec. 1999, pp. 209-214
[4] Rehtanz, C.: "Visualisation of voltage stability in large electric power systems", IEE Proc. Generation, Transmission and Distribution, vol. 14, no. 4, Nov. 1999

[5] Rehtanz, C., Leder, C.: "Stability Assessment of Electric Power Systems using Growing Neural Gas and Self-Organizing Maps", Proc. of 8th European Symposium on Artificial Neural Networks, Bruges, Belgium, 2000, pp. 401-406
[6] Kohonen, T.: "Self-Organizing Maps", Springer-Verlag, Berlin, 1995
[7] Jolliffe, I.T.: "Principal Component Analysis", Springer-Verlag, Berlin, 1986

Neural Networks for Contingency Evaluation and Monitoring in Power Systems

F. García-Lagos[1], G. Joya[1], F. J. Marín[2], and F. Sandoval[1]

[1] Dpto. Tecnología Electrónica. E.T.S.I. Telecomunicación.
Universidad de Málaga. Campus Teatinos s/n. 29071 Málaga, Spain
{lagos, joya, sandoval}@dte.uma.es
[2] Dpto. Electrónica. E.T.S.I. Informática.
Universidad de Málaga. Campus Teatinos s/n. 29071 Málaga, Spain
marin@ctima.uma.es

Abstract. In this paper an analysis of the applicability of different neural paradigms to contingency analysis in power systems is presented. On one hand, unsupervised Self-Organizing Maps by Kohonen have been implemented for visualization and graphic monitoring of contingency severity. On the other hand, supervised feed-forward neural paradigms such as Multilayer Perceptron and Radial Basis Function, are implemented for severity numerical evaluation and contingency ranking. Experiments have been performed with successfully result in the case of Kohonen and Multilayer Perceptron paradigms.

1 Introduction

The main objective of a Power System Security Assessment is to keep the system operation point into certain security range, thus guaranteeing a continuous and acceptable electrical energy supply. The energy market conditions have evolved due to new circumstances such as deregulation, increasing energy consume, and economical, social and ecological constrains in the building of new grids. These conditions force the system to work near its security limits, producing less and less conservative operation points. Consequently, it is necessary a continuous system monitoring to detect dangerous situations as soon as possible. In particular, the Contingency Analysis task must inform whether the current state is secure, critical or insecure with respect to a possible fault in a particular system component. The contingency analysis is conditioned by two facts: thousands of partially known and noised variables are required to model the system, and its behavior is highly non linear with respect them. Thus, although advances in telecommunication and computation allow a more and more detailed understanding of the system, the classical analysis techniques based on the solution of load flow problems are unable for producing real time solutions.

Neural Network Techniques present interesting features such as extracting classification criteria from an unsupervised analysis of the pattern set, complex non-linear function approach, or fault tolerance. These features seem to convert neural networks in methods appropriated to face this Contingency Analysis task.

However, at the moment, most reported contingency analysis methods are based on Fast Decoupled Load Flow calculation [3]. The main drawback of these last methods is the amount of AC load flow calculations needed to determine the line flows and bus voltages for each contingency, making the methods inapplicable for on-line purposes. Also, because these methods always use fast converging load flow algorithms (e.g. fast de-coupled Newton-Rhapson), the convergence is not guaranteed in heavily loaded power systems. In [5] a module for contingency screening and ranking using the Reactive Support Index (RSI) and an iterative filtering process is developed. Although the results are very promising when using the RSI in combination with the iterative filtering process, the method RSI by itself misclassifies some local contingencies. On the other hand, the iterative filtering process needs the solution of a load flow problem, requiring extra time to be carry out. Most recently, solutions based on artificial neural networks has been presented. Thus, in [3] Radial Basis Function Networks (RBFN) are used for contingency analysis in planning studies. The system uses different RBFN for active power flow contingencies and for bus voltage contingency, improving the learning capacity of the network. The output of the RBFN is an index that determines the capacity of a power system to support a peak demand under a contingency. The method, even though accurate and efficient for planing purposed, is not feasible for real time operation. In [4] a Multilayer Perceptron is used to calculate two different performance index. Each input vector is composed by the network power injections plus a number indicating the outaged element, this input vector is preprocessed using the Fast Fourier Transform (FFT). There is an only network, which evaluates all contingencies. We think that the use of heterogenous input patterns, containing both electrical measurements and qualitative information difficulties a correct function approximation. On the other hand, the use of an only neural network reduce the possibility of finding out an appropriate hidden weight set, which must be common to all functions. In this paper, we study the application of three neural paradigms to contingency analysis in its different specifications: On one hand, Self-Organized Maps by Kohonen are used to allow, in a visual way, the monitoring of the evolution of the system with respect to its security level. On the other hand, Feed-forward supervised neural networks such as Backpropagation and Radial Basis Functions are used to obtain a numerical evaluation of the security level of each contingency, allowing so building a contingency ranking. This numerical evaluation is based on approaching some determinate Performance Index functions: Real and Reactive Power Performance Indexes in our case.

The rest of this paper is organized as follows. In section 2 we explain the process of generation of the patterns used as neural network input; as well as the mathematical expression of the performance indexes used for contingency evaluation. In section 3 we detail the Self-Organizing Maps used for severity contingency visualization. All simulation parameters are indicated and results are commented. Section 4 is devoted to study the applicability of Feed-Forward Supervised Neural Networks to numeric evaluation of contingency severity. Finally, conclusions are related in section 6.

2 Generation of Input Patterns and Performance Indexes

For test purposes we have used the standard network IEEE 14 bus. Two different indexes are used taking advantage of the decoupling principle between active and reactive power: PI_p (active power performance index) and PI_v (reactive power performance index). These indexes are defined by equations (1) and (2), respectively [4]:

$$PI_p = \sum_\alpha w_{pL}(P_L/P_{L,lim})^2 \tag{1}$$

$$PI_v = \sum_\beta w_{vi}(|V_i - V_{i,lim}|)/V_{i,lim} \tag{2}$$

where P_L is the active power flow on line L, $P_{L,lim}$ is the active power flow limit on line L, w_{pL} is the severity weight of the line L, V_i is the voltage magnitude at bus i, $V_{i,lim}$ is the voltage magnitude limit at bus i, w_{vi} is the severity weight of voltage magnitude of bus i, and α and β are the sets including only lines whose active power flows is greater than their active power limits, and buses whose voltage magnitudes are out of their voltages limits, respectively. Hereafter, we indicate with $PI_{p,i}$ and $PI_{v,i}$ the respective performance index of the particular contingency i.

The input patterns for the neural networks applied to study the mentioned indexes are composed by active power flows and reactive power flows, respectively. From the standard network base case, we simulate a 41 days hourly load profiles, which is obtained by means of the following procedure. Each hourly base load is multiply by a variable coefficient between 0.5 and 1.8, thus simulating a very near real load pattern. This global load is distributed among all load buses according to their predefined load capacity. Finally, a normally distributed noise with 0 mean and 2% standard deviation is applied to the resultant bus load. From this load pattern, a complete load flow solution is carry out obtaining our active and reactive flow patterns.

3 Contingency Visualization with Self-Organizing Maps

Self-organizing Maps (SOM) by Kohonen are characterized by their capacity to classify a set of complex pattern in an unsupervised way, extracting no obvious classification criteria. Consequently, their use appear promising in contingency monitorization tasks, where sometimes it is more important to know the relative position of the current system state in a contingency ranking than to know the exact value of its performance index for this contingency.

In this study, we use a SOM of 10x10 neurons to classify the previously mentioned patterns composed by the active power flows of each line of the network IEEE 14. In parallel, an identical SOM is used for classifying the reactive power flow patterns. The results are equivalent in both cases, therefore, we focus our comments on the first. The update weight rule is given by equation 3:

$$w_i(t+1) = w_i(t) + l_r h_{iv}(x - w_i(t)) \tag{3}$$

where x is an input vector, w_i is the weight vector of neuron i, and $h_{iv}()$, the neighborhood function, determines the weight increment of each neuron as a function of proximity to the winner neuron. In our case, the neighborhood area is determined by a square centered in the winner neuron whose side diminishes until zero along the training. l_r is a dynamic learning rate, which evolves along the learning according to equation 4:

$$l_r(t) = l_{r0}/(1 + \frac{ct}{nn}) \qquad (4)$$

being l_{r0} the initial learning rate (0.3), c a constant (0.2), t the current iteration and nn the number of neurons [1].

After training, it is obvious that the obtained classification incorporates the severity level criterium for each contingency. Thus, as particular cases, figures 1 and 2 show the pattern classification respect to contingencies 1 (line 1-2 outaged) and 13 (line 6-13 outaged), respectively. Symbol *dot* represents states producing a performance index zero. Symbols *circle, plus, box, triangle down, diamond, pentagon* and *hexagon* represents patterns producing a performance in the intervals $[0.58, 0.97]$, $[0.98, 1.46]$, $[1.47, 1.9]$, $[2, 2.44]$, $[2.45, 3]$, $[3, 5]$ and $[5.1, \infty]$ respectively. One can see that the distribution of classes along one map diagonal follows, in a continuous way, the ascending order of the performance index. This fact suggests the possibility of using a linear SOM for this problem. This possibility has been confirmed by simulating a linear SOM of 10 neurons. It is necessary highlight that although each contingency produces a different set of classes, the trained SOM is one.

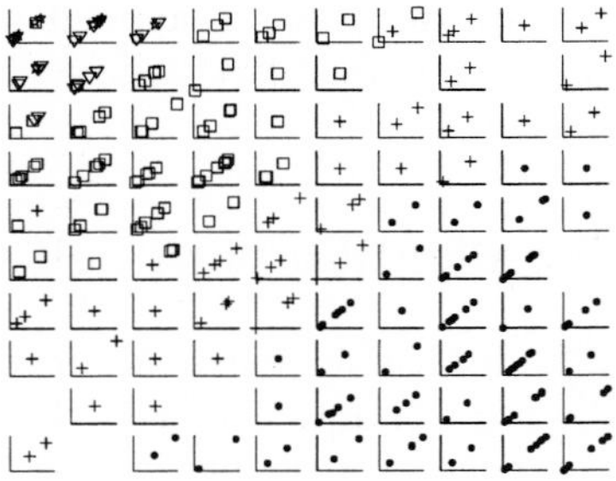

Fig. 1. Patterns distribution and classification for contingency 1 in a 10x10 SOM.

The obtained results show high applicability of this neural paradigm to two important contingency analysis subtask: on one hand, a rapid visualization of the severity level of the current system state respect to a particular contingency, and, on the other hand, a very early forecast of the tendency of the system to a possible dangerous operation point respect to a particular contingency.

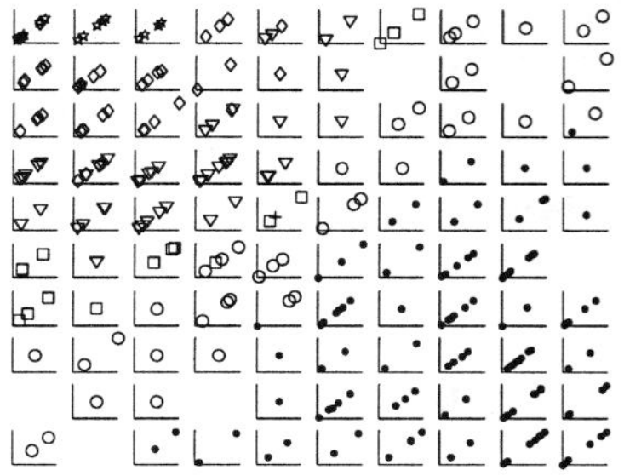

Fig. 2. Patterns distribution and classification for contingency 13 in a 10x10 SOM.

4 Numerical Contingency Evaluation and Ranking with Feed-Forward Neural Networks

Another contingency analysis subtask consists on obtaining a numerical evaluation of the severity of a particular contingency and its later ranking. The objective is to dispose of a selection criterium to prepare the system to firstly face the most dangerous outages. Normally, this numerical evaluation is obtained from calculating the value of a prefixed performance index.

Feed-forward supervised neural networks such as Multilayer Perceptron with Backpropagation and Radial Basis Functions are characterized by their ability for approaching complex non-linear functions, with a possible relatively long training time but a real time response in operation mode. We have just applied both paradigms to the problem of calculating the previously defined $PI_{p,i}$ and $PI_{v,i}$ performance index for every contingency of the IEEE 14 standard network.

Some reported applications use only one neural network to obtain the PIs of all contingencies. Heterogeneous input patterns, containing both electrical measurements and qualitative information about each particular contingency, are then needed. We think that this procedure is not adequate by two reasons: first, these neural paradigms have many problems to work with variables of very difference nature; moreover, the fact of representing logical variables in a numerical format may produce serious mistakes during the learning process. Second, the approximation of a lot of functions by means of the same neural network force to find out a set of hidden weight which must be used to implement these different functions. This set may be not find out, or, in the best case, may need a very long training process. Consequently, in this work a feed-forward network is used for calculating PI of each contingency.

Another consideration about the input pattern must be done. Input patterns of very high dimension will difficult the learning process convergence, specially if these patterns contain a high percentage of no significance information. To avoid this problem, we have made a principal component analysis among all active (reactive) flow components. As a result, the original forty component patterns have been reduced to six component patterns. The six selected components explain more than 99.9% of the variability of the original pattern set.

Next we show the simulation results for the application of multilayer perceptron and radial basis functions to obtain the $PI_{p,i}$ for the contingency set of the IEEE 14 network. A similar behavior is obtained for $PI_{v,i}$.

4.1 Multilayer Perceptron

A multilayer perceptron (MLP) has been used for calculating each $PI_{p,i}$. Its structure is composed by a six neuron input layer, a twenty neuron hidden layer and one output neuron. The activation function for hidden neurons is the hyperbolic tangent and the activation functions for output neurons is linear. 900 input patterns have been used for training and 100 different patterns have been used for testing. A Levenberg-Marquardt [2] training algorithm have been used with a cross validation strategy. Thirty training epochs are needed to obtain the best results.

Table 1. Mean relative training and testing errors (in percentage)(MAPE) for problematic contingencies of IEEE 14 network. The results are referred to MLP and RBF.

Contingency	1	2	3	8	10	13	14	16	19
MLP (training)	1.0	1.4	1.08	0.01	2.38	2.01	0.001	0.01	1.04
MLP (testing)	2.1	1.3	2.22	0.3	3.0	3.0	0.019	0.3	1.98
RBF (training)	1.8	2.01	1.0	4.1	1.7	4.02	0.001	0.23	1.98
RBF (testing)	4.3	4.1	10.7	12.0	12.56	14.30	0.008	1.78	10.28

Table 1 contains the mean relative training and testing errors (in percentage) for those contingencies producing a $PI_{p,i}$ greater than zero for some input pattern. Figure 3 illustrates in a graphical mode the high matching between

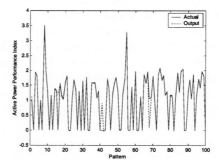

Fig. 3. Actual and MLP calculated performance index values for testing pattern set for contingency 1 (line 1-2 outaged).

real and calculated performance index for the testing simulation for contingency 1 (line 1-2 outaged). The results obtained for the rest of contingencies are in the same order. These reduced errors confirm the excellence of this paradigm for constructing a contingency ranking based on the numeric evaluation of a performance index.

4.2 Radial Basis Function Network

The radial basis function (RBF) networks stand out by their local learning capacity. In these networks each particular piece of the objective function is approaches by a reduced neuron set, theoretically allowing a better adjustment of the function. We have tested this hypothesis in the context of the contingency performance index calculation.

A radial basis network is used for calculating the $PI_{p,i}$ of each contingency. Its structure is composed of a six neuron input layer, a variable number of radial function neurons in the hidden layer (in our study this number varies from 68 to 215 depending on the contingency analyzed) and one linear output neuron.

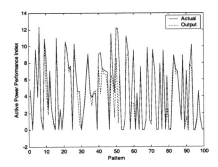

Fig. 4. Actual and RBF calculated performance index values for testing pattern set of contingency 13 (line 6-13 outaged).

Mean relative training and testing errors for $PI_{p,i}$ obtained with RBF appear in Table 1, and figure 4 shows, as illustration, the testing results for contingency 13 (line 6-13 outaged). Both table and figure results highlight the poor generalization capacity of this paradigm in the context of our particular problem. In addition, this network presents another drawback: obtaining the appropriate value of the spread parameter is a very laborious and time expensive task, making this paradigm less flexible than the multilayer perceptron.

5 Conclusions

In this work a study have been carry out about the applicability of different neural paradigms to the contingency analysis task, in its two different aspects:

visual monitorization of the system state evolution with respect to the severity level of a particular contingency, and the contingency ranking by mean of the numerical evaluation of some performance indexes.

Respect to the visual monitorization, we show the viability of the Kohonen's linear self-organizing map to correctly classify the severity of all contingencies from an unique network trained with a set of active (reactive) power flow patterns. Simulation results show that this paradigm is very useful to both a rapid visualization of the severity level of the current system state respect to a particular contingency, and a very early forecasting of the tendency of the system to a possible dangerous operation point respect to a particular contingency.

Respect to the contingency ranking, the simulation results show that multilayer perceptrons are a very interesting solution to obtain a numerical evaluation of the contingency severity, by mean of calculating some particular performance index. We have used the active power flow performance index (PI_p) and the reactive power flow performance index (PI_v). Additionally, we have shown that input patterns dimensionality may be drastically reduced. In the particular case of the IEEE 14 network, the original forty component vector has been reduced to a six component vector.

Radial basis function networks have been analyzed in this context, too. However, results are not so promising as those obtained by the multilayer perceptron. Thus, radial basis function network shows a poor generalization capacity and a greater difficulty for its construction.

Acknowledgments. This work have been partially supported by the Spanish Comisión Interministerial de Ciencia y Tecnología (CICYT), Project No. TIC98-0562.

References

1. M. Cottrell and P. Letremy. Classification et analyse des correspondances au moyen de L'Algoritme de kohonen: Application à l'ètude de données socio–économiques. *Prépublication du SAMOS,*, (42), 1995.
2. M. Hagan and M. Menhaj. Training feedforward networks with the marquardt algorithm. *IEEE Transactions on Neural Networks*, 4(6):989–993, 1994.
3. M. M. Refaee, J.A. and H. Maghrabi. Radial basis function networks for contingency analysis of bulk power systems. *IEEE Transactions on Power Systems*, 14(2):772–778, May 1999.
4. T. S. Sidhu and L. Cui. Contingency screening for steady-state security analysis by using FFT and artificial neural networks. *IEEE Transactions on Power Systems*, 15(1):421–426, February 2000.
5. E. Vaahedi, C. Fuchs, W. Xu, Y. Mansour, H. Hamadanizadeh, and G. Morison. Voltage stability contingency screening and ranking. *IEEE Transactions on Power Systems*, 14(1):256–265, February 1999.

Hybrid Framework for Neuro-Dynamic Programming Application to Water Supply Networks

M. Damas, M. Salmerón, J. Ortega, G. Olivares

Department of Computer Architecture and Computer Technology, University of Granada.
Facultad de Ciencias. Campus Fuentenueva s/n. E-18071, Granada, Spain
E-mail:{mdamas,moises,julio,gonzalo}@atc.ugr.es

Abstract. A hybrid method, based on evolutionary computation, Monte Carlo simulation, and neural networks for functional approximation and time series prediction, is proposed to reduce the high computational cost usually required by dynamic programming problems, that appear in complex real applications. As an example of application a scheduling problem related with the control of a water supply network is considered.

1 Introduction

Problems of planning and sequential decision making under uncertainty appear in a wide range of real applications and were considered to be intractable for a long time [1-3]. In these problems, the outcome of each decision is not fully predictable but can be estimated to some extent before making the next decision. Each decision has an *immediate cost* associated but it also affects the state in which future decisions are to be made thus having influence in the cost of future periods of time (*cost-to-go*). Thus, a low cost solution at the present must be balanced against the undesirability of possible high costs in the future. Examples of these problems are the dynamic scheduling problems, where the scheduler has to be able to react to external events in order to find solutions to the problems resulting from the changes produced by those events.

These problems can be considered from the point of view shown in Figure 1. There is an agent that interacts with its environment (everything outside the agent, including the system to control), and selects actions applied to this environment, which responds to these actions and presents new patterns to the agent. The agent has the goal of optimizing (over the long run) a cost which is defined from the outputs and states observed in the environment at discrete time steps, $i*t$ ($i=0,1,2,...$). In Figure 1, the agent performs some actions, Y, over a system and receives information about its inputs, W, and its outputs and/or states, X. The inputs to the system, and its states and outputs define the representation of the environment state. Thus at each time step, the agent has to determine a mapping, called policy, from the environment state space to the space of actions.

Dynamic Programming (DP) [1,2] has been the usual procedure to determine optimal policies in these kind of problems, although it requires a *perfect model* of the environment, that provides a complete description of all possible states and costs and their probabilities (there is still uncertainty in the state transitions appearing, since the external input values and perturbations are not previously known and are uncontrollable). Thus, as DP assumes a perfect model, its practical utility is limited cause the high computational requirements for real applications and other approaches, commonly referred as reinforcement learning (RL) [15,17] or neuro-dynamic programming (NDP) [3], have been proposed as ways to approximate the DP results.

In this paper we propose a framework that can be considered as a neuro-dynamic procedure. It applies neural networks for functional approximation and time series prediction, and also evolutionary computation and Monte Carlo simulation to optimize the behavior of the agent.

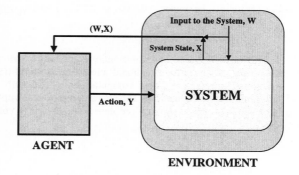

Fig. 1. Interaction between the agent and its environment

The proposed hybrid framework we propose and its relation with DP and NDP are considered in Section 2, while Section 3 provides some details of the application to water supply networks. Finally, Section 4 gives the experimental results and Section 5 summarizes the conclusions of the paper.

2 Description in the Context of Dynamic Programming, and Reinforcement Learning

Dynamic Programming uses a *cost-to-go function* to structure the space of possible agent decisions. Thus, it provides the required formalization of the tradeoff between immediate and future costs (values of the cost-to-go function) by considering a discrete-time dynamic system whose states change according to transition probabilities that depend on the control signal applied (decision made), $Y(i*t)$. If the system is in the state $X(i*t)=a$, and the control $Y(i*t)=u$ is applied, it moves to state $X((i+1)*t) = b$ with probability $p_{ab}(u)$. The transition from a to b when u is applied has an immediate cost $g(a,u,b)$. To rank the possible next state b, the optimal cost (over all remaining stages), $J^*(b)$, is used starting from each state b. The function $J^*(b)$ is the so called *optimal cost-to-go* of state b and satisfies the Bellman's equation [2]:

$$J^*(a)= min_u\, E[g(a,u,b) + J^*(b) \mid a,u\,]= min_u\{ \bullet_b\, (p_{ab}(u)[g(a,u,b) + J^*(b)]) \}. \qquad (1)$$

where $E[.\mid a,u]$ refers to the expected value with respect to b, for given a and u. Thus, for each state a, the optimal control signal u is the minimum of the previous expression, and the goal of dynamic programming is to determine the optimal cost-to-go function J^* and an optimal policy u for each state a. Two alternatives are commonly used to do that: *policy iteration*, and *value iteration* [17]. An important drawback of

DP algorithms is that they perform operations that use the entire state space. If this space is large, a search would require a high amount of computational work.

NDP provides sub-optimal optimization methods based on the evaluation and approximation of the optimal cost-to-go function J^*. In these methods the transition probabilities are not explicitly estimated, but instead the cost-to-go function, $J(a)$, of a given policy, •, applied from state a, is progressively calculated by generating several sample system trajectories and their associated costs. Monte Carlo methods [3,4,13], Temporal Difference Learning [16,17], and the procedure here presented are examples of NDP procedures. In Monte Carlo methods, the value of $J(a)$ is estimated by using several samples of trajectories from state a and using the policy •. Thus a complete trajectory is required to determine the value of cost-to-go function for this trajectory. Instead, Temporal Difference methods use a sample of the immediate cost $g(a,u,b)$ for state a, and action $u=•(a)$ and an estimation of $J(b)$, and they do not require to wait for a complete trajectory to be determined but only to wait for the next stage.

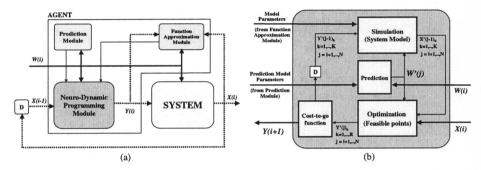

Fig. 2. (a) Modules of the proposed hybrid procedure. (b) Outline of procedure implemented by the Neuro-Dynamic Programming Module

The framework we propose in this paper comprises a set of modules shown in Figure 2. There is a Prediction module that provides the parameters of a model that allows the estimation of the system input in the next stage. It is also possible to use a Function Approximation module that will provide a description that makes possible an easier simulation of the system. The prediction model provided by the Prediction module, and the model of the system provided by the Function Approximation module are used by the NDP module to determine a set of feasible functioning points, and select the best one to be applied at the next stage.

Thus, at stage i, the agent has to determine the control vector $Y(i+1)$ for the next stage. To do this, the vector $W'(i+1)$, obtained by a model determined by the Prediction module, and the output $X(i)$ are used to define the cost function $Cost(i)$ whose minima correspond to the feasible control vector $Y(i+1)$ for the next stage. These feasible values for vector $Y(i+1)$ also allow the determination of the next state, $X(i+1)$ satisfying all the established restrictions. Once a set of feasible control vectors is obtained for a given stage $(i+1)$, $\{Y(i+1)_j, j=1,...,M_{i+1}\}$, it is necessary to select among them the best one to be used in the next control stage. The rank of $Y(i+1)_j$ $(j=1,...,M_{i+1})$ is determined by taking into account the immediate cost associated with its application, and the future evolution of the system that can be estimated by simulation. This simulation uses the predictions of the input vectors for the following stages, $W'(i+2),...,W'(N-1),W'(N)$, from the prediction model available at stage i. With these predicted input vectors it is possible to estimate sets of feasible control values (by simulation and using the predicted values W') for the following stages $\{Y'(i+2)_j, j=1,...,M_{i+2}\},..., \{Y'(N)_j, j=1,...,M_N\}$, and to generate sample trajectories (as in Figure 3) in order to approximate the cost-to-go function at stage $i+1$.

An approximation for the cost-to go function, noted as $J'(X(i+1)_j)$, can be:

$$J'(X(i+1)_j) = (1/K)(d(X(I+1)_j,1)+d(X(i+1)_j,2) +...+ d(X(i+1)_j,K)). \qquad (2)$$

where K is the number of simulated trajectories and, for the m-th trajectory, and

$$d(X(i+1)_{j1},m) = g(X(i+1)_{j1},Y'(i+2)_{j2},X'(i+2)_{j2}) + g(X'(i+2)_{j2},Y'(i+3)_{j3},X'(i+3)_{j3})+ ... \\ ...+ g(X'(N-1)_{j(N-i-1)},Y'(N)_{j(N-i)},X'(N)_{j(N-i)})$$

is the cumulative cost up to reaching the final state at the last stage of this trajectory. It is assumed that different simulated trajectories are statistically independent. It is possible to determine $J'(X(i+1)_j)$ iteratively starting with $J'(X(i+1)_j)=0$, and by using [3]:

$$J'(X(i+1)_j) = J'(X(i+1)_j) + G_m * (d(X(i+1)_j,m) - J'(X(i+1)_j)). \qquad (3)$$

with $G_m=1/m$, $m=1,2,..,K$. Once the values for $J'(X(i+1)_j)$ are computed for $j=1,2,..,M_{i+1}$, they are ranked, and the control $X(i+1)_j$ ($j=1,..,M_{i+1}$) with the lowest value for the approximate cost-to-go function is selected for the next stage.

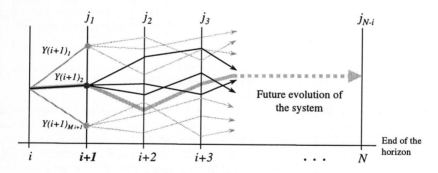

Fig. 3. Generation of sample trajectories

Thus, the procedure here presented can use the simulation of the full sample control trajectories (as in Monte Carlo methods) or only some stages (as Temporal Difference). Moreover, the procedure reduces the space of possible trajectories to sample by using a prediction procedure that anticipates the future values for the inputs to the system.

3 Application on Water Supply Networks

In this section we illustrate the use of the previously described framework in a dynamic scheduling problem that appears in a water supply system. More specifically, the flow that the drinkable water treatment station (*ETAP* in Spanish) has to be distributed among the different tanks of the system (as in Figure 4), such that demand is satisfied and that the tanks neither overflow nor fall bellow the established minimum water volume, while valve movements are minimized and the flow from the *ETAP* is kept constant.

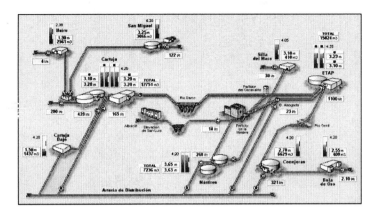

Fig. 4. Scheme of the Water Distribution System of Granada (Spain)

In the application, $x[i][j]$ is the volume of water in tank j (1...n) in the time interval i (1...N, where N is the horizon equal to 24 hours); $w[i][j]$ is the volume of water expected to be consumed during time interval i from tank j; and $y[i][j]$ is the volume of water supplied to tank j during time interval i. The feasible functioning points for the time interval i (i.e. the values of $x[i][j]$), are the minima of a cost function $Cost[i]$, obtained by summing four terms:

$$Cost(i) = Cost1(i) + Cost2(i) + Cost3(i) + Cost4(i). \qquad (4)$$

corresponding to the following conditions: (a) the sum of water quantities entering each tank should be similar to the total water supply provided by the *ETAP*; (b) the volume of water within each tank never exced its maximum capacity; (c) the volume of water within each tank should be greater than zero (thus it never empties completely); and (d) the water level in each tank should remain close to preset levels to allow the system to react appropriately in cases of very high or very low demand.

The procedure that customize the framework described in Section 2 is showed in Figure 5. The feasible points (i.e. the minima of expression (4)) are obtained, in lines (4) and (18) of Figure 5, by an hybrid genetic algorithm that includes a local search procedure applied after the mutation operator and is described in [6].

The maximum number of trajectories used (determined by R and r), is limited by the maximum computing time available, which is determined by the time required to apply the control action and to let the water levels gets their corresponding values. As can be well understood, as more trajectories are processed, the performance of the agent improves.

```
1)  For each i (i = 1,2,..,N)  {
2)     Obtain W'[i][j]  (j=1,2,.., n);
3)        Determine Cost[i];
4)        Compute R minima of Cost[i];
5)        For each k (k=1,..,R){
6)            Trajectory_generation(i,k,r);
7)        }
8)        Select S such that J'(S)=min{J'(1),...,J'(R)}
9)  }
10) Trajectory_generation(i,k,r) {
11)    Obtain W'[i+1][j]  (j=1,2,...,n)
12)    Generate Cost[i+1];
13)    Determine r minima of Cost[i+1];
14)    For each p (p=1,..,r) {
15)        For each s (s=i+2,..,N) {
16)            Obtain W'[s][j]  (j=1,2,..,n)
17)            Generate Cost[s];
18)            Determine a minima of Cost[s];
19)        }
20)    }
21)    Evaluate J'(k);
22) }
```

Fig. 5. Description of the procedure for tanks scheduling

In Figure 5, lines (2) and (11) indicate that during each stage i, the vector $W'(i+1)$ represents the water consumption prediction for the following interval $i+1$ is required. The prediction of these values is done by using a neural model obtained from an hybrid technique based on RBF networks [7] and classical orthogonal transformations such as SVD and QR [8,9] that allow not only the computation of the parameters of the RBFs, as in the majority of the neural procedures, but also the automatic determination of the number of inputs and RBFs that define the structure of the network.

As it is well known, an RBF network consists of a linear combination of gaussian activation functions with centres c_i and radius σ_i for the RBF i-th. The equation that defines the output, $W'(i+1)$, from an input vector that comprises lagged values from the series of water consumption. The learning process for the different parameters involved can be performed by means of relatively simple rules, such as the LMS rule or any similar rule, which enables us to use it *on-line* as successive data become available.

Certain matrix factorization techniques collectively known as *orthogonal techniques*, namely the *Singular Value Decomposition* (SVD) and the *QR factorization with column pivoting* (QR-cp), have traditionally been used [8] to solve *variable subset selection* problems in multiple regression and similar applications. In the context of neural networks based on multilayer perceptrons, the potential of these tools has been employed by means of an *off-line* procedure that starts with an overdimensioned network and then determines the relevant inputs and nodes by setting up the corresponding activation values in matrix form and calling the SVD and QR-cp procedures [9].

The prediction procedure we have developed constitutes a recent extension of that scheme, in which SVD and QR-cp are incorporated into the *on-line* learning procedure of an RBF network [10,11]. By creating overdimensioned matrices **A** of order $2D \times D$ with lagged values from the series, where D denotes the prediction window size (that is, the number of past samples considered for prediction of a future value), the SVD and QR-cp routines determine which of these regressors are fundamental for prediction, and the neural network takes as its inputs only those selected by this method. Moreover, SVD and QR-cp have been proposed, also in [11], as highly useful tools to perform a *pruning* operation on the neurons or RBFs that might lose predictive capacity during the learning process.

4 Experimental Results

To predict the water demand from each tank, the RBF was trained using a window of size equal to $D=8$; thus samples from the tank demand observed during the previous eight hours were used to predict the next hour's demand. The relative error, taken in absolute value, is at the 5% level and the NRMSE [10] for these 24 samples is approximately equal to 0.40. In Figure 6 the predicted demand values for one of the three tanks during a complete day are compared to their real counterparts; this gives some indication about the accuracy of the method.

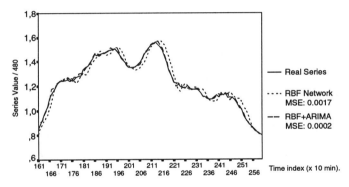

Fig. 6. Comparison of the real and the used predicted demands

Figure 7 shows the curves corresponding to the evolution of water levels in a given tank when the controller proposed here is used (controller) and when different manual controls (Man1, Man2, Man3) are used. It can be seen the solution obtained by the controller is feasible and presents a slower degree of variation, with no extreme maximum and minimum levels, than the curves corresponding to the manual control of the system.

The parallel procedure corresponding to the Neuro-Dynamic Programming module was implemented in a cluster of PCs (Pentium II, 333 MHz). Figures 8.a and 8.b compare the changes in the volumes of the considered (three) tanks, $X(t)$, and the

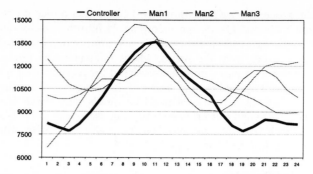

Fig. 7. Comparison between different manual control (Man1,Man2,Man3) and the controller obtained with the procedure described

amount of water that enters each tank during one hour, $Y(t)$, when using only one of the processors (Figure 8.a) and when using eight processors (Figure 8.b) to search the feasible solutions in a given amount of time (each processor processes a trajectory).

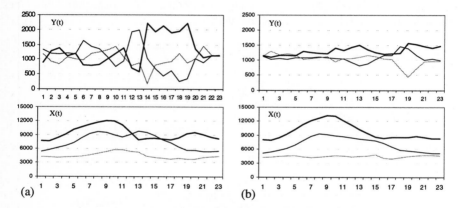

Fig. 8. a) Results for 1 trajectory, and b) Results for 8 trajectories

The solution determined when using several processors presents a smaller change in the volumes of water entering the tanks. To give an idea of the computing times and the speedups obtained by the parallel implementation of the procedure, the time required to compute the control actions for the next hour using eight trajectories is 42 min when using one processor, 14 min for four processors, and 8 min for eight processors. Thus, the speedup grows linearly with the number of processors [18].

5 Conclusions

A hybrid method based on Neural Networks, Monte Carlo simulation, and Evolutionary Computation has been presented and applied to solve dynamic programming problems. The method can be used to solve complex planning and control problems in which decisions are made in stages, and the states and control belong to a continuous space. The evolutionary computation procedure determines the feasible operation points at each stage, by using a predicted value of the inputs. The cost-to-go functions for each feasible state are approximated by Monte Carlo simulation, thus including the influence of the future cost in the selection of the control for the next stage. The hybrid method has been applied in the automatic control of water distribution network. As it has been demonstrated in the section 4 the obtained results are satisfactory.

Also, this procedure has been implemented in parallel in a cluster of computers. Where parallel processing allows us to calculate the greatest number of trajectories at a given time, thus improving the effectiveness of the control procedure. Moreover, it can be used to reduce the time required to process a given number of trajectories.

Acknowledgements. This work has been financed by project TIC2000-1314 (Spanish *Ministerio de Ciencia y Tecnología*); we are also grateful to the water supply company of the city of Granada, EMASAGRA Corp.

References

1. Bertsekas, D.P.: "Dynamic Programming and Optimal Control". *Athena Scientific*, Beltmont, Massachusetts, 1995.
2. Bellmann, R.; Dreyfus, S.:"Applied Dynamic Programming". *Princeton University Press*, Princeton, N.J., 1962.
3. Bertsekas, D.P.; Tsitsiklis, J.N.: "Neuro - Dynamic Programming". *Athena Scientific*, Belmont, Massachusetts, 1996.
4. Tesauro, G.; Galperin, G.R.: "On-line policy improvement using Monte-Carlo search". In *Advances in Neural Information Processing*, 9, pp.1068-1074, MIT Press, 1996.
5. Watkins, C.J.C.H.: "Learning from Delayed Rewards". PhD Thesis, Cambridge University, 1989.
6. Renders, J.M.; Flasse, S.P.: "Hybrid Methods using Genetic Algorithms for Global Optimization". *IEEE Trans. on Systems, Man, and Cybernetics (Part B)*, Vol.26, No.2, pp.243-258. Abril, 1996.
7. Moody, J., Darken, C.: "Fast Learning in Networks of Locally-Tuned Processing Units". *Neural Computation* 1 (2), pp. 281-294, 1989.
8. Golub, G.H., Van Loan, C.F.: "Matrix Computations". *Johns Hopkins University Press*, Baltimore, MD, 1989.
9. Kanjilal, P.P., Banerjee, D.N..: "On the Application of Orthogonal Transformation for the Design and Analysis of Feedforward Networks". *IEEE Transactions on Neural Networks*, 6 (5), pp. 1061-1070, 1995.
10. Salmerón, M.; Ortega, J.; Puntonet, C.G.: "On-line optimization of Radial Basis Function Networks with Orthogonal Techniques". In *Foundations and Tools for Neural Modeling* (Mira, J.; Sánchez-Andrés, J.V. Ed.), Lecture Notes in Computer Science, 1606, pp.467-477, Springer-Verlag, Berlin, 1999.
11. Salmerón, M.; Ortega, J.; Puntonet, C.G.; Prieto, A.:"Improved RAN sequential prediction using orthogonal techniques". Neurocomputing (in press).
12. Whitley, D.; Dominic, S.; Das, R.; Anderson, C.: "Genetic Reinforcement Learning for Neurocontrol Problems". *Machine Learning*, 13, pp.259-284, 1993.
13. Tesauro, G.J.; Galperin, G.R.:"On-line policy improvement using Monte-Carlo search". Advances in Neural Information Processing Systems, 9, MIT Press, pp.1069-1074, 1997.
14. Tsitsiklis, J.N.; Van Roy, B.:"An analysis of temporal-difference learning with function approximation". IEEE Trans. on Automatic Control, Vol.42, No.5, pp.674-690. May, 1997.
15. Crites, R.H.; Barto, A.G.:"Improving elevator performance using reinforcement learning". Advances in Neural Information Processing, 8, MIT Press, pp.1017-1023, 1996.
16. Zhang, W.; Dietterich, T.G.:"High-performance job-shop scheduling with a time-delay TD(lambda) network". Advances in Neural Information Processing, 8, MIT Press, pp.1024-1030, 1996.
17. Sutton, R.; Barto, A.G.:"Reinforcement Learning: An Introduction". MIT Press, Cambridge, MA, 1998.
18. Damas, M.; Salmerón, M; Fernández, J.; Ortega, J.; Olivares, G.: "Parallel Control of Water Supply Networks in Clusters of Computers". 4th World Multiconference on Systemics, Cybernetics and Informatics (SCI'2000), pp.538-543, ISBN: 980-07-6693-6. Orlando (USA), 2000.

Classification of Disturbances in Electrical Signals Using Neural Networks

Carlos León[1], Antonio López[1], Juan C. Montaño[2], and Íñigo Monedero[1]

[1]Departamento de Tecnología Electrónica
Escuela Universitaria Politécnica
Universidad de Sevilla
C/ Virgen de Africa 7
41011 Sevilla (Spain)
cleon@cica.es
phone: 34-954552836 fax: 34-954552833

[2]IRNAS (Spanish Research Council-CSIC)
Campus Universitario Reina Mercedes
P.O. Box 1052
41080 Sevilla (Spain)
montaño@irnase.csic.es
phone: 34-954624711

Abstract. This paper describes a currently project accomplished by the authors in the area of Power Quality (PQ) using artificial neural networks (ANN). The efforts are oriented to obtain a product (Power disturbances monitor for three-phase systems) that permits a real time detection, automatic classification, and record process of impulsive or oscillatory voltage transients, long term disturbances, and waveform distortions in electrical three-phase AC signals. To classify the electrical disturbances, we consider using a fully connected feedforward ANN with a backpropagation learning method based on Generalized Delta Rule. In order to select the best alternative more than 200 network architectures were tested. Long-term disturbances, like swells or long-duration interruptions, have been detected using a method based on the test of the RMS value of the signal. Short-term disturbances, like sags, are detected by sampling a cycle of the electrical signal, and waveform distortions are detected using the main harmonics of the signal. To train the ANN we have developed a three-phase virtual generator of electrical disturbances. In order to compress the ANN input data we use the Wavelet Transform.

1. Introduction

The term *power quality* is applied to a wide variety of electromagnetic phenomena in power systems. In recent years, the use of electronic equipment has increased the interest in power quality and several terms have been developed in order to describe power systems disturbances. There can be completely different definitions for power quality, depending on the point of view of utilities, manufacturer of load equipment, or customer. For instance, utilities may define power quality as reliability and show statistics demonstrating that the system is almost 100 percent reliable. The manufacturer of load equipment may define quality power as those characteristics of the power supply that enable the equipment to work properly. However, power quality is ultimately a customer-driven issue and the customer's point of reference must take precedence. Therefore, a power quality problem can be defined as *any power problem manifested in voltage, current, or frequency deviations* that results in failure or malfunctioning of customer equipment.

The subject of this paper is the analysis of the voltage quality for sensing any deviation of the voltage waveform out of certain limits. Alternating current power systems are designed to operate at a sinusoidal voltage of a given frequency (typically 50 or 60 Hz) and magnitude. Any significant deviation in the magnitude, frequency, or purity of waveform is a potential power quality problem. These deviations must be fast detected for further actions or storing for classification and statistical studies.

Software procedures have been developed, applying the FFT for analyzing these disturbances [1]; however, due to the great amount of stored data and the time of required processing, such procedure is slow and not very efficient. To minimize storage space, we need to represent signals by as few bits as possible. This is the data compression problem, and has been studied for several decades especially for image compression [2]. Likewise, the noise reduction has been dealt with from a statistical point of view [3-5].

Continuous and discrete wavelet transform (DWT) have been used in analysis of non-stationary signals and several papers [6-8] have proposed the use of wavelets for the analysis of power systems. They are able to remove noise and achieve high compression ratios because of the "concentrating" ability of the wavelet transform. If a signal has its energy concentrated in a small number of wavelet coefficients, this signal will be relatively large compared to any other signal or noise that has its energy spread over a large number of coefficients. This means that thresholding the wavelet transform will remove the low amplitude and undesired coefficients in the wavelet domain and reconstructs the signal with little loss of information. Wavelet thresholding has important applications in statistic. Donoho and Johnstone [9] propose to start with a wavelet decomposition of the data set, thresholding later the coefficients, and then use the wavelet reconstruction as an estimate function. This is the model of the fast and efficient algorithm for data-compression that we consider in this paper.

2. Power Quality Measurement for Three-Phase Systems

The power quality measurement system or power quality monitor (PQM) described in this paper has been specifically developed for the analysis of three-phase line voltages. It stores data by sampling the three phase-to-neutral voltages simultaneously. Then, an efficient measurement algorithm, based on the power-frequency data estimation obtained from three equidistant samples of a sinusoidal signal, calculates the instantaneous frequency for the synchronization of the voltage and sampling periods. It allows the estimated power frequency to be defined from the pondered mean of the estimation performed in each phase R-S-T.

Thus, the detected R-phase of the voltage signal can be processed to construct a perfect three-phase system for being used as reference. The PQM detects individual events, at the time of occurrence, by comparing the monitored signals to the reference three-phase voltages. When a threshold parameter is exceeded, a disturbance event is detected. Threshold parameters are adjustable over a specified range to accommodate different monitoring circumstances.

- *Instantaneous voltage*. Instantaneous voltage amplitude measured with respect to the power-frequency sine wave. Short-duration voltage variations such as impulsive and oscillatory transients, waveform distortion and voltage fluctuations can be detected. The measurement interval for short-duration voltage variation is from 500ns to 50ms.
- *AC rms voltage*. With rms sensing, according to the above strategy of synchronization, the measurement interval is an integral number of cycles of the fundamental power frequency. Harmonic content, swell, long-term interruptions and voltage unbalance events can be detected. The measurement interval is from 1 cycle to 1 min.

Having detected the disturbance event, the digitized samples are stored in memory. As subsequent processing, measurement, and reporting of the disturbance event will be based entirely upon the stored samples, the PQM retain two-cycles data from before and after the detection point to accurately reconstruct the entire disturbance event.

Furthermore, the digitized data is formatted to provide a compressed and detailed graphic representation of the disturbance waveform. Therefore, the PQM includes two algorithms: one for calculating the harmonic spectrum of the incoming voltage data, using the discrete fourier transform (DFT), and other algorithm for filtering and compressing the collected disturbance data using the discrete wavelet transform (DWT). The two algorithms are applied concurrently.

The conventional DFT is applied to the original digitized samples, $f(n)$, getting the set of fourier coefficients and the first 50 harmonics in phasor form. The second algorithm consists of the following steps.

A. *Wavelet decomposition*. $f(n)$ samples are transformed in order to generate a set of signal coefficients. The DWT used (Daubechies family Db4) is applied to $f(n)$, getting signals $a_j(n)$ and $d_j(n)$, where j is the index level. Family Db4 is particularly appropriate for detecting disturbances of high frequency (transients), as it is more localized in time than other members of the same family are.

B. *Threshold wavelet estimators and reconstructed signal* A process of comparison between the input signal and the reconstructed signal $a_j(n)$ begins. This process stops when the difference between the two signals is less than the set threshold. One of the goals of the present work is to reach a high compression ratio. This expresses the minimum amount of data necessary for recovering the original signal.

In a first phase, the algorithm of coefficient filtering performs a comparison of signals a_j with the original signal $f(n)$ to obtain the error signal ξ_j. In a second phase, the absolute maximum value of ξ_j is compared with a fixed threshold λ. If the magnitude of the error signal is less than λ, then the signal resulting (reconstructed signal) is the new reconstructed signal $f(n)^*$. In both phases, the *optimal relative error* between the original $f(n)$ and the reconstructed $f(n)^*$ signal is used for measuring the quality of the estimator.

These recording mechanisms make the PQM most suitable for automatic classifying of disturbance waveforms and analyzing complex power-quality problems when properly applied by the expert user.

2.1. Three-Phase Arbitrary-Function Generator

We are developing a three-phase arbitrary-function generator (Fig. 1) that simulates all kinds of electrical disturbances in line voltages such as oscillatory transients, waveform distortion, voltage fluctuations, sag, swell, interruptions and voltage unbalance. Generated signals simulate those obtained at low-voltage level by line voltage transducers.

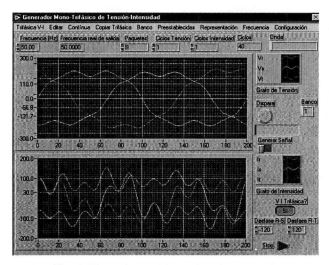

Fig. 1. Three phase arbitrary-function generator

A wide range of parameter settings and combination possibilities make the instrument an appropriate tool for training artificial neural networks in the classification process of electrical disturbances. Local operation via PC-control, using the LabView program running under Windows, makes the unit user-friendly during test parameter set-up.

The instrument enables tests to be performed in accordance with EN-50160 and the other common standards of the European Union (EU). Tests can be pre-programmed and stored for being recalled at any arbitrary time at the touch of a button. Continuously varying values for three-phase voltages and dropout time can be defined to occur autonomously. These tests can also be programmed to run in an endless loop.

At present, three arbitrary functions are generated after completion of the parameter settings using the initial program. It superimposes three sinusoidal signals with combined harmonic content (up to the 50th harmonic). Power frequency variations, total harmonic distortion and voltage imbalance of the three-phase voltage signals can be initially selected too. Fig. 2 shows the screen for parameter settings in case of the voltage waveforms of Fig. 1.

Unlike a natural environment, however, where disturbance events are unpredictable, the instrument allows the user to develop controlled, repeatable simulations. Results from simulation testing can be used to validate in real time a complete system of power disturbance analysis, including data capture, recording, classification, and reporting the results. These waveforms can be used to train an ANN too. The system can also generate three analogue sinusoidal signals to be utilized as AC threshold or, for example, as reference in the disturbance classification training processes of the ANN.

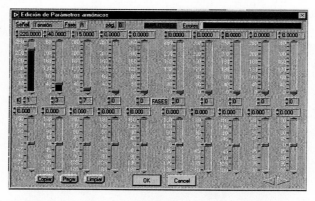

Fig. 2. Parameter settings showing the harmonic content of the voltage waveform

3. PQ Disturbances Classification Using Neural Networks

In power engineering, the analysis of PQ problems is not focused only to the detection of electrical disturbances. Far more important is the ability to classify various types of disturbances as well. As an alternative to classify the PQ disturbances, we consider using an artificial neural network (ANN). This technology has been widely used in Power Systems management [10-14]. After to evaluate several alternatives (ART2, LVQ, Counterpropagation, etc), authors selected a fully connected feedforward ANN with a backpropagation learning method [15,16] based on Generalized Delta Rule. This type of network can resolve the function approximation problem (to find the unknown function that relates a set of training patterns) [17,18]. To simulate ANNs we used simultaneously a set of five Intel PIII 450Mhz computers.

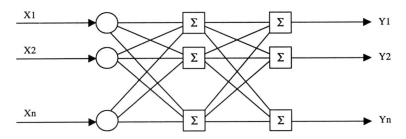

Fig. 3. Feedfordward Neural Network

The feedforward neural network type has an input layer, one or more hidden layers, and an output layer, as shown in Fig. 3.

The neurons in different layers are connected by means of weights. When a training pattern p is presented, the input for a neuron j is the sum of the weighted input signal i_{pi} :

$$net_{pj} = \sum_i w_{ij} i_{pi} \qquad (1)$$

where w_{ji} is the weight of ith input at j^{th} node. The output o_{pj} from the neuron is given by

$$o_{pj} = f_j \times net_{pj} = \frac{1}{1 + e^{-net_{pj}}} \qquad (2)$$

when a sigmoid activation function is used.

Training process is carried out using the backpropagation algorithm. This learning method updates the interconnection weights using the Generalized Delta Rule. In this method, the error at a given output node o_{pj}, when a training pattern p is presented, is:

$$\delta_{pj} = (t_{pj} - o_{pj}) f_j' net_{pj} \qquad (3)$$

where t_{pj} = target value for j^{th} output node produced by input pattern p; and f'' = first derivative of activation function used by node j. The error at a given non-output node j, when training pattern p is presented, is:

$$\delta_{pj} = f' net_{pj} \sum_k (\delta_{pk} w_{kj}) \qquad (4)$$

in which k = number of neurons in next layer; and δ_{pk} = already computed for the kth neuron in the next layer. Weights may be modified after each training pattern is presented. So, the weight change applied to weight w_{ji}, after pattern p has been presented, is:

$$\Delta_p w_{ji} = \eta(\delta_{pj} o_{pi}) + \alpha \Delta_{p-1} w_{ji} \tag{5}$$

where η = learning rate; and α = momentum factor. Effectiveness and convergence of the learning algorithm depend on the value of and. If the selected learning rate is too high, the network tends to oscillate avoiding the learning process of the correct mapping from the input to the target. If the value of is very small, the network can take a very long time to learn. Momentum factor is used to speed network training. Proper selection of a momentum factor can prevent network from oscillating.

Mean Square Error (MSE) is the measure of how well the network output matches target output. MSE is calculated at the end of each training cycle. MSE is summed over each output node for each pattern. The MSE at the end of a given cycle is:

$$MSE = \frac{1}{NT} \sum_{p=1}^{N} \sum_{j=1}^{T} (t_{pj} - o_{pj})^2 \tag{6}$$

in which N = number of patterns; and T = number of outputs nodes.

There are not specific rules to select an optimal ANN architecture [19]. In the PQ problem the number of input neurons is 113, clustering in three groups:

- 48 time inputs, sampling an electrical signal period of 20 ms, in order to detect impulsive or oscillatory voltage-transients.
- 50 RMS inputs, one for each electrical signal cycle during 2 second, in order to detect long-duration disturbances like overvoltages or undervoltages.
- 15 main harmonics of the signal inputs, to detect waveform distortions.

The number of output neurons is 8: one for a global evaluation of the voltage quality (PQ) and 7 for each disturbance: power frequency variations, transients, sags, short duration interruptions, swells, long duration interruptions and waveform distortions.

To select the other variables related with the ANN design and training is usually very complex. For example, the number of hidden neurons and the number of hidden layers to use are difficult to be determined. If the architecture is too small, the network may not have enough degrees of freedom to learn the process correctly. On the other hand, if the network is too large, the solution may not converge during training or the network may overlearn the data.

In order to select the best alternative, we have developed a three phase heuristic method. The first phase began testing the transfer functions for each neuron layer (linear, sigmoid or tanh), weight initialization (randomly or used specified values), learning rule (Generalized or Cumulative Delta Rule), and the order of training patterns presentation. A second phase sets the number of hidden neurons (10 to 140), the number of hidden layers (3 to 5), the learning rate (0.1 to 0.5) and momentum factor (0.1 to 0.5). Other objective of this design phase is to establish the utility of using partially connected networks (each ANN output only depends on a part of the inputs). In this second phase 200 network architectures were tested. The third phase objective is to test the addition of Gaussian noise to input training pattern and little variations on hidden neurons number.

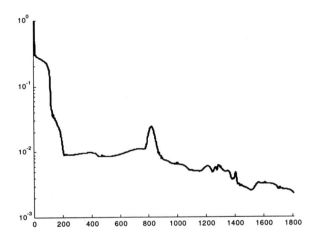

Fig. 4. Variation of MSE during the training process (1800 epochs)

Another problem is to select the training patterns. In the PQ problem we generate 1100 patterns by using the three-phase generator of electrical disturbances described above. Authors selected 900 patterns to use during training, 110 validation patterns to avoid that the ANN overlearns the training patterns, and 90 testing patterns used for evaluate the performances of the network. The training, validation and testing patterns range values are between -1 and 1 (inputs are scaled by the maximum value of the patterns). Authors employed a total of 5,000 training cycles. Fig. 4 shows the reduction of MSE during the training process for one architecture. The results obtained after training the ANN for each possibility with different architectures are shown in Table 1. The best test result is a MSE of 0.0554 (94.5% of correct PQ disturbances classification).

Table 1. Results of the ANN selection heuristic process

Phase	Parameter	Best results
Ph.1	Transfer function per layer	Linear (first layer), Sigmoid (other layers)
	Weight initialization	Randomly between -1 and 1
	Learning rule	Generalized Delta Rule
	Training patterns presentation	Randomly
Ph.2	Architecture	113 - 100 - 100 - 9 partially connected
	Learning rate	0.5
	Momentum factor	0.5
Ph.3	Gaussian input noise	Irrelevant
	Variation on hidden neurons	113 - 100 - 100 - 9 partially connected

If we analyze the error obtained for each output (Table 2) we can see that the worst figure is the output number 8 (waveforms distortions problems), but even in this case, the test MSE is minor than 8%. The only other output MSE upper 5% is the MSE for output 1 (that offers a global evaluation of the signal PQ). As conclusion, the errors obtained are satisfactory to detect and classify PQ disturbances. The next phase in our project is to integrate the ANN into the power disturbance monitor for three phase systems describes above.

Table 2. MSE for each ANN output.

ANN Output	MSE value
Unit 1: Global voltage quality (PQ)	0.051
Unit 2: Power frequency variations	0.009
Unit 3: Transients	0.026
Unit 4: Sags	0.005
Unit 5: Short duration interruption	0.001
Unit 6: Swells	0.032
Unit 7: Long duration interruption	0.001
Unit 8: Waveform distortion	0.075

4. Conclusions

This paper describes the developing of a power disturbance monitor for three-phase systems. This system classifies and stores short-term and long term disturbances, and waveform distortions in electrical three-phase AC signals, using a fully connected feedforward neural network with a backpropagation learning method. Wavelet transform is used for compress data. An arbitrary function generator has been developed for training the ANN. Preliminary tests show that the system obtains good results in the classification of electrical PQ incidences. The next project phase is to integrate the ANN into the power disturbance monitor.

References

[1] J.C. Montaño, M. Castilla, J. Gutierrez, A.López, Sistema de medida y vigilancia de la calidad del suministro de potencia eléctrica, 5as Jornadas Hispano-Lusas de Ingeniería Eléctrica, Salamanca, Spain, pp. 1643-1650, July 1997.

[2] B.V.Brower, Low-bit rate image compression evaluation, SPIE, Orlando, FL, April 1994. SPIE

[3] H. Guo, M. Lang, J. E. Odegard and C. S, Burrus. "Nonlinear Shrinkage of Undecimated DWT for noise reduction and data compression". Proceedings International Conference on Digital Signal Processing. June 1995. Limasol, Cyprus.

[4] D. L. Donoho."De-noising by Soft-thresholding". IEEE Trans. Inform. Theory, 41(3), pp. 613-627. May 1995.

[5] D. L. Donoho, I. M. Johnstone. Ideal denoising in an ortonormal basis chosen from a library of bases, .Dept. of Statistics, Stanford Univ.,1994.

[6] D. C. Robertson, O. I. Camps, J. S. Mayer, W. B. Gish. "Wavelets and electromagnetic power system transients". IEEE Trans. Power Delivery, Vol.11, N°2.April 1996.

[7] D. Borrás, M. Castilla, J.C. Montaño."Wavelet and neural structure: a new tool for diagnostic of power system disturbances", .IEEE SDEMPED'99, pp. 375-380. Sept. 1999. Gijón (Spain).

[8] M. Castilla, D. Borrás, JC. Montaño."Wavelet and Neural Network Structure for Analyzing and Classifying Power System Disturbances", EPE 99, Paper 488. ISDN:90-75815-04-2, Sept.1999,Lausanne (Switzerland)

[9] D. L. Donoho, I. M. Jonhstone, Ideal Spatial Adaptation by Wavelet Shrinkage, Stanford Statistics Dept. Report TR-400, July 1992.

[10] C. Liu, D. A. Pierce y H. Song, Intelligent System applications to Power Systems, IEEE Computer Application in Power, Oct. 1997.

[11] T. Dillon, Neural Net Applications in Power System, CRL Publishing, 1996.

[12] I. Dabbaghchi, R. D. Christie, G. W. Rosenwald, y C. Liu, AI Applications Areas in Power System, IEEE Expert, Jan-Feb. 1997.

[13] M. A. El-Sharkawi, Neural Network´s Power, IEEE Potentials, Jan. 1997.

[14] M. Kezunovic y I. Rikalo, Detect and Classify Faults Using Neural Nets, IEEE Computer Applications in power, Oct. 1996.

[15] J. J. Buckley y T. Feuring, Fuzzy and Neural: Interactions and Applications, Physica-Verlag, 1999.

[16] J. A. Freeman y D. M. Skapura, Redes Neuronales: Algoritmos, aplicaciones y técnicas de programación, Addison-Wesley, 1993.

[17] D. R. Tveter, The Pattern Recognition Basis of Artificial Intelligence, IEEE Computer Society Press, 1998.

[18] S. Haykin, Neural Networks. A Comprehensive Foundation, Prentice Hall, 1999.

[19] M. H. Choueiki, C. A. Mount-Campbell and S. C. Ahalt, Building a quasi optimal neural network to solve the short-term load forecasting problem, IEEE Trans. on Power Systems, 12 (4), 1432-1439, 1997.

Neural Classification and « Traditional » Data Analysis: An Application to Households' Living Conditions

Sophie Ponthieux[1] and Marie Cottrell[2]

[1] INSEE, Division "Conditions de vie des ménages",
sophie.ponthieux@insee.fr
[2] SAMOS-MATISSE, CNRS UMR 8595, Université Paris 1
cottrell@univ-paris1.fr

Abstract. The description, classification and « measurement » of living conditions present many difficulties. A very important one comes from the qualitative nature of the data, and the large number of characteristics that may be taken into account. For this reason, it is difficult to obtain a description that could give an overall view of the arrangements between the modalities, and be usable to breakdown the observations into a reasonable number of classes. In this paper, we propose several examples of the use of neural network techniques, precisely the Kohonen algorithm, to classify a population of households according to their characteristics in terms of living conditions.

1 Introduction

Since the 1970s in United-Kingdom, more recently in France, poverty is analyzed both in terms of income and in terms of living conditions, with a multi-dimensional approach. Living conditions include a great number of domains. Dickes (1994) lists ten of them: dwelling, durable, food, clothing, financial resources, health, social relations, leisure, education and work. Not all the existing studies include this complete set of domains. The choice of including or not one of those domains may be based on two arguments: in the first one, the main hypothesis is that the subjects are rationales in their behavior, which leads to select only the domains where privations are assumed to decrease or disappear when the financial resources are increased (Mack & Lansley, 1984); the second one is based on the notion of « standard » (Townsend, 1989), and leads to consider any domain as soon as all the subjects are, at least potentially, involved, i.e. following Dickes (1994, p.184), « all the households, whatever their composition or their situation in the life cycle ».

The main difficulty is that we have to deal with a great quantity of information that is mainly qualitative. We propose here to focus on two main objectives, taking into account only a relatively small number of domains (accommodation – in two parts: convenience, problems -, environment, durable and deprivations):

- obtain a good description of the characteristics: how they are combined, *i.e.* what are the most frequent associations between modalities, and in turn how the different sets of modalities are organized together,

- obtain an operational grouping of the observations when using only their characteristics in the domains of the living conditions.

We compare two techniques: « traditional » data analysis (multiple correspondence analysis and clustering), and neural classification (Kohonen algorithm). The methods and data are shortly presented in *section 2*. Then we present in *section 3* two descriptions of the arrangement between the characteristics, the first one resulting from a Multiple Correspondence Analysis (MCA), the second one obtained with the Kohonen algorithm[1]. In *section 4* we compare different classifications of the observations: first a simple « score », then several classifications, obtained successively from a hierarchical clustering, then from Kohonen classification.

Due to the small space allowed for the papers, most of the graphs and detailed tables of results are gathered in an Appendix available on request by E-mail.

2 Method and Data

We suppose that the reader is familiar with the Kohonen algorithm. See for example Kohonen (1984, 1993, 1995), Kaski (1997), Cottrell, Rousset (1997) for an introduction to the algorithm and to its applications to data analysis. For our work, the main property of the Kohonen algorithm is the so-called topology conservation property. After convergence, in the resulting classification, similar data are grouped into the same class or into neighbor classes. This feature allows to represent the proximity between data, as in a projection, along the Kohonen map. As a further treatment, the Kohonen classes can be clustered into a reduced number of macro-classes (which only contain neighbor Kohonen classes) by using a classical hierarchical classification.

The data source used for this paper is the French part of the European Community Households Panel, here in its third wave (year 1996). It provides detailed information, both at the individual and the household level, about incomes (which allows to define an indicator of monetary poverty), and living conditions (material living conditions: dwelling, environment, durable goods, deprivations, but also financial living conditions: whether the monetary resources allow to live from "very comfortably" to "with great difficulty").

In what follows, the observations (households) are described according to their answers to questions covering the different domains of living conditions (*cf. infra*) and by a set of general characteristics: type of household, average age of the adults (persons aged 17 years and over), number of children under 17 years, type and location of the dwelling, financial living conditions, score for the material living conditions, indicator of monetary poverty. Only the observations with no missing variable for all theses descriptors are kept for the analysis, that is 6458 households. Only the variables describing the material living conditions are used in the classifications; we have kept all the information available[2]. The other variables are

---

[1] The SAS programs used for Kohonen classifications are due to Patrick LETREMY (SAMOS/MATISSE, Université de Paris 1)
2 This is rather different from what Dickes (1994) recommends ; according to this author, the choice of the

used to compare the different classifications. Finally, living conditions are described by 10 items about dwelling (5 about convenience and 5 about problems), 4 items about environmental topics, 6 items for the durable and 6 items about deprivations, a total of 26 dummy variables[3], that is to say 52 modalities.

This information is not always used under the same form: when we classify the characteristics, the inputs are a response table; when we classify the observations, the input is either the partial scores (score by domain) or the coordinates (obtained after a MCA, Multiple Correspondence Analysis) of the observations.

3. Description of the Characteristics: Classifications of the Modalities

The modalities ended by 0 indicate the absence of problem (the households has a bath or a shower, has a separate kitchen, and so on). The suffix 1 corresponds to a problem (nor bath neither shower, no separate kitchen, and so on).

3.1. Results from a Multiple Correspondence Analysis (MCA)

In all the representations (graph 1), most of the « positive » modalities appear in a very tight location. On the "negative" side, a group of modalities appears systematically isolated. It corresponds to what could be called « absence of a minimum set of conveniences in the dwelling » and combines no running hot water, no bath or shower, no indoor toilet. The other modalities are organized in several groups, often associating problems in the dwelling, bad quality of the environment, and scarcity of durable goods. It is interesting to notice that the « modern » durable (micro-wave, VCR) and the « traditional » ones (telephone, TV) appear to form two different sub-groups. It seems also that some deprivations (holidays and furniture) may have a particular status.

Finally, the MCA suggests more or less 4 groups of modalities:
- a first one grouping all the positive modalities (all the conveniences, no problem in the dwelling or in the environment, all the durable, no deprivation) with maybe one exception in the case of holidays and furniture;
- a second one where are associated absence of problems in the dwelling, bad quality of the environment, no holidays, and no possibility to replace worn-out furniture;
- and a third – possibly also a fourth group -, around the « absence of a minimum set of conveniences in the dwelling », added to no telephone and no TV set, combined (when representing the axis 3 and 4) with food deprivation (cannot buy meat/chicken/fish every second day if wanted).

---

items must be based on a consensus about their necessity. This criteria is very often reduced to their frequency... because there is not a lot of information allowing to control for a consensus.
3 In the case of the durable, the answer distinguishes between "not having by choice" and "not having because cannot afford". Here, we have grouped « not having by choice » and « having », under the assumption that in this case there is no deprivation.

3.2 Results from a Kohonen Classification

The modalities are classified first on a 10 x 10 grid (graph 2), then along a five classes string (graph 3). At first glance, it appears that the « positive » modalities are grouped at the top and on the right of the grid, and the « negative » ones at the bottom and on the left. The class in the bottom left cell corresponds to the « absence of a minimum set of conveniences in the dwelling » already mentioned with the MCA. It is here interesting to notice that the "very" negative modalities have no immediate neighbors, and are located on the grid quite on the exact opposite from their "positive" appearance. If we group the classes according to their closeness using a hierarchical classification, we obtain 3 sets of modalities, which are consistent with those resulting from the MCA. The classification along a string (one-dimensional Kohonen map) gives a very synthetic view of the associations between the modalities, and confirms the particular status of the holidays and replacement of furniture. A study of the profiles suggests a neat gradation of the negative modalities, which is an interesting result in that it could be usable for a "weighting" of the items.

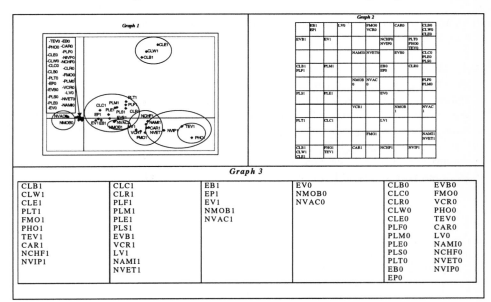

4. « Measuring » Living Conditions: Classifications of the Observations[4]

The simplest way to classify the observations according to their living conditions

---

[4] Table 1 provides a summary of the results : it gives the proportion of the super-classes obtained with each classification, and the concentration indicators (for an interpretation in terms of over-representation) of the different characteristics. We have added some general characteristics (not used for the classifications) in order to have a broader view of the population in the classes (type of dwelling, location, type of household).

consists in calculating a score of « bad points » for the whole set of items. The problem is there to determine the threshold to be used. This question is discussed in Lollivier & Verger (1997); they propose to use the rate of monetary poverty to define, by comparing this rate and the cumulative frequencies of the score, a threshold for living conditions "poverty". One problem with this method is that it does not allow to compare across the time the evolutions of these two measures of poverty, because, by construction, the two rates will be very close. Another problem is that it "cuts" the population into one group having "good" living conditions and one group having "bad" living conditions only on the criteria of one "bad point", whatever the item considered; therefore, even though it would be difficult to weight the items, it is clear that they are not all at the same level of "seriousness". So we have tried in what follows to obtain classes defined using the qualitative dimension of the information and independently from any exogenous threshold. The results are summarized in Table 1.

4.1 Classifications Using the Partial Scores

The information used in this part is the scores obtained by each observation for each of the 5 domains.

4.1.1 Hierarchical Classification
Ten clusters are then grouped in 3 super-classes; it gives about 15 % of the observations as living in the least favorable conditions (class 1), about 25 % in less unfavorable conditions (class 2), and 60 % in better living conditions (class 3). This classification appears to be consistent with the poverty rates (in monetary terms) in the three classes: about 30 % in class 1, 14 % in class 2, and only 1,6 % in class 3 (to be compared with 10,7 % for the whole sample). The three classes are differentiated also in terms of financial living conditions, neatly better when going from class 1 to class 3.

4.1.2 Kohonen Classification
Here the Kohonen algorithm is utilized to obtain 10 classes, in turn grouped in 3 super-classes. Given the type of input (scores by domain), we have chosen to classify the observations along a string, in order to illustrate the *continuum*. The "progression" along the string appears neatly from best to worse living conditions. The proportions obtained in the three super-classes is slightly different from that obtained with the hierarchical classification: about 11 % of the observations appear as living in the least favorable conditions (class 3), again 11 % in less unfavorable conditions (class 2), and 78 % in better living conditions (class 1). In terms of monetary poverty, the classes are less different than in the previous classification. An interesting result is that some differences appear between classes 2 and 3: the scores in the domains of dwelling problems and bad quality of the environment are higher in class 2, while class 3 is characterized by higher scores in the domains of dwelling convenience, durable and deprivations.

4.2 Kohonen Classifications Using the Observations Coordinates (After a Multiple Correspondence Analysis)

The inputs are now the coordinates of the observations, resulting from a « traditional » MCA. So it corresponds to a transformation of the responses into quantitative values, (we use all the qualitative dimensions of the initial information). We use the Kohonen algorithm to classify the observations first on a grid, then along a string. Each one of these classifications is then grouped in 3 final super-classes.

4.2.1 Grid
One Kohonen classification is used here to build a 8 X 8 grid. The 64 classes obtained have then been grouped into 10, then 3 super-classes. The result is rather similar to that obtained in the classification of the modalities. The grouping into 10 and into 3 classes illustrates rather well the property of neighborhood that is one interest of this type of classification.

The final 3 super-classes give a distribution that is rather different from that obtained when using the partial scores: 15,2% (class C), 14,2% (class B) and 70,6% (class A). If class A appears clearly as the group having the « best » living conditions, the situations are not neatly different at first glance between classes B and C. This could be an interesting effect of taking into account the information at a very detailed level. Comparing classes B and C is particularly interesting: as for the partial scores, these two classes differentiate mainly in the domains of dwelling (in both terms of conveniences and problems) with higher scores in class C, while the scores are higher for durable and deprivations in class B. In terms of total score, poverty and financial living conditions, class B appears to be more disadvantaged than C.

So what we obtain here is a classification that suggests that the situations may be different in terms of living standards and in terms of living conditions, the difference coming mainly from the characteristics of the dwelling.

4.2.2 String (One Dimensional Kohonen Map)
Here, the observations are classified in 10 Kohonen classes along a string, then grouped in 3 super-classes.

Super-classes 1 to 3 represent respectively 70%, 14% and 16% of the observations; this is the classification that gives the higher proportion of unfavorable living conditions (if we add classes 2 and 3). It is also the classification that differentiates the least the classes in terms of monetary income, even though the poverty rate increases from class 1 to 3 (this indicating a greater heterogeneity of the incomes in class 3 which combines a higher income and a higher poverty rate than class 2). The global score increases from class 1 to 3, and between classes 2 and 3, the same differences appear for the domain of dwelling.

4.3. Comparison of the Classifications

At first, and it is not very surprising, the proportion of households that can be said to have unfavorable living conditions appear to be rather variable. Secondly, some neat differences appear according to the form of the information in input:

- classifications using as inputs the partial scores show generally a neat gradation in living conditions, ranking –by construction- from « worse » to « better »; but the Kohonen classification shows some differences between the domains that the hierarchical classification does not recreate.

- classifications based upon the observations coordinates show most interesting results: especially, the absence of minimum convenience in the dwelling distinguishes almost always one class, which is not systematically the most « underprivileged » in terms of global living conditions, poverty rate and financial living conditions. These classifications, because they use the qualitative dimension of the information, show some associations or specificities that do not appear if we « measure » living conditions with a « score ». It is particularly interesting in the case of dwelling convenience; an explanation could be that, given the high proportion of households having a minimum set of conveniences in the dwelling, the situation of not having this minimum set discriminates in itself a small proportion of households even though it were their only "negative" characteristic.

As for the general characteristics of the households (type of dwelling, location, type of household), the distributions appear rather consistent over all the classifications. Generally, unfavorable living conditions are more often than on average observed among households living in large structures, and among persons living alone and lone parents.

References

Cottrell, M. & Ibbou, S. (1995) : Multiple correspondence analysis of a crosstabulation matrix using the Kohonen algorithm, *Proc. ESANN'95*, M.Verleysen Ed., Editions D Facto, Bruxelles, 27-32.
Cottrell, M., Fort, J.C. & Pagès, G. (1998) : Theoretical aspects of the SOM Algorithm, *Neurocomputing*, 21, p. 119-138.
Cottrell, M. & Rousset, P. (1997) : The Kohonen algorithm : a powerful tool for analysing and representing multidimensional quantitative et qualitative data, *Proc. IWANN'97*, Lanzarote.
Dickes, P (1994), *Ressources financières, bien-être subjectif et conditions d'existence*, in F.Bouchayer (ed.) Trajectoires sociales et inégalités, Eres.
Kaski, S. (1997) : Data Exploration Using Self-Organizing Maps, *Acta Polytechnica Scandinavia*, 82.
Kohonen, T. (1984, 1993) : *Self-organization and Associative Memory*, 3°ed., Springer.
Kohonen, T. (1995) : *Self-Organizing Maps*, Springer Series in Information Sciences Vol 30, Springer.
Lollivier, S & Verger, D (1997), *Pauvreté d'existence, monétaire ou subjective sont distinctes*, Economie & statistique n°308/309/310 « Mesurer la pauvreté aujourd'hui ».
Mack, J & Lansley, S (1984), *Poor Britain*, Allen & Unwin.
Mayer, S.E & Jencks, C (1989), *Poverty and the distribution of material hardship*, Journal of Human Resources vol.24.
Townsend, P (1989) : *Deprivation*, Journal of Social Policy, vol.16.

Table 1 – Summary of the results

		Scores			Classes based on partial scores									Classes based on coordinates					
					Hierarchical				Kohonen			Grid+3 super-classes			String+3 super-classes			
		0	1		1	2	3	1	2	3	A	B	C	1	2	3
Distribution (%)		89.2	10.2		60.7	24.6	14.8	77.6	10.9	11.5	70.6	14.2	15.2	70.0	13.8	16.1
		Concentration indicators (proportion –or mean– for a given class / proportion –or mean – for the whole sample)														
Score for the material living conditions	Total (all domains)	0.7	3.2		0.5	1.2	2.5	0.5	1.8	3.9	0.3	1.0	4.4	0.7	2.3	1.1
	Dwelling, convenience	0.8	2.9		0.8	0.6	2.5	0.4	4.3	2.0	0.7	1.5	1.7	0.8	1.2	1.5
	Dwelling, problems	0.9	2.2		0.9	0.8	1.7	0.8	2.2	1.2	0.9	1.4	1.1	0.6	1.1	2.6
	Environment	0.7	3.7		0.2	1.6	3.4	0.7	1.2	2.7	0.6	2.9	1.2	0.7	1.3	1.9
	Durables	0.7	3.2		0.1	1.7	3.4	0.4	0.8	5.6	0.4	3.4	1.7	0.8	1.3	1.8
	Deprivations	0.7	3.2		0.4	1.2	2.9	0.7	1.9	2.4	0.6	2.2	1.5	0.7	1.3	1.9
Monthly income per c.u.(a)		1.0	0.6		1.2	0.8	0.6	1.1	0.8	0.6	1.1	0.7	0.9	1.0	0.8	0.9
Monetary poverty	poor(b)	0.7	3.2		0.4	1.3	3.1	0.6	1.4	3.2	0.6	2.3	1.8	0.9	1.1	1.6
	non poor	1.0	0.7		1.1	1.0	0.7	1.0	1.0	0.7	1.1	0.8	0.9	1.0	1.0	0.9
Financial living conditions	very difficult	0.5	5.4		0.1	0.9	4.8	0.5	1.3	4.1	0.4	3.7	1.4	0.7	0.7	2.5
	difficult	0.8	2.3		0.4	1.5	2.4	0.8	1.4	2.0	0.8	2.1	1.0	0.9	1.3	1.4
	rather difficult	1.0	1.1		0.7	1.5	1.1	0.9	1.3	1.2	0.9	1.2	1.1	0.9	1.4	1.0
	rather comfortable	1.1	0.2		1.3	0.7	0.2	1.1	0.8	0.4	1.1	0.4	0.9	1.1	0.8	0.7
	comfortable and very c.	1.1	0.1		1.5	0.4	0.1	1.2	0.6	0.2	1.2	0.2	0.8	1.2	0.4	0.8
Type of dwelling	House, isolated	1.0	0.6		1.1	0.9	0.7	1.1	0.9	0.7	1.1	0.7	0.9	1.1	0.9	0.7
	House, in a neighborhood	1.0	1.0		1.0	1.1	1.0	1.0	1.1	1.0	1.0	1.0	1.0	1.0	1.1	1.0
	Structure <10 units	0.9	1.4		0.9	1.0	1.3	0.9	1.5	1.4	1.0	1.2	1.4	0.9	1.1	1.2
	Structure >=10 units	1.0	1.4		0.9	1.0	1.3	0.9	1.3	1.2	0.9	1.4	1.0	0.9	1.0	1.4
	Other	1.0	1.3		1.0	1.1	1.1	0.9	1.1	2.0	0.8	0.7	2.3	1.0	1.2	0.8
Location	Rural town	1.0	0.9		1.0	1.1	1.0	1.0	0.9	1.1	1.0	0.9	1.1	1.1	1.0	0.6
	City <10000 inh	1.0	0.8		1.1	1.0	0.8	1.0	0.9	0.8	1.0	0.9	1.0	1.0	0.8	1.0
	10000 to <100000 inh	1.0	1.0		0.9	1.1	1.0	1.0	1.0	1.0	1.0	1.1	1.0	1.0	1.1	1.0
	100000 to <2000000 inhh	1.0	1.2		1.0	0.9	1.1	1.0	1.0	1.1	1.0	1.1	0.9	0.9	1.0	1.3
	Paris area	1.0	0.9		1.1	0.8	0.9	1.0	1.3	0.9	1.1	0.9	0.8	0.9	1.1	1.3
Type of household (children taken into account if <25 years old)	Person living alone	0.9	1.5		0.8	1.2	1.4	0.9	1.0	1.6	0.9	1.2	1.4	0.9	1.3	1.1
	Couple without child	1.1	0.5		1.1	0.9	0.6	1.1	0.8	0.6	1.1	0.7	0.9	1.0	0.8	0.9
	Couple with child(ren)	1.0	0.8		1.1	0.8	0.8	1.0	1.1	0.6	1.1	0.9	0.8	1.1	0.8	0.9
	Lone parent family	0.9	1.9		0.6	1.3	2.0	0.9	1.1	1.9	1.1	1.9	0.8	0.8	1.3	1.5
	Other	1.0	0.9		1.0	1.1	1.0	1.0	0.9	1.3	1.0	1.0	1.1	1.1	0.9	0.6

(a) Equivalence scale : 1 – 0.5 – 0.3
(b) Poverty threshold at 50 % of the median income per c.u.

Nonlinear Synthesis of Vowels in the LP Residual Domain with a Regularized RBF Network

Erhard Rank[1] and Gernot Kubin[2]

[1] Institute of Communications and Radio-Frequency Engineering,
Vienna University of Technology,
Gusshausstrasse 25/E389, A–1040 Vienna, Austria
erank@nt.tuwien.ac.at

[2] Institute of Communications and Wave Propagation,
Graz University of Technology,
Inffeldgasse 16c, A–8010 Graz, Austria
g.kubin@ieee.org

Abstract. In this paper we present a speech analysis/synthesis coder based on a combination of linear prediction with nonlinear modeling of the residual using a regularized radial basis function (RBF) network. The model has been applied to synthesis of sustained vowel signals and has been found to preserve the dynamics and spectra of the original speech signal. While several nonlinear speech models reportedly suffer from high-frequency losses in the synthesized speech due to system inherent low-pass behavior, our approach achieves good speech signal reproduction even in the higher frequency ranges. The decomposition of the speech signal by linear prediction analysis supports processing during synthesis such as pitch modifications while the nonlinear modeling provides the means for adequate reproduction of the fine-grained dynamic characteristics of speech.

1 Introduction

Synthesis of speech signals with nonlinear system models working in the phase space domain is a growing field of interest in the area of nonlinear signal processing. Synthetic signals incorporating the natural dynamical properties of speech can be generated. Speech re-synthesis with nonlinear systems is based on a phase space embedding of the time-domain signal, and on modeling of the system dynamics by a nonlinear function, e.g., with an artificial neural network (ANN).

Phase space embedding is used to find a representation of an unknown autonomous dynamical system that fully captures the dynamics of the system, i.e., for an unknown mapping $\Theta : R^d \to R^d$ that describes the evolution of a trajectory on a d-dimensional attractor in the system's phase space, we can find an equivalent mapping $F : R^D \to R^D$, with reconstruction dimension D, from which the output of the unknown system can be reconstructed exactly. Furthermore, for a discrete-time system, prediction of the next sample of a scalar signal like speech can be done by a mapping $f : R^D \to R$ with

$$x[n+1] = f(\boldsymbol{x}[n]).$$

The D-dimensional state vector $x[n]$ can be constructed from a scalar time signal by a time delay embedding

$$x[n] = \{x[n], x[n-M], \ldots x[n-(D-1)M]\}^T,$$

with a properly chosen delay parameter M. Using a reconstruction dimension $D \geq 2d + 1$, it is generally possible to capture and reproduce the dynamics of the original system [1]. However, a good choice of the delay M—e.g., at the first minimum of the mutual information function—facilitates a good reconstruction for an embedding space of even lower dimensionality [2].

The system dynamics is captured by estimating the function $f()$ from observations of $x[n+1]$ and $x[n]$. ANNs provide a non-parametric approach for this function approximation problem. Two different structures have been considered as ANN realizations:

- Multilayer perceptrons (MLPs) have been proven to be *universal approximators*, however, no explicit constraint for the actual structure and complexity needed for a specific approximation accuracy can be given [3]; nevertheless, MLPs of reasonably small size give good approximation in many cases. Training of MLPs is a time consuming step-by-step procedure with the caveat of getting stuck with suboptimal solutions (local minima of error function). Interpretation of the network parameters is not self-evident.
- Radial basis function (RBF) networks have been designed for *data interpolation in multi-dimensional space* and achieve universal approximation as well. Approximation accuracy is determined by the density of RBF centers. Network training can be performed in a one-step procedure by matrix inversion under certain conditions. Network weights are related to the RBF center positions and thus to a region in the signal embedding phase space.

The long-range objective of our work is to come up with a general synthesis model including the modeling of speech dynamics, like transitions between phonemes, and fundamental frequency modifications. Therefore, we favor the application of an RBF network with their clear relation between network parameters and position in phase space.

The source-filter model of speech production [4] is considered as the basis for many speech synthesis and coding algorithms: the influence of the vocal tract filter on the glottis signal is estimated from the speech signal (inverse filtering), usually by linear prediction (LP) analysis [5]. The task is to find a vocal tract representation as all-pole LP filter with system function $1/A(z)$ and filter the sampled speech signal $s[n]$ with the inverse filter (with system function $A(z)$) to get a residual signal $x_R[n]$ with maximally flat (white) spectrum [5]. The residual signal of voiced phonemes is related to the derivative of the glottis signal. It includes information about fundamental frequency, timing and voice quality as well as speaker characteristics and can be processed, e.g., for fundamental frequency modifications in speech synthesis, or efficiently coded independently of the LP filter parameters. The spectral envelope of the speech signal, related to the phonemic content, is captured by the vocal tract filter parameters and is superimposed on the processed residual in the re-synthesis or decoding stage.

Application of a combination of linear prediciton and the nonlinear oscillator model can be questioned since it can be shown that nonlinear prediction alone can always outperform a linear predictor when applied to signals originating from a nonlinear system (like speech production) in terms of mean square prediction error [6]. However, for speech synthesis models—in particular for voiced speech—the direct relation between the linear prediction filter and the influence of the vocal tract on the acoustic glottis signal justifies the choice of a linear filter model. The combination of linear prediction and the nonlinear oscillator model also shifts the nonlinear system modeling task from modeling of the full speech signal to modeling of the output of the physical oscillator in speech production, the glottis signal.

A general overview on nonlinear time-series modeling can be found in [6,7], the application of ANNs to time series modeling is described in [8]. Re-synthesis of the full speech signal based on radial basis function (RBF) networks is described in [6, 9–13]. A model that includes linear prediction into a nonlinear synthesis model using a back-propagation MPL with two hidden layers is described in [14]. They propose a nonlinear function model working with a phase space embedding of a low-pass filtered linear prediction residual. However, due to the non invertible low-pass filtering by a Butterworth filter the system suffers from a lack of high-frequency components in the residual signal, which are not regenerated in the synthesis stage and thus are also missing in the synthetic speech signal. The very problem has also been encountered for synthesis of the full speech signal with a nonlinear model [9, 10, 12, 13]. Due to the necessary application of regularisation or singular value decomposition in the signal analysis stage the nonlinear system model always generates a smoothed, i.e., low-pass version of the original signal.

Here we present a system involving an RBF network for modeling the filtered linear prediction residual and its capability to reproduce sustained vowel signals with high accuracy in the spectral content. In our model the problem of high frequency losses is tackled by the application of an invertible integration filter for estimation of the glottis signal from the LP residual and modeling the glottal oscillation pattern with a nonlinear model. During re-synthesis an inverse (differentiation) filter is applied to the synthetic glottis signal. Thus, the negative effect of smoothing by the model is less significant than in the case where the model is applied to the full speech signal or when estimating the glottis signal with a non-invertible filter which cannot be compensated during re-synthesis. By this means, the original spectrum is well preserved in the synthetic speech signal in the perceptually relevant range up to 5 kHz and, above this frequency, the spectral envelope is reproduced with reasonably small energy losses.

As our overall goal is a flexible speech synthesis structure, processing of the linear prediction residual supports the application of prosodic manipulations, like pitch modification, by the nonlinear glottis oscillator model while the nonlinear modeling itself is capable of reproducing the natural dynamics of the speech production process. In the following, our complete synthesis model is described including the necessary steps for applying regularized RBF networks, and an example for analysis/re-synthesis of vowel sounds will be given (section 2). We will conclude with a short discussion and an outlook to future work (section 3).

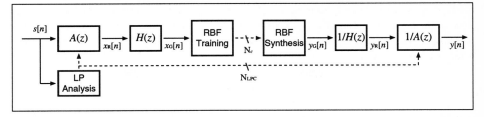

Fig. 1. Synthesis model. The original speech signal $s[n]$ is LP analyzed and filtered, resulting in a residual signal $x_R[n]$, then filtered by a low-pass (integration) filter to yield an estimate of the glottis signal $x_G[n]$. This signal is used to train an RBF network as a one-step ahead predictor. In the synthesis part, the RBF network is used as an autonomous oscillator to reproduce a synthetic glottis signal $y_G[n]$, which is in turn filtered by the complementary synthesis filters to get the synthetic full speech signal $y[n]$. Solid lines denote signal paths, while dashed lines denote parameter paths.

2 Synthesis Model

The synthesis model (figure 1) consists of LP analysis and filtering of the speech signal $s[n]$ by the inverse filter $A(z)$, followed by low pass filtering of the residual signal $x_R[n]$ with a simple one-pole low-pass filter $H(z)$. The resulting glottis signal estimate $x_G[n]$ is applied to the RBF network for learning. The input vector $x[n]$ for the RBF network is constructed from delayed samples of $x_G[n]$ with a delay of M samples and the network is trained for one-step ahead prediction (*teacher forcing*) [15]: $x[n+1] = f(x[n])$.

The radial basis function (RBF) network is defined by its input-output map

$$f(x) = \sum_{i=1}^{N_c} w_i \, \varphi(\|x - c_i\|),$$

with the basis function $\varphi(\cdot)$, the function centers c_i, weighting coefficients w_i, the number of centers and weighting coefficients N_c, and the input vector x. The Gaussian basis function is

$$\varphi(\|x - c_i\|) = \exp\left(-\frac{\|x - c_i\|^2}{2\sigma_i^2}\right),$$

where σ_i^2 is the variance of the Gaussian.

The modeling properties of the network are strongly influenced by the choice of the centers for the RBFs which can be fixed or adapted to the data. If the radial basis functions and centers c_i are fixed a priori (e.g., Gaussians with fixed variance and centers on a hyper-lattice) network training affects only the weighting coefficients w_i which can be optimized in a one step procedure by matrix inversion with an optional regularisation term. We determine the least squares solution of

$$(G^T G + \lambda G_0) w = G^T z,$$

with $\boldsymbol{G}$ a N_s (number of training samples) by N_c matrix composed of the output of the radial basis functions for the training data, $\boldsymbol{w}$ a column vector composed of the weighting coefficients w_i, $\boldsymbol{z}$ a column vector of output targets (i.e., of $x_G[n+1]$ for the training data), λ the regularisation factor, and $\boldsymbol{G}_0$ a N_c by N_c matrix composed of

$$G_{0_{i,j}} = \varphi(\|\boldsymbol{c}_i - \boldsymbol{c}_j\|),$$

which leads to a smoothing (regularisation) of the estimated input-output map for a priori fixed center positions $\boldsymbol{c}_i$ [13] of the network.

In the re-synthesis stage the predicted sample $y_G[n+1]$ is fed back to the input delay line and the RBF network is free running as an autonomous oscillator which re-generates an artificial glottis signal $y_G[n]$. To obtain the synthetic full speech signal $y[n]$, the artificial excitation $y_G[n]$ is filtered by $1/H(z)$ and by the LP synthesis filter $1/A(z)$.

This model has been applied to sustained vowel signals and, assuming stationarity, both LP analysis and network training were performed over the whole duration of the input signal.

As an example, figure 2 shows the signals $s[n]$, $x_R[n]$, $x_G[n]$, $y_G[n]$, $y_R[n]$, and the synthetic speech signal $y[n]$ first as time-domain waveforms, next in a three-dimensional phase-space representation, and finally as the corresponding spectra, all for the vowel /i/ produced by a female speaker. In figure 3 a direct comparison between the time signals and low-frequency part of the spectra of input $s[n]$ and synthetic speech signal $y[n]$ is given. The sampling rate was chosen as $f_s = 22050$ Hz and the total signal length for LP analysis and network training is $N_s = 3000$ samples (136 ms). LP analysis is performed using the auto-correlation method, with filter order $N_{\text{LPC}} = 22$. The single pole of the low-pass filter is set to $z_p = 0.98$. The time delay between the input vector components of the RBF network and for the phase space representation in figure 2 is $M = 10$ samples (0.45 ms). The RBF network uses $D = 4$ delayed samples as input and has its centers placed on the cross points of a fixed hyper-lattice with $K = 5$ levels in each dimension, comprising $N_c = 625$ centers. Regularisation is applied during network training with a regularisation factor of $\lambda = 0.05$. During re-synthesis, the delay line of the RBF network is initialized with initial samples copied from $x_G[n]$, but thereafter the network operates strictly as an autonomous oscillator. We observe the output of the system during steady-state operation after several hundred samples have been discarded from the beginning of the output signal.

By comparing each pair of natural and synthetic signals in figure 2, one can see that the dynamical properties are well preserved by the model. The fundamental frequency and the formant structure of the synthetic signal closely match those of the original signal. However, as can be seen best in the phase space representation (middle row in figure 2), the synthetic signal $y_G[n]$ exhibits a much smoother trajectory in comparison with $x_G[n]$. While the synthetic signal trajectories are not always as smoothly bundled as in this example and the small-scale variations inherent to nonlinear dynamical systems are still reasonably well reproduced, a certain amount of smoothing is a general and inevitable characteristic caused by the approximation of the nonlinear mapping $f(\boldsymbol{x})$. Generically, this smoothing results in a loss of high-frequency components which can be identified in the spectra beyond 5 kHz. However, in the frequency band below 5 kHz

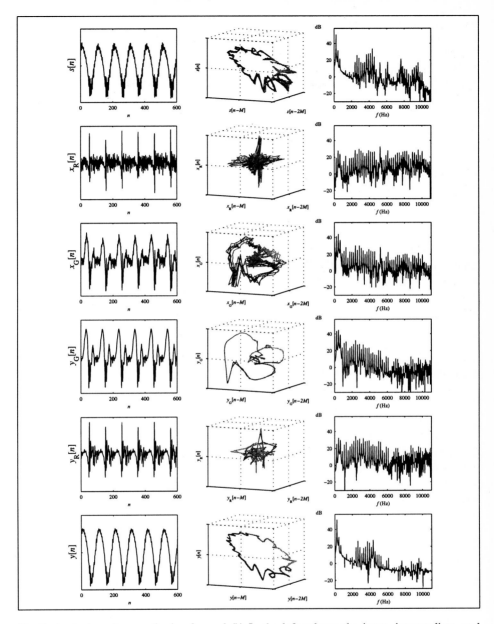

Fig. 2. Analysis and re-synthesis of vowel /i/. In the left column the input, intermediate, and output signals of the system are plotted as time-domain waveforms, the middle column shows a three-dimensional phase-space embedding with delay $M = 10$ (the same delay is used as embedding delay for the nonlinear model), and in the right column the corresponding spectra are displayed. Note the similarity of the natural (first three rows) and the synthetic (last three rows) signals. Moreover, albeit the fact that the synthetic signals are smoother than the original—which is clearly visible in the phase space representation—the high-frequency losses are small and the spectral envelope is reproduced well over the whole frequency range.

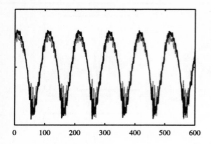 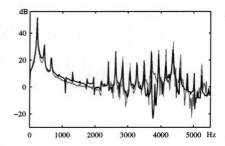

Fig. 3. Direct comparison of the time-domain signals and the low-frequency parts of the spectra of the synthetic speech signal $y[n]$ (black) and the original signal $s[n]$ (grey).

(see figure 3) which carries most of the perceptually relevant information for voiced speech the spectrum is faithfully reproduced and in the higher frequency range—in spite of some energy loss—the spectral envelope is preserved. This is considered an notable improvement towards natural nonlinear synthesis, since nonlinear oscillator design for speech synthesis has been reported to suffer from high-frequency loss in most implementations.

It has to be noted that the stability of the system and the quality of the output signal are highly dependent on the selection of tuning parameters. Crucial parameters are the delay M, the low-pass filter pole location z_p, and the regularisation factor λ. The density of the basis function lattice (parameter K) should not be reduced either, although it was found that the number of network centers could be pruned significantly for a given vowel. A higher dimension of the embedding phase space $D > 4$ does *not* yield improved stability in general; on the contrary, even $D = 3$ may suffice for stable synthesis in some cases. This supports the finding that the production of vowels can be characterized as a low-dimensional, nonlinear deterministic process [16].

Finally, the system has been tested on speech signals at different sampling rates, with the parameters N_{LPC}, M, and z_p adjusted according to the sampling rate. For sampling rates above 30 kHz, unstable system behavior was observed frequently. On the positive side, with parameter values as stated above and 22 kHz sampling rate, we generically obtain stable system behavior for vowels from female speakers[1].

3 Discussion and Outlook

We have described a speech analysis/synthesis model combining linear prediction and nonlinear modeling of the low-pass filtered residual signal. In comparison to previous approaches such as nonlinear synthesis of the full speech waveform or combined LP and nonlinear modeling with a non-invertible low-pass filter, our new system offers improved reproduction of speech at little additional complexity. Particularly the effect

---

[1] Male voices require a slightly different parameter set, in particular an integration filter pole closer to the unit circle (i.e., longer integration time constant) and higher embedding lag M.

of smoothing by the nonlinear system model resulting in poor synthesis in the high-frequency range is relieved by applying the inverse filter of the low-pass during re-synthesis.

Due to the high number of parameters (the total number of system parameters is 8 constants plus—in our example— $N_{\text{LPC}} + N_c = 647$ signal-dependent parameters), finding a stable regime is difficult, but can be based on knowledge from previous studies such as appropriate values for the dimension D and the delay M of the embedding space [16, 2]. Both the application of standard LP models and the fact that turbulent noise sources along the vocal tract are not included in our model explicitly lead to a certain quality impairment. These shortcomings will be addressed in our future work by more elaborated LP algorithms, by analysis of the nonlinear modeling errors and introduction of an adequate compensation technique during synthesis. Furthermore, the application of the model to non-stationary speech signals will be investigated.

References

1. Floris Takens, "On the numerical determination of the dimension of an attractor," in *Dynamic Systems and Turbulence*, D. Rand and L. S. Young, Eds., vol. 898 of *Warwick 1980 Lecture Notes in Mathematics*, pp. 366–381. Springer, Berlin, 1981.
2. Hans-Peter Bernhard, *The Mutual Information Function and its Application to Signal Processing*, Ph.D. thesis, Vienna University of Technology, 1997.
3. Kurt Hornik, M. Stinchcombe, and H. White, "Multilayer feedforward networks are universal approximators," *Neural Networks*, vol. 2, pp. 359–366, 1989.
4. Gunnar Fant, *Acoustic Theory of Speech Production*, Mouton, The Hague, Paris, 1970.
5. John D. Markel and Augustine H. Gray, Jr., *Linear Prediction of Speech*, Springer, Berlin, Heidelberg, New York, 1976.
6. Gernot Kubin, "Nonlinear processing of speech," in *Speech Coding and Synthesis*, W. Bastiaan Kleijn and K. K. Paliwal, Eds., pp. 557–610. Elsevier, Amsterdam, 1995.
7. José Principe, Ludong Wang, and Jyh-Ming Kuo, "Nonlinear dynamic modeling with neural networks," in *The first European Conference on Signal Analysis and Prediction*, 1997.
8. Simon Haykin, "Neural networks expand SP's horizon," *IEEE Signal Processing Magazine*, vol. 13, no. 2, pp. 24–49, Mar. 1996.
9. Martin Birgmeier, "A fully Kalman-trained radial basis function network for nonlinear speech modeling," in *Proc. of the IEEE ICNN'95*, Perth, Australia, 1995, pp. 259–264.
10. Martin Birgmeier, *Kalman-trained Neural Networks for Signal Processing Applications*, Ph.D. thesis, Vienna University of Technology, 1996.
11. Gernot Kubin, *Signal Analysis and Prediction*, chapter Signal Analysis and Speech Processing, pp. 375–394, Birkhaeuser, Boston, 1998.
12. Iain Mann and Steve McLaughlin, "Stable speech synthesis using recurrent radial basis functions," in *Proc. of EuroSpeech'99*, Budapest, Hungary, 1999.
13. Iain Mann, *An Investigation of Nonlinear Speech Synthesis and Pitch Modification Techniques*, Ph.D. thesis, University of Edinburgh, 1999.
14. Karthik Narasimhan, José C. Principe, and Donald G. Childers, "Nonlinear dynamic modeling of the voiced excitation for improved speech synthesis," in *Proc. of ICASSP'99*, 1999.
15. Simon Haykin, *Neural Networks. A Comprehensive Foundation*, Macmillan College Publishing Company, New York, Toronto, Oxford, 1994.
16. Hans-Peter Bernhard and Gernot Kubin, "Detection of chaotic behaviour in speech signals using Fraser's mutual information algorithm," in *Proc. of 13th GRETSI Symposium on Signal and Image Processing*, Juan-les-Pins, France, Sept. 1991, pp. 1301–1311.

Nonlinear Vectorial Prediction With Neural Nets*

Marcos Faúndez-Zanuy

Escola Universitària Politècnica de Mataró
Universitat Politècnica de Catalunya (UPC)
Avda. Puig i Cadafalch 101-111, E-08303 Mataró (BARCELONA) SPAIN
faundez@eupmt.es

Abstract. In this paper we propose a nonlinear vectorial prediction scheme based on a Multi Layer Perceptron. This system is applied to speech coding in an ADPCM backward scheme. In addition a procedure to obtain a vectorial quantizer is given, in order to achieve a fully vectorial speech encoder. We also present several results with the proposed system

1 Introduction

Most of the speech coders use some kind of prediction. The most popular one is the scalar linear prediction, but several papers have shown that a nonlinear predictor can outperform the classical LPC linear prediction scheme. On the other hand, vectorial prediction schemes have also been proposed. In this paper we propose and study a vectorial nonlinear prediction scheme based on a MLP neural net. We apply the nonlinear predictor inside an ADPCM scheme for speech coding. This predictor replaces the LPC predictor in order to obtain an ADPCM speech encoder scheme with nonlinear prediction. Figure 1 summarizes the scheme. If the quantizer is a vectorial quantizer, this scheme is also known as Predictive Vector Quantization [1].

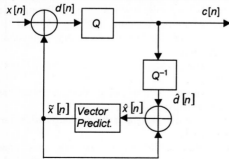

Fig. 1. Vectorial ADPCM scheme.

---

* This work has been supported by the CICYT TIC2000-1669-C04-02

Classical ADPCM block-adaptive waveform coders compute the predictor coefficients in one of two ways:
a) backward adaptation: The coefficients are computed over the previous frame. Thus, it is not needed to transmit the coefficients of the predictor, because the receiver has already decoded the previous frame and can obtain the same set of coefficients.
b) forward adaptation: The coefficients are computed over the same frame to be encoded. Thus, the coefficients must be quantized and transmitted to the receiver. In [2] we found that the SEGSNR of forward schemes with unquantized coefficients is similar to the classical LPC approach using one quantization bit less per sample. On the other hand with this scheme the mismatch between training and testing phases is smaller than in the previous case, so the training procedure is not as critical as in backward schemes, and the SEGSNR are greater.

The SEGSNR is computed with the expression: $SNRSEG = \frac{1}{K}\sum_{j=1}^{K} SNR_j$, where SNR_j is the signal to noise ratio (dB) of frame j : $SNR = \frac{E\{x^2[n]\}}{E\{e^2[n]\}}$, and K is the number of frames of the encoded file.

In [3] we found that the quantization of the predictor coefficients (forward scheme) is not a trivial question. In [4] we proposed another training approach of the neural net, in order to improve the performance of the backward scheme. The goal is to obtain similar results with backward ADPCM nonlinear prediction and forward (unquantized coefficients) schemes. We will work with the backward scheme. Figure 2 shows the network architecture for scalar and prediction.

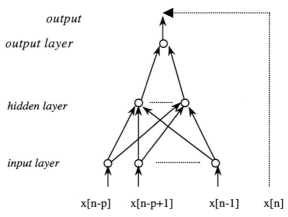

Fig. 2. Neural net training for scalar prediction

In our previous work we fixed the structure of the neural net to 10 inputs, 2 neurons in the hidden layer, and one output. The selected training algorithm was the Levenberg-

Marquardt, that computes the approximate Hessian matrix, because it is faster and achieves better results than the classical backpropagation algorithm. We also apply a multi-start algorithm with five random initializations for each neural net. In [4] we studied the following training schemes:
a) The use of regularization.
b) Early stopping with validation.
c) Bayesian regularization.
d) Validation and bayesian regularization.
e) Committee of neural nets [5].
The combination between Bayesian regularization with a committee of neural nets increased the SEGSNR up to 1.2 dB over the MLP trained with the Levenberg-Marquardt algorithm, and decreases the variance of the SEGSNR between frames.

1.1 Conditions of the Experiments

The experimental results have been obtained with an ADPCM speech coder with an adaptive scalar quantizer based on multipliers [7]. The number of quantization bits is variable between Nq=2 and Nq=5, that correspond to 16kbps and 40kbps (the sampling rate of the speech signal is 8kHz). We have encoded eight sentences uttered by eight different speakers (4 males and 4 females). The neural net has been trained with the Levenberg-Marquardt algorithm combined with bayesian regularization [7-8]

2. Vectorial Prediction

In this paper we propose a different scheme, that consists on a vectorial neural net prediction.
Previously to the vectorial prediction scheme we will study the classical scalar prediction scheme, using a neural net training with hints. The proposed scheme consists on the neural net architecture shown on figure 3, where the second output is only used for training purposes. That is, x[n+1] is used as a hint for a second neural net output during training, but this output is ignored during the speech coding process.
Table 1 compares the obtained results for one and two outputs using the best approach found in [4], that consists on bayesian regularization with a committe of 5 neural nets for the prediction process.
Table 1 shows the obtained segmental signal to noise ratio SEGSNR and the standard deviation of the SEGSNR (σ).
The results of table 1 are similar to the best results that we obtained in [4] (using the scheme of figure 1) for the best situation. Table 2 reproduces these results for comparison purposes.
The next experiment consists on an ADPCM speech coding with vectorial prediction and scalar adaptive quantization (same quantizer used in the previous experiments). Table 3 shows the obtained results with a prediction of two samples simultaneously.

Table 1. SEGSNR for several ADPCM schemes with the nnet of figure 2 and scalar prediction.

Epoch	combination	Nq=2 SEGSNR	σ	Nq=3 SEGSNR	σ	Nq=4 SEGSNR	σ	Nq=5 SEGSNR	σ
6	Mean	14.2	5.1	20.4	5.8	25.6	6.3	30.4	6.7
50	mean	13.9	5.5	20.8	6.0	26	6.6	30.9	6.9
6	median	14.7	5.1	20.8	6.0	25.7	6.4	30.4	6.7
50	median	14.0	5.6	21	6.1	26.2	6.7	31.1	7.0

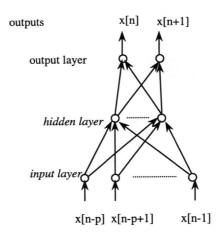

Fig. 3. Neural net training for vectorial prediction

Table 2. SEGSNR for several ADPCM schemes with the nnet of figure 1

Epoch	combination	Nq=2 SEGSNR	σ	Nq=3 SEGSNR	σ	Nq=4 SEGSNR	σ	Nq=5 SEGSNR	σ
6	Mean	14.5	4.9	20.5	5.9	25.6	6.5	30.5	6.8
50	mean	14.6	5.6	21.3	6.4	26.5	6.8	31.4	7.2
6	median	14.9	5.2	21	6	25.9	6.5	30.7	6.9
50	median	14.3	5.5	21.1	6.2	26.4	6.8	31.2	7.1

Table 3 shows that the behavior of the vectorial prediction scheme is worst than the scalar scheme. We believe that this is due to the quantizer, because the experiments of table 2 show a good performance with the same neural net scheme. It is interesting to observe that the vectorial scheme is more suitable with a vectorial quantizer. In order to check this assert we have plot the vectorial prediction errors on figure 4. It is interesting to observe that the vectors to be quantized do not lie in the whole space. Most of the vectors are around y=x that means that the error vector presents the following property: $\vec{e}[n] = (e_{1n}, e_{2n})$, $e_{1n} \cong e_{2n}$. Thus, a vectorial quantizer can exploit this redundancy and improve the results.

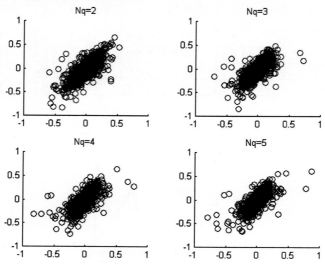

Fig. 4. Vectorial prediction errors for several quantization bits

Table 3. SEGSNR for several ADPCM schemes with the nnet of figure 2 and vectorial pred.

Epoch	combination	Nq=2		Nq=3		Nq=4		Nq=5	
		SEGSNR	σ	SEGSNR	σ	SEGSNR	σ	SEGSNR	σ
6	Mean	12.58	4.4	18.3	4.9	23.0	5.3	27.5	5.4
50	mean	12.9	5	18.9	5.4	23.7	5.5	28.3	5.8
6	median	12.9	4.5	18.6	5.2	23.2	5.3	27.7	5.5
50	median	12.9	4.8	18.8	5.3	23.6	5.5	28.1	5.7

Non-linear predictive vector quantization

A special case of vector quantization is known as predictive vector quantization (PVQ) [1]. Basically, PVQ consists on an ADPCM scheme with a vector predictor and a vectorial quantizer. Obviously, if the vector predictor is nonlinear the system is a NL-PVQ scheme. We propose to use the NL-vector predictor described in previous sections, and a vectorial quantizer. In order to design the vectorial quantizer we use the residual signal of the vectorial prediction and the scalar quantizer based on multipliers applied as many times as the dimension of the vectors. Thus, we have used the residual errors of the first speaker (with Nq=3) and the generalized Lloyd algorithm [1] in order to create codebooks ranging from 4 to 8 bits. The initial codebook has been obtained with the random method.

It is interesting to observe that the optimization procedure must be a closed loop algorithm, because there are interactions between the predictor and the quantizer. Thus, the system can be improved computing again the residual errors with the actual quantization scheme

Figure 5 shows the obtained codebooks and table 4 the SEGSNR for several quantization bits per sample (this value is obtained by the ratio between the number of bits of the vectorial quantizer and the dimension of the vectors). The committee of neural nets is obtained with the *median{}* function and 50 epochs. It can be seen that the Vector Quantizer results of table 4 outperform the scalar quantizer of table 3 in nearly 2 dB for Nq=4.

Table 4. SEGSNR for NL-PVQ.

Nq=2		Nq=2.5		Nq=3		Nq=3.5		Nq=4	
SEGSNR	σ	SEGSNR	σ	SEGSNR	σ	SEGSNR	σ	SEGSNR	σ
10.4	8.3	15.8	6.9	19	5.9	21.4	6.1	25	5.7

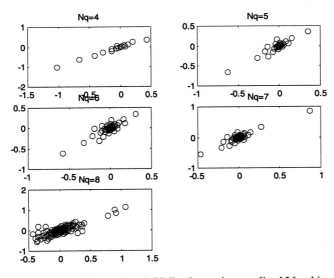

Fig. 5. Codebooks with a random initialization and generalized Lloyd iteration.

Scalar linear prediction

The AR modeling of order P is given by the following relation:

$$x[n] = \sum_{i=1}^{P} a_i x[n-i] + e[n]$$

where $\{a_i\}_{i=1,\ldots P}$ are the scalar prediction coefficients. Their value is usually obtained with the levinson-durbin recursion [1].

Vectorial linear prediction

The AR-vector modeling of order P is given by the following relation:

$$\vec{x}[n] = \sum_{i=1}^{P} A_i \vec{x}[n-i] + \vec{e}[n]$$ where $\{A_i\}_{i=1,\ldots P}$ are the $m \times m$ matrices equivalent to the prediction coefficients of the classical scalar predictor, and m is the dimension of the vectors. The prediction matrices can be estimated using the Levinson-Whittle-Robinson algorithm, that has been previously applied to speaker verification in [5]. The algorithm is the following:

Computation of the correlation matrices R_i:

$$R_i = \sum_{n=0}^{N-i-1} \vec{x}[n+i](\vec{x}[n])^T$$

where $(\vec{x}[n])^T$ means the transpose of the vector $\vec{x}[n]$.

for i=0:P-1,

$$F_i = \sum_{j=0}^{i} A1_j^i R_{i+1-j}$$

$$K1_i = -F_i (D2_i)^{-1}$$

$$K2_i = -F_i^T (D1_i)^{-1}$$

$$D1_{i+1} = (I - K1_i K2_i) D1_i$$

$$D2_{i+1} = (I - K2_i K1_i) D2_i$$

$$A1_0^{i+1} = A1_0^i$$

$$A2_0^{i+1} = K2_i A1_0^i$$

$$A1_{i+1}^{i+1} = K1_i A2_i^i$$

$$A2_{i+1}^{i+1} = A2_i^i$$

for j=1:i,

$$A1_j^{i+1} = A1_j^i + K1_i A2_{j-1}^i$$

$$A2_j^{i+1} = A2_{j-1}^i + K2_i A1_j^i$$

end

end

for i=0:P,

$$A_i = A2_i^P$$

end

Conclusions and Future Work

A PVQ system implies a good predictor and a good quantizer. In this paper we have first proposed a MLP training with output hints, in order to check that the proposed scheme for vectorial prediction is suitable. The next step has been the evaluation of the system with scalar and vectorial quantizers, and we have found that the SEGSNR is smaller than the achieved one with the scalar system. we believe that this is because of the quantizer. Thus, the next step will be the evaluation on the PVQ system proposed in this paper with an improved quantizer in two ways:
a) The VQ will be computed with a better algorithm, like the Linde-Buzzo-Gray (LBG) one. Looking at figure 5 it can be seen that the centroids are not well distributed along the vectorial space.
b) The VQ is computed in an open loop design, and it is used in a closed loop system. An iterative algorithm [10] must be applied in order to jointly optimize the predictor and the quantizer.

Future work will include the study of vectorial prediction of order higher than 2. In addition, the PVQ scheme will be the first step towards a CELP speech coder.

References

[1] A. Gersho & R. M. Gray "Vector Quantization and signal compression". Ed. Kluwer 1992.
[2] M. Faúndez, F. Vallverdu & E. Monte , "Nonlinear prediction with neural nets in ADPCM" International Conference on Acoustic , Speech & Signal Processing, ICASSP-98 .SP11.3.Seattle, USA, May
[3] O. Oliva, M. Faúndez "A comparative study of several ADPCM schemes with linear and nonlinear prediction" EUROSPEECH'99 , Budapest, Vol. 3, pp.1467-1470.
[4] M. Faúndez-Zanuy, "Nonlinear predictive models computation in ADPCM schemes". Vol. II, pp 813-816. EUSIPCO 2000, Tampere.
[5] C. Montacié & J. L. Le Floch, "Discriminant AR-Vector models for free-text speaker verification", pp.161-164, Eurospeech 1993.
[6] S. Haykin, "neural nets. A comprehensive foundation", 2on edition. Ed. Prentice Hall 1999.
[7] N. S. Jayant and P. Noll "Digital Coding of Waveforms". Ed. Prentice Hall 1984.
[8] D. J. C. Mackay "Bayesian interpolation", Neural computation , Vol.4, N° 3, pp.415-447, 1992.
[9] F. D. Foresee and M. T. Hagan, "Gauss-Newton approximation to Bayesian regularization", proceedings of the 1997 International Joint Conference on Neural Networks, pp.1930-1935, 1997.
[10] V. Cuperman, A. Gersho "Vector predictive coding of speech at 16 kbits/s". IEEE trans. on Commun. vol. COM-33, pp.685-696, July 1985.

Separation of Sources Based on the Partitioning of the Space of Observations

Manuel Rodríguez-Álvarez, Carlos G. Puntonet, Ignacio Rojas

Departamento de Arquitectura y Tecnología de Computadores.
Facultad de Ciencias. Universidad de Granada.
Campus Universitario Fuentenueva. Av. Fuentenueva s/n.
18071 Granada (Spain).
e-mail: mrodriguez@atc.ugr.es ; cgpuntonet@atc.ugr.es ; irojas@atc.ugr.es

Abstract. The techniques of Blind Separation of Sources (BSS) are used in many Signal Processing applications in which the data sampled by sensors are a mixture of signals from different sources, and the goal is to obtain an estimation of the sources from the mixtures. This work shows a new method for blind separation of sources, based on geometrical considerations concerning the observation space. This new method is applied to a mixture of two sources and it obtains the coefficients of the unknown mixture matrix A and separates the unknown sources, S_o. Following an introduction, we present a brief abstract of previous work by other authors, the principles of the method and a description of the algorithm, together with some simulations.

Keywords. Blind Separation of Sources, Mixture Matrix, Linear Mixture of Sources, Space of Observations.

1   Introduction

When *p* different signals (called sources) propagating through a real medium have to be captured by sensors, these sensors are sensitive to all sources $s_{oi}(t)$ and thus the signal, $e_k(t)$, observed at the output of sensor k, is a mixture of source signals. The solution of the problem of separation consists of retrieving the unknown sources $s_{oi}(t)$ from just the observations $e_k(t)$, eliminating the effect introduced by the medium. To achieve this it is necessary to apply the following hypotheses:

1. The sources $s_{oi}(t)$ and the mixture matrix A are unknown.
2. The number of sensors is equal to the number of sources, *p*.
3. The observed signals $e_k(t)$, satisfy:

$$e_k(t) = \sum_{i=1}^{p} a_{ki} \cdot s_{oi}(t) , \ k=1,\ldots,p , \ a_{ki} \in R \tag{1}$$

In matrix notation equation (1) can be written:

$$E(t) = A \cdot S_o(t) \ ; \ E, S_o \in R^p , \ A \in R^{p*p} \tag{2}$$

Hypothesis 1 establishes the conditions for a Blind Separation of Sources. Equation (1) shows an instantaneous linear mixture. E(t) and $S_o(t)$ denote column vectors with components $e_k(t)$ and $s_{oi}(t)$ respectively, which are included in expression (2). A is a $p * p$ matrix (called mixture matrix), which we assume to be regular. This matrix A is introduced in order to model the effect of the real medium. To resolve the problem of separation of sources a matrix W is defined, such that:

$$S(t) = W^{-1}(t) \cdot E(t) = W^{-1}(t) \cdot A \cdot S_o(t) \quad (3)$$

The procedure is intended to obtain W^{-1} such that $W^{-1} \cdot A = D \cdot P$, that is, a diagonal matrix D modified by a permutation matrix P.

2 Previous Work

A great diversity of approaches have been proposed to solve the problem of Blind Separation of Sources, most of which use some kind of statistical analysis. Initially, Jutten et al. [1] proposed a solution based on a neural network (Herault-Jutten's network). Other authors have developed algorithms based on higher-order statistics [2], or based on the contrast function concept [3], Maximum Likelihood (ML) technique [4], neural networks for separate non-stationary signals [5] or the entropy concept [6]. In our previous work ([7], [8], [9]) we have proposed a geometrical procedure to separate digital or analog signals with a linear or non-linear mixture.

3 Principles of the Method

For $p = 2$ and $(s_{o1}, s_{o2}) \in \mathbb{R}^+$, with bounded values $s_{o1} \in [0, s_{o1}^M]$ and $s_{o2} \in [0, s_{o2}^M]$, and for two uniform noises, the observed signals $(e_1(t), e_2(t))$ form a parallelogram in the (e_1, e_2) space, as shown in Figure 1. In this special case we have demonstrated [8] that, through a matrix transformation, the coefficients of the W matrix coincide with the slopes of the parallelogram.

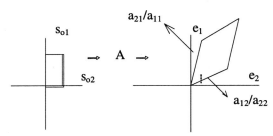

Figure 1 : Representation of $(s_{o1}$ vs. $s_{o2})$ and $(e_1$ vs. $e_2)$ when $p = 2$ and $s_{oi} \in \mathbb{R}^+$.

Nevertheless, if $(s_{o1}, s_{o2}) \in \mathbb{R}$, then the parallelogram is not strictly positive. In this case, it is possible to translate the parallelogram by detecting translation vector T and to translate the space of observations with this vector T. This translation is now equivalent to considering the signals $(s_{o1}, s_{o2}) \in \mathbb{R}^+$. As shown in Figure 1, the linear application represented for the mixture matrix A transforms the original form in the space of sources (s_{o1}, s_{o2}), into a parallelogram in the space of observations (e_1, e_2).

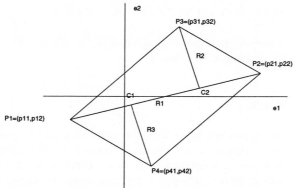

Figure 2: Space of observations: Representative points and straight lines.

From Figure 2 and it can seen the space of observations (e_1, e_2), that, for random uniform sources, the parallelogram is geometrically bounded for the segments between the points $P_1 = (p_{11}, p_{12})$, $P_2 = (p_{21}, p_{22})$, $P_3 = (p_{31}, p_{32})$ and $P_4 = (p_{41}, p_{42})$. The slopes of these segments give the coefficients of the mixture matrix W, as shown in Figure 1. In order to obtain these segments, it is necessary to know the coordinates of the points $P_1 = (p_{11}, p_{12})$, $P_2 = (p_{21}, p_{22})$, $P_3 = (p_{31}, p_{32})$ y $P_4 = (p_{41}, p_{42})$.

In other case, for example for speech signals, the form of the figure in the space of observations is very different, and it can seen in Figure 3

Than this method proceeds as follows:

1.- First of all, the procedure detects if the mixture proceeds from an two uniform sources or from a non-uniform sources. To make this, the procedures computes the kurtosis of the two observations and the coefficient of correlation between the two observations. If the kurtosis of both of them is positive, the procedure looks for the zones of maximum density of points. In other case, the procedure loos for the border of the figure in the space of observations.

2.- The procedure makes a lattice of the space of observations (e_1, e_2), with N-rows and M-columns (lattice of N by M) as shown in Figure 4. Then, the procedure computes through a threshold (TH) the number of cells in the lattice in which the number of points inside it is greater than the threshold TH. These cells are marked with a point and it represents each one.

3.- The procedure then substitutes the cells in the lattice in which the number of points inside it is greater than the threshold TH by a point that represents the cell (see

Figure 4). This point is not located exactly in the centre of the cell because it is weighted by the density of points (e_1, e_2) in this cell. This step greatly reduces the complexity of the algorithm, because the greatest number of points that the procedure needs to compute is N * M (i.e. if the space of observations has 10^5 points (e_1, e_2) and N = M = 20 then the procedure only requires N * M = 400 points).

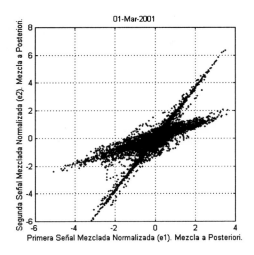

Figure 3: Space of observations for two speech signals.

4.- To further reduce the number of points, the next step of the procedure finds the points which form the border or the zone of maximum density of points in the figure in the space of observations.

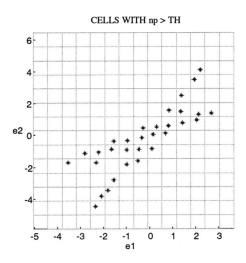

Figure 4: Lattice of the space of observations in cells.

To do this, the algorithm looks for cells that have a neighbourhood (for a cell, the neighbourhood is formed by the cells that are above, below, to the left and to the right of it) in which there is no cell containing a representative point (such cells have fewer points than the threshold TH). Then these cells without a complete neighbourhood form the border of the figure, which have NR points in the space of observations.

5.- The algorithm then calculates the coordinates of the points $P_1 = (p_{11}, p_{12})$ and $P_2 = (p_{21}, p_{22})$. The space of observations has been reduced in step 4 to NR points which are pairs of values (e_1, e_2). In this set of NR points, there are P_1 and P_2 with the greatest euclidean distance (see equation (4)) in the reduced space of observations :

$$d(P_1, P_2) = \max\ d(P_i, P_j) \quad \forall \quad i, j \in \{1, 2, ...\ NR\} \tag{4}$$

6.- Once points P_1 and P_2 are obtained, the algorithm calculates the equation of the straight line R_1 which passes through points P_1 and P_2 (see equation (5)) :

$$A\ e_1 + B\ e_2 + C = 0 \tag{5}$$

where $A = (P_{22} - P_{12})$
$B = (P_{11} - P_{21})$
$C = (P_{21} \cdot P_{12}) - (P_{22} \cdot P_{11})$.

7.- After this, the algorithm obtains the coordinates of the points $P_3 = (p_{31}, p_{32})$ and $P_4 = (p_{41}, p_{42})$ as follows: the straight line R_1 divides the reduced space of observations (e_1, e_2) into two subspaces where R_1 is the border between them. The points which lie within one of these subspaces (for example, above the straight line R_1), give a negative result in equation (5). There is one point $P_3 = (p_{31}, p_{32})$ that provides the most negative value of all the possibilities in (5), and which is also the point at the greatest Euclidean distance from the straight line R_1 in the subspace above R_1. In the same way, points in the other subspace (below the straight line R_1), give a positive result in equation (5). There is one point $P_4 = (p_{41}, p_{42})$ that provides the most positive value of all the possibilities in (5), and which is also the point at the greatest Euclidean distance from the straight line R_1 in the subspace below R_1. In both cases, the algorithm calculates the Euclidean distance from a generic point $P_i = (p_{i1}, p_{i2})$ in the reduced space of observations (e_1, e_2) to the straight line R_1 by equation (6) :

$$d(P_i, R_1) = \frac{AP_{i1} + BP_{i2} + C}{\sqrt{A^2 + B^2}} \tag{6}$$

where A, B and C are the coefficients of the straight line R_1 given in equation (5).

8.- Once the characteristic points of the parallelogram are obtained, the algorithm computes the slopes of the segments (P_1P_3 and P_1P_4) or (P_2P_4 and P_3P_2) in order to obtain the coefficients of the calculated matrix W as in equation (7) (see Figure 5).

$$\begin{pmatrix} a_{12} \\ a_{22} \end{pmatrix}^{-1} = \frac{P_{32} - P_{12}}{P_{31} - P_{11}} \quad ; \quad \begin{pmatrix} a_{21} \\ a_{11} \end{pmatrix} = \frac{P_{42} - P_{12}}{P_{41} - P_{11}} \tag{7}$$

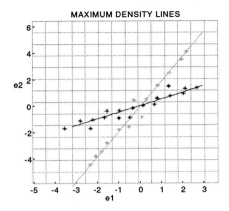

Figure 5: Obtention of the straight lines of the figure.

9.- Using the coefficients of the calculated mixture matrix W, the procedure computes the inverse matrix and reconstructs the signals ($s_i(t)$) (see equation (3)).

4 Simulations and Results

4.1 Simulation 1

The original signals are two uniform noises. The original mixture matrix (A) and the matrix (W) obtained for the procedure are:

$$A = \begin{pmatrix} 1 & 0.5 \\ -0.5 & 1 \end{pmatrix} \quad ; \quad W = \begin{pmatrix} 1 & 0.499 \\ -0.501 & 1 \end{pmatrix}$$

The procedure uses a lattice of 20 by 20 cells to divide the space of observations (N = 20, M = 20). The total number of points for each signal is 10000. The threshold TH is set at 15 points. The Crosstalk figures obtained for each signal after reconstruction of the estimated signals $s_i(t)$ were :

- Crosstalk 1 = - 61.50 dB
- Crosstalk 2 = - 59.09 dB

4.2 Simulation 2

The original signals are two words. The original mixture matrix (A) and the matrix (W) obtained for the procedure are:

$$A = \begin{pmatrix} 1 & 0.5 \\ 0.5 & 1 \end{pmatrix} \; ; \; W = \begin{pmatrix} 1 & 0.52 \\ 0.52 & 1 \end{pmatrix}$$

The procedure uses a lattice of 15 by 15 cells to divide the space of observations (N = 15, M = 15). The total number of points for each signal is 10000. The threshold TH is set at 25 points. The Crosstalk figures obtained for each signal after reconstruction of the estimated signals $s_i(t)$ were :

- Crosstalk 1 = - 19.95 dB
- Crosstalk 2 = - 20.13 dB

4.3 Simulation 3

The original signals are $s_{o1} = \cos^2(t \cdot \cos^2 t)$ and $s_{o2} = 1 + \sin(t \cdot \cos t)$. The original mixture matrix (A) and the matrix (W) obtained for the procedure are:

$$A = \begin{pmatrix} 1 & 0.75 \\ -0.25 & 1 \end{pmatrix} \; ; \; W = \begin{pmatrix} 1 & 0.73 \\ -0.24 & 1 \end{pmatrix}$$

The procedure uses a lattice of 10 by 10 cells to divide the space of observations (N = 10, M = 10). The total number of points for each signal is 10000. The threshold TH is set at 25 points. The Crosstalk figures obtained for each signal after reconstruction of the estimated signals $s_i(t)$ were :

- Crosstalk 1 = - 32.78 dB
- Crosstalk 2 = - 40,72 dB

5 Conclusions

In this work we have shown a new geometry-based method for blind separation of two sources. In comparison with other methods this new one has a low computational cost and it is very intuitive in terms of computer application. Furthermore, this method could be used to detect the perimeter or outlines in simple two-dimensional figures.

6 Future Research

In the future we will intend to implement this method for more than two signals (in general for a p-dimensional space with p mixed signals). We are currently developing a system to make real mixtures with two microphones in real time and with one interface in order to apply this algorithm. We are also thinking of developing a Blind Separation of Sources system based on a digital signal processor (DSP).

Acknowledgments

This work has been supported by the Spanish CICYT Project TIC98-0982 "ALYASS" (Algorithms and Architectures for Source Separation).

References

1. Jutten, C., Hérault, J., Comon, P., Sorouchiary, E.: Blind separation of sources, Parts I, II, III. Signal Processing, 24 (1), 1991, 1-29.
2. P. Comon, Independent Component Analysis, A new concept?, *Signal Processing*, 36 (3), 1994, 287-314.
3. E. Moreau and O. Macchi, New self-adaptive algorithms for source separation based on contrast functions, *Proceedings of IEEE Signal Processing Workshop on Higher-Order Statistics*, Lake Tahoe (USA), 1993, 215-219.
4. M. Gaeta and J. L. Lacoume, Source separation without a priori knowledge: the maximum likelihood solution, *Proceeding of EUSIPCO' 90*, Barcelona (Spain), 1990, 621-624.
5. K. Matsuoka, M. Ohya and M. Kawamoto, A neural net for blind separation of nonstationary signals, *Neural Networks*, 8 (3), 1995, 411-419.
6. A. J. Bell and T. J. Sejnowski, An information-maximisation approach to blind separation and blind deconvolution, *Neural Computation*, 7, 1995, 1129-1159.
7. C. G. Puntonet, A. Prieto, C. Jutten, M. Rodríguez and J. Ortega, J, Separation of sources: a geometry-based procedure for reconstruction of n-valued signals, *Signal Processing*, 46 (3), 1995, 267 - 284.
8. C. G. Puntonet, M. Rodríguez-Alvarez and A. Prieto, A geometric method for blind separation of sources.. *Proceedings of the IASTED International Conference on Artificial Intelligence and Soft Computing*. Banff (Canada), 1997, 372-375.
9. C. G. Puntonet, M. Rodríguez-Alvarez, A. Prieto and B. Prieto, Separation of speech signals for nonlinear mixtures. *Proceedings of the 5th International Work-Conference on Artificial and Natural Neural Networks (IWANN'99)* Alicante (Spain), 1999, 665 - 673.

Adaptive ICA with Order Statistics in Multidimensional Scenarios

Yolanda Blanco, Santiago Zazo, Jose M. Paez-Borrallo

ETS Ingenieros Telecomunicacion. Universidad Politecnica de Madrid
ETSIT Ciudad Universitaria S/N. 28040. Madrid. Spain.
yolanda@gaps.ssr.upm.es

Abstract. In this paper we propose an alternative statistical Gaussianity measure whose optimization provides the extraction of one non-gaussian independent component at each stage of an ICA procedure; this measure is based on the Cumulative Density Function (cdf) instead of traditional distribution distances over Probability Density Functions (pdf's). Additionally, a novel multistage-deflation algorithm is proposed in order to perform ICA in multidimensional scenarios very efficiently; although this approach can be applied to any multistage ICA method, we have developed it to speed up our ICA procedure based on Order Statistics (OS). The algorithm consists on a gradient learning rule plus an orthonormalization projection technique that decreases the vector space dimension progressively [1].

1 Introduction

The goal of BSS (Blind Source Separation) is to extract N unknown independent sources from a set of mixtures of them. The generic model is shown in Fig.1 where it can be appreciate that a useful spatial decorrelation preprocessing is performed (*for more detail see* [10]); afterwards Independent Component Analysis (ICA) is applied in order to find a linear transformation $\mathbf{w} = \mathbf{Bz}$ so that the components w_i are a consistent estimation of the sources (although they might appear in an arbitrary order and multiplied by scale factors).

Most of the ICA methods consist on extracting all the independent components simultaneously by means of a maximization of an independence measure between the outputs signals. The original independence measure is the Mutual Information (MI) [5] as the Kullback - Leibler divergence between the joint pdf and the product of marginal pdf's. Many methods are deduced from MI: let us mention the Maximum Likelihood approach [9] and fourth cross cumulants methods [5],[4].

More recently a new ICA approach line has appeared: the independent components (IC's) are extracted one by one by means of the optimization of a Gaussianity measure at each single output channel [3],[7],[6]. One of the main

---
[1] Work partly supported by national projects 07T/0032/2000 and TIC2000-1395-C02-02 .

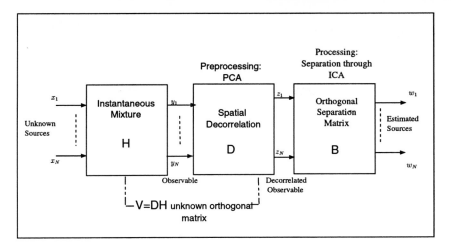

Fig. 1. Generic model of BSS

advantages in front of traditional ICA is that it decreases computational cost because the degrees of freedom are reduced. Also it's highly recommended when N increases because methods like JADE ([4]) has high computational memory requirements.

In the next section we deduce a new Gaussianity measure in terms of cdf's instead of extended pdf's. It reports some interesting advantages like the robustness against outliers. *Section 3* will concern with the adaptive processing of the new cost function and *section 4* will explain a more efficient deflation procedure for the multidimensional case. Some successful comparative results conclude the paper.

2 Gaussianity measure based on Order Statistics

The Central Limit Theorem states that the linear combination of independent random variables always increases the Gaussianity of the combined distribution. Therefore ICA must simply decrease the Gaussianity of the output channel to extract one of the original non-Gaussian independent components. Gaussianity is an appropriate distance between the analyzed signal w_i and the equivalent Gaussian distribution g with the same power.

Negentropy is the original Gaussianity measure; it was defined as the Kullback Leibler distance between both pdf's: f_{w_i}, f_g. Kurtosis ([7]) arised like an estimation of Negentropy when the pdf is considered as $A_4 \exp(-a_4|w_i|^4)$; recently a family of Gaussiniaty measures (built with non-linearities) has been proposed in [6] like an extension of the Kurtosis measure.

By our own we recall that distance between two distributions can be equally evaluated using cdf's instead of pdf's. Specifically let us define a family of distributions distances through the norm concept applied to the difference between both cdf's. For ICA proposal the distance between the analyzed signal w_i and its equivalent Gaussian distribution would be in terms of the Infinite Norm ($L^\infty Norm$):

$$d(F_{w_i}, F_g) = \max_w |F_{w_i}(w) - F_g(w)| \qquad (1)$$

Furthermore, it can be appreciate that the Infinite norm in *equation* (1) is exactly the well known Kormogoroff - Smirnov test [8] to evaluate if two distributions are or not coincident.

The estimation $\widehat{F}_{w_i}(w)$ is necessary in the evaluation of previous distance but it would require a high computational cost if it is performed through accumulated histograms; fortunately, an equivalent distance easy to estimate can be established in terms of inverse cdf's: $Q_{w_i} = F_{w_i}^{-1}, Q_g = F_g^{-1}$; more indeed the only difference with (1) is that the distance is taken over the horizontal axis instead of the vertical one.

$$D(Q_{w_i}, Q_g) = \max_w |Q_{w_i}(w) - Q_g(w)| \qquad (2)$$

The estimation of Q_{w_i} can be performed very robustly in a simple practical way reordering the n temporal discrete samples from $\{w_i[1], ...w_i[n]\}$ and obtaining the set of Order Statistics $w_{i(1)} < w_{i(2)} < ... < w_{i(n)}$:

$$w_{i(k)} = \widehat{Q}_{w_i}(\frac{k}{n}) \Leftrightarrow \widehat{F}(w_{i(k)}) = \frac{k}{n} \qquad (3)$$

Consequently the estimation of equation (2) can be expressed using Order Statistics notation:

$$D(Q_{w_i}, Q_g) = \max_k |w_{i(k)} - Q_g(\frac{k}{n})| \qquad (4)$$

More indeed for symmetric distributions the couple of symmetric Order Statistics :$w_{i(k)}$, $w_{i(l)}$ (with *l=n-k)* provides the same information than eq. (4) for the appropriate values of *(k,l)* :

$$D(Q_{w_i}, Q_g) = |w_{i(k)} - Q_g(\frac{k}{n})| + |w_{i(l)} - Q_g(\frac{l}{n})| \qquad (5)$$

The previous expression has been evaluated for both kind of distributions (sub and super-Gaussians), resulting the following Gaussianity distance in both cases. Let us call $J(w_i)$ this distance which denotes the cost function [2] to be maximized in order to extract a non-Gaussian symmetric source in the analysis channel w_i:

$$J(w_i) = D(Q_{w_i}, Q_g) = |w_{i(k)} - w_{i(l)} + 2Q_g(\frac{l}{n})| \qquad (6)$$

Let us remark that Infinite-Norm in equation (6) is a modified Kormogoroff-Smirnov (K-S) estimator that allows evaluating how much closer both distributions are [8]. In practice $k=75\%$ can be chosen like the optimum value to yield the K-S test in eq.(6), (see [10]).

---

[2] Let us give the numerical deduced constant values $Q_g(1\%) \cong -3.8245$, $Q_g(25\%) \simeq -0.8$, they are corresponding to the Gaussian distribution $(0, \sigma^2 = 1)$. The output channel power is always 1 because $E[\mathbf{zz}^T] = \mathbf{I}$ and **B** is unitary .

3 Adaptive Formulation

According to the previous section separation requires the maximization of the cost function at equation (6) at the i-stage of a multistage procedure.

Taking account that $w_i = \mathbf{b}_i^T \mathbf{z}$, where $\mathbf{b}_i^T$ is the i-row of the separation matrix $\mathbf{B}$, the goal is to update $\mathbf{b}_i$ properly at each stage. Let us apply an adaptive algorithm based on stochastic gradient rules. Let us remark that the gradient expression of a similar cost function was deduced in [2] in terms of a rotation angle for the simplest scenario where $N=2$. The goal of the present work is to generalize the gradient algorithm in [2] to the multidimensional problem:

$$\mathbf{b}_i[t+1] = \mathbf{b}_i[t] + \mu \nabla J|_{\mathbf{b}_i[t]} \quad (7)$$

Where the gradient is the derivative of $J(w_i(\mathbf{b}_i))$ in equation (6) respect to $\mathbf{b}_i$, it results:

$$\nabla J|_{\mathbf{b}_i[t]} = S \left. \frac{d(w_{i(k)} - w_{i(l)})}{d\mathbf{b}_i} \right|_{\mathbf{b}_i[t]} \quad (8)$$

where $S = sign(w_{i(k)} - w_{i(l)} + 2Q_g(\frac{l}{n}))_{\mathbf{b}_i[t]}$

It results after applying the chain rule:

$$\left. \frac{d(w_{i(k)} - w_{i(l)})}{d\mathbf{b}_i} \right|_{\mathbf{b}_i[t]} = \left(\left. \frac{dw_{i(k)}}{dw_i} \right|_{\mathbf{b}_i[t]} - \left. \frac{dw_{i(l)}}{dw_i} \right|_{\mathbf{b}_i[t]} \right)^T \mathbf{z} \quad (9)$$

The derivative of any r-order statistic respect to w_i was obtained by means of the derivative of a vector in [2] and it finally gives:

$$\left. \frac{dw_{i(r)}}{dw_i} \right|_{\mathbf{b}_i[t]} = \mathbf{d}_r = [0, 0, ..0, 1, 0..0]^T;$$

where

$$\mathbf{d}_r[j] = \left\{ \begin{array}{l} 1 \ if \ w_i[j] = w_{i(r)} \\ 0 \ rest \end{array} \right\}_{j=1...n} \quad (10)$$

In order to extract several IC's a multistage procedure is proposed in [6], which is based on the following fact: the decorrelation preprocessing implies that the linear ICA transformation is an orthogonal matrix transformation; more indeed one could exploits orthonormality of $\mathbf{B}$ matrix in order to obtain the vectors $\mathbf{b}_i$ successively. The procedure would be for every i-stage:

1-Updating $\mathbf{b}_i$ according to equation (7) until the maximum is reached (when gradient of J is null).

2-Normalize $\mathbf{b}_i$ to norm one.

3-Projecting $\mathbf{b}_i$ onto the space NxN orthonormal to the previous vectors $\{\mathbf{b}_j\}_{j<i}$, in order to guarantee the independence of w_i with the previously extracted sources.

Previous method can be described by means of the block diagram in Fig.2, valid for every i-stage as a parallel deflation procedure:

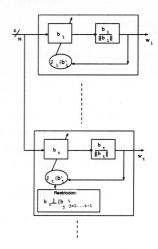

Fig. 2. Parallel "ICA with OS" procedure.

4 Effective Implementation based on serial Orthogonal Projections

Previous method in *section 3* performs fairly well because it consists basically as a maximization technique and the projection of the separation vector onto the orthonormal full-space NxN spanned by the previous separation vectors. However, let us recall that any stage has to deal with the same dimension **z** vector where, from an intuitive point of view, it is expected that when a source is extracted, the problem can also be reduced in one dimension.

A more efficient implementation can be developed constraining the search of any separation vector $\mathbf{b}_i$ (of dimension $(N-i+1)$) to the orthonormal subspace $\mathbf{B}_i$ (of dimension $(N-i)$x$(N-i+1)$) spanned by only one separation vector which is the previous one $\mathbf{b}_{i-1}$; therefore the dimension of the problem is reduced at each stage. An schematic view can be observed in *figure* (3).

Let us observe that there is an evident relationship with the GSLC (Generalized Side Lobe Canceler) procedure where the blocking matrix $\mathbf{B}_i$ is blocking previous separation vectors $\mathbf{b}_1, \mathbf{b}_2 ... \mathbf{b}_{i-1}$. Let us remark that now the maximization algorithm is unconstrained and also with faster converge because the vector $\mathbf{b}_i$ is previously forced to belong to the orthonormal subspace $\mathbf{B}_i$.

The calculus of matrix $\mathbf{B}_i$ is based on the Gram-Schmidt (GS) procedure:

$$\{\mathbf{b}_{i-1}, \mathbf{e}_1, ..., \mathbf{e}_{N-i}\} \Rightarrow \{\mathbf{u}_1, ..., \mathbf{u}_{N-i}\} \qquad (11)$$

where $\{\mathbf{e}_k\}$ are the unitary vectors to complete an $N-i$ dimension basis and $\{\mathbf{u}_k\}$ are the orthonormal-basis components through the GS procedure. Therefore, the blocking matrix is $\mathbf{B}_i = [\mathbf{u}_2 \, \mathbf{u}_3 \, ... \, \mathbf{u}_{N-i}]$. Let us observe that $\mathbf{B}_i$ is orthonormal to guarantee that vector $\mathbf{z}_i$ maintains the spatial decorrelation feature.

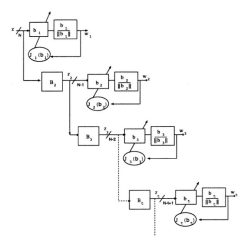

Fig. 3. Serial "ICA with OS" procedure.

5 Simulations

A random mixture of $N=6$ independent sources has been separated by means of both multistage procedures (*parallel and serial*) where the gradient rule is the explained in *section 3*. Excellent Amari's performance index [1] around -18 dB has been obtained in both procedures. Quality of the separation that has been carried out by the serial deflation procedure is shown in *figure* (4). On the other hand, *figure* (5) shows clearly the faster and improved convergence of the serial method described in *section 4*.

Finally, the new Gaussianity measure has been compared with Kurtosis and the non-linear Gaussianity measure proposed in [6]. They have been applied to the separation of several scenarios of mixtures of two sources, outliers (with power $=4\sigma_{x_i}^2$) have been added in 1% of the samples. Amari's index [1] in dB is shown in 5 for 1000 processed samples, being lower with the OS measure. In table 5 it's observed that the new measure improves the behavior against outliers when the OS are (75%,25%), it is because the contaminated samples are filtered by the procedure towards the extreme OS.

6 Conclusion

We propose a new Gaussianity measure to solve the ICA problem that offers certain advantages:

1. J must be always maximized, allowing a simple procedure to separate hybrid mixtures.
2. Implementation of a gradient rule for J optimization is really efficient and easy based only on samples ordering.

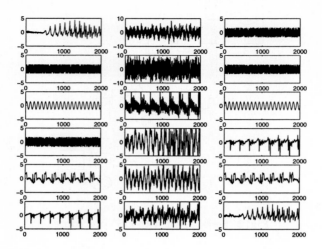

Fig. 4. Column 1: Original signals. Column 2: Observed mixed signals. Column 3: Extracted signals.

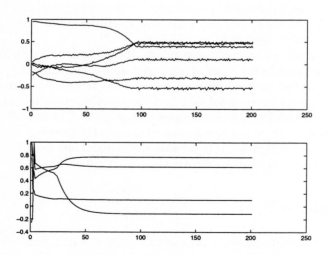

Fig. 5. Convergence of the coefficients of b_3 at *3-stage: Up:* parallel procedure. *Bottom:* serial deflation procedure.

Table 1. Comparative performance index in presence of outliers for several Gaussianity measures: Kurtosis; based on non-linearity : $E\{\log \cosh(w_i)\}$; based on OS: $|w_{1(75\%)} - w_{1(25\%)}|$.

	Kurtosis	Non-lin.	OS
2 Laplacian	-11.90	-14.58	-15.48
2 Uniform	-10.06	-8.93	-10.89
1 Lap.,1 Unif.	-10.80	-10.11	-11.44
1 Unif, 1Gaussian	-9.00	-8.00	-14.68
1 Lap., 1 Gaussian	-11.01	-10.86	-14.34

3. It is more robust against outliers . Other side, an efficient implementation for multidemensional mixtures is provided constraining the search of any separation vector to the subspace orthonormal to the previous extracted vector. This solution reduces computational complexity and speeds up convergence.

References

1. S. Amari , A. Cichocki, H. H. Yang. A New Learning Algorithm for Blind Signal Separation. *Proc. of Neural Information Processing Systems, NIPS 96*, vol. 8, pp 757-763.
2. Y. Blanco, S. Zazo, JM Paez-Borrallo. Adaptive Processing of Blind Source Separation through 'ICA with OS' *Proceedings of the International Conference On Acoustics And Signal Processing ICASSP'00* . Vol I, 233-236.
3. Y.Blanco, S. Zazo, J.C. Principe. Alternative Statistical Gaussianity Measure using the Cumulative Density Function. *Proceedings of the Second Workshop on Independent Component Analysis and Blind Signal Separation: ICA'00* . pp 537-542
4. JF Cardoso, A. Soulimiac. Blind beamforming for Non Gaussian Signals. *IEE Proceedings-F* , vol 140, n 6, pp 362-370, December 1993.
5. P.Common . Independent Component Analysis, A New Concept. *Signal Processing*, n 36, pp 287-314. 1992.
6. A. Hyvrien, E. Oja. A Fast Fixed Point Algorithm for Independent Component Analysis *Neural Computation*, vol 9 (n 7), pp 1483-1492.1997.
7. S.Y. Kung, C.Mejuto. Extraction of Independent Components from Hybrid Mixture: Knicnet Learning Algorithm and Applications. In *proc. International Conference On Acoustics And Signal Processing ICASSP'98* . vol II, 1209,1211.
8. A. Papoulis. Probability and Statisitics. Prentice Hall International, INC. 99
9. D. T. Pham, P. Gharat, C. Jutten. Separation of a Mixture of Independent Sources through a Maximum Likelihood Approach. *Proc. EUSIPCO 1992*771-774
10. S. Zazo, Y. Blanco, J.M. Paez-Borrallo. Order Statistics: a New Approach to the problem of Blind Source Separation. *Proc. of the first Workshop on Independent Component Analysis and Blind Signal Separation: ICA'99*. pages 413-41

Pattern Repulsion Revisited

$Fabian J. Theis^1$, $Ch. Bauer^{1,2}$, $C. Puntonet^2$, $Elmar W. Lang^1$

1 Institute of Biophysics
University of Regensburg, D-93040 Regensburg, Germany
email: elmar.lang@biologie.uni-regensburg.de
2 Departamento de Arquitectura y Tecnologia de Computadores
E.T.S. Ingenieria Informatica, Universidad de Granada, E-18071 Granada, Spain

Abstract. Marques and Almeida [9] recently proposed a nonlinear data seperation technique based on the maximum entropy principle of Bell and Sejnowsky. The idea behind is a pattern repulsion model alluding to repulsion of physical particles in a restricted system which leads to a uniform distribution, hence a maximum entropy, under certain conditions to the energy function. In this paper, we want to revisit their model and give a rigorous mathematical framework in which this algorithm indeed converges to a demixing function.

1 Introduction

In the square case of **independent component analysis** (ICA), a random vector $X : \Omega \to \mathbb{R}^N$ called **sensor signal** is given. It is assumed to result from an independent random vector $S : \Omega \to \mathbb{R}^N$, which will be called **source signal**, via a mixing tranformation through an invertible **nonlinear mixing function** $\mu : \mathbb{R}^N \longrightarrow \mathbb{R}^N$, ie. $X = \mu \circ S$. Only the sensor signal is usually known, and the task is to recover μ and therefore $S = \mu^{-1} \circ X$.

In the linear case, where μ is linear, many algorithms have been proposed with the Bell-Sejnowski maximum entropy algorithm being the most popular and also most widely studied. In the general nonlinear case however, hardly any solution exists. Without any further assumption the separation problem is ill-posed and an infinite number of solutions exist.

In this paper we study an ICA algorithm called **pattern repulsion** introduced by Marques and Almeida [9]. The underlying idea is the fact that in a closed restricted physical system, where particles are subject to a force decreasing faster than the Coulomb force, these particles will eventually arrive at a uniform distribution. This in turn means that the entropy of the system is maximized and therefore its mutual information minimized, as will be shown.

2 Mathematical Preliminaries

In the following the square case of independent component analysis (ICA) will be considered only, i.e. the number of sources n is assumed to equal the number of sensors m.

2.1 A contrast function

Now for $m, n \in \mathbb{N}$ let $M(m \times n)$ be the $\mathbb{R}$–vectorspace of real $m \times n$ matrices, and $\mathrm{Gl}(n) := \{W \in M(n \times n) \mid \det(W) \neq 0\}$ be the general linear group in $\mathbb{R}^n$. Given then a **transformation space**

$$\mathcal{T} \subset \{f : \mathbb{R}^N \longrightarrow \mathbb{R}^N \mid f \text{ measurable }\}$$

and a **contrast function**

$$\kappa : \mathcal{T} \longrightarrow \mathbb{R}$$

, the goal of any nonlinear ICA algorithm is to minimize κ.
Consider for example the **canonical contrast function** $\kappa(\tau) := I(\tau \circ X)$, $\tau \in \mathcal{T}$, where $I(Y)$ denotes the **mutual information** of the random vector Y, minimizing κ means minimizing the mutual information. Hence at a minimal τ_0, the transformed random vector $\tau_0 \circ X$ is as independent as possible. This solution thus corresponds to a **minimal mutual information**(MMI) of the components of the random output vector $\tau_0 \circ X$.
An alternative contrast function is the negative entropy $\kappa(\tau) := -H(\tau \circ X)$ (sometimes a limiting function is added in the formula in order to render the entropy finite). The induced minimization method, denoted by **maximum entropy** (ME), is often easier to implement. Further, it can be shown [10] to also extract, under mild restricting assumptions, independent components of any given sensor signals, hence is, at least locally, equivalent to MMI even in a nonlinear setting.

2.2 A minimization procedure

If it is assumed that the inverse of the mixing function lies already in the transformation space ($\mu^{-1} \in \mathcal{T}$), then the global minimum of the canonical contrast function has value $\kappa = 0$, so indeed the global minimum corresponds to a truely independent random vector. Of course it cannot be hoped that μ^{-1} can be found in general, as uniqueness in this general setting cannot be achieved [8] — in contrast to the linear case ($\mathcal{T} \subset \mathrm{Gl}(N)$) [5].

Obviously the setting above is far too general to yield useful computational results, because it is not possible to do minimization if it is not clear how to describe elements of $\mathcal{T}$. Therefore a means of describing $\mathcal{T}$ by a finite set of real-valued parameters is needed. An attractive possibility is to consider output functions of neural networks, see [2] and [6]. This has the advantage that in neural networks, being adaptive systems, it is known how to algorithmically minimize a given energy function for example by using any version of the standard gradient descent method. Moreover, more general functions can then be learned approximately as sufficiently complex neural networks are so called universal approximators [7].

Fixing notation, for any neural net Γ with N inputs and N outputs, $A(\Gamma) : \mathbb{R}^N \longrightarrow \mathbb{R}^N$ denotes its output function. For fixed Γ, $\Gamma(w)$ is the net Γ, where the weights have been replaced with the new weights $w \in \mathbb{R}^n$. So one obtains

$$\mathcal{T} \subset \{A(\Gamma(w)) \mid w \in \mathbb{R}^n\}$$

and $\overline{\kappa}(w) := \kappa(A(\Gamma(w)))$ is to be minimized.

2.3 Basics from information theory

In the following use will be made of an important lemma from information theory:

Lemma 1. *Let $A \subset \mathbb{R}^n$ be measurable of finite Lebesgue measure $\lambda(A) < \infty$. Then the maximum of the entropy of all n-dimensional Lebesgue-continuous random vectors X, whose corresponding density functions have support in A and for which $H(X)$ exists, is obtained exactly at the random vector X_* being uniformly distributed in A.*

In other words: For the random vector X_* with the density

$$f_* := \frac{1}{\lambda(A)}$$

holds: All X as above with density $f_X \neq f_*$ satisfy $H(X) < H(X_*) = \log \lambda(A)$.

Proof. Let X be as above with density f_X. The Gibbs-Inequality for X and X_* then shows:

$$H(X) \leq -\int_A f_X \log f_* dx = -\log\left(\frac{1}{\lambda(A)}\right) \int_A f_X dx = \log \lambda(A) = H(X_*)$$

and equality holds iff $f_X = f_*$. □

3 Pattern Repulsion

3.1 The particle repulsion analogue

Let $\Theta : \mathbb{R}_+ \longrightarrow \mathbb{R}_+$ be a given continuously differentiable positive function with compact support $[0, \epsilon]$; we call Θ **particle energy function**. So Θ vanishes on $(\epsilon; \infty)$. Note, however, that in real world applications the particle energy function is often not explicitly assumed to have compact support, instead it is mostly chosen to be integrable and to be 'small' outside a small spherical region around the origin. We however want to restrict ourselves to the case of compact support in order to omit an additional approximation constant.

In the physical picture, a particle or a density distribution around the point $\hat{x} \in \mathbb{R}^N$ subjects the particle $x \in \mathbb{R}^N$ to the **repulsive force**

$$F(\hat{x}, x) := \frac{\partial \Theta(|\hat{x} - x|)}{\partial x} = \Theta'(|\hat{x} - x|)\frac{\hat{x} - x}{|\hat{x} - x|} \in \mathbb{R}^N,$$

if $|.|$ denotes the euclidean norm on $\mathbb{R}^N$. Hence, the particle x in the field generated by $\hat{x}$ has the potential energy $\Theta_{\hat{x}}(x) := \Theta(|\hat{x} - x|)$.

3.2 Conditions on pattern density distributions

In the pattern repulsion model, the density distribution corresponds to the density ρ_w of the random vector $Y := A(\Gamma(w)) \circ X$ which is the vector X transformed with the neural network output function $A(\Gamma(w)) \in \mathcal{T}$. We assume now that ρ_w has support in $K \subset \mathbb{R}^N$ compact for all $w \in \mathbb{R}^n$; note that this is always the case for the usual activation functions having a bounded image. Moreover, we assume ρ_w to be square integrable, that is $\rho_w \in \mathcal{L}^2$ for all $w \in \mathbb{R}^n$. A further assumption to ρ_w is that ρ_w is not allowed to vary too much with respect to the particle energy function support width ϵ. This is to say there exists a $\delta > 0$ such that for all $w \in \mathbb{R}^n$ and all $x, \hat{x} \in \mathbb{R}^N$ with $|x - \hat{x}| < \epsilon$ the relation $|\rho_w(x) - \rho_w(\hat{x})| < \delta$ holds.

The ICA-problem then corresponds to the algorithmic computation of a $w \in \mathbb{R}^n$ such that $\rho_w : \mathbb{R}^N \longrightarrow \mathbb{R}$ is uniform in K; this will lead, according to lemma 1, to the maximization of the entropy (ME) and (under mild assumptions) hence to a minimization of the mutual information (MMI), as will be discussed at the end of this paper [10].

With these assumptions we now can define the **total energy** for $w \in \mathbb{R}^n$ by

$$\Omega(w) := \frac{1}{2} \int_{\mathbb{R}^N} \int_{\mathbb{R}^N} \Theta(|x - \hat{x}|) \rho_w(x) dx \rho_w(\hat{x}) d\hat{x},$$

where ρ designates the density of the random vector X.

3.3 Estimation of the total pattern energy

In actual computations, the density of the random vector X is not known, however. It is only represented through a collection of samples x_λ, which result from i.i.d. sampling according to X for $\lambda \in \Lambda$. These samples can be used to approximate the total energy as follows:

Lemma 2. *For the number of samples* $|\Lambda| \to \infty$,

$$\Omega^\Lambda(w) := \frac{1}{2|\Lambda|^2} \sum_{\lambda, \mu \in \Lambda} \Theta(|A(\Gamma(w))(x_\lambda) - A(\Gamma(w))(x_\mu)|)$$

converges almost surely to the total energy $\Omega(w)$.

Proof. Let X_λ be the random vector according to which x_λ has been sampled for $\lambda \in \Lambda$. Using the theorem of large numbers,

$$\frac{1}{|\Lambda|} \sum_{\lambda \in \Lambda} \Theta(|A(\Gamma(w))(x_\lambda) - A(\Gamma(w))(\hat{x})|)$$

converges for fixed $\hat{x} \in \mathbb{R}^N$ almost surely to the expectation

$$E\left(\Theta(|A(\Gamma(w)) \circ X - A(\Gamma(w))(\hat{x})|)\right) = \frac{1}{|\Lambda|} \sum_{\lambda \in \Lambda} \int_{\mathbb{R}^N} \Theta(|x - A(\Gamma(w))(\hat{x})|) \rho_w(x) \, dx.$$

Application of the theorem of large numbers once more to the second last equation yields almost sure convergence of

$$2\Omega^A(w) = \frac{1}{|A|^2} \sum_{\lambda,\mu \in A} \Theta(|A(\Gamma(w))(x_\lambda) - A(\Gamma(w))(x_\mu)|)$$

to

$$E\left(\frac{1}{|A|} \sum_{\lambda \in A} \Theta(|A(\Gamma(w))(x_\lambda) - A(\Gamma(w)) \circ X|)\right)$$

so altogether to $2\Omega(w)$. □

3.4 The density uniformization method

Now let $C := \int_{\mathbb{R}^N} \Theta(|x|)dx$. Then using affine and spherical coordinate transformations we calculate for all $\hat{x} \in \mathbb{R}^N$:

$$C = \int_{\mathbb{R}^N} \Theta(|x - \hat{x}|)dx$$
$$= \int_{\mathbb{R}_+} \Theta(r)r^{n-1}dr \operatorname{vol}(S^{n-1}),$$

where $\operatorname{vol}(S^{n-1})$ denotes the volume of the $n-1$-sphere.

The following two lemmata now give the desired density uniformization method:

Lemma 3. *The total energy can be written as*

$$\Omega(w) = \frac{C}{2}\int_K \rho_w^2(x)dx + \frac{C}{2}\int_K \delta(x)\rho_w(x)dx,$$

where $\delta : \mathbb{R}^N \longrightarrow \mathbb{R}$ has support in K and

$$\sup_{x \in K} |\delta(x)| < \delta.$$

Proof. First note that ρ_w is integrable and positive in accord with the assumptions. $\Theta_{\hat{x}}(x) := \Theta(|\hat{x} - x|) : \mathbb{R}^N \longrightarrow \mathbb{R}$ is continuous so that application of the mean value theorem of integral calculus to the integral

$$\int \rho_w|\overline{B}_\epsilon(\hat{x})\Theta_{\hat{x}}|\overline{B}_\epsilon(\hat{x})d\hat{x},$$

where $\overline{B}_\epsilon(\hat{x})$ denotes the closed sphere in $\mathbb{R}^N$ centered at $\hat{x}$ with radius ϵ, yields the following: There exists a $\xi_{\hat{x}} \in \mathbb{R}^N$ with $|\xi_{\hat{x}} - \hat{x}| < \epsilon$ such that

$$\int_{\overline{B}_\epsilon(\hat{x})} \Theta_{\hat{x}}(x)\rho_w(x)dx = \rho_w(\xi_{\hat{x}})\int_{\overline{B}_\epsilon(\hat{x})} \Theta_{\hat{x}}(x)dx.$$

Since $\Theta_{\hat{x}}$ has support in $\overline{B}_\epsilon(\hat{x})$ this means

$$\int_{\mathbb{R}^N} \Theta_{\hat{x}}(x)\rho_w(x)dx = C\rho_w(\xi_{\hat{x}}).$$

So, if we define for $\hat{x} \in \mathbb{R}^N$

$$\delta(\hat{x}) := \frac{1}{C}\left(\int_K \Theta_{\hat{x}}(x)\rho_w(x)\,dx\right) - \rho_w(\hat{x}),$$

then δ has support in K and fulfills the first equation from the lemma. Furthermore, due to the above

$$|\delta(x)| = |\rho_w(\xi_{\hat{x}}) - \rho_w(\hat{x})| < \delta,$$

because of the assumptions to the densities ρ_w. This proves the lemma. $\square$

Lemma 4. *Assume there is a $w_0 \in \mathbb{R}^n$ such that ρ_{w_0} is of the form*

$$\rho_{w_0} = \frac{1}{\mathrm{vol}\,K}\chi_K,$$

where χ_K denotes the characteristic function of K. Then the $\mathcal{L}^2$-norm of ρ_w i.e. the mapping

$$w \longmapsto \|\rho_w\|_2 := \left(\int_K \rho_w^2(x)dx\right)^{1/2}$$

is minimal in $w \in \mathbb{R}^n$ iff ρ_w is uniform in K i.e. if $\rho_w = \rho_{w_0}$.

Note that in the case of using neural networks for the inverse (demixing) transformations it is not possible to construct a non-continuous density of the form $\frac{1}{\mathrm{vol}\,K}\chi_K$. The above lemma then has to be restated in a form that $w \longmapsto \|\rho_w\|_2$ converges to an infimum iff ρ_w converges uniformly to $\frac{1}{\mathrm{vol}\,K}\chi_K$.

Proof. For $\rho_{w_0} = \frac{1}{\mathrm{vol}\,K}\chi_K$, we have

$$\|\rho_{w_0}\|_2^2 = \int_K \frac{1}{\mathrm{vol}^2\,K}\chi_K = \frac{1}{\mathrm{vol}\,K}.$$

Since density functions ρ_w have the $\mathcal{L}^1$-norm 1, it follows

$$0 \leq \|\rho_w - \rho_{w_0}\|_2^2$$
$$= \int_K (\rho_w - \rho_{w_0})^2$$
$$= \int_K \rho_w^2 - 2\frac{1}{\mathrm{vol}\,K}\int_K \rho_w + \int_K \rho_{w_0}^2$$
$$= \|\rho_w\|_2^2 - \frac{1}{\mathrm{vol}\,K}$$

so that
$$\|\rho_w\|_2 \geq \|\rho_{w_0}\|_2.$$
Hence ρ_{w_0} is the global minimum of $w \longmapsto \|\rho_w\|_2$. Vice versa, if $\|\rho_w\|_2 = \|\rho_{w_0}\|_2$, then using the above calculation, we deduce
$$\|\rho_w - \rho_{w_0}\|_2 = 0,$$
and therefore $\rho_w = \rho_{w_0}$, as claimed. □

3.5 Marques-Almeida Theorem and statistical independence

If δ is sufficiently small, the above two lemmata show that the total energy is minimal at $w \in \mathbb{R}^n$ where the density ρ_w is uniform. But this means according to lemma 1 that $H(Y) = H(A(\Gamma(w)) \circ X)$ attaines its global maxima exactly at the global minima of the total energy, so that we have shown the following theorem:

Theorem 1 (Marques, Almeida). *For sufficiently small δ the global minima of the total energy $\Omega(w)$ are solutions $A(\Gamma(w)) \in \mathcal{T}$ to the ICA problem in the maximum entropy (ME) model.*

If the source random vector has zero mean, one can show that maximum entropy induces minimal mutual information at least locally [10]. This then shows the next theorem:

Theorem 2. *For sufficiently small δ the transformed random vectors at global minima of the total energy $\Omega(w)$ are statistically independent.*

3.6 Gradient of the total pattern energy function

If we use the approximated total energy Ω^Λ from lemma 2, then the following lemma shows how to calculate the gradient of the total energy. This can be used in an optimization algorithm to minimize the energy.

Lemma 5. *The gradient of the approximated total energy can be calculated as:*
$$\frac{\partial}{\partial w}\Omega^\Lambda(w) = \frac{1}{|\Lambda|^2} \sum_{\lambda,\mu \in \Lambda} F\left(A(\Gamma(w))(x_\lambda), A(\Gamma(w))(x_\mu)\right) \frac{\partial}{\partial w} A(\Gamma(w))(x_\lambda)$$

Proof. Write $y_\lambda := A(\Gamma(w))(x_\lambda)$. First calculate for fixed $\lambda' \in \Lambda$:
$$\frac{\partial}{\partial y_{\lambda'}} \sum_{\lambda,\mu \in \Lambda} \Theta(|y_\lambda - y_\mu|) = \sum_{\lambda,\mu \in \Lambda} (-\delta_{\lambda'\lambda} + \delta_{\lambda'\mu}) F(y_\lambda, y_\mu)$$
$$= -\sum_\mu F(y_{\lambda'}, y_\mu) + \sum_\lambda F(y_\lambda, y_{\lambda'})$$
$$= \sum_\mu F(y_\mu, y_{\lambda'}) + \sum_\lambda F(y_\lambda, y_{\lambda'})$$
$$= 2\sum_\lambda F(y_\mu, y_{\lambda'})$$

Using the chain rule, it follows

$$\frac{\partial}{\partial w}\Omega^\Lambda(w) = \frac{1}{2|\Lambda|^2}\sum_{\lambda'}\frac{\partial}{\partial y_{\lambda'}}\left(\sum_{\lambda,\mu\in\Lambda}\Theta(|y_\lambda - y_\mu|)\right)\Bigg|_{y_{\lambda'}=A(\Gamma(w))(x_\lambda)}$$
$$\frac{\partial}{\partial w}A(\Gamma(w))(x_\lambda)$$
$$= \frac{1}{|\Lambda|^2}\sum_{\lambda,\mu\in\Lambda}F\left(A(\Gamma(w))(x_\lambda), A(\Gamma(w))(x_\mu)\right)\frac{\partial}{\partial w}A(\Gamma(w))(x_\lambda)$$

□

Note that the partial derivatives in the above formula can be calculated by a neural network using the backpropagation learning rule.

4 Conclusions

We have presented a rigorous mathematical formulation to the Marques-Almeida pattern repulsion model of nonlinear ICA. Using lemmata 2 and 5, it is possible to calculate the total energy and its gradient (approximately). This can be used to apply minimization techniques in order to minimize the total energy. According to the theorems above this then gives a solution to the ICA problem.

References

1. L. B. ALMEIDA, *Multilayered Perceptrons.* Handbook of Neural Computation by E. Fiesler and R. Beale, Eds. Institute of Physics (1997), Oxford University Press, available at http://www.oupusa.org/acadref/ncc1_2.pdf
2. M. ANTHONY - P. L. BARTLETT, *Neural Network Learning: Theoretical Foundations.* Cambridge University Press.
3. H. BAUER, *Maš- und Integrationstheorie.* Walter de Gruyter, Berlin – New York (1990).
4. H. BAUER, *Wahrscheinlichkeitstheorie.* 4. Auflage, Walter de Gruyter, Berlin – New York (1990).
5. P. COMON, *Independent component analysis - a new concept?* Signal Processing, (36), pp. 287-314 (1994).
6. S. HAYKIN, *Neural Networks.* Macmillan College Publishing Company (1994).
7. K. HORNIK - M. STINCHCOMBE - H. WHITE, *Multilayer feedforward networks are universal approximators.* Neural Networks, (2), pp. 359-366 (1989).
8. A. HYVÄRINEN - P. PAJUNEN, *Nonlinear Independent Component Analysis: Existence and Uniqueness Results.* Preprint.
9. G. C. MARQUES - L. B. ALMEIDA, *Separation of nonlinear mixtures using pattern repulsion.* in J. F. Cardoso, Ch. Jutten, Th. Loubaton EDS. ICA '99, pp. 277-283.
10. F. J. THEIS, CH.BAUER, E. LANG, *Comparison of ME and MMI in a nonlinear setting* preprint (2001).
11. F. TOPSØE, *Informationstheorie.* Teubner Studienbücher Mathematik (1973).

The Minimum Entropy and Cumulants Based Contrast Functions for Blind Source Extraction

Sergio Cruces[1], Andrzej Cichocki[2], and Shun-ichi Amari[2]

[1] Signal Processing Group, Camino Descubrimientos, 41092-Seville, Spain
sergio@cica.es,
WWW home page: http://viento.us.es/~sergio
[2] Brain Science Institute, RIKEN, 2-1 Hirosawa, Wako-shi, Saitama, 351-0198 Japan
{cia,amari}@brain.riken.go.jp,
WWW home page: http://www.bsp.brain.riken.go.jp/

Abstract. In this paper we address the problem of blind source extraction of a subset of "interesting" independent sources from a linear convolutive or instantaneous mixture. The interesting sources are those which are independent and, in a certain sense, are sparse and far away from Gaussianity. We show that in the low-noise limit and when none of the desired sources is Gaussian, the minimum entropy and cumulants based approaches can solve the problem. These criteria, with roots in Blind Deconvolution and in Projection Pursuit, will be proposed here for the simultaneous blind extraction of a group of independent sources. Then, we suggest simple algorithms which, working on the Stiefel manifold perform maximization of the proposed contrast functions.

1 Introduction

In the recent years the criteria for blind source separation (BSS) of independent and non-Gaussian sources have been an active field of research [15, 8, 4, 16, 6, 3]. The blind separation problem considers the case where a certain number of sources is linearly combined to give the observations and only from these observations we try to recover all the possible sources.

More recently, blind source extraction, the problem of recovering or extraction of only a subset of the most "interesting" independent sources from the linear mixture has gained increasing attention due to its practical applications in communications [22] and in biomedical engineering [20].

The first approaches on blind source extraction can be traced back to the work of Donoho and many others [13, 21] in the single input single output blind deconvolution problem where one is interested in blindly recovering a non-minimum phase filtered non-Gaussian signal which is independent and identically distributed (i.i.d). Independently, but nearly at the same time, another field coined under the name of Projection Pursuit [17] was interested in the spatial counterpart of this problem. The motivation here was to find "interesting" low-dimensional projections of high-dimensional data sets. Both fields arrived to the common conclusion that the pursuit of the most non-Gaussian projection allows to extract one of the independent sources. More recently, but with the same aim, other related criteria and algorithms has been developed in the context of blind source separation and independent component analysis [12, 7, 18, 22, 19]. However, in most cases the developed methods allow only the extraction of the sources one by one or all of them simultaneously.

The main objective of this paper is to extend the concepts and approaches proposed by Amari et al. [1, 2] and Cruces et al. [11] to the case of the simultaneous extraction of several "interesting sparse" sources without the need of employing a deflation procedure. The organization of the paper is as follows. Section 2 discuss mixture signal model and some basic but useful results. Section 3 presents the minimum entropy contrast for blind source extraction while section 4 presents a contrast function based on higher order cumulants. Section 5 extends some of the previous results to the problem of blind deconvolution and section 6 presents family of the algorithms that can perform the optimization of the proposed contrast functions. Finally, section 7 summarizes the conclusions of the paper.

2 Signal Model and Basic Results

Let us consider a vector of N unknown statistically independent source signals $\mathbf{s} = [s_1, \cdots, s_N]^T$ with zero mean and normalized covariance $Cov(\mathbf{s}) = \mathbf{R}_{ss} = E[\mathbf{s}\mathbf{s}^T] = \mathbf{I}_N$. These signals are linearly mixed by unknown nonsingular mixing matrix $\mathbf{A}$ as

$$\mathbf{x} = \mathbf{A}\mathbf{s}, \tag{1}$$

where $\mathbf{x}$ is the available vector of observations (sensor signals).

Without loss of generality we assume that the unknown mixing matrix $\mathbf{A}$ is orthogonal. Note, that the orthogonality of the mixing matrix ($\mathbf{A}\mathbf{A}^T = \mathbf{I}_N$) can be always enforced by simply performing pre-whitening of the original observations.

In order to extract $E < N$ sources, the observations will be further processed by an $E \times N$ semi-orthogonal separating matrix $\mathbf{U}$ satisfying relationship $\mathbf{U}\mathbf{U}^T = \mathbf{I}_E$ which yields to the outputs vector (or estimated sources)

$$\mathbf{y} = \mathbf{U}\mathbf{x} = \mathbf{G}\mathbf{s} \tag{2}$$

where $\mathbf{G} = \mathbf{U}\mathbf{A}$ is also the semi-orthogonal $E \times N$ global transfer matrix of mixing-separating system.

It is well known [8, 5] that for Gaussian stationary sources, independence is not a sufficient condition to separate them. This result is a direct consequence of the Darmois-Skitovich theorem which is presented below.

Theorem 1 (Darmois-Skitovich). *Let* $\mathbf{s} = [s_1[n], s_2[n], \ldots, s_N[n]]^T$ *be an N-dimensional random vector ($N \geq 2$) whose components are mutually independent and consider the two outputs obtained from a linear combination of these sources*

$$\begin{aligned} y_1[n] &= G_{1,1}s_1[n] + G_{1,2}s_2[n] + \ldots + G_{1,N}s_N[n] \\ y_2[n] &= G_{2,1}s_1[n] + G_{2,2}s_2[n] + \ldots + G_{2,N}s_N[n] \end{aligned} \tag{3}$$

Assuming that $y_1[n]$ and $y_2[n]$ are independent, if for any index i $G_{1,i} \neq 0$ and $G_{2,i} \neq 0$ hold, then $s_i[n]$ will have a Gaussian distribution.

Thus, the blind source separation of $E = N$ sources can be identifiable, up to the arbitrary scaling and ordering indeterminacies, if and only if there is at most one Gaussian source in the mixture. This result is easily extended to the case of blind extraction for $E < N$ in the following corollary.

Corollary 1. *For a linear mixture of N stationary and independent sources, the extraction from the observations of any non-Gaussian subset with $E < N$ can be performed, up to the arbitrary ordering and scaling, if and only if, there are at most $N - E_{max}$ (where $E \leq E_{max}$) Gaussian sources in the mixture.*

Another useful result is the entropy power inequality [10] which provides a lower bound on the differential entropy of a sum of two random variables in terms of their individual differential entropies.

Theorem 2 (Entropy power inequality). *If a and b are independent continuous random variables then*

$$2^{2h(a+b)} \geq 2^{2h(a)} + 2^{2h(b)} \tag{4}$$

where $h(a) = -\int p_a(a) \log p_a(a) da$ is the differential entropy of a continuous random variable a with probability density function $p_a(a)$.

3 The Minimum Entropy Contrast for Blind Extraction of Groups of Sources

The central limit theorem tells us that the linear mixture of N independent signals will became asymptotically Gaussian (as N grows towards ∞). This has suggested that the pursuit of non-Gaussianity can be a separating process and, since the Gaussian distribution for a fixed variance maximizes the entropy, one can intuitively think that by minimizing the entropy of the outputs while keeping the variance constant leads to separation of sources.

Projection Pursuit approaches [17] have shown that, indeed, such idea is correct and several algorithms for the blind extraction of a single source and for the simultaneous blind source separation of the whole set of sources from the mixture have been developed [14]. The following theorem will extend these results to consider of the simultaneous extraction of a specific subset of $E \leq E_{max} \leq N$ sources from the linear mixture.

Theorem 3. *Let us assume that the sources can be ordered by increasing value of the differential entropy (or uncertainty) as*

$$h(s_1) \leq \ldots \leq h(s_E) < h(s_{E+1}) \leq \ldots \leq h(s_N) . \tag{5}$$

If s_E is not Gaussian distributed, then the following objective function

$$\Psi_{ME}(\mathbf{y}) = -\sum_{i=1}^{E} h(y_i) \quad \text{subject to} \quad Cov(\mathbf{y}) = \mathbf{I}_E \tag{6}$$

is a contrast function whose global maxima correspond to sources with the smallest value of entropy (i.e., the least uncertain sources of the mixture), i.e., $\mathbf{y} = [s_1, \ldots, s_E]^T$ up to an arbitrary reordering or permutation.

Proof. The proof of this theorem is based on the entropy power inequality. First note that $Cov(\mathbf{y}) = \mathbf{G}\mathbf{G}^T$. Taking into account that $y_i = \sum_{j=1}^{N} G_{ij} s_j$, and applying theorem 2 we can see that

$$2^{2h(y_i)} \geq \sum_{j=1}^{N} 2^{2h(G_{ij}s_j)} \qquad (7)$$

$$= \sum_{j=1}^{N} G_{ij}^2 \, 2^{2h(s_j)} \qquad (8)$$

$$= [\mathbf{V}]_{ii} \qquad (9)$$

where $\mathbf{V} = \mathbf{G}\mathbf{\Lambda}_1\mathbf{G}^T$ and $\mathbf{\Lambda}_1$ is the diagonal matrix with elements $[\mathbf{\Lambda}_1]_{ii} = 2^{2h(s_i)}$.

Now, expressing marginal entropies of the outputs in terms of the diagonal elements of matrix $\mathbf{V}$, and taking into account that $\sum_{j=1}^{N} G_{ij}^2 = 1$, we can rewrite the sum of marginal entropies as

$$\sum_{i=1}^{E} h(y_i) = \sum_{i=1}^{E} \frac{1}{2} \log(V_{ii}) \qquad (10)$$

$$\geq \operatorname{trace}\{\mathbf{G}\mathbf{\Lambda}\mathbf{G}^T\} \qquad (11)$$

where $\mathbf{\Lambda}$ is a diagonal matrix with elements $[\mathbf{\Lambda}]_{ii} = h(s_i)$, and the inequality between (10) and (11) follows from the concavity of the logarithm. Note that, from the semi-orthogonality of $\mathbf{G}$, the equality is attained if and only if $\mathbf{V}$ is diagonal. Then, as a consequence of the Poincaré's separation theorem of matrix algebra we obtain that

$$\min_{Cov(\mathbf{y})=\mathbf{I}_E} \sum_{i=1}^{E} h(y_i) = \min_{\mathbf{G}\mathbf{G}^T=\mathbf{I}_E} \operatorname{trace}\{\mathbf{G}\mathbf{\Lambda}\mathbf{G}^T\} = \sum_{i=1}^{E} h(s_i) \qquad (12)$$

Thus the global maxima of the contrast function $\Psi_{ME}(\mathbf{y})$ are achieved for such matrices $\mathbf{G}$ of which rows are orthogonal vectors that span the same subspace of the eigenvectors associated with the E lowest eigenvalues of $\mathbf{\Lambda}$, what enforces that $G_{ij} = 0 \; \forall j \geq E$.

The necessary and sufficient condition for the equality between (10) and (11) ($\mathbf{V}$ being diagonal) will trivially hold for any subset of decorrelated Gaussian sources. However, from the given hypotheses and due to ordering the sources according increasing entropy, the first E sources are non-Gaussian and have the lowest possibly entropy. Thus, the condition for $\mathbf{V}$ to be a diagonal matrix implies that each row of $\mathbf{G}$ should contain only one nonzero element $+1$ or -1. Then, $\mathbf{G}$ is any generalized permutation matrix which extracts the E sources with the lowest individual differential entropy, i.e., $\mathbf{G}$ can be reduced by row permutations to the form $[\mathbf{I}_E, \mathbf{0}]$. □

The following lemma brings us another interpretation of the minimum entropy contrast which reveals, in a more explicit fashion, the pursuit of non-Gaussianity.

Lemma 1. *Let be* $\mathbf{g} = [g_1, \ldots, g_E]^T$ *a vector of normalized Gaussian random variables. The maximization of*

$$\Psi_{ME}(\mathbf{y}) = -\sum_{i=1}^{E} h(y_i) \quad \text{subject to} \quad Cov(\mathbf{y}) = \mathbf{I}_E \tag{13}$$

is equivalent to the maximization of the following quasi-distance of the outputs marginal distributions from the Gaussianity

$$\sum_{i=1}^{E} KL(p_{y_i} \| p_{g_i}) \quad \text{subject to} \quad Cov(\mathbf{g}) = Cov(\mathbf{y}) \tag{14}$$

where $KL(p_{y_i} \| p_{g_i}) = \int p_{y_i} \log \frac{p_{y_i}}{p_{g_i}} dy_i$ *denotes the Kullback-Leibler divergence (relative entropy) between the involved (marginal) probability density functions.*

The proof of the lemma easily follows from the decomposition of the Kullback-Leibler divergence in terms of the individual differential entropies of the outputs $KL(p_{y_i} \| p_{g_i}) = h(p_{g_i}) - h(p_{y_i})$ when the constraint $Cov(\mathbf{g}) = Cov(\mathbf{y})$ applies.

4 Contrast Function Based on Higher Order Cumulants

In the preceding section we have observed how the negative of the differential entropy of the outputs can give us an index of non-Gaussianity. However, it is well known that this is not the only function we can use for this task. In particular, the absolute value of the higher order autocumulants $C_s^r = Cum(\underbrace{s, \ldots, s}_{\times r})$ with $r > 2$ can also measure the departure from Gaussianity. This result is summarized in the following theorem.

Theorem 4. *When the sources signals are normalized such that* $cov(\mathbf{s}) = \mathbf{I}_N$ *and for a decreasing order arrangement of them with regard to the absolute values of their* $(1+\beta)$-*order autocumulants*

$$|C_{s_1}^{1+\beta}| \geq \ldots \geq |C_{s_E}^{1+\beta}| > |C_{s_{E+1}}^{1+\beta}| \geq \ldots \geq |C_{s_N}^{1+\beta}|, \tag{15}$$

if $|C_{s_E}^{1+\beta}| \neq 0$, *the following function*

$$\Psi_{Cum}(\mathbf{y}) = \frac{1}{1+\beta} \sum_{i=1}^{E} |C_{y_i}^{1+\beta}| \quad \text{subject to} \quad cov(\mathbf{y}) = \mathbf{I}_E \tag{16}$$

is a contrast whose global maxima lead to the extraction of the first E *sources with the largest* $(1+\beta)$-*order autocumulants.*

Proof. We will only sketch the proof which is parallel to that of the minimum entropy contrast. After some straightforward simplifications and subject to the semi-orthogonality of $\mathbf{G}$ we obtain that

$$\sum_{i=1}^{E} |C_{y_i}^{1+\beta}| \leq \sum_{j=1}^{N} |C_{s_j}^{1+\beta}| \sum_{i=1}^{E} G_{ij}^2 \qquad (17)$$

Then, we can apply the Poincaré's separation theorem to observe that the maximum of (17) is given by

$$\max_{cov(\mathbf{y})=\mathbf{I}_E} \Psi_{Cum}(\mathbf{y}) = \frac{1}{1+\beta} \sum_{j=1}^{E} |C_{s_i}^{1+\beta}| \qquad (18)$$

Here, the bound is only attained for that matrices $\mathbf{G}$ that can be reduced by row permutations to the form $[\mathbf{I}_E, \mathbf{0}]$, i.e., $\mathbf{G}$ is the extraction matrix of the first E sources. □

5 Multichannel Blind Deconvolution

The previous results for blind source extraction can be also extended to solve the Multichannel Blind Deconvolution problem as is shown in the following theorem.

Theorem 5. *Consider N source random processes which are mutually independent and temporally i.i.d. (independent and identically distributed) and with cross-covariance $Cov_{\mathbf{s},\mathbf{s}}[n] = \delta[n]\,\mathbf{I}_E$. Let the function $\psi(\cdot) = \{\psi_{ME}(\cdot), \psi_{Cum}(\cdot)\}$ where $\psi_{ME}(\cdot) = \frac{1}{2}\log(2\pi e) - h(\cdot)$ and $\psi_{Cum}(\cdot) = |C_{(\cdot)}^{1+\beta}|$ with $\beta > 1$. Assuming that the sources can be ordered by decreasing value of the function $\psi(\cdot)$ of their random variables as*

$$\psi(s_1[n]) \geq \ldots \geq \psi(s_E[n]) > \psi(s_{E+1}[n]) \geq \ldots \geq \psi(s_N[n]) \,, \qquad (19)$$

and if $\psi(s_E[n]) > 0$, the function

$$\Psi(\mathbf{y}[n]) = \sum_{i=1}^{E} \psi(y_i[n]) \quad \text{subject to} \quad Cov_{\mathbf{y},\mathbf{y}}[n] = \delta[n]\,\mathbf{I}_E \qquad (20)$$

is a contrast whose global maxima are found at the extraction of the, in a certain sense, less Gaussian sources of the mixture, i.e., $\mathbf{y}[n] = [s_1[n], \ldots, s_E[n]]^T$ up to a permutation and arbitrary individual delays.

It is interesting to note that theorem 5 includes several special cases already known in the literature. For $E = N = 1$ equation (20) includes the minimum entropy contrast for blind deconvolution (whose optimality has been analyzed by Donoho in [13]) and also includes the cumulants based contrast function proposed by Shalvi and Weinstein in [21]. For $E = N > 1$ equation (20) is the contrast for blind deconvolution proposed by Comon in [9]. Furthermore, Tugnait [22] and Inouye et al. [19] have analyzed similar cumulants based criteria to (20) in the case of $E = 1$ and $N > 1$.

6 Blind Source Extraction/Deconvolution Algorithms

When the observations are decorrelated ($Cov_{\mathbf{x},\mathbf{x}}[n] = \delta[n]\,\mathbf{I}_E$) and of zero mean, one of the possible methods to maximize the proposed contrast $\Psi(\mathbf{y}[n])$ is to employ the natural Riemannian gradient ascent in the Stiefel manifold [1] which preserves the outputs decorrelation constraint. The desired gradient, which is given by

$$\tilde{\nabla}_{\mathbf{U}[n]}\Psi = \nabla_{\mathbf{U}[n]}\Psi(\mathbf{y}[n]) - \mathbf{U}[n] * (\nabla_{\mathbf{U}[n]}\Psi(\mathbf{y}[n]))^T * \mathbf{U}[n], \qquad (21)$$

where $*$ denotes the convolution operator, leads to Amari's gradient algorithm[1]

$$\mathbf{U}^{(k+1)}[n] = \mathbf{U}^{(k)}[n] + \mu \left(\mathbf{R}^{(k)}_{\varphi,x}[n] - \mathbf{R}^{(k)}_{y,\varphi}[n] * \mathbf{U}^{(k)}[n] \right) \qquad (22)$$

where $\mathbf{R}^{(k)}_{\varphi,x}[n] = E[\varphi(\mathbf{y}^{(k)}[l])\,(\mathbf{x}[l-n])^T]$ and $\varphi(\mathbf{y}[n]) = [\frac{d\Psi}{dy_1[n]}, \ldots, \frac{d\Psi}{dy_E[n]}]^T$. The exact expressions of $\varphi(\mathbf{y}[n])$ depend on the used criteria. When $\psi(\cdot) = \psi_{ME}(\cdot)$ approximations to these derivatives can be found in [8] and in [23] where the marginal p.d.f. of the outputs are truncated at low orders of the Edgeworth or Gram-Charlier expansions, respectively. By using the contrast function expressed by cumulants $\psi(\cdot) = \psi_{Cum}(\cdot)$, the learning algorithm takes a specific form

$$\mathbf{U}^{(k+1)}[n] = \mathbf{U}^{(k)}[n] + \mu \left(\mathbf{S}_y \mathbf{C}^{\beta,1}_{y,x}[n] - \mathbf{C}^{1,\beta}_{y,y}[n] \mathbf{S}_y * \mathbf{U}^{(k)}[n] \right) \qquad (23)$$

where $\mathbf{S}_y$ is the diagonal matrix with entries $[\mathbf{S}_y]_{ii} = \text{sign}(C^{1+\beta}_{y_i}[0])$ and $\mathbf{C}^{\beta,1}_{y,x}[n]$ is the $(1+\beta)$-order cross-cumulant matrix whose elements are given by $\left[\mathbf{C}^{\beta,1}_{y,x}[n]\right]_{ij} = Cum(y_i[l], \ldots, y_i[l], x_j[l-n])$, similarly $\left[\mathbf{C}^{1,\beta}_{y,y}[n]\right]_{ij} = Cum(y_i[l], y_j[l-n], \ldots, y_j[l-n])$. Note that the stochastic versions of these algorithms can be easily obtained.

7 Conclusions

The minimum entropy and cumulants based approaches for blind source extraction/deconvolution of temporal i.i.d. sources has been extended to allow the simultaneous recovery of groups with the least Gaussian sources (in a certain specific sense) from a linear mixture. The connections of this approach with other criteria have been shown and Amari's gradient algorithm has been suggested for the optimization of the proposed contrast functions.

References

1. S. Amari, "Natural gradient learning for over- and under-complete bases in ICA," *Neural Computation*, vol. 11, pp. 1875–1883, 1999.

---

[1] Algorithm (22) was originally proposed in [1, 2] to solve the blind separation problem for memoryless mixtures.

2. S. Amari, A. Cichocki, Y.Y. Yang, *Blind Signal Separation and Extraction*, chapter 3 in "Unsupervised Adaptive Filtering" Volume I, edited by S. Haykin, Wiley, 2000.
3. S. Amari, A. Cichocki, "Adaptive blind signal processing – neural network approaches," *Proceedings of the IEEE*, vol. 86, no. 10, pp. 2026–2048, 1998.
4. A.J. Bell, T.J. Sejnowski, "Blind separation and blind deconvolution: An information-theoretic approach," in *ICASSP*, 1995.
5. X-R. Cao, R-W. Liu, "General Approach to Blind Source Separation," in *IEEE Transactions on Signal Processing*, vol. 44(3),pp. 562-571, March 1996.
6. J. F. Cardoso, "Blind signal separation: Statistical principles," *Proceedings of the IEEE*, vol. 86, no. 10, pp. 2009–2025, 1998.
7. A. Cichocki, R. Thwonmas, S. Amari, "Sequential blind signal extraction in order specified by stochastic properties," *Electronics Letters*, vol. 33, no. 1, pp. 64–65, 1997.
8. P. Comon, "Independent component analysis, a new concept?," *Signal Processing*, vol. 3, no. 36, pp. 287–314, 1994.
9. P. Comon, "Contrasts for Multichannel Blind Deconvolution," *IEEE Signal Processing Letters*, vol. 3, no. 7, pp. 209–211, 1996.
10. T. M. Cover, J. A. Thomas, *Elements of Information Theory*, Wiley series in telecommunications. John Wiley, 1991.
11. S. Cruces, A. Cichocki, L. Castedo, "Blind source extraction in gaussian noise," in *proc. of the 2nd International Workshop on Independent Component Analysis and Blind Signal Separation (ICA'2000), Helsinki, Finland*, June 2000, pp. 63–68.
12. N. Delfosse, P. Loubaton, "Adaptive blind separation of independent sources: A deflation approach," *Signal Processing*, vol. 45, pp. 59–83, 1995.
13. D. Donoho, *On Minimun Entropy Deconvolution*, Applied Time Series Analysis II, D. F. Findley Editor, Academic Press, New York, 1981.
14. M. Girolami, C. Fyfe, *Negentropy and kurtosis as projection pursuit indices provide generalized ICA algorithms*, pp. 752–763, Boston, MA: MIT Press, 1996.
15. C. Jutten, J. Herault, "Blind separation of sources, part i: An adaptive algorithm based on neuromimetic architecture," *Signal Processing*, vol. 24, pp. 1–10, 1991.
16. J. Karhunen, E. Oja, L. Wang, R. Vigario, J. Koutsensalo, "A class of neural networks for independent component analysis," *IEEE Transactions on Neural Networks*, vol. 8, no. 3, pp. 486–503, May 97.
17. P.J. Huber, "Projection pursuit," *Annals of Statistics*, vol. 13, pp. 435–525, 1985.
18. A. Hyvarinen, E. Oja, "A fast fixed-point algorithm for independent component analysis," *Neural Computation*, vol. 9, pp. 1483–1492, 1997.
19. Y. Inouye, T. Sato, "Iterative algorithms based on multistage criteria for blind deconvolution," *IEEE Transactions on Signal Processing*, vol. 47, no. 6, pp. 1759–1764, June 1999.
20. T-P. Jung S. Makeig, A. Bell, T.J. Sejnowski, *Independent Component Analysis of Electroencephalographic Data*, vol. 8, pp. 145–151, M. Mozer et al., Cambridge, MA: MIT Press, 1996.
21. O. Shalvi, E. Weinstein, "New criteria for blind deconvolution of nonminimun phase systems (channels)," *IEEE Transactions on Information Theory*, vol. 36, no. 2, pp. 312–321, 1990.
22. J. K. Tugnait, "Identification and deconvolution of multichannel linear non-Gaussian processes using higher order statistics and inverse filter criteria," *IEEE Transactions on Signal Processing*, vol. 45, no. 3, pp. 658–672, 1997.
23. H. H. Yang, S. Amari, "Adaptive on-line learning algorithms for blind source separation – maximum entropy and minimum mutual information," *Neural Computation*, 1997.

Feature Extraction in Digital Mammography: An Independent Component Analysis Approach

Athanasios Koutras, Ioanna Christoyianni, Evangelos Dermatas, and George Kokkinakis

WCL, Electrical & Computer Engineering Dept., University of Patras
26100 Patras, Hellas
koutras@giapi.wcl2.ee.upatras.gr

Abstract. In this paper we present a new feature extraction technique for digital mammograms. Our approach uses Independent Component Analysis to find the source regions that generate the observed regions of suspicion in mammograms. The linear transformation coefficients, which result from the source regions, are used as features that describe the observed regions in an effective way. A Principal Component Analysis preprocessing step is used to reduce dramatically the features vector dimensionality and improve the classification accuracy of a Radial Basis Function neural classifier. Extensive experiments in the MIAS database using a very small dimensioned feature vector gave a recognition accuracy of 88.23% when detecting all kinds of abnormalities which outperforms significantly the accuracy of the commonly used statistical texture descriptors.

1 Introduction

Breast cancer is a leading cause of fatality among all cancers for women. Still, there is no known way of preventing breast cancer but early detection allows treatment before it is spread to other parts of the body. Currently, X-ray mammography is the single most effective, low-cost, and highly sensitive technique for detecting small lesions [1] resulting in at least a 30 percent reduction in breast cancer deaths. The radiographs are searched for signs of abnormality by expert radiologists but complex structures in appearance and signs of early disease are often small or subtle. That's the main cause of many missed diagnoses that can be mainly attributed to human factors [1,2]. Since the consequences of errors in detection or classification are costly, there has been a considerable interest in developing methods for automatically classifying suspicious areas of mammography tissue, as a means of aiding radiologists by improving the efficacy of screening programs and avoiding unnecessary biopsies.

Among the various types of breast abnormalities, which are visible in mammograms, clustered microcalcifications (or "calcifications") and mass lesions are the most important ones. Masses and clustered microcalcifications often characterize early breast cancer [3] that can be detected in mammograms before a woman or the physician can palp them. Masses appear as dense regions of varying sizes and properties and can be characterized as circumscribed (Figure 1a), spiculated (Figure 1b), or ill defined (Figure 1c). On the other hand, microcalcifications (Figure 1d) appear as small bright arbitrarily shaped regions on the large variety of breast texture background. Finally, asymmetry (Figure 1e), and architectural distortion (Figure 1f), are also very important and difficult abnormalities to detect. The great variability of the mass appearance along with the other abnormalities in digital mammograms is the main obstacle of building a unified mass detection method.

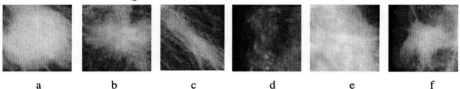

    a   b   c   d   e   f

Fig. 1. Types of breast cancer

Neural network computer-aided diagnosis (CAD) for detecting masses or microcalcifications in mammograms has already been used [4-7]. However, the use of neural networks for detecting all kinds of abnormalities is still very limited. One of the main points that should be taken under serious consideration when implementing a neural classifier for recognizing breast tissue is the selection of the appropriate features that describe and highlight the differences between the abnormal and the normal tissue in an ample way, especially when building a robust diagnostic system. Currently, various types of features are used for detecting abnormal regions in mammograms [7,8]. In this study, we propose a novel technique for feature extraction from regions of suspicion in mammograms using Independent Component Analysis (ICA).

ICA is a signal processing technique employed in various signal processing applications [9] the purpose of which is to find a linear non-orthogonal coordinate system in multivariate data. The directions of the axes are determined not only by the data's first and second order statistics, but also by higher order statistics. In our approach we consider the normal and the abnormal regions of mammograms to be generated by a set of independent images, namely the source regions that are estimated using standard ICA techniques. The coefficients of the linear combination of the independent source regions are the features that are fed into a Radial Based Function Neural Network (RBFNN) classifier. Additionally, a preprocessing step implementing Principal Component Analysis (PCA) is presented for reducing the dimensionality of these features without affecting the classification accuracy. Extensive experiments have shown a greater accuracy of 88.23% in recognizing normal and abnormal breast tissue in mammograms, which is a significant improvement, compared to other feature descriptors based on the statistical textural analysis (accuracy less than 85%) [8].

The structure of this paper is as follows: In the next section, a detailed description of the features extracted from mammograms using ICA techniques is given. Additionally, the implemented technique for reducing the dimensionality of the extracted features is presented. In section 3, a brief description of the RBFNN classifier is given. In section 4 we present the data set and the experimental results and finally, in section 5 some conclusions are drawn.

2 Feature Extraction from Mammograms

2.1 ICA based Feature Extraction

The aim of the proposed feature extraction technique is to find a set of descriptors that can be used to describe the healthy and tumorous regions of mammograms in an effective way. To this direction, we assume that the observed regions of mammograms are generated by a linear combination of an unknown set of statistically independent source regions according to the equation

$$X = A \cdot S, \qquad (1)$$

where A is the mixing matrix that generates the X observed regions from the S independent source regions, with the coefficients used in the linear combination being in the rows of A. These coefficients are employed as feature descriptors that describe uniquely the abnormal and the normal regions in X, in a most fitting way. The source regions are estimated using standard ICA techniques as follows:

Let us consider a set of N regions of mammograms used for the training procedure, containing normal and abnormal tissue with dimensions KxL pixels. The regions are first converted to one-dimensional vectors with length $D=\{KxL\}$. These vectors form the rows of the observation matrix X_{train} with dimensions NxD and are fed into the ICA neural network seen in Figure 2.

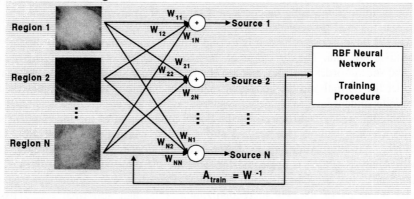

Fig. 2. The ICA based feature extraction scheme for the training procedure.

The source regions are estimated by: $S = W \cdot X_{train}$, where W is the NxN ICA separating matrix. To estimate this matrix in an unsupervised manner, we have applied the Maximum Likelihood Estimation criterion (MLE). The log-likelihood of the observed regions is given by:

$$L = \log(p_X(X_{train};W)) = \log(|W|) + \log(p_S(S)) \qquad (2)$$

The weights of the ICA network are estimated using the stochastic gradient of L with respect to the matrix W:

$$\frac{\partial L}{\partial W} = \left[W^{-1}\right]^T - \Phi(S) \cdot X_{train}^T , \qquad (3)$$

where $\Phi(S) = -\left[\dfrac{p_1'(s_1;W)}{p_1(s_1;W)}, \cdots, \dfrac{p_N'(s_N;W)}{p_N(s_N;W)}\right]^T$ and $p_i(s_i;W)$ is the probability density function of the ith source region. The marginal pdf of the source regions was chosen experimentally to follow the hyperbolic cosine distribution with $p_i(s_i;W) \propto 1/\cosh(s_i)$, so $\Phi(s_i) = \tanh(s_i)$. From equation (3) and using the natural gradient approach, we conclude to the following weight adaptation rule for W:

$$\Delta W = -n \frac{\partial L}{\partial W} W^T W = n\left[I - \Phi(S) S^T\right] \cdot W \qquad (4)$$

By applying the above learning rule, we can find the separating matrix W with dimensions $N \times N$ and the N independent source regions in the matrix S with dimensions $\{K \times L\}$. The features that are used to describe the observed regions of mammograms are contained in the rows of the inverse of the separating matrix $A_{train} = W^{-1}$.

For the testing procedure seen in Figure 3, each observed ROS is generated by a linear combination of the learned source regions S using the coefficients $A_{test} = X_{test} \cdot S^{\#}$, where the operator # denotes the pseudoinverse of a matrix. The feature vectors in A_{test} are then fed into the RBFNN classifier and the final decision is made whether the tested ROS is normal or abnormal.

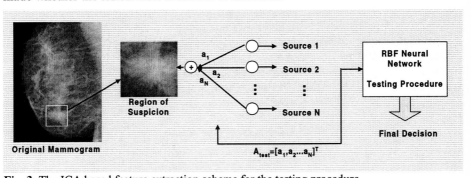

Fig. 3. The ICA based feature extraction scheme for the testing procedure.

2.2 Dimensionality reduction

From the above, it can be easily seen that the dimensionality of the extracted features depends strongly on the number of the regions used in the training procedure. In case of a large training set, the extracted features present big dimensionality, which complicates the task of the implemented classifier. On the other hand, when using a small training set, the performance of the ICA network deteriorates and the independent source regions cannot be estimated correctly.

In order to face the problem of dimensionality reduction, we have added a PCA preprocessing step. In this case, instead of performing ICA on the N observed regions, we perform ICA on a subset of K linear combinations where $K<N$. This technique does not affect the ICA network's performance, as the initial regions have been replaced with another linear combination. The use of the PCA preprocessing step does not destroy the higher order relationship between the initial regions. These relations still exist in the data and are not separated.

PCA can be implemented using eigenvalue decomposition on the covariance matrix of the observed regions in X_{train}. Let P be the matrix DxN with the N principal components in its columns, sorted by descending order with respect to their variances. By taking the first K more significant principal components and performing ICA on the data in P^T, we find the K independent source images in the rows of S. The new feature vectors of the observed regions in X_{train} are determined as follows:

The representation R_m of X_{train} based on the principal components in P is defined as $R_m = X_{train} \cdot P$. The regions in X_{train} can be approximated using the minimum squared error [10] as $X_{rec} = R_m \cdot P^T$. By applying the rule in equation (4) on the first K principal components we estimate the matrix W such that $S = W \cdot P^T$, therefore $P^T = W^{-1} \cdot S$. Using the above equation we find that $X_{rec} = R_m \cdot P^T = R_m \cdot W^{-1} \cdot S$. From this, we can deduce that the rows of the transformation matrix

$$B_{train} = R_m \cdot W^{-1} \qquad (5)$$

contain the coefficients of the linear combination of the statistically independent regions in S that generate the observations in X_{rec}, therefore they can be used as feature vectors with reduced dimensionality K to describe the observed regions in a compact and more efficient way.

For the testing procedure, we process each region in X_{test} with the principal components P_m estimated from the training procedure $R_{test} = X_{test} \cdot P_m$ and the feature vector is calculated by

$$B_{test} = R_{test} \cdot W^{-1}. \qquad (6)$$

3 Neural Network Classifier

Neural networks have been widely used in situations where the expert knowledge is not explicitly defined and cannot be described in terms of statistically independent rules. In digital mammography, artificial neural networks, a massively interconnected computing architecture of simple processing units, have been widely used to mimic the computational power and the perception capabilities of the human brain. In our work a radial-basis-function neural network (RBFNN) is employed as proposed in

[4,8]. Our network's input layer handles the ICA based features extracted from the training and the testing set. Two output neurons denote the presence or absence of a lesion. The hidden layer performs non-linear transformation of the input vector without bias elements.

The radial-basis-function neural networks (RBFNN) have the advantage of fast learning rates and have been proved to provide excellent discrimination in many applications. In this paper we present results of the classification capabilities of the RBF neural network in successfully recognizing any type of cancer in digital mammography.

4 Experiments

4.1 The MIAS Data set

In our experiments the MIAS MiniMammographic Database [11], provided by the Mammographic Image Analysis Society (MIAS), was used. The mammograms are digitized at 200- micron pixel edge, resulting to a 1024x1024-pixel resolution.

In the MIAS Database there is a total of 119 ROS containing all kinds of existing abnormal tissue from masses to clustered microcalcifications. The smallest abnormality extends to 3 pixels in radius, while the largest one to 197 pixels. These 119 ROS along with another 119 randomly selected sub-images from entirely normal mammograms were used throughout our experiments.

The MIAS database provides groundtruth for each abnormality in the form of circles; an approximation of the center and the radius of each abnormality. Since the abnormal tissues are rarely perfectly circular, and the MIAS policy was to err on the side of making the groundtruth circles completely inclusive rather than too small, these regions often contain a substantial amount of normal tissue as well. The groundtruth for each abnormality was used to crop and scale the abnormal regions to 35x35 pixels based on the center of each abnormality as our main interest was primarily focused on the detection of the core of the cancerous regions. For the subsequent analysis, the rows of each region were concatenated to produce 1x1125 dimensional vectors.

4.2 Classification Results

From a total number of 238 ROS included in the MIAS database, we used 119 regions for the training procedure (*jackknife* method): 59 groundtruthed abnormal regions along with 60 randomly selected normal ones. For the evaluation we used the remaining 119 regions that contain 60 groundtruthed abnormal regions together with 59 entirely normal regions. Therefore, no ROS was used both in the training and testing procedure.

In order to reduce the dimensionality of the extracted feature vectors, we implemented the PCA preprocessing step as explained earlier. The principal components were found by calculating the eigenvectors of the covariance matrix of the training set. ICA was then performed successively on the first 5 to 35 most significant of these eigenvectors, which resulted in 5 to 35 dimensional feature vectors and 119 independent source regions with 1125 pixels length. These features were used to train the implemented RBFNN classifier.

For the testing procedure, the remaining 119 ROS were first preprocessed with the eigenvectors estimated from the previous step and then their features were extracted using equation (6). The extracted features were used in the RBFNN classifier, the results of which are shown in Table 1. Specifically, it can be seen that the implemented feature extraction technique, resulted in features that can describe effectively the healthy and tumorous regions achieving high recognition accuracy above 80% in most of the cases. The best results were obtained when using only 5 principal components in the PCA preprocessing step, which resulted in an extremely low-dimensionality feature vector and a low complexity neural classifier, achieving a total recognition accuracy of 88.23%. In detail, the RBFNN classified correctly 53 out of 60 normal regions and 52 out of 59 abnormal ones, showing an overall accuracy of 105 out of 119 correctly classified regions. The recognition results showed that the proposed features perform better than techniques which use standard statistical feature descriptors which present a recognition accuracy of 84.87% as shown in [8], are easier and faster in the implementation, and they are able to recognize all kinds of abnormalities in mammograms in a more effective way.

Table 1. Recognition accuracy for the ICA features

components	ABNORMAL	NORMAL	TOTAL
35	76.66%	88.13%	82.23%
30	81.66%	83.05%	82.23%
25	81.66%	83.05%	82.23%
20	73.33%	83.05%	78.15%
15	76.66%	83.05%	79.83%
10	78.33%	84.74%	81.51%
5	**88.33%**	**88.13%**	**88.23%**

Further experiments were performed using more than 35 components in the PCA preprocessing step, but the results showed no further improvement of the recognition accuracy of the RBFNN classifier. Results when using all the 119 dimensional feature vectors without the PCA preprocessing step showed a recognition accuracy of 82.23%, far below that achieved when only the first 5 principal components were used.

6 Conclusion

In this paper we investigated the performance of an ICA based feature extraction technique used for recognizing breast cancer in digital mammograms. Our approach consists of finding the independent source regions that generate the observed ones using standard ICA techniques. The coefficients of the linear transformation of the source regions are the features that describe them effectively. A principal component analysis (PCA) preprocessing step is implemented to reduce the extracted features dimensionality without decreasing the classification accuracy. Extensive experiments have shown that the derived features present a greater class discriminability than the commonly used statistical texture descriptors and lead to a better performance when used to detect all kinds of abnormalities in mammograms.

References

1. Martin, J., Moskowitz, M. and Milbrath, J.: Breast cancer missed by mammography. AJR, Vol. 132. (1979) 737
2. Kalisher, L.: Factors influencing false negative rates in xero-mammography. Radiology, Vol.133. (1979) 297
3. Tabar, L. and Dean, B.P.: Teaching Atlas of Mammography.2nd edition, Thieme, NY (1985)
4. Christoyianni, I., Dermatas, E., and Kokkinakis, G.: Fast Detection of Masses in Computer-Aided Mammography. IEEE Signal Processing Magazine, vol. 17, no 1. (2000) 54-64
5. Meersman, D., Scheunders, P. and Dyck, Van D.: Detection of Microcalcifications using Neural Networks. Proc. of the 3rd Int. Workshop on Digital Mammograph, Chicago, IL (1996) 97-103
6. Dhawan, P.A., Chite, Y., Bonasso, C. and Wheeler K.: Radial-Basis-Function Based Classification of Mammographic Microcalcifications Using Texture Features. IEEE Engineering in Medicine and Biology & CMBEC (1995) 535-536
7. Doi, K., Giger, M., Nishikawa, R., and Schmidt, R. (eds.): Digital Mammography 96. Elsevier Amsterdam (1996)
8. Christoyianni, I., Dermatas, E., and Kokkinakis, G.: Neural Classification of Abnormal Tissue in Digital Mammography Using Statistical Features of the Texture. IEEE International Conference on Electronics, Circuits and Systems, vol 1. (1999) 117-120
9. Lee, Te-Won: Independent Component Analysis: Theory and Applications. Kluwer Academic Publishers (1998)
10. Bartlett, M., Lades, M., Sejnowski, T.: Independent component representation for face recognition. Proc. SPIE Symposium on Electronic Imaging: Science and Technology (1998)
11. http://s20c.smb.man.ac.uk/services/MIAS/MIASmini.html

Blind Source Separation in Convolutive Mixtures: A Hybrid Approach for Colored Sources

Frédéric Abrard and Yannick Deville

Laboratoire d'Acoustique Métrologie Instrumentation, Université Paul Sabatier,
118 route de Narbonne, 31062 Toulouse Cedex, France.
abrard@cict.fr, ydeville@cict.fr

Abstract. This paper deals with the blind source separation problem in convolutive mixtures when the sources are colored processes. In this case, classical methods first extract the innovation processes of the sources and then color them, which yields two successive filter approximations. On the contrary, we propose here a new concept allowing to directly extract estimates of the colored sources in one step.

1 Introduction

In this paper, we focus on the 2-source to 2-sensor blind source separation problem, as shown in Fig. 1. The two colored sources X_1 and X_2 result from two white non-Gaussian signals P_1 and P_2, colored by two ARMA filters D_1 and D_2. They are transferred through a mixing matrix which consists of filters $A_{ij}(z)$. To simplify the notations, all the filters $D_i(z)$ and $A_{ij}(z)$ are supposed to be MA. According to this configuration, we can write the two measured signals Y_1

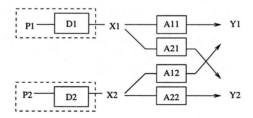

Fig. 1. Source generation and mixing matrix.

and Y_2 as:

$$Y_1(t) = A_{11}(t) * X_1(t) + A_{12}(t) * X_2(t)$$
$$Y_2(t) = A_{21}(t) * X_1(t) + A_{22}(t) * X_2(t) \qquad (1)$$

where $*$ denotes the convolution operator.

Blind source separation (BSS) consists in estimating the two sources X_1 and X_2 from the two observed signals Y_1 and Y_2 without any knowledge of the mixing filters or sources properties, except that the sources are assumed to be independent. It is well known that classical methods such as kurtosis maximization allow to extract the white process P_1 or P_2 even if the sources are colored [1], [2], [3], [4].

This may easily be seen by deriving from the $\mathcal{Z}$-transform of (1) that:

$$[Y(z)] = [A(z)] [D(z)] [P(z)] \tag{2}$$

which becomes:

$$[Y(z)] = [V(z)][P(z)] \tag{3}$$

We see in (3) that there is no distinction between the coloration and the propagation stages. Moreover, each linear combination of several filtered i.i.d signals makes the result closer to the Gaussian density than the signal whose normalized kurtosis has the highest absolute value among all the original signals thus filtered and combined [5], [2]. Therefore, maximizing the absolute value of the output normalized kurtosis by means of the separation system used in [3] will directly give the innovation process P_i whose normalized kurtosis has the highest absolute value.

Some authors [3], [4] worked on this problem and proposed a solution which consists in first estimating the white process and then building a post-processing filter which will try to artificially color the signal in order to restore the contribution of one source on one sensor. This method has the main drawback to consist of two steps and to combine two successive approximations: The first inverse filters and the coloration filter.

We here propose a new approach which directly estimates the different contributions of the two sources on one sensor in one step.

2 Preliminary Version

To extract the sources we need a separation system which will recompose the two observations in order to cancel one source. This could be performed by the direct separation system showed in figure 2. We use higher-order statistics in our

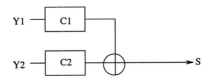

Fig. 2. Preliminary separation system.

separation criterion. We choose to maximize the absolute value of the normalized kurtosis of the output S of the separation system. This parameter is expressed as:

$$|k(S)| = \left| \frac{\text{cum}_4(S)}{[\text{cum}_2(S)]^2} \right| \qquad (4)$$

It is well known that by maximizing the expression in (4) vs the coefficients of C_1 and C_2 we extract a white process P_i up to a scale factor and delay [3]. The output is then:

$$S(z) = \alpha_i z^{-p_i} P_i(z) \qquad (5)$$

with α_i constant, z^{-p_i} delay operator and i source index.

Our idea came from the observation that maximizing the kurtosis leads to:

- the separation of two non-Gaussian signals, provided they are statistically independent at order four,
- the fourth-order whiteness of the extracted signals.

We here aim at avoiding the above whitening effect in order to extract the colored sources (or the resulting observed signals, which each consist of the contribution of only one source on a sensor). According to this idea, we add another filter, whose transfer function is denoted B, which will artificially whiten the output S. The maximization of the kurtosis of the resulting whitened signal U is then intended to restore $U = P_1$ or $U = P_2$, which allows S to become equal to X_1 or X_2, or to the observed versions of these signals. The new separation system thus obtained is showed in Fig. 3.

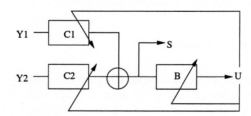

Fig. 3. Separation system including a whitening filter.

We adapt the three filters C_1, C_2 and B by maximizing $|k(U)|$ vs the respective coefficients of the different filters.

But this system is not constrained enough to guarantee that it converges to filter values such that its output S extracts a source signal itself, or its observed version contained in a sensor signal. We can show this by writing in $\mathcal{Z}$-domain:

$$\begin{aligned} S(z) = & \; (C_1(z).A_{11}(z) + C_2(z).A_{21}(z)).X_1(z) \\ & + (C_1(z).A_{12}(z) + C_2(z).A_{22}(z)).X_2(z) \end{aligned} \qquad (6)$$

Supposing that $S(z)$ contains only $X_1(z)$, i.e. the maximization of the kurtosis cancelled the contribution from $X_2(z)$, then:

$$C_1(z).A_{12}(z) + C_2(Z).A_{22}(z) = 0 \tag{7}$$

which means

$$C_2(z) = -\frac{C_1(z).A_{12}(z)}{A_{22}(z)} \tag{8}$$

Combining this filter value with (6) yields:

$$S = \left(C_1(z).(A_{11}(z) - \frac{A_{12}(z).A_{21}(z)}{A_{22}(z)})\right).X_1(z) \tag{9}$$

There is no other constraint here to fix the value of $C_1(z)$ since, in practice the whiteness effect is achieved by the filter $B(z)$. S is then a version of X_1 transferred through an arbitrary time varying filter, which yields an uncontrolled frequency distortion for this source. On the contrary, we want this filter to take a specific value, such that S is equal to X_1 itself or to the contribution of X_1 in one of the sensor signals. To this end, we now introduce a modified version of this approach.

3 Proposed Approach

3.1 First Idea

In the particular case when the signal S only contains a contribution from one source transferred through an unknown filter H, we can write $S(t) = H(t)*X_1(t)$. Combining this expression with (1) yields:

$$Y_1(t) - S(t) = (A_{11}(t) - H(t)) * X_1(t) + A_{12}(t) * X_2(t) \tag{10}$$

and therefore:

$$E\left[(Y_1(t) - S(t))^2\right] = E\left[((A_{11}(t) - H(t)) * X_1(t))^2\right] + E\left[(A_{12}(t) * X_2(t))^2\right] \tag{11}$$

The expression in (11) is minimized when $H(t) = A_{11}(t)$. In this case, we extract $S(t) = A_{11}*X_1(t)$, which is the contribution of the source X_1 measured on sensor 1. In other words, we thus get an adequately colored version of the innovation process of this source.

We then choose to adapt the coefficients of C_1 and C_2 by maximizing the new criterion defined as:

$$\Gamma_{C_1,C_2} = |k(U)| - \lambda E\left[(Y_1(t) - S(t))^2\right] \tag{12}$$

with $\lambda > 0$, while the whitening filter B is adapted so as to maximize $|k(U)|$.

This aims at combining two effects i.e:

- separating the sources by kurtosis maximization,
- extracting the signal $S(t) = A_{11} * X_1(t)$ by minimizing $E\left[(Y_1(t) - S(t))^2\right]$. Note that the other source may then easily be derived, as $Y_1(t) - S(t) = A_{12}(t) * X_2(t)$.

But in this intuitive approach we considered that the output S only contains one source and that whatever happens no other source could appear, which is not guaranteed here.

When both sources are present in S, the function $E\left[(Y_1(t) - S(t))^2\right]$ is clearly minimized vs C_1 and C_2 when $S = Y_1$, which may be far from what we expected before. We see here the danger to reach the solution $S = Y_1$. However, in this case the value of the kurtosis $|k(B*Y_1)|$ is not maximum because Y_1 still includes the two sources.

Therefore, we now provide a more general theoretical analysis, which shows in which conditions the proposed approach works.

3.2 Theoretical Considerations

Considering a point defined by fixed values of (C_1, C_2, B), for any signal S the condition for our cost function to be lower at that point than at the desired convergence point may be expressed as:

$$|k(B*S)| - \lambda E\left[(Y_1 - S)^2\right] < |k(P_1)| - \lambda E\left[(A_{12} * X_2)^2\right] \quad (13)$$

Then we have two different cases:

1. If $E\left[(A_{12} * X_2)^2\right] - E\left[(Y_1 - S)^2\right] < 0$ then (13) yields the following constraint on λ:

$$\lambda > \frac{|k(P_1)| - |k(B*S)|}{E\left[(A_{12} * X_2)^2\right] - E\left[(Y_1 - S)^2\right]} \quad (14)$$

The numerator $|k(P_1)| - |k(B*S)|$ is always positive, so the expression (14) is true for all $\lambda > 0$.

2. If $E\left[(A_{12} * X_2)^2\right] - E\left[(Y_1 - S)^2\right] > 0$ then (13) yields the constraint:

$$\lambda < \frac{|k(P_1)| - |k(B*S)|}{E\left[(A_{12} * X_2)^2\right] - E\left[(Y_1 - S)^2\right]} \quad (15)$$

The condition (15) means that, for some positive values of λ, there may exist points where the considered cost function takes higher values than at the desired convergence point. The latter point then does not correspond to the maximum of this function. This results from the hybrid nature of this cost function (12): whereas its term $|k(U)|$ alone is maximized exactly at the desired convergence point, the additional term $-\lambda E\left[(Y_1(t) - S(t))^2\right]$ that we introduced in this function shifts the position of the global maximum of the overall resulting function. It may be shown that the value of λ controls the magnitude of this shift, and therefore the accuracy of the extraction of the source signals. The behavior of the proposed separation system then depends as follows on λ:

1. For $\lambda = 0$ the maximum of the cost function coincides exactly with the desired convergence point. However, $\lambda = 0$ is not acceptable because the system is then underdetermined, as explained above.
2. For a very small λ, the above underdetermination disappears but the system is still ill-conditioned.
3. For intermediate values of λ, the system becomes well-conditioned and the maximum of its cost function is still close to the desired convergence point. This is the range of λ to be used.
4. If λ is further increased, the position of the maximum may shift significantly from the desired position. Especially, if λ is very high, (12) shows that the cost function has almost the same behavior as $-\lambda E\left[(Y_1(t) - S(t))^2\right]$, so that its maximum almost corresponds to $S(t) = Y_1(t)$.

The proposed approach thus provides an original alternative to classical solutions: whereas the latter methods perform an approximation by using an artificial coloring step, the solution proposed in this paper avoids this approximation by extracting the colored sources directly, but the specific cost function defined above entails another type of approximation, i.e. in the position of the convergence point.

A quantitative assessment of the above-defined ranges of values of λ and of the accuracy of the proposed approach is provided in the next section.

4 Experimental Results

We here present several results obtained with the following settings:

Two binary white processes with $|k(P1)| = 1.9987$ and $|k(P2)| = 2$.
The impulse responses of the coloration filters correspond to the following coefficient arrays: $D_1 = [1\ 0.8\ 0.5\ 0.3\ 0.1]$ and $D_2 = [1\ 0.9\ 0.7\ 0.45\ 0.4]$
Similarly, the mixing matrix is built as:

$$\begin{array}{l} A_{11} = [\ 0.5\ -0.9\ 0.4]\ A_{12} = [-0.6\ 0.6\ 0.3] \\ A_{21} = [\ 0.4\ -0.7\ 0.3]\ A_{22} = [\ 0.7\ 0.9\ 0.5] \end{array} \quad (16)$$

The length of the separating and whitening filters is set to 9.
We adapt the coefficients with an extended version of the modified gradient ascent algorithm described in [6] in the case of linear instantaneous mixtures. This algorithm here reads:

$$\begin{aligned} B^j(n+1) &= B^j(n) + \mu tanh\left(\text{sign}[k(U)]\frac{\partial k(U)}{\partial B^j}\right) \\ C_1^j(n+1) &= C_1^j(n) + \mu tanh\left(\text{sign}[k(U)]\frac{\partial k(U)}{\partial C_1^j} - \frac{\partial \lambda E\left[(Y_1 - S)^2\right]}{\partial C_1^j}\right) \\ C_2^j(n+1) &= C_2^j(n) + \mu tanh\left(\text{sign}[k(U)]\frac{\partial k(U)}{\partial C_2^j} - \frac{\partial \lambda E\left[(Y_1 - S)^2\right]}{\partial C_2^j}\right) \end{aligned} \quad (17)$$

Denoting $\hat{e}_l$ the lth estimated source and e_l the contribution of this source on the first sensor, we evaluate the accuracy of the separation by the criterion:

$$E(\hat{e}_l) = \frac{\Sigma_n |\hat{e}_l(n) - e_l(n)|^2}{\Sigma_n |e_l(n)|^2} \quad (18)$$

The following results are obtained after 500 iterations, where each iteration includes an update of all three filters. The cumulants values are computed on a 300-sample data window. With several values of λ we get:

value of λ	source restored	accuracy $E(\hat{e}_1)$	accuracy $E(\hat{e}_2)$
3	no	0.779	0.4118
2.5	yes (after erratic convergence)	0.2407	0.1272
2	yes	0.1587	0.0839
1	yes	0.0817	0.0432
0.85	yes	0.0824	0.0435
0.5	yes	0.093	0.0486
0.2	yes	0.1276	0.0674
.05	no, without any convergence	2.2639	1.1966

We see that, as predicted in the previous sections, with λ over 2.5 we do not extract a source. Moreover, our experimental tests show that for all $\lambda > 2.5$ the output we get is equal to Y_1. Besides, as we mentioned, when λ is below a certain value we do not extract the right solution nor Y_1: the filters then keep on evolving but do not converge. Fig. 4 to 7 show the results obtained with $\lambda = 0.85$.

5 Conclusions and Future Work

In this paper, we introduced a new approach for extracting colored sources from their convolutive mixtures. This approach consists of a separating structure and adaptation algorithms for its filters. This structure makes it possible to directly extract a colored source, whereas classical methods use a two-step approach including an artificial "recoloring" step, which is likely to yield only an approximate extraction of the sources. The proposed adaptation algorithms yield specific convergence properties, which entail another type of (slight) approximation. In our future investigations, we will aim at improving these convergence properties by developing modified versions of these algorithms . Anyway, the numerical results reported above show that this first version of our approach already succeeds in separating the considered signals.

References

1. J. A. CADZOW. Blind deconvolution via cumulant extrema. *IEEE signal processing magazine*, pages 24–42, May 1996.
2. J. S. SEO. *Blind fault detection and source identification using higher order statistics for impacting systems*. PhD thesis, University of Southampton, Faculty of Engineering and Applied Science, Institute of Sound and Vibration Research, November 2000.

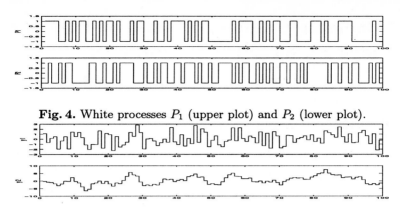

Fig. 4. White processes P_1 (upper plot) and P_2 (lower plot).

Fig. 5. Sensor signals Y_1 (upper plot) and Y_2 (lower plot).

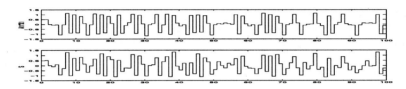

Fig. 6. Signals to be compared: $A_{11} * X_1$ (upper plot) and S (lower plot).

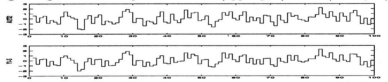

Fig. 7. Comparison between $A_{12} * X_2$ (upper plot) and $Y_1 - S$ (lower plot).

3. J. K. TUGNAIT. Identification and deconvolution of multichannel linear non-gaussian processes using higher order statistics and inverse filter criteria. *IEEE Trans. on Signal Processing*, 45(3):658–672, March 1997.
4. C. SIMON PH. LOUBATON C. VIGNAIT C. JUTTEN G. D'URSO. Blind source separation of convolutive mixtures by maximization of fourth-oder cumulants: the non-iid case. *Proc. Asilomar*, November 98.
5. D. DONOHO. On minimum entropy deconvolution. *In Applied Time series Analysis II ed. D. Findley, Academic Press, New York*, pages 556–608, 1981.
6. F. ABRARD Y. DEVILLE M. BENALI. Numerical and analytical solution to the differential source separation problem. In *Proceedings of EUSIPCO*, Tampere, Finland, 2000.

A Conjugate Gradient Method and Simulated Annealing for Blind Separation of Sources

Rubén Martín-Clemente[1], Carlos G. Puntonet[2], and José I. Acha[1]

[1] Área de Teoría de la Señal y Comunicaciones, Universidad de Sevilla, Avda. de los Descubrimientos s/n., 41092-Sevilla (Spain)
e-mail: ruben@cica.es, acha@viento.us.es

[2] Departamento de Arquitectura y Tecnología de Computadores, Universidad de Granada, E-18071, Granada (Spain)
e-mail: carlos@atc.ugr.es

Abstract. This paper presents a new procedure for Blind Separation of non-Gaussian Sources. It is proven that the estimation of the separating system can be based on the cancellation of some second partial derivatives of the output cross-cumulants. The resulting method is based on a conjugate gradient algorithm on the Stiefel manifold. Simulated annealing is used to obtain a good initial value and improve the rate of convergence.

1 Introduction

Blind Source Separation (BSS) is a problem that has recently attracted a great deal of attention. It consists in recovering a set of n unknown source signals from the observation of another set of linear mixtures of the sources, where the 'blind' qualification emphasizes that no information is assumed to be available about the coefficients of the mixture. In matrix form, the mixture model is represented by the equation:

$$\mathbf{x}(t) = A\mathbf{s}(t) \qquad (1)$$

where $\mathbf{s}(t) = [s_1(t),...,s_n(t)]^T$ is a vector of n source signals, $A = (a_{ij})$ is an unknown $n \times n$ invertible *mixing matrix* and vector $\mathbf{x}(t)$ collects the observed signals, being the only data available. The aim of BSS is to determine a $n \times n$ *separating matrix* $B = (b_{ij})$ such that:

$$\mathbf{y}(t) = B\mathbf{x}(t) = G\mathbf{s}(t) \qquad (2)$$

is an estimate of the source signals, *i.e.*, the *global matrix* $G = (g_{ij})$ has one and only one non-zero coefficient per row and column, where $G = A \cdot B$. In this case, G is said to be a *'generalized permutation matrix'*. The only assumptions are that the sources are statistically independent, stationary, unit-variance and zero-mean. In addition, *at most* one source is gaussian-distributed.

Since the inverse system has to be learned with the sole knowledge of the observed mixtures, this problem can be termed "unsupervised" or "self-learning".

It has many applications in diverse fields, like communications, speech separation (the 'cocktail party problem'), processing of arrays of radar or sonar signals, in biomedical signal processing (e.g., EEG, MEG, fMRI analysis), etc.

In the past ten years, numerous solutions to this problem have been proposed, starting from the seminal work of Jutten and Herault [11]. In the Independent Component Analysis (ICA) approach, which is closely related to BSS, the goal is to linearly transform the data such that the transformed variables are as statistically independent as possible [2,3,10]. Other approaches are concerned with batch (off-line) algorithms based on higher-order statistics [6]. The properties of the vectorial spaces of sources and mixtures also lead to geometrical methods for BSS [15,16]. For further reading, see the review paper [4] and the references therein.

In this paper, we first introduce a new set of necessary and sufficient conditions for BSS which lead to a new dependence measure. Then, this measure is minimized by using a conjugate gradient algorithm that naturally preserves the geometry of the problem. In addition, a simulated annealing technique is explored in order to provide fast initial convergence.

2 A New Cumulant-based Criterion

By using the multi-linearity property of the cumulants, the output cross-cumulant $cum_{31}(y_i(t), y_j(t)) = cum(y_i(t), y_i(t), y_i(t), y_j(t))$ can be expanded as follows:

$$cum_{31}(y_i(t), y_j(t)) = \sum_{l=1}^{n} g_{il}^3 \, g_{jl} \, \kappa_l \tag{3}$$

where κ_l is the kurtosis (fourth-order cumulant) of the l-th source. By setting (3) to zero for all $i \neq j$, we obtain a set of equations with respect to the coefficients of the matrix B. These equations are satisfied when G is a *generalized permutation matrix*. However, spurious roots are also possible depending on the source statistics.

Nevertheless, that (3) should vanish when G is a *generalized permutation matrix* is due to the fact that all the terms '$g_{il}^3 \, g_{jl} \, \kappa_l$' that appear in (3) take the value of zero but not to the fact that the coefficients 'g_{il}' are raised to the power of three. On the other hand, cumulant (3) can be simplified by differentiation, without losing any fundamental property. Thus, let us introduce the following second-order derivatives of (3):

$$\Delta_{ij} \triangleq \frac{1}{6} \frac{\partial^2 c_{31}}{\partial b_{ij}^2} = \sum_{l=1}^{n} g_{il} \, g_{jl} \, a_{jl}^2 \, \kappa_l, \quad (i \neq j) \tag{4}$$

Then, by using $cum(x_l, x_m, x_j, x_j) = \sum_p a_{lp} a_{mp} a_{jp}^2 \kappa_p$ (which is a consequence of the multi-linearity property), it is possible to express (4) as:

$$\Delta_{ij} = \sum_{l=1}^{n} \sum_{m=1}^{n} b_{il} b_{jm} cum(x_l, x_m, x_j, x_j), \tag{5}$$

We maintain that the set of equations $\Delta_{ij} = 0$ for all $i \neq j$ provide us with a set of *necessary* and *sufficient* conditions to asssure the source separation, i.e., spurious roots are not possible, provided that at most one source is gaussian (e.g. for mixtures of two sources, a direct solution can be obtained in agreement with previous results [13]).

Proof. Trivial non-separating solutions, in which any of the output signals vanishes, occur when either the mixing or the separating matrix is not full row rank. These solutions are avoided by assuming that the signals in $\mathbf{x}(t)$ are uncorrelated and have unit variance since, in this case, A, B and G must necesarily be orthogonal matrices. This assumption can be easily accomplished by means of the so-called *whitening* or *sphering* of the original mixtures, in which uncorrelated linear combinations of the observed signals are obtained [7].

Then, observe that (4) can be written compactly in a matrix form:

$$\Delta_{ij} = \mathbf{g}_i^T \Gamma_j \mathbf{g}_j \qquad (6)$$

where $\mathbf{g}_k^T$ is the k-th row of G and Γ_j is the diagonal matrix whose (l,l)-entry is equal to $a_{jl}^2 \kappa_l$. If at most one source is gaussian (i.e., its kurtosis equals zero), we can also assume without loss of generality that each diagonal element of Γ_j is different (i.e. Γ_j has no repeated eigenvalues) since A (and thus Γ_j) can be properly adjusted by multipliying the observations $\mathbf{x}(t)$ by any orthogonal matrix.

Since both A and B are orthogonal matrices, it follows that $\mathbf{g}_i^T \mathbf{g}_j = \delta_{ij}$, where δ_{ij} stands for the Kronecker delta. Now, we prove our main result:

(*Necessity*). If B is a separating matrix, then each vector $\mathbf{g}_i$ is a different canonical vector. Therefore, it follows from (6) that $\Delta_{ij} = 0$ for all i, j ($i \neq j$).

(*Sufficiency*). Let us assume that $\Delta_{ij} = 0$ for all $i \neq j$. The vector $\Gamma_j \mathbf{g}_j$ can be expressed as a linear combination of the vectors $\mathbf{g}_i$ since the set $\{\mathbf{g}_1, \mathbf{g}_2, \ldots, \mathbf{g}_n\}$ forms an orthonormal basis of the space. Therefore:

$$\Gamma_j \mathbf{g}_j = \Delta_{jj} \mathbf{g}_j + \sum_{i \neq j} \Delta_{ij} \mathbf{g}_i \qquad (7)$$

where $\Delta_{ij} = \mathbf{g}_i^T \Gamma_j \mathbf{g}_j$ as was defined before and $\Delta_{jj} = \mathbf{g}_j^T \Gamma_j \mathbf{g}_j$. Since, $\Delta_{ij} = 0$ for all $i \neq j$, we obtain that $\Gamma_j \mathbf{g}_j = \Delta_{jj} \mathbf{g}_j$, which implies that each $\mathbf{g}_i$ is an eigenvector of a diagonal matrix, i.e., a canonical vector. As a consequence, B is a separating matrix.

3 Blind Separation based on a Conjugate Gradient Algorithm

According to the previous Section, we propose to minimize the following cost function:

$$f(B) = \sum_{i \neq j} \Delta_{ij}^2, \qquad (8)$$

where $\mathbf{x}(t)$ satisfies $E\{\mathbf{x}^T(t)\mathbf{x}(t)\} = I$ (see above), being I the identity matrix, and B is constrained to the set of matrices such that

$$B^T B = I \tag{9}$$

In the literature, this constraint surface is known as the Stiefeld manifold. Edelman et al [8] have shown that the gradient of $f(B)$ at B on the Stiefeld manifold is given by:

$$\nabla f(B) = \partial f(B) - B\left(\partial f(B)\right)^T B \tag{10}$$

where $\partial f(B)$ is the $n \times n$ matrix of partial derivatives of $f(B)$ with respect to the elements of B, i.e.,

$$(\partial f(B))_{ij} = \frac{\partial f(B)}{\partial b_{ij}} \tag{11}$$

In view of (10), some comments are provided in order:

1. Consider small deviations of B in the direction $\nabla f(B)$, as follows:

$$B \to \tilde{B} \equiv B - \mu \nabla f(B) \tag{12}$$

where $\mu > 0$. If matrix B is orthogonal, then it is straightforward to check that the translation (12) preserves the orthogonality constraint, in the sense that $\tilde{B}\tilde{B}^T = I + o(\mu^2)$.

2. The first order Taylor expansion of $f(B)$ gives

$$f(B + \Delta B) = f(B) + <\partial f(B)|\Delta B> + o(\Delta B) \tag{13}$$

where $<M|N> = \text{Trace}[M^T N]$ stands for the standard inner product of matrices. If B is modified into $\tilde{B}$, it follows from (12) and (13) that

$$f(\tilde{B}) = f(B) - \mu <\partial f(B)|\nabla f(B)> + o(\mu) \tag{14}$$

Then, taking into account that

$$<\partial f(B)|\nabla f(B)> \equiv \frac{1}{2} <\nabla f(B)|\nabla f(B)> \tag{15}$$

which is always *positive*, it follows that $f(B)$ is decreased by the translation (12). Identity (15) is readily obtained by expanding its right-hand side and using that $\text{Trace}[M]=\text{Trace}[M^T]$ and $\text{Trace}[B^T M B] = \text{Trace}[M]$, where M is any matrix and B is an orthogonal matrix.

3. Thus, the steepest descent method for minimizing $f(B)$ is given by:

$$B_{t+1} = B_t - \mu \nabla f(B_t) = B_t - \mu H(B_t) B_t \tag{16}$$

where $\mu > 0$ and $H(B_t) \equiv \partial f(B_t) B_t^T - B_t \partial f(B_t)^T$. We would point out that algorithm (16) is an EASI-type method [5] for the serial update of matrix B.

Different algorithms can be chosen to minimize $f(B)$. For the sake of computational simplicity, we propose a *conjugate gradient method* [9], which is an improved version of the abovementioned steepest descent method. On the Stiefeld manifold, it is as follows[8]:

1. Let B_0 be the initial point and set $G_0 = \nabla f(B_0)$ and $F_0 = -G_0$.
2. For $t = 0, 1, \ldots$
   2.1 Let $B_{t+1} = B_t \exp(\mu B_t^T F_t)$, where $\mu > 0$. When μ is small enough, observe that it can be replaced with $B_{t+1} \simeq B_t + \mu F_t$.
   2.2 Compute $G_{t+1} = \nabla f(B_{t+1})$
   2.3 Compute the new search direction
   $$F_{t+1} = -G_{t+1} + \gamma_t F_t \exp(\mu B_t^T F_t),$$
   where $\gamma_t = \frac{<G_{t+1}|G_{t+1}>}{<G_t|G_t>}$ (which gives a Fletcher-Reeves conjugate gradient formulation [8]).
   2.4 Reset $F_{t+1} = -G_{t+1}$ (the steepest descent direction) if $t+1 \equiv 0 \bmod n(n-1)/2$.

4 Simulated Annealing For Obtaining Initial Conditions

In order to obtain a good starting point for the conjugate gradient algorithm, we have explored the 'simulated annealing' technique. Simulated annealing is a stochastic algorithm which provides a fast solution to some combinatorial optimization problems [1]. In a recent paper [17], this technique has been shown to be useful for Blind Source Separation purposes.

To make use of the method, we must provide the following elements:

1. A cost function ('energy') to minimize, i.e., the same function $f(B)$ that was defined in (8).
2. A control parameter T ('temperature') and a cooling schedule, which, in order to provide fast convergence, is given by[17]:

$$T(t) = \frac{T_0}{(1+t)^2} \tag{17}$$

where T_0 is the initial temperature.

The method works by performing a sequence of *random* updates $B_{t+1} \leftarrow Q_t B_t$ with the property that B_{t+1} is almost always 'more separating' than its predecessor. More specifically [1], if $f(B_{t+1}) < f(B_t)$, then the update is accepted; otherwise, the update will be rejected and matrix B will remain unchanged, i.e., $B_{t+1} \leftarrow B_t$, unless $\exp(-\Delta f/T(t)) > r$, where r is a random number between 0 and 1 and $\Delta f = f(B_{t+1}) - f(B)$. From [1], it is conjectured that the algorithm will tend to a global minimum of $f(B)$.

The generator of random updates would be inefficient if 'uphill' moves were almost always proposed. In view of our simulations, the problem is reduced if we consider a 'pairwise' approach similar to that used in [7]. In our method, each step involves: 1. choosing and index pair (i,j) $(1 \leq i < j \leq n)$ and 2. generating a *random* cosine-sine pair (c, s) and a Givens $n \times n$ rotation matrix Q which verifies $(Q)_{ii} = (Q)_{jj} = c$ and $(Q)_{ij} = -(Q)_{ji} = s$. Then, the above update is given by $B \leftarrow Q^T B$. Observe that it affects just two rows of B, preserves the orthogonality of this matrix and, last but not least, is computationally simple, since requires only $6n$ flops.

5 Links of the Cost Function with other Methods

First, note that (5) can be written in a matrix form as:

$$\Delta_{ij} = \mathbf{b}_i^T C_j \mathbf{b}_j \tag{18}$$

where $\mathbf{b}_k = [b_{k1}, \ldots, b_{kN}]^T$ and C_j is the $N \times N$ matrix whose (n, m)-entry equals $cum(x_n, x_m, x_j, x_j)$. Hence, the same argument that led to eqn. (7) now gives the identity:

$$C_j \mathbf{b}_j = \Delta_{jj}\mathbf{b}_j + \sum_{i \neq j} \Delta_{ij}\mathbf{b}_i = \Delta_{jj}\mathbf{b}_j \tag{19}$$

where $\Delta_{jj} = \mathbf{b}_j^T C_j \mathbf{b}_j$, which means that $\mathbf{b}_j$ is also an *eigenvector* of C_j. Interestingly, following a different approach, Cardoso and Souloumiac proposed the minimization of the sum of squares of non-diagonal entries of matrices $B^T C_j B$ in JADE [6]. Clearly, (8) comes from a *sufficient* subset of those equations which are solved by JADE. Consequently: *i.)* a significant computational cost saving is expected if only the cost function $f(B)$ is minimized and *ii.)* since JADE satisfies more statistical equations, it may also be numerically more robust. The experiments are in accordance with these intuitions.

6 Simulations

The validity of the procedure has been corroborated by many simulations with different number of sources. Figure 1 shows the mean *Signal to Noise Ratio* of the estimated sources averaged over 10 experiments for a simple mixture of three uniform sources, where the cumulants were estimated over 1000 samples. It is found that the separating solution is obtained remarkably earlier when simulated annealig is used to provide initial conditions. In addition, simulated annealing involves only about the 2% of the computational burdem.

7 Conclusions

In this Communication, we present a set of conditions for BSS which satisfy: (1) absence of spurious solutions and (2) Δ_{ij} is a quadratic function of the coefficients of B whereas $cum_{31}(y_i, y_j)$ depends on as high as fourth-order powers

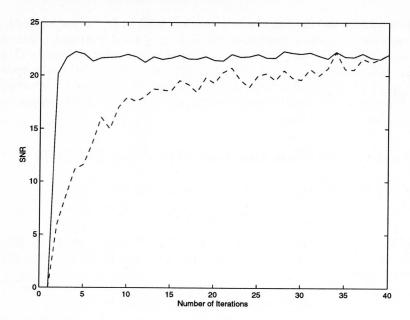

Fig. 1. Signal to Noise Ratio of the estimated sources. (a) Solid Line: with annealing. (b) Dashed Line: without annealing.

of these coefficients. It was shown that they are a sufficient subset of those conditions presented in JADE.

Imposing the orthogonality constraints in (9) is problematic in practice. Alternating projection methods have been widely used for this purpose[14]. Nevertheless, it is difficult to prove their convergence[12]. On the other hand, we use a conjugate gradient method on the Stiefel manifold which naturally preserves the constraint.

Simulated annealing seems to be a promising technique which can improve the rate of convergence of the algorithms. Further work concerning this subject will be presented in a subsequent paper.

References

1. E. Aarts and J. Korst, "Simulated Annealing and Boltzmann Machines", *John Willey and Sons*, 1989.
2. S.Amari, A. Cichocki and H.H. Yang "A new Learning Algorithm for Blind Source Separation", in *Advances in Neural Information Processing 8*, Cambridge, MA:MIT Press, 757-763, 1996.
3. A.J.Bell and T. Sejnowski "An Information-Maximization Approach to blind separation and blind deconvolution", *Neural Computation*, vol. 7, pp.1129-1159, 1995.
4. J.-F. Cardoso "Blind Signal Separation: Statistical Principles", *Proc. of the IEEE*, vol. 86, No. 10, pp.2009-2025, Oct.1998.

5. J.-F. Cardoso and B. Laheld "Equivariant Adaptive Source Separation", *IEEE Transactions on Signal Processing*, vol. 45, No. 2, pp.433-444, 1996.
6. J.-F. Cardoso and A. Souloumiac "Blind Beamforming for Non Gaussian Signals", *IEE Proc.-F*, vol. 140, No. 6, pp.362-270, Dec. 1993.
7. P. Comon, "Independent Component Analysis - A New Concept?", *Signal Processing*, vol. 36, No. 3, pp.287-314, 1994.
8. A. Edelman, T. A. Arias and S. T. Smith "The Geometry of Algorithms with Orthogonality Constraints", *SIAM Journal on Matrix Analysis Applications*, vol. 20, No. 2, pp.303-353, 1998.
9. P. E. Gill, W. Murray and M. H. Wright, "Practical Optimization", *Academic Press*, 1981.
10. A. Hyvarinen and E.Oja "A fast fixed-point algorithm for independent component analysis", *Neural Computation*, vol. 9, pp.1483-1492, 1997.
11. C. Jutten and J. Herault, "Blind Separation of Sources, Part I: an adaptive algorithm based on neuromimetic architecture", *Signal Processing*, vol. 24, pp.1-10, 1991.
12. R. J. Marks II, "Alternating projections onto convex sets", in *Deconvolution of Images and Spectra*, P. A. Jansson, Ed., Academic, 1997.
13. R. Martín-Clemente and J.I.Acha, "Blind Separation of Sources using a New Polynomial Equation", *Electronics Letters*, Vol.33, No.1, pp.176-177,1997.
14. E. Oja "Nonlinear PCA Criterion and Maximum Likelihood in Independent Component Analysis", in *First Intern. Workshop on Independent Component Analysis and Signal Separation*, Aussois, France, pp. 143-148, Jan. 1999.
15. C. G. Puntonet, A. Prieto, C.Jutten and J. Rodriguez-Alvarez, "Separation of Sources: A Geometry-Based Procedure for reconstruction of N-Valued Signals", *Signal Processing*, vol. 46, No. 3, pp.267-284, 1995.
16. A. Prieto, C. G. Puntonet and B. Prieto, "A Neural Learning Algorithm for Blind Separation of Sources based on Geometrical Properties", *Signal Processing*, vol. 64, No. 3, pp.315-331, 1998.
17. C. G. Puntonet, M. Rodríguez-Álvarez, C. Bauer, E. W. Lang, "Simulated Annealing and Density Estimation for the Separation of Sources", *Proceedings of the IASTED Int. Conf. on Signal Processing and Communications*, pp. 347-352, Marbella, Spain, Sept. 2000.

The Problem of Overlearning in High-Order ICA Approaches: Analysis and Solutions

Jaakko Särelä[1], Ricardo Vigário[1,2]

[1] Neural Network Research Centre, Helsinki University of Technology,
FIN - 02015 HUT, Finland
[2] GMD - FIRST, Kekuléstrasse 7, D-12489 Berlin, Germany
{Jaakko.Sarela, Ricardo.Vigario}@hut.fi

Abstract. We consider a type of overlearning typical of independent component analysis algorithms. These can be seen to minimize the mutual information between source estimates. The overlearning causes spike-like signals if there are too few samples or there is a considerable amount of noise present. It is argued that if the data has flicker noise the problem is more severe and is better characterized by bumps instead of spikes. The problem is demonstrated using recorded magnetoencephalographic signals. Several methods are suggested that attempt to solve the overlearning problem or, at least, diminish/reduce its effects.

1 Introduction

Let us consider a general linear blind source separation problem (BSS), where the observed n signals are weighted sums of m underlying source signals: $\mathbf{x}(t) = \mathbf{A}(t)\mathbf{s}(t) + \mathbf{n}(t)$. The weighting or mixing matrix $\mathbf{A}$ is assumed constant. The origin of the noise term $\mathbf{n}$ can lay in the observing device or in the fact that the linear instantaneous model cannot explain the observations to their full extent, e.g. the relation between the sources and mixtures can be non linear.

Solving the BSS problem consists then in estimating both unknowns $\mathbf{A}$ and $\mathbf{s}$ from the observed $\mathbf{x}$. One way to do this is through the use of the independent component analysis (ICA) [7, 1], where the sources $\mathbf{s}$ are assumed mutually independent. Usually two restrictions are made: it is assumed that there are at least as many mixtures as sources i.e. $m \geq n$, and that the noise is negligible. The abovementioned model becomes:

$$\mathbf{x} = \mathbf{As}. \qquad (1)$$

Typically ICA algorithms search for an inverse linear mapping from $\mathbf{x}$, $\hat{\mathbf{s}} = \mathbf{Bx}$, so that $\hat{\mathbf{s}}$ minimize the *mutual information* between the source estimates, rendering them statistically independent. This strategy works only for non-Gaussian sources, since any rotation of independent Gaussian gives another set of independent Gaussians. Most high order statistics ICA algorithms use this strategy (see [4]). A pre-whitening stage is often required, which simplifies the search for independence, and reduces $\mathbf{B}$ to a simple orthogonal transformation.

In [5] it has been shown that insufficient data samples leads to an overlearning of ICA algorithms that take the form of estimates with a single spike or bump and practically zero everywhere else, no matter what the data **x** is. In the paper it was suggested that the problem is caused by insufficient (efficient) amount of samples. In this paper we further argue that the problem may arise no matter how large finite data sets we have, if there is some part in the system that cannot be modeled by **As**, but there is structured time-dependent noise present.

In the experiments FastICA [2] is used as a representative ICA algorithm. The absolute value of the kurtosis is used as the objective function because of its convenient analytical properties. The results are straightly applicable to all other ICA algorithms as demonstrated in [5].

Illustrations with magnetoencephalographic (MEG) recordings [3] are used for three main reasons: 1) The linear and instantaneous model is plausible within reasonable assumptions [3]). 2) The signal to noise ratio in MEG is very poor. In particular, the power spectrum of the noise is best characterized by an $1/f$-curve. Hence the name $1/f$-noise or 'flicker' noise. 3) Despite the previous, ICA has been used successfully in the analysis of MEG signals (see e.g. [9,6]).

2 Setting the problem

2.1 Spikes maximize super-Gaussianity

The overlearning problem becomes easily understandable, when we consider a case where the sample size T equals the dimension of the data m. Then, by collecting the realizations of **x** in (1) into a matrix, we get: $\mathbf{X} = \mathbf{AS}$, where all the matrices are square. Now, by changing the values of **A**, we can give any values for **S**. This phenomenon is similar to the classical overlearning in the case of linear regression with equal number of observations and parameters. In the present case then, whatever our principle for choosing **S** is, the result depends little on the observed data **X**, but is totally determined by our estimation criterion. For ICA, we have the following propositions (proves in [8]):

Proposition 1. *Denote by* $\mathbf{s} = (s(1),...,s(T))^T$ *a T-dimensional (nonrandom) vector that could be interpreted here as a sample of a scalar random variable. Denote by H the set of such vectors (samples) **s** that have zero mean and unit variance, i.e.* $\frac{1}{T}\sum_t s(t) = 0$ *and* $\frac{1}{T}\sum_t s(t)^2 = 1$.

Then the kurtosis of **s**, *defined as* $\sum_t s(t)^4 - 3$, *is maximized in H by spike-like signals, with just one significantly non-zero component.*

Proposition 2. *Under the conditions of Proposition 1, also the absolute value of kurtosis is maximized by vectors of previous form, provided that T is sufficiently large.*

Thus, in the extreme case where we only have as many samples as dimensions, our estimate for the source matrix **S** is roughly a scaled permutation matrix. To demonstrate this, we generated 500 mixtures from three artificial signals of 500 samples Fig. 1a. Their original kurtoses were, respectively, 8.65, -1.50, 5.75. Some noise was added to the mixtures to render full rank to the covariance matrix. A representative sample of the noisy mixtures is shown in Fig. 1b. Some of the resulted components by FastICA are shown in Fig. 1c. Note that the kurtoses of these components are now 463.32, 484.11, 483.06.

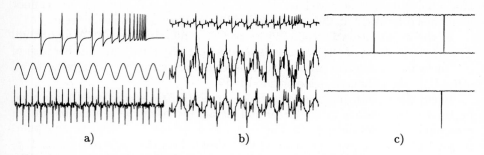

Fig. 1. *a) Three artificial signals with zero mean and unit variance. b) Three of the 500 mixtures generated by those signals. c) Three estimated independent components by FastICA.*

2.2 Are we safe if $T > m$?

In a rule of thumb, drawn for parameter estimation (e.g. linear regression), the number of samples should be at least five times the number of free parameters. In the case of ICA, we need to estimate the unmixing matrix **B**. If we assume that the data has been pre-whitened, the number of free parameters is $n^2/2$, where n is the number of sources[1]. Thus we should have $T > 5 \times \frac{n^2}{2}$ samples.

In the case of infinite samples of random variables, the *central limit theorem* ensures that the kurtosis of the linear projection $\hat{s}_i = \mathbf{w}_i^T \mathbf{x}$ is at most the kurtosis of the most kurtotic independent source generating **x**. Unfortunately, this does not hold for non-Gaussianity measures of finite sample sizes! Consider a finite sample of a Gaussian i.i.d random vector. It is easy to produce a spike by enhancing one time instance and dampening all the others: one time instance is used as our unmixing vector, i.e. $\hat{s}_i = \mathbf{x}(t_0)^T \mathbf{x}$, where $\mathbf{x}(t_0)$ corresponds to the realisation of **x** at t_0. Some spikes, generated in this way, are shown in Fig. 2a. Note that although small, their kurtoses are now clearly non-zero (see table 1).

Hence spikes are generated if their kurtoses (or any other measure of mutual information used by the specific algorithm) are greater than the ones of the actual independent sources.

2.3 Bumps are generated, if low frequencies dominate

What happens if the sources are not Gaussian i.i.d? We can try to generate spikes in any kind of data using the strategy mentioned above. In Fig. 2b there are five bumps generated from a random vector that has a $1/f$ frequency content. The

---
[1] If $m > n$, the spare dimensions are eliminated during whitening.

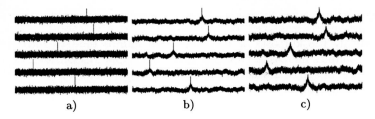

Fig. 2. a) Artificially generated spikes from 80 dimensional Gaussian data with 16000 samples ($T = 5 \times n^2/2$). b) Artificially generated bumps with spikes from 80 dimensional data with 16000 samples having $1/f$ power spectra. c) Artificially generated smooth bumps.

frame / sig	sig 1	sig 2	sig 3	sig 4	sig 5
a)	0.18	0.52	0.30	0.40	0.56
b)	2.61	1.76	2.08	1.21	3.06
c)	3.94	3.45	4.65	2.05	5.01

Table 1. Kurtoses of signals in Fig. 2.

spike is still visible, but now it points out from the middle of a bump, because close samples in time are strongly correlated. Note that the kurtoses have now increased. An even stronger effect occurs if the mixing vector is taken to be a weighted average around some time point. Then the bump is described by

$$\hat{s}(t) = \mathbf{b}^T \mathbf{x}(t), \quad \mathbf{b} = \sum_{t=0}^{L} w(t) \mathbf{x}(t_0 - L/2 + t). \quad (2)$$

$w(t)$ is a windowing function, normalized so that $\sum w(t) = 1$, and $L+1$ is the width of the window, at best also the width of the bump. Bumps generated with this method don't have a spike and their kurtoses are even greater. Some bumps using a triangular window of length $L = 101$ are presented in Fig. 2c.

Note that the kurtoses of the generated smooth bumps are now greater than 2 (table 1), which is the theoretical maximum of the absolute value of the kurtosis of sub-Gaussian signals. This renders very hard to identify any sub-Gaussian source from MEG data (or any other with a $1/f$ characteristics).

3 Attempts to solve the problem

The key question in solving the problem of spikes and bumps is: How can we ensure that the kurtoses of spike or bump artifacts are smaller than the kurtoses of the desirable components? The next sections will review a few possible ways to achieve this goal. With the first two we can increase the number of samples per free parameter in ICA estimation. The third increases the *effective* number of samples: by filtering out the low frequencies, we emphasize the information on the higher frequencies. Other approaches are suggested in Sec. 3.4.

3.1 Acquiring more samples

Asymptotically the kurtoses of the spikes and bumps tend to zero, when the number of samples increases. How fast does that happen? Consider, once again, the generation of spikes in case of Gaussian i.i.d. data. The kurtosis of the spike can be easily approximated assuming that T is so large that the rest of the source estimate $\mathbf{x}(t_0)^T \mathbf{x}(t)_{t \neq t_0}$ still has zero mean and unit variance. We then get

Proposition 3. *Let $s(t)$ be a spike generated from $\mathbf{x}(t)$ by $s(t) = \mathbf{x}_0^T \mathbf{x}(t)$, where $\mathbf{x}_0 = \frac{\mathbf{x}(t_0)}{\|\mathbf{x}(t_0)\|}$ is the unmixing vector and t_0 is the time instance at which the spike occurs. Let $\mathbf{x}(t)$ be a Gaussian i.i.d. signal. Then the kurtosis of the spike $s(t)$ is approximately*

$$kurt(s(t)) \approx \frac{\|\mathbf{x}(t_0)\|^4}{T} \propto \frac{1}{T}. \tag{3}$$

If gathering more data is possible, this should solve the problem. Often this is not the case, and gathering more data may be expensive and/or time consuming.

A similar reasoning to proposition (3) can be drawn to the relation of the kurtosis of a bump with the sample size T. Experimental evidence shows that the width of a bump grows roughly proportional to $\log T$. Hence the kurtosis evolves closer to $kurt(s(t)) \propto \frac{\log T}{T}$, which decreases with T much slower than $kurt(s(t)) \propto \frac{1}{T}$. Furthermore the kurtoses of the bumps tend to be bigger than the kurtoses of the spikes with the same T. Thus even bigger amounts of samples are required for the bumps to have kurtoses smaller than the desired signals.

The effect of acquiring more samples, in case of bumps, was experimentally illustrated using MEG data containing strong artifacts. ICA results for this data were first reported in [10]. Three optimally isolated sources were chosen as target signals, shown in Fig. 3a. They consist of cardiac, eye activity and a digital watch

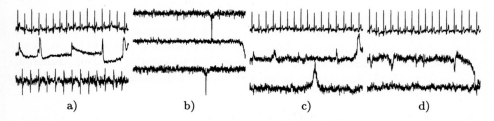

Fig. 3. *a) original target signals, b) ICA results using 2000 points, c) 5000 points and d) 17000 data points.*

components. ICA was performed to data of different lengths, ranging from 2000 points to 17000 points. A sample of the results is presented in Fig. 3b- 3d. For comparison purposes, only the first 2000 points are shown in each frame. The bump of the last component in Fig. 3d is out of this chosen range.

We conclude that acquiring more samples can, in some cases, be used to solve the problem of spikes, but is not enough for solving completely the problem of bumps.

3.2 Reducing dimension

Consider once again the spikes generated from Gaussian i.i.d. data. The kurtosis of the spike does not depend only on the amount of samples, but is also highly dependable on the dimension of $\mathbf{x}(t)$. It can be seen that this dependence is of the form: $\text{kurt}(s(t)) \propto m^2$. Furthermore, experimental results show that the width of a bump is not a function of the dimension size. The quadratic relation is therefore valid for both spikes and bumps.

The important factor to avoid overlearning is not the absolute amount of samples or dimensions, but rather the ratio between them. Hence, decreasing the number of dimensions seems a much more efficient way to avoid spikes and bumps than acquiring more samples. Additionally, it is also easier to leave behind something we already have than to gather some new information.

How should we choose this reduction optimally? If $m > n$, i.e. we have more sensors than sources, this is straightforward: if (1) holds, $m - n$ eigenvalues of the covariance matrix are zero, thus a projection to the non-zero eigenvectors loses no information. If there is some amount of independent sensor noise, we can still do this effectively using principal component analysis (PCA), which guarantees optimal reconstruction in mean squared error sense. Then the sensor noise is concentrated in the components having the smallest variances.

The risk in doing the aforementioned dimension reduction is that some relevant, but weak, portion of a source may get discarded from the processed data. The choice of the retained components is therefore of great importance. These limitations are even more relevant if the number of underlying sources is very high (e.g. comparable to the number of observations). Although ICA would no longer be strictly valid, we can expect that the decomposition found helds some meaningful sums of the original independent components.

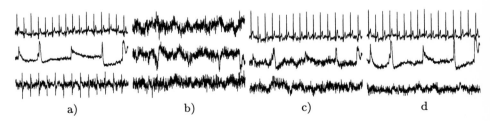

Fig. 4. a) *original target signals, b) ICA results using 127, c) 117 and d) 77 dimensions.*

Figure 4 shows the results of applying dimension reduction to the data in Fig. 3. Note that, as the dimension reduces, the structured targets start to show. Although 77 is not yet the ideal dimension, we are on the correct direction.

3.3 Pre-filtering of the data

High-pass filter A bump can be considered as a set of multiple spikes in nearby vicinity, or as a low pass filtered version of a single spike. In fact, a bump is mainly dominated by low frequencies. An obvious way to reduce its influence is to high-pass filter the data, prior to any BSS decomposition. This technique, used in MEG recordings [10], enabled the detection of a source signal otherwise impossible to detect (the digital watch at the bottom of Fig. 3a).

One problem in performing a simple high-pass filtering of the data is the determination of its cut off frequency. Data driven techniques should enable to find the best suited filtering parameters, in a more robust manner.

Use singular spectrum analysis Singular spectrum analysis (SSA, for a recent review see e.g. [11]) consists in a PCA decomposition of an estimate of the correlation matrix that is based on a number of lagged copies of each time series. The eigen filters generated in this fashion differ in general from the sines and cosines typical for a Fourier analysis. Yet, ordered in a decreasing order of the respective eigenvalues, they present increasing frequency contents. It is reasonable to expect that the suppression of the principal component(s), with its low frequency, leads to a reduction in the probability of generation of bumps from the data, in much the same way as stated in the beginning of this section, with the advantage that the design of the correct filters is completely data-adaptive.

Solve ICA for innovation processes Another data driven filtering technique can be performed if we consider the sources as the sum of an auto-regressive process (modeling all temporal structures in the data), and an innovation process: $\mathbf{s}(t) = \mathbf{s}_{AR}(t) + \mathbf{s}_{INN}(t)$. Similarly, the observed data could be described as: $\mathbf{x}(t) = \mathbf{x}_{AR}(t) + \mathbf{x}_{INN}(t)$.

Then, if the innovation processes are independent and non-Gaussian, we can solve the mixing matrix $\mathbf{A}$ from the innovation process only: $\mathbf{x}_{INN} = \mathbf{A}\mathbf{s}_{INN}$. The estimated mixing matrix can then be used to find the complete sources. This approach has been successfully used in BSS of correlated images [4].

3.4 Some other approaches

If the source of overlearning is fully characterized, a *regularizing term* can be added to the cost function, that penalizes that sort of solutions. In this particular ICA problem this could correspond to allow a search for higher kurtosis while penalizing the extremely sparce solutions, or a combination of low frequency contents and sparsity.

The introduction of *prior information* in the search for the independent components would lead to avoiding the type of overfitting mentioned in this paper. As an example, limiting the search for signals produced by a dipolar source can be useful in the analysis of neuromagnetic evoked responses. Another example is given by the modeling of strong dynamical information in the time series.

4 Conclusion

We have reviewed and characterized a form of overfitting present in most high order statistics ICA approaches, when used to solve the BSS problem. It consist in the generation of estimates of the source signals that are zero everywhere except for a single spike or bump. A series of preliminary solutions were as well outlined.

5 Acknowledgements

This work has been partially funded by EC's - # IST-1999-14190 BLISS project. The authors would like to thank Dr Aapo Hyvärinen for the proves of propositions 1 and 2.

References

1. P. Comon. Independent component analysis – a new concept? *Signal Processing*, 36:287–314, 1994.
2. The FastICA MATLAB package. Available at http://www.cis.hut.fi/projects/ica/fastica/, 1998.
3. M. Hämäläinen, R. Hari, R.J. Ilmoniemi, J. Knuutila, and O.V. Lounasmaa. Magnetoencephalography—theory, instrumentation, and applications to noninvasive studies of the working human brain. *Reviews of Modern Physics*, 65:413–497, 1993.
4. A. Hyvärinen, J. Karhunen, and E. Oja. *Independent component analysis*. Wiley, 2001. in press.
5. A. Hyvärinen, J. Särelä, and R. Vigário. Spikes and bumps: Artefacts generated by independent component analysis with insufficient sample size. In *Proc. Int. Workshop on Independent Component Analysis and Blind Signal Separation (ICA'99)*, Aussois, France, 1999. To appear.
6. O. Jahn and A. Cichocki. Identification and elimination of artifacts from MEG signals using efficient independent components analysis. In *Proc. of the 11th Int. Conf. on Biomagnetism (BIOMAG-98*, Sendai, Japan, 1998.
7. C. Jutten and J. Herault. Blind separation of sources, part I: An adaptive algorithm based on neuromimetic architecture. *Signal Processing*, 24:1–10, 1991.
8. Proves of the propositions in the paper. Available at http://www.cis.hut.fi/jaakko/proves/IWANN01.ps, 2001.
9. R. Vigário, J. Särelä, V. Jousmäki, M. Hämäläinen, and E. Oja. Independent component approach to the analysis of eeg and meg recordings. *IEEE transactions on biomedical engineering*, 47(5):589–593, 2000.
10. R.N. Vigário, V. Jousmäki, M. Hämäläinen, R. Hari, and E. Oja. Independent component analysis for identification of artifacts in magnetoencephalographics recordings. In *Advances in Neural Information Processing 10 (Proc. NIPS'97)*, pages 229–235. MIT Press, Cambridge MA, 1997.
11. Pascal Yiou, Didier Sornette, and Michael Ghil. Data-adaptive wavelets and multiscale singular-spectrum analysis. *Physica D*, 142:254 – 290, 2000.

Equi-convergence Algorithm for Blind Separation of Sources with Arbitrary Distributions

L.-Q. Zhang, S. Amari, and A. Cichocki

Brain-style Information Systems Research Group
RIEKN Brain Science Institute
Wako shi, Saitama 351-0198, JAPAN
http://www.bsp.brain.riken.go.jp

Abstract. This paper presents practical implementation of the equi-convergent learning algorithm for blind source separation. The equi-convergent algorithm [4] has favorite properties such as isotropic convergence and universal convergence, but it requires to estimate unknown activation functions and certain unknown statistics of source signals. The estimation of such activation functions and statistics becomes critical in realizing the equi-convergent algorithm. It is the purpose of this paper to develop a new approach to estimate the activation functions adaptively for blind source separation. We propose the exponential type family as a model for probability density functions. A method of constructing an exponential family from the activation (score) functions is proposed. Then, a learning rule based on the maximum likelihood is derived to update the parameters in the exponential family. The learning rule is compatible with minimization of mutual information for training demixing models. Finally, computer simulations are given to demonstrate the effectiveness and validity of the proposed approach.

1 Introduction

Blind source separation or independent component analysis [14, 12] introduces a novel paradigm for signal processing and has attracted considerable attention in signal processing society. A number of neural networks and statistical methods [14, 12, 11, 5, 16, 9, 10, 17, 18] have been developed for blind signal separation. There are a number of factors which are likely to affect separation results in applications, such as the number of active sources, the distributions of source signals, time-variable mixtures and noise.

It is the main purpose of this paper to develop a learning algorithm with certain uniform convergence in the sense that all components converge to true solution at the same rate regardless of what types of the distributions the source signals have. Such an equi-convergent algorithm has been proposed in [4]. However, because the algorithm includes unknown activation functions and certain statistics of source signals, its practical implementation still remains open. The problem attributes how to estimate the statistics of source signals. If the activation functions are not suitably chosen, the learning algorithm will not converge to the true solution. Thus the online estimators of the statistics of the source signals might not be accurate enough to realize the equi-convergence algorithm.

Stability of learning algorithms [4, 10], is critical to successful separation of source signals from measurements. One way is to estimate those statistics adaptively [17, 13]. Some other statistical models, such as the Gaussian mixture model [8, 15] are also employed to estimate the distributions of source signals. Usually, such methods are computing demanding.

In contrast to the previous works on the estimation of the distributions for source signals, this paper attempts to avoid to estimate source distributions, but rather to online adapt activation functions for source signals. The adaptation of activation functions has two purposes: one is to modify the activation functions such that the true solution is the stable equilibrium of the learning system; the second is to estimate the sparseness of source signals. This simplification makes it easy to estimate the parameters in the generative models and reduce computing cost. Usually, the adaptation of activation functions needs only a very few parameters for each component. There are some advantages by using the exponential family to estimate activation functions. It is easy to reveal the relation between the distributions and activation functions (score functions). And also we can easily construct a linear connection of the score functions for the exponential family if we want to separate signals with specific distributions. Another advantage of the approach is its compatibility, i.e. both the updating rules for the demixing model and for the free parameters in the distribution models make the cost function decrease to its minimum, if the learning rate is sufficiently small.

Assume that source signals are stationary zero-mean processes and mutually statistically independent. Let $\mathbf{s}(k) = (s_1(k), \cdots, s_n(k))^T$ be the vector of unknown independent sources and $\mathbf{x}(k) = (x_1(k), \cdots, x_m(k))^T = \mathbf{A}\mathbf{s}(k) + \mathbf{v}(k)$ a sensor vector, which is a linear instantaneous mixture of the sources with additive noises $\mathbf{v}(k)$. The blind separation problem is to recover original signals $\mathbf{s}(k)$ from observations $\mathbf{x}(k)$ without prior knowledge on the source signals and the mixing matrix, but the assumption of mutual independence of source signals. The demixing model here is a linear transformation of the form

$$\mathbf{y}(k) = \mathbf{W}\mathbf{x}(k), \qquad (1)$$

where $\mathbf{y}(k) = (y_1(k), \cdots, y_m(k))^T, \mathbf{W} \in \mathbf{R}^{m \times m}$ is the demixing matrix to be determined during training. Assume that $m \geq n$, i.e. the number of the sensor signals is larger than the number of the source signals. We intend to train the demixing model $\mathbf{W}$ such that n components are designed to recover the n source signals and the rests correspond to the zeros or noise.

2   Learning algorithms

The estimation of demixing model $\mathbf{W}$ can be formulated in the framework of a semiparametric statistical model [3]. The probability density function of $\mathbf{x}$ can

be expressed as
$$p_X[\mathbf{x}, \mathbf{W}, p(\mathbf{s})] = |\det(\mathbf{W})| p(\mathbf{W}\mathbf{x}), \qquad (2)$$
which depends on two unknowns separating matrix $\mathbf{W} = \mathbf{A}^{-1}$ and pdf $p(\mathbf{s})$. Here, the statistical model (2) includes not only an unknown matrix $\mathbf{W}$ to be estimated but also an unknown pdf function $p(\mathbf{s})$ which we do not need to estimate. Such a model is said to be semi-parametric. The semi-parametric model theory suggests to use an estimating function to estimate demixing matrix $\mathbf{W}$. In blind source separation, the estimating function is a matrix function $\mathbf{F}(\mathbf{y}, \mathbf{W})$, which does not depend on $p(\mathbf{s})$, provided it satisfies certain regularity conditions which can be found in [6]. Amari and Kawanabe [6] proposed an information geometric theory of estimating functions by extending the differential geometry of statistics [1], and gave the set of all the estimating functions. It is also proved that the effective part of the off-diagonal elements of estimating functions F_{ij} for $(i \neq j)$ is spanned by the functions of the form $\varphi(y_i)y_j$ and $\varphi(y_j)y_i$ and the diagonal part by $f(y_i) = \psi(y_i)y_i - 1$, where φ and ψ are arbitrary functions. The optimal one for φ and ψ is to choose $\varphi_i(y) = -\frac{d}{dy_i}\log p_i(y_i) = -\frac{\dot{p}_i(y_i)}{p_i(y_i)}$, if we can estimate the true source probability distribution $p_i(y_i)$ adaptively. From the theory of estimating functions, we have the general form of estimating functions [3]
$$\mathbf{F}(\mathbf{y}, \mathbf{W}) = \mathcal{K}(\mathbf{W}) \circ [\mathbf{I} - \varphi(\mathbf{y})\mathbf{y}^T]\mathbf{W}, \qquad (3)$$
where $\varphi(\mathbf{y}) = [\varphi_1(y_1), \ldots, \varphi_m(y_m)]^T$ is an arbitrary vector function and $\mathcal{K}(\mathbf{W})$ is an arbitrary linear operator which maps matrices to matrices. An online learning algorithm based on the estimating function is described by
$$\mathbf{W}(k+1) = \mathbf{W}(k) - \eta(k)\mathbf{F}(\mathbf{y}, \mathbf{W}). \qquad (4)$$

Different linear operator $\mathcal{K}(\mathbf{W})$ leads to different existing algorithms such as the natural gradient algorithm [5]. It should be noted that different algorithms have different stability regions. Therefore, the choice of the nonlinear activation functions and the linear operator $\mathcal{K}(\mathbf{W})$ is vital to successful separation of source signals.

2.1 Equi-convergence Learning Algorithm

Given an estimating function, learning algorithm (4) may have different convergence rate for different component $y_i(k)$ of $\mathbf{y}(k)$. It is the purpose of this paper to present an algorithm such that all the components of the output $\mathbf{y}(k)$ have the same convergence rate in the sense of
$$\frac{\partial E[\mathbf{F}(\mathbf{y}, \mathbf{W})]}{\partial \mathbf{W}} = \mathcal{I}, \qquad (5)$$
where $\mathcal{I}$ is the $n^2 \times n^2$ identity matrix, that is, the Hessian of the learning algorithm is the identity. This implies that algorithm (4) converges equally well in

any directions. In paper [4], a universally convergent learning algorithm was developed with an estimating function $\mathbf{F}(\mathbf{y}, \mathbf{W}) = \mathbf{G}(\mathbf{y})\mathbf{W}$, where $\mathbf{G}(\mathbf{y}) = [g_{ij}(\mathbf{y})]$ is given by

$$g_{ij} = \frac{1}{\sigma_i^2 \sigma_j^2 \kappa_i \kappa_j - 1} \{\varphi(y_j)y_i - \sigma_j^2 \kappa_i \varphi(y_i)y_j\}, \quad i \neq j \quad (6)$$

$$g_{ii} = \frac{1}{m_i + 1} \{\varphi(y_i)y_i - 1\}, \quad (7)$$

where $\sigma_i^2 = E[y_i^2]$, $\kappa_i = E[\dot{\varphi}_i(y_i)]$, $m_i = E[y_i^2 \dot{\varphi}_i(y_i)]$, where $\dot{\varphi} = d\varphi/dy$. However, the estimating function includes the unknown activation functions and a number of unknown statistics σ_i, κ_i, m_i, which depend on the source signals. How to estimate these statistics becomes the key problem to realize the equi-convergence learning algorithm. In this paper, we suggest to estimate both the activation functions and the statistics σ_i, κ_i, m_i. The motivation is if the activation functions are not well chosen, the algorithm may not converge to the true solution and the online estimators for the statistics will be not accurate enough to realize the equi-convergence algorithm.

3 Exponential Family

In order to model the probability distributions of the source signals, we suggest to use an exponential family [7], which is expressed in term of certain functions $\{C(y), \psi_1(y), \cdots, \psi_N(y)\}$ as $p(y, \boldsymbol{\theta}) = \exp\left[-C(y) - \sum_{i=1}^{N} \theta_i \psi_i(y) + \mathcal{N}(\boldsymbol{\theta})\right]$, $\boldsymbol{\theta}$ is the vector of free parameters, and $\mathcal{N}(\boldsymbol{\theta})$ is a normalization term such that the integral of $p(y, \boldsymbol{\theta})$ over the whole interval $(-\infty, \infty)$ is equal to one. There are some good properties in the exponential family, such as *flatness*, as a statistical model. Refer to [7] for further properties of the exponential family.

We provide here a feasible way to construct the exponential family for blind source separation. First, we define a parametric family for activation functions $\varphi(y, \boldsymbol{\theta}) = \sum_{i=1}^{N} \theta_i \varphi_i(y)$, where $\boldsymbol{\theta} = (\theta_1, \cdots, \theta_N)^T$ are the parameters to be determined adaptively and φ_i are certain activation functions. From the definition of activation functions, $\varphi(y, \boldsymbol{\theta}) = -\frac{\partial \log p(y, \boldsymbol{\theta})}{\partial y}$, the probability density function is equivalently given by

$$p(y, \boldsymbol{\theta}) = \exp\left\{-\boldsymbol{\theta}^T \boldsymbol{\psi}(y) + \mathcal{N}(\boldsymbol{\theta})\right\}, \quad (8)$$

where $\mathcal{N}(\boldsymbol{\theta})$ is the normalization term and $\boldsymbol{\psi}(y) = (\int_0^y \varphi_1(\tau)d\tau, \cdots, \int_0^y \varphi_N(\tau)d\tau)$.

In blind separation, it is well known that the hyperbolic tangent function $\varphi(y) = tanh(y)$ is a good activation function for super-Gaussian sources and the cubic function $\varphi(y) = y^3$ is a good activation function for sub-Gaussian sources. Thus, we define the exponential family by linearly combining the three activation functions in the form of $\varphi(y, \boldsymbol{\theta}) = \theta_1 \alpha \, tanh(\alpha y) + \theta_2 4\beta \, y^3 + \theta_3 \, y$, where $\alpha = \frac{\pi}{2}$, $\beta = (\Gamma(0.75)/\Gamma(0.25))^2$. The purpose of such choice of α and β is to ensure the distributions have the uni-variance. The corresponding exponential family is expressed as

$$p(y, \boldsymbol{\theta}) = \exp\left(\theta_1 \log sesh(\alpha y) - \theta_2 \beta y^4 - \theta_3 \frac{y^2}{2} + \mathcal{N}(\boldsymbol{\theta}))\right). \quad (9)$$

In this exponential family, $\boldsymbol{\psi}(y) = (-\log sesh(\frac{\pi}{2}y), \beta y^4, \frac{y^2}{2})^T$ and $\mathcal{N}(\boldsymbol{\theta})$ is the normalization term. This exponential family includes three typical distributions: the super-Gaussian, sub-Gaussian and Gaussian distributions. In particular, the exponential family covers the homotopy family $p(y, \theta, 1 - \theta, 0)$ which connects the following two density functions $\frac{sech(\alpha y)}{2}$ and $exp(-\beta y^4 + \mathcal{N}(0, 1, 0))$. Figure 1 shows the waveform of the homotopy family varying θ from -1 to 1. This family will be used to model the distributions of source signals in this paper.

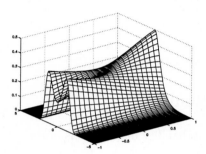

Fig. 1. The waveform of the homotopy family varying θ from -1 to 1

4 Adaptation of Activation Functions

Assume that $q_i(y_i, \boldsymbol{\theta}_i) = \exp\left\{-\boldsymbol{\theta}_i^T \boldsymbol{\psi}(y_i) + \mathcal{N}(\boldsymbol{\theta}_i)\right\}$ is a model for the marginal distribution of y_i, $(i = 1, \cdots, m)$. Various approaches such as entropy maximization and minimization of mutual information lead to the following cost function,

$$l(\mathbf{y}, \boldsymbol{\theta}, \mathbf{W}) = -\log(|det(\mathbf{W})|) - \sum_{i=1}^{m} \log q_i(y_i, \boldsymbol{\theta}_i), \quad (10)$$

where $\boldsymbol{\theta} = (\boldsymbol{\theta}_1^T, \boldsymbol{\theta}_2^T, \cdots, \boldsymbol{\theta}_m^T)^T$ is the parameters determined adaptively. The main purpose of the adaptation of activation functions is to modify the activation function such that the true solution is the stable equilibrium of learning dynamics.

4.1 Natural Gradient Learning

By minimizing the cost function (10) with respect to $\boldsymbol{\theta}$ by using the gradient descent approach, we derive learning algorithms for training parameters $\boldsymbol{\theta}$. When the parameter space has a certain underlying structure, the ordinary gradient of a function does not represent its steepest direction. The steepest descent direction in a Riemannian space is given by the natural gradient [2], which takes the form of

$$\tilde{\nabla}_\theta l(\mathbf{y}, \boldsymbol{\theta}, \mathbf{W}) = \mathcal{G}^{-1} \nabla_\theta l(\mathbf{y}, \boldsymbol{\theta}, \mathbf{W}) \tag{11}$$

where $\mathcal{G}$ is the Riemannian metric of the parameterized space, $\nabla_\theta l(\mathbf{y}, \boldsymbol{\theta}, \mathbf{W}) = (\frac{\partial l}{\partial \boldsymbol{\theta}_1}^T, \cdots, \frac{\partial l}{\partial \boldsymbol{\theta}_m}^T)^T$ and $\frac{\partial l}{\partial \boldsymbol{\theta}_i} = -\boldsymbol{\psi}(y_i) + \mathcal{N}'(\boldsymbol{\theta}_i)$. The Riemannian structure of the parameter space of statistical model $\{q_i(y_i, \boldsymbol{\theta}_i)\}$ is defined by the Fisher information [1]

$$\mathcal{G}_i(\boldsymbol{\theta}_i) = E\left[\frac{\partial \log q_i(y_i, \boldsymbol{\theta}_i)}{\partial \boldsymbol{\theta}_i} \frac{\partial \log q_i(y_i, \boldsymbol{\theta}_i)^T}{\partial \boldsymbol{\theta}_i}\right] \tag{12}$$

in the component form. The learning algorithm based on the natural gradient descent approach is described as

$$\boldsymbol{\theta}_i(k+1) = \boldsymbol{\theta}_i(k) - \eta(k) \mathcal{G}_i^{-1}(\boldsymbol{\theta}_i(k)) \frac{\partial l(\mathbf{y}, \boldsymbol{\theta}, \mathbf{W})}{\partial \boldsymbol{\theta}_i}. \tag{13}$$

From the cost function (10), we see that the minimization of mutual information is equivalent to maximizing the likelihood for parameters $\boldsymbol{\theta}_i$ because the first term in (10) does not depend on $\boldsymbol{\theta}_i$. Thus it should be noted that the above learning rule is actually equivalent to the maximum log-likelihood rule for each component. Both learning rules for updating parameters $\boldsymbol{\theta}$ and the demixing model $\mathbf{W}$ make cost function $L(\boldsymbol{\theta}, \mathbf{W}) = E[l(\mathbf{y}, \boldsymbol{\theta}, \mathbf{W})]$ smaller and smaller, provided the learning rate is sufficiently small.

5 Implementation of Equi-Convergence Algorithm and Simulation

In this section, we present how to implement the equi-convergence algorithm and give computer simulations to demonstrate the effectiveness and the performance of the proposed approach.

In order to implement the equi-convergence learning algorithm, we need to estimate the statistics of the output signals, which estimate source signals. The statistics σ_i, κ_i, m_i are evaluated by the following iterations

$$\kappa_i(k+1) = (1-\mu)\kappa_i(k) + \mu \dot{\varphi}(y_i(k), \boldsymbol{\theta}_i(k)), \tag{14}$$

where μ is a learning rate. The other statistics are estimated in a similar way.

We choose $\mathbf{F}(\mathbf{y}, \mathbf{W}) = \mathbf{G}(\mathbf{y})\mathbf{W}$, defined by (6) and (7), as the estimating function. Therefore, the equi-convergent algorithm (4) is realized by the following two stages: the first stage is to train demixing matrix $\mathbf{W}$ using the natural gradient algorithm (4), to train $\boldsymbol{\theta}_i$ using algorithm (13) and simultaneously, to estimate statistics σ_i, κ_i, m_i adaptively. After certain times iteration, we shift the learning algorithm to the equi-convergence algorithm (4). In order to remove the fluctuation caused by the non-stationarity of source signals, we use the averaged version of the algorithm (4) over certain time windows.

A large number of simulations have been done to show the validity and performance of the proposed algorithm. Here, we give one simulation example. The five source signals consist of two super-Gaussian, two sub-Gaussian and one Gaussian signals. We set the initial values $\theta = 0.5$ for all the five components. The mixing matrix is randomly generated by computer. For each batch iteration, we take 20 sample data as the output signals $\mathbf{y}(k)$. Figure 2 shows on-line estimation of $\mathbf{y}(k)$ and $\boldsymbol{\theta}$ by using the equi-convergent learning algorithm. It can be proved that the parameter θ converges to 1 for the super-Gaussian signal with distribution $sech(\alpha\ y)/2$ and to 0 for the sub-Gaussian signal with distribution $exp(-\beta|y|^4 + \mathcal{N}(0,1,0))$.

The computer simulations show that the proposed approach can separate both super-Gaussian and sub-Gaussian source signals simultaneously without presuming any knowledge on the source signals.

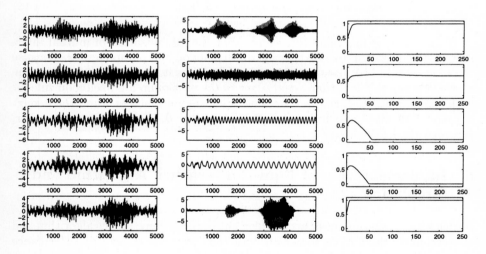

Fig. 2. a) The sensor signals; b) Online estimation of $\mathbf{y}(k)$; c) Online estimation of $\theta(k)$

6 Conclusion

In this paper, we present a new approach to realize the equi-convergence learning algorithm for blind source separation. An exponential family is employed as a model for the distributions of the source signals. A adaptation rule is developed for updating the parameters in the model of distributions of source signals. The equi-convergence algorithm is implemented by updating the demixing matrix and the parameters in the distributions simultaneously. The equi-convergence algorithm has two favorite properties: the isotropic convergence and universal convergence. Computer simulations verify such properties.

References

1. S. Amari. *Differential-geometrical methods in statistics, Lecture Notes in Statistics*, volume 28. Springer, Berlin, 1985.
2. S. Amari. Natural gradient works efficiently in learning. *Neural Computation*, 10:251-276, 1998.
3. S. Amari and J.-F. Cardoso. Blind source separation- semiparametric statistical approach. *IEEE Trans. Signal Processing*, 45:2692-2700, Nov. 1997.
4. S. Amari, T. Chen, and A. Cichocki. Stability analysis of adaptive blind source separation. *Neural Networks*, 10:1345-1351, 1997.
5. S. Amari, A. Cichocki, and H.H. Yang. A new learning algorithm for blind signal separation. In G. Tesauro, D.S. Touretzky, and T.K. Leen, editors, *Advances in Neural Information Processing Systems 8 (NIPS*95)*, pages 757-763, 1996.
6. S. Amari and M. Kawanabe. Information geometry of estimating functions in semiparametric statistical models. *Bernoulli*, 3(1):29-54, 1997.
7. S. Amari and H. Nagaoka. *Methods of Information Geometry*. AMS and Oxford University Press, 2000.
8. H. Attias. Indepedent factor analysis. *Neural Computatation*, 11(4):803-851, 1999.
9. A.J. Bell and T.J. Sejnowski. An information maximization approach to blind separation and blind deconvolution. *Neural Computation*, 7:1129-1159, 1995.
10. J.-F. Cardoso and B. Laheld. Equivariant adaptive source separation. *IEEE Trans. Signal Processing*, SP-43:3017-3029, Dec 1996.
11. A. Cichocki and R. Unbehauen. Robust neural networks with on-line learning for blind identification and blind separation of sources. *IEEE Trans Circuits and Systems I : Fundamentals Theory and Applications*, 43(11):894-906, 1996.
12. P. Comon. Independent component analysis: a new concept? *Signal Processing*, 36:287-314, 1994.
13. S. Douglas, A. Cichocki, and S. Amari. Multichannel blind separation and deconvolution of sources with arbitrary distributions. In *Proc. of NNSP'97*, pages 436-445, Florida, US, September 1997.
14. C. Jutten and J. Herault. Blind separation of sources, Part I: An adaptive algorithm based on neuromimetic architecture. *Signal Processing*, 24:1-10, 1991.
15. T. Lee and M. Lewicki. The generalized gaussian mixture model using ica. In P. Pajunen and J. Karhunen, editors, *Proc. ICA'2000*, pages 239-244, Helsinki, Finland, June 2000.
16. E. Oja and J. Karhunen. Signal separation by nonlinear hebbian learning. In M. Palaniswami, Y. Attikiouzel, R. Marks II, D. Fogel, and T. Fukuda, editors, *Computational Intelligence - A Dynamic System Perspective*, pages 83-97, New York, NY, 1995. IEEE Press.
17. W. Lee T, M. Girolami, and T. Sejnowski. Independent component analysis using an extended infomax algorithm for mixed sub-gaussian and super-gaussian sources. *Neural Computation*, 11(2):606-633, 1999.
18. L. Zhang, A. Cichocki, and S. Amari. Natural gradient algorithm for blind separaiton of overdetermined mixture with additive noise. *IEEE Signal Processing Letters*, 6(11):293-295, 1999.

Separating Convolutive Mixtures by Mutual Information Minimization

Massoud Babaie-Zadeh[1,2], Christian Jutten[1], and Kambiz Nayebi[2]

[1] Institut National Polytéchnique de Grenoble (INPG), Laboratoire des Images et des Signaux (LIS), 46 Avenue Félix Viallet, Grenoble, France
[2] Sharif University of Technology, Tehran, Iran

Abstract. Blind Source Separation (BSS) is a basic problem in signal processing. In this paper, we present a new method for separating convolutive mixtures based on the minimization of the output mutual information. We also introduce the concept of joint score function, and derive its relationship with marginal score function and independence. The new approach for minimizing the mutual information is very efficient, although limited by multivariate distribution estimations.

1 Introduction

Blind Source Separation (BSS) is a basic problem in signal processing, which has been considered intensively in the last decade. In the linear instantaneous case, the mixture is supposed to be of the form:

$$\mathbf{x} = \mathbf{A}\mathbf{s} \qquad (1)$$

where $\mathbf{s}$ is the source vector, $\mathbf{x}$ is the observation vector, and $\mathbf{A}$ is the (constant) mixing matrix. The separator system, $\mathbf{B}$, tries to estimate the sources via:

$$\mathbf{y} = \mathbf{B}\mathbf{x} \qquad (2)$$

For linear mixtures, it can be shown that independence of the components of $\mathbf{y}$, is a necessary and sufficient condition for achieving the separation (up to a scale and a permutation indeterminacy) [3].

The convolutive case, too, has been addressed by a few authors [7, 6, ?, ?, 2, 4]. In that case, the mixing and separating matrices are linear time invariant (LTI) filters, *i.e.* the mixing system is:

$$\mathbf{x}(n) = [\mathbf{A}(z)]\,\mathbf{s}(n) \qquad (3)$$

and the separating system is:

$$\mathbf{y}(n) = [\mathbf{B}(z)]\,\mathbf{x}(n) \qquad (4)$$

For these mixtures too, it has been shown that the independence of the outputs is a necessary and sufficient for signal separation (up to a filtering and a permutation indeterminacy) [7]. However, in convolutive mixtures, the independence of two random processes y_1 and y_2, cannot be deduced from the solely independence of $y_1(n)$ and $y_2(n)$, but required the independence of $y_1(n)$ and $y_2(n-m)$, for all n and all m.

Several methods have been proposed for satisfying the above condition. Most of them are based on higher (than 2) order statistics : cancellation of cross-spectra [7], of higher order cross-moments [6], of higher order cross-cumulants [6, 2], or more generally on a contrast function [4].

In this paper, we introduce a method based on minimizing the output mutual information. This method is inspired by the method proposed by Taleb and Jutten [5] for the instantaneous mixtures, but contains a few new points. Section 2 contains preliminary results on stochastic process independence and the definition and a few properties of joint score function. Estimating equations are derived in Section 3. The algorithm and experiments are presented in Section 4 and 5, respectively.

2 Preliminary Issues

2.1 Independence in the Convolutive Context

In convolutive mixtures, $y_1(n)$ and $y_2(n)$ can be independent, while y_1 and y_2 are not [1]. For example, if the sources s_i are iid, and:

$$\mathbf{B}(z)\mathbf{A}(z) = \begin{bmatrix} 1 & z^{-1} \\ 0 & 1 \end{bmatrix} \tag{5}$$

then the outputs are:

$$\begin{cases} y_1(n) = s_1(n) + s_2(n-1) \\ y_2(n) = s_2(n) \end{cases} \tag{6}$$

It is obvious that in this case, $y_1(n)$ and $y_2(n)$ are independent for all n, but y_1 and y_2 are not, and thus the source separation is not achieved.

To check the independence of two random variables x and y, one can use the mutual information:

$$I(x,y) = \int_{x,y} p_{xy}(x,y) \ln \frac{p_{xy}(x,y)}{p_x(x)p_y(y)} dxdy \tag{7}$$

This quantity is always non-negative, and is zero if and only if the random variables x and y are independent.

However, $I(y_1(n), y_2(n)) = 0$ is be a separation criterion. Conversely, one can use $I(y_1(n), y_2(n-m)) = 0$ for all m. But, this criterion, for all m is practically untractable. Thus, we restrict ourselves to a finite set, say $m \in \{-M, \ldots, +M\}$. For example, Charkani [2] and Nguyen and Jutten [6] considered the independence of $y_1(n)$ and $y_2(n-m)$ for $m \{0, \ldots, +M\}$, where M is the maximum degree of the FIR filters of the separating structure.

---

[1] Recall that, by definition, two stochastic processes X_1 and X_2 are independent if and only if the random variables $X_1(n)$ and $X_2(n-m)$ are independent for all n and all m.

2.2 JSFs versus MSFs

In this subsection, we introduce the concepts of Joint Score Function (JSF) and Marginal Score Function (MSF).

Definition 1 (Score Function) *The score function of the scalar random variable x, is the log derivative of its distribution, i.e.:*

$$\psi(x) = \frac{d}{dx} \ln p_x(x) = \frac{p'_x(x)}{p_x(x)} \tag{8}$$

For the N dimensional random vector $\mathbf{x} = (x_1, \ldots, x_N)^T$, we define two score functions:

Definition 2 (MSF) *The Marginal Score Function (MSF) of $\mathbf{x}$, is the vector of score functions of its components, i.e.:*

$$\boldsymbol{\psi}_{\mathbf{x}}(\mathbf{x}) = (\psi_{x_1}(x_1), \ldots, \psi_{x_N}(x_N))^T \tag{9}$$

Note that the ith element of $\boldsymbol{\psi}_{\mathbf{x}}(\mathbf{x})$ is:

$$\psi_i(\mathbf{x}) = \frac{d}{dx_i} \ln p_{x_i}(x_i) \tag{10}$$

where $p_{x_i}(x_i)$ is the PDF of x_i.

Definition 3 (JSF) *The Joint Score Function (JSF) of $\mathbf{x}$, is the vector function $\boldsymbol{\varphi}_{\mathbf{x}}(\mathbf{x})$, such that its ith component is:*

$$\varphi_i(\mathbf{x}) = \frac{\partial}{\partial x_i} \ln p_{\mathbf{x}}(\mathbf{x}) = \frac{\frac{\partial}{\partial x_i} p_{\mathbf{x}}(\mathbf{x})}{p_{\mathbf{x}}(\mathbf{x})} \tag{11}$$

where $p_{\mathbf{x}}(\mathbf{x})$ is the mutual PDF of $\mathbf{x}$.

Generally, MSF and JSF are not equal, but we have the following theorem:

Theorem 1 *The components of the random vector $\mathbf{x}$ are independent if and only if its JSF and MSF are equal, i.e.:*

$$\boldsymbol{\varphi}_{\mathbf{x}}(\mathbf{x}) = \boldsymbol{\psi}_{\mathbf{x}}(\mathbf{x}) \tag{12}$$

For a proof, refer to appendix.

Definition 4 (SFD) *The Score Function Difference (SFD) of $\mathbf{x}$ is the difference between its JSF and MSF, i.e.:*

$$\boldsymbol{\beta}_{\mathbf{x}}(\mathbf{x}) = \boldsymbol{\varphi}_{\mathbf{x}}(\mathbf{x}) - \boldsymbol{\psi}_{\mathbf{x}}(\mathbf{x}) \tag{13}$$

As a consequence of Theorem 1, SFD is an independence criterion.

3 Estimating Equations

3.1 Estimating MSF and JSF

For estimating the MSF, one must simply estimate the score functions of its components. In [5], the following theorem is used for estimating the score function of a scalar random variable:

Theorem 2 *Consider a scalar random variable x, and a function f with a continuous first derivative, satisfying:*

$$\lim_{x \to \pm\infty} f(x) p_x(x) = 0 \qquad (14)$$

then :

$$E\{f(x)\psi(x)\} = -E\{f'(x)\} \qquad (15)$$

Note that the condition (14) is not very restrictive, since most of densities $p_x(x)$ vanishes as x tends towards infinity.

Now, consider the score function estimate equal to a linear combination of some kernel functions $k_i(x)$, *i.e.*:

$$\hat{\psi}(x) = \sum_i^L w_i k_i(x) = \mathbf{k}^T(x) \mathbf{w} \qquad (16)$$

where $\mathbf{k}(x) = (k_1(x), \ldots, k_L(x))^T$ and $\mathbf{w} = (w_1, \ldots, w_L)^T$. We estimate $\mathbf{w}$ for minimizing the mean square error $E\left\{[\psi(x) - \hat{\psi}(x)]^2\right\}$. Applying the orthogonality principle, and using Theorem 2, $\mathbf{w}$ is obtained by:

$$E\left\{\mathbf{k}(x)\mathbf{k}^T(x)\right\} \mathbf{w} = E\left\{\mathbf{k}(x)\psi(x)\right\} \qquad (17)$$
$$= -E\left\{\mathbf{k}'(x)\right\} \qquad (18)$$

This method can be easily generalized to multivariate pdf. First, we prove the generalization of Theorem 2:

Theorem 3 *Consider a random vector $\mathbf{x} = (x_1, \ldots, x_N)^T$, and a multivariate scalar function f with continuous derivatives with respect to x_i, satisfying:*

$$\lim_{x_i \to \pm\infty} \int_{x_1, \ldots, x_{i-1}, x_{i+1}, \ldots, x_N} f(\mathbf{x}) p_\mathbf{x}(\mathbf{x}) = 0 \qquad (19)$$

then:

$$E\{f(\mathbf{x})\varphi_i(\mathbf{x})\} = -E\left\{\frac{\partial}{\partial x_i} f(\mathbf{x})\right\} \qquad (20)$$

For a proof, refer to appendix.

Now, suppose we model $\varphi_i(\mathbf{x})$, the ith element of JSF as a linear combination of the kernel functions $k_1(\mathbf{x})$, ..., $k_L(\mathbf{x})$, i.e. $\hat{\varphi}_i(\mathbf{x}) = \mathbf{k}^T(\mathbf{x})\mathbf{w}$ where $\mathbf{k}(\mathbf{x}) = (k_1(\mathbf{x}), \ldots, k_L(\mathbf{x}))^T$ Following similar computations than those used for developing (18), it can be shown:

$$E\left\{\mathbf{k}(\mathbf{x})\mathbf{k}^T(\mathbf{x})\right\}\mathbf{w} = -E\left\{\frac{\partial}{\partial x_i}\mathbf{k}(\mathbf{x})\right\} \quad (21)$$

3.2 Gradient of the Mutual Information

Suppose the separating system consists of FIR filters whose the maximum degree is M. Hence, the separating system writes as:

$$\mathbf{B}(z) = \mathbf{B}_0 + \mathbf{B}_1 z^{-1} + \cdots + \mathbf{B}_M z^{-M} \quad (22)$$

For developing a gradient-based algorithm, we must estimate the derivative of the mutual information with respect to each matrix $\mathbf{B}_k$.

Theorem 4 *If the separating system $\mathbf{B}(z)$ satisfies (22), then:*

$$\frac{\partial}{\partial \mathbf{B}_k} I\left(y_1(n), y_2(n-m)\right) = E\left\{\boldsymbol{\beta}^{(m)}(n)\mathbf{x}^T(n-k)\right\} \quad (23)$$

where I denotes the mutual information, and $\boldsymbol{\beta}^{(m)}(n)$ is defined by:

$$\boldsymbol{\beta}(n) = \boldsymbol{\beta}_{y_1(n),y_2(n-m)}(y_1(n), y_2(n-m)) \quad (24)$$

$$\boldsymbol{\beta}^{(m)}(n) = \begin{bmatrix} \beta_1(n) \\ \beta_2(n+m) \end{bmatrix} \quad (25)$$

where $\beta_{\mathbf{x}}$ denote the SDF of the random vector $\mathbf{x}$.

For a proof refer to appendix.

In other words, for computing $\boldsymbol{\beta}^{(m)}(n)$, one must first shift forward the second component of $\mathbf{y}$, then compute its SFD to obtain $\boldsymbol{\beta}(n)$, and then shift back its second component to obtain $\boldsymbol{\beta}^{(m)}(n)$.

Note that the algorithm stops when $\boldsymbol{\beta}(n) = 0$, which is equivalent to the independence of the outputs.

4 The Algorithm

The steepest descent algorithm has been used for achieving the output independence, *i.e.* in each iteration, all of the $\mathbf{B}_k$s are updated according to:

$$\mathbf{B}_k = \mathbf{B}_k - \mu \frac{\partial}{\partial \mathbf{B}_k} I(y_1(n), y_2(n-m)) \quad (26)$$

where μ is a small positive constant, the derivative is computed following (23), and SFD is computed using (16), (18), (21). For estimating the MSFs, we have chosen the 4 kernels ($L = 4$):

$$k_1(x) = 1 , \ k_2(x) = x , \ k_3(x) = x^2 , \ k_4(x) = x^3 \quad (27)$$

For estimating the JSFs, we used the 7 kernels ($L = 7$):

$$\begin{aligned} k_1(x_1, x_2) = 1 , \ k_2(x_1, x_2) = x_1 , \ k_3(x_1, x_2) = x_1^2 , \ k_4(x_1, x_2) = x_1^3 \\ , \ k_5(x_1, x_2) = x_2 , \ k_6(x_1, x_2) = x_2^2 , \ k_7(x_1, x_2) = x_2^3 \end{aligned} \quad (28)$$

For tractability of the algorithm, the value of m is randomly chosen from the set $\{-M, \ldots, M\}$ at each iteration, implying a stochastic implementation of the independence criterion . Note that this algorithm is not equivariant [1], and consequently its performance is not independent of the mixing filter.

5 Experimental Results

For measuring the separation performance, we define the output SNR. Assuming no permutation, and denoting $\mathbf{C}(z) = \mathbf{B}(z)\mathbf{A}(z)$, the output SNR on the first channel is:

$$\text{SNR}_1 = \frac{E\left\{[y_1(n)]^2\right\}}{E\left\{\{[C_{12}(z)]\,s_2(n)\}^2\right\}} \quad (29)$$

Hence a high SNR_1 means that the effect of the second source in the first output is negligible. However, the first output is not necessarily equal to the first source, it can be a filtered version of the first source (see Sect. 2.1).

In the first experiment, we have chosen two sinusoids (500 samples), with different frequencies, and mixed them with the $\mathbf{A}(z)$:

$$\begin{bmatrix} 1 + 0.2z^{-1} + 0.1z^{-2} & 0.5 + 0.3z^{-1} + 0.1z^{-2} \\ 0.5 + 0.3z^{-1} + 0.1z^{-2} & 1 + 0.2z^{-1} + 0.1z^{-2} \end{bmatrix} \quad (30)$$

We thus used a 2-degree FIR separating filter, $\mu = 0.3$, and $M = 4$.

Figure 1 shows the separation results, after 13000 iterations (we have only drawn 200 samples). The output SNRs are 48.64dB and 48.61dB. Figure 5, shows the time variation of output SNRs.

In the second experiment, we mixed two uniform white noises, with zero means and unit variances. We choose the same mixing system and parameters as in the first experiment. Figure 5 shows the output SNRs in dB.

6 Conclusion

In this paper, we proposed a new method for separating the convolutive mixtures, based on a stochastic implementation of the minimization of delayed output mutual informations. Moreover, each mutual information term is minimized using

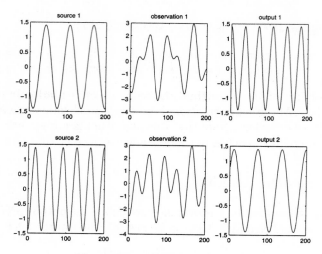

Fig. 1. Separating two sinusoids

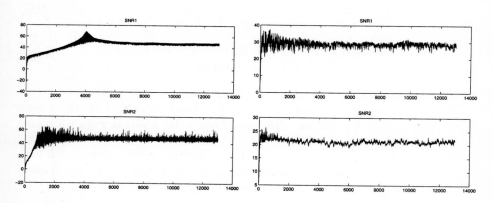

Fig. 2. Output SNRs, in separating two sinusoids

Fig. 3. Output SNRs, in separating two uniform white noises

Marginal and Joint score functions. The experiments show its efficiency. The main restriction of this new method is related to JSF estimation which requires large samples, and whose implementation will be difficult for more than 3 or 4 sources.

A Appendix

Proof of Theorem 1: The proof is given in the two dimensional case. Its generalization to higher dimensions is obvious.

If the elements of $\mathbf{y}$ are independent, then (12) can be easily obtained. Conversely, suppose that (12) holds, then we prove that the elements of $\mathbf{y}$ are independent. Following (12), we have $\frac{\partial}{\partial y_1} \ln p_\mathbf{y}(y_1, y_2) = \frac{\partial}{\partial y_1} \ln p_{y_1}(y_1)$. Integrating both sides of this equation with respect to y_1, leads to:

$$\ln p_\mathbf{y}(y_1, y_2) = \ln p_{y_1}(y_1) + g(y_2) \Rightarrow p_\mathbf{y}(y_1, y_2) = p_{y_1}(y_1) h(y_2) \qquad (31)$$

By integrating both sides of this equation with respect to y_1 from $-\infty$ to $+\infty$, we have $h(y_2) = p_{y_2}(y_2)$ thus the result holds. □

Proof of Theorem 3: Without loss of generality, let $i = 1$. We have:

$$E\{f(\mathbf{x})\varphi_1(\mathbf{x})\} = \int f(\mathbf{x})\varphi_1(\mathbf{x}) p_\mathbf{x}(\mathbf{x}) d\mathbf{x}$$
$$= \int_{x_2,\ldots,x_N} \int_{x_1} f(\mathbf{x}) \frac{\partial}{\partial x_1} p_\mathbf{x}(\mathbf{x}) dx_1 dx_2 \cdots dx_N \qquad (32)$$

Using integration by parts for the inner integral and (19) leads to the desired relation. □

Proof of Theorem 4: Here, because of the limited space, we only prove the theorem for $m = 0$. The generalization to the case $m \neq 0$ is straightforward, but contains somewhat complicated calculations.

Let $\mathbf{B}(z)$ satisfy (22), and $b_{ij}^{(k)}$ denote the ijth element of $\mathbf{B}_k$, then:

$$\frac{\partial H(\mathbf{y}(n))}{\partial b_{ij}^{(k)}} = -E\left\{\frac{\partial}{\partial b_{ij}^{(k)}} \ln p_\mathbf{y}(\mathbf{y})\right\} \qquad (33)$$

Among the y_1 through y_N, only y_i depends on $b_{ij}^{(k)}$, hence:

$$\frac{\partial H(\mathbf{y}(n))}{\partial b_{ij}^{(k)}} = -E\left\{\frac{\partial \ln p_{\mathbf{y}(n)}(\mathbf{y}(n))}{\partial y_i(n)} \cdot \frac{\partial y_i(n)}{\partial b_{ij}^{(k)}}\right\}$$
$$= -E\{\varphi_{\mathbf{y},i}(n) x_j(n-k)\} \qquad (34)$$

Consequently:

$$\frac{\partial H(\mathbf{y}(n))}{\partial \mathbf{B}_k} = -E\{\varphi_\mathbf{y}(n) \mathbf{x}^T(n-k)\} \qquad (35)$$

We now compute the marginal entropy derivatives:

$$\frac{\partial}{\partial b_{ij}^{(k)}} \sum_i H(y_i(n)) = \frac{\partial}{\partial b_{ij}^{(k)}} H(y_i(n))$$

$$= -E\left\{ \frac{\partial}{\partial b_{ij}^{(k)}} \ln p_{y_i(n)}(y_i(n)) \right\}$$

$$= -E\left\{ \frac{\partial \ln p_{y_i(n)}(y_i(n))}{\partial y_i(n)} \cdot \frac{\partial y_i(n)}{\partial b_{ij}^{(k)}} \right\}$$

$$= -E\left\{ \psi_{y_i(n)}(n) x_j(n-k) \right\} \tag{36}$$

Hence:

$$\frac{\partial}{\partial \mathbf{B}_k} \sum_i H(y_i) = -E\left\{ \boldsymbol{\psi}_\mathbf{y}(n) \mathbf{x}^T(n-k) \right\} \tag{37}$$

Combining (7), (35) and (37) proves the theorem. □

References

[1] J.-F. Cardoso and B. Laheld. Equivariant adaptive source separation. *IEEE Trans. on SP*, 44(12):3017–3030, December 1996.
[2] N. Charkani. *Séparation auto-adaptative de sources pour des mélanges convolutifs. Application à la téléphonie mains-libres dans les voitures.* Thèse de l'INP Grenoble, 1996.
[3] P. Comon. Independent component analysis, a new concept? *Signal Processing*, 36(3):287–314, 1994.
[4] U. A. Lindgren and H. Broman. Source separation using a criterion based on second-order statistics. *IEEE Trans. on SP*, 5:1837–1850, 1998.
[5] C. Simon. *Séparation aveugle des sources en mélange convolutif.* PhD thesis, l'université de Marne la Vallée, Novembre 1999. (In French).
[6] A. Taleb and C. Jutten. Entropy optimization, application to blind source separation. In *ICANN*, pages 529–534, Lausanne, Switzeland, October 1997.
[7] H.L. Nguyen Thi and C. Jutten. Blind sources separation for convolutive mixtures. *Signal Processing*, 45:209–229, 1995.
[8] S. Van Gerven and D. Van Compernolle. Signal separation by symmetric adaptive decorrelation: Stability, convergence and uniqueness. *IEEE Trans. on SP*, 43:1602–1612, 1995.
[9] D Yellin and E. Weinstein. Criteria for multichannel signal separation. *IEEE Trans. Signal Processing*, pages 2158–2168, August 1994.

Author Index

Abarbanel, H.D.I.   I-490
Abels, C.   II-328
Abraham, A.   I-269, II-679
Abrard, F.   II-802
Acevedo Sotoca, M.I.   I-482
Acha, J.I.   II-810
Ackermann, G.   II-328
Affenzeller, M.   I-594
Aguilar, P.L.   II-208
Aisbett, J.   I-783
Aizenberg, I.   II-219, II-254
Aizenberg, N.   II-254
Aknin, P.   I-637
Aleksander, I.   I-765
Aler, R.   I-799
Alonso-Betanzos, A.   I-293, I-301, II-393
Alvarez, J.L.   II-644
Álvarez-Sánchez, J.R.   I-772, II-459
Álvaro Fernández, J.   II-246
Amari, S.-I.   I-325, II-786, II-826
Amirat, Y.   II-436
Ammermüller, J.   I-55
Andina, D.   II-111, II-385
Andreu, E.   I-14
Angulo, C.   I-661
Aoki, H.   II-369
Apicella, G.   II-484
Arulampalam, G.   I-410
Ascoli, G.A.   I-30
Aso, H.   I-215
Astilleros Vivas, A.   II-184

Babaie-Zadeh, M.   II-834
Bachiller, M.   II-319
Balsi, M.   II-484
Barahona da Fonseca, I.   I-184
Barahona da Fonseca, J.   I-184
Barakova , Emilia I.   II-55
Barbi, M.   I-87
Barro, S.   I-21, II-418
Bäumler, W.   II-328
Bauer, C.   II-328, II-778
Belanche, L.A.   I-243
Bellei, E.   II-611

Benítez, R.   I-522
Benmahammed, K.   II-227
Berechet, I.   II-619
Berlanga, A.   I-799
Bermejo, S.   I-669
Bernier, J.L.   II-136
Berzal, J.A.   I-166
Bhattacharya, J.   I-108
Biondi Neto, L.   II-353
Blanco, A.   I-285
Blanco, Y.   II-770
Blasco, Xavier   I-457, I-466
Bluff, K.   I-807
Bode, M.   I-363
Bolea, J.A.   I-55
Bologna, G.   I-791
Bongard, M.   I-55
Botía, J.A.   II-547
Bourret, P.   II-80
Bouzerdoum, A.   I-410
Breithaupt, R.   I-363
Buldain, J.D.   I-223, I-253
Burbidge, R.   I-653
Burian, A.   II-311
Butakoff, C.   II-254
Buxton, B.   I-653

Cabestany, J.   I-669
Calderón-Martínez, J.A.   II-235
Campoy-Cervera, P.   II-235
Camps, G.   II-345
Canca, D.   I-749
Canedo, A.   I-21
Cañas, A.   II-136
Carmona, E.   II-401
Carrasco, R.   II-587
Carreira, María J.   II-628
Casañ, G.A.   II-671
Castaño, M.A.   II-671
Castedo, L.   II-603
Castellanos, J.   I-621
Castillo, E.   I-293, I-316
Castillo, P.A.   I-506, I-629
Català, A.   I-661
Chang, C.   II-508

Author Index

Chiang, S.   I-196
Chillemi, S.   I-87
Chinarov, V.   I-333
Christensen, J.   I-63
Christoyianni, I.   II-794
Cichocki, A.   II-786, II-826
Cofiño, A.S.   I-308
Cooper, L.N.   II-696
Cortés, P.   I-749
Costa, J.F.   I-158
Cotta, C.   I-474, I-709
Cottrell, M.   II-738
Courant, M.   II-492
Cruces, S.   II-786

Damas, M.   I-530, II-719
Dapena, A.   II-603
Deco, G.   II-64
Delgado, A.E.   I-38, I-772
Delgado, M.   I-285
Dermatas, E.   II-794
Deville, Y.   II-802
Dorado, J.   I-717
Dorronsoro, J.   I-427
Douence, V.   II-31
Drias, H.   I-586
Dunmall, B.   I-765
Duro, R.J.   I-207

El Dajani, R.   II-377
Engelbrecht, A.P.   I-386

Faúndez-Zanuy, M.   II-754
Feliú Batlle, J.   I-394, II-47
Fellenz, W.A.   I-677
Feng, J.   I-47, I-418
Fernández López, P.   II-96
Fernández, E.   I-55
Fernández, M.A.   II-660
Fernández-Caballero, A.   II-660
Ferrández, J.M.   I-55
Fessant, F.   I-637
Fischer, J.   I-363
Fiz, J.A.   II-361
Fontenla-Romero, O.   I-293, I-301, II-393
França, Felipe M.G.   I-435
Fukuda, F.H.   II-353

Galindo, J.   II-587

Galindo Riaño, P.   II-88, II-152
Gallardo Caballero, R.   II-72
Galván, I.M.   I-514
Di Garbo, A.   I-87
Garcia-Bernal, M.A.   I-355
Garcia Dans, O.   II-393
García Báez, P.   II-96
García Feijó, J.   II-319, II-401
García Orellana, C.J.   I-482, II-72
García de Quirós, F.   I-55
García, J.M.   I-749
García-Lagos, F.   II-711
García-Pedrajas, N.   I-645
Garrido, S.   I-612
Gaussier, P.   II-516
Ghennam, S.   II-227
Gibbon, G.   I-783
Gómez, F.J.   II-660
Gómez-Moreno, H.   I-685
Gómez-Ruiz, J.A.   I-355, II-168
Gómez-Skarmeta, A.F.   II-547
González Velasco, H.M.   I-482, II-72
González, A.   I-427
González, E.J.   I-579, II-104
González, J.   I-498, I-530, I-538, II-136
Gonzalo, I.   I-150
Graña, M.   II-563
Grittani, G.   II-500
Guerrero Vázquez, E.   II-88, II-152
Guerrero, F.   I-749
Guidotti, D.   II-595, II-611
Guijarro Berdiñas, B.   I-293, I-301, II-393
Guinot, C.   II-119
Gupta, J.N.D.   I-741
Gutiérrez, G.   I-514
Gutiérrez, J.M.   I-308

Hadi, A.S.   I-316
Hamami, L.   II-336
Hamilton, A.F.   II-104
Hayasaka, T.   II-176
Hernández Ábrego, G.   II-279
Herreras, O.   I-1
Herrero, J.C.   I-814
Herrero, J.M.   I-466
Hervás-Martínez, C.   I-645
Holden, S.   I-653
Horiba, I.   II-303
del-Hoyo-Alonso, R.   II-531

Huerta, R. I-490
Hülse, M. II-410

Ibarz, J.M. I-1
Inui, T. I-126
Isasi, P. I-514, II-262
Ishii, N. I-561

Jacoby, N. I-602
Jarabo Amores, P. II-652
Jaramillo, M.A. II-246
Jia, X. II-476
Jiménez, N.V. II-345
Joya, G. II-711
Juez, J. II-571
Jutten, C. II-834

Kanniah, J. II-468
Karagianni, H. II-127
Kim, D.-J. I-450
Kim, I.-Y. I-450
Kim, S.I. I-450
Kitano, H. I-95
Kokkinakis, G. II-794
Kosugi, Y. II-369
Koutras, A. II-794
Kubin, G. II-746
Kuosmanen, P. II-311
Kurematsu, A. II-287
Kwok, T. I-733

Lacruz, B. I-316
Lago-Fernández, L.F. II-64
Lang, E.W. II-295, II-328, II-778
Lara, B. II-410
Lassouaoui, N. II-336
Lázaro, M. I-347
Leder, C. II-704
Ledezma, A. I-799
Lee, J.-M. I-450
León, C. II-728
Lewis, J.P. I-442
Li, S. I-757
Lima, P.M.V. I-822
Linaje, M. II-208
de Lope Asiaín, J. II-451
López Aligué, F.J. II-72, II-184
López Coronado, J. I-394, II-47
López Ferreras, F. I-685, II-652

López, A. II-728
López, A.M. I-725
López, H. I-725
López-Aguado, L. I-1
López-Rubio, E. I-355, II-168
Lourens, T. I-95
Luengo, S. II-484

Macías Macías, M. I-482, II-184
Madani, K. II-200, II-619
Maldonado-Bascón, S. I-685
Mange, D. II-1, II-39
Maraschini, A. II-484
Maravall Gómez-Allende, D. II-160, II-451
Marichal, R.L. I-579, II-104
Marín, F.J. II-539, II-711
Mariño, J. I-21
Martín-del-Brío, B. II-271, II-531
Martín-Clemente, R. II-810
Martín, J.D. II-345
Martín-Vide, C. I-621
Martínez, M. I-457, I-466
Martínez, P. II-208
Martínez, R. I-772
Le Masson, G. II-31
Mata, J. II-644
Medraño-Marqués, N.J. II-271, II-531
Menzinger, M. I-333
Merelo, J.J. I-506, I-629
Mérida, E. I-522
Midenet, S. I-637
Mies, C. II-328
Miftakov, R. I-63
Milligan, G. I-442
Minami, T. I-126
Miquel, M. II-377
Mira, J. I-38, I-772, II-319, II-401, II-660
Mitrana, V. I-621
Miyake, S. I-215
Miyamoto, Y. I-378
Moga, S. II-516
Molina Vilaplana, J. I-394
Molina, J.M. I-514
Monedero, Í. II-728
Montanero, J.M. II-246
Montaño, J.C. II-728
Montaser Kouhsari, L. I-72

Monte, E.   II-361
Montesanto, A.   II-444
Moraga, C.   I-142
Del Moral Hernandez, E.   I-546
Morano, D.   II-426
Moreno, L.   I-579, II-104
Moreno, L.   I-612
Morton, H.   I-765
Moya, E.J.   II-571
Mucientes, M.   II-418
Mulero, J.I.   II-47
Muñoz García, D.F.   II-235
Muñoz-Pérez, J.   I-355, I-522, II-168
Muruzábal, J.   I-701
Myasnikova, E.   II-219

Nagata, A.   II-369
Nakamura, M.   I-561
Nakano Miyatake, M.   II-192, II-287
Nakauchi, S.   II-176
Nayebi, K.   II-834
Negri, S.   I-243
Neskovic, P.   II-696
Nogueras, R.   I-709
Normann, R.A.   I-55

Ohnishi, E.   II-176
Okada, H.   I-370, I-378
Okuno, H.G.   I-95
Olivares, G.   II-719
Omlin, C.W.   I-339, II-579
Omori, T.   I-370
Oota, M.   I-561
Ortega, J.   I-498, I-570, I-136, II-144, II-719
Ortiz-Boyer, D.   I-645
Oukhellou, L.   I-637
Ozeki, T.   I-325

Pacheco, M.A.C.   I-174
Padilla, A.   II-547
Páez-Borrallo, J.M.   II-770
Pantaleón, C.   I-347
Park, H.   I-325, I-402
Pasemann, F.   II-410
Patricio Guisado, M.A.   II-160
de la Paz López, F.   II-459
Pegalajar, M.C.   I-285
Pelletier, B.   II-80
Peña, D.   II-235

Penas, M.   II-628
Penedo, M.G.   II-628
Perán, J.R.   II-571
Pereda, E.   I-108
Perez-Meana, H.   II-192, II-287
Pérez, O.M.   II-539
Pérez, R.M.   II-208
Pérez-Uribe, A.   II-492
Petacchi, R.   II-595, II-611
Piñeiro, J.D.   I-579
Pizarro Junquera, J.   II-88, II-152
Plaza, A.   II-208
Pomares, H.   I-498, I-530, I-538
Pomares, R.   I-14
Ponthieux, S.   II-738
Pontnaut , J.   II-436
Porras, M.A.   I-150
Preciado, J.C.   II-208
Preciado, V.M.   II-636
Prieto, A.   I-570, I-629, II-144
Puente, J.   I-693
Puliti, P.   II-444
Puntonet, C.G.   II-328, II-762, II-778, II-810

Quero, G.   II-508
Quoy, M.   II-516

Rabinovich, M.I.   I-490
Rabuñal, J.R.   I-717
Raducanu, B.   II-563
Rajimehr, R.   I-72
Ramdane-Cherif, A.   II-436
Ramírez, J.   II-500
Rank, E.   II-746
Raptis, S.   II-127
Rehtanz, C.   II-704
Renaud-Le Masson, S.   II-31
Requena, I.   I-285
Restrepo, H.F.   II-39
Restum Antonio, E.   II-353
Reyneri, L.M.   II-14, II-426, II-595, II-611
Rincón, M.   II-319
Riquelme, J.C.   II-644
Rivas, V.M.   I-506
Rivera, A.J.   I-570
Rizzi, I.   II-611
De Roberto Junior, V.   II-353
Rodrigues, P.   I-158

Author Index

Rodrigues, R.J.   II-687
Rodríguez Álvarez, M.   II-328, II-762
Rodríguez, F.   I-490
Rojas, I.   I-498, I-530, I-538, I-570, II-762
Romero, G.   I-629
Rosa Zurera, M.   II-652
de la Rosa, M.G.   II-401
Rousset, P.   II-119
Rubel, P.   II-377

Saadia, N.   II-436
Saarinen, J.   II-311
Şahin, E.   II-524
Saïghi, S.   II-31
Sainz, G.I.   II-571
Salcedo, F.J.   II-144
Salinas, L.   I-142
Salmerón, M.   I-538, II-595, II-719
Saltykov, K.A.   I-81
Samsonova, M.   II-219
Sanchez-Andres, J.V.   I-14
Sanchez-Perez, G.   II-192
Sánchez Maroño, N.   II-393
Sánchez, E.   II-418
Sánchez, E.   I-21
Sánchez, L.   I-725
Sánchez-Montañés, M.A.   I-117
Sanchis, J.   I-457
Sanchís, A.   I-514
Sandoval, F.   II-711
Santa Cruz, C.   I-427
Santamaría, I.   I-347
Santos, A.   I-717
Santos, J.   I-207
Sanz-González, J.L.   II-111
Sanz-Tapia, E.   I-645
Särelä, J.   II-818
Satoh, S.   I-215
Scorcioni, R.   I-30
Segovia, J.   II-262
Sempere, J.M.   I-621
Senent, J.   I-457, I-466
Serrano, A.J.   II-345
Serrano Pérez, A.   II-184
Setiono, R.   I-277
Shevelev, I.A.   I-81
Shirazi, M.N.   I-134
Siegelmann, H.T.   I-158
Sigut, J. F.   I-579, II-104

Simancas-Acevedo, E.   II-287
Simões da Fonseca, J.   I-184
Smith, K.A.   I-733, I-741
Snyders, S.   I-339
Solé i Casals, J.   II-361
Solsona, J.   II-484
Sopena, N.   II-361
Soria, B.   I-14
Soria, E.   II-345
de Souza, F.J.   I-174
Stauffer, A.   II-1
Steinberg, D.   II-679
Steinmetz, U.   II-410
Suárez Araujo, C.P.   I-184, II-96
Sugie, N.   II-303
Suzuki, K.   II-303

Tannhof, P.   II-200
Tascini, G.   II-444
Tempesti, G.   II-1
Tereshko, V.   I-554
Teuscher, C.   II-1
Theis, F.J.   II-778
Thomé, A.C.G.   II-687
Tobely, T.E.   I-235
Torres Sánchez, I.   II-279
Toscano-Medina, K.   II-192
Trelles, O.   II-539
de Trémiolles, G.   II-200
Trotter, M.   I-653
Troya, J.M.   I-474
Tsuruta, N.   I-235
Tzafestas, S.   II-127

Usui, S.   II-176

Valdés, M.   II-547
Varela, R.   I-693
Varona, P.   I-1, I-490
Vega Corona, A.   II-385
Vela, C.   I-693
Velasco, M.   II-262
Vellasco, M.M.R.   I-174
Vianna, G.K.   II-687
Vigário, R.   II-818
Vila, A.   II-587
Vilasís-Cardona, X.   II-484
Vilela, I.M.O.   I-822

Wallace, J.G.   I-807

Wang, L.   I-757
Watanabe, E.   II-369
Weir, M.K.   I-442
Wickert, I.   I-435

Yamakawa, H.   I-370, I-378
Yamauchi, K.   I-561
Yañez Escolano, A.   II-88, II-152
Yang, Y.   II-468, II-476
Yoshiki, Y.   I-235

Yue, T.-W.   I-196
Yun, T.S.   II-555

Zayas, F.   II-246
Zazo, S.   II-770
Zhang, L.-Q.   II-826
Zhou, C.   II-468, II-476
Ziegaus, C.   II-295
Zufiria, P.J.   I-166
van Zyl, J.   II-579

Lecture Notes in Computer Science

For information about Vols. 1–1990
please contact your bookseller or Springer-Verlag

Vol. 1991: F. Dignum, C. Sierra (Eds.), Agent Mediated Electronic Commerce. VIII, 241 pages. 2001. (Subseries LNAI).

Vol. 1992: K. Kim (Ed.), Public Key Cryptography. Proceedings, 2001. XI, 423 pages. 2001.

Vol. 1993: E. Zitzler, K. Deb, L. Thiele, C.A.Coello Coello, D. Corne (Eds.), Evolutionary Multi-Criterion Optimization. Proceedings, 2001. XIII, 712 pages. 2001.

Vol. 1994: J. Lind, Iterative Software Engineering for Multiagent Systems. XVII, 286 pages. 2001. (Subseries LNAI).

Vol. 1995: M. Sloman, J. Lobo, E.C. Lupu (Eds.), Policies for Distributed Systems and Networks. Proceedings, 2001. X, 263 pages. 2001.

Vol. 1997: D. Suciu, G. Vossen (Eds.), The World Wide Web and Databases. Proceedings, 2000. XII, 275 pages. 2001.

Vol. 1998: R. Klette, S. Peleg, G. Sommer (Eds.), Robot Vision. Proceedings, 2001. IX, 285 pages. 2001.

Vol. 1999: W. Emmerich, S. Tai (Eds.), Engineering Distributed Objects. Proceedings, 2000. VIII, 271 pages. 2001.

Vol. 2000: R. Wilhelm (Ed.), Informatics: 10 Years Back, 10 Years Ahead. IX, 369 pages. 2001.

Vol. 2001: G.A. Agha, F. De Cindio, G. Rozenberg (Eds.), Concurrent Object-Oriented Programming and Petri Nets. VIII, 539 pages. 2001.

Vol. 2002: H. Comon, C. Marché, R. Treinen (Eds.), Constraints in Computational Logics. Proceedings, 1999. XII, 309 pages. 2001.

Vol. 2003: F. Dignum, U. Cortés (Eds.), Agent Mediated Electronic Commerce III. XII, 193 pages. 2001. (Subseries LNAI).

Vol. 2004: A. Gelbukh (Ed.), Computational Linguistics and Intelligent Text Processing. Proceedings, 2001. XII, 528 pages. 2001.

Vol. 2006: R. Dunke, A. Abran (Eds.), New Approaches in Software Measurement. Proceedings, 2000. VIII, 245 pages. 2001.

Vol. 2007: J.F. Roddick, K. Hornsby (Eds.), Temporal, Spatial, and Spatio-Temporal Data Mining. Proceedings, 2000. VII, 165 pages. 2001. (Subseries LNAI).

Vol. 2009: H. Federrath (Ed.), Designing Privacy Enhancing Technologies. Proceedings, 2000. X, 231 pages. 2001.

Vol. 2010: A. Ferreira, H. Reichel (Eds.), STACS 2001. Proceedings, 2001. XV, 576 pages. 2001.

Vol. 2011: M. Mohnen, P. Koopman (Eds.), Implementation of Functional Languages. Proceedings, 2000. VIII, 267 pages. 2001.

Vol. 2012: D.R. Stinson, S. Tavares (Eds.), Selected Areas in Cryptography. Proceedings, 2000. IX, 339 pages. 2001.

Vol. 2013: S. Singh, N. Murshed, W. Kropatsch (Eds.), Advances in Pattern Recognition – ICAPR 2001. Proceedings, 2001. XIV, 476 pages. 2001.

Vol. 2014: M. Moortgat (Ed.), Logical Aspects of Computational Linguistics. Proceedings, 1998. X, 287 pages. 2001. (Subseries LNAI).

Vol. 2015: D. Won (Ed.), Information Security and Cryptology – ICISC 2000. Proceedings, 2000. X, 261 pages. 2001.

Vol. 2016: S. Murugesan, Y. Deshpande (Eds.), Web Engineering. IX, 357 pages. 2001.

Vol. 2018: M. Pollefeys, L. Van Gool, A. Zisserman, A. Fitzgibbon (Eds.), 3D Structure from Images – SMILE 2000. Proceedings, 2000. X, 243 pages. 2001.

Vol. 2019: P. Stone, T. Balch, G. Kraetzschmar (Eds.), RoboCup 2000: Robot Soccer World Cup IV. XVII, 658 pages. 2001. (Subseries LNAI).

Vol. 2020: D. Naccache (Ed.), Topics in Cryptology – CT-RSA 2001. Proceedings, 2001. XII, 473 pages. 2001

Vol. 2021: J. N. Oliveira, P. Zave (Eds.), FME 2001: Formal Methods for Increasing Software Productivity. Proceedings, 2001. XIII, 629 pages. 2001.

Vol. 2022: A. Romanovsky, C. Dony, J. Lindskov Knudsen, A. Tripathi (Eds.), Advances in Exception Handling Techniques. XII, 289 pages. 2001

Vol. 2024: H. Kuchen, K. Ueda (Eds.), Functional and Logic Programming. Proceedings, 2001. X, 391 pages. 2001.

Vol. 2025: M. Kaufmann, D. Wagner (Eds.), Drawing Graphs. XIV, 312 pages. 2001.

Vol. 2026: F. Müller (Ed.), High-Level Parallel Programming Models and Supportive Environments. Proceedings, 2001. IX, 137 pages. 2001.

Vol. 2027: R. Wilhelm (Ed.), Compiler Construction. Proceedings, 2001. XI, 371 pages. 2001.

Vol. 2028: D. Sands (Ed.), Programming Languages and Systems. Proceedings, 2001. XIII, 433 pages. 2001.

Vol. 2029: H. Hussmann (Ed.), Fundamental Approaches to Software Engineering. Proceedings, 2001. XIII, 349 pages. 2001.

Vol. 2030: F. Honsell, M. Miculan (Eds.), Foundations of Software Science and Computation Structures. Proceedings, 2001. XII, 413 pages. 2001.

Vol. 2031: T. Margaria, W. Yi (Eds.), Tools and Algorithms for the Construction and Analysis of Systems. Proceedings, 2001. XIV, 588 pages. 2001.

Vol. 2032: R. Klette, T. Huang, G. Gimel'farb (Eds.), Multi-Image Analysis. Proceedings, 2000. VIII, 289 pages. 2001.

Vol. 2033: J. Liu, Y. Ye (Eds.), E-Commerce Agents. VI, 347 pages. 2001. (Subseries LNAI).

Vol. 2034: M.D. Di Benedetto, A. Sangiovanni-Vincentelli (Eds.), Hybrid Systems: Computation and Control. Proceedings, 2001. XIV, 516 pages. 2001.

Vol. 2035: D. Cheung, G.J. Williams, Q. Li (Eds.), Advances in Knowledge Discovery and Data Mining – PAKDD 2001. Proceedings, 2001. XVIII, 596 pages. 2001. (Subseries LNAI).

Vol. 2037: E.J.W. Boers et al. (Eds.), Applications of Evolutionary Computing. Proceedings, 2001. XIII, 516 pages. 2001.

Vol. 2038: J. Miller, M. Tomassini, P.L. Lanzi, C. Ryan, A.G.B. Tettamanzi, W.B. Langdon (Eds.), Genetic Programming. Proceedings, 2001. XI, 384 pages. 2001.

Vol. 2039: M. Schumacher, Objective Coordination in Multi-Agent System Engineering. XIV, 149 pages. 2001. (Subseries LNAI).

Vol. 2040: W. Kou, Y. Yesha, C.J. Tan (Eds.), Electronic Commerce Technologies. Proceedings, 2001. X, 187 pages. 2001.

Vol. 2041: I. Attali, T. Jensen (Eds.), Java on Smart Cards: Programming and Security. Proceedings, 2000. X, 163 pages. 2001.

Vol. 2042: K.-K. Lau (Ed.), Logic Based Program Synthesis and Transformation. Proceedings, 2000. VIII, 183 pages. 2001.

Vol. 2043: D. Craeynest, A. Strohmeier (Eds.), Reliable Software Technologies – Ada-Europe 2001. Proceedings, 2001. XV, 405 pages. 2001.

Vol. 2044: S. Abramsky (Ed.), Typed Lambda Calculi and Applications. Proceedings, 2001. XI, 431 pages. 2001.

Vol. 2045: B. Pfitzmann (Ed.), Advances in Cryptology – EUROCRYPT 2001. Proceedings, 2001. XII, 545 pages. 2001.

Vol. 2047: R. Dumke, C. Rautenstrauch, A. Schmietendorf, A. Scholz (Eds.), Performance Engineering. XIV, 349 pages. 2001.

Vol. 2048: J. Pauli, Learning Based Robot Vision. IX, 288 pages. 2001.

Vol. 2051: A. Middeldorp (Ed.), Rewriting Techniques and Applications. Proceedings, 2001. XII, 363 pages. 2001.

Vol. 2052: V.I. Gorodetski, V.A. Skormin, L.J. Popyack (Eds.), Information Assurance in Computer Networks. Proceedings, 2001. XIII, 313 pages. 2001.

Vol. 2053: O. Danvy, A. Filinski (Eds.), Programs as Data Objects. Proceedings, 2001. VIII, 279 pages. 2001.

Vol. 2054: A. Condon, G. Rozenberg (Eds.), DNA Computing. Proceedings, 2000. X, 271 pages. 2001.

Vol. 2055: M. Margenstern, Y. Rogozhin (Eds.), Machines, Computations, and Universality. Proceedings, 2001. VIII, 321 pages. 2001.

Vol. 2056: E. Stroulia, S. Matwin (Eds.), Advances in Artificial Intelligence. Proceedings, 2001. XII, 366 pages. 2001. (Subseries LNAI).

Vol. 2057: M. Dwyer (Ed.), Model Checking Software. Proceedings, 2001. X, 313 pages. 2001.

Vol. 2059: C. Arcelli, L.P. Cordella, G. Sanniti di Baja (Eds.), Visual Form 2001. Proceedings, 2001. XIV, 799 pages. 2001.

Vol. 2062: A. Nareyek, Constraint-Based Agents. XIV, 178 pages. 2001. (Subseries LNAI).

Vol. 2064: J. Blanck, V. Brattka, P. Hertling (Eds.), Computability and Complexity in Analysis. Proceedings, 2000. VIII, 395 pages. 2001.

Vol. 2066: O. Gascuel, M.-F. Sagot (Eds.), Computational Biology. Proceedings, 2000. X, 165 pages. 2001.

Vol. 2068: K.R. Dittrich, A. Geppert, M.C. Norrie (Eds.), Advanced Information Systems Engineering. Proceedings, 2001. XII, 484 pages. 2001.

Vol. 2070: L. Monostori, J. Váncza, M. Ali (Eds.), Engineering of Intelligent Systems. Proceedings, 2001. XVIII, 951 pages. 2001. (Subseries LNAI).

Vol. 2072: J. Lindskov Knudsen (Ed.), ECOOP 2001 – Object-Oriented Programming. Proceedings, 2001. XIII, 429 pages. 2001.

Vol. 2073: V.N. Alexandrov, J.J. Dongarra, B.A. Juliano, R.S. Renner, C.J.K. Tan (Eds.), Computational Science – ICCS 2001. Part I. Proceedings, 2001. XXVIII, 1306 pages. 2001.

Vol. 2074: V.N. Alexandrov, J.J. Dongarra, B.A. Juliano, R.S. Renner, C.J.K. Tan (Eds.), Computational Science – ICCS 2001. Part II. Proceedings, 2001. XXVIII, 1076 pages. 2001.

Vol. 2077: V. Ambriola (Ed.), Software Process Technology. Proceedings, 2001. VIII, 247 pages. 2001.

Vol. 2081: K. Aardal, B. Gerards (Eds.), Integer Programming and Combinatorial Optimization. Proceedings, 2001. XI, 423 pages. 2001.

Vol. 2082: M.F. Insana, R.M. Leahy (Eds.), Information Processing in Medical Imaging. Proceedings, 2001. XVI, 537 pages. 2001.

Vol. 2083: R. Goré, A. Leitsch, T. Nipkow (Eds.), Automated Reasoning. Proceedings, 2001. XV, 708 pages. 2001. (Subseries LNAI).

Vol. 2084: J. Mira, A. Prieto (Eds.), Connectionist Models of Neurons, Learning Processes, and Artificial Intelligence. Proceedings, 2001. Part I. XXVII, 836 pages. 2001.

Vol. 2085: J. Mira, A. Prieto (Eds.), Bio-Inspired Applications of Connectionism. Proceedings, 2001. Part II. XXVII, 848 pages. 2001.

Vol. 2089: A. Amir, G.M. Landau (Eds.), Combinatorial Pattern Matching. Proceedings, 2001. VIII, 273 pages. 2001.

Vol. 2091: J. Bigun, F. Smeraldi (Eds.), Audio- and Video-Based Biometric Person Authentication. Proceedings, 2001. XIII, 374 pages. 2001.

Vol. 2092: L. Wolf, D. Hutchison, R. Steinmetz (Eds.), Quality of Service – IWQoS 2001. Proceedings, 2001. XII, 435 pages. 2001.

Vol. 2097: B. Read (Ed.), Advances in Databases. Proceedings, 2001. X, 219 pages. 2001.